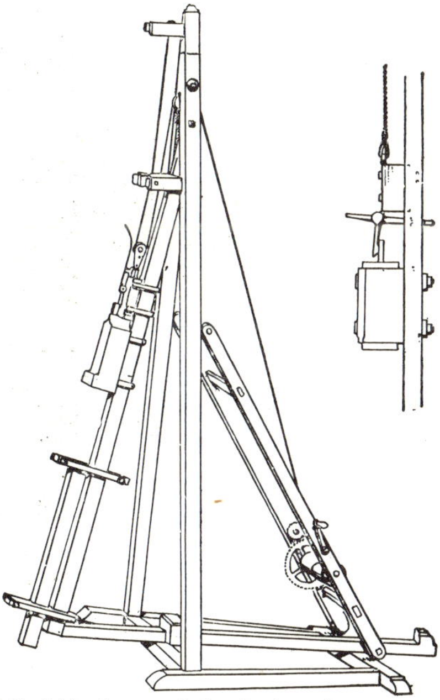

Adjustable driver with fixed release trip

Another form of trip release

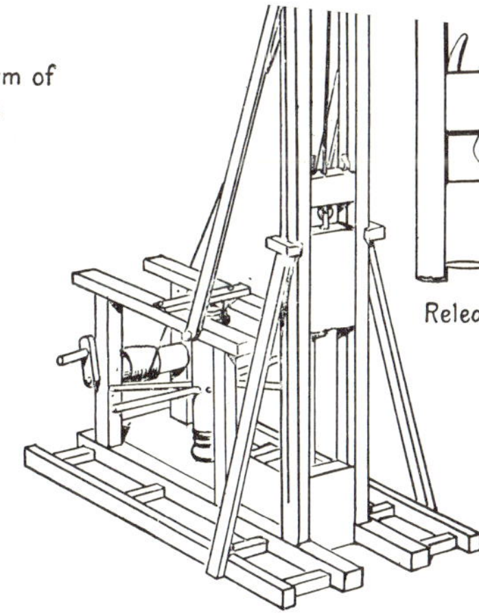

Driver with horizontal and vertical winches

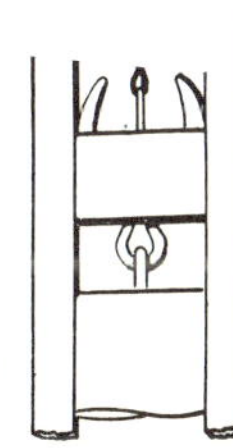

Release

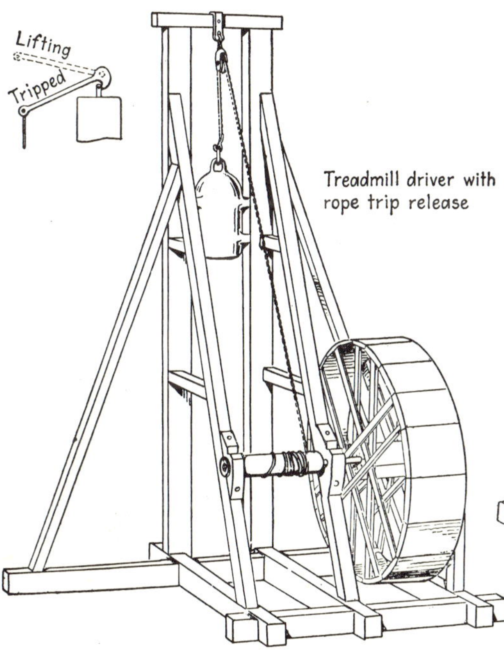

Treadmill driver with rope trip release

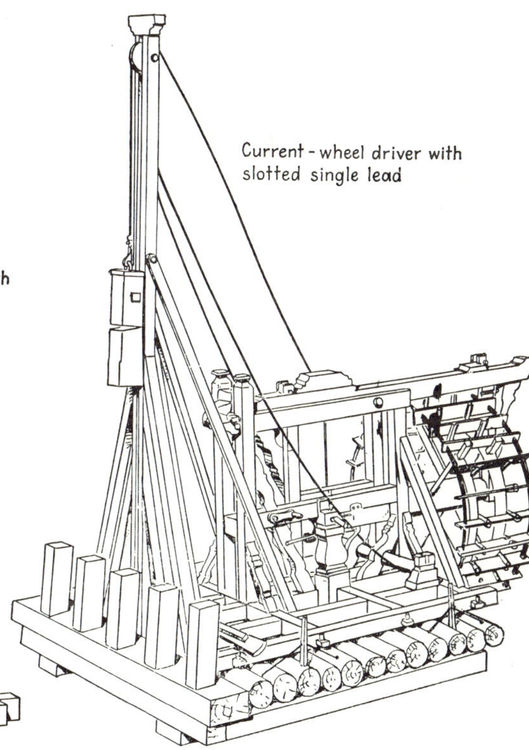

Current-wheel driver with slotted single lead

Medieval Pile Drivers

PILE FOUNDATIONS

PILE FOUNDATIONS

ROBERT D. CHELLIS, B.S., C.E.

Structural Engineer Consultant, Stone & Webster Engineering Corp.
Fellow, American Society of Civil Engineers

SECOND EDITION

McGRAW-HILL BOOK COMPANY

New York Toronto London

1961

PILE FOUNDATIONS

 Library of Congress Catalog Card Number: 60-53032

11 12–MAMM–7 5

ISBN 07-010751-3

PREFACE

This new edition has been made desirable because of the great advances, since 1951, in many of the aspects of foundations treated in this book. Some subjects were still in their infancy at that time. Literature in the fields covered has become increasingly voluminous and instructive and contains much information of great value; such results have been incorporated in the text and references added in the bibliography. Engineers, contractors, and students should avail themselves of this new knowledge. Some of the principal aspects affected are mentioned below.

Among phases of the subject amplified or added are new uses for piles, beginnings of successful application of the longitudinal-wave theory for pile driving, approximate relation between numbers of blows on the sampling spoon and types of soils, formulas for determination of end-bearing values of single piles or groups, cone-penetration tests, vane-shear tests, and natural frequencies of soils.

Diesel hammers have been developed and are described and their properties shown. Relative merits of types of hammers are discussed. A number of new types and sizes of steam or air hammers have become available, and their properties are included. Other items amplified or added include driving piles by vibration, and loosening or tightening the friction between piles and soil by means of electric currents.

Factors in developing comparative estimated costs of pile types and rigs are considered.

Still other subjects added or amplified include piles in permafrost, pile group efficiencies, and column formulas for unsupported lengths of piles. The section on earthquakes has been rewritten.

The subject of ship impact and wave forces on piles and wharves has been expanded and design charts added, thus making much previously disjointed design information available in one place. Numerous types of fendering systems that have been developed are illustrated.

The most recent practice in designing for ice thrusts is given.

A study and resumé of design factors for lateral resistances of piles have been made.

A number of new types of piles and caissons are described, including prestressed piles. Information on design and practice for prestressed concrete piles was given by Mr. N. L. Scott of the Prestressed Concrete Institute; Mr. H. K. Preston of John Roebling's Sons; Mr. R. H. Singer of Ben C. Gerwick, Inc.; Mr. S. N. Clary of the Virginia Department of Highways; Mr. Harry Palmbaum of Blakeslee Prestress, and Mr. W. E. Dean of the Florida State Road Department, to whom thanks are due.

New tables of properties of piles have been added. The sections on drilled piles and drilled caissons have been rewritten to include much progress in these methods, and thanks are expressed to Mr. C. C. Gauntt of Case Foundation Co. for information on these subjects. The sections on corrosion and cathodic protection of steel piles have been greatly extended. Additional methods for encasing piles are included.

Later methods of soil solidification have been included, and criteria shown. Appreciation is due to Dr. A. M. Swift and Mr. R. H. Karol of the American Cyanamid Co. for review of this new material.

Standard specifications have been revised to current issues and new standards included. Instructions for pile inspectors have been amplified and record forms revised.

Gratitude is expressed to the many engineers who have contributed information on various phases of the subject during verbal discussions or by correspondence. In some instances, much time and effort were expended and have contributed to the scope and accuracy of the book. In particular, thanks are due to Messrs. E. A. Smith, F. M. Fuller, C. E. Simmons, and Dr. R. L. Nordlund of Raymond International, Inc., for detailed reviews of the entire book and for information on the wave equation and its application using electronic computers. The text and tabular values for new material on British piles and hammers were reviewed by Mr. S. Packshaw of the British Steel Piling Co., Ltd., who also obtained information from various British firms other than his own, which was of great value to the book.

Appreciation is expressed to the donors of new photographs and to those who granted permission to reproduce sketches and tables, for which credits appear in the text.

Indebtedness is also acknowledged to the many authors of books and articles to which references appear in the bibliography. This field is too broad to be covered fully by the experience of any one person, and only by digesting a wide selection of published material could its scope be indicated and the most pertinent material correlated so that more convenient use could be made of the experiences and recommendations of the best authorities.

Robert D. Chellis

CONTENTS

TABLES

APPENDIX

CHAPTER 1

BASIC PRINCIPLES OF PILE FOUNDATIONS

Basic Principles

This book is confined to the design of pile foundations, including selection of types of piles and driving equipment, methods of driving or placing, determination of capacities, and means of obtaining longevity.

The need for piles under a foundation may be self-evident in certain cases, or it may appear as the result of studies in soil mechanics to determine the characteristics and predicted action of the various strata.

A prediction of the expected action of the soil may sometimes be made from information about the action of previous structures on the site or structures on adjoining sites of which the soil character is definitely known to be the same. Unless such data are available, good engineering practice requires ample borings, taken in a manner which will provide samples capable of showing the information necessary to enable the engineer or the soil-mechanics laboratory, if tests are required, to determine the capabilities of the various strata.

The design of piles rests upon three equally important basic considerations: first, a consideration of the geology of the site and the study of boring results; second, the study of pile types and equipment by means of a dynamic pile-driving formula; and third, the study of pile carrying capacities by the static formula. Test-pile and test-load results should be combined with these studies in many cases. The capacity of the soil below the piles should be considered.

Geology and Subsurface Explorations

The first consideration is the probable relative depths, characters, consistencies, and load-carrying capacities of the various strata, to determine the type and probable lengths of piles. Only the inexpert are courageous enough to proceed without this information.

Selection of Site. Geologic and subsurface conditions should be investigated sufficiently prior to purchase of a site to permit reasonably accurate comparative estimates to be made. The cost of foundations adequate to prevent detrimental settlements may be prohibitive on some sites. Considerable savings may be made at the most advantageous site.

In the case of large or heavy structures in regions where ample information is not obtainable, an engineering geologist should be consulted.

Types of Soils. *Glacial soils* in the United States lie north of a line through northern New Jersey and Pennsylvania, then along the Ohio and Missouri rivers, into Kansas, then northwesterly a little west of the Missouri River to the Rocky Mountains, then westerly, south of the Canadian border to the Pacific Ocean. In this region, foundation conditions can be very irregular, and results at one boring may not represent the adjacent soil. Peat, inorganic silt, and soft clay often occur above the rock and may be overlain by other soils that are of better quality but which may be rendered unsuitable for heavy loads by the presence of the poor materials below. In the northeastern section of the United States sharp changes occur, with numerous glacial ponds filled in with heterogeneous unconsolidated materials such as peat and muck, surrounded by shores of hard glacial drift (Fig. 16.1). Some of these soft spots are of considerable depth, well over 100 ft of peat having been encountered.

Alluvial soils occur in present or former flood plains and deltas, often to great thickness. They are deposited in thin layers and may be quite variable.

Residual soils have been formed by decay of rock. They are generally clayey and grade down to disintegrated rock at moderate depths and are often encountered south of the glacial line outside of river valleys in the United States.

Borings. Adequate borings are invaluable, and it is poor practice, and may prove poor economy, to skimp on the number, length, or quality of borings. A real knowledge of the materials through which the piles are driven and upon which they are founded is the most essential part of pile-foundation design. Misleading opinions may be formed by generalizing from a small number of borings of insufficient depth. The number and spacing of borings should depend upon the character of the underground. It is suggested that widely spaced borings be drilled initially, to bound the site and explore the center. It can then be seen whether conditions appear to be uniform and if intermediate borings are warranted. If the underground is variable, additional sets of intermediate borings, or rock soundings if it has been determined by this time that piles are to be end-bearing, should be made until uncertainties have been removed. Sometimes it has been found necessary to make borings or soundings at every footing.

Borings should be made to a depth of 100 ft or more. Unless rock has been encountered by this depth, or unless it has been verified that a hard stratum occurs at a higher level and that it is not underlain by compressible strata, at least one boring should be continued to rock or to a depth $1\frac{1}{2}$ times the width of the structure.

The character of the borings is important. Wash-water samples give no indication of natural soil characteristics other than that the soil contains some fine material capable of being washed out. For general initial identification, possibly to be followed later by more accurate samples, it is often desirable to take dry samples by a pipe or split sampler, driving into the ground below wash holes. For compressible soils of any appreciable extent, *undisturbed samples* of larger diameter are required. These need be taken only in the questionable strata. Thin-walled tubes 2 in. in diameter (Shelby tubes) are suitable for general examination and tests.

To retain the natural water contents of samples, containers should always be sealed with paraffin until inspections, regardless of the method of sampling.

A discussion of subsurface-exploration necessity and methods is contained in "Pile Foundations and Pile Structures," *ASCE Manuals of Engineering Practice, No. 27.*[*,1h]

Space will not permit detailed description of sampling methods to be included in the scope of this book. This is an important phase of all foundation engineering and should be studied elsewhere.[†]

Selection of Pile and Driving Equipment

The second consideration in the design of pile foundations is a study, by means of an adequate dynamic pile-driving formula, of the suitable types of piles and driving equipment. Such computations will readily determine, for any selected type, size, and length of pile, the carrying capacity of the pile and the stress in the pile during driving, for any given amount of penetration per blow. With these data, ample-sized hammers can be chosen to secure the desired amount of embedment in load-carrying strata with reasonable sets and without damage to the pile. It is desirable to study this problem in conjunction with driving test piles, if possible.

Study of Static Friction

The third consideration in pile-foundation design is a study of the static-friction values required to be developed, in the strata selected for load carrying, based on the embedded surface areas of pile, after having driven the pile to satisfactory tip grade by means of the dynamic-

* Superior numbers (with or without letters) refer to the Bibliography, at the end of the book, in which the material is organized by subject.

† For a full description of all types of borings and boring apparatus, see "Subsurface Exploration and the Sampling of Soils for Civil Engineering Purposes," the final report of the American Society of Civil Engineers, by M. Juul Hvorslev.

FIG. 1.1. Vulcan No. 1 single-acting steam hammer driving Monotube piles. (*Courtesy of Union Metal Mfg. Co.*)

formula results. Before investigating static-friction values, the amount allowable for end bearing should be deducted.

Test Piles

Study of the above considerations will be greatly facilitated by data obtained from continuous records of test-pile driving and by loading and pulling tests, if possible.

The history of a soil deposit is made up of interactions of many geologic processes and physical forces under which it has come to equilibrium and includes weight of present or past overburden, capillary pressure, ground-water levels, hydrostatic pressure, permeability, void ratio, and grain shape and structure. Forcing a pile into the soil, whether or not accompanied by displacement, jetting, or vibration, causes a change in these stress conditions. Renewal of conditions similar to the original may take place quickly, as in the case of some well-graded cohesionless materials, or it may take a long time. The temporary conditions may be worse than conditions will become later, or they may be better.

Driving several test piles on the site, adjacent to boring locations, so that driving records may be studied in conjunction with boring data, is most desirable. Some piles should be driven to a deeper tip grade than appears probable for the building piles, to determine load-carrying capacity below expected tip grades.

Redriving of several test piles after a period of time will provide valuable information as to the increase or decrease in resistance. Such

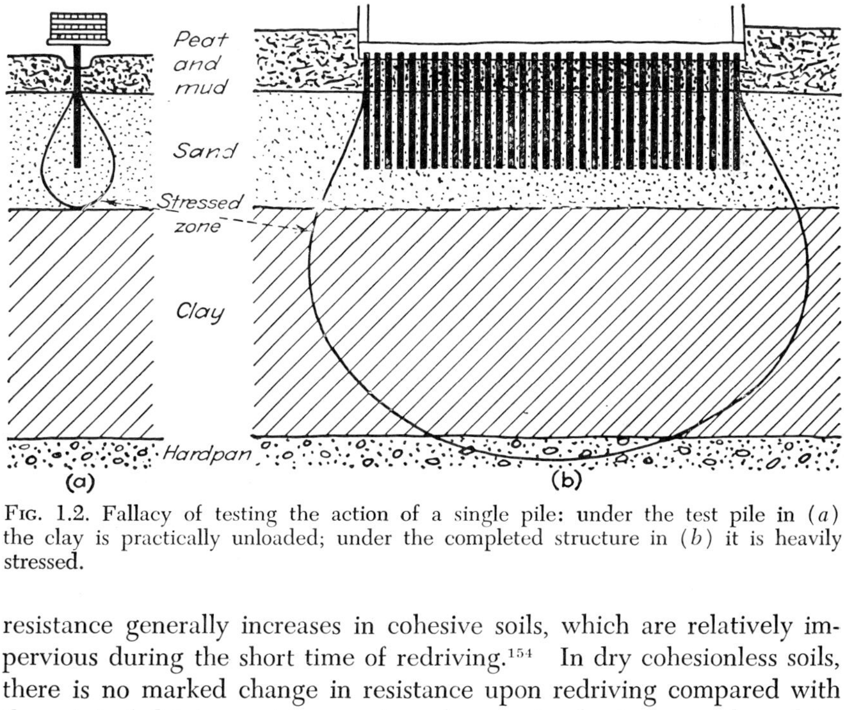

FIG. 1.2. Fallacy of testing the action of a single pile: under the test pile in (*a*) the clay is practically unloaded; under the completed structure in (*b*) it is heavily stressed.

resistance generally increases in cohesive soils, which are relatively impervious during the short time of redriving.[154] In dry cohesionless soils, there is no marked change in resistance upon redriving compared with the original driving resistance. In submerged cohesionless soils, redriving resistance depends upon grain size and density as follows: (*a*) if the soil is dense, either fine or coarse-grained, a considerable decrease;[25,154] (*b*) if the soil is loose and coarse-grained, little change from the original; and (*c*) if the soil is loose and fine-grained (silts and very fine sand), showing small resistance to original driving, a marked increase in the number of blows, for a few blows, until the increase in pore-water pressure, which results in a quicksand effect, is again created.

It has been a general conception that the short, simple *Engineering News* formula quickly gave the bearing value for each and all piles under a structure, but that as an added precaution, sometimes taken by ultraconservative engineers, the driving of one or more single test piles would give a safe bearing value per pile. Modern concepts of soil mechanics have changed the picture and, although test piles have their value, the interpretations to be drawn from the results may bear no relation to those which might formerly have been drawn. The assumption that the action of an individual pile under load will indicate the behavior of large groups cannot be defended; in many instances the settlement of the structure as a whole bears no relation whatever to the movement of an individual pile when tested under working load[17] (Fig. 1.2).

The capacity of a single pile—involving its ability to transmit load to the soil in friction and end bearing, and the carrying value of the strata when affected by the stressed zone formed in the soil by a single pile—are, however, determinable by load tests.[92,96] This information has its value and a definite place in the design of pile foundations. It does not complete the design, according to modern concepts, as it did in the past.

Pile Test Loadings

Information Obtainable from Test Loadings. Test loadings carried to failure, if possible, will provide friction values, including end resistance. One of the most important factors conducive to misleading results from the test loading of a single pile is disregard of the time of testing. Common practice has been to test-load single piles to 150 to 200 per cent of the design load, leaving the load on the pile 24 to 48 hr. If the settlement has not exceeded a specified maximum (often 0.01 in. per ton of applied load), the test has been considered satisfactory. Such results may vary widely from the final value of a single pile, since great changes often occur in the soils after pile driving and before or after the load tests, in accordance with the type of soil.

Effect of Type of Soil. If piles are driven in a coarse-grained saturated pervious soil where losses in resistance may reach over 40 per cent during the 24 hr following driving,[25] the load tests should not be made until several days have elapsed.

If piles are driven in some submerged, uniform, fine-grained sands which are so loose that the jarring from pile driving makes the sand temporarily quick the piles will drive easily, but since the condition ceases after the blow, the stratum will be capable of supporting a much larger static load than indicated by the driving resistance. A load test will determine the true capacity.

If piles have been driven in an impervious type of soil which heaves during driving, in which case the setup may amount to several hundred per cent in the weeks following, it would seem advisable to defer the load tests until results representative of the permanent capacity of the soil could be obtained. If tests on various piles are made at different time intervals after driving, results may not be comparative unless experience indicates the length of time required for the temporary soil-distortion effects to dissipate.

Pulling Tests. Pulling several piles will also determine friction values, without end resistance, and the end resistance may be determined by comparing the total pulling force with a test loading.

Test Locations. To permit ordering piles to proper lengths, the borings and test piles should be located in positions that will cover the site and permit the drawing of contours or cross sections of the strata.

Transfer of Load to Soil. The distribution of the transfer of load from the pile to the soil and pile tip can easily be checked in the cases of precast concrete piles, hollow piles such as hollow concrete, pipe, and Union Metal Monotube shells, and even in H piles, by means of measurements of downward movement, under the test loads, at the tip, quarter points, and mid-point of a pile, or of several piles in a group. This can be accomplished in the case of precast concrete piles by casting closed-end pipes into the concrete, in the case of hollow piles by inserting closed-end pipes, and in the case of H piles by welding the pipes to the inner portion of the pile, with the bottoms of the pipes located at the various heights at which the movement is to be measured. Rods can be inserted in these pipes and the elevations of the tops of the rods taken. The load test should be applied to the pile so as to leave the rods free standing. This has been done in the case of a single rod in a pipe to the tip of a precast concrete pile, by Boonstra and Franx in the Netherlands,[13,14] who found that an appreciable load was required before any load reached the tip of the pile. Hansen and Kneas[23] made tests using rods set in pipes embedded in or attached to Raymond, Monotube, pipe, and H piles, to various points of their heights, and with rods set in pipes in the ground outside the piles. These tests confirm the theoretical conclusions given later regarding rate and character of distribution of load from pile to soil. This method has also been suggested by Solomon.[15] Carlson and Baldwin strain meters have been installed in concrete and Monotube piles and axial-stress readings at various heights taken without difficulty.[*,119,124]

By recording the settlements of the pile at the tip and intermediate points as well as in the usual manner at the head, the elastic shortening of the pile throughout its length may be obtained by subtraction, thus enabling determination of the distribution of the load to the various strata.

[*] *Bearing Pile Investigation for Sepulveda Dam,* U.S. Engineer Office, Los Angeles, Calif., April, 1940.

End-bearing capacity might be tested separately from friction-load-bearing capacity by driving a closed-end pipe pile inside a pipe shell, or a pipe shell and core with a common plate shoe in the manner of a MacArthur pile, and load-testing the core. The outer shell could be pulled to give frictional resistance alone. A closed-end-pile load test would give combined end-bearing and frictional resistance. By selection from these methods, the desired separation of values could be obtained. The time element involved in redistribution of the load to the different strata may also be studied if the test is left on as long as it should be.

A test arrangement for determining the settlement of a pile under load has been used as shown in Fig. 1.3, consisting of an access shaft and tunnel extending under the pile points, where a recording level arm is stationed.[5f] This permits movement of the pile tip to be measured, as well as settlement at the head, under loads. By comparing the results of this test, which permits determining friction values alone, with corresponding tests on a pile having tip bearing, the separate values of friction and end bearing can be determined. This type of test is usable only in soil that permits safe tunneling, such as that in Los Angeles, in which pile holes are customarily bored.

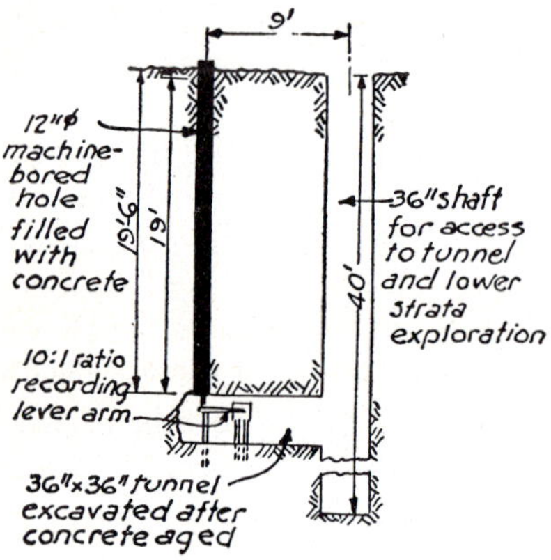

Fig. 1.3. Test arrangement for determining settlement of cast-in-place concrete piles under load.

The settlement of the pile head is made up of pile shortening, plus soil consolidation adjacent to the pile as the soil is dragged down by the pile, plus movement of the pile through the soil in case the shearing value of the soil is exceeded. Sometimes the friction grip between soil and pile exceeds the soil shearing value. Shearing values of soils may be tested in the laboratory on "undisturbed" samples, although compaction due to the pile may have increased the actual value.

Load distribution is complex, but a method of analysis follows. Results depend upon character of strata, magnitude of test-load increments, time interval between application of increment loadings, and rapidity of application of loads. The time-settlement relationship would be different for a 10-ton increment applied over a period of 60 min, from the same load applied instantly by a jack.

Stress distribution along piles may be studied by considering the diagrams in Fig. 1.4, according to Solomon.[15]

Stress in the soil quickly increases from zero at the surface to a fairly constant value in homogeneous soil, the shearing value of the soil being the limiting value, until sufficient resistance has been mobilized in an upper section of pile length denoted as l_1. The pile shortening is denoted by Δ_1. Increasing the load to $R_1 + R_2$ reduces the value of l_1 by $\Delta + \Delta_{12}$ and also causes a new hydrodynamic-stress condition in the soil, which adjusts itself with the escape of entrained water in the course of hours or days. In order to reach equilibrium again, soil in the length l_2 will also be stressed. The length l_2 will be shortened by Δ_2, and l_1 will be shortened an additional amount Δ_{21}. The unit stress p_1 will be increased to p_2. This process will be continued under successive increments of loading, until load reaches the tip.

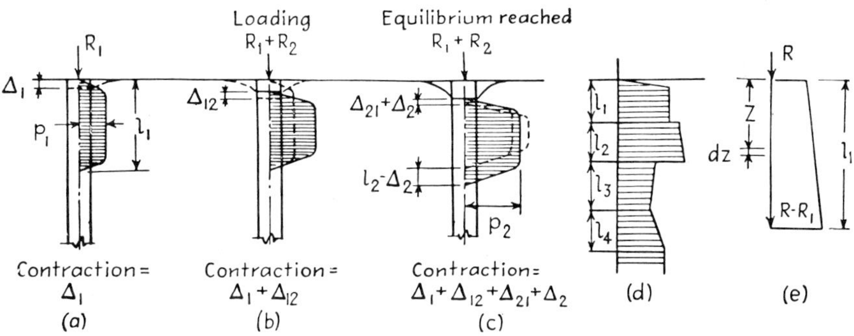

FIG. 1.4. Stress distribution along height of cylindrical piles.

To check the load distribution, the test loads should be removed when about one-third to one-half the failure load, and just before test loads have reached the tip of the pile.

Time Factor in Tests. When removing test loads, it is desirable to continue settlement readings, for at least as long a period as the loading time. The data thus obtained may, in certain types of soils, be valuable in determining relative elasticities of pile and soil.

While short-time test loads on piles may not permit the loads time to consolidate upper layers sufficiently to reach the intended lower bearing strata for permanent loads, such short-time tests may simulate certain live loading conditions, such as for train, wind, or impact loads, and thus be of much more use than for determination of permanent load values.

Testing Proper Strata. It is important that the stratum tested be the one which has been selected for permanent load-carrying purposes. In a recent series of load tests, made on long piles driven through firm clay, then softer clays and silts, and a sufficient number of feet into firm sand to remove the load in end bearing and friction in the sand, it was observed that test loads of approximately the same magnitude as the net driving resistance to the hammer blow caused only about one-third as much elastic compression, measured at the pile cap, as was computed by the formula Pl/AE, even using l only to the assumed center of driving resistance. The load was left in place for several hours and when no further measurable settlement occurred in such a short period, the pile was considered to have been safely tested to the applied load. Actually, what had happened was that the upper, firm clay stratum had been tested for its capacity to carry the load for a few hours, and the intended bearing stratum had not been tested at all. The clay and silts would not permit the escape of the entrained water, in a few hours, at a measurable rate, after the initial settlement, but over a period of years considerable settlement would have occurred, as the load would have been transferred gradually to the pile as a column, and thence to the sand stratum. In this case, the load test tested the stratum which the piles were intended to by-pass, and the test-load results had no relationship to the dynamic-driving results or to the actual load-carrying capacity of the pile. The tests were of value for transient loads but would not have been of value for the dead-load portions of the loads. This illustrates the study which must be given to planning, making, and interpreting load tests.

Group Reduction Value. When selecting a working value for pile loads, it should be remembered that piles in groups may not, unless they are strictly end bearing, have the full value of the sum of the individual pile values multiplied by the number of piles in the group. This may be visualized by considering the bulb of pressure developed around and below a single pile and the effects of adding other similar bulbs that overlap. This subject is discussed in Chap. 6.

It would be desirable, particularly in the case of a large pile-driving operation for an important structure to be located upon friction-load-carrying strata, to test a typical pile group. Length of pile should be provided to give several feet of projection above the ground, and strain-gage readings taken on each pile.

Value of Load Tests. The information from load tests will provide information to enable a sound engineering decision to be made as to the foundation design. The usual type of load test, using only a single pile, in which only the head settlement is measured under fairly rapid application of loads, partial unloadings are not performed, the total load is left on the pile an insufficient length of time, and the load is removed quickly instead of by increments over a period of time corresponding to the loading time, provides such doubtful information that a correct interpretation is most difficult if not entirely impossible.

An adequate load test provides the means of determination of the correct lengths of pile to order. A brief description of typical pile test loading and pulling apparatus is given in Chap. 15 and in Appendix VI.

Uses of Piles

Piles are most commonly effective in one of four ways: first, transferring the load through soft upper strata to end bearing on a hard substratum; second, as friction piles in their lower portions in transferring the load through soft upper strata into stiffer strata below, which are considered to have adequate distributing value; third, as pure friction piles for their full length; and fourth, occasionally in compacting the soil (Fig. 1.5).

A special case of the use of piles as columns is to transfer loads through soils which possibly may be eroded or scoured away but which otherwise

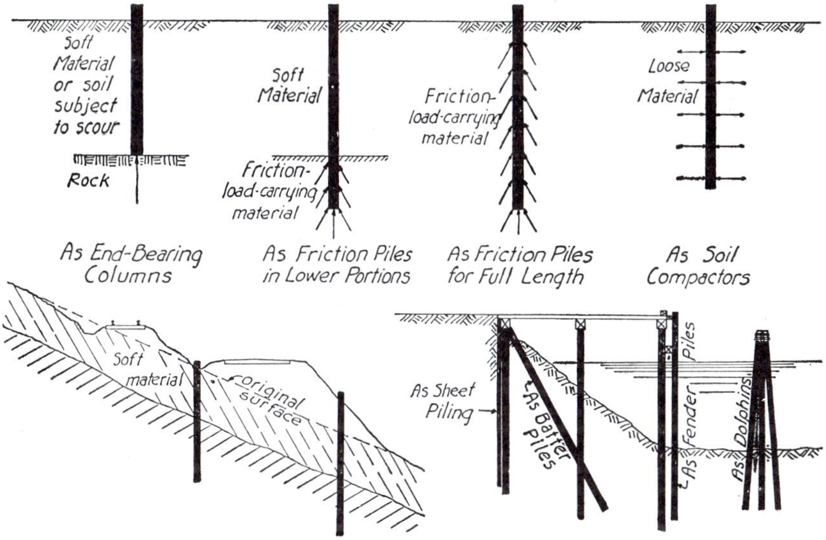

FIG. 1.5. Uses of piles.

would be suitable foundation strata, to rock or to other strata not subject to erosion or scour.

Piles may also be used to stabilize sliding strata in banks (Fig. 1.5), but in such cases they should be of a drilled or other type which will not cause a slip by jarring the bank beyond the point of stability.

Piles are also used for anchors against horizontal forces; as means of assisting in stabilizing structures with regard to the earth against earthquake forces; as fenders for wharfs, dolphins, etc.; and to resist uplift or lateral forces.

Bearing Capacity of Pile Foundations

Factors Involved. The design of pile foundations falls into two general fields: first, the selection and design of the piles and the selection of the driving equipment to be used; and second, the study of the soils to which the load is transmitted by the piles, by the principles of soil mechanics.

The briefest possible discussion of the second of these fields, dealing with the general principles of the bearing capacities of pile foundations as dependent upon the principles of soil mechanics, will be given since a detailed study of this matter falls outside the scope of this book. The subject is covered to this extent, however, in order to place the specific problem of design and driving in its proper perspective when considered as merely one phase of the much broader subject of the ultimate bearing power of foundations involving piles.

Piles in End Bearing. The first use of piles is in end bearing on bedrock or other firm stratum capable of carrying the additional load caused by the building without objectionable settlement. The building load is then considered as applied at the level of the pile tips, and consideration should be given to the effects on the underlying strata, if compressible, from a load at this level. The pressure distribution on soil below the tip of a single pile may be assumed to take the form shown in Fig. 1.6*a*, as computed by means of the Boussinesq equations,* the curved lines representing points of equal intensity of vertical pressure. The general method of plotting these lines is described below.

Although it is true that overlapping of bulbs of pressure from adjacent piles indicates that the intensity of stress in the soil is increased in those areas, the only damaging overlaps are those resulting in additive stresses capable of producing damaging deformations. If the inner bulbs in Fig. 1.6*b* represent lines of 10 per cent intensity and the outer bulbs 1 per cent, it is evident that an overlapping of the outer bulbs is of no practical moment. A spacing of 3 to 3½ pile diameters is usually sufficient to avoid detrimental overlaps from adjacent piles which might cause local settlements. The merging of the individual bulbs of pressure into one large one under a group of piles, as shown in Fig. 1.6*c*, indicates stresses carried into deeper underlying strata, however, and these must be investigated for capacity to carry the load, regardless of whether the individual bulbs overlap. This theory of the separation of bulbs of pressure may be further extended to apply to the bulbs under groups, as shown in Fig. 1.6*d*, in which the load applied through the hard stratum to the softer strata below is spread by battering all piles under the substructure mat.

Combined End-bearing and Friction Piles. The second use of piles, in carrying load through overlying strata not considered capable of supporting the additional building load to firmer adequate strata

* For these equations see any modern text on soil mechanics. Tables of values and graphs are also available.

in which the lower portions of the piles act in friction, involves consideration of the probable friction values in the lower material and a study of the effects on the underlying strata from a load applied at these firmer strata. This condition is a combination of the first and third uses, and the conditions down from the top of the actual friction-load-carrying

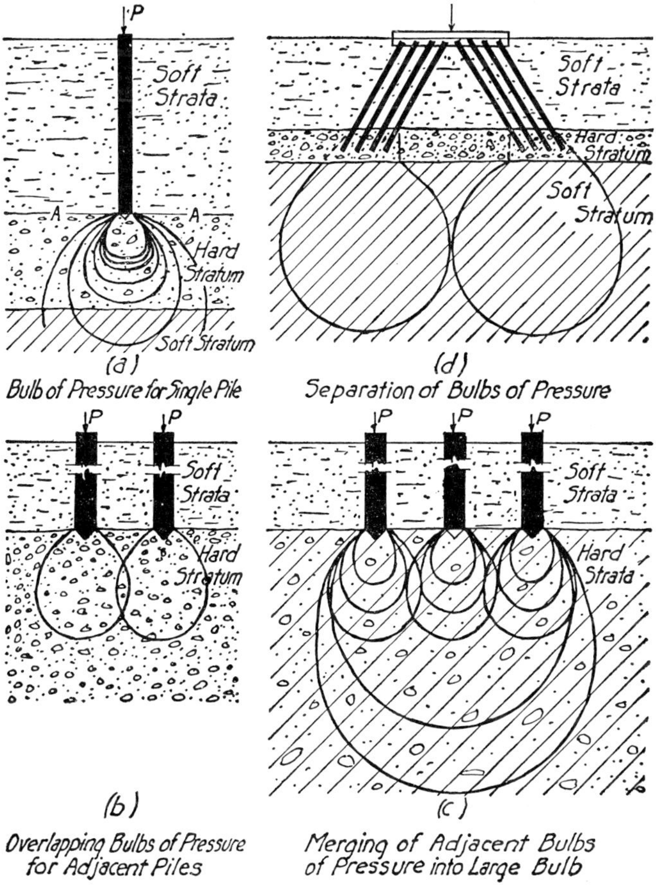

FIG. 1.6. End-bearing-pile pressure distributions in soil.

strata are similar to those described below under the third use of piles, providing due account is taken of the weight of the non-load-carrying overburden.

Friction Piles. The third use of piles, as pure friction piles for their entire length, driven into deep strata of fairly uniform consistency, also involves consideration of safe friction values and a study of the probable consolidation effects in friction-load-carrying strata and in strata below pile tips or bulbs of pressure. Pressure distribution in the soil varies widely between single piles and groups of piles, and also with the shape of structure and relation between its width and the length of piles. Figures 1.7*a*, 1.7*b*, and 1.7*c*, from an article by Morrison,[16] show the type

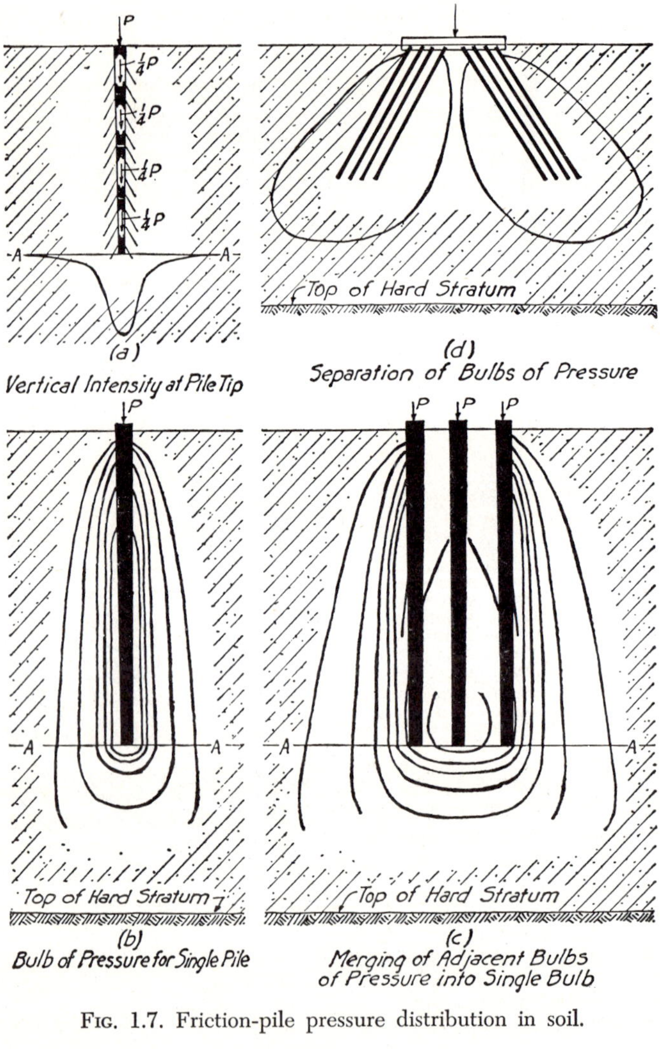

FIG. 1.7. Friction-pile pressure distribution in soil.

of soil action, as computed from the Boussinesq equations. In Fig. 1.7*a*, one-quarter of the total load is taken as acting on the soil at each section, and vertical intensity of pressure at the pile tip is shown. By computing these intensities at different levels and drawing lines through points of equal pressure, a typical bulb of pressure for a single pile is obtained, as shown in Fig. 1.7*b*. When a group of piles are driven in a footing, bulbs of pressure overlap and are blended into a bulb for the entire group, as shown in Fig. 1.7*c*. If the piles are all loaded as fully as a single pile, this results in a much higher intensity of loading on plane *A-A* than would be the case if the pile spacing were such that the bulbs did not merge. In the group shown in Fig. 1.7*c*, each pile should carry one-third of the load, although the middle pile will settle more, and the group will settle more than will an equally loaded single pile; but if the footing is rigid, more load is thrown on the outer piles, thus accelerating settlement.

Use of friction piles in the upper part of a very deep deposit of fairly uniform consistency is for the purpose of reducing the intensity of pressure acting at the ground level and shifting the zone of maximum stress to the lower portions, where less settlement will be caused, and is governed by two basic principles: Settlement produced by a uniform load increases in proportion to the diameter of the loaded area for cohesive soils, whereas in cohesionless soils it is more nearly independent of size. Settlement under a unit load decreases with increasing depth of foundation. However, in addition to depending upon the depth of the foundation, settlement also depends upon the ratio of depth to diameter of the loaded area, so that for equal potential settlement reductions, the depth-to-diameter ratios should be kept equal. In cohesionless materials the effect of this ratio on settlement is less than in more cohesive soils. This principle indicates that the value of the piles may be greatly affected by the relation of their lengths to the width of the loaded area. Under a narrow structure, every effort should be made to keep the piles longer than the width of the structure, so that a lowered bulb of pressure will occur. This will reduce settlements. Figure 1.8, from an article by Terzaghi, in which the pressure distributions were computed by the Boussinesq equations, illustrates the two conditions. In Fig. 1.8, loading *A*, in which the building foundation is narrower than the length of the piles, the piles are seen to be effective by means of enlarging the bulb of pressure and lowering its location very markedly. In Fig. 1.8, loading *B*, in which the building foundation is several times the pile length, the top portion of the bulb of pressure is lowered just as much as in the base of loading *A*, but owing to the larger dimensions of the bulb, soil pressures in the strata in which the lower portions of the bulb occur are changed little or not at all. Therefore, in the case of loading *B*, piles might be of little or no value in carrying the load or reducing settlements, unless the soil immediately beneath the footing is unsatisfactory. In fact, they might even be detrimental, because of remolding of the soil. In the case of an individual pile, the bulb of pressure spreads the load out over a large area, thus reducing soil pressure; in loading *A*, a great deal of this same effect still remains; in loading *B*, the piles spread the load out very little, and the effect of the pile foundation on the soil is practically the same as that of a raft foundation without piles. In this last case, the total bearing value of the piles in the foundation bears no relation to the carrying value of an individual pile by itself.

It is possible to reduce loading intensity from friction piles by driving the piles on batters, as shown in Fig. 1.7*d*, creating separate bulbs of pressure instead of a single bulb having greater stress intensities. Bulbs can be allowed to overlap somewhat, if the overlapping portions are of

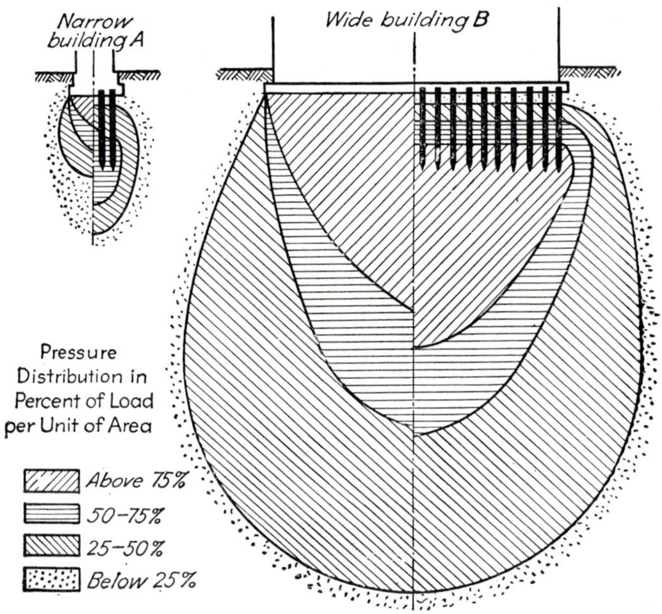

FIG. 1.8. Effect of relation between foundation width and pile length on pressure distribution.

low stress intensities. This method is useful where the soil is incapable of supporting the intensity of load resulting from use of vertical piles but has sufficient cohesion and friction to prevent the piles from bending out of axis. This scheme has been in general use in Denmark where this type of soil condition is prevalent.[33]

Supporting forces on H piles are shown in Fig 1.9.[11a] The effect of boxing an upper section in a cohesionless stratum is shown in Fig. 1.9*c*.

For piles depending upon friction for their support, displacement piles and tapered piles are generally more effective for the same length of embedment in the strata selected for carrying the load, or shorter piles may be used. However, taper in portions of piles passing through inadequate upper strata is not desirable if too great a portion of the driving energy is resisted in such material. Sometimes lower portions are tapered and upper portions straight-sided.

Piles as Soil Compactors. The fourth use of piles is as compactors of the soil. It may be found more economical to compact a soft layer above a fairly firm stratum by driving pile tips down only to the firm stratum than to drive the piles into the firm material to support the load in friction. In this case, piles would be driven in rows at large intervals,

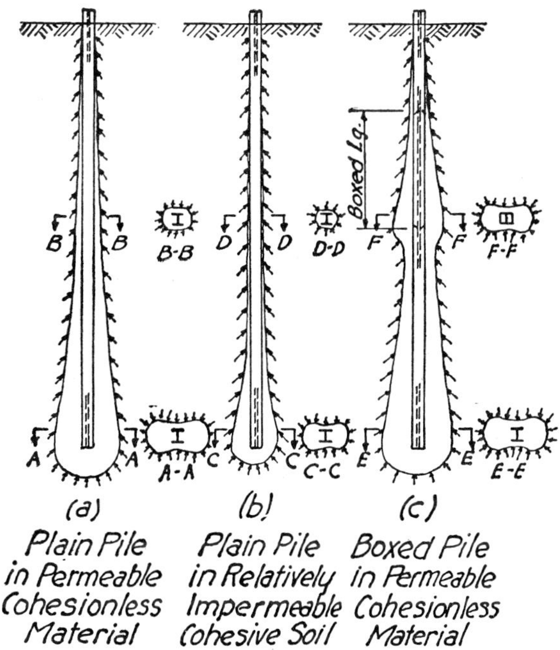

FIG. 1.9. Diagrams of supporting forces and zones of compacted soil for H piles.

and intermediate piles driven at closer and closer intervals until the desired degree of driving resistance would be obtained. It is almost impossible to consolidate clay soils by compaction in this manner. For soils which can be compacted but yet have some degree of moldability, care may be required to avoid driving so many piles that the ground becomes remolded to such an extent that final bearing capacity is reduced rather than increased.

Piles are also used for some other purposes. They may serve to resist uplift in foundations of structures having overturning moments, such as transmission or radio towers. They are sometimes relied upon to resist lateral forces, although they must be used with caution and batter piles are more efficient. They also find use in marine structures, such as dolphins and wharf fenders.

When to Use Piles. *Unless piles serve one of these purposes, they should not be used.* It has been a common belief that piles were never harmful, but this is not the case. Piles may be without value in some instances, and under some conditions their use may be actually very harmful. For example, in the case of a fairly thick layer of reasonably firm clay over a deep bed of soft clay, soil studies might show the firm clay to have adequate carrying value, and it could act as a distributing mat over the soft clay to spread the load, whereas the driving of piles might destroy this advantage of spreading the load and thus decreasing unit pressure on the soft material, with the result that increased settlements would be experienced in the soft underlying stratum, and total settlement might be further aggravated by a kneading or remolding of the firm clay above by the action of pile driving.

Soil Bearing

Every stratum must support the load from all soils above, including the weight of any structures they carry. Therefore, a safe pile foundation is not automatically ensured if satisfactory driving resistance and friction values are secured, even in strata which are satisfactory in themselves for this purpose. The capability of the strata underlying the pile tips to support additional loads without detrimental failure must be investigated. This involves a consideration or study of the degree of preconsolidation of the strata to be subjected to additional load. A decision must be reached that the settlements from underlying strata will be permissible, or else the piles should be extended to reach acceptable firm support. Occasionally it is possible to compensate, to a sufficient degree for building load added, by excavation of present earth load, providing basement space.

In order to obtain the distributing value from a firm stratum in which the pile tips rest, when located above a soft stratum through which it is not feasible to carry the piles but upon which it has been decided to float the entire load, pile tips should be specified not to punch through the distributing stratum but to be stopped at as high a distance from the underside of this stratum as permitted by friction and driving-resistance considerations. In no case should such pile tips approach closer than 3 ft, and preferably a minimum of 5 ft, to the underside of the distributing layer. Punching and shear values should be investigated.

Average soil pressures at or below the level of individual pile points, under any group of piles or under the piles in each entire footing, and under the entire structure, should in no case exceed the bearing value of the soil, regardless of the values obtained for individual piles by the pile-driving formula or by applying a probable square-foot unit skin-friction capacity to the embedded length of pile to obtain a total value. When computing average pressure under an entire group of piles or structure, the bounding area of the group or structure often may be increased by projecting each side a distance, depending on amount of friction embedment of the load-carrying friction length of the pile. The Boussinesq equations may often be used to develop the stresses in the bulbs of pressure. Stresses at and below pile tips have been studied by this method.[15]

It is possible to reduce the intensity of loading on a deep soft stratum underlying a pile-supporting good distributing stratum below an upper soft stratum by battering the piles, as shown in Fig. 1.6*d*. Use of batter piles permits separation of the portions of the bulbs of pressure having sufficient intensity to be objectionable when overlapping. The firm intermediate stratum should be one which is capable of resisting the horizontal reactions from the piles.

The Boussinesq equations strictly apply only in the case of a vertical point load acting on the horizontal surface of a semi-infinite solid. With increasing depth below the surface, Mindlin's equations, which give stresses due to a force acting at a point in the interior of a semi-infinite solid, approach more nearly to the case.* Ruderman's solution covers the case of a friction pile, giving the intensity of stresses in the ground due to shearing resistance along the pile.† Frequently both friction and end bearing provide support, however, and a combination of these solutions is needed. The solutions all result in the typical bulb of pressure, although the magnitude of the values would be affected by use of the incorrect solution. These solutions are quite complicated, and it is not expected that they will be solved numerically by the pile engineer for the usual cases of pile driving. They are described and illustrated because an understanding of the types of action occurring in soils under various conditions of soils, pile lengths, spacings, and groupings is necessary to the proper design of pile foundations. Should it be thought necessary to solve such problems numerically, in conjunction with studies of possible future total or differential settlements, the matter becomes one where the services of a soil-mechanics expert are desirable.

Objects to Be Accomplished by Use of Pile Formulas

Experience and judgment are of great value in pile-driving operations, and any pile-driving formula should be looked upon as a tool in aiding

* R. D. Mindlin, "Force at a Point in the Interior of a Semi-infinite Solid," *Physics*, vol. 7, No. 5, pp. 195–202, May, 1936.

† J. Ruderman, *Stress Distribution around a Loaded Pile*, Thesis No. 500, Department of Civil Engineering, Columbia University.

judgment. Pile-driving experience and developed judgment are the principal considerations in constructing pile foundations. The experience of one person is a plant of slow growth. Many engineers do not have the opportunity to develop it and yet are faced with the responsibility for a pile foundation. Principles and data in this book should be applicable to any driven pile-foundation problem, even if experience is lacking, thus permitting the engineer to form an opinion of his own. This book is based upon experience of many engineers, put in usable form by other engineers, and the theoretical data in connection with the dynamic formulas have been checked by repeated field observations.[96] (See Appendix IV for typical examples of driving with various types and sizes of hammers and piles and comparisons between load tests and computed pile capacities.)

The use of the procedure outlined is intended to promote four principal objects, namely: *uniformity of driving results* with different kinds of piles and types of hammers; *economies* of pile lengths due to such uniformities by avoiding excessive driving in some cases; *avoidance of failure* to drive the piles to an adequate depth into the bearing strata to obtain proper permanent safe bearing values on the soil; and guarding against obtaining *apparent friction values,* in the case of friction-load-carrying or uplift piles, which are in excess of the actual values.

Historical Background of Pile-driving Formulas

Types of Formulas. The pile formulas which have been proposed fall into four general types: *empirical formulas; static formulas,* which equate the soil resistance (both end bearing and skin friction) to the bearing value; *dynamic formulas,* which equate the soil resistance to the energy of the blow, thus obtaining a bearing value; and formulas based upon the *theory of longitudinal impact* on a rod. These types of formulas will be discussed in the following paragraphs.

Empirical Formulas. Various empirical formulas have been proposed, usually based on the results of tests for limited conditions. Such is the Wilcoxen formula. There are other empirical formulas based on the strengths of posts, but these have no relation to driving conditions; they include the formulas of Sankey, Adams, Brereton, and one by Rankine. These formulas are now little, if at all, in use.

Static Formulas. Many static formulas have been proposed, and some have been quite widely used. Formulas such as those of Vierendeel, Patton, Griffith, and Howe, are based on Rankine's theory of conjugate stresses in the earth. Another more empirical approach to a static formula is represented by those formulas, such as Arrol's and another by Patton, which add the tip and frictional resistances to obtain a bearing capacity. This latter method appears to be more useful, in the light of present-day knowledge of soil mechanics, and is discussed more fully hereafter.

Dynamic Formulas. A great number of dynamic pile-driving formulas have been proposed and used. Their basic assumption is that the ultimate carrying capacity is equal to the dynamic driving force, and the principle behind them is that the weight of the ram multiplied by the stroke may be equated to the driving resistance multiplied by the set* of the tip of the pile. The simplest dynamic formula, that proposed by Major Sanders in 1851, is exactly this equation, with a factor of safety of 8. Merriman proposed the same formula, with a factor of safety of 6, in his *American Civil Engineers' Pocket Book.* Goodrich[2] proposed a factor of safety of 3.6 for wood piles driven by a drop hammer with a fall of 15 ft and a penetration of 1 in.

The next most simple group consists of dynamic formulas which contain a fixed coefficient, designed to compensate to some degree for the factors present but not expressed by variable terms in the formulas. To this group belong the *Engineering News,* Wellington, Vulcan, and Bureau of Yards and Docks (not now used by the Navy) formulas.

Another group of formulas attempts to cover the variables by using expressions for efficiency of applied energy by including the relative weights of the pile and hammer. The Dutch, Ritter, and Benabencq formulas are of this type.

There are other similar formulas, which endeavor to compensate for the variables by using both fixed coefficients and expressions for the relative weights of the pile and ram, such as the Eytelwein and Navy-McKay (not now used by the Navy) formulas.

The next group of these formulas contains either all or part of a series of terms designed to represent impact loss and various elastic losses in the driving cap, pile, and soil during driving. Formulas of this class are one by Rankine and those proposed by Redtenbacher,[3] Hiley,[7] and Schenk.[1m]

The pile-loading formula which has been in most general use in the past in this country is the so-called *Engineering News* formula,[1] while Redtenbacher's formula[3] has been used quite extensively in Europe. A history and comparison of formulas in use in this country and abroad, such as the Eytelwein, Navy-McKay, Bureau of Yards and Docks, Dutch, Redtenbacher and Rankine formulas, has been done by various writers,[4,6,8,1c,5a] although in Appendix I their interrelationship is briefly traced as a matter of informative and historical interest. Some of these formulas have the advantage of simplicity but neglect entirely many

* The term "set" is used, as in British practice, to define the net penetration per blow, thus avoiding confusion with the word "penetration" as used when referring to the total depth of embedment of the pile in the ground.

widely variable factors and use fixed coefficients for other variables. The factors of safety actually obtained by the use of some of these formulas may vary widely, often being much greater or much less than advisable because of the attempt to provide a very short formula which will cover a wide range of conditions with a factor of safety adequate to fit the worst cases. By comparison with formula (2.1), the factor of safety obtained by use of the *Engineering News* formula may vary from as low as ½ to as high as 16 or more. Unduly high capacities are indicated with small sets in particular. This also results in the phenomenon of obtaining widely varying results when using some of these formulas with different types or sizes of hammers on the same kind of piles. It is desirable to design piling to a more uniform factor of safety and to a more uniform length, and this can be done in a simple manner.

Qualifications to the use of the *Engineering News* formula accompanied it when it was proposed by Wellington.[1] He stated that the formula was offered "for *safe working loads for piles under all ordinary conditions, to be reduced under exceptional conditions* [as notably with irregular penetration] *but never exceeded* unless the pile is known to rest on rock, and act as a column. [The formula is] assumed to be sensible at an approximately uniform rate." If these qualifications were borne in mind, together with the fact that it was designed for driving wood piles, which are necessarily light in weight, and that double-acting hammers and steel and concrete piles were unknown in those days, many of the unduly high results now obtained by the use of this formula would not occur. The qualifications indicate that the formula was intended for use with friction piles, rather than with those fetching up sharply to small sets. Appendix IV contains examples of results of the use of this formula for all types of piles and magnitudes of sets.

The *Engineering News* formula was derived to be used with drop hammers. In those days rams of two to three thousand pounds were usually dropped 20 or 30 ft. This resulted in large sets or damaged piles. This formula included as the last term in the denominator a figure of 1.0 in the case of drop hammers, which was stated by its author to represent the excess of initial friction, caused by settling of the earth around and into irregularities of the pile since the last blow, that must be overcome during the first inch of the penetration due to each blow, over the assumed uniform friction acting during the rest of the blow. When using a single-acting hammer, the author of the formula reduced this term from 1.0 to 0.1, on account of the shorter interval between blows which presumably allowed less chance for setup between the soil and the pile.[1] Extensive tests using the two types of hammer have been made by the Texas Highway Department, who state that the inherent inaccuracy of these formulas was easily demonstrated by discrepancy of the results. It was invariably found that the steam-hammer formula gave a higher apparent bearing value than the drop-hammer formula, a difference which became more pronounced with increased driving resistance, until the steam-hammer values became five or six times as great as the drop-hammer figures. The *Engineering News* formula has also been observed to give different values of driving resistance when using different drops, which evidently cannot all be correct.

In the more comprehensive formulas discussed below, the term in the formulas following the value of the set s represents the sum of the average elastic compressions of the pile cap, pile, and ground. Observations as well as computations indicate that these values vary enormously with the size of hammer and material, area and length of pile. The concept of this term was different in the *Engineering News* formula, where, being a constant, it does not serve to replace the several variables in the comprehensive formulas.

A dynamic pile-driving formula, which will adequately reconcile the varying factors introduced by the wide range of hammers and piles now used, may be obtained by taking advantage of the fact that the temporary elastic compression can be observed and that the driving resistance must be a definite amount to cause this compression, thus utilizing the spring of the pile itself as a gage of carrying capacity.

The dynamic pile-driving theories are based on assuming instant propagation of force through the pile, whereas the force actually travels in a wave. The amount of side friction and proportion of plastic to elastic yield of the soil have a great effect on these waves. In order to apply the theory, it is necessary to assume all motion at the top as elastic, and this is done by assuming that the equivalent elastic movement equals the observed elastic movement plus twice the plastic movement.

Formulas (2.1*a*) and (2.1*b*) were suggested by the work of Hiley,[7] and are stated in a form which will permit accomplishment of the object stated in the preceding paragraph. The basic theory behind these formulas is contained in several much older pile-loading formulas, certain of which include some or most of the principal energy losses caused by elastic compressions and impact. Redtenbacher, Rankine, and Weisbach assumed perfectly elastic piles, and Eytelwein assumed perfect inelasticity. Krapf (1906) seems to have made the assumption of imperfect elasticity in the pile for the first time, and Stern developed his formula based upon this same assumption.[79] Kafka, in Vienna, seems to have been the first to propose a theoretical dynamic formula that considered the elasticity of the soil as well as of the pile.[80,81] The more complete of these formulas are often not in convenient form, and the necessary data and information to permit their use have generally not been readily available. The Hiley type of pile-driving formula is in common use in England and is coming into more frequent use in this country, having already appeared in one form or another in several building codes. In such cases, however, certain variables generally have been either omitted or given in approximate form, without sufficient information to enable the engineer to obtain a full understanding of the theory. Of course, it is not within the province of such codes to explain the theory behind their stated requirements. However, lack of ready availability of this information from other sources is a serious handicap to the engineer.

Wave Equations. De St. Venant and Boussinesq developed a theory of end impact on rods, which has been found to give stresses that compared well with test results for concrete piles.[21] The general wave equation for bearing value on a rod is, when complicated by actions of the ram, cap block, pile, and ground, too complex for manual solution. Electronic digital computers can produce solutions for end-bearing or frictional resistances.[121] Computer programming may not be practicable for small jobs, and use of charts covering different combinations of variables may be necessary.

Results from Hiley-type dynamic formulas agree fairly well with those from wave equations, except that the former may tend to underestimate the capacities of long, heavy piles and mandrels. When a short pile is hit, the entire length of pile and ground are in compression at the moment of maximum compression, as assumed in the Hiley formula; when a long, heavy pile is hit, part of the pile and ground are in tension because of return waves when the cap block and the upper part of the pile are in compression. This indicates that less energy is being absorbed by the pile and ground at this instant than assumed by dynamic formulas and that more energy is available for penetration. For long, heavy piles and mandrels, investigation of capacities by use of wave-equation charts, if available, is desirable for comparison with results of dynamic and static formulas and load tests.

Development of Suggested Pile-driving Formulas for Use

While further research combining theory with field observations may lead to a more accurate pile bearing-value formula than any now available, or to other methods of approach to the subject, it is meanwhile necessary to continue designing and constructing pile foundations to the best of our abilities. With this thought in mind the various pile-driving formulas have been studied, with the result that formulas (2.1*a*) and (2.1*b*) are suggested for use, as giving realistic results, as being capable of reconciling the many variables in piles, and as giving comparable driving results under different conditions and an understanding of the effects of different variables. For acceptable short cuts the International Conference Uniform Building Code formula (Appendix I, page 563) may be used without serious error (see Fig. A.4), although the results sometimes may be more conservative than necessary. Any of these formulas of the Hiley type are preferable to other commonly used formulas, because they shut out unduly high results otherwise obtained with various small sets, heavy piles, or light hammer–pile weight ratios. The Hiley type of formulas will provide adequate uniformity of computed driving resistance, or carrying capacity, with various types and sizes of piles and hammers. After relative uniformity has been attained, the matters of relationship of driving resistance to bearing capacity, as indicated by test loads or friction values, and of factor of safety will be considered. It is useless to consider these matters without first having obtained uniformity of driving results, for the problem then becomes involved in a maze of contradictions, as engineers are well aware. It should be constantly borne in mind also that pile-driving formulas must be used as only one of several tools on each problem, the others being boring results, static values, and test-loading results.

To make formulas (2.1*a*) and (2.1*b*) readily usable, tabulations have been made of various weights and dimensions of pile-driver parts, caps, and piles, of hammer energies, and of the several temporary elastic compressions required in the formula, so that all necessary data will be available in one source, to enable the engineer in the office, by entering these variable values in the formula, to produce quickly a set–bearing value graph, such as is shown in Fig. 2.1 or Fig. 2.2, applicable throughout the entire range of sets, for use in the field.

While the Hiley-type formulas have been recommended for general use, they appear to underestimate the capacities of very heavy or very long piles. Measured compressions are often less than computed, not only because of friction, but because of nonuniformity resulting from wave action in the length of pile. The wave equations indicate that stresses in the pile may somewhat exceed those obtained by the Hiley-type formulas. If means have become available to obtain comparative results by wave equations, this should be done for heavy, long piles.

CHAPTER 2

PILE-DRIVING ANALYSIS

DYNAMIC FORMULA

Functions of Dynamic Formula

The theory of dynamic impact, namely that the ultimate carrying capacity is the same as the ultimate driving resistance, results in a so-called "dynamic" pile-driving formula. Such a formula for a single pile bears no relationship to the capacity of the soil below the pile tips to carry the widespread load of a structure; it takes no account of the reduced value of a friction pile when in a group; and it takes no account of changes in soil structure and hydrostatic conditions induced temporarily during driving, or other short- or long-time adjustments in bearing value. Therefore, the driving resistance is only one item of design information that must be considered with other conditions in order to design the pile foundations intelligently, safely, and economically.

Such a formula can apply only in the case of cohesionless strata, such as sand, gravel, or permeable fill; in these strata the resistances acting while the pile is being driven bear a reasonably close relationship to those acting on a pile carrying a static load—although it should be borne in mind that in coarse-grained cohesionless soils cases have been reported where piles driven in such saturated pervious material have shown up to 50 per cent decrease in resistance to driving resumed in 24 hr, presumably caused by readjustment of the positions of the sand grains to relieve internal stresses,[25] and that piles driven in submerged uniform fine-grained sands which are so loose that they become quick temporarily under the blow may show much less resistance than will occur under a static load. The criterion for safe use of the dynamic formula is that redriving after rest should not differ too widely from the ultimate driving resistance during the original driving.

In the case of driving piles in plastic material such as soft clay or fine-grained silt, relations between the temporary resistance to driving and the permanent resistance to the applied load on the pile are uncertain. In materials of this second class, the friction action during driving is very much less than the friction which occurs after a period of time, but the resistance to a dynamic blow is far greater than the resistance to a static load carried over a long period. This phenomenon is caused by the fact that in order to make room for the pile it is necessary for the water in the voids of the clay to escape, which is a process requiring considerable time and one which cannot be accomplished by a sharp blow. This inability of the water to escape rapidly through the clay causes water to be present against the surface of the pile during driving, thus lubricating the surface. After cessation of driving, hydrostatic excess pressure in the clay will gradually relieve itself, and this material will close in against the pile, increasing the friction grip on the pile, with the result that more load will be carried in friction and less by end bearing. The difference between temporary and permanent bearing values in permeable or nonsaturated soils and saturated clays may be shown graphically in Table 2.1.

TABLE 2.1. DISTRIBUTION OF LOAD BETWEEN FRICTION AND END BEARING

Soil	Operation	Bearing[a]
Permeable, and nonsaturated materials......	During driving	- - - - - - ———
	Redriving after rest	- - - - - - ———
	Static	- - - - - - ———
Saturated clays..........................	During driving	- - ———
	Redriving after rest	- - - - ———
	Static	- - - - - - - - - - - -

[a] - - - = friction, ——— = end bearing.

It is an error to use a pile-driving formula to obtain carrying capacity in materials of the second class. The stress in the pile during driving can be computed by means of the pile-driving formulas when driving in cohesive soils, however, provided the temporary driving resistance thus computed is not considered as having any relation to the permanent load-carrying capacity of the pile. This fact is of value in determining allowable sets and types and sizes of piles and hammers.

Many soils partake of the characteristics of both classes, and judgment should be used as to extent of dependence on a formula of this type. The need for accurate knowledge of the strata, obtained from adequate borings, is emphasized.

Some codes restrict the use of a dynamic formula to soils in which the dynamic resistance to redriving after a 24-hr rest shows an increase of not over 25 per cent.

Practical Use of Dynamic Formula

The ultimate carrying capacity R_u (considered as ultimate resistance to driving) of each pile would be obtained from the relation $R_u s = W_r h$ if it were not for energy losses due to various sources. Formulas (2.1*a*) and (2.1*b*), which consider energy losses due to the principal causes (efficiency of hammer, impact, temporary compression in pile cap and head, temporary compression in pile, temporary compression or quake of ground), will give the ultimate carrying capacity R_u. They are applicable for the case of $W_r > W_p e$, which is the usual case and in which conditions favorable to driving prevail.

*Formula for use with drop hammers, single-acting steam hammers:**

$$R_u = \frac{e_f W_r h}{s + \frac{1}{2}(C_1 + C_2 + C_3)} \times \frac{W_r + e^2 W_p}{W_r + W_p} \tag{2.1a}$$

*Formula for use with double-acting, differential-acting steam, and diesel hammers:**

$$R_u = \frac{12 e_f E_n}{s + \frac{1}{2}(C_1 + C_2 + C_3)} \times \frac{W_r + e^2 W_p}{W_r + W_p} \tag{2.1b}$$

In (2.1*a*) and (2.1*b*),

R_u = ultimate carrying capacity of pile (considered as ultimate resistance to driving), in pounds, before applying any factor of safety.

W_r = weight of falling mass, in pounds (see Table IV) (usually the ram, although the casing cylinder in many European hammers).

W_c = weight of casing, in pounds, for double-acting and differential-acting steam hammers (see Table IV).

E_n = rated energy of hammer per blow, in foot-pounds, as published by manufacturers and listed in Table IV. The hammers should run at the speeds listed, if possible, to obtain the greatest over-all efficiency. In the cases of Vulcan-California double-acting and Super-Vulcan differential-acting steam hammers, this is taken by the manufacturer as the sum of the energies obtained by multiplying the weight of the ram by the stroke plus the product of the piston area times the published steam pressure at the hammer, with the maximum value equal to the sum of the weights of the ram plus casing times the stroke, $E_n = (W_r + W_c)(h/12)$. This is based on the theory that the steam pressure cannot exceed the weight of the casing without causing the casing to lift from the pile. In the case of National and Union Iron Works double-acting steam hammers, the rated energies are also obtained in this same manner.

* For derivation, see Appendix I, pp. 559–561.

In the case of the McKiernan-Terry double-acting hammers the value of E_n is based on indicator diagram readings, confirmed by high-speed moving-picture readings of the velocity of the ram at impact, and although these values agree reasonably well with the theory of limiting the steam pressure by the weight of the casing, the variations are enough so that it is recommended that the rated energies be used, since the hammers act most efficiently at these figures.

h = height of free fall of ram, in inches, for drop hammers; normal (shortest) stroke of ram, in inches, for single-acting steam hammers; and normal stroke of ram for double-acting and differential-acting steam hammers (see Table IV).

e_f = efficiency.* The following percentages† are suggested for use in computing bearing capacities (when computing fiber stresses in piles it is well to increase values under 100 per cent by 10 per cent):

100 per cent for drop hammers released by trigger.

75 per cent for drop hammers actuated by rope and friction winch, but bearing in mind that this figure may decrease when the drop is small or the drag considerable, and increase somewhat if the drop is very large or the drag not great. The haul on the line and drum, the friction in the guides, and the friction band effect (if the operator does not fully release it or if he "picks up" the hammer before the fall is complete in order to keep the line taut), all reduce the energy of the blow.

85 per cent‡ for McKiernan-Terry single-acting steam hammers. (The manufacturer recommends 90 per cent as the lowest value of e_f for these hammers, even when

* Manufacturer's rated energies for European hammers and for other makes not stated should be investigated to determine whether they are gross or net after losses in efficiency have been deducted.

† All of the percentages have been intentionally given as slightly on the low side for all types of hammers, compared to the values which might be used provided the hammers were all in excellent condition, in order to be reasonably certain of obtaining at all times approximately the intended energy as a minimum, since possible slight overdriving is preferable to underdriving. However, it is very likely that there are hammers in use today which do not deliver more than 50 to 60 per cent of their rated energies. Whereas for computing carrying capacities the efficiencies should be taken on the low side, for investigating driving stresses they should be taken on the high side.

‡ Includes 10 per cent to cover poor condition of hammer, wear, improper adjustment of valve gear, poor lubrication, unusual weather conditions causing condensation, unusually long hose, restricted areas at hose connections, minor hose leaks, unduly tight packing in types of hammers having manual adjustment of packing, unnoted minor drops in steam pressure which will reduce stroke, binding in guides, etc.

operated under unfavorable circumstances, pointing out that the hammer is totally enclosed and designed to keep friction at a minimum; that the stroke is constant since the blow is on an anvil block on the pile; that the actual stroke is longer than the normal listed by an amount sufficient to compensate for the slight back pressure; and that the hammers are designed to develop the full rated energy.)

80 to 85 per cent* for Raymond single-acting and differential-acting hammers. (These are manufactured and used solely by the Raymond Concrete Pile Co., which claims that improved designs and maintenance result in high efficiencies.)

75 per cent* for Warrington-Vulcan single-acting steam hammers.

65 per cent* for Vulcan-California double-acting hammers.

85 per cent* for McKiernan-Terry Series B double-acting hammers. Since the rated energies of these hammers are based on indicator diagrams, losses caused by back pressure, preadmission of steam before completion of the downstroke, expansion losses owing to drop from entering pressure of steam, mean effective pressure in the cylinder, wire drawing of steam, and losses in valves and ports have been deducted before obtaining the rated energies, leaving only mechanical losses such as those caused by piston-ring friction and tight packing to be covered by the value of e_f. (The manufacturer recommends 90 per cent as the lowest value of e_f for these hammers, even when operated under unfavorable circumstances.)

85 per cent* for Industrial Brownhoist, National and Union double-acting hammers.

75 per cent* for McKiernan-Terry double-acting hammers, sizes 0 to 7.

75 per cent* for differential-acting steam hammers.

Since the rated energies for these hammers are based on the product of the entering steam pressure times the areas of the piston, the value of e_f should be such as to cover losses caused by wire drawing of steam, piston-ring friction, back pressure resulting from preadmission of steam just prior to impact, losses in valves and ports, tight packing, and other mechanical losses. It is claimed

* See footnote ‡ on page 29.

that the nonexpansive use of steam in the steam cycle in this type of hammer obviates a drop from the entering steam pressure to mean effective pressure. (The manufacturer recommends 84 per cent for the value of e_f for the large- and medium-sized hammers and 80 per cent for small-sized hammers, in first-class condition and operated under favorable circumstances.)

100 per cent for diesel hammers. (The rated energies are intended by the manufacturers to be the available energies. However, engineers generally downgrade these ratings because of the difficulties and variable methods used in determinations. The McKiernan-Terry diesels are the least subject to reduction because the method of computation and field observation is simplest and conservative. For discussion see page 76.)

80 per cent* for BSP semiautomatic single-acting steam hammers. (The manufacturer recommends 90 per cent for hammers in normal condition.)

W_p = weight of pile, in pounds, including shoe and driving cap for drop hammers and single-acting steam hammers; weight of pile including shoe and weight of anvil in case of double-acting and differential-acting steam hammers (see Tables IV and V). Note that steel sheet piles are sometimes driven in pairs, and the weight of both should be included. Also includes weight of follower if used. In the case of driving to refusal in end bearing on rock, use only half the actual weight in the formula. The weight of earth wedged into the spaces between the flanges of H piles, or into open-end pipe piles, during driving should also be included. It is probable that the weight of earth clinging to the pile should also be included, which may be assumed as ½ to 1 in. thick, in the absence of definite information, and may equal 50 to 100 per cent of the weight of a wood pile. (In European hammers that obtain the blow from the fall of the casing, the weight of the hammer, less the weight of the casing, should also be included.) Weights are confined to driven portions and include driving cores or mandrels.

l = length of pile, in inches, measured from head to center of resistance to driving.

L = length of pile, in feet, measured from head to center of resistance to driving.

e = coefficient of restitution:

* See footnote ‡ on page 29.

= 0.80* for micarta cushion when driving Raymond piles. (This type withstands 400 to 500 blows of heavy driving and therefore gives more uniform results.)
= 0.55† for no cushion. Steel on steel when driving pipe piles.
= 0.50* for oak cap blocks when driving Raymond piles.
= 0.50† for well-compacted cushion when driving pipe piles.
= 0.50‡ for ram of double-acting hammers striking on steel anvil and driving steel piles or precast concrete piles.
= 0.40† for medium-compacted wood cushion when driving pipe piles.
= 0.40‡ for ram of double-acting hammers striking steel anvil and driving timber piles, also for striking steel helmet containing wood and driving steel piles.
= 0.40‡ for ram of single-acting or drop hammers striking directly on head of precast concrete piles not fitted with driving cap.
= 0.32‡ for ram of single-acting hammers striking on steel plate cover of wood cap of steel piles.
= 0.25‡ for fresh wood cushion when driving pipe piles.
= 0.25‡ for ram of single-acting or drop hammers striking on well-conditioned wood cap of driving cap in driving precast concrete piles or directly on wood pile heads.
= 0.0‡ for deteriorated condition of heads of timber piles or of wood cap and for excess packing in driving cap.

s = final set of pile (using average of last five blows for drop hammers and of last 20 blows for other types), in inches.

C_1 = temporary compression allowance for pile head and cap, in inches (see Table I).

C_2 = temporary compression of pile, in inches (see Table II).

C_3 = temporary compression allowance for ground for average cases where pile is driven into penetrable ground, in inches (see Table III).

A = average of cross-sectional areas of pile at butt and at center of resistance to driving, in square inches (shell only, in case of steel-pipe and Monotube piles, shells and core for cast-in-place concrete piles, or mandrel) (see Table V). In the case of end-bearing piles, the center of resistance is at the tip. In the case of reinforced-concrete piles, the area of reinforcing should be

* Values from tests made by the Raymond Concrete Pile Co. However, studies by wave equation do not indicate such vastly different effects, although the difference is considerable. Values as high as 0.80 should be used with caution.

† Values of *e* from *Seamless Steel Pipe Piles,* National Tube Co.[11n]

‡ Values of *e* from experiments by Hiley.[7] These values agree closely with those given by E. Noë and L. Troch.[1a]

transformed into equivalent concrete area and included.

E = modulus of elasticity for pile material.

p_1 = stress per square inch on driving cushion, or on pile head if no cushion is used.

$$p_1 = \frac{R_u}{\text{area of pile head}} \tag{2.2}$$

p_2 = stress per square inch on average cross section of wood or concrete piles, or on shell area of tubular steel pile shells, or on area of steel sheet piling, or on average area of mandrels, or on net areas of shell and core for cast-in-place piles.

$$p_2 = \frac{R_u}{A} \tag{2.3}$$

p_3 = stress per square inch on horizontal projection of pile tip, including driving points under steel pipe casings, and on bounding area under H piles (for end-bearing piles and piles of constant cross section.)

$$p_3 = \frac{R_u}{\text{area of tip}} \tag{2.4a}$$

p_3 = stress per square inch on gross area of pile at ground surface in case of tapered friction piles (for tapered friction piles).

$$p_3 = \frac{R_u}{\text{gross area at ground surface}} \tag{2.4b}$$

In so far as use of formulas (2.1*a*) and (2.1*b*) are concerned, it should be noted that the driving of pipe piles, Union Metal Monotube pile shells, and steel mandrels consists of the driving of steel and not concrete, since the concrete is poured later as a filling in the pipe or casing.

Set–Bearing Value Graph

Set versus Resistance Graphs. Graphs, as shown in Figs. 2.1[80,81] and 2.2[92], should be prepared for studying the problem and for field use by means of formulas (2.1*a*) and (2.1*b*), in which final sets are plotted against ultimate driving resistances or against working loads at any selected factor of safety. One of these graphs is very convenient for the pile inspector, because he can pick the pile value corresponding to any set directly from the curve and enter it upon his report.

The curves in Fig. 2.1 can be plotted readily by assuming several different driving resistances and solving for the corresponding values of s. This is simpler than assuming values for s and solving for R_u, owing to the fact that the value of R_u enters into the value of C_2, and would make

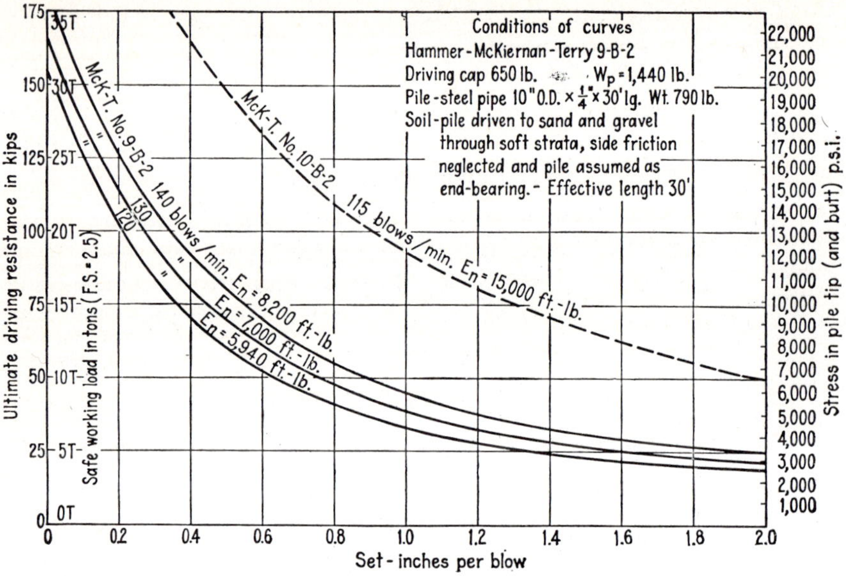

FIG. 2.1. Set-bearing value curves.

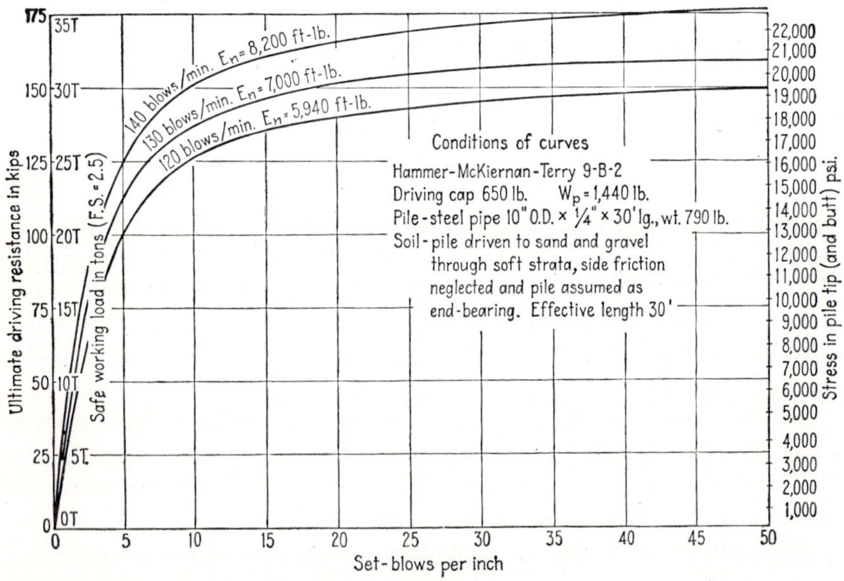

FIG. 2.2. Set-bearing value curves.

the solution in this case a matter of cut and try. Points in Fig. 2.2 can be plotted by reading values from the curves in Fig. 2.1. For an example, see Appendix II, pages 568, 569.

The curves in Fig. 2.2 show graphically the tip resistance to be expected for the entire range of values of s, and are especially interesting as they show what may be expected as driving becomes harder and approaches refusal. The flattening of the curve as the number of blows per inch increases indicates that beyond a certain point a considerable increase in the number of blows per inch has slight effect on the carrying capacity. If the *Engineering News* formula is plotted on this graph, the curve will continue ascending at a steep slope instead of flattening, thus graphically illustrating one of its weaknesses. The curves in Fig. 2.2 predict maximum ultimate bearing values of the piles, with assumed driving equipment.

Pile Fiber Stresses. The fiber stresses in the piles can be computed by dividing the ultimate driving resistance by the area of the pile at the point under consideration, and showing scales of values for different tip sizes, as on the right-hand sides of Figs. 2.1 and 2.2, for comparison with the maximum fiber stress or yield point of the pile material. By drawing horizontal lines across the graph at the limiting values selected for maximum allowable fiber stresses during driving, the minimum permissible sets can be read at the intersections with the curves. A brief table of these limiting sets sometimes is sufficient information for field use.

Analysis of Energy Losses and Check of Computations

Analysis of Energy Losses. After obtaining the value of R_u, it is recommended that the amounts of the total applied energy lost due to various causes when driving the pile be ascertained in order to observe the efficiency of the equipment being considered and as a check upon the computations, by substituting the value of R_u in one of the formulas given in Table 2.2.

Selection of Proper Hammer. The preceding formulas are of assistance in selecting the proper weight of hammer to use, as the proportion of useful energy to wasted energy may be observed. For economical and efficient driving, a reasonable proportion of the applied energy should remain available for driving. Furthermore, if the percentage of remaining useful energy is very small, slight uncertainties in the assumptions may be as great, numerically, as the value of the net energy remaining for driving, indicating that the sensitivity of the formula is too great to place too much reliance on the computed results in such cases. For reasons of economy and in the interests of reliance on the computed results, a hammer of efficient size should be used. It is better to select a hammer which may be on the heavy side rather than one on the light side. In general, the hammer should be as large as can be safely used without damaging the pile, and computed stresses in the pile should be compared with the yield point of the pile material, reduced by a reasonable factor of safety, to be sure that such a value is not exceeded.

TABLE 2.2. ANALYSIS OF DRIVING-ENERGY LOSSES

Net effective energy available for driving = ultimate resistance to driving × final penetration under last blow	Total kinetic energy applied by hammer	Loss in energy due to impact	Loss in energy due to imperfectly elastic compression of		
			Pile head and cap	Pile	Soil

For use with drop hammers, single-acting steam hammers, and diesel hammers

$$R_u s \quad = e_f W_r h \quad - e_f W_r h W_p \frac{(1 - e^2)}{W_r + W_p} - \frac{R_u C_1}{2} - \frac{R_u{}^2 l}{2AE} - \frac{R_u C_3}{2} \qquad (2.5a)$$

For use with double-acting and differential-acting steam hammers

$$R_u s \quad = 12 e_f E_n \quad - 12 e_f E_n W_p \frac{(1 - e^2)}{W_r + W_p} - \frac{R_u C_1}{2} - \frac{R_u{}^2 l}{2AE} - \frac{R_u C_3}{2} \qquad (2.5b)$$

Temporary Compressions. The denominator $s + \frac{1}{2}(C_1 + C_2 + C_3)$ of the first term of formulas (2.1*a*) and (2.1*b*) represents the total average downward movement under the blow of the ram. The average temporary compression is covered by the term $\frac{1}{2}(C_1 + C_2 + C_3)$ and the permanent residual set by the term s. Field measurements indicate that the total amount of temporary compression in the pile and ground often is not as great as computed when using the full length of the pile for L. This is due to the fact that friction on the side of the pile absorbs considerable energy before it can reach the tip. If reduced values of L and l were used in certain cases, a closer agreement would be obtained. Rankine's formula uses one-half the pile length, but it would seem better generally to take more than this, depending on the relative amounts of friction which it is judged the various strata can exert. This depends on the material, density, and depth of the various strata, and on the relative amounts of resistance which it is judged are obtained from friction and from end bearing. The value of L should be taken as the distance from the head of the pile to the center of driving resistance. Until driving records are available, the value may be estimated from an inspection of the boring log and blows on the sampling spoon in each stratum. After driving records are available, the foot-by-foot sets may be translated into driving resistances from a set–bearing value graph such as shown in Fig. 2.1. For an example, see Fig. A.2. When estimating this value, the amount assumed to act as tip resistance should be considered. For straight-sided piles, this procedure should give very good results. For tapered piles, the friction on the upper portions keeps increasing during driving and thus raising the center of driving resistance. This effect can be estimated, probably with sufficient accuracy. In some cases, it will be noted that a portion of the pile or mandrel projects considerably aboveground at final penetration, and this should be taken into account.

The graph of set values plotted against depths may show a large increase in number of blows per foot when sets become small, greatly disproportionate to the increase in driving resistance R_u. Graphs of sets obtained with hammers of different weights or energies will be entirely different, and convey entirely different impressions of the strength of the soil. This is shown in Fig. A.2. The exact center of driving resistance is not important, because an error of several feet in a fairly long pile would evidently affect C_2 little, C even less, relatively, and $s + (C/2)$ even less than that.

Field Measurements of Compressions. Field measurements can readily be taken of the sum of C_2 and C_3 and, if possible, it is desirable to take these on the first piles driven to serve as a check on the assumptions used. By holding a piece of paper, or, preferably, bristol board, on the pile just prior to the completion of driving (Fig. 2.3*a*) and moving a pencil horizontally continuously across the edge of a board, a graph will be obtained which will show the total downward movement of the top of the pile under each blow, and the amount of upward spring, represented in the formula by $C_2 + C_3$. The computed curves in Figs. 2.1 and 2.2 can be checked by taking these field graphs on several points of the pile as it passes the board, in order to record various combinations of rebounds and permanent sets as the driving resistance varies. Since the readings will be made at points below the head of the pile, account should be taken of the elasticity of the portion of the pile above the location at which the graph was secured. (For a typical field graph, see Fig. 2.3*b*.) Field measurements have been taken of C_2 alone, when driving steel H piles to rock through material having very little friction value and without using a driving cap, which check as closely as the graph can be made, the values of C_2 computed by the following formula on which Table II is based:

$$C_2 = \frac{R_u l}{AE} \tag{2.6}$$

In this manner the pile itself can be considered as a spring gage to check the value of R_u, since R_u varies directly with C_2. C_1 is more difficult to measure and, since a difference in this figure would produce comparatively little change in the answer, it is not usually desirable to attempt to measure it, except in the case of drop hammers as described hereafter. The values given in Table I are the results of field observations by Hiley[7] and Goodrich.[2] In the case of anvils having a cup to receive a driving block, practice is sometimes to throw in short pieces of wood from time to time, which may compact to a hard layer, with the result

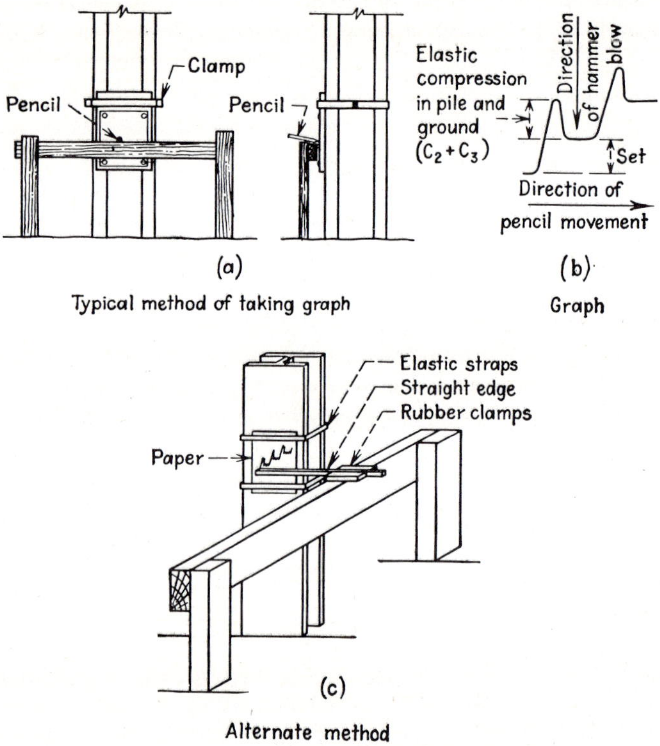

FIG. 2.3. Typical and alternative methods of taking graphs.

that the value C_1 may vary widely. This tends to confuse comparisons of hardness of driving for the various piles. It should never be done when the pile is in the region where it should take up its final resistance. Every effort should be made to obtain consistency of driving at final penetration, to approximate the value of C_1 used in the equation.

The value of A should be taken as the average of the areas at the butt and at the center of driving resistance instead of the average of the butt and tip areas unless the pile is practically end bearing. The use of this larger area will also tend to reduce the computed elastic compression, as does the use of l taken only to the center of driving resistance, and thus bring the computations more into line with field observations.

Mechanical and electrically operated recorders have been used.[38,106,112]

A method for determining the proportion of the applied energy which results in permanent set of the pile and that which is lost owing to various sources is applicable for field observations on driving with a drop hammer; it consists of dropping the hammer from various heights, when within the range of the final desired set, and of plotting the drop heights against the sets obtained.[8] By prolonging this graph, which should be a straight line, until it intersects the height axis, an intercept is obtained which gives the maximum height of fall h' for no set. This is a measure of all the energy losses, represented by the term $\frac{1}{2}(C_1 + C_2 + C_3)$, for that height of fall, the only additional losses being the impact losses for any greater heights of fall, represented by the last term of formulas (2.1*a*) and (2.1*b*). A typical graph is shown in Fig. 2.4. If field measurements of $C_2 + C_3$ are also taken as described above, this result may be subtracted from the sum of $C_1 + C_2 + C_3$, and values of C_1 determined for future reference. The condition of the head of the pile or filler in the driving cap should always be noted when taking the readings for this graph, in order to obtain data useful for future work.

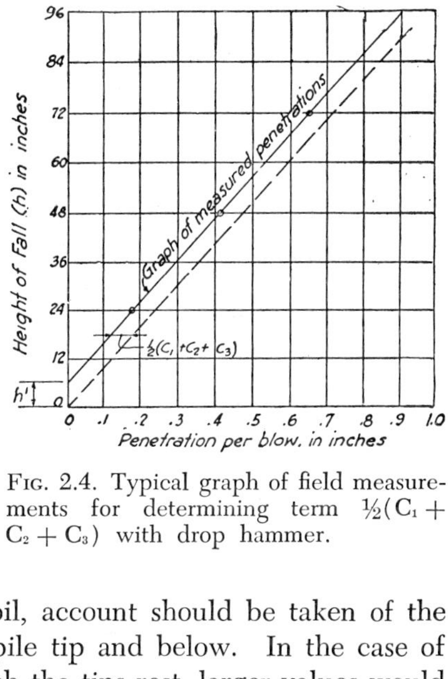

FIG. 2.4. Typical graph of field measurements for determining term $\frac{1}{2}(C_1 + C_2 + C_3)$ with drop hammer.

When selecting the value of C_3 for temporary compression of the soil, account should be taken of the compressibility of the strata at the pile tip and below. In the case of a soft bed under the stratum in which the tips rest, larger values would be used than if hard material were present, possibly as much as double those in Table III. This indicates the desirability of obtaining field measurements of rebound on the pile.

Schenk[1m] has proposed the values of C_2 and C_3 from load-test diagrams, instead of from rebounds measured during driving, by taking their value as the tangent of the angle between the horizontal and the slope of the elastic-rebound line found upon removal of load near the failure point or beyond.

Investigation of Load Capacity from Known Set

In the case of an investigation, where the set is known and it is desired to find the corresponding value of R_u, a curve should be plotted such as in Fig. 2.1, and from this curve the exact values sought.

Batter Piles

When driving batter piles with drop, single-acting, or diesel hammers, the height h in the formulas is reduced, and friction also occurs in the

FIG. 2.5. Raymond special rig capable of driving both in and out batters, shown driving a Raymond step-taper pile. (*Courtesy of Raymond Concrete Pile Co.*)

guides. Taking the coefficient of friction at 0.1 when θ is the angle between the batter and the vertical, the effective value of drop h' to be used in the formulas in place of h may be taken as follows for drop, single-acting steam, and diesel hammers:

$$h' = h(\cos\theta - 0.1 \sin\theta) \tag{2.7a}$$

When driving batter piles with double-acting or differential-acting steam hammers, the gravity drop of the ram is reduced, but not the effective steam pressure on the piston. The effective value of energy

FIG. 2.6. Union marine pile driver driving precast concrete batter pile. (*Courtesy of Union Iron Works.*)

E'_n, in foot-pounds, to be used in the formulas in place of E_n may be taken as follows for double-acting and differential-acting steam hammers:

$$E'_n = E_n - W_r h \frac{(1 - \cos \theta)}{12} \tag{2.7b}$$

STATIC FORMULA

Uses of Static Formula

Before ordering piles, it is necessary to consider total lengths of embedment in load-resisting friction strata and resulting carrying capacities strictly from a friction point of view, after deducting such percentage for

TABLE 2.3. FRICTION BETWEEN PILE AND SOIL[a]

Material	Identification	Ordinary range of values per square foot of bounding area of pile[b]	Blow counts per linear foot on sampler, N, N', or N''[c] from standard penetration tests		
			Standard Raymond sampler 2 in. o.d., 1.375 in. i.d., 140-lb hammer, 30-in. drop	New York City Code sampler 2.5 in. o.d., 300-lb hammer, 18-in. drop	Corps of Eng., N.E. Div. sampler 3.0 in. o.d., 300-lb hammer, 30-in. drop
Fine-grained soils:					
Very soft clay	Fist penetrates several inches easily	125 ± 125	0–2	0–2	0–8
Mud		250 ± 200			
Silt		300 ± 200	—	0–2[d]	8–16[d]
Soft clay	Thumb penetrates several inches easily	400 ± 200	2–4		
Silty clay		600 ± 200	2–4	3–10[d]	
Sandy clay		600 ± 200	—	—	16–55[d]
Medium clay	Thumb dents several inches, moderate effort	700 ± 200	4–8	5–10[d]	11–30
Medium stiff clay		—	—	—	16–55
Sandy silt		800 ± 200			
Firm clay	Thumb dents several inches, great effort	900 ± 200	—	Over 30[d]	55–110[d]
Dense silty clay		1,200 ± 300			
Stiff clay	Thumb dents readily, great effort to penetrate	1,500 ± 400	8–15	11–30	
Stiff to medium-hard clay		—	—	—	55–110
Very stiff clay	Thumbnail indents readily	3,000 ± 1,000	15–30		

Hard clay	Thumbnail indents with difficulty	Over 4,000	Over 30	Over 30	
Very hard clay		—	—	—	Over 110[d]
Coarse-grained soils:					
Very loose sand and silt or clay		100 ± 50[d]	0–6[d]	0–15[d]	0–10[d]
Medium sand and silt or clay		500 ± 100[d]	6–30[d]	16–50[d]	10–15[d]
Dense sand and silt or clay		700 ± 100[d]	30–50[d]	Over 50[d]	55–110[d]
Very dense sand and silt or clay		900 ± 100[d]	Over 50[d]	—	Over 110[d]
Very loose sand		—	0–4[e]	—	0–8
Loose sand		—	4–10[e]	0–15	8–16
Medium sand		1,200 ± 500[f]	10–30[e]	—	16–55
Dense sand		—	30–50[e]	16–50	55–110
Very dense sand		—	Over 50[e]	Over 50	Over 110
Sand and gravel		2,000 ± 1,000			
Gravel		2,500 ± 1,000			

[a] See reference 99 for analysis of these values.

[b] The ± figures indicate a range governed by the character of the soil. Not all soils falling in the same general category have equal properties.

[c] N is number of blows reported by driller and shown on the boring logs. N' represents value of N corrected for depth below ground to adjust approximately for energy changes due to various lengths of drilling rods, as shown on Fig. 2.7. (Values shown for N' are for Raymond standard sampler rod; values will be less for heavier, larger rods.) If the sand is very fine or silty and is below ground water, the value of N' shall be reduced to N'' as follows:

$$N'' = 15 + \tfrac{1}{2}(N' - 15) \tag{2.8}$$

[d] Values from N. L. Liver, Jr., and R. C. Jennings, *The Prepakt Reporter*, October–November, 1958.

[e] N values from R. B. Peck, W. E. Hanson, and T. H. Thornburn, *Foundation Engineering*, John Wiley & Sons, Inc., 1953, p. 109.

[f] If not micaceous, muddy, or under hydrostatic pressure or vibration.

point resistance as it is felt will be present, as a rough check upon the designed lengths. Examination of a large number of pile test records indicates that the soil friction values in Table 2.3 show the order of magnitude that probably may be assumed for preliminary studies.[99]

Another extensive tabulation of test results for cohesion and adhesion of soils around piles is also available.*

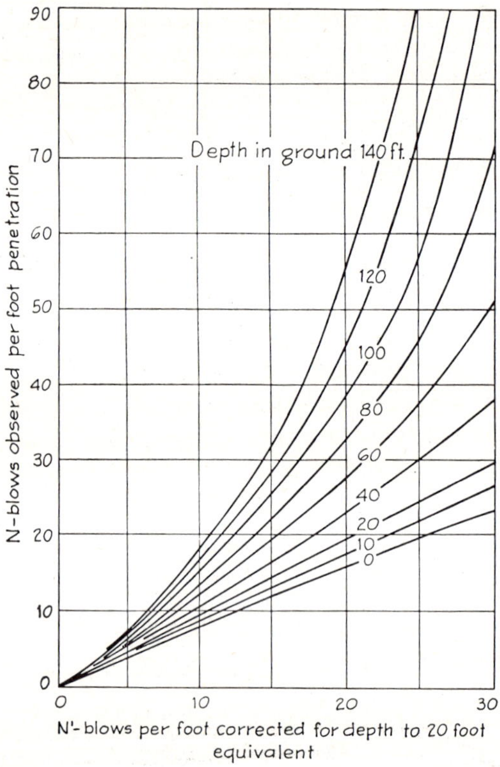

FIG. 2.7. Chart for correcting *N* values for depth in standard penetration test.

Such values as those in Table 2.3 can only be average ranges. Caution should be exercised in using these figures for large projects; there can be no substitute for test information. For small projects, it may be cheaper to provide additional pile length than to make tests.

This static formula may be expressed in the form

$$f_u = \frac{R_u - R_t}{A_s} \tag{2.9}$$

* M. J. Tomlinson, "The Adhesion of Piles Driven in Clay Soils," *Proc. Fourth Intern. Conf. Soil Mech. and Foundation Eng.*, London, vol. 2, 1957, pp. 66–71.

where f_u = ultimate friction value, in pounds per square foot;
R_t = amount of ultimate load assumed carried on tip, in pounds;
A_s = surface area of portion of pile acting in friction, in square feet.

The percentage of load carried by end bearing varies with the soil, pile, and method of placing. It varies with the amount of load, and may not act unless the load is sufficiently large. Hard clay may provide considerable point resistance, and soft clay practically none.

The following semiempirical formula has been presented:*

$$R_u = \pi r_p^2(1.3cN_c + \gamma D_f N_q + 0.6\gamma r_p N_\gamma) + 2\pi r_p D'_f s \qquad \text{(for round piles)} \qquad (2.10a)$$

$$R_u = 4r_p^2(1.3cN_c + \gamma D_f N_q + 0.8\gamma r_p N_\gamma) + 8r_p D'_f s \qquad \text{(for square piles)} \qquad (2.10b)$$

where r_p = round-pile radius or half side of square pile, in feet;
γ = effective weight of soil, in pounds per cubic foot;
D_f = depth of tip below ground surface, in feet;
D'_f = friction lengths in load-carrying strata;
N_c, N_q, N_γ = dimensionless bearing-capacity factors depending on value of ϕ, for which see reference shown* (ϕ may be assumed as at least 30 deg in cohesionless material unless otherwise determined);
c = cohesion, in pounds per square foot (term $1.3cN_c$ is used only in cohesive soil); and
s = average ultimate shearing value of soil, in pounds per square foot.

Values of N (or N' or N'') from the standard penetration test can be fairly reliably correlated with values of ϕ, N_γ, and N_q.

Study of this subject has been extended to include the influence of compaction and prestressing of cohesionless soils during pile driving by Meyerhoff.[206] Results have been expressed by an equivalent angle of internal friction. The theory indicates that the ultimate bearing capacity of a pile in the compacted soil may be as much as twice that without soil compaction and up to six times for a displacement caisson. A method is suggested to estimate the settlement and allowable loads of single piles and displacement caissons. Theoretical results show that load per pile, when in a group, decreases considerably with larger size of the group and is least under a raft. It has been estimated that settlement increases with group width and pile spacing. Such methods are useful for estimating but should not replace load tests.

* K. Terzaghi and R. B. Peck, *Soil Mechanics in Engineering Practice*, John Wiley & Sons, Inc., 1948.

Ground Testing before Pile Driving

It is an advantage to obtain accurate information as to pile load-carrying capacity of the ground before pile driving commences on a large scale. Methods of making tests of the ground without the delay and expense of driving full-size piles would be very useful. Several such procedures have been proposed, and used to a larger extent in Europe and Great Britain than in the United States to date.

Penetration Tests. In Great Britain, methods for predetermining pile needs are described in detail in the *Specification for Concrete Pile-driving* of the Institution of Structural Engineers,[1e] and are performed by firms specializing in this work. These methods consist of driving rods or pipes and taking frequent readings of the driving or pulling resistances, which are calibrated to the results of full-size piles. The proportions caused by end bearing and friction are segregated, and the values of setup or negative downward drag considered.

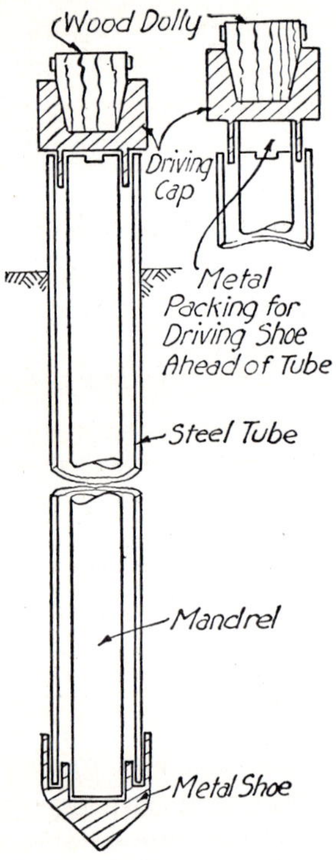

FIG. 2.8. Test pile.

The specification of the Institution of Structural Engineers permits a 4½- to 16-in. diameter tube to be driven as a test, with a shoe, and a mandrel which can be driven separately from the tube at times by insertion of a driving block on the mandrel head, thus enabling the end resistance to be obtained independently from the total resistance. The residue is friction. The frictional resistance may be checked by jacking the tube a short distance. All resistances are calculated by means of dynamic formulas. A test pile which would meet this specification is shown in Fig. 2.8. The fact that the outside diameter of the shoe is greater than that of the tube would affect the friction value in some types of soil, and it would be desirable to have a flush surface if this could be obtained. The tube may be coupled together from short sections, and driven with a ½-ton drop hammer operated from a light pile frame 20 to 25 ft high, with a gasoline-driven winch, possibly mounted on a truck.

The British Steel Piling Co., Ltd., uses 4½-in. tubes with 6½-in.-o.d. shoes. Pre-Piling Surveys Ltd. also uses small-scale driving equipment, with the outside diameters of the shoe and tube the same. Operators of these processes claim correspondence of results with driving and test data from full-size pile driving and test records, by means of proper calibration. The plant employed is generally patented.

Dutch Cone-penetration Tests.[198,199] This method, when used in soft soils, consists of the use of a small cone having a base area of 10 sq cm, on the bottom of a rod housed in a pipe, the cone and pipe being pushed down alternately at a slow constant rate while the pressures are recorded. The resistance of a cone in a homogeneous soil is believed to have a uniform unit value.[199] The assumption is also made that the dimensions of the shear surface are proportional to the pile diameter. Such results have been correlated with the shearing strengths of Dutch soils, and empirical rules developed. It is claimed that the resistance of large pile points can be predicted with an accuracy of 50 per cent, more or less.

Cone-penetration tests for investigating relative densities of sands and gravels can be made.[206] Static application of load is preferable, but in dense soils large reactions are required, so that dynamic tests are cheaper and are fairly reliable in cohesionless soil. A cone is fitted loosely into the bottom of a series of pipe sections, all driven into the soil by a drop hammer, using a constant drop. The numbers of blows per foot of depth are recorded. The cone diameter is slightly larger than that of the pipe, to minimize friction. These tests are more rapid and cheaper than standard borings; however, they provide no samples.

Some standard penetration tests in adjacent locations are required, in any event, to allow the soil to be identified and tested in the laboratory and to enable cone-test results to be roughly correlated with standard N values, relative densities, and angles of internal friction. Meyerhoff[206] and Huizinga[201] state that, on the average,

$$q_c = 4N \tag{2.11}$$

where q_c = static cone resistance, in tons per square foot;
N = standard penetration resistance, in number of blows per foot. (Use $N' = 15 + \frac{1}{2}(N - 15)$ for $N > 15$, in saturated very fine or silty sands.)

The observed point resistance of piles q_p varies from about $\frac{2}{3}$ to $1\frac{1}{2}$ times the static cone resistance q_c, and on the average

$$q_p = q_c \tag{2.12}$$

and

$$q_p = 4N \tag{2.13}$$

where q_p is in tons per square foot.

Field loading tests on driven displacement piles[201] have shown that observed unit skin friction f_s of piles varies from about 1¼ to 3 times the static skin friction f_c on the shaft of a penetrometer, and on the average

$$f_s = 2f_c \tag{2.14}$$

For small displacement piles such as H piles,

$$f_s = f_c \tag{2.15}$$

Other useful relationships among cone and standard penetration tests, bearing, and shear values are also set forth by Meyerhoff.[206]

Vane Shear Tests. The vane apparatus for shear testing of soils in place consists of four vertical rectangular blades welded at right angles to a vertical shaft. The vane is pushed into the soil and then twisted until the soil is ruptured in cylindrical form. Shear strength is computed from the maximum moment needed for rupture and dimensions of the soil cylinder.[120] The device appears to have been developed in Germany and Sweden about 1928 and began to come into use by the Swedish Geotechnical Institute in 1947. A symposium of papers on vane testing appears in reference 130, and an annotated bibliography in reference 207.

Laboratory Testing of Soil Samples

Any method whereby friction-pile capacities could be determined from undisturbed or unconsolidated undrained triaxial compression tests of soil samples would be useful because it would avoid the expense of placing a test-pile-driving rig on the job at a considerable time before the start of actual driving. Such a method has been proposed.*

Fairly good correspondence has been found between shear values of cohesive soils determined by field and laboratory tests in which the shear is taken as one-half the unconfined compressive strength.

The bearing or uplift capacity of friction piles in clay approximately equals the surface area times the shearing strength of the clay. It may be assumed that

$$f_u = c = \frac{q_u}{2} \tag{2.16}$$

where c = cohesion of soil, in pounds per square foot; and
q_u = unconfined compressive strength of clay, in pounds per square foot.

Dynamic Pile-bearing Formula with Static Supplement

The application of coefficients varying with pile characteristics and soil strata to a formula resembling the Hiley, but with a static supplement,

* Dames and Moore, Los Angeles and San Francisco, Calif. See W. W. Moore, "Experiences with Predetermining Pile Lengths," *Trans. ASCE, Paper* 2372, vol. 114, pp. 351–385, 1949.

has been used by the State of Ohio Department of Highways.* A factor is used which is the product of four factors: soil, pile type, pile length, and pile cross section. The soil factor is selected with regard to soil classification and water content. A factor of safety of 2 is used for static loads and 3 for vibrating loads. Results agree well with those obtained from load tests and by means of formulas (2.1*a*) and (2.1*b*).

SKIN FRICTION

Friction

Factors Affecting Friction Value. Friction values depend on type of soil, depth in the ground, degree of natural consolidation and saturation, shape of the pile, amount of compaction by pile, surface texture of pile, and sometimes on the time interval between driving and testing. Published data on the relative effects of these conditions on friction values are scarce.[96,99] In many instances, data to enable these various factors to be considered were not reported.

The relative amount of a load compared with failure load is a factor. Upper strata pick up the smaller loads, and no load may reach the lower strata or tip until large loads are applied, possibly beyond the range of working loads. Duration of load is a factor in cohesive soils. Changing conditions may have an effect.

To study the factors affecting friction between piles and soil, the first essential is full boring logs and set of soil samples for examination.

Effect of soil type. *Unit values of skin friction for a pile in sand* and of point bearing increase with increasing depth. Tabular values shown for rough estimating purposes and most load- or pulling-test results are *average* values, obtained by dividing the total load by the total pile surface area. In dense sand, driving resistance may reach refusal in a very few feet. In loose sand, piles may be driven long distances.

Unit value of skin friction for a pile in soft clay depends upon the properties of the clay. Point resistance is negligible in soft clay. While driving resistance may remain small and be fairly constant with depth, skin friction that will develop varies, in general, with depth. The rate of transfer of load to soil is low for shallow depths.[155]

Unit value of skin friction in soft silt is low during driving, because of liquefaction, but within a few days or weeks the silt apparently regains its original strength. In Shanghai, piles driven in silty clay varying to sandy silt were observed to have constant skin-friction values up to lengths of 45 ft, after which a slight decrease was noted.[34]

* W. H. Rabe, "Dynamic Pile-bearing Formula with Static Supplement," *Eng. News-Record,* vol. 137, No. 26, pp. 868–871, Dec. 26, 1946.

The unit value of skin friction for a pile in clay may vary widely for the same clay, depending upon the method used in placing the pile. Driving may have remolded the soil to such an extent that the original structure has broken down and the clay has become more plastic around the pile—as compared with a pile poured in a cored hole, for example. Hydrostatic pressure may prevent, temporarily at least, bond between the pile and ground. Therefore, because a certain type of pile has been tested to a given friction value in a particular soil, it does not follow that a pile placed by some other method would sustain an identical load, even if the period of setup were to be the same. The time factor is important because bearing capacity is likely to increase with time, as the water dissipates itself and the clay structure re-forms itself. Therefore, some load tests might be made as long as possible after driving.

Friction on piles driven through soft cohesive material to a firm stratum should not be included in the carrying capacity.

Distribution of Friction. An idea of amounts of friction being obtained from various strata may be gained by translating foot-by-foot penetration records, taken continuously during driving of a pile, into driving resistances by means of formula (2.1*a*) or (2.1*b*) and a graph such as shown in Fig. 2.1. By deducting an allowance for end bearing, the friction capacity of the various strata may be noted, and by deducting these amounts occurring above any point from the driving resistance at a lower point, an approximation of the friction value of the lower stratum alone may be obtained.

The set for the last five blows for drop hammers or the last 20 blows for other types is not a good measure of the bearing capacity of the pile unless the prior rate of driving has been such, in suitable material, that sufficient friction can be considered as capable of sustaining load, or unless the end-bearing material reached is capable of carrying a very high unit pressure. If driving has been fairly hard, for instance, but through material not judged suitable for carrying permanent loads, it is hardly justifiable to stop driving after 5 or 20 blows in good material, for then the pile becomes practically end bearing, and both the soil pressure beneath the tip and the friction on the sides of the pile in the good material would be very large and further settlement might occur. The static formula should be applied in such cases as a check upon the dynamic driving formula.

Initial versus Permanent Friction. Of the total amount of friction, all of the amount occurring in sands and gravel, except in quick conditions, and, according to a suggestion by Hiley,[7] one-half of the amount occurring in soils containing an appreciable amount of materials of clays and fine silts, may be taken as being operative during driving, and consequently included in the driving resistance.

It has been observed[25] that if piles are driven in a saturated coarse-grained pervious soil, they may lose, upon redriving, up to 40 to 50 per cent of their resistance in 24 hr. This is probably a consequence of the compaction of the sand during driving, after which the sand has a chance to absorb water and readjust itself. An example has been reported* in which 40-ft wood piles driven 22 ft into wet sand gave a good immediate test; but on the next day all resistance had disappeared and it was necessary to drive another 10 ft.

On the other hand, sometimes piles drive very easily in loose submerged fine uniform sand strata, which, however, are firm under steady pressure and are capable of sustaining large quiescent loads both in end bearing and friction. In fine-grained sedimentary deposits along the Gulf Coast, it is not uncommon for piles to indicate low loading capacity in terms of dynamic driving formulas but later show excellent supporting values under static load tests.[2aw] These observed conditions agree with the theory of critical density.† The jarring action of the blow temporarily causes a quick condition close to the pile; this condition rapidly disappears. The capacities of such piles cannot be ascertained by redriving, because the same condition recurs. The capacities of such piles may not always be ascertainable by redriving, although the first few blows may show increased resistance before the quick effect reestablishes itself. A load test is the surest way to be certain of the bearing capacity and to avoid the expense of unnecessary extra length of piles. If this condition is known or is found to exist when driving close to existing structures, it is possible that settlement will occur since sands as loose as this can decrease in volume when jarred; means of avoiding the vibrating effects of pile driving, particularly in the quick stratum, should be considered fully. In the vicinity of heavy vibrating machinery, use of piles extending into sands should be avoided if possible. Trouble also may be experienced in earthquake regions.

Pile driving in cohesive soil breaks down the soil structure and remolds it so that it loses the larger part of its compressive strength and shear value temporarily. Soft clay reconsolidates itself after cessation of driving quite rapidly at first and continuing for some days or weeks. This effect may be noted from redriving piles, load tests, vane-shear tests, or laboratory unconfined-compression tests of soil samples. Load tests made soon after driving will show lower values than will tests made after a longer interval. For comparative purposes, time intervals before testing should be similar, unless the period is long in each case. Dynamic pile-driving formulas are misleading when applied to piles in clay

* Perrin, *J. Inst. Engrs. Australia*, vol. 4, No. 3.

† A. Casagrande and R. E. Fadum, "Application of Soil Mechanics in Designing Building Foundations," *Proc. ASCE*, vol. 68, No. 9, pp. 1487–1520, November, 1942.

and should not be used for determining carrying capacity. For a discussion of the action of soft clay along friction piles, reference 155 may be consulted. It must be borne in mind that friction capacity alone is not the full solution of the problem; after the load is safely transmitted to the clay, the settlement due to consolidation of the clay during the life of the structure must be computed by use of recognized standard soil-mechanics principles.

Back pressure from soil compaction increases the coefficient of friction. In certain clays, unbalanced pressure may not be dissipated for years, possibly not entirely during the life of the structure, although this factor is so uncertain that little reliance can be placed upon it and it should be considered merely as an additional factor of safety.

Seepage. Seepage around piles driven in unwatered excavations may reduce the skin friction to less than the hydrostatic uplift on the piles, and they may rise. Considerable lifts have been observed. After normal hydrostatic balance has been restored, the skin friction again can act. Redriving or load tests may be necessary to determine if sufficient unseating effect has occurred to reduce the bearing capacity of the pile.

Surface Texture. Surface-texture effect on friction values was observed in one test to result in a precast pile having only six-tenths the value of a cast-in-place pile.[25] The shaft of a compressed concrete cast-in-place pile secures improved friction values owing to the roughness of the surface formed by forcing concrete into the interstices of the soil, and variations in shaft diameter on account of differences in firmness of the soils. In the case of Franki piles, with their rough and warty surfaces, it is stated that the coefficient of friction of soils on concrete is the same as of soil on soil, the soil being in effect under shear.

When casings are required for concrete cast-in-place piles on account of the softness of the soil, a casing having the greatest possible coefficient of friction is desirable, and for this purpose driven corrugated shells have been found effective, since the soil is densified around the pile and engages the corrugations and places the soil in shear. If the shell is dropped into a pipe casing later withdrawn, the advantages of increased compaction in improving the shearing value in the annular soil ring are largely lost.

When driving H piles, voids appear between the upper portions of the flanges, owing to subsidence of the soil gripped.* Most H piles, when pulled, show the space between the flanges compactly wedged with soil from tip to 5 to 15 ft below the ground surface, and the effective displacement volume and friction perimeter of the lower portion of the pile seems

* *Bearing Pile Investigations for Sepulveda Dam,* U.S. Engineer Office, Los Angeles, Calif., April, 1940.

to lie between the rectangle enclosing the pile and a circle circumscribing it. The outer faces of the flanges and of the wedged earth have been noted to be covered with layers of laminated clayey material as thick as ½ in. when the H pile has been pulled, thus corroborating laboratory experiments which show that the friction which can be developed between a pile and clay loam is greater than the shearing strength of the soil under corresponding pressures. It has also been observed that H piles slide out of the ground without disturbing the soil surrounding the adhering envelope while pulling tests on other types of piles might cause radial ground cracks extending several feet from the pile.

Observations have also shown that when untreated timber friction piles are withdrawn from soil after having been loaded, they are generally coated with a thin film of material, indicating excellent adhesion between the pile and the soil, and that failure occurred by soil shearing.[15,2j]

The reason that the value of skin friction between plastic soil and the pile is greater than the shearing value of the soil lies in the fact that the plastic material consolidates again following removal of the remolding force and will have a smaller void ratio and higher shearing strength than the partly remolded material located at a little distance from the pile. Because of this, piles driven in plastic soils are observed to have a coating of ½ to 1 in. of soil adhering when pulled. Consequently, the use of smooth casings should often be as satisfactory as the use of corrugated casings, from a skin-friction standpoint, in some soils.

Shape of Pile. The shape of the pile affects the unit skin-friction value. It has been found by tests in silty clay that the skin-friction value is larger per square foot for round than for square piles where the diameter of the round pile equals the side dimension of the square pile, the ratio being approximately 4 to 3.[34] The tests also indicated that skin-friction values increased with diameters of piles.

Taper. Taper increases the apparent friction on the side of the pile. The load creates both horizontal and vertical reactions in the soil as the tapered surface is pushed down. For vertical-sided piles, all horizontal compaction of the soil is done during driving, before working load is applied, but for a tapered pile to settle, some additional side compaction must occur. The taper also creates the effect of vertical bearing throughout the pile length, as well as at the tip. These are valuable effects in the strata selected for permanent load-carrying qualities, but conversely are undesirable in strata that are to be by-passed with load during driving, testing, and throughout the life of the structure.

Sequence of Driving. Sequence of driving piles in groups may affect lengths and resistances because of tightening of the ground, and should be noted on pile inspector's reports if possible.

Liquefaction of Soil. Liquefaction of the soil occurs in some regions, including the Middle Western portion of the United States. During much of the year a river bed may be dry, and if piles were driven at such a season reliance might be placed on both dynamic- and static-friction values based on the distance below the ground line. When the river fills, it has been noted that the river bottom may liquefy to a distance equal to the depth of water. It has been found that even shale bottoms may liquefy. This effect is distinct from scour, for the material is not removed, but it loses its pile load-carrying capacity and lateral supporting power as well.

Scour. Scour may greatly reduce the friction area of a pile. This may result from currents, floods, or ship-propeller action. In the southwestern part of the United States, a rule of thumb is to consider that scour may extend twice as far below the river bottom as the depth of floodwater in the channel.[48] Occasionally a figure of three times has also been found advisable. Jetting may be required to obtain the required penetration so that adequate embedment will remain after scour.

Avoidance of Friction. During driving, it is possible to avoid effects of friction from hard upper strata by driving through a temporary pipe sleeve which can be driven with a mandrel that completely fills the pipe when driving it. After the pile is driven, the temporary casing is pulled and used for the next pile. It is also possible to avoid expending energy in a hard upper layer of cohesive soil by spudding or coring a hole through it prior to driving each pile. In both of these ways, the driving energy of the hammer can be expended in effective driving at lower levels.

Negative Friction. Negative friction, in which soil pulls down the pile instead of supporting load, occurs often in regions of incompletely consolidated soft clay, silt, and organic soil. Natural consolidation may provide this effect, which is accelerated by fills and sedimentation. Additional load from a loose cohesionless top stratum may be caused by compaction of the stratum from vibrating equipment, traffic, or natural settlement.

Where a pile is driven through a hard upper layer, then through a soft layer and into a sand or gravel stratum, the friction of the upper layer might, as the soft layer settles, cause the upper layer to put additional load on the pile. In the case of piles driven through deep upper strata consisting of very soft material (which can neither exert much friction during driving, nor set up to cause much load by friction after driving) and only driven a relatively short distance into a firm stratum, piles act essentially in end bearing and may be computed by the dynamic formula, uncomplicated by friction. It is often sufficient, particularly on small jobs where the amount of difference which might be involved in total pile footage is inconsiderable, merely to bear in mind the possible effects of friction when selecting the factor of safety.

FACTOR OF SAFETY

Ratio of Working to Ultimate Load

After the total ultimate carrying capacity, including such adjustments for friction as deemed proper, has been determined, this value should be divided by a suitable factor of safety to obtain the total working or safe design load per pile. A factor of safety of 2 or 2½ is often satisfactory for use with the dynamic formula given in formulas (2.1*a*) and (2.1*b*) and with other ground-testing methods outlined in this chapter. *Ultimate* driving resistance is obtained from the Hiley formula as shown; the *Engineering News* formula contains a built-in factor of safety purported to be 6.

The factor of safety used with ultimate friction values is often lower than that used with the dynamic formula. If the ultimate value is known, as from a pulling test, or even from a loading test after making allowance for end bearing, the absence of other variables permits a closer approach of the working value to the ultimate. There has been a great tendency to neglect static-friction values and drive by dynamic formula alone, and to stop driving after a satisfactory set has been reached; this often causes opposition to requirements calling for deeper penetrations that may involve hard slow driving. The factor of safety used for static friction should be kept as low as is consistent with safety, or else considerable difficulties and great expense, all out of proportion to the additional factor of safety gained, may be needlessly caused. However, the designer should not allow himself to be forced below whatever factor he has finally concluded is actually required by good engineering judgment. A factor of safety against static friction of 1.5 seems to be about as low as it is desirable to go generally, and has been used on a number of projects. This factor also conforms to the requirements of some engineers as to load tests, which are often specified to be carried to 1½ to 2 times the design load. The value may be governed somewhat by the character of the soil, with less reliance placed on yielding soils than on sands and gravels.

Uniform settlement may not be serious under some structures. Differential settlements may or may not be detrimental.

Under heavy bridge piers or under turbine foundations, sheet-glass polishing machinery, large printing presses and similar expensive equipment which must be kept in alignment, and vibrating equipment, conservative values should be selected.

Higher pile values may be used under light manufacturing buildings, structures articulated for differential settlement, light bridges, wharves, piers, and trestle bents. Some structures can be arranged for shimming or jacking at necessary intervals, thus saving a large expense in installing fully nonyielding foundations.

Ratio of Live to Dead Load

When selecting the factor of safety the ratio of live to dead load should be borne in mind. Conservative pile values should be used under tall or heavy buildings for which the ratio of dead to live load is large and the intensity of bearing is high. In cases of end bearing or where the load-carrying strata surrounding the pile and below the tip consist of cohesionless materials, no appreciable further settlement will take place, and no distinction need be made between live and dead loads. When load from the pile is transferred into cohesive strata around and below the pile, however, a distinction should be made and the amount and duration of the live load considered, owing to the considerable length of time required for the applied load to squeeze the water out of the clay. The following examples are illustrative of this problem:

Wharf decks must necessarily be designed for a large live load per square foot to take care of local loadings. Such loads may reach only a small group of piles, and thus cause less intensities of soil loading than would the same unit load on the entire wharf. Furthermore, wharf live loads are usually of short duration.

Loads from gantries and other cranes are usually of short duration.

Live loads for which building floors must be designed are seldom reached over the entire structure at any one time. Many building codes specify allowable reductions for live loads over larger areas and for multistoried structures.

Live loads from trains and highway bridges are intermittent.

Wind loads, in such cases where these are a large enough factor to enter into the design of the structure, are usually of short duration.

Loads from earthquakes or earth tremors are of short duration.

Hydrostatic uplift loads, although they may last several days or longer, are yet of relatively short duration.

Effect of Number of Piles in Group

Pile grouping in individual footings may reduce friction-pile values, the corner pile having the lowest value. The end and corner pile groups under the entire building, if the building is stiff and the soil cohesive, will be required to carry more load per pile than piles in the central groups unless the number of such piles is increased. These considerations may appreciably affect the factor of safety and should be borne in mind unless already considered when the pile working loads were chosen.

Effect of Changing Soil Conditions

Lateral Movement. In earthquakes, oscillating lateral movements of the soil relative to buildings supported on piles may cause the soil around the upper portions of the piles not to remain in contact with the piles throughout their full lengths, and should be considered before placing full reliance on the friction value for the upper portions.

Liquefaction of the Soil and Scour. If either of these occur, they will reduce or remove the friction supporting the pile; allowance for these effects should be made in selection of length of embedment.

Vibration and Critical Density

Vibration. The effect of vibration from structures supported on piles, particularly in noncohesive materials, is a problem about which little has been published. In general, vibration tends to compact noncohesive soils, the extent of compaction and resulting settlement depending upon the initial density and grading, the energy and period of the vibrating force, and the length of time. During the pile-driving operations, the displacement volume of displacement piles usually more than compensates for the decrease in soil volume in the immediate area around the pile.

Sands are generally moderately dense in nature, and vibrations transmitted to the soil are usually small for most industrial installations, so that vibration settlements are usually small and occur over a long period of time. This is not always the case, and heavy hammers, large blanking presses, heavy shaking screens, or earthquake shocks may cause serious settlement even when the sand or sand and gravel appear to be moderately dense.

Susceptibility of the pile and soil to settlement under continued vibration may be investigated by loading the pile to working load and applying a vibrator. In one case, a shaking screen was set on a top platform. In another, a 12-in.-diameter steel disk 1 in. thick, mounted with 1 in. eccentricity, was located on the loading beam.[87,100] In both cases, the piles depended for support upon underlying submerged very fine sands or silts, and were intended to carry vibratory loads. Tests were run for several weeks and, in both instances, piles showed continued settlements when vibration was in effect and none when it was stopped. This was repeated for several cycles. In one case, the building site was changed and in the other, piles were extended to rock. The eccentrically mounted disk was driven by a variable-speed electric motor, and it was found that resonance periods were obtainable in all types of piles. Tests were run at resonance speeds. If the permanent-equipment frequencies approached the resonance period determined, or a harmonic, more serious results might be anticipated.

Natural Frequency. There is a natural frequency of vibration of the soil-and-pile system.* By varying the speed of the vibrator this may be observed; at this time the rate of settlement is apt to be many times that occurring when the frequency of the applied vibrations is higher or lower. There appears to be a critical range in which accelerated settlements take place, lying between about 0.5 and 1.5 times the natural frequency of the system. This critical range seems to be quite independent of the vibrator size, so that if test settlements are noted it is more than likely that settlements may occur after the structure is built, if vibrations in this range take place. In a test it was noted that settlement was greater, when the pile test load was greater, with vibration applied, thus indicating that vibration may lessen friction on the pile.

There are no definite natural frequencies which can be assigned to soils, but ranges of values have been observed experimentally, as shown in Table 2.4.

A method of sinking piles rapidly by exciting resonance to applied vibration has been invented.†

Critical Density. Very loose sands are occasionally encountered and, if these are also saturated, rather spectacular failures or settlements may result from pile-driving operations, vibrating equipment, or earthquake shocks. Experiments have shown that under shearing forces very loose sands decrease in volume, whereas dense sands increase in volume. The density at which no volume change occurs is termed the "critical density."‡ Since the sand grains do not change in volume, the change in volume of the soil mass due to shearing forces is actually a change in volume of voids. When soils below critical density are suddenly subjected to shearing forces or severe vibration, there is a rapid decrease in volume of soil mass. For sands above ground water this decrease results in appreciable settlements. If such sands are below ground-water level, such decrease in volume can occur only if water is expelled. Until this excess water has left the soil mass, individual grains falling into new positions are supported by the water rather than by contact with other sand grains, and the entire mass is temporarily liquefied in a manner similar to quicksand. This is the principle upon which the concrete vibrator works

* For a discussion and derivation of formulas, as applicable to loaded soil areas, see K. Terzaghi, *Theoretical Soil Mechanics,* John Wiley & Sons, Inc., 1943, pp. 434–454.

† Tests being made by C. L. Guild Construction Company, Inc.

‡ A. Casagrande and R. E. Fadum, "Application of Soil Mechanics in Designing Building Foundations," *Proc. ASCE,* vol. 68, No. 9, pp. 1487–1520, November, 1942.

TABLE 2.4. NATURAL FREQUENCIES OF SOILS

Supporting soil or rock	Natural frequency, cps	Safe bearing pressure, tons per sq ft	Remarks
Peat, 6 ft, overlying sand	12.5		From H. Lorenz, "Neue Ergebnisse der dynamischen Baugrunduntersuchung," *Z. Ver. deut. Ingr.*, vol. 78, 1934. Values measured using oscillation.
Medium sand, with peat remnants, old, 6 ft	19.1	1.05	
Gravelly sand with clay lenses	19.4		
Dense old slag fill, compacted by traffic	21.3	1.6	
Loamy sand fill, very old, dense, well compacted	21.7	1.9	
Clay, tertiary, moist	21.8		
Clay, Lias, moist	23.8	2.7	
Sand, medium, uniform, fairly dense	24.1	3.25	
Sand, fine, with 30 per cent medium	24.2	1.5	
Sand, coarse, uniform, dense	26.2	4.35	
Sand, nonuniform, very dense	26.7	4.9	
Gravel, pea, dense	28.1	4.9	
Clay, compact	28.1	4.9	
Limestone, soft, undisturbed	30.1		
Sandstone, undisturbed	34.0		
Granite, undisturbed	34.0		
Peat	7.5		From Andrews and Crockett, *Large Hammers and Their Foundations, Institute of Structural Engineers*, London, 1945. Values measured using oscillator. Pressures are engineer's estimates.
Silt, estuarine, waterlogged	10.0	0.8	
Clay, soft	12.0	1.05	
Sand, light, waterlogged	15.0	1.6	
Clay, medium	15.0	2.17	
Sand and hardish peat layers, wood	17.0	2.17	
Clay, stiff	19.0	3.25	
Silt and sand, mixed	23.3	3.25	
Sand and rubble, loosely compacted	23.5	3.25	
Limestone	30.0	5.25– 26.25	
Granite	40.0	10.5 –105.0	

SOURCES: References 159 and 160.

and upon which the vibroflotation process of soil compaction is based. It can be demonstrated in the laboratory by allowing a slow upward flow of water through a filter to loosen a container of uniform fine sand, while stirring, then allowing the water and sand to settle until their levels are the same, and placing a small weight on the surface of the saturated sand. A light tap on the container will cause liquefaction, and the weight will sink instantly into the almost liquid mass. The water standing on top of the sand will represent the reduction in volume occupied by the soil.

Vibration or impact from equipment reaching a sand of lower than critical density, which is supporting a load causing shearing stresses, may result in disastrous settlement. Where pile driving is done adjacent to a structure supported on such material, damage may be expected.

Pile driving in submerged sands below critical density shows rapidly decreasing resistance as soon as the tip enters such material underlying other strata, and the impression gained at such a site is that driving could continue almost indefinitely. If piles have been ordered to lengths based upon classification of boring samples, the lack of resistance will come as a shock, and the lengths available will be insufficient to reach a satisfactory dynamic resistance. Actually, there is no relation between driving resistance and static load in such soils, and if driving is stopped when adequate penetration is secured in the sand, computed from a static formula, it may be assumed that the sand will settle into a position around the pile as soon as the quick condition induced by the blow has dissipated itself. A load test is advisable, to satisfy critics.

A section through a site in which sand below critical density was encountered is shown in Fig. 16.1. Upon reaching the lowest stratum, very fine rounded and uniform clean sand, pile driving resistance immediately fell off to a small amount, although above this point it had been considerable.

Deposits of sand appreciably looser than critical density are not common. Natural modes of deposition usually result in well-graded sands or sands and gravels at or above critical density. This does not hold for sands of a uniform grain size. Where loose uniform sands and especially clean rounded fine or very fine sands are encountered, the possibility of the material being below critical density and the possible effects of pile driving and vibration should be considered.

Methods of protecting against settlements resulting from pile driving, vibration, or earthquake shocks, affecting noncohesive soils below critical density, may consist of grouting the soil or soil compaction. Cases of settlement are described in Chap. 16, and methods of soil solidification are given in Chap. 14. It is recommended that the services of a soil-mechanics engineer be secured when operations of this character, involving possible settlements of structures, are encountered.

CHAPTER 3

HAMMER SPEEDS, STROKES, AND DRIVING STRESSES

Speed of Hammer

Double-acting and differential-acting hammers should be run at full listed speeds, since the *net* available energy at the pile tip falls off rapidly at lesser speeds, particularly sharply in the case of undersized hammers. The number of strokes per minute occurring when taking final penetration readings should always be noted on pile-driving reports, particularly if for any unavoidable reason the speed is less than the maximum specified, for if the speed has fallen off and is not noted, the energy is reduced, and the smaller penetrations obtained will indicate falsely high bearing values. Strokes must be counted when driving a pile, and not with the hammer running free, as the hammer speeds up at the end of driving as the penetrations become small and resistance increases, owing to the greater rebound of the hammer helping on the upstroke. Frequently, it is impossible to place any faith in old pile-driving reports, because of lack of notations regarding hammer speeds; it is, and probably has been more so in the past, fairly common practice to pay very little attention to the hammer speed or its variations. Sometimes steam pressures have been noted at the boiler or in the line near the hammer. The difficulty of translating such a pressure reading into terms of energy with any degree of accuracy may be observed in the comments given under the definition of the term for efficiency e_f. Many pile inspectors are not fully aware of the theory behind pile driving; therefore, the specifications should emphasize this point.

It may be assumed that if a differential-acting hammer is operating at or near the full stroke and if the body of the hammer is rising noticeably with each blow, thus indicating that steam pressure is adequate, the hammer is delivering a full blow so far as practical pile driving is concerned.

Stroke

In the case of single-acting steam hammers, the *actual stroke* should be measured at the time close to final penetration. Strokes frequently are less than the nominal figures indicate, and because of the sensitivity of the formula, when the net force at the tip is small compared with the applied force, a reduction of a few inches in the stroke, with consequent reduction of applied energy, may make a large percentage reduction in the net energy after deducting losses during driving. Cases have been observed where the hammers have been rebuilt, with strokes considerably shorter than nominal figures.

The inspector should note the stroke on his reports and, if less than normal, should immediately call the matter to the attention of the engineer.

Stresses in Piles

Normal Stresses. It is important to compute stresses in the pile, with the hammer and speed being used, or proposed, at the head, tip, and mid-point of the pile, or at points of entering harder strata, and to compare these values with the allowable stress in the material being driven. It is suggested that stresses be computed by means of formulas (2.2), (2.3), and (2.4); that is, by dividing the ultimate driving resistance R_u corresponding to the applied energy and the set by the net area of the pile at that point. An example of the computation of stresses in a pile by this method is shown in Appendix II (pages 569–575).

Overdriving. Overdriving of wood piles is indicated by (*a*) bending of the pile; (*b*) bouncing of the hammer; (*c*) cutting of driving plate into head of pile; (*d*) separation of wood along annual growth rings near the surface; and (*e*) brooming of the head. However, damage may still occur, in the case of tapered piles, before any of these symptoms appear, since in tapered piles the stress at the tip is usually greater than at the head, owing to the much smaller cross-sectional area of the pile at this point. The stress at some intermediate point may be greatest, depending on the pile taper and the relative heights of the friction strata carrying the load. Therefore, it is *not a sufficient warning of damage to the pile to observe incipient brooming or crushing at the head.* Stresses computed from driving resistances obtained by some of the brief formulas in common use may often be several hundred per cent from the figures thus obtained.

To avoid damage to the pile, the movement or value of the set s must be of sufficient magnitude to keep the stress in the pile within allowable limits, unless the elasticity of the pile is sufficient to absorb all of the applied energy when the value of s is zero.

The rapidly increasing rate of stress increase for small sets is shown by upper portions of the typical curves in Fig. 2.1. In this particular example, the allowable fiber stresses are not exceeded with the No. 9-B-2 McKiernan-Terry hammer even at very small sets, owing to the small size of the hammer; if a No. 10-B-2 McKiernan-Terry hammer had been used, it can be seen how the upper end of the curve would have ascended to much higher values of fiber stress with small sets, and overdriving would have caused stresses in the pile to reach approximately the

FIG. 3.1. Typical effects of overdriving upon wood piles. Piles exhumed along Norfolk & Western Railway. (*Courtesy of Raymond Concrete Pile Co.*)

yield point of the steel. For this reason overdriving should be avoided as possibly doing much more harm than good. It is better to have an unbroken or undamaged pile extending to a certain tip grade than to have a broken or broomed pile, possibly with the damage unknown, driven apparently to a slightly lower head grade. In some cases, it may be necessary to resort to jetting to reach the necessary tip grade without overdriving.

It would be well to select speeds or drops which would have some factor of safety with respect to the maximum allowable fiber stress, when driving at the required ultimate resistance. This factor should be largest in the case of wood piles to allow for weakening because of slight kinks, bends, knots, turpentining, etc. Treatment may weaken piles. Tip stress may govern, particularly in the case of tapered piles or end-bearing piles, and therefore the use of small tips should be avoided with materials having low strengths in compression. The use of large tips in wood piles is generally advisable. Governing stress in non-end-bearing piles of constant cross section is usually at the head.

FIG. 3.2. McKiernan-Terry 9-B series double-acting steam hammer being used in demolition work, illustrating possibilities of overdriving and damage to pile. (*Courtesy of Spencer, White & Prentis, Inc.*)

In borderline cases, particularly when using wood piles, it is desirable to pull one pile or more to observe the condition of the wood. With steel shells, there is the advantage that a light may be lowered into each shell and conditions inspected for the full height.

Tension Failures. Tension failure during driving conventional precast piles takes the form of many minute cracks, which are either not indicated or are considered unimportant, because they heal with time and under load. However, there have been some bad failures with prestressed piles. In this case, the point of failure may be much more concentrated.

Tension failure tends to occur during the first few blows when the point of the piles can shoot ahead rapidly and has to be held back by the reinforcing steel. In cases where prestressed piles were inadequately reinforced, it has been necessary to reduce the hammer blow during the early part of driving and resume full power when the pile met serious resistance. Double-acting and differential-acting hammers can be throttled to reduce the blow. Single-acting hammers require an adjustable valve control.

Use of new packing or cushion on the head of each pile at the start of driving will greatly reduce tension failures by producing soft blows. By the time the packing has fully consolidated, the pile point has reached firmer resistance.

Head Failure. Head failure may often be prevented in wood or concrete piles by banding the head. In precast concrete piles, the head stress caused by impact and lack of care in the setup of the cap, etc., is often close to the crushing strength of the concrete, and the best way of avoiding trouble during driving is to take care to provide a strong mix, adequate damp curing, sufficient aging, and a casting method which will ensure flat smooth heads cut at a normal to the axis of the pile. Furthermore, the ill effects of slight irregularities may be reduced by the use of a thin layer of soft packing material against the pile head. The Building Research Board has found[21] that the results of a series of impact tests on concrete blocks indicate that the impact strength of concrete may run as low as 50 per cent of the cube-crushing strength.

Setting Piles

Failure due to excessive stresses caused by lack of care in the setup of the driving equipment may be caused by the incidence of conditions not intended to occur, such as much fuller consolidation of the packing material in the driving cap than usual, tilting of the driving cap, nonaxial blows, unsymmetrical placing of the packing in the cap, uneven head on the pile, etc. Driving caps should not have undue clearance, since this may allow the packing to move off center of the pile head to such an extent that it fails to cover it fully. Sharp edges in cap packing on the head of piles is also a cause of high local stress.

It is important to set the piles vertically or on the axis which they are to follow and to be sure that the hammer is centered over the pile so that the blow will be entirely axial and the pile will not be driven out of line.

Eccentric blows may stress the pile greatly and may cause damage. It is very important that the hammer strike a square blow on the pile head. If the head is not square, it becomes damaged and the effective energy of the blow may be much reduced; also if energy is consumed in springing or bending the pile at each blow, the effective energy is reduced.

Although a certain amount of force may be used to pull a pile back into line, too much of a bend may cause eccentric forces which will break the pile when struck. If a pile deviates too far from its proper axis, it should be cut off and abandoned and a new pile driven adjacent if the design of the structure will permit, or it should be pulled and reused.

Templates or gates may be laid on the ground and braced in position, or extended as elevated frames. A floating template for use on water has been employed to align piles in bents. Multiple leads or templates for setting considerable groups of piles at one time have been used,[65,74,3y] sometimes mounted on barges. A cantilever jig has been used for driving piles in a trestle, fitting over the last two pairs of piles and projecting forward, at two elevations, with openings to guide the next pair of piles.

Inclined piles should be held firmly in position so that the hammer blows will remain axial.

FIG. 3.3. Template on ground for guiding piles when driving without leads (note bracing). An upper frame, as well as a hole bored in the ground, may also be necessary. (*Courtesy of Portland Cement Association.*)

Banding

Concrete Piles. Concrete pile heads may be banded to prevent spalling or shattering, for piles which otherwise appear capable of being driven satisfactorily. Two or three bands may be used, the first being placed 1 in. clear of the head and the remainder at centers about equal to half the over-all diameter of the pile. To ensure that the top ring shall not work loose it should be either anchored into the concrete or roughened on the inside to provide good bond. When the heads of precast concrete piles are removed after driving, the bands can be reused.

Wood Piles. Brooming of heads should be avoided since a great deal of effective energy is lost this way. The head should be recut if broomed. Wood pile driving caps usually are recessed to receive the chamfered head of the pile; if not, it is often advisable to place a steel ring around the head of a wood pile or to wrap it with wire in order to prevent splitting the head portion of the pile. A small-diameter cable with a clamp may be used. The use of butt protection is particularly necessary when driving with a drop hammer in hard soil. One railroad wraps all wood pile butts with a 3-in. band of No. 12 annealed iron wire fastened in place with 1½-in. fence staples, keeping the top of the band 12 in. below the pile butt. The butt is cut square, chamfered for a standard follower block, and wrapped at the treating plant, after seasoning and immediately before treatment, to avoid the possibility of the wire loosening because of shrinkage during the seasoning period.[43] Wrought-steel bands may be purchased from hammer manufacturers; ordinarily ¾ by 3 in. is used. Banding is inferior to the use of pile caps as a means of butt protection. Banding should not be considered as increasing the allowable fiber stress in the wood but merely as a means of avoiding local damage from such causes as incipient checking and hammer blows struck slightly out of the line of the axis of the pile.

Driving Shoes

The use of driving shoes on the tips of wood or concrete piles may be advisable to assist in cutting through some hard overlying stratum which is to be penetrated. However, the use of such shoes should not lead to the conclusion that the pile will not be damaged. Cases have been observed where the shoes, which do not generally have as large an overall diameter as the pile tip, have been driven up into the pile. The stress at the pile tip corresponding to the energy applied and the penetration obtained should be investigated as described to determine the possibility of damage. By pulling a pile, the actual condition may be observed. It apparently makes little difference whether shoes are open or closed at the end. Typical driving shoes are shown in Fig. 3.4.

Cobi pile tips provide protection and a cutting edge while avoiding the tendency of pointed tips to drive out of line. They are available in malleable iron for wood piles with 6-, 7-, and 8-in. tips, and as inverted paraboloids of forged or cast steel for precast concrete and pipe piles from 8 to 20 in. diameter. Metal in the paraboloid type is in tension and thus allows ¼-in. metal to be used and weight to be saved. Either inside or outside fits for pipe piles may be obtained. For use on precast piles a welded-on collar is provided. Larger special sizes may be obtained. They are made in the United States, and are controlled by the Albert Pipe Supply Co., Inc., Brooklyn, N.Y.

The Pilot timber pile boot is a heat-treated casting tapered to fit various pile diameters and has a flat bottom plate. Four external fins flare out toward the base. These lugs act as a drill, penetrating coarse sand, hardpan, and decomposed rock. Boulders may be broken or pushed aside. These boots are made by Associated Pipe & Fitting Co., Inc.*

Shoes may be made on the job by bending a piece of strap steel of less width than the tip diameter and sharpening the pile to a wedge as wide as the strap instead of to a point.

A piece of pipe about 12 in. long, into which the lower end of the pile is fitted, makes a shoe that prevents tip damage. The pipe shoe is set

* Also manufacturers of many types of fittings for pipe, shell, wood, and composite piles.

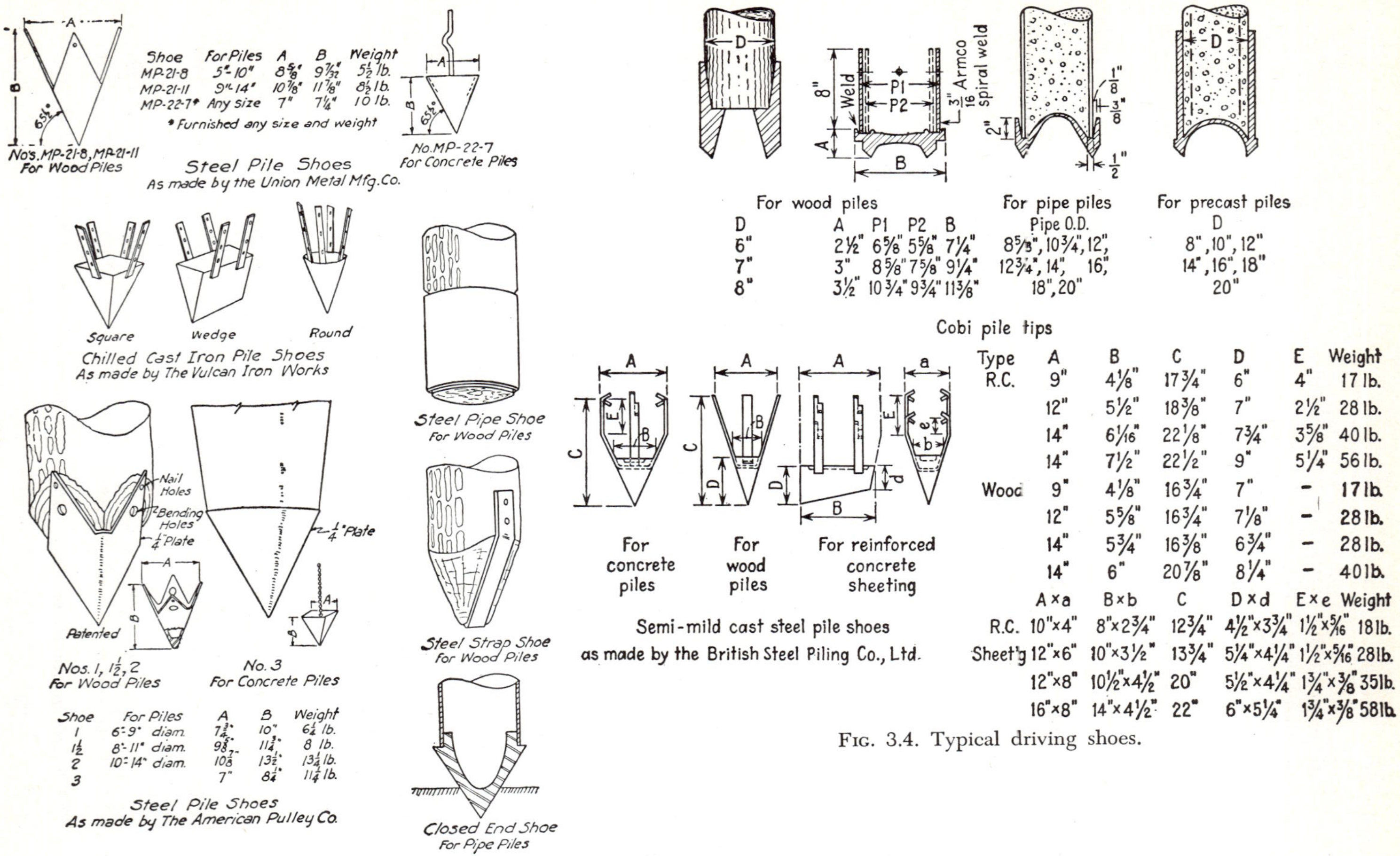

Steel Pile Shoes, as made by the Union Metal Mfg. Co.

Shoe	For Piles	A	B	Weight
MP-21-8	5"-10"	8⅝"	9 7/32"	5½ lb.
MP-21-11	9"-14"	10⅞"	11⅞"	8½ lb.
MP-22-7*	Any size	7"	7¼"	10 lb.

* Furnished any size and weight

Steel Pile Shoes, as made by The American Pulley Co.

Shoe	For Piles	A	B	Weight
1	6"-9" diam.	7¾"	10"	6¼ lb.
1½	8"-11" diam.	9⅝"	11¾"	8 lb.
2	10"-14" diam.	10⅞"	13½"	13¼ lb.
3		7"	8¼"	11¼ lb.

For wood piles

D	A	P1	P2	B
6"	2½"	6⅝"	5⅝"	7¼"
7"	3"	8⅝"	7⅝"	9¼"
8"	3½"	10¾"	9¾"	11⅜"

For pipe piles — Pipe O.D.: 8⅝", 10¾", 12", 12¾", 14", 16", 18", 20"

For precast piles — D: 8", 10", 12", 14", 16", 18", 20"

Cobi pile tips

Type	A	B	C	D	E	Weight
R.C.	9"	4⅛"	17¾"	6"	4"	17 lb.
	12"	5½"	18⅜"	7"	2½"	28 lb.
	14"	6 1/16"	22⅛"	7¾"	3⅝"	40 lb.
	14"	7½"	22½"	9"	5¼"	56 lb.
Wood	9"	4⅛"	16¾"	7"	–	17 lb.
	12"	5⅝"	16¾"	7⅛"	–	28 lb.
	14"	5¾"	16⅜"	6¾"	–	28 lb.
	14"	6"	20⅞"	8¼"	–	40 lb.

Type	A×a	B×b	C	D×d	E×e	Weight
R.C.	10"×4"	8"×2¾"	12¾"	4½"×3¾"	1½"×5/16"	18 lb.
Sheet'g	12"×6"	10"×3½"	13¾"	5¼"×4¼"	1½"×5/16"	28 lb.
	12"×8"	10½"×4½"	20"	5½"×4¼"	1¾"×⅜"	35 lb.
	16"×8"	14"×4½"	22"	6"×5¼"	1¾"×⅜"	58 lb.

Fig. 3.4. Typical driving shoes.

so that the bottom edge is flush with the bottom end of the pile, and the pile taper prevents the sleeve from moving up the pile. The pipe sleeve may be open-ended. The pile is not sharpened.

Piles with square ends are more easily kept in line while driving and provide better bearing for end-bearing piles. Points add little, if any, to the rate of penetration.

Yield-point Stresses in Piles

Yield-point Stresses in Wood. Maximum allowable stresses at failure for wood are given in Table VI. The great increase in allowable stresses for air-seasoned over green wood and the considerable reduction in allowable stresses for wood seasoned by steaming or boiling under vacuum should be noted. In case it is not convenient to test a short section to determine the allowable stress which it will stand, a rough approximation of the amount may be made by determining the weight per cubic foot and thus estimating the degree of seasoning by comparing this weight with green and air-seasoned weights given in Table VI. The exact volume of the short piece of timber can be determined accurately by measuring the volume of water which it will displace.

FIG. 3.5. This pile hit a block of broken concrete fill.

Yield-point Stresses in Steel. The yield point of the steel used in Union Metal Monotube shells is approximately 50,000 psi. The yield point of structural steel is generally taken at approximately 36,000 psi.

Yield-point Stress in Concrete. The ultimate crushing strength of concrete depends on the mix, age, and method of curing. Control-test specimens should be made when piles are cast, cured under the same conditions as the pile and not under laboratory conditions, and tested, when pile driving is ready to take place, for ultimate compressive strength per square inch and value of E. Values of E for concrete may range from a low of 2,000,000 to a high, in exceptionally good and well-cured mixes, of 6,000,000.

Field Determination of Stresses in Piles

The investigation of the Department of Scientific and Industrial Research[21] in England has concerned itself with the driving stresses in precast concrete piles of uniform cross section, and the results have been put in the form of charts. Although the results which will be obtained by the method of stress determination outlined below will often show considerable variation from those charts, the charts are not readily available and are limited to that type of pile, not frequently driven in the United States.

Because the unit fiber stress in a tapered pile may be larger at some point below ground than at the butt, nonoccurrence of damage at the butt is no assurance that damage is not taking place at some lower point.

The following brief method is suggested as a simple means of ready field determination of the approximate magnitude of the fiber stress at any point in a pile during driving, in case curves such as shown in Figs. 2.1 and 2.2 are not available.

The proposed method of stress determination is not useful when driving thin shells with heavy mandrels, since the compression of the blocking in the driving cap is large as compared with the elastic compression of the mandrel. However, it is not with this type of piles but with wood, steel, or concrete piles of relatively small net area or reasonably high fiber stress that we are concerned in regard to the possibility of damage.

The proposed method of stress determination is based on the use of the pile itself as a spring gage measuring its own stress. This may be done by assuming the validity of Hooke's law as applied to the pile and soil. The simple basic relation $W_r h = R_u s$ would hold true were it not for losses in efficiency in the hammer, impact loss, and elastic losses in the pile cap, pile, and soil. The above equation, when written to represent conditions at the top of the driving cap on the head of the pile where it must also take account of movement at that point, becomes $W_r h = R_u(s + C)$, in which C represents the average elastic movement of pile, cap, and soil. Inclusion of all these losses results in formulas (2.1*a*) and (2.1*b*).

In the field determination of stress, the terms C_1, C_2, and C_3 (which equal C) are not computed, but the rebound of the pile, measured on the pile near the head, is measured by the field graph made as shown in detail in Fig. 2.3. Since the need for determining pile stresses occurs only during such hard driving that the values of $C_2 + C_3$ are relatively large, it is permissible and slightly on the safe side to ignore C_1, the temporary compression in the pile cap, which is relatively small and also difficult to measure, and to measure $C_2 + C_3$ for use in the determination of stress. With wood piles there is often no cap in any event and C_1 would be zero. Also, with the types of piles under consideration, which exclude heavy mandrels, the use of zero for the coefficient of restitution e causes very little change in the results and, for the purpose of determining fiber stresses in these types of piles during driving, the accuracy resulting from the inclusion of this term is not warranted.

The field procedure is to measure from the graph the rebound of the pile close to the head, as well as the net set. The observed rebound is $C_2 + C_3$. If the term $\frac{1}{2}(C_1 + C_2 + C_3)$ in formulas (2.1*a*) and (2.1*b*) is replaced by the term $B/2$, which is one-half the observed bounce, representing the average temporary compression, and e equals zero, then formulas (2.1*a*) and (2.1*b*) become, for use with drop hammers, single-acting steam hammers, and diesel hammers:

$$R_u = \frac{e_f W_r h}{S + B/2} \times \frac{W_r}{W_r + W_p} \tag{3.1a}$$

and for use with double-acting and differential-acting steam hammers:

$$R_u = \frac{12 e_f E_n}{S + B/2} \times \frac{W_r}{W_r + W_p} \tag{3.1b}$$

In the case of end-bearing piles, since the value of the driving resistance R_u is constant throughout the length of the pile regardless of whether or not the pile is tapered, it is possible to divide the value of R_u by the net area A_t of the tip of the pile and obtain the maximum unit fiber stress p in the pile. This relationship may be stated in the following formula for end-bearing piles:

$$p = \frac{R_u}{A_t} \tag{3.2a}$$

where p = the unit fiber stress in the pile.

In the case of friction piles, or piles resisting the net driving force partly by end bearing and partly by friction, the point of maximum stress may be at some point above the tip. The fiber stress at any point may be computed by deducting from R_u the proportion of total driving resistance which it is judged has been removed by friction above that point.

The amounts of force resisted by the tip and removed by the different strata may be quite closely judged from a foot-by-foot driving record of the number of blows per foot of the pile as driven. If some doubt is felt as to the reasonably close selection of these proportions, it is possible to assume upper and lower limits for these values, thus bracketing conditions, with good assurance that the actual value lies within, or close to, this range. The formula for fiber stress then becomes, for friction or combined friction and end-bearing piles,

$$p = \frac{R_u - R_f}{A_p} \tag{3.2b}$$

where A_p = net area of the pile at any point considered.

Although the inapplicability of a dynamic driving formula to pile driving in cohesive or clayey soils cannot be too strongly emphasized as regards the determination of permanent load-carrying values, this type of formula gives the amount of temporary resistance to the applied force of driving in this type of soil as well as in noncohesive soils. And in investigating the stress in the pile during driving, this fact becomes useful and the above formulas are entirely applicable. Since the driving force is generally several times larger than the working load on the pile, it is the stress during driving which requires investigation. The stress in the pile under working load was presumably considered during the selection of the pile and is within safe working limits.

Driving should not stop to permit placing the paper for the graph on the pile, since this might result in unduly hard driving to loosen the pile and give results not in line with normal driving. On the other hand, if for some reason it has become necessary to stop driving before having reached the desired depth, a graph will afford means of checking any unusually high stresses which may occur during the loosening of the piles. In such cases, it should be borne in mind that the center of driving resistance is often raised to quite a high point in the pile, owing to temporary setup of the ground around the pile, although, after the pile has become loosened, driving may become easier and the center of resistance fall to the position which it would have had if no stoppage had occurred. This affects the area of the pile cross section to be considered in the case of tapered piles.

The efficiency values previously given for e_f were intentionally given on the low side, in order to be reasonably sure of securing the net energy expected. When investigating possible fiber stresses in the pile during driving, however, the efficiencies should be taken as high as may reasonably be expected, in order to be fairly sure that stresses will not be greater than expected. For this reason, it is suggested that 10 per cent be added to all values tabulated for e_f when using formulas (3.1*a*) and (3.1*b*).

For a numerical example of this field method of investigating pile stresses during driving, see Appendix III, pages 577–579.

The wave equation indicates that stresses in the pile are considerably higher in some cases than those given by the Hiley-type formula. The stresses are due to force required to overcome ground resistance and to accelerate the pile very rapidly.

CHAPTER 4

DRIVING EQUIPMENT

Hammers

Drop Hammers. These are still frequently used. The weight is raised by a rope running over the top of a framework and extending back to a drum or geared shaft. It is released by tripping it to drop free of the rope or by releasing the drum to allow the rope to unwind. The drag of the rope and drum reduces efficiency.

FIG. 4.1. Vulcan No. 1 single-acting steam hammer driving 18-in. pipe piles 184 ft long to rock for Starrett-Lehigh Building, New York City. (*Courtesy of Spencer, White & Prentis, Inc.*)

Single-acting Steam Hammers. Steam or air raises the movable mass of the hammer which then drops by gravity alone. Hammers of this type, in which the ram is the movable part, are always employed in leads. In European countries, some single-acting steam hammers employ the casing as the movable part, the piston being stationary and resting on the pile by means of a hammer rail and moving down with the pile. The characteristics of the blow are a low striking velocity, owing to the low fall, and a heavy striking weight. The blows are much more rapidly delivered than for a drop hammer.

Double-acting Steam Hammers. These employ steam or air to raise the striking part and also to impart additional energy during the down-stroke. The downward acceleration of the ram owing to gravity is increased by the acceleration due to steam pressure. The energy of the blow is given by the term $E_n = W_r v^2/2g$. They run at greater speeds than do single-acting hammers.

Differential-acting Steam Hammers. These also employ steam or air to raise the striking part and impart additional energy during the down-stroke. It is claimed that the nonexpansive use of steam in the steam cycle obviates a drop from the entering steam pressure to mean effective pressure. They have been made in both open and closed* types, and run at greater speeds than do single-acting hammers. The closed type is no longer made.

Internal-combustion Hammers. These are operated essentially by internal combustion and are not diesel hammers.

Diesel Hammers.[68] These consist of a cylinder, ram, anvil block, and simple fuel-injection system. Principal makes on the United States market are the Delmag, McKiernan-Terry, and Link-Belt Speeder. Each has its own modifications and advantages. To start, a cable lifts the ram and a trip drops it. The falling ram actuates a fuel-metering pump which injects the fuel into the cylinder. In the Delmag and McKiernan-Terry hammers, the fuel is squirted into a cup in the anvil block, atomized by the impact of the ram, and ignited by the heat of compression. These two hammers are stopped by pulling a rope to disengage the fuel-pump cam. In the Link-Belt hammer, a high-pressure fuel-injection system atomizes the fuel as it is squirted into the chamber, permitting the operator to control the energy of the blow and to stop the hammer from his cab.

FIG. 4.2. Link-Belt Speeder Model 105 diesel pile hammer. (*Courtesy of Link-Belt Corp.*)

While the Delmag and McKiernan-Terry hammers are open at the top, permitting the rising ram to project, the Link-Belt cylinder head is closed, and as the ram rises it compresses air in the head and two bounce chambers, shortening the upstroke and speeding the downstroke, which increases the blow rate. The Link-Belt hammer has a battery-operated glow plug to help ignition in cold weather and with soft driving.

* Manufacture of closed type has been discontinued.

The length of stroke is proportional to the pile resistance; therefore the harsher the driving, the more effective the hammer. This is particularly true of the Delmag and McKiernan-Terry hammers, which depend entirely on ram impact for ignition; in this fact also lie the greatest weaknesses of diesel hammers, namely, difficulty of rating the energy and of operating in soft driving.

Diesel hammers weigh considerably less than steam hammers, are free of the need of boilers or compressors, are more mobile, require less moving time, use only a gallon or two of fuel per hour, have less maintenance, operate at temperatures as low as $-18°$ F, and have no water, steam, or air hoses to disconnect and drain each night nor boilers to keep hot overnight or weekends or to start up again. Diesel hammers, being basically single-acting, depend on a long drop. Lengths of Delmag and McKiernan-Terry diesels compare favorably with correspondingly rated steam-air hammers but are larger than double-acting hammers and also require space for the rising of rams. Link-Belt diesels are somewhat shorter. Long hammer lengths reduce length available for piles in the leads. Diesels are not fast, although Link-Belts are considerably faster than Delmag or McKiernan-Terry diesels. Because pile resistance is a requisite for diesel hammers, some of them will not continue to strike a broken pile or continue to operate if a pile falls out from under the anvil; this protects workmen and equipment.

Accurate energy determination for diesels is difficult. McKiernan-Terry rate their diesels by multiplying the ram weight by the measured ram stroke; Delmag computes the energy of fuel injected; Link-Belt considers the above factors and others, including the effect of the bounce chamber. McKiernan-Terry diesels carry a scale showing the height of ram at each stroke, and while there are losses due to friction and compression of air in the combustion chamber, they are probably more than offset by the larger explosion energy. Delmag contends that the energy per blow is constant, regardless of stroke length, after the fuel valve is set and the amount of fuel for each cycle is constant; whatever energy is not used to push the pile down lifts the piston and becomes available in the form of kinetic energy; heat and friction losses are neglected. Link-Belt diesel energy determination is complex and academic because the operator can change it at will by controlling the fuel valve; the combustion rate of the atomized fuel is also dependent upon temperature and pressure. Engineers often downgrade the manufacturer's rated diesel energies, the McKiernan-Terry least because it is simply observed and conservatively rated.[68] Energy reductions for diesel hammers are indeterminate and are not specifically evaluated in definitions of the efficiency factor in the driving formula, but should be selected by judgment and experience.

Whereas steam hammers start by opening a valve, the diesel requires an assist from the crane. Extremely soft driving can be a problem. Inability to reactivate itself in the absence of pile reaction is one of the diesel's greatest weaknesses. This could result in dropping and lifting the hammer by the crane like a drop hammer as long as soil was soft.

FIG. 4.3. McKiernan-Terry No. 7 double-acting steam hammer turned upside down and rigged for pulling a doubtful pile. (*Courtesy of Engineering News Record.*)

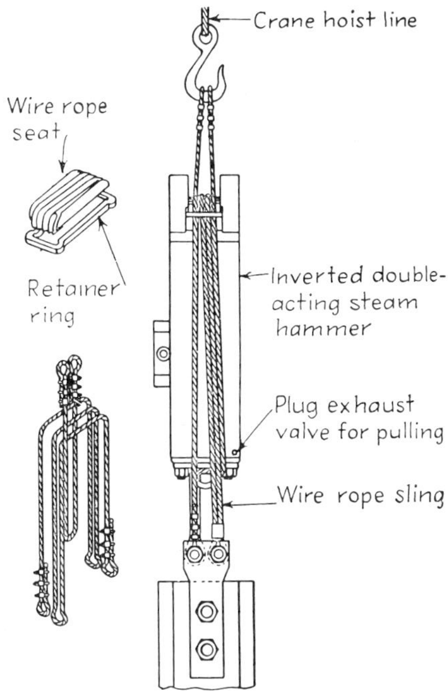

FIG. 4.4. Double-acting steam hammer rigged for extracting piles.

The amount of resistance required to reactivate a diesel varies in proportion to its size, and a small hammer will sometimes do a good job where a larger machine will not run.

Extractors

Extractors are made particularly for pulling piles. Double-acting hammers may also be used for this purpose. Properties are given in Table IV.

A steady pull may be obtained by using a multiple-part line on a drum cable. A 14-part line was used for pulling old piles on Chicago subways, by providing a post support to the top of which a seven-sheave block was connected, with a 14-part line connected to a suspended seven-sheave block. The main hoist line led from the bottom block, and an auxiliary hoist line held the post erect.[83] Such devices may be required to start piles or may be used to supplement extractor blows.

A rig employing four 35-ton hydraulic rams has been built to pull wood piles.[2bm,2bo] The unit is self-contained and requires only a truck crane to assist in its operation. A gripping collar with four levers having detachable shoes is used to hold the pile. As the rams move up, the shoes dig into the pile and move it up the 6 or 8 in. necessary to loosen it. Then the truck crane with a cable sling pulls the pile out. Mobile hydraulic jack rigs for pulling wood piles without jetting have been used.[2bm,2bo]

Marine piles can usually be pulled by a 6- to 10-part line supported over a gallows frame on the forward end of a barge and operated by a hoist engine at the other end. The pull is often limited by the flotation of the barge. Surging the barge may help. Jetting will reduce the resistance.

Steel sheet piling has been pulled by a Heppenstall tong suspended from an A frame mounted on a skid rig. Pulls of 50-ton rated and 100-ton actual values have been used with little damage to the pile faces. Percentages of recovery were high.*

Driving Rigs

Typical Rigs. Various typical pile-driving rigs are shown in diagram in Fig. 4.6 and in photographs. Rigs comprise framework and platforms to support the engine, boiler, winches, and drivers. Both standard and special rigs are made by pile-hammer manufacturers, and their catalogues may be consulted. Very heavy rigs for handling exceptionally long piles are built to suit special conditions on large projects. Frameworks for drop hammers are usually made from wood. The simpler types of frameworks for other hammers may be either wood or steel, whereas the more elaborate and portable rigs use steel.

Land Rigs. Drop-hammer rigs using hand winches, horsepower, or small gasoline engines for lifting are still available. Examples are Township and Joe Heaver drivers.

Standard contractors' pile drivers, intended for use with the heavier sizes of drop hammers, may be provided with or without extension sills to carry the engine. Where small rollers are shown under sills, rigid

* *Construction Methods and Equipment,* vol. 41, No. 3, March, 1959, pp. 109, 110.

roller bearings can be used, to permit the use of two 10-in. pipe rollers. Spool-roller pile drivers roll sideways and forward and back, the flanged spools running on two longitudinal pipes. Batter leader drivers, or pendulum rigs, drive piles at an angle, such as required for trestles, and will also drive vertically. They are also often mounted on barges.

FIG. 4.5. Universal driver capable of driving both in and out and side batters up to 1:1. Note motor-driven pile spotter for moving leads in and out and moonbeam guide which permits side motion. Shown driving composite piles at Naval Air Base, Alameda, Calif. (*Courtesy of Raymond Concrete Pile Co.*)

Swiveling pile drivers, or turntable rigs, are among the most useful and efficient drivers where large numbers of piles are driven in a small area, for piles can be driven anywhere within the area reached by the leads, when swinging to either side, rolling forward or backward, and sliding sideways. The ability to use telescope leads with this rig adds to its value. Tilt rigs also provide one of the most economical rigs for large compact operations, as they are rolled forward or back or slid sideways on the long cross-pipe rollers by quick pulls on cables attached to drums

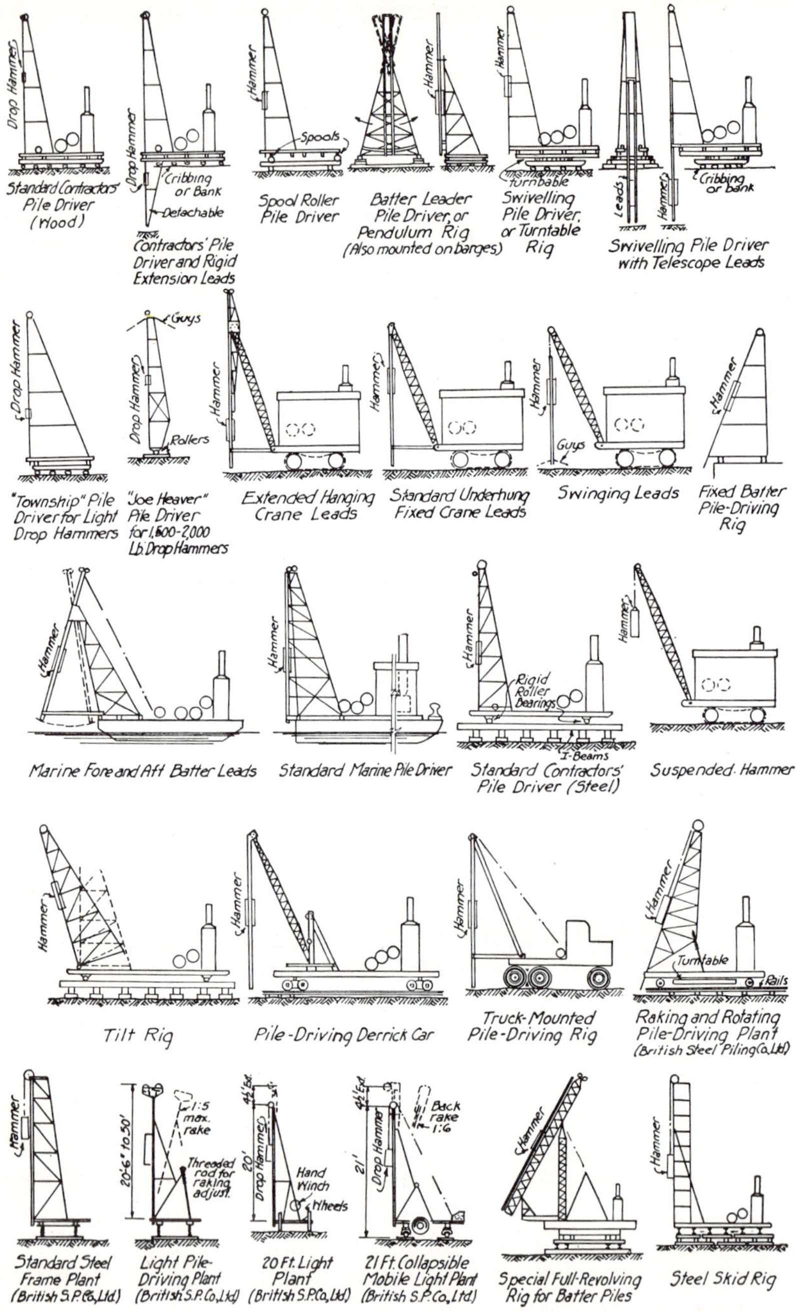

FIG. 4.6. Typical pile-driving rigs.

on the platform and powered by the boiler. This rig also permits driving batter piles with the same driver. Pendulum rigs and fore-and-aft rigs may be mounted on barges, cars, trucks, or skid rigs, and drive vertical as well as batter piles.

Light steel-framed pile-driving plants, particularly designed for use with light sections of steel piling, are available in Great Britain. These

FIG. 4.7. Pendulum rig driving batter piles at Ford Motor Co. plant, Richmond, Calif. (*Courtesy of Western Foundation Corp.*)

plants are operated by hand for the smaller, and by petrol or diesel winches for the larger, sizes. These rigs have an adjustment to allow driving on a batter. The frames are mounted on a flat base, on rail wheels, or rollers, or on pneumatic tires for use on loose sand or beach. There is also a 21-ft frame which folds up on a two-wheel trailer.

Locomotive, Caterpillar, and Whirley Cranes. These are used for pile driving when equipped with a boom for handling pile leads and steam hammer, and a drum or winch head having a line to the head of the boom for handling the pile. The steam hammer is handled by a fall from the head of the boom. The locomotive crane operates on railway tracks.

When driving from a crane, it has been common practice to use fixed, or hanging or swinging leads, but with the growing practice of using templates or gates, leadless hammer suspensions by cables from cranes are becoming more popular for bridges, trestles, etc., on account of the portability and flexibility of the equipment. For groups containing a large number of piles, leads are usually more economical; however, sheeting braces and changes in the level of the excavation seriously interfere with the free movement of the equipment and, in such cases, the use of derricks or cranes is found to be more advantageous.

FIG. 4.8. Union railroad pile driver No. 1½-A. (*Courtesy of Union Iron Works.*)

Railway Pile Drivers. These have drop or steam hammers mounted on special flatcars. Steel leads fold inside railway clearances and are usually mounted on a turntable.

Floating Pile Drivers. These have drop or steam hammers mounted on a pontoon or barge which may be towed but which is moored or anchored during driving.

Leads. Drop hammers require the use of fixed leads or leaders, which form tracks in which the hammer is engaged for the full length of travel. Such leads are built as part of the tower structure. In the United States the hammer and pile are placed between the leads. In Great Britain and Australia the hammer and pile are generally placed on the face of the leads, permitting use of large butts. Single-acting hammers also require the use of leads, but these may

FIG. 4.9. Driving 24-in. octagonal precast concrete piles below deck with locomotive crane without leads and without followers, where clearance was sufficient for hammer, on Chicago & North Western Railroad. Note guide frame extending aboveground. Vulcan No. OR single-acting steam hammer. (*Courtesy of Portland Cement Association.*)

FIG. 4.10. Union pile-driving and derrick car with No. 3A double-acting steam hammer. (*Courtesy of Union Iron Works.*)

be fixed, or swinging or hanging. Double-acting and differential-acting hammers may use fixed, or swinging or hanging leads, or may be suspended by cable over the pile and set directly on it. Small sizes of sheeting hammers are held in place by hand.

Extension leads or telescope leads below fixed rigs permit running the hammer down below the level of the rig, into excavations, trenches, or water. Telescope leads are most convenient, being lowered or raised by a separate line.

Box leads are useful in maintaining alignment of double-acting hammers suspended from crane cables, when driving precast concrete piles.[5t] These may consist of a light boxlike framework of angles riveted to the sides of the hammer, extending about 5 ft down over the pile.

FIG. 4.11. Osgood Type 20, Model 200, pile driver, with 2,000-lb drop hammer. Adjustable braces hold leads vertical. Capstan head with manila line raises piles into driving position. A great number of these were used by the Corps of Engineers all over the face of the globe in World War II in isolated spots where steam was not available. (*Courtesy of The Osgood Co.*)

Steam Generators.[104] A portable steam-generating unit may be used instead of a boiler. Such a unit may weight only 5,000 lb, be contained in a box 5 by 7 by 6 ft, evaporate 3,000 lb per hr, develop up to 300 psi in 2 min, and save two-thirds of the fuel. Steam pressure may be kept up on hammers and waiting time is reduced.

Vibrator Drivers.[127,157,158,174] Steel pipe piles, caisson piles, reinforced-concrete piles, sand piles, and short piling have been sunk economically in the Soviet Union, China, and West Germany by longitudinal vibrations and the weight of pile and vibrator. The principle is that high-frequency longitudinal vibrations of proper amplitude and frequency destroy friction and the soil behaves as a viscous medium. A rigid connection is used between pile and vibrator. Vibrators have been of an electromechanical type, but could be electromagnetic. Vibrator size, weight, amplitude, frequency, resonance, types of soil, soil density, friction, and end bearing are discussed in detail in the references. High point resistance can be overcome by jetting while vibrating. Vibrators are useful in cold regions since no steam or condensate are present. Vibrators have been found to reduce placing time materially and result in large decreases in power consumption and labor costs. These vibrators largely reduce the pull required when used as extractors.

A method of sinking piles rapidly by resonant vibration, called Infra Sonic Piledriving, has been studied (by C. L. Guild Construction Co., Inc.).

FIG. 4.12. Driving Monotube piles, 40 to 50 ft long, using Vulcan No. 1 single-acting steam hammer with McDermid base, for water tower, Lake, Minn. Leads suspended over and extending down into excavation, with driver on bank. *Engineer:* W. Darby, City Engineer. *Pile-driving contractor:* E. E. Gillen Co. *General contractor:* Pittsburgh-Des Moines Steel Co. *Pile manufacturer:* Union Metal Mfg. Co. (*Courtesy of Union Metal Mfg. Co.*)

Type of Hammer and Rig to Select

Choice of hammer type must be made between drop, single-acting, double-acting, differential-acting, and the new diesel. No particular type is preeminent for all classes of work, and for some kinds of driving more than one, or possibly all types, may be suitable.

When driving heavyweight piles such as precast concrete, and when driving into dense strata such as stiff clay, heavy gumbo, compact gravel, hardpan, and shale, a heavy blow with a heavy ram, and with a fairly short stroke and low velocity at moment of impact, has been found most satisfactory. This allows more energy to be transferred into motion of the pile and reduces the impact and shattering effect on packing or pile head of a high-velocity light weight striking a heavy mass. Such a type of blow is obtained with single-acting or differential-acting hammers, or by a short stroke of a heavy drop hammer.

When driving light or average-weight piles or casings in materials of average consistency, the manufacturers of double-acting hammers mention that the rapidity with which the blows follow one another keeps the pile in motion and reduces inertia, friction, and point resistance.

Differential-acting hammers are claimed by the manufacturers to combine the advantages of both single- and double-acting hammers. The range of ram weights is the same as for single-acting hammers, while the numbers of blows a minute approach those for double-acting hammers.

When steam passed out of general construction use, it survived for standard pile hammers. Development of large portable air compressors did away, to some extent, with expensive, single-purpose steam equipment, but added cumbersome hose. The ideal solution is a small, light, self-contained, and self-activated pile hammer. The diesel hammer is bidding for this field.

Impact effect between different types of hammers and piles may be observed in the last term of formulas (2.1*a*) and (2.1*b*).

Tension values may become high in precast concrete piles if the natural period of the rebound synchronizes with the speed of the hammer, but this will occur only rarely. There is no good way of predicting this occurrence, and the use of a heavier hammer or softer packing in the cap may be tried as remedies.[21]

Damaging stresses in thin shells, such as may occur when driving fairly thin pipe or Monotube pile shells, may sometimes happen when the set becomes small before the desired depth has been obtained. To alleviate the situation, a wider spacing of piles or the use of a lighter hammer requiring a longer driving time may avail. If an unduly large hammer has been used, crushing of the shell may occur because of soil compaction. Short waits of a few minutes may give the soil time to readjust itself so that driving to the desired depth can proceed further without damage.

The time required for driving is an important factor in all but the smallest pile jobs. For example, a drop hammer may strike very few blows a minute compared with other types, but owing to the high drops possible, it may exert more energy and thus drive the pile further per blow; on the other hand, the energy of the drop hammer may require a limitation on the drop when harder strata are reached, in order to avoid overstressing the pile, thus putting the slower drop hammer at a disadvantage. This saving in time may or may not be significant to either a contractor or the purchaser, depending upon the required completion date and the form of the contract, whether lump-sum, unit-price, or rental-of-equipment-and-crew basis. If a test pile has been driven by one type of hammer and foot-by-foot driving records have been kept, set–bearing value curves such as shown in Fig. 2.1, prepared both for

FIG. 4.13. Special lightweight Raymond skid rig, for use on small jobs where moving is not an important item. Rig is portable and easily assembled. (*Courtesy of Raymond Concrete Pile Co.*)

the hammer used and any other alternate one considered, will provide means of translating the number of blows required by one hammer into the number needed if using the other. Then, knowing the speeds per minute of each type of hammer, the comparative times of driving may be obtained. Since the size of hammer affects the time needed for driving, it is advisable to use as heavy a hammer as possible without overstressing the pile.

The most suitable type of pile-driving plant depends much on conditions such as the number, length, spacing, and weight of piles; conditions of ground; presence of batter piles; and depth of excavation, cribbing, or water. On level ground, a frame resting on skid beams, spools, rollers, or rails is commonly used. On rough ground or inaccessible site, a crane mounted on a caterpillar or truck is preferable. The use of crane rigs for other locations is growing also, with the increasing use of swinging or hanging leads and gates or templates. Driving from a barge should be avoided if possible. If this method is used, a double-acting or differential-acting hammer to avoid the surging caused by use of a heavy falling weight should be employed. For driving a cofferdam, a centrally located jib crane carrying a hammer sometimes avoids moving a pile-driving rig.

FIG. 4.14. Driving Monotube piles 23 ft long with McKiernan-Terry No. 9-B-3 double-acting steam hammer, using driving rig on pneumatic-tired truck for bridge over Mojave River, Barstow, Calif. *Engineers:* State of California, Division of Highways. *General contractor:* A. S. Vinnell Co. *Pile-driving contractor:* Kiss Crane Service. *Pile manufacturer:* Union Metal Mfg. Co. (*Courtesy of Union Metal Mfg. Co.*)

Availability and cost of purchase or rental may be a factor in selection of hammer and rig.

Steam pressure available may limit selection of the hammer.

Headroom clearance may govern the choice of rig; hammer and stroke lengths vary, a drop hammer requiring considerable overhead clearance.

Accessibility may limit the type of rig. In some cases, only swinging leads can be used.

Underwater driving may be required and will limit the hammer selection to certain types, unless followers are to be used.

Batter or raking piles, if to be driven with the same rig as vertical piles, will affect choice of hammer and rig.

Driving Caps

The following comments will outline briefly common uses of types of hammer bases, caps, and anvils.

For Drop Hammers. Typical Vulcan driving caps for drop hammers are shown in Fig. 4.15.

Pile caps for wood piles (Fig. 4.15*a*) are interposed between the hammer and the pile head. The cap is composed of a toggling element and a cushion against which the hammer strikes. The toggling element is an iron casting having jaws corresponding to those on the hammer so that it may engage the same leaders. In the bottom face of the casting is a deep conical recess for the tapered head of the pile, and on the top surface is a shallow recess into which is set a short round wood cushion block. A steel band is pressed over the top end of the block to prevent

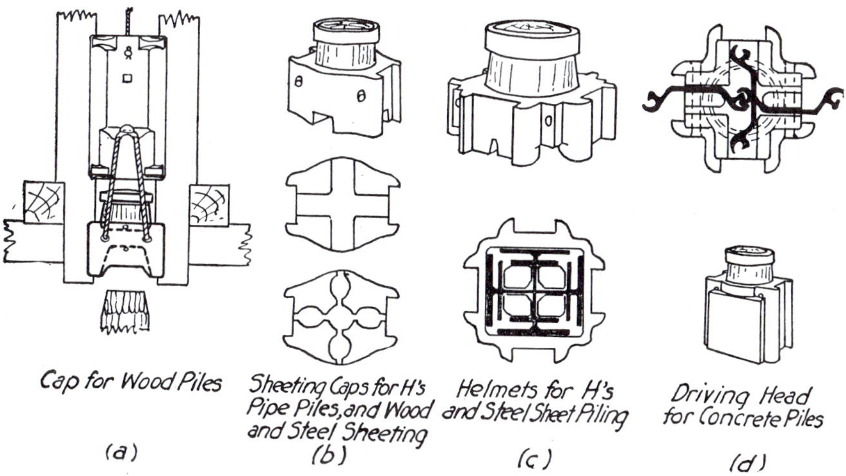

FIG. 4.15. Typical Vulcan caps, helmets, and driving heads for use with drop hammers.

it from splitting. After driving, the cap is attached by rope slings to hooks on the hammer so it can be lifted.

Sheeting caps for wood sheeting (Fig. 4.15*b*) are of cast iron and are similar to pile caps for round wood piles except that the bottom recess is designed to receive rectangular wood piling. The grooves run both ways and can suit the plank thickness.

Sheeting caps for H's, pipe piles, and steel sheeting (Fig. 4.15*b*) are of cast iron and are similar to wood pile caps except that grooves to suit the shape of the pile are used in the lower face. For steel sheeting, they can be used only with certain sections, principally the straight-web and light. Deep web sections can be driven only in one direction, as grooves at right angles cannot be provided. Grooves for H's can be provided at right angles.

Helmets for H's and steel sheeting (Fig. 4.15*c*) serve the same purpose as sheeting caps but are not subject to the same limitations and can be arranged for almost any style or combination of piles, corners, Z piling, obsolete piles, etc. They are made of cast steel and are practically indestructible.

The helmet has grooves in the bottom face to suit the shape of piling. This feature permits toggling and assures alignment of hammer and pile. Either one or two piles can be driven at a time. There are jaws on all four sides so that the driver may be placed parallel to or at right angles to the line of piling. On the upper surface is a shallow recess to receive a wood cushion block with steel band, as for pile caps.

Because of the great variety of helmets needed, all kinds may not be found in stock, so that time for making may be required. In Fig. 4.15*c* are shown the bottom face of a helmet for driving four sizes of H's, and of a typical helmet for driving steel sheeting.

Driving heads for concrete piles (Fig. 4.15*d*) provide a means for toggling the pile. They are made of cast steel. The *plain* type is for use when reinforcing rods do not project. The wood blocks and bands are the same as those used for pile caps. The bottom recess is very deep and is made to fit the pile closely, with a layer of planking to be interposed directly on the head of the pile to distribute the blow evenly and prevent spalling due to uneven contact. The *pedestal type* is used when reinforcing rods project.

For Vulcan Hammers. Typical driving caps for Vulcan steam hammers, both single-acting and differential-acting, are shown in Fig. 4.16.

Standard bases (Fig. 4.16*a*) have a conical recess and, in conjunction with suitable accessories, permit driving any kind of pile. Round wooden piles can be driven by allowing the end to enter the base until the side of the recess rests on top of the chamfered pile end. The pile must not project beyond the shoulder at the top of the conical recess. To prevent this from occurring when driving wood piles having a small butt diameter, either the *plate for standard bases* (Fig. 4.16*d*) or *dished cap for standard bases* (Fig. 4.16*e*) should be used. They are also advantageous in the case of large soft piles or when difficult driving is encountered.

Driving heads for concrete piles (Figs. 4.16*h* and 4.16*k*) are available. The *plain type* (Fig. 4.16*k*) is commonly used in driving precast concrete piles when the reinforcing rods do not project above the top of the concrete. The top ring is completely filled with a short wood block to receive the blow. Coils of wire rope or a stack of thin steel plates are also sometimes used. The deep skirt can be obtained to fit square, round, octagonal, or other piles. The bottom recess is sufficiently deep to permit placing a layer of plank capping between the top of the pile and the cap to equalize the force of the blow in case of any unevenness. This type of cap can be used only with a standard base (Fig. 4.16*a*).

The *pedestal type* (Fig. 4.16*h*) is intended for use on concrete piles where the reinforcing rods extend above the face of the concrete. The ring containing the cushion block is raised above the rods by a pedestal, which is integral for short lengths or separate from the block for long lengths. This type of cap also requires a standard base (Fig. 4.16*a*).

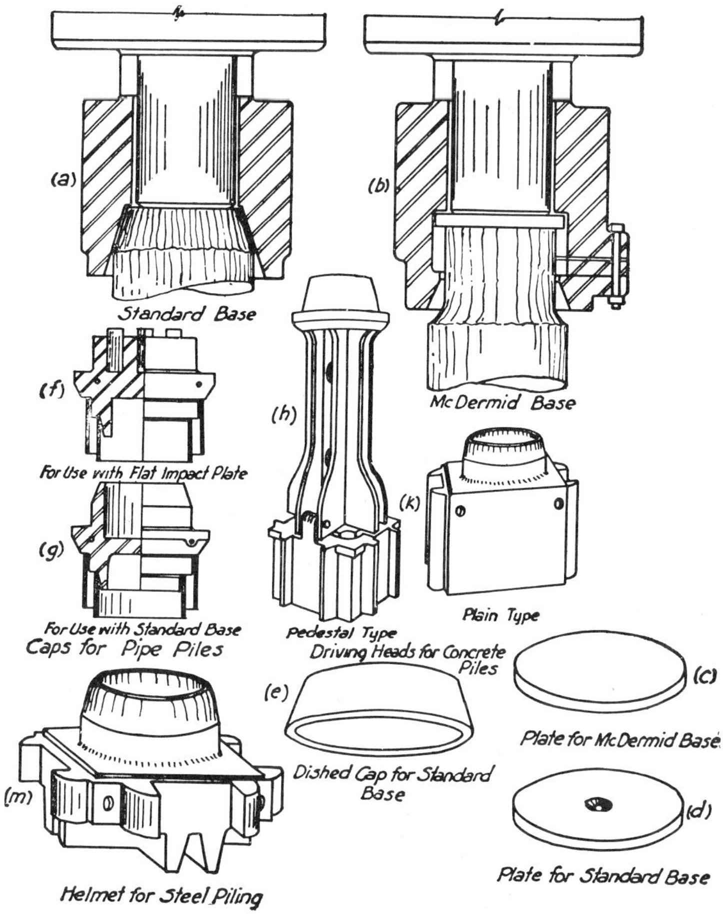

FIG. 4.16. Typical driving caps for Vulcan steam hammers.

Pipe caps (Figs. 4.16*f* and 4.16*g*) are designed for use with a standard base (Fig. 4.16*a*) when driving pipe piles. A double skirt forms a recess for the pipe shell. The size of pipe must be specified. The top recess is filled with a wood cushion block (Fig. 4.16*g*) or a series of wood plugs which head over and form a pad between the cap and the impact plate with hammers having flat impact plates (Fig. 4.16*f*).

Helmets for steel piling (Fig. 4.16*m*) are provided for use with standard bases when driving sheet steel piling or H piles. The upper ring is filled with a wood cushion block. Because of the great variety of such steel sections, these helmets usually are made to order. Such helmets can be arranged to drive steel sheeting singly, in pairs, in four pieces forming a cross, two pieces forming a corner, etc.

McDermid bases (Fig. 4.16*b*) are intended only for driving round wood piles. When the pile is soft or driving exceptionally hard, this

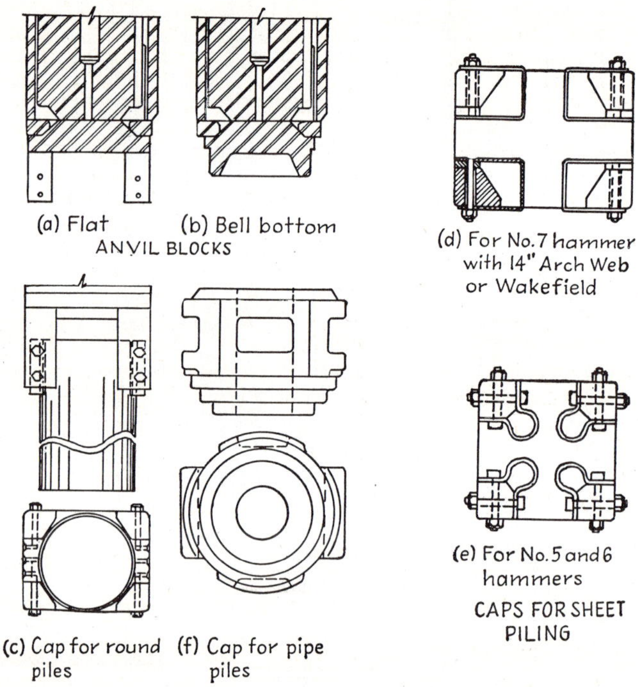

FIG. 4.17. Typical anvil blocks and driving caps for McKiernan-Terry double-acting hammers.

form of base obviates the necessity of placing a plate or dished cap on top of the pile before driving, as is necessary with the standard base. The McDermid base is interchangeable with the standard base on Vulcan hammers. Piles or wood cushion blocks having steel bands around the top should not be driven when using this form of base.

For McKiernan-Terry Double-acting Hammers. Typical anvil blocks and driving caps for McKiernan-Terry hammers are shown in Fig. 4.17.

Anvil blocks may be flat (Fig. 4.17*a*) or bell bottom (Fig. 4.17*b*).

Pipe caps (Fig. 4.17*f*) are furnished for sizes of pipe from 12 to 18 in. in diameter.

Round pile caps (Fig. 4.17*c*) may be bolted to the bottom lugs of the hammer.

Sheet piling caps may be bolted to the bottom lugs of the hammer, for driving steel sheeting (Figs. 4.17*d* and 4.17*e*) or Wakefield sheeting (Fig. 4.17*d*).

For McKiernan-Terry Single-acting Hammers. Typical driving caps for McKiernan-Terry single-acting hammers are shown in Fig. 4.18. Piles larger than those mentioned require special anvil blocks, wider than

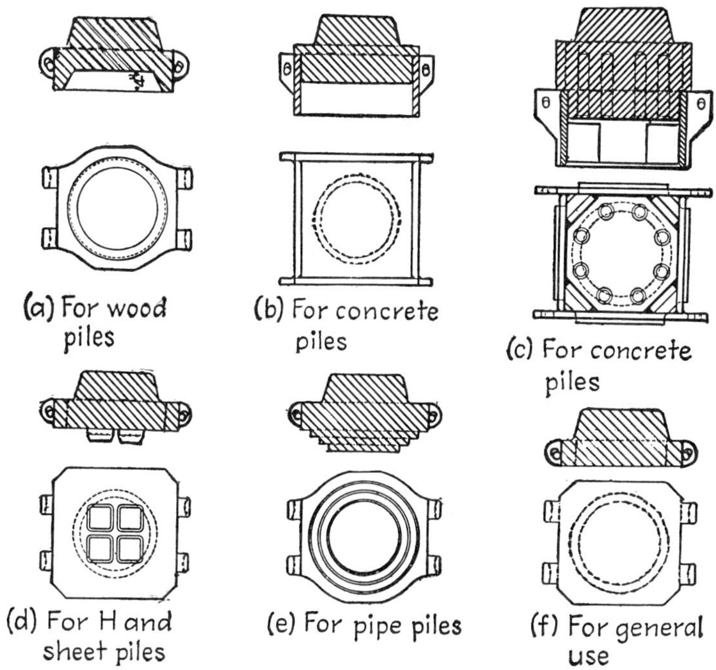

FIG. 4.18. Anvil blocks for McKiernan-Terry single-acting hammers.

the width of the hammer, and leads must be wide enough to fit these special anvil blocks.

Wood-pile anvil blocks (Fig. 4.18*a*) provide the largest diameters at the top of the recess as 13, 16, 18, 22, and 28 in., respectively, for the S3, S5, S8, S10, and S14 hammers. The bottom diameter of the recess is 4 in. greater in all cases except for size S14.

Concrete-pile anvil blocks are shown in Figs. 4.18*b* and 4.18*c*. Figure 4.18*b* shows an anvil block for concrete piles without projecting rods. These blocks are furnished to suit either square or octagonal piles, and the maximum standard sizes are 16, 20, 22, 26, and 32 in., respectively, for the S3, S5, S8, S10, and S14 hammers. Figure 4.18*c* shows an anvil block for concrete piles with rods projecting; the number, size, and length of holes can be made to suit.

H- and steel-sheet piling anvil blocks (Fig. 4.18*d*) have grooves made to suit the piles. For H piles, the maximum sizes are 24, 27, 30, 36, and 36 in., respectively, for S3, S5, S8, S10, and S14 hammers.

Concrete-sheet piling anvil blocks are made to suit the piling.

Pipe-pile anvil blocks (Fig. 4.18*e*) are made in all sizes to suit pile diameters. The maximum sizes are 20, 24, 26, 30, and 32 in., respectively, for S3, S5, S8, S10, and S14 hammers.

Flat anvil blocks (Fig. 4.18*f*) are for general use, and the maximum sizes are 20, 24, 26, 30, and 36 in., respectively, for S3, S5, S8, S10, and S14 hammers.

For Use with Union Metal Monotube Steel Shells. A full series of special driving caps designed for use when driving different sizes of Monotube steel shells with various types and sizes of hammers is available with the piles. Steel knockout plates, hickory cushions, and a laminated type of cushion block consisting of alternate thicknesses of wood or fiber fillers between steel plates are available. The range of such caps available for use with Vulcan, McKiernan-Terry, and drop hammers is shown in Table IV.

For Use with H Piles. Cushions are not generally necessary between the heads of H piles and the hammer, although with very hard driving and large hammers it may be desirable to use some form of head to prevent damage to the pile from too sudden fetching up or from driving necessary to reach a definite depth. Driving heads may be of the conventional cast-steel type (Fig. 4.15), or of the follower type consisting of a short length of H pile of the same section number as the pile but of considerably greater weight (Fig. 4.19 *h* and Table V.9).

Followers

When followers are used, they should be treated in the same manner as a pile, computing an additional value of C_2 for elastic loss in them. A few typical followers are shown in Fig. 4.19. Some may be purchased complete, parts for making others may be obtained, and many are homemade.

Follower Caps for Piles (Fig. 4.19*d*). Piles are sometimes required to be driven below the ground surface 20 ft or more. When driven to the end of the leads, a follower must be used for the remaining distances. The cast-iron follower cap is recessed on the bottom, the same as a pile cap, to fit over the pile. In its upper end is inserted and bolted a wood pile section of proper diameter and length, with its upper end trimmed

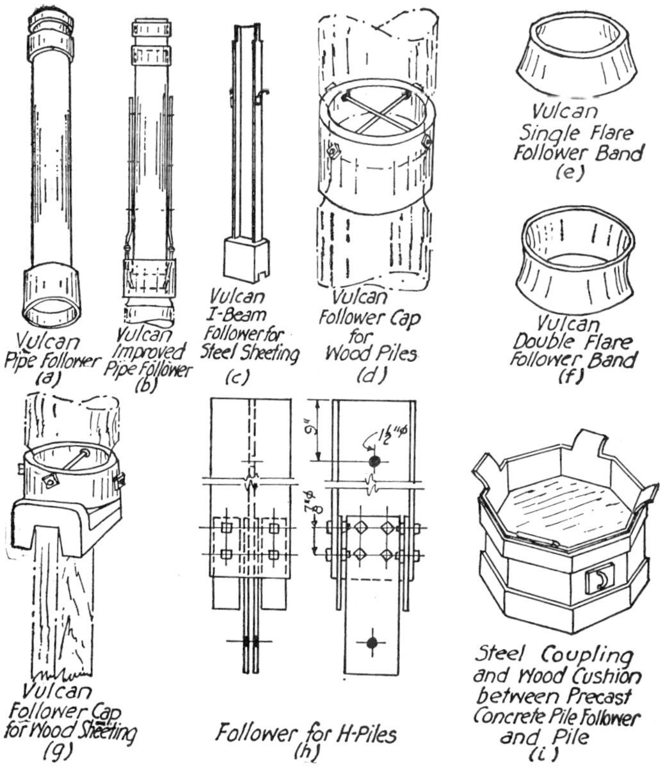

FIG. 4.19. Typical follower details.

to fit into the pile cap or steam hammer. These follower caps are furnished by the Vulcan Iron Works in the following sizes:

Size	Largest diameter of pile, inches	Diameter of upper recess, inches	Over-all length, inches
A	10	10	10½
B	12	12	12
C	16	14	15

Follower Caps for Wood Sheeting (Fig. 4.19*g*). These are similar to the follower cap shown in Fig. 4.19*d* but are grooved on the lower face. Different sizes are obtainable.

Follower Bands (Figs. 4.19*e* and 4.19*f*). When there is not room for a cast-iron cap, a wrought-steel band may be used. They are obtainable with single or double flare, in any size. Twelve-inch inside diameter with 7-in. face, or 14-in. diameter with 8-in. face, will meet all ordinary needs. The diameter is measured in the narrowest part of the band.

Pipe Followers (Figs. 4.19*a* and 4.19*b*). These are used for driving round piles below the surface of the ground or water, in the same way as for wooden followers with iron caps, but this form is much more satisfactory. These are made by Vulcan in the *plain* and *improved* (Kearn) *types.* Both styles consist of a cast-iron cap, having a recess for the pile head, that is cast on the lower end of a heavy steel pipe of suitable length. Shrunk around the upper end is a thick steel band forming a recess for a cushion block. This cushion block rests directly on a turned oak or gumwood filler, furnished with the follower, that must be driven into the pipe, for its full length, in the field. The cushion block and band for it are furnished with the follower. The improved type has, in addition, two small pipes leading to the space between the pile head and top of the recess. These pipes are attached by driving staples through holes in the pipe into the wood filler. By admitting steam or compressed air, the follower may be released after driving, if necessary. For piles 10 to 14 in. in diameter, 10-in. extra-strong pipe is used. A 12-in. extra-strong pipe is used for larger piles, or where the length warrants.

Fig. 4.20. Follower for driving 24-in. octagonal precast concrete pile below deck, made from a length of pile, for Chicago & North Western Railway. A steel coupling (*right*) containing a wood cushion connects the follower to the pile being driven. Used where clearance did not permit hammer to pass deck. Vulcan No. OR single-acting steam hammer. (*Courtesy of Portland Cement Association.*)

I-beam Followers (Fig. 4.19*c*). These are suitable for driving steel sheeting below the ground surface. The upper end is prepared for insertion into helmets or sheeting caps. The lower recess fits over the sheeting. The I beams are cast in the base. These followers are made in *plain form* for light work as shown in the figure and in *reinforced form* for heavy work. They are made for various sizes of sheeting.

H-pile Followers (Fig. 4.19*h*). These are furnished by steel manufacturers. Table V.9 shows standard sizes, dimensions, and weights.

Couplings for Precast Followers (Fig. 4.19*i*). The steel coupling with wood cushion is used between the head of a precast concrete pile and a precast follower section of the same dimensions. Chains attached to lugs on the coupling permit raising the coupling and follower.

Underwater Driving

When driving to specified set values under water with suitable types of hammers such as double-acting steam hammers, closed-type differential-

FIG. 4.21. Number 11-B McKiernan-Terry double-acting steam hammer driving wood piles to an average cutoff of 55 ft below water. (*Courtesy of McKiernan-Terry Corp.*)

acting steam hammers, or McKiernan-Terry single-acting or Vulcan Mariner type steam hammers, adequate compensation must be made for the buoyant effect of the water by hanging sufficient equivalent dead weight on the hammer casing; otherwise the energy of the blow will be reduced.

Driving can be done to 80 ft under water. It is only necessary to supply compressed air to the bottom cylinder of the McKiernan-Terry hammers and to the top of the Super-Vulcan hammers to provide pressure to keep out the water and to carry the exhaust pipe or hose above water level, the quantity of air delivered by a ½-in.-diameter hose being 60 cfm at a pressure of ½ psi for every foot of submergence. Steam or compressed air can be used, but the latter is preferable. For the differential-acting hammers, it is advisable to reduce the air leakage by sealing the oil and drain holes in the guides with wooden plugs, to interpose a thin rubber gasket between the cylinder heads and top of housing, and to keep the hammer plumb under water to assure even contact of the anvil and seat. Small McKiernan-Terry double-acting hammers, Nos. 1 and 3, can also be used to 10 ft under water, with the modification that the exhaust be carried above water level, although the power of the blow is somewhat reduced. If steam is used for motive power when driving under water, it will be very wet as an effect of condensation, and a special grade of cylinder oil recommended by the hammer manufacturer should be used. If air is used, the heating effect of the steam will be entirely missing, and a still different grade should be used. The proper grade of oil should be used to prevent damage to the hammer, clogging, or failure to secure the expected energy.

It is quite customary to avoid underwater driving, with its possible losses of effective energy, by using followers. Loss of energy in compressing the follower must be deducted. Use of shorter piles made possible by driving under water may conserve labor, materials, and expense.

Noise Reduction

Noise from pile driving may be undesirable in congested areas and near hospitals or hotels. If economic reasons dictate the use of driven piles, if public authorities do not forbid the noise, and if public-relations considerations do not govern, piles may be driven with the usual hammers.

Noise-reduction methods are not looked on generally with favor by pile-hammer manufacturers or contractors in the United States, and have been tried in only a few cases. Noise is from two sources, the first being in the area around the striking plate, particularly noticeable when driving steel piles. This type of noise may be more or less muffled by canvas hoods and other such devices, none very practicable. The second type of noise is from the exhaust, to which a muffler might possibly be applied.

The British Steel Piling Co., Ltd., states that a simple and effective silencer can be provided. It consists of a large tube fitted with baffle plates. The worst noise is said to be from exhausts, and it can be cured by silencers.*

A homemade muffler was used throughout three continuous shifts when driving piles for the Canadian Pacific Railway in the heart of the city of Vancouver after hotels had attempted to stop the noise by injunction.[41]

The performance of the muffled hammer should be calibrated to that of an unmuffled hammer.

Noise may often be avoided by placing piles by other methods than driving, such as drilling, jetting, jacking, loading, washing out, or wet rotary preexcavation.

Demolition and Rock Breaking

Double-acting hammers can be adapted easily for breaking up rock, concrete, old masonry, or ore (Fig. 3.2). Types suitable for underwater driving can be used to break submarine rocks. A chisel consisting of a sharpened steel bar is fitted under the anvil block and held in place by a framework of steel channels bolted to the body of the hammer.

Cutting Off Piles under Water

Wood Piles. It is often necessary to have the cutoff grade under water. If pile heads are to be embedded in concrete, exact accuracy is unnecessary, but if framing rests on them, considerable care is required in cutting off the heads exactly level and smooth. A diver can operate a power-driven chain saw while standing either on the bottom or on a platform suspended at the proper depth. If the number of piles warrants, a track can be rigged above water, attached to the next row of uncut piles or other special supports. Upon this a framework extending down to the proper elevation can be run, with a power saw attached to the bottom of the framework. For a rectangular area containing a sufficient number of piles to warrant the expense, a traveling bridge having a cross-traveling trolley (in the manner of a traveling crane), arranged to carry a horizontal circular saw, can be constructed. Numerous similar devices are used.

Dynamite has been used successfully to cut off piles under water, particularly in remote regions where other equipment is inaccessible. In one stream, piles were cut off cleanly 10 ft below water level by placing half a stick of dynamite in 3 ft of old fire hose, plugging the ends with

* O. Faber, *Proc. Inst. Civil Engrs.* (*London*), vol. 226, pp. 327–331, discussion on Paper 4652, 1929.

clay, wrapping the hose around the pile, and sinking it to the proper elevation.[2v] It was necessary to obtain permission from the game warden to kill the fish, however. In other instances, three half sticks of dynamite (with a percussion cap in each) fastened to a wire around the pile, and a fuse long enough to reach the cutoff elevation and connected to a battery, have given clean cutoffs.

Steel Piles. Steel H piles, steel pipe piles, or steel sheet piles can be cut under water by a diver using an oxyacetylene or oxyhydrogen flame and special burner, or a fully insulated arc-oxygen underwater cutting torch and tubular steel electrodes. These procedures present no difficulties and are often preferable to splicing. No special skill is needed by the diver other than the ability to use ordinary equipment on the surface. For flame cutting, the flame may be ignited on the surface or by a battery spark under water. For arc-oxygen cutting, it is only necessary for the diver to call for "current on," strike the arc, and drag the electrode across the work. Progress depends on visibility, accessibility, and absence of current in the water. Pipe piles have been cut off at great depths under water by an internal expanding rotary cutter.[3ac]

CHAPTER 5

SELECTION OF PILE AND METHODS OF DRIVING

Building Codes

In localities where building codes apply, their requirements must be met. In every case, however, the problem should be fully investigated without regard to code-allowable designs, and if proper design indicates that construction in excess of that called for by the code is required, it should be used. Building-code pile values are generally conservative, and most public authorities will permit the use of higher values if careful and accurate tests demonstrate that these are justified. Some codes set forth the rules under which such tests should be conducted.

Type of Pile to Select

Factors Affecting Choice. When new projects are financed, the type of construction which gives the required facilities at the lowest first cost will be demanded. This leads to a consideration of untreated wood piles which are generally among the most available and plentiful materials, are lightweight and easy to handle and drive, but which should not be selected without consideration of the various factors involved, since they may not prove ultimately to be the economical choice.

The solution should be obtained by considering broadly the various materials, destructive influences, economic life, and investment justified. More than one solution may sometimes be suitable; on the other hand, only certain types may be capable of fitting conditions of the site.

It would be unavailing to provide protection against almost all destroying influences and yet overlook one. For example, the most carefully constructed concrete jackets, extending a short distance below the mud line on wood piles, provided to protect the wood piles against collapse from marine-borer attack, would be useless if propeller wash caused scour below the bottoms of the jackets, thus exposing unprotected surfaces.

Temporary Piling. In harbors where borer attack is known to be light or in which the seasonal period of heavy attack has just passed, or in locations subject to decay or insect attack, temporary piles may economically be of untreated wood. If it is practicable to pull the piling after it has served its purpose, salvage of unattacked creosoted piling may be economically sound.

Semipermanent Piling. Somewhat longer life of wood piles, at less expense than that involved by more permanent piles, may be obtained by leaving the bark on, provided abrasions and knots are covered with metal of life expectancy equal to that of the wood, if the fact that the inner skin of the bark may decompose and become slippery is of no consequence. Other and better methods of obtaining somewhat longer life (for one or two seasons) are to paint the pile with proven preservatives or to sheath it with creosoted battens nailed in place over the danger zone.

Permanent Piling. This may require treating or armoring. Creosote treatments for wood piles are usually efficient protection for piles in the ground, piles not fully submerged, and piles in temperate waters free from *Martesia, Sphaeroma,* and *Limnoria.* Such piles are widely used and are generally economical. In tropical waters sheathing may also be required. Sheathing or armoring probably provides the longest protection for wood piles, and this may be done with sheet metal, precast concrete, gunite, vitrified pipe, cast-iron pipe, etc. Steel piles and concrete piles also are used when permanent piling is necessary, using water-line protection where advisable. In the case of concrete piles in sea water, protection such as asphalt impregnation has been tried.

Sea Water. This presents additional problems due to the presence of marine borers, wave action, salt spray, drift, and abrasion.

Current Movement. Strong water currents accelerate the leaching of preservative, scour, deterioration of paint coatings, damage caused by the movement of boats, and marine-borer attack which is generally heaviest on the side facing the current.

Abrasion of tubular steel piles at the mud line may be only a fraction of that of H piles. Circular shields can be used on H piles.

Wave Action. Spray and alternate wetting and drying have deteriorating effects.

Chemical Attack. Chemical attack, and the action of some alkali soils on concrete piles, occasionally are important.

Electrolysis. This may be a factor with steel piles in salt water, particularly where some other metal is in the water adjacent or where direct electrical currents may be connected to the piles or induce current in them. Electrolysis is not commonly a factor, however, particularly where the tops of the piles are insulated by concrete caps. Muntz-metal sheathing is subject to electrolysis if not homogeneous.

Character of the Ground. A thorough knowledge of the materials through which the piles are to be driven and upon which they are to be founded is the most important factor when selecting the type of pile. If a pure-friction pile is desired in a very uniform soil, or in soils such as clays or silts, or soils containing even a moderate percentage of clay, probably a displacement pile, tapered, would drive most readily and give good results. Because of the larger butt diameter of tapered piles, a greater proportion of load per foot of pile will be delivered to the soil near the upper portion of the pile. On the other hand, to reach a load-carrying stratum by driving through a considerable bed of fairly firm clay, it would appear that a pile which had the greatest frictional area and end bearing in the carrying strata, and by which the least energy would be expended in the non-load-carrying upper strata, would be advisable, such as a straight-sided pile in preference to a tapered pile. In such a case it is possible that a pile shell having a tapered lower section and a vertical upper section might be most desirable. Recent patented tapered enlargements on bottom sections for use in firm strata underlying soft strata have resulted in shorter, cheaper piles in some cases. H piles are very efficient when driven to refusal in hard material or to a high resistance in sand or gravel.

Long timber piles driven through a relatively soft stratum, or water, into a hard stratum are subject to breakage, and it may be advisable in such cases to use a stronger material.

Penetration to rock with the least wasted energy may indicate an H pile, open-end pipe pile, or box pile in preference to displacement or tapered piles, although, if the driving is easy, the difference may not be important. To aid in determining the most effective weight of piles to drive to firm bearing, light and heavy weights may be compared by observing the relative effects of the C_2 term and the W_p terms in formulas (2.1*a*) and (2.1*b*). While heavy piles absorb much energy in impact, piles that are long and light absorb much energy in spring action. When required resistance is large, a heavy pile is usually most efficient.

Penetration into rock is sometimes necessary to secure stability. This can usually be done best with an H pile or a Drilled-In Caisson. Penetrations of H piles for 4 ft into decomposed mica schist such as occurs around Philadelphia, for 8 to 17 ft into soft shale, and for 15 to 20 ft into coquina rock have been obtained, and give an idea of driving possibilities. Some limestone rock is so soft that refusal cannot be obtained, as at Bethlehem, Pa. Sufficient penetration to obtain good seating can be obtained in hard rock. The hardness of the rock is no barrier to Drilled-In Caissons. In the past, it has been the practice to blast holes in shale rock to take wood piles and set the piles in the holes, but the shale crushes and the holes become filled with mud, whereas if H piles are driven into the rock they do not lose contact with the rock.

Soupy strata may require concrete piles to be cased cast-in-place or precast.

Firmer strata may permit the use of uncased concrete piles, although if too firm, pile shafts may be damaged by vertical or lateral soil displacement.

Hard clay strata may require the use of drilled or cored holes to permit penetration or avoid lateral movement. The "clays" of tropical regions are likely to be laterite rather than true clay, and since it is not practicable to drive wood or concrete piles into fairly dry laterite, as they shatter after a penetration of 6 to 12 ft, pits may be dug or bored and then filled with concrete.*

Lateral or vertical movement of the soil may occur at previously driven piles in firm or cohesive soils which are relatively incompressible and may cause distortion, displacement, necking, or pulling apart of uncased piles. Under such conditions, this would make the use of uncased poured-in-place concrete piles inadvisable, and the use of strong metal shells desirable. In order to obtain the greatest coefficient of friction on the soil, corrugated metal is indicated, although, as discussed elsewhere, it seems that in some soils the soil outside the pile shears before the bond of the pile to the soil is broken. In such soils, drilled or cored holes may be used to remove the soil and thus avoid displacement.

Pedestal piles are designed to retain permanently the good resistance effects of the temporary bulbs of pressure formed during driving, by forming a concrete bulb which hardens. They may be advantageous in transferring footing loads through soft material to a firmer distributing stratum capable of bearing the load thus spread over it. These are particularly useful in case the bearing stratum is hard to penetrate to sufficient depth to obtain adequate frictional area, or in case it is so thin that a friction pile might be in danger of punching through to softer material beneath. Piles with large pedestals, such as Franki piles, are often capable of sustaining very large loads and are so used in Europe and Canada although such loads would not be permitted under many building codes in the United States.

Dredging. Future dredging may change conditions of embedment, earth thrust, lateral support and unsupported length, scour, and protection of encasements.

Scour. Scour[48] or liquefaction of the soil may increase the unsupported length of piles, may create horizontal forces or permit them to act on piles, or may expose lower unprotected sections to attack from marine borers, if these factors are not taken into account in the original design.

* C. B. Andrews, "Engineering Problems in the Tropics," *Civ. Eng.*, vol. 1, No. 13, pp. 1184–1188, October, 1931.

Length. This may limit the choice of types. Pine piles over 80 ft long are available only in limited quantities. While Douglas fir piles 125 ft long have been used, wood piles generally are considerably shorter but can be used by driving in clusters and capping with a concrete caisson if in water. Wood piles with gunite splices have been tested and used.[2bl] The ability to splice, the amount of lateral support or lateral strength needed, and the weight of the pile should be considered.

Character of Structure. Some structures are of a type that will not be appreciably damaged by settlement, whereas settlements in other types would cause great damage and expense.

Character of Loading. For instance, piles in soil which could not sustain a permanent load can often satisfactorily support a considerable temporary load. Special conditions of loading may be caused by ice, uplift, earthquake, soil pressure, etc. Flexibility may be more desirable than rigidity, as in the case of ferry slips and some wharfs where the piles are subject to great impact. The softer the soil below the mud line, the greater the flexibility and the smaller the liklihood of overstress.

Availability. Availability of pile materials often governs price and time of delivery and may limit the choice. Selection may sometimes be limited to types which can be driven with available hammers.

Adjacent Structures. Proximity to adjacent structures may require use of a pile type that can be placed without vibration and without displacement of the soil. Loss of ground into the pile may be serious in some soils and must be prevented. Additional expense of cantilever foundations may be avoided by the use of a type of pile which can be installed tangent or close to an existing wall.

Ground-water Level. Ground-water level is a vital factor in selection of pile type. Untreated wood will resist decay indefinitely if constantly submerged between low and high tides, unless attacked by beetles or marine borers. The elevation of ground water cannot be considered a fixed point as has often been done in past years. In urban regions, the construction of subways and deep basements and sewers has often lowered ground water enormously and is likely to do so in the future. In some sections, industrial users have lowered the water level by artesian wells. Even in some rural, agricultural districts this has happened. Multitudinous cases of damage or destruction from this cause have occurred. With our present-day awareness of the possibility of the construction of works that may affect ground-water levels and the availability of various alternatives types of piles, there is far less reason for damage to occur to new structures from this cause than there was in the past.

Wood piles must be treated above any possible ground-water level in order to be lasting, or concrete, steel, or composite piles must be used.

Change in ground-water level may alter the bearing capacity of a pile. If piles were to be driven in strata drained by wells or well points during construction, probably considerably higher driving resistances would be encountered than would be present after the water level rose. Load tests should not be made on soil under conditions different from those that will be present during life of the structure.

Economics

Probable and justifiable costs are an integral phase of engineering and must be considered along with technical merits of various possible types of suitable piles. Probable life of the pile should be related to economic life of the structure.

Foundations are considered from an economic aspect in reference 1*o*.

Relative Costs of Piles and Driving Rigs. Presentation of unit costs per linear foot for various types of piles might be misleading. A discussion will indicate the reasons, as well as methods of reaching the desired comparisons.

The cost comparison really required is cost per ton of load carried for the entire substructure, including piles, excavation, pumping, forms, concrete, reinforcing, backfill, and any other items required. The sizes of pile caps or thicknesses of mats vary with the numbers and working loads of the piles, and foundation costs other than piles may readily be estimated from preliminary designs.

Items for pile costs include rig rental with crew, transportation and living expenses of foremen, cost of rigs "on and off" site, fuel, tools, supplies, rental of auxiliary equipment, costs of pile materials, inspection service for untreated and treated wood piles, freight on pile materials, handling of pile materials to and from storage, wastage, cutoffs, shoes, caps, splicing, jetting, driving through obstructions, redriving to test for or overcome heaving, down time for maintenance and repairs, allowance for rainy days and holidays, taxes, duties and exchange involved with foreign materials or locations, insurance, overhead, and profit. Overtime, multiple shifts, and winter work must also be included, if required.

The portion of pile cost due to the driving rig varies greatly with the number and length of piles, and the costs of rigs have a wide spread, depending upon their sizes and capacities to handle light, heavy, short, or long piles. The cost of a rig "on and off the site" includes freight, unloading, assembling, and the reverse. A caterpillar rig driving in its home city would cost far less than a heavy skid rig shipped a long distance. Not all types of rigs are equally available at all spots.

Accessibility of the site also affects both rig and pile costs; a railroad siding may enter a flat site where a large number of closely spaced piles are to be driven, or the site may be many miles by truck over construction roads where scattered groups of piles are needed. Storage space and congestion affect costs. Helicopters have been used to carry 90-ft-long wood piles to transmission-tower footings in mountainous rough country otherwise inaccessible except over narrow, winding, steep mountain roads.[215]

The cost of driving per linear foot of pile is affected materially by pile length, since a considerable portion of the time per pile is consumed in setting it in the leads, and the fewer settings for a given footage, the cheaper per foot. Unit cost also depends upon hardness of driving, and the total number of blows depends upon the hammer size and speed, energy per blow, and allowable pile fiber stress. Piles may drop through water, sink long distances under the hammer weight, or require scores of blows per foot. Times may be required to penetrate obstructions, even by other methods than driving. Jetting will expedite driving, but its own cost must be included.

The construction schedule may affect pile costs. If overtime is required, payroll rates increase; however, the rig will be required for fewer days than on a straight-time basis, and supervisory costs and field office expenses will run for a shorter period. Unit costs will be less on large jobs requiring several rigs because such expenses can be distributed over more rigs and pile footage. From 6 to 30 piles per shift may be driven, depending upon variables; a common range for medium conditions may be from 16 to 20.

Availability of piles or pile materials affects costs. If unspliced wood piles are required on the East Coast, of longer length than available in southern pine, Douglas fir must be shipped from the West Coast at increased freight cost, which depends upon the length of haul and use of water or rail shipments. Cost of creosote treatments varies with pile lengths, kind of wood, amount of treatment, and relation of the location of the treating plant to source of supply and destination; in some less populous areas, treating plants may be distant or capacities may be limited. The effects of freight or hauling rates are quite different on H piles, or light pile shells, or precast concrete piles, or concrete ingredients.

The unit cost of wood pile material increases more or less directly with length and appears in *Engineering News-Record.* Shell and uniform-diameter poured-in-place concrete material unit costs are relatively unaffected by length, and H-pile material unit costs do not vary until length extras become effective.

Prices of H piles may be secured from steel companies or from importers of foreign sections. Creosoting estimates may be secured from pile contractors or treating plants, a list of which is available in the annual *Proceedings* of the American Wood-Preservers' Association.

Cost indices vary from year to year. Varying percentages of materials and labor, each having its own index factors, may affect costs of the various types of piles differently. Cost indices also vary considerably with location.

It is evident that there are no such things as standard unit prices for pile types. Any spreads given would have to be so wide as to be largely valueless. Approximate comparative estimates can be made by evaluating the various cost items. Closer values may be reached by consultation with pile-driving contractors. Sometimes it is impossible to tell from preliminary estimates or experience which of several suitable pile types will cost least, and bids should be requested on several types and combined with foundation estimates if the foundation varies. In the final result, the unit price that will have to be paid will be that resulting from cost estimates by the successful bidder. Much depends upon the bidder's eagerness for work, availability of equipment, and size and attractiveness of the job.

Ordered Lengths of Piles

Customary methods of determining the lengths of piles are the examination of the boring results and the driving of test piles. Sometimes only the latter step is taken, but it is difficult to be sure of safe results if the character of the strata along the pile and below the point are not known. Shearing values may be estimated or determined from laboratory tests on undisturbed samples.

Occasionally it is possible to order piles, of the types where the lengths must be predetermined, to the shortest likely length and place extensions on the heads where necessary,[2ac] to avoid cutting off the excesses.

Precast concrete jackets were used to make up for unequal top grades of precast piles ordered to the lengths of the shortest piles required.[2ac] These jackets were hollow reinforced-concrete boxes with 4-in.-thick side walls, having hooked stubs projecting from the top to anchor into the concrete deck. The jackets were slipped down over the piles until the tops were at the proper grades and then filled with concrete. Concrete projections cast on the insides of the jackets provided a ½-in. grouting space between the pile and the jacket. The jackets, being in the tidal range, were coated with asphalt mopped on hot over a primer consisting of a 70 per cent distillate and 30 per cent roofing asphalt, and gave useful protection against abrasion as well as good seats and connections for the deck, and ready economical adjustment of unequal lengths.

In tropical regions, where piles may be driven through mud, marl, sand, or gravel to coral-reef rock, very pockety conditions may be encountered, with some pockets filled with loose material, making the proper length of each pile a matter for individual determination.

In limestone regions, the cavernous nature of the rock may result in break-throughs when driving piles.

Sectional piles, or pipe piles that can be burned off or have pieces welded on, are useful where lengths are indeterminate in advance.

Uplift Piles

Piles are sometimes required to resist uplift as part of the static design or for occasional or emergency conditions.

Causes of Uplift. *Static uplift* may be present at all times in designs where it is felt to be safe to rely upon tension on piles, without expectation of undue differentials. Retaining walls on piles and fixed cranes are examples. *Intermittent static uplift* may occur from such conditions as certain positions of revolving cranes, the added weight of ice on wires, or breakage of wires.

Hydrostatic uplift may occur if the weight of the structure is not sufficient to resist flotation. Empty graving dry docks, basements of heavy structures subjected to floodwater levels before superstructure weight has been added, and floor construction between column reactions are examples.

Wind uplift may occur from pressure against such structures as chimneys, tall buildings, raised bascule bridges, and ice-coated wires.

Earthquake uplift may cause rocking action of the structure.

Ice uplift may result from the grip of ice or frozen ground on the pile. Ice grip may exceed the holding-down value of the pile. In tidal waters constant change in elevation does not permit a thick ice sheet to grip a pile, but on the Great Lakes, where the ice forms undisturbed for weeks, temperatures are low, and the ice adheres tenaciously. Then a storm may raise the water level on a weather shore, drawing piles out of the ground. Rivers above tidal influence, unless confined by dams, are affected by rising water caused by melting snow or rains, which lifts the ice. When the temperature is below freezing, adhesion of ice to timber equals its cohesion. One set of model experiments indicated this figure to be about 30 psi.* Late tests indicate that values may reach 100 psi with thick sheets.[113] Sun on piles rapidly weakens adhesion. Swedish observations of the uplift exerted on the face of a dam by an ice sheet of considerable thickness, in consequence of a rise in the water level, indicated that the force might approach 10,000 lb per lin ft. Lifting ice may exert damaging forces on batter piles or bracing.

Frost uplift has been observed in a number of instances.[6]

Uplift due to lateral forces may occur from such causes as ice thrust behind retaining walls, impact from ships or floating objects, pressure

* J. W. James, *Proc. Inst. Civil Engrs.* (*London*), vol. 41, Part 2, pp. 191–201, Paper 1374, 1875.

from high water, and pressure from velocity of current. Uplift on dolphin and ferry-slip wall piles may be avoided by leaving the tops free to rub against each other, so that the individual piles will act in bending and not produce an overturning couple in the structure.[1h]

Heaving may cause uplift, particularly when driving piles in incompressible clays. Concrete piles are liable to fracture from this source.

Means of Resisting Uplift. Piles should be sufficiently long to result in safe friction values for the various strata. Pulling tests will provide this information, and the friction values given in Table 2.3 will give rough indications of the ranges of ultimate values which may be expected.

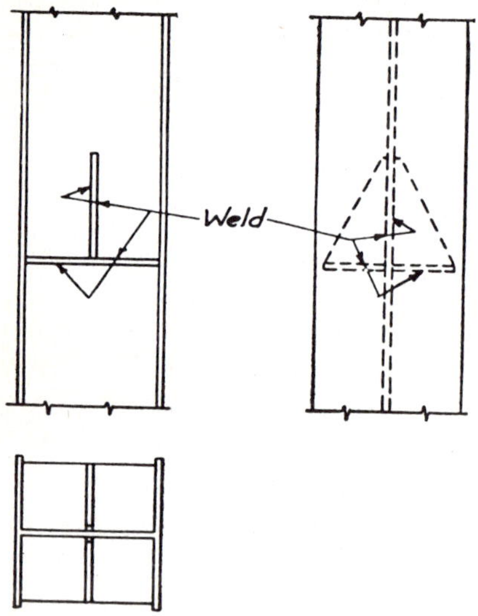

FIG. 5.1. Core stopper.

Factors of safety need not be large; they depend somewhat upon the importance of the structure and expected frequency and duration of uplift.

Time factors should be considered. Some soils grip the pile better after a time, and some less well after driving. Tests should be deferred until some time after driving in such soils.

Variable resistance may occur in soils intermittently submerged.

Weight of earth clinging to the piles should be considered since, regardless of the magnitude of the friction value on a single pile, the group of piles cannot pick up more earth than is adjacent and above their tips.

Transmission of uplift from structure to piles must be considered. Details of anchorages between wood piles and concrete caps are discussed on pages 229–232.

Tension in piles must be covered in pile design, particularly in concrete.

Type of Pile to Resist Uplift. Generally, the type of pile with the largest perimeter should be chosen to resist uplift, if a friction pile is used for this purpose. Sometimes core stoppers (Fig. 5.1) are used in the webs of H piles, but there is contention as to whether or not their use adds to the pulling value of the pile. It is generally assumed that the effective friction surface of an H pile is equal to the outside bounding dimensions, and that the soils in the core become solidly wedged in place, in which case no added value would be figurable from core stoppers. Tests have also indicated that piles which displace all of the earth within their bounding perimeters have a much greater friction value than piles which do not. This fact should be borne in mind when using friction values.

Tests have indicated uplift values of friction H piles to be from 75 to 90 per cent of the load-carrying capacity. Straight-sided piles show no great reduction of holding power after small upward movement; step-taper piles fall in this class. Uniformly tapered piles may not fill the hole after upward movement if the soil is at all cohesive, and uplift value may be lost after the initial movement.

Certain types of piles with enlarged tips, such as button-bottom piles or bulb piles in which bulbs of concrete are forced out at the tip, may be used to develop large uplift values with shorter piles than would otherwise be used. For bulb piles, the weight of a cone of earth having a side slope of possibly 30 deg with the vertical might be considered, taking due account of overlapping cones from adjacent piles. Buoyancy should be considered. Provision must be made for transmitting the additional tension through the pile.

Wood piles occasionally have been driven butt end down, to avoid uplift which would otherwise act against the taper of the pile. Such cases might occur because of ice or frost action, or from the elasticity of vegetation mats in swampy ground. Wood piles were driven butt end down through 25 ft of peat to rest in a 5- to 10-ft sand stratum over clay, when tip-end-down piles punched through the sand.[2bs] Internal or external pipe-thread protectors have been used on the small ends of wood piles, to stop brooming and splitting when the butt end was driven down.

Jetting

Uses. Jetting is not required on the majority of jobs, but may be of importance and be found to be a tricky operation. It should be used particularly in the case of wood piles, if necessary to reach final full desired embedment in good condition. Hard driving in sand or mixed strata may result in damaging stresses in the pile. Even if the butts are banded to hold the tops together, it is probable that the tips and lower sections are broomed, split, crumpled, or broken off. Heavy gravel or boulders may have gouged a treated wood shell so that it has lost its value. Jetting should be done when more than a few feet of penetration into hard sand is desired. Insufficiency of water for jetting may require that coring or drilling be used instead. Jarring of structures in the vicinity may be avoided by jetting, although care must be taken to avoid loss of soil. Loss of friction or lateral support due to forming a quagmire should be avoided.

Final sets should be obtained by driving for several feet above final tip grade without jetting. Provided this is done, dynamic formulas may be used for jetted piles.

Jetting is seldom needed for steel sheet piles, but is more often used with wood or concrete sheet piles, to relieve driving stresses, secure desired penetration, and save time.

Type of Soil. Jetting is usually effective and practical in almost all soils except very coarse and loose gravel, hardpan, and rock, although it is difficult to jet in clay without plugging the jets. Occasionally jet pipes can be extended through clay to reach sand strata below. In some soils, jetting is so effective that ordinary driving can nearly be omitted, but if there is considerable ground water and the material disturbed by the jet cannot escape, jetting is not so successful. Wet rotary equipment for "jetting" in cohesive soils is available.[209]

It is claimed that the puddling action causes the soil to pack closely around the pile after cessation of jetting, and that equally good or better friction values are obtained than would be the case without jetting. Jetting is inadvisable if it washes the fine material from the coarse, since this may reduce the bearing value.

In silty soils, jetting may loosen the soil around piles already driven, and such soil may not settle into place as firmly as before. If such a condition is considered possible, piles should be redriven after all jetting within 25 ft has been completed, or a load test applied, to ascertain the existence of such a condition and to correct it.

Water-jetting Methods. Water jetting is done by displacing the soil by means of a jet discharging a proper volume of water under sufficient pressure to allow the discharge to come up around the pile to the surface. It is not generally necessary or desirable to fasten the jet pipe to the pile or drive it into the ground, since it can usually be pushed down quite easily by hand.

Lubricating action generally contributes more to ease of driving than does the size of the pocket washed out at the tip. To obtain a greater lubricating effect, it is often desirable to move the pipe up and down during driving, particularly when the water pressure or water volume is low.

When using a single jet, the pile tends to move toward the jet, and this fact is sometimes of use when it is necessary to straighten piles. In order to avoid driving off line, the jet may be put down first on one side of the pile, then on the other, but this is usually too expensive a procedure to adopt. Instead, two or three jets can be used. With two jets, one should be kept slightly ahead of the tip, and the other slowly raised and lowered along the pile to keep a clean flow alongside. If rigid standard leads are used, or if the pile is driven at least one-third of its distance before jetting starts, it is generally found that a single jet pipe on one side will not cause the pile to deviate from its true course.

Holes for 19-in.-diameter precast piles were spudded in tight marl in the Savannah River, using 110 ft of H as a spud, with a cutting edge. A 3-in. jet pipe with 1¼-in. nozzle used 325 psi pressure.

Steel sheet piling 110 ft long for the cutoff under Garrison Dam was driven using a hairpin jet, an inverted U of extra-heavy 4-in. pipe kept 5 ft above the pile tip, to flush both sides of the sheet. Jets were off for the last 5 ft.[107]

Spade or multiple jetting is effective for driving sheet piling.[101] The jet is formed by fastening a pipe along the arch of a length of sheet piling, leading to a manifold chamber at the tip. Steel teeth beyond the manifold and a line of jet holes in the bottom chamber give a chopping and flushing action as the jet is raised and lowered. At Garrison Dam the pipe was 3½ in. in diameter reducing to 2½ in. 6 ft from the tip, entering a manifold chamber 10 in. from the tip, formed by welding plates to the pile. Five chisel-pointed teeth 1.6 in. wide spaced across the 16-in. pile extended 5 in. below the manifold. Jet water escaped though ⅝-in. holes between the teeth. Two more holes through the interlocks formed horizontal jets to reduce edge friction. The spade jet was threaded into the interlock of the last driven sheet pile, jetting with water at 125 psi recirculated from a pool in the trench. After a hole was jetted, a sheet pile was dropped into it. After the line of piles was set, they were seated in impervious underlying material by driving. In extremely difficult going, the jet may be used alongside a pile during driving.

FIG. 5.2. Raymond step-taper pile being driven with the assistance of two water jets. (*Courtesy of Raymond Concrete Pile Co.*)

Fixed versus Movable Jets. Some precast concrete piles and concrete sheet piles are made with jet pipes and jets cast into them, thus avoiding off-center or unsymmetrical jetting and some of the consequent difficulties of keeping plumbness and alignment. However, casting jet pipes into concrete piles is more expensive, and only a movable jet will permit correcting plumbness. Two jets give the most rapid penetration and the best control over the path of the pile.

Equipment. Centrifugal pumps having two, three, or four stages, driven by gasoline or diesel engines, are widely used for jetting.

Pumps should be bronze-fitted if salt water is to be pumped, and such pumps are desirable in any event since the water used often contains materials harmful to steel and iron.

Jet pipes are often 1½, 2, or 2½ in. in diameter, running sometimes to 4 in., but with small nozzles, usually ¾ to 1½ in. in diameter.

Pressures of about 100 to 300 psi are frequently used, but they may run to 700 psi. Selection of volume and pressure is very important. Often both are underestimated, or too much emphasis is laid on pressure and too little on volume. The volume should be large enough to allow discharged water to come to the surface along the sides of the pile.

Hose should be about ½ in. greater in diameter than the jet pipe and should have a protecting jacket of canvas, cotton, or steel wire. The length of hose should be as short as possible on account of the friction losses. The full size of pump discharge should be piped to the hose, if possible, for this reduces friction loss and also permits the use of more small jets or one large one.

Design data regarding pump sizes, discharges, and loss of pressure in jet pipes and hoses are given in Table VIII. These values were furnished by the Griffin Equipment Corp.

Air-jetting Methods.[61] Air jets in conjunction with water are being increasingly used and are covered in some specifications. In water jetting, the problem is to secure an upward flow along the sides of the pile. Even upward jets will not always do this because the water may turn off in some other direction, and compaction of the soil due to withdrawal of water may actually make driving harder. Air travels upward more readily than water and cannot have this compacting action.

Use of compressed-air jets with water jets discharging at about the same elevation indicates that at least some water tends to follow the air up. When double water jets and heavy driving cannot secure the desired penetration, the addition of air jets may easily obtain the result.

Combined air and water jets are often made by attaching a small air pipe to the outside of the water-jet pipe, and are satisfactory if the length is fixed. For shipment or other reasons requiring variable lengths, it has been found that sections are useful. In this case the air pipe is inside the water pipe, with the air jet discharging just inside the water-jet nozzle.

Air jets alone, without water jets, have been found effective for shallow depths for probing or for friction reduction. Used alone, they have not been found suitable for deep penetrations.

Soil Lubrication

When hard driving through film clay not desired to be used for final load-carrying purposes occurs, occasionally lubrication by a stream of water at the surface around the head of the pile will ease driving and make more energy available at lower strata.

Occasionally, a useful expedient for obtaining deeper penetration in the soil is to wet the point and shaft of the pile before placing it in the hole. On one project the piles slipped 4 to 5 ft deeper when wet.[5t]

Effect of Electric Current on Pile Friction[122]

Tests have shown that a weak direct current flowing along a metal pile ionizes the surrounding clay minerals, causing them to move away from it and immediately reduce driving friction. The negative side of a direct-current circuit is attached to several places around the butt, while the positive side of the circuit is grounded several feet away. When the desired depth is reached, reversing the current removes the water from the surrounding soil and increases adhesion.

Coring or Drilling Prior to Driving

Soil augers or churn drills may be used where jetting is impracticable. By the use of these methods or by jetting, it should be possible to place undamaged wood piles, and to preserve intact the preservation shell on treated piles. A soil auger should be about 2 in. greater in diameter than the pile. Drills can bore gravel, clay, frozen ground, caliche, hard gravel conglomerate, and hardpan. Drills are truck- or car-mounted, and can operate on batters from 10 to 15 deg fore and aft, and from 10 to 45 deg sideways, depending upon the model. Sand or grout can later be placed in the annular space to provide friction value if necessary.

Holes can be drilled from 6 to 72 in. in diameter up to 200 ft deep. One tractor-mounted rig with a pantograph arrangement permits drilling holes on banks 15 ft above or below the tractor level, and 15 ft distant.

A churn drill may be used where the ground is too wet, loose, or hard for a soil auger. A churn drill of 10-in. diameter can be swung from the pile-driver leads and operated on a ½-in. cable from one of the hoist drums while the hammer is held as high in the leads as possible.[48] Loosening the ground with the drill alone is sometimes enough, but more often it is necessary to sink a 10- or 12-in. casing with the drill and withdraw it just before placing the pile. If boulders or gravel fall into the hole, it may be necessary to use a 16-in. casing and drive the pile through it and then pull the casing. A 6- or 8-in. tongue bailer is useful in removing mud and material from inside the casing instead of depending upon displacement by the drill. Sometimes in sand it is necessary to fill the casing with clay slurry to prevent caving in when the pipe is withdrawn. A groove in the pile, extending up from the tip to the top of the drilled hole, is advisable to permit water and mud to be forced out by the piston action of the pile.

Wet rotary preexcavation equipment for jetting and drilling in cohesive soils is available. A clay slurry is left in the hole to hold it open. The purpose is to eliminate displacement of clay with horizontal and vertical soil movements. This method has been used to depths of 100 ft in Detroit clay and 210 ft in Boston clay.[209]

Driving through Obstructions

Sometimes, in order to drive a pile through obstructions such as timbers, boulders, riprap, and thin stone strata, or to avoid damage to the permanent pile, special methods are adopted. These practices may permit much lighter piles to be used than would otherwise be the case.

On sites having buried boulders, these may be located in advance of pile driving by means of rod soundings or water jets, or diamond-drill rigs if boulders are so numerous that one is apt to be found below another. This will determine the need for special procedures for any pile location.

Pilot piles, or spuds, are sometimes used before driving wood or concrete piles or sheeting, and are useful in breaking or cutting through obstructions. I beams, H-pile sections, and cylindrical mandrels have been used for this purpose. A hollow steel spud with hinged jaws at the bottom has been developed that also will permit the installation of longer piles in the leads.[98]

Mud- or sandblasting will sometimes shatter boulders sufficiently so that the pile can be driven through. *Diamond drilling* into a boulder followed by blasting is sometimes more effective. *Jack-hammer drilling* into boulders through pipes can be tried. Blasting by lowering a stick with a can of dynamite attached into the casing of a cased concrete pile, when it reaches a boulder, has been tried with fair results.

Smaller pipe sections can sometimes be driven through the obstructions after blasting and be made into composite piles.

Explosives have been used for loosening holes for piles, and explosive charges have been set off at the foot of steel sheet piling to enable it to be driven deeper.* The use of explosives to form enlargements is part of the basic procedure in placing Wilhelmi (French) piles.

Spudding a hole by alternately raising and dropping the pile itself has been done in the case of heavy piles, such as large, long, precast concrete piles, when little progress was being made with a hammer. The pile design should be investigated before spudding.

* *Eng. News-Record,* vol. 130, No. 22, pp. 806–810, 811, June 3, 1943.

The special *jet pile* shown in Fig. 5.3 was developed to aid in driving wood piles on 7-ft centers into a sea bottom composed of mud underlain by sand and rock which was friable and rather loose but full of lumpy boulderlike knurls which caused a pile to jump aside.[40] This jet was swung from the hammer leads by the handle and cut its way by its own weight, except that, where lumps were encountered, the handle was laid down and the hammer dropped on the jet head. The teeth cut through the rock and the jet remained plumb. When the jet was withdrawn, the hole filled with mud and sand, but since it was clear of rock, the wood pile followed the hole accurately and successfully.

Bulling through is sometimes less expensive than taking special measures. Where using steel casings, it may be necessary to double the thickness and use Stellite or Stoodite rod for cutting edges.

Sand filling of pipe piles has been used successfully to minimize their crimping when driving through boulders or riprap. The sand was later blown out and concrete placed.[125]

Button-bottom piles were driven through 16 ft of dumped large rock and 16 in. of lean concrete, using a No. O Vulcan hammer, for Stuyvesant Town, in New York City, with breakage of few concrete points and little delay.[5w]

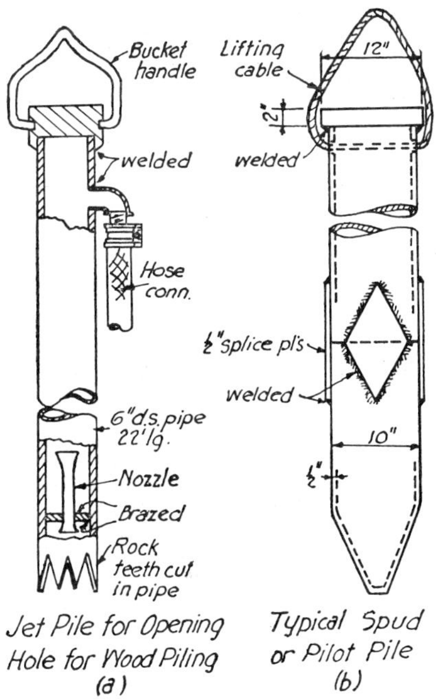

FIG. 5.3. Equipment used for driving through obstructions.

When sheet piling is being driven and an obstruction is encountered that might tend to displace the pile, it is sometimes possible to defer driving this pile but continue driving on either side of it. The obstruction may be dislodged by the other piles striking it more effectively; in any case the pile will then be guided by the piles on each side, and will be better able to dislodge the obstruction without being deflected.

Driving in Short Lengths and Restricted Headroom

Where clearance makes it necessary, some types of piles can be driven or jacked in short lengths, and the joints welded or otherwise satisfactorily spliced. Jacking piles can be placed by jacking against a portion of the structure above, if the structure is of sufficient weight and strength so that the jacking force will not damage it. Piles placed in sections are well suited to jacking. A hydraulic jack should be used with a direct-reading load gage, or a pressure gage from which the total load can be obtained by multiplying the pressure by the piston area. Jacking avoids vibration from hammer driving.

Fig. 5.4. Driving 10-in. open-end pipe pile in short sections with McKiernan-Terry No. 9-B-2 double-acting steam hammer, using tower leads, in restricted headroom, Staten Island, N.Y. (*Courtesy of Spencer, White & Prentis, Inc.*)

Short leads will permit driving 2- to 5-ft-long pile sections in headroom as low as 10 ft or less, and rigs can enter through low doorways. A McKiernan-Terry No. 7 double-acting hammer has been mounted between the elevated forks of a fork-lift truck, with short leads attached and braced to the forks.

Driving Piles Longer than Leads

When available leads will not receive full-length piles, a type of pile must be chosen that can be placed sectionally or else provision made for setting the tip in an excavated or spudded hole or unfilled shell. Pipe and H piles can be welded in sections as driven. Precast concrete and wood piles should be driven in one piece, and it may also be economical to avoid splicing the H or pipe piles.

A heavy square steel casing may be driven with bottom flaps closed, of sufficient length to permit placing the pile in the leads after the casing is driven.[98] The flaps at the bottom of the casing are forced open by the pile. After the pile is driven, the casing is pulled for reuse. The same casing can be used to penetrate near-surface obstructions.

A method of driving button-bottom concrete piles longer than the leads is described in Chap. 9.

FIG. 5.5. Driving 10-in.-diameter by ⅜-in. shell pipe piles in restricted headroom with McKiernan-Terry No. 7 double-acting steam hammer suspended from chain hoist. (*Courtesy of Spencer, White & Prentis, Inc.*)

Extended Leads and Cribbing

Piles may be driven in excavations or below the edges of banks by the use of extended leads, which should be securely attached to the main leads. The driving rig can also be run out over an excavation on cribbing.

If telescopic leads are not used, it is possible to install ordinary steel angle guides on each side of the hammer, extending upward and bracing them together with welded struts, thus allowing the hammer to follow the pile down many feet below the leads.[71]

Heaving of Ground

Heaving of earth around piles will sometimes be noted when driving piles in a group. Grades should be taken on each pile as driven, and if the piles show raising, they should be redriven unless the piles are friction piles in a uniform stratum. Care should be taken not to lose the grade markings if cutting off piles before completing the driving of the group. Heaving may be as much as several feet, and extend several piles away.

Checking on heaving of piles driven to cutoffs below the ground level, as in cases where it is more convenient to drive all piles from ground level and excavate around the pile groups later for the footings, is generally neglected and, in the case of uncased poured-in-place concrete piles, would be very difficult.

The use of uncased poured-in-place concrete piles driven using a temporary pipe casing should be avoided in dense or cohesive soils because driving of this casing can cause considerable lateral and vertical movement of the soil. These effects may critically damage adjacent

FIG. 5.6. Driving Monotube shells with No. 1 Vulcan single-acting steam hammer on turntable rig, with extended leads to permit driving 15 ft below underside of rig, Cleveland, Ohio. (*Courtesy of Western Foundation Corp.*)

piles previously poured by causing "necking" or pulling apart of the shaft concrete, distorting the shaft so that reduced or eccentric sections result, or displacing a pedestal bulb so that it will be eccentric. Uncased poured-in-place piles do not lend themselves to redriving after heaving.

Much can be done to control heaving by the use of wider pile spacing, as discussed in the next chapter, and by selecting the proper sequence of driving, such as working from the center of a group outward and working back and forth in long rows along a hard bank, instead of toward it.

In the cases of composite or spliced piles, care should be taken to ascertain that heaving has not pulled apart the joints or splices.

With piles having open casings (such as corrugated shells driven in sections, and pipe piles) an interior inspection can be made. Tip elevations of corrugated shells should be checked after withdrawing the mandrel, to be sure that the shell has not crumped longitudinally and raised the tip; use of a heavier-gage metal may be required.

In heaving, there is always the possibility that the upper section will break away from the lower, if the piles are of uncased concrete or of cased concrete having only thin metal shells, and it is recommended that all such piles have a 5/8-in.-diameter rod, anchored top and bottom.[5d]

If heaving is causing damage to the completed piles and expensive redriving is required, it may be advisable to use a cored-out type if in clay, or to use an open-end pipe pile and blow out the material. The Raymond Concrete Pile Company's wet rotary preexcavation method can also be used to eliminate heaving.[209]

Shrinkage of Ground

The driving of piles in fine-grained saturated sand may cause subsidence of the surface. In such cases, when loading piles one or more days later, a considerable gain over the indicated driving resistance may be observed in static-load-carrying capacity.

Weaving of Piles

When piles are driven on close centers, weaving of piles is apt to result if they are displacement piles and if the material is not removed. This will result in several inches lateral movement of the heads. Whether or not corresponding movement occurs throughout the length of the pile depends somewhat upon pile taper and character of the strata. If the piles are forced out of plumb, cracks may develop which may permit corrosion of reinforcement, shear value against lateral forces may be impaired, tension resistance may be reduced, and stresses due to eccentricity from axial loads may be caused. Bad effects may be minimized or avoided by the use of blown-out, cored-out, drilled, or H piles or by wider spacing.

Torsion during Driving[5au]

Both conventionally reinforced and prestressed concrete piles have occasionally been observed to crack across the pile at about one-third the distance from the pile head. This is probably due to a combination of tension and torsion stresses. Puffing of dust from the crack and then spalling appear. Continuance of driving results in more cracks, then failure. Piles frequently rotate while driving, as much as 90 deg having been recorded. This damage does not seem to be a function of hardness of driving. It has been remedied by filling the driving head so that it laps down only 3 or 4 in. over the pile butt and by casting or chipping the top 12 in. of the pile to a round shape. Apparently the corners of square or octagonal piles gouged into the inside faces of the driving head and the resistance of the driving rig to rotation caused torsional stresses.

Backfilling around Piles

Backfilling of the annular space between the dropped-in inner shell and the soil, left when the outer casing is pulled, may be necessary to develop required friction values or to prevent subsidence of the ground. This is usually done by sand grouting. The same procedure may be necessary after the withdrawal of temporary pipe sleeves.

Effects of Driving on Existing Structures

It may be necessary to select a design which keeps new piles at a distance from existing structures in order to avoid temporary jars, or to avoid causing possible consolidation of the ground under the existing foundations. Steam hammers may cause more settlement from vibration than drop hammers.

When driving piles near retaining walls, thought should be given to the character of the ground behind the walls. If it is a material which can be settled by vibration, the pressure on the wall may be doubled, according to the wedge theory. If the material is noncompressible, the earth may be forced against the wall, thus greatly increasing the overturning moment. Heaving of ground caused by driving displacement piles in dense soils may also damage adjacent structures.

Driving in submerged uniform fine-grained cohesionless soils may sometimes cause a great deal of settlement of the ground and adjacent structures owing to the temporary quick action induced by the displaced water, permitting closer rearrangement of the soil grains upon relief from the temporary water pressure.

Jetting may have a detrimental effect on the soil under existing structures, or it may be considered advisable to jet in certain soils in order to reduce vibration from the hammer blows.

Jacking piles down may be a means of avoiding effects of vibration from pile driving. Bored pilot holes may also avoid vibration.

To avoid displacement of ground close to existing structures, it may be desirable to use H piles, or open-end pipe piles that may be washed out or blown out. Bored pilot holes may also be used. The wet rotary preexcavation method may be used.[209]

Heaving of ground may also be a cause of displacement of existing adjacent structures and may affect choice of location and type of piles.

Level readings, photographs, observations, and sketches of structures that may be subject to damage from vibration, settlement, or heaving of ground due to new adjacent pile driving are advisable. Legal aspects of damage caused by pile-driving questions are available elsewhere.*

FIG. 5.7. Driving 10-in. open-end steel pipe piles with No. 9-B-2 McKiernan-Terry double-acting steam hammer, in limited side clearance at existing building, Staten Island, N.Y. (*Courtesy of Spencer, White & Prentis, Inc.*)

Installation of Piles under Existing Buildings during Demolition[11t]

Foundations such as Tuba piles, caissons, or Pretest cylinders can be installed under existing buildings during demolition, to save time.

Installation of Piles during Building Construction[11t]

First developed for underpinning, the Pretest cylinder method has been used under new buildings to save cost and time under suitable conditions. Cylinders are concreted in short sections in pits, temporary short wood posts set on top, and footings poured. As erection of the building adds load, various cylinders are jacked down against this load reaction, permitting building construction to proceed without waiting for

* I. V. Werbin, "Notes on Legal Decisions—Damages from Blasting and Pile Driving," *Eng. News-Record*, vol. 147, No. 12, Sept. 20, 1951, pp. 33, 44, and *Legal Guide for Contractors, Architects, and Engineers*, McGraw-Hill Book Company Inc., 1952.

pile driving. Each cylinder may be pretested by jacking one at a time to the exact depth required to obtain resistance to the particular load to be carried.

Lagging

Lagging consists of timbers bolted to the sides of a pile, and is occasionally used on wood or H piles when driving in very soft ground. The lagging is usually much shorter than the pile, and consists of two, three, or four timbers symmetrically located. The lagging has about the same effect as using a larger diameter pile and gives an increased bearing area and friction surface. The friction surface should probably be considered as based on the bounding perimeter. Lagging often consists of 6- by 8-in. or similar-size timbers.

Boxed H piles are, in effect, the same as lagged piles and are formed by welding plates across the grooves to the flange toes, with the spaces at the bottoms closed by sloping plates to form a wedge shape pointing down (Fig. 5.8). The same effect may be had by installing the sloping bottom plates and pouring concrete between the flanges for the extent which it is desired to box.

The value of lagging is a subject of considerable discussion. It seems best to approach the problem as one in soil mechanics rather than as one in pile driving. Short lagging creates a somewhat rounded bulb of pressure, whereas longer lagging creates a more pear-shaped bulb around the pile. Since lagging is used in soft soils, creation of these bulbs spreads the load and reduces settlement in the vicinity of the piles, but no different effect should be expected in soil underlying the bulbs of pressure. For this reason, lagging might be much less effective over a long period of time if a deep bed of soft material were below the piles than would be the case if a shallow bed were present. When test loads are placed on lagged piles, much greater resistances for the same settlement are to be expected from lagged piles because test loads are generally left on the piles only a short time. For long-continued loading, the increase would be less noticeable and might eventually be negligible. In the case of piles for a wharf carrying a railroad track, lagging might be quite effective to prevent undue settlement if the train loads were only there for a day or two, whereas if the wharf were used for heavy storage, settlement might be expected. Vibrating loads also might cause continued settlement. The spacing of the piles affects the overlapping of the bulbs of pressure, and should be borne in mind.

In soft ground, lagging should preferably extend for the full length of penetration to avoid loss of carrying capacity in the upper part of the pile. If a pile is driven through alternate soft and firm strata, the lagging should be located so as to come to rest in a firm stratum but some

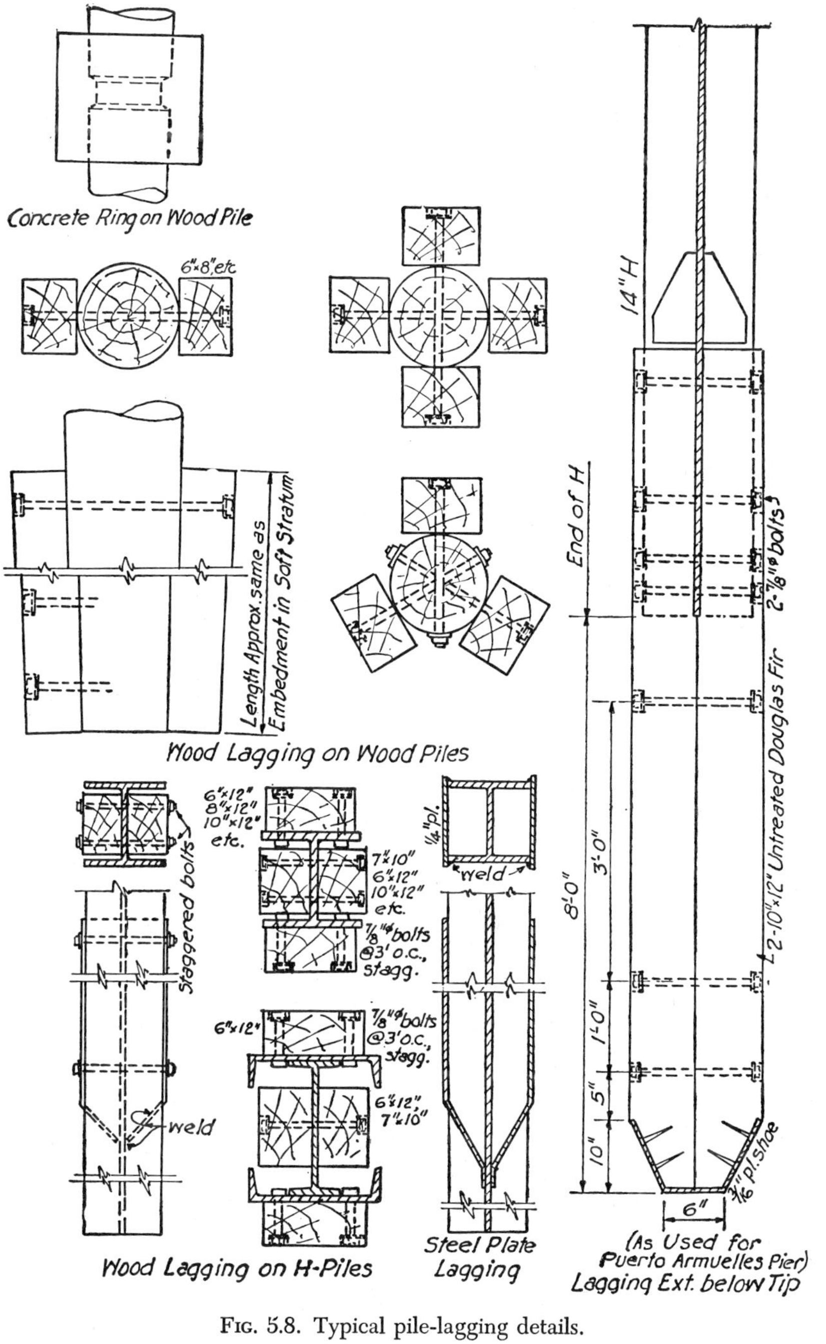

FIG. 5.8. Typical pile-lagging details.

distance from the underside of it. In sand, lagging should be about 6 to 10 ft long, located just above the middle of the pile penetration[11a,11b] (see Fig. 1.9*c*).

Lagging occasionally makes unnecessary the use of long composite piles. It is of much more value on individual piles than on group piles. Lagging may be useful on piles subject to temporary tension in soft soil.

Lagging is sometimes used to secure increased bearing on soft or coral rock, thus controlling the lengths of the piles. At a pier in deep water at the U.S. Training Base at San Clemente, Calif.,[25] successive layers of softer rock and sandy soil overlaid bedrock. To reach bedrock, long and heavy piles would have been required, while it would not have been safe to stop narrow steel pile points on the softer rock. Tests showed the feasibility of lagging the piles with wooden fillers and stopping on one of the higher soft-rock strata. The use of lagging does not preclude the use of encasement protection against marine borers, since the encasement can extend to a point a short distance below the mud line, with lagging on the necessary portions of the embedded length.

Ground Water during Driving

Presence of ground water has an effect on choice of pile type, precluding uncased poured-in-place piles where the soil is too soft and soupy to permit proper forming. Holes have been drilled in alternating layers of sand and clay, a pipe dropped into the hole, sealed in a bottom clay stratum, and pumped out.[7f] Well points have been used to lower ground water temporarily, thus permitting use of less expensive uncased piles. If total unwatering is too costly, it may be that partial lowering of the head will be of value. Ground water on one large project was lowered by installing drain pipes from the bottoms of sand piles.[9a]

Sometimes ground water is found under pressure below some firm stratum which it is necessary to penetrate. On one project where corrugated shells were used, water under pressure was found in sand and gravel underlying clay, hardpan, peat, and fill.[5d] Test borings became open wells, and loose pile casings filled and overflowed rapidly. The hardpan, as well as the lower bed of sand and gravel, was particularly destructive to the shells, allowing water to rush in so it was necessary to drive the points with an extra casing of plain steel. Sometimes it was necessary to drive a second, third, or fourth casing inside the original casing before the water flow was stopped. It was found essential that the casings be practically dry, as water under head will bubble up through a large head of concrete, washing laitance to the top and leaving uncemented gravel below. It is sometimes necessary to cut off some of the head to expose this condition.

In another instance, a large 30-ft-deep pit for an underground structure was excavated in clayey silt below ground water. It was known that water-bearing sand, into which H piles were to be driven for support, was present at some distance below the bottom of the excavation and it therefore had been specified that the pit be kept flooded during driving and that followers be used, with the thought that if water channels were not permitted to form along the sides of each pile during driving because of the head present when puncturing the clayey bed, the water could be pumped out after the piles had been driven, and forms could be built. However, a pile was driven in the unwatered clay, whereupon water shot up into the pit, cutting a considerable hole around the pile, and it was necessary to flood the pit and place tremie concrete around the pile before work could proceed as planned.

Ground water may also affect driving results if present in uniform fine-grained sands, which may temporarily become quick around the pile because the pressure from driving forces away the displaced water.

Horizontal Driving

Horizontal driving of piles[123] or tie rods may be accomplished by providing leads to support the hammer in a horizontal position. Double-acting hammers have been used for this purpose, but if the resistance is calculated, the rated energies should not be used since there is no gravity pull on the ram; the steam pressure at the cylinder times the area of the piston should be computed. Steel angles bolted to the hammer support it in the leads and allow it to slide forward. The cradle formed by the leads can be used over land or water, and swung by a derrick into the new position. If over water, temporary pile supports may be used.

Tie rods may be driven successfully.[45] The tendency of rods to sag in the leads may be prevented by using U-bolt hangers from cross yokes, the yokes having lugs to hold them from slipping sideways off the leads but being free to move longitudinally. To protect ends of threaded rods, short sections may be coupled on each end, that on the driven end being removed temporarily for insertion of the next rod section. To keep the rod from dipping, a horizontal fin 6 in. wide by 3 ft long welded to the top of the first section has been found satisfactory. To guard against rotation, chalk marks along the top of the projecting portion of the rod may be watched by a man with a wrench. Wedge points and conical points are not generally as satisfactory. An upward slope of leads of about 6 in. in 100 ft has been found desirable. By these means 3¼-in. rods have been driven in strings of 20-ft sections over 100 ft under existing stuctures for attachment to deadmen or anchor piles, with sufficient accuracy to hit a 2-ft diameter target, using a McKiernan-Terry 9-B-3 hammer.

As driving progresses, the hammer should be pulled forward in the leads by a manila rope, using two tackle blocks and a three-part line.

Winter Work

Pile-driving operations have been conducted in severe cold. In very soft ground, added expense of winter work is somewhat offset by savings from being able to carry the rig on frozen ground.

Where it is desired to drive piles, unslaked lime spread 4 in. thick, sprinkled with water or snow, and covered with a tarpaulin on which snow is also placed, has been found to melt through 3 ft of frost in 12 hr. Frost up to at least 2 ft thick can be broken up successfully by driving a heavy casing, or by a heavy pilot pile. Driving through for the first pile in a group usually shatters the frozen ground so that the remainder drive readily.

Earth drills are capable of drilling holes for piles through frozen ground. Some types can be used on batters.

Fresh concrete can be kept from freezing even in the coldest weather by heating the aggregate and covering the pile with straw.

To prevent forming ice films caused by condensation of exhaust on oily exposed moving parts, a sprinkling of table salt may be used.

In cold weather a steam hammer should be warmed up gradually, by admitting steam slowly and allowing the ram to strike a few gentle blows, to avoid bursting the cylinder.

When using air in the hammer in cold weather with high humidity, the air exhaust may freeze because of the rapid expansion of compressed air, forming pellets that may be expelled at high speed and damage eyesight. Use of an air-line lubricator of equal parts of alcohol and ice-machine oil will usually prevent exhaust air from freezing.

Driving through Permafrost

When the Russians build on frozen ground in the arctic, they sink piles deep into permanently frozen soil, melting holes with steam jets. The piles are wrapped in tar paper and greased, so that the topsoil, freezing and thawing with the seasons, cannot stick to them and heave them.

Ground has been thawed 40 ft deep by pumping cold water through points driven in the ground, the water having sufficient caloric heat to soften the ground in 10 to 15 days.

Pipe piles required for firm foundations and uplift have been set in drilled holes, removing drilled material by circulating mud.[105] The mud was then forced out by air and the piles filled with sand. Piles quickly froze in place. Tests showed a smooth steel-soil bond of 75 psi. After driving, the final foot of frozen excavation was chipped out, and 12 in. of sand, 4 in. of fiber glass, and 6 in. of sand were placed to insulate the frozen ground against heat of hydration in the footing concrete. Well-points were also found necessary for a time to prevent heaving until the ground refroze.

Refrigerating coils have also been placed around wood piles, and the backfill slurry quickly frozen.[150] Wood piles have been placed with the butt end down, to resist uplift. Adfreezing values of 20 psi on wood piles, 26 psi on concrete piles, and 23 to 29 psi on steel piles have been reported.[210] An adhesive value of soil to pile of 20 psi appears to be a safe working value except for soils of natural density less than 85 per cent of optimum or with a moisture content of less than 80 per cent of optimum; this excludes highly organic or very dry well-drained sand or gravel.[151] Methods for determining heat flow down a pile from the structure in summer, and length of embedment, are available.[152]

If upper strata are melted by heat from a structure, negative friction load may result. If the permafrost contains ice inclusions, melting may cause destructive unequal settlements. This can be avoided by insulation or a ventilated space or by thawing to a maximum depth to which thawing might proceed and then settling the piles in stable permafrost materials. When roofs of cavities over melting ice subside, negative friction may result as the soil consolidates.[152]

Experiments indicate that the piles should be sunk into permafrost to twice the depth of the active zone to prevent frost uplift. The use of piles in permafrost is treated comprehensively in references 152, 205, 210 and 214.

Piles are scarce in the arctic, and U.S. Army engineers have had success with use of thick insulating mats to keep the frozen topsoil from thawing, then omitting piles.

Increasing Capacity of Existing Piles

Sometimes it is desired to increase the capacity of existing piles in order to arrest settlement, support additional load, or replace lost capacity due to decay or corrosion, where it is not feasible to drive additional piles on account of space limitations, inaccessibility, or interference with operations. One method of doing this is by installing concrete encasements in steel shells around the pile, under air pressure. Another method is to solidify the ground. Both methods are discussed in detail elsewhere in this book. Possibly the load can be decreased, or use made of hydrostatic uplift.

Reuse of Piles

Piles remaining after demolishing structures should not be used for the support of a new load unless evidence shows them to be adequate. The new load-carrying capacity is limited to 75 per cent of the rated capacity of the piles by the New York building laws, which also limits the loads on additional piles under such structures to 75 per cent of their rated loads.

Pulling Existing Piles

Piles have been pulled and found safe for reuse in new locations in some instances. Jetting may be necessary to loosen the soil. Extractors are often used for this purpose. Concrete piles are apt to be damaged in pulling, but steel piles can be pulled for reuse quite well.

Electroosmosis has been used to decrease the resistance of steel piles or sheet piling to pulling. Where ground water is present, a direct-current voltage applied to the steel for a short time, using the members to be pulled as cathodes, will draw water to the surface of the steel and provide lubrication and decrease the skin friction. An electric welding machine has been used as a source of current. Chapter 14 contains more information on electroosmosis.

Vibrators have been attached to piles, in the Soviet Union, and used successfully for pulling piles.

Possible Use of Short Piles instead of Spread Footings

Spread footings instead of piles are usually installed under light loads, without further estimates of their cost as compared with short piles. Tests on spread footings and cast-in-place piles poured in auger-bored holes were made in rather loose soils, of low to moderate bearing value, near Los Angeles. These tests showed that less concrete was used in the piles, that the costs were less for piles than for spread footings, and that both total and differential settlements of the piles were less than for the spread footings, under equal loads.[5f] Furthermore, the piles appeared to gain carrying capacity with time. An added advantage obtained in this region by the use of piles was the avoidance of difficult determinations of proper footing depths owing to the varied stratified nature of the soils which contained thin bedded layers of soft clays to firm gravels at different depths.

Similar advantages have been found in using short precast or poured-in-place concrete piles or posts under load-carrying cross members, instead of poured-in-place concrete sleepers, for supporting bands of piping in refineries.

Precast concrete piles are used in the San Antonio, Tex., area for use with light loads such as for residence foundations.[5h] Shale, used for support of heavy structures, is 40 to 60 ft deep, and the upper soil is of such a nature that weathering causes heaving. The driven-pile theory was used to avoid heaving by carrying the loads in skin friction at depths sufficient to be unaffected by this action. Pilot holes 2 to 3 ft deep are drilled by hand to assure alignment and spacing; then 9-in.-diameter reinforced concrete piles 10 to 16 ft long are driven with an 1,800-lb double-acting compressed-air hammer mounted on a truck. It is usual to drill test holes to aid in estimating pile lengths. The use of piles provides a large factor of safety, but the price per foot of pile is low, and the house foundation is very inexpensive, with all excavation and caving avoided. About 20 to 25 piles are used under a five-room residence, and are driven in 4 to 6 hr.

In tropical regions where the "clays" are usually laterites, which are very hard when fairly dry but expand greatly during wetting and are characterized by causing differential settlements of spread footings, the use of concrete piles in dug, cored, or drilled holes, to carry the load to deeper strata, is advisable. Precast piles driven into fairly dry laterite usually shatter.*

Cast-in-place concrete piles, poured in drilled holes from 18 to 36 in. in diameter in soils which would stand for a short time without caving, have been found economical, in some cases, for uplift footings for transmission towers.[5g]

Unusual Uses of Piles

Piles may often be used in somewhat unusual ways to accomplish economical results.

An interesting way of reusing existing piles was observed in New York when it was desired to rebuild old wooden piers using much heavier fireproof structures with concrete decks. The old piles were only good for about one-quarter of the new weight, so a hollow concrete box substructure was set on the old piles, which were cut off below water. The hydrostatic uplift reduced the net load to the available pile capacities.

Both wood and metal piles have been used to support highway slabs over poor ground, forming pavement bridges.[216] Creosoted wood-pile bents under concrete roads through peaty swamps have been used by the Illinois Division of Highways.†

* C. B. Andrews, "Engineering Problems in the Tropics," *Civ. Eng.*, Vol. 1, No. 13, pp. 1184–1188, October, 1931.

† *Proc. AWPA.*, vol. 36, 1940, pp. 145–149, 156.

FIG. 5.9. Monotube piles 20 to 65 ft long used to support roadway slab over muck pocket over stretch of 583 ft on U.S. Highway 31 near Kokomo, Ind. (Ground shown cut away to show piles.) *Engineer:* Paul Sawyer. *Contractors:* Gradle Bros., Inc. *Pile Manufacturer:* Union Metal Mfg. Co. (*Courtesy of Union Metal Mfg. Co.*)

FIG. 5.10. Raymond cylinder piles, 36 and 54 in. o.d., with maximum distance of 72 ft from mud line to cutoff. High-level bridge on U.S. Route 11, Pass Manchac, La. *Pile contractor:* Raymond Concrete Pile Co. *Owner:* Louisiana State Highway Department. (*Courtesy of Raymond Concrete Pile Co.*)

CHAPTER 6

PILE GROUPING AND SPACING

INTERACTION BETWEEN PILES AND SOIL

Effects of Pile Spacing

The usual standards of pile spacing in the United States are generally too close for the loads applied on piles in plastic soils. If proper spacing is used, taking into account the length, size, shape, and surface texture of the piles, and soil characteristics, loads can be carried with less danger of settlement. A wider spacing of piles will materially reduce heaving and possible uplifting of the pile, damage by tension caused by heaving, and the possibility of crushing thin shells. A wider spacing will also permit the tips of the later piles driven in the group to reach more nearly to the desired tip grade, whereas, in case of close spacing, it is often noted that the later piles are unable to obtain the desired embedment.

Tip elevations are often selected to obtain end bearing and friction in a chosen stratum having properties that will permit it to carry the loads without shear failure or excessive settlement; in such cases, the driving sequence should be such as not to increase compaction unequally of the overlying material to be penetrated. Unsymmetrical patterns of tip elevations might be a source of future unequal settlements or tilting of a footing.

In driving piles through relatively incompressible strata, great horizontal forces are set up and, if piles are too closely spaced, damage may be done to piles already driven. This might occur to a corrugated or fluted steel shell either empty or filled with green concrete, or to an uncased concrete pile after the casing had been withdrawn and before the concrete had hardened. Lifting forces may cause necking of green concrete or separation of unreinforced set concrete. On this account, a minimum center-to-center distance of 2½ times the pile diameter should be a limiting value. One advantage of the open shells is that the interiors can be, and always should be, inspected before filling with concrete. It is a good rule to require that no shell be filled until all piles within a radius of 5 ft have been driven. The destructive effects on thin casings have been observed with sufficient frequency to warrant such delays in pouring as this may cause. In three-row pile clusters driven at 2.5 ft on centers, for instance, this means that concreting would follow a minimum of five piles behind driving, with all intervening piles open for inspection. It was noted on one project that the lateral pressure from a hardpan stratum over softer soil through which the piles penetrated overcame the resistance of the corrugated shells filled with wet concrete and destroyed the piles without surface evidence of such destruction.[5d] Occasionally, a shell would suffer crushing upon withdrawal of the mandrel. Frequently several piles in a cluster would be in place before any would crush in, and then one or more would yield. The immediately adjacent pile is not always the one to crush. Dummy mandrels in the middle piles of groups of three were used in one case. Alternate rows were driven in this manner, filled, and allowed to set prior to driving the intermediate rows. Pipe can be inserted temporarily in thin shells of constant diameter, to serve the same purpose.

It is claimed that freshly poured concrete pile shafts in contact with the ground are less compressible than the ground and that, therefore, the shafts are not constricted in section owing to driving from adjacent piles; however, some tendencies toward reductions have been reported.[25] A greater danger to this type of pile than the danger from close spacing is that of lateral displacement of a section of the pile, in cases where a relatively incompressible stratum occurs between two compressible strata, although this type of pile would have been a poor choice in such conditions.

Delayed progressive collapse of fluted shells with time has been noted in a very few instances, some of the shells in a group suddenly collapsing in toward the centers of the piles in the middle portions which were in the middle of a clay bed, with no failure occurring until after completion of the entire group. After this piles failed in the order of driving, although not all were crushed. If the shells had been filled immediately after driving and inspection, the collapse might have been prevented, or if not, the overflow of fresh concrete should have given indication. It is held by some engineers to be better to specify immediate filling of shells than to specify that they shall be left open until all adjacent piles within a certain distance have been driven.

If piles are too closely spaced, the carrying capacity of the soil upon which the group acts may be less than the sum of the capacities of the soils surrounding the individual piles. For a given number of piles in a group of friction piles, the value of the group may be reduced if the spacing is decreased. In one case piles driven in a muddy bottom under water settled one-third more under test load when spaced 2½ ft on centers than when spaced 3½ ft on centers.[*]

[*] C. W. Staniford, *Eng. News*, vol. 74, p. 76, July 8, 1915.

A spacing of 3 ft 6 in. to 5 ft on centers for friction piles is recommended. Although this will require larger foundation caps, the gain in individual pile-bearing capacities in the group is great, and the piles can be put down to the desired grades. It is better to begin pile driving on the center piles of a group and work toward the edge, in this way avoiding a great deal of center compaction. Starting at the edge makes the piles more and more difficult to drive and results in a one-sided bearing group. Close spacing results in a smaller bounding perimeter of the group, and often in much more rapid settlement of the structure than would occur with a group having a larger bounding area but containing the same number of piles. This would indicate that it might be poor economy to crowd the piles. A rough check which can be applied to groups of friction piles is to be sure that the bounding perimeter of the group is not shorter than the sums of the perimeters of the individual piles, thus assuring as much shear value in the soil around the group as in the soil around the piles.

When driving on close centers for a small group, such as three or four piles, piles driven off centers by even a few inches may result in considerable overloading of one pile because of the relatively great eccentricity on the group. This may sometimes require going back to drive another pile. Driving of additional piles if eccentricities cause loads over 110 per cent of the allowable bearing-load capacity is required by the New York building laws. It may be more economical to use wider spacings and larger caps.

Piles under marine structures subject to wave action should be spaced to a minimum of 5 diameters apart to reduce eddying and abrasion.[3ae]

Building-code Restrictions on Pile Spacing

There have been numerous rules for the spacing of piles. Many of these were established before the actions as explained by soil mechanics were understood and have been incorporated in building codes. This fact, the desire to keep footings as small as possible, and the wish not to increase footing sizes for one type of pile over those required for another have resulted in the general use of the specified minimums as maximums. These figures are usually low, such as 2 ft or 2 ft 6 in. on centers, or 12 in. clear between piles.

Effect of Grouping on Pile Bearing Values for Friction Piles

Reduction of Pile Value in Group. Both theory and tests have shown that the total bearing value of a group of friction piles, particularly in clay, may be less than the product of the bearing value of an individual pile multiplied by the number of piles in the group. The reduction in value per pile depends on the size and shape of the pile group and the size, spacing, and length of the piles. No reduction due to grouping occurs when the piles are end-bearing piles, and for groups which partake of both actions, only the portion taken in friction is reduced.

The reason for the bearing value of a group of friction piles being less than that of the sum of the individual piles is illustrated in Figs. 1.7*b* and 1.7*c*, which show how the zones of pressure spread out with depth and overlap. It appears that the zones spread out more with longer piles, and thus that the length of pile is a factor in selecting spacing, and probably a much more important one than the diameter of the pile. Model tests have shown that overlapping of pressure zones and reduced carrying capacities have resulted from a spacing of even 4 diameters. The Geotechnic Commission of the Swedish State Railways made tests in 1930 on piles 15 m long with 30-cm butts and 15-cm tips in clay, in which single piles carried 19.2 tons, piles spaced 0.7 m carried 12.0 tons, and piles spaced 1.2 m carried 18.5 tons. These observations indicate that a spacing of at least 10 per cent of the length was required to avoid group action.[67] However, piles presumably should not be driven for bearing purposes in soft clay. Probably the lengths in resisting strata only should be considered, weighted according to the relative stiffness of the soils.

Group action of friction piles presents a soil problem which should be considered in a manner similar to the case of a spread footing. Shear occurs in the soil around a spread footing, whereas with a pile group the shear is transferred to a lower level and bounds the pile group. In the spread footing there is compression under the footing, and with a pile footing there is compression in the soil around and below the piles. In the case of relatively nonelastic soils, the problem of reduced pile values would seem to be quite different from that in elastic soils where loads and settlements are dependent upon the effects of single or overlapping bulbs of pressure. No distinctions generally seem to be stated as to the type of soil with which the various group-reduction formulas are to be used. It would not appear logical to use formulas derived from the Boussinesq elastic-bulb-of-pressure theory in sands.

Several "efficiency formulas" are in use for assigning reductions to carrying capacity of piles in groups. The wide range of values for a given group is shown on Fig. 6.3. Some of these formulas are based on shielding actions of adjacent piles based upon relative spacing and pile diameters; these disregard pile lengths; the condition that horizontal sections at different elevations vary because the stressed zones in bulbs of pressure vary at different heights; varying diameters of tapered piles with height, variation of soil properties with depth and stratification and ground-water level; and rapidly decreasing effect of zones of pressure with distance from the pile.

Converse-Labarre Method. One method of assigning a reduced bearing value to a group of piles is by the use of formula (6.1), which is contained in the Uniform Building Code of International Conference of Building Officials and specifications of the American Association of State Highway Officials. The curves in Fig. 6.1 have been plotted by

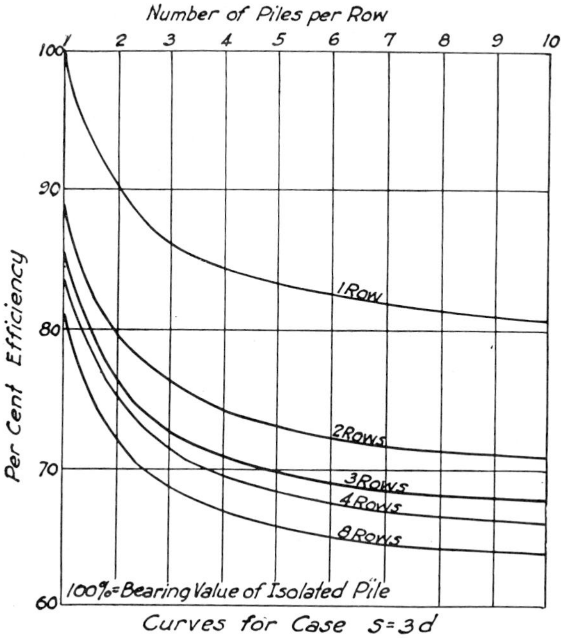

$$\text{Efficiency} = 1 - \phi \left[\frac{(n-1)m + (m-1)n}{90mn} \right] \qquad (6.1)$$

Formula contained in Uniform Building Code of Pacific Coast Building Officials Conference and A.A.S.H.O. Specifications

m = number of rows
n = number of piles in a row
ϕ = d/s, in which ϕ is numerically equal to the angle whose tangent is d/s, expressed in degrees
s = spacing center to center of piles
d = pile diameter

FIG. 6.1. Efficiency of supporting value of piles depending upon friction when driven in groups—Converse-Labarre method.

the use of this formula. Lengths of piles or distances embedded in friction strata do not appear.

Los Angeles Group-action Method. This is a more recent variation of the above method, using radians instead of degrees, and is as follows:

$$\text{Efficiency} = 1 - \frac{D}{\pi S m n} [m(n-1) + n(m-1) + \sqrt{2}(m-1)(n-1)] \qquad (6.2)$$

Masters Method. Another method of determining reduced bearing values, which may give considerably greater reductions, particularly in the case of long piles, is that proposed by Frank M. Masters,[15] in which the vertical shear values on the soil around each pile in the group are computed and their effects on each of the other piles in the group calculated by means of the Boussinesq equations. These data were developed in conjunction with the Morganza Floodway project, and the theoretical results checked by tests of pile groups of various shapes and numbers of piles. In this method the lengths of piles are taken into account. By length, however, should be meant the length embedded in friction-load-carrying strata. The discussions of this method should be studied before attempting to use it, as exceptions are taken to the use of Boussinesq equations for the transfer of load to soil in a vertical line, as with a pile, the equations being held to be applicable only to conditions for which they were derived, namely, application of load to a plane boundary of a semi-infinite elastic isotropic solid. Furthermore, the driving of piles remolds and compresses the soil, thus greatly changing its properties. Groups of piles destroy the continuity of the elastic medium, the piles in effect acting as reinforcing bars in the soil.[15] This makes doubtful the validity of applying the principle of superposition of effects from adjacent piles, which is part of the procedure used by Masters. In Masters' article, the ratio of shearing stress at the butt to that at the tip was taken as a constant, but it has frequently been observed[13,14,15,23] that this ratio varies with the amount of load on the pile, and that load does not progress down the pile until the applied load is sufficient magnitude to overcome the shearing resistance of the upper strata, so that this ratio is a variable. In Fig. 19 of reference 15 it can be seen that the settlement of the pile-point curve does not form a straight line as it would have done had the ratio of shearing-stress distribution at butt and tip been a constant. Considerable load was necessary to obtain permanent movement of the tip. It is evident that at smaller loads, the upper strata are doing the work, and that as the load increases, the lower strata are called more and more into action.

The value of the group, relative to a single pile, is dependent upon the shearing value of the soil bounding the group as well as on the bearing areas on the strata receiving load from the group. Masters' tables give values for both efficiencies, for the length of pile selected by him as an illustration.

Feld Method. A rule-of-thumb method of obtaining the efficiency value for piles in groups, used by Feld,[15] consists of reducing the value of each pile in the group by one-sixteenth, on account of the effect of the nearest pile in each diagonal or straight row of which the pile in question is a member. The application of this rule is shown in Fig. 6.2. The rule is based on using minimum pile centers. A pile fully enclosed in the center of a group has a value of 50 per cent of a single pile. Note that omitting the center pile of the nine-pile group has no effect on the bearing value of the cluster. This rule requires assigning different loads to various piles, whereas the other methods assign an equal reduced value to each pile and all piles in the group can be lengthened to effect the reduction. It would not appear advisable to use piles of different lengths in one group as a result of applying this rule.

Seiler and Keeney Method. Seiler and Keeney[42] consider the assumptions upon which the Converse-Labarre formula and Masters' tabular values are based, and divergencies of results. The Seiler-Keeney formula is empirical, the proponents desiring a formula which would give curves following the general shape of the Converse-Labarre curves but would

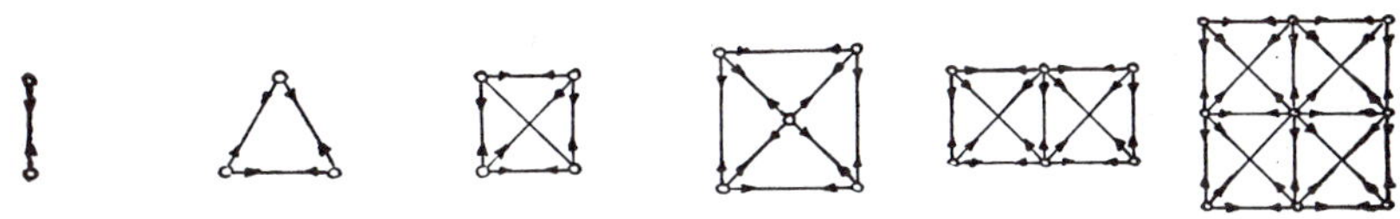

2 Piles	3 Piles	4 Piles	5 Piles	6 Piles	9 Piles
@ $\frac{15}{16}$	@ $\frac{14}{16}$	@ $\frac{13}{16}$	4 @ $\frac{13}{16}$	4 @ $\frac{13}{16}$	4 @ $\frac{13}{16}$
94%	87%	82%	1 @ $\frac{12}{16}$	2 @ $\frac{11}{16}$	4 @ $\frac{11}{16}$
	(Similarly for other groups)		80%	77%	1 @ $\frac{8}{16}$
					72%

FIG. 6.2. Efficiency of pile groups—Feld rule.

fall in the range of Masters' values for the smaller groups, in view of the satisfactory agreement found by him for test values of such groups with his theory. This Seiler-Keeney formula is as follows:

$$\text{Efficiency} = \left[1 - \frac{11S}{7(S^2 - 1)} \times \frac{m + n - 2}{m + n - 1}\right] + \frac{0.3}{m + n} \tag{6.3}$$

where S = spacing of piles center to center, in feet.

Pressure-area Formula.[42] Some recent building codes recognize the limitations of group piles and provide for the maximum load that can be transmitted to an appropriate area occupied by and immediately surrounding the group, depending upon the bearing power of the soil. The average individual pile capacity is then determined by dividing the capacity of the total area over which the load is distributed by the number of piles in the group. The efficiency of the pile group in bearing on the underlying soil may be computed by the pressure-area formula, if the distance beyond the center line of the outer row of piles over which the load acts is assumed (Fig. 6.4). This formula expresses the proportions of the total area per pile for the group, compared with the area

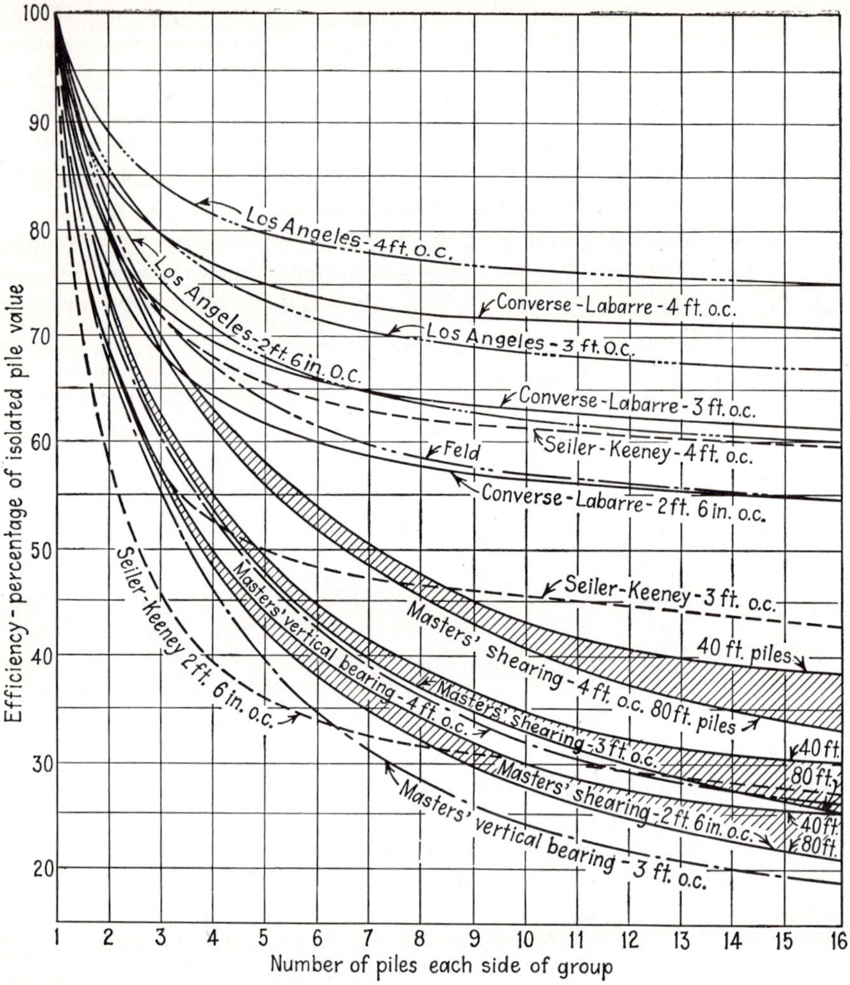

Fig. 6.3. Efficiencies of square pile groups by various formulas.

under a single pile with the pressure distributed over an area having the same marginal distance, and is

$$\text{Efficiency} = \frac{1 - (2k + n)[2 - (2k + n)]}{(2kn)^2} \tag{6.4}$$

where $k = S'/S$ (see Fig. 6.4), and

$n = m$.

Values from this formula may be computed and compared with those in Fig. 6.3.

Pretest Method.[11t] When Pretest cylinders are installed so that their bulbs of pressure overlap, it is customary to test them in groups in addition to testing them singly. This removes the possibilities of overloading the cylinders and of releasing the load on any cylinder by subsequent jacking down of an adjacent one. This method has been widely used by Spencer, White and Prentis, Inc.

Cylindrical-pier Method. Terzaghi and Peck[1n] approximate an upper limiting value of a group of piles which should not be exceeded by efficiencies obtained by other methods. A pile group and the enclosed soil are assumed to act as a cylindrical pier having a total value composed of bearing and perimeter shear given by the following formula:

$$Q_c = Q_{br} + 2\pi r_g D'_f s \tag{6.5}$$

where Q_c = ultimate carrying capacity of pile group, in pounds.

Q_{br} = ultimate bearing capacity of base of a cylinder having a perimeter equal to that of the boundary of the pile group, in pounds. [For a rectangle with a and b as sides, $2(a + b) = 2\pi r_g$]. The value of Q_{br} is the same as that given by formula 2.10, using Q_{br} in place of R_u and r_g in place of r_p.

r_g = radius of periphery of pile group, in feet.

If this method is used, a pile group may be considered safe if the number of piles times the design load does not exceed $Q_c/3$, except with friction piles in soft clay or end-bearing piles on a thin, firm stratum over soft clay. Load tests on model groups have confirmed the value of this approach.[37,156]

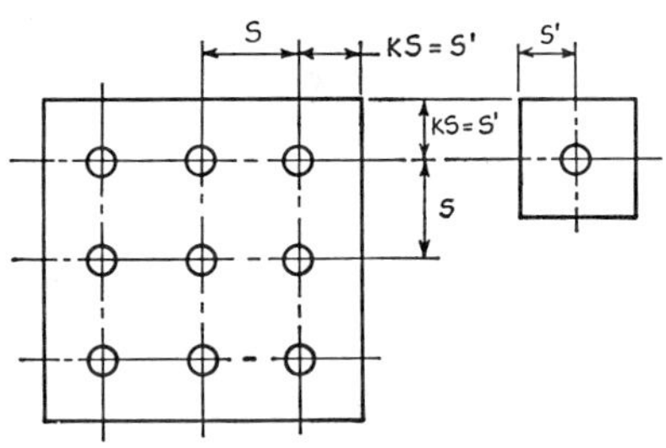

FIG. 6.4. Distribution of vertical pressure under single pile and group.

For long piles in small groups, the bearing resistance afforded by the soil below the level of the pile tips and bounded by a line enclosing the pile group is small compared with the shearing strength of the soil on this line for the length of the piles. On the other hand, as the number of piles increases, the bearing area under the group increases much faster than the shearing surface around the group. The optimum pile spacing, which utilizes the capacity of each pile, can be found by trial. The sum of the shearing and bearing capacities of the group of piles must be at least equal to the capacity of a single multiplied by the number of piles in the group. No benefits will accrue from driving additional piles within the boundary of the group if the spacing is equal to, or less than, that determined by this criterion. Use of this method is recommended.

Direct Measurements. Direct measurements of the loads being taken by the various piles in a group, by means of a strain gage applied to portions of the piles projecting above ground, will probably give more accurate information than theoretical or empirical assumptions at the present time.

Hansen and Kneas[23] report observed settlements under test load of various points along the height of four piles spaced in a square 3 ft 3 in. on a side, and touch upon the relative amounts of downward drag caused on the three piles, compared to the settlement of the fourth pile under load, which in the particular case under test was in the ratio of 1:10. These piles were approximately 56 ft long, driven through fill and silt into sand and gravel.

Swiger[24] describes an investigation of the behaviors of groups of piles as compared with single piles. Groups of three were tested, loads on individual piles being determined by jacking against a heavily loaded overhead platform, to predetermined settlements, using calibrated hydraulic jacks. This loading simulated that on a group of piles connected by a very rigid cap so that the settlement of each pile would be the same. For the particular conditions of the test, comprising wood piles 17 ft long, having 8-in. tips and 10-in. butts, driven in the test pit approximately 10 to 12 ft into moderately compact fine sand, the load carried by the center pile was slightly less than 70 per cent of that carried by an outer pile, this figure reducing to slightly less than 60 per cent when correction was made for elastic deformations of the piles. For this group of three piles, the group efficiency under approximately working loads, considering each of the two outer piles as 100 per cent and the central pile as 70 per cent, was indicated as 90 per cent. It was found that the efficiency increased 2 or 3 per cent when test loads of about 1½ times the range of working loads were applied. For the particular conditions of this test, the results obtained agreed closely with the result obtained by the use of formula (6.1).

A useful detailed account of determination of pile loads in a group and of distribution of friction in the group and over lengths 50 ft in clay confirmed formula (6.1).[119,124]

Effect of Shape of Pile Group. Friction piles in rectangular or circular groups will fail in shear at a larger load per pile than a square group.[15] A rectangular group will have smaller vertical pressures, and a circular group will have larger vertical pressures on the underlying strata than a square group.

Effect of Pile Length. The usual manner of providing for reduced values of piles in groups has been to provide more piles, thus reducing the working loads on group piles below those for individual piles, and attempting to maintain a uniform factor of safety. However, under any given footing, the shearing area of soil in the bounding perimeter of the group can be increased by using longer piles, thus reducing the unit shearing value. At the same time, deeper penetrations will result in smaller unit bearing values on lower strata owing to wider horizontal distribution of load. Furthermore, at increased depths, the increment of load attributable to the footing, compared with original soil load, is less. There is also the possibility that the longer piles will penetrate into better strata. Therefore, it appears advantageous to use fewer but longer piles, at greater spacing and greater load per pile. This should result in a cost saving, particularly for piles which are friction piles only in the lower portions, with the upper portions merely acting as columns to by-pass a soft stratum, since the added footage on the bottoms of the piles would be much less than the footage required to drive additional piles. It is not suggested that lengths vary in the same group, but there is no reason why separated footings should not have different tip grades.

Effect of Pile Grouping under Structure

The arrangement of pile footing groups in a building is analogous to the arrangement of individual piles in a group, and both from this analogy and from consideration of the dish-shaped settlement to be expected under the structure, in accordance with soil-mechanics theory, it is evident that the corner and outer footing groups are required to take more load than central groups, the amount depending upon the rigidity of the building compared to the depressed surface which the bearing strata would tend to form under normal column reactions. In stiff multistory buildings, the beam action of the building causes settlements to be more nearly uniform by throwing more load on the outer pile groups.

DISTRIBUTION OF DESIGN LOADINGS ON PILES

Distribution of Load between Vertical and Batter Piles

The distribution of load between vertical and batter piles under retaining walls may be determined graphically. One of the simplest methods, known as *Culmann's method*, is shown in Fig. 6.5.* Let the resultant force acting on the foundation mat be represented by R, and replace each group of similar piles by an imaginary pile at the center of the group, as in (b). Groups a and b must resist the axial forces R_a and R_b. Group c must resist an axial pull R_c. The resultant must pass through both points I and II. A force polygon can therefore be drawn, as in (c). If R is taken per lineal foot of wall, then the forces are per lineal foot of

* K. Culmann, *Die Graphische Statik*, Zurich, 1866.

wall, and the spacing of piles may be determined on the basis of the allowable axial load per pile.

Because different piles in a group are not all the same distance from the resultant, they probably are not all called upon to resist the same amount of force. The *trapezoidal method,* which determines a force

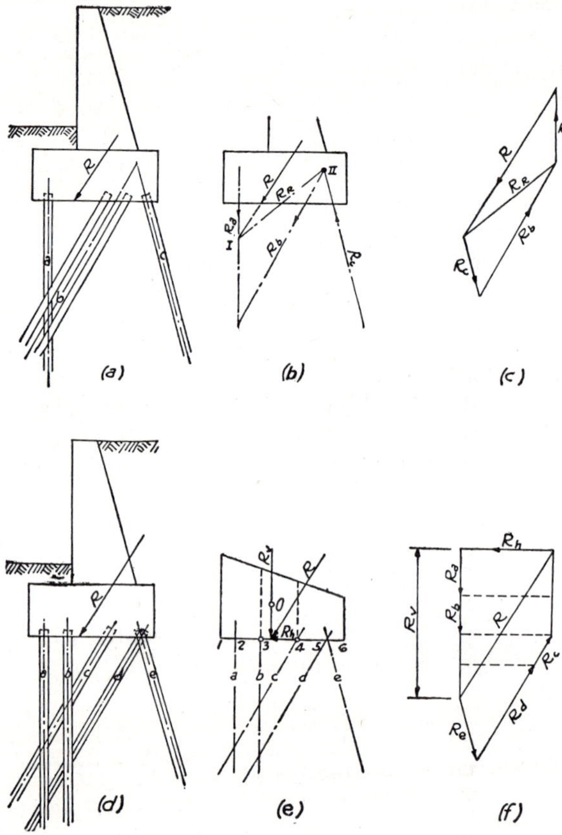

FIG. 6.5. Determination of pile reactions. (*a*), (*b*), and (*c*) Culmann's method (*d*), (*e*), (*f*) Trapezoidal method.

for each individual pile, is that of Brennecke-Lohmeyer, shown in Fig. 6.5.* For the vertical component R_v of the resultant force, substitute the trapezoidal loading which will have its center of gravity O on the line of action of R_v, as shown in (*e*). The vertical forces on piles at points 2, 3, 4, and 5 are computed from the trapezoidal loadings 1–3, 3–4, and

* L. Brennecke and E. Lohmeyer, *Der Grundbau,* vol. 2, 4th ed., W. Ernst und Sohn, Berlin, 1930.

4–6, assuming pinned connections at 3 and 4. The axial forces on the piles are determined from the force polygon as shown in (*f*).

Eccentric Loadings on Vertical Piles from Vertical Loads

Eccentric loadings from vertical loads on footings supported by vertical piles may be necessary owing to space limitations, changes in loadings caused by transient loads or structural alterations, or temporary loadings during the construction periods. Loads on individual piles may be obtained by the following methods.

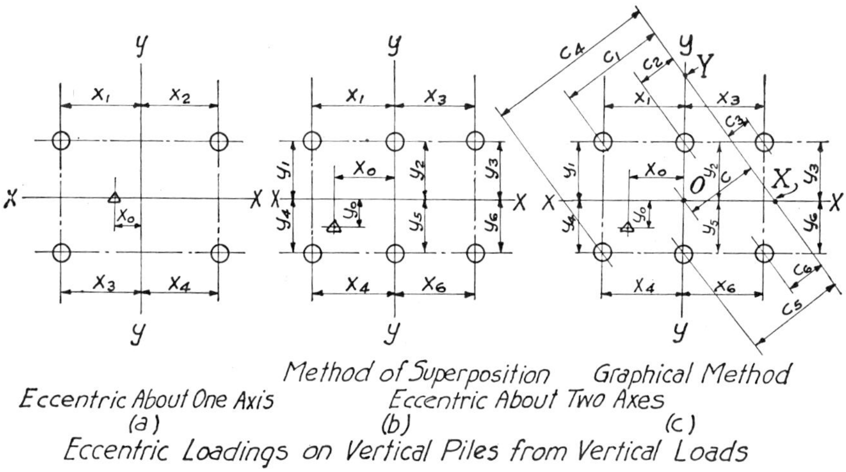

FIG. 6.6. Eccentric loadings on vertical piles from vertical loads.

Eccentric about One Axis. (Fig. 6.6*a*). When the eccentricity is about one axis only,

$$R_p = \frac{P}{n} \pm \frac{P x_0 x_n}{I_g} \tag{6.6}$$

where R_p = load on any pile, in pounds;

P = resultant of all vertical loads on pile group, in pounds;

n = number of piling in footing;

x_0 = distance from center of gravity of applied loads to center of gravity of pile group, in feet;

x_n = distance from center of gravity of the pile group to the line of pile, measured parallel to x_0, in feet (x_1, x_2, etc.); and

I_g = moment of inertia of pile group, measured about axis normal to direction of eccentricity.

To obtain I_g, use the formula

$$I_g = Ax_1^2 + Ax_2^2 + \cdots + Ax_n^2 \tag{6.7}$$

where $x_1, x_2, \ldots, x_n$ are the distances from the center of gravity of the pile group to the line of each pile, measured parallel to x_0, in feet. Since all piles in the group are assumed to be identical, A may be taken as unity in each case.

Eccentric about Two Axes. Where the eccentricity is about both axes, the individual pile loads may be computed by the method of superposition, or determined graphically.

Method of Superposition (Fig. 6.6*b*). The individual pile loads may be computed by determining separately the effects of eccentricity by the above method, and then the results may be added algebraically. The formula in this case is

$$R_p = \frac{P}{n} \pm \frac{Py_0y_n}{I_x} \pm \frac{Px_0x_n}{I_y} \tag{6.8}$$

where the terms are as defined above, with the following additional definitions:

y_0 = distance from center of gravity of applied loads to center of gravity of pile group, in feet;
y_n = distance from center of gravity of the pile group to the line of pile, measured parallel to y_0, in feet (y_1, y_2, etc.);
I_y = moment of inertia of pile group, measured about axis normal to x_0; and
I_x = moment of inertia of pile group, measured about axis normal to y_0.

To obtain I_y and I_x, use the following formulas:

$$I_y = Ax_1^2 + Ax_2^2 + \cdots + Ax_n^2 \tag{6.9}$$
$$I_x = Ay_1^2 + Ay_2^2 + \cdots + Ay_n^2 \tag{6.10}$$

In these formulas, $x_1, x_2, \ldots, x_n$ and $y_1, y_2, \ldots, y_n$ are the distances from the center of gravity of the pile group to the line of each pile, measured parallel to x_0 and y_0, respectively.

Graphical Method (Fig. 6.6*c*). The position of the neutral axis is determined mathematically, then plotted, and the moment arms to the various piles scaled from this axis. The intercepts on the x and y axes, determining the location of the neutral axis, may be obtained by the following formulas:

$$OY = \frac{I_x}{ny_0} \quad \text{(in feet)} \tag{6.11}$$
$$OX = \frac{I_y}{nx_0} \quad \text{(in feet)} \tag{6.12}$$

The following terms are defined:

c = distance from center of gravity of pile group to neutral axis, measured normal to neutral axis, in feet; and
c_n = distance from neutral axis to any pile, measured normal to neutral axis, in feet (c_1, c_2, etc.).

After scaling the values of c_1, c_2, etc., the load on any pile is determined by using the following expression:

$$R_p = \frac{c_n}{c} \times \frac{P}{n} \tag{6.13}$$

The values on the two sides of the neutral axis have opposite signs, provided the neutral axis does not fall entirely outside the pile group. That is, piles on one side will be in compression, and on the other side in tension.

Eccentricity Computations by Electronic Computer. Electronic computers have been used economically for rapid determination of revised pile loads in groups because of eccentricities resulting from piles being driven off center.[128] They may also be used to determine individual pile reactions in initial group designs.

Varying Spacings of Piles for Equal Pile Reactions

Under loads eccentric to the foundations, these may be obtained by the following method, based on the theories of arithmetical progression.[37] This avoids the sometimes tedious cut-and-try methods.

Analytical Method. 1. *Eccentric about One Axis when Resultant Falls within Middle Third* (Fig. 6.7*a*). Assume the width and location of the foundation with respect to the resultant applied load, and select suitable edge distances *a* and *b*. Then make

$$x = \frac{6z - 2D}{n - 1} \tag{6.14}$$

and

$$d = \frac{6D - 12z}{(n - 1)(n - 2)} \tag{6.15}$$

where P = the resultant load, in pounds ($P = Rn$);
R = the working load on each pile, in pounds;
D = distance center to center of front and rear piles, in feet;
z = distance from resultant to front pile, in feet;
n = number of piles in row considered;
x = distance between front and second piles, in feet; and
d = increments in spacing for succeeding piles, in feet.

If the value of x obtained is too close to permit pile driving, the following principles will aid in making proper changes in assumptions. It is evident that the value of x decreases as either D or n increases. The remedies are to (1) decrease D by increasing b; (2) increase z by extending at a as dotted and moving the front pile to the left; (3) use fewer piles per row by using more rows; (4) use fewer piles of higher bearing value per pile. These revisions may be easily tried by quick substitutions in formula (6.14), and guesswork avoided.

The load on a foundation is often variable and there may be many values of z and P, with two limiting values. It is not difficult to design for both limits and select an average or probable value for the final arrangement.

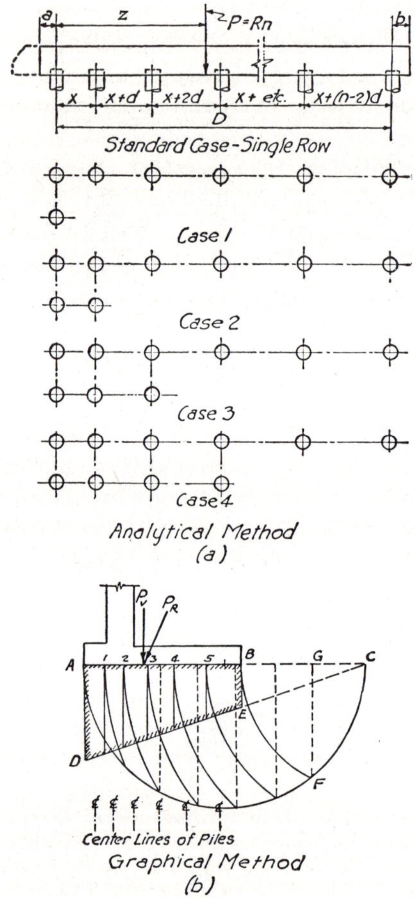

FIG. 6.7. Spacing of vertical piles for equal reactions under eccentric vertical loads.

The values of x and d will generally be fractional, but no significant error will result from using a balanced adjustment and easily measured dimensions.

The derivation of the above formulas will be given, since some of the preliminary formulas will be of use in special conditions, as discussed below. The variation in spacing is assumed to start from x and increase in arithmetical progression. The summation of spaces is

$$D = (n - 1)x + \frac{(n - 1)(n - 2)}{2} d \tag{6.16}$$

By taking moments about the center of the left-hand pile,

$$Pz = Rnz = Rn\frac{(n-1)x}{2} + Rn\frac{(n-1)(n-2)d}{6} \tag{6.17}$$

This gives

$$(n-1)(n-2)d = \left[z - \frac{(n-1)x}{2}\right]6$$

The equation for D gives

$$(n-1)(n-2)d = [D - (n-1)x]2$$

Therefore

$$6z - 3(n-1)x = 2D - 2(n-1)x$$

and

$$x = \frac{6z - 2D}{n-1} \tag{6.14}$$

as above. The value of d in formula (6.15) is obtained by substituting this value of x in formula (6.16).

2. *Eccentric about One Axis when Resultant Falls outside Middle Third* (Fig. 6.7*a*). When z is equal to or less than $D/3$, formulas (6.14) and (6.15) do not apply. In this case, if the distance between rows makes it feasible to drive an intermediate pile in the front row, the number of piles in formulas (6.16) and (6.17) may be increased by one, and formula (6.17) becomes, for *Case* 1, *one pile in the front row,*

$$Pz = R(n+1)z = Rn\frac{(n-1)x}{2} + Rn\frac{(n-1)(n-2)d}{6} \tag{6.18}$$

and new values are determined for x and d, as follows:

$$x = \frac{6z(1 + 1/n) - 2D}{n-1} \tag{6.19}$$

$$d = \frac{6D - 12z(1 + 1/n)}{(n-1)(n-2)} \tag{6.20}$$

In extreme cases, it may be necessary to use more than one intermediate row of piles. For *Case* 2, *two piles in the intermediate row,* formula (6.17) becomes

$$Pz = R(n+2)z = Rx + Rn\frac{(n-1)x}{2} + Rn\frac{(n-1)(n-2)d}{6} \tag{6.21}$$

For *Case* 3, *three piles in the intermediate row,* formula (6.17) becomes

$$\begin{aligned} Pz &= R(n+3)z \\ &= R(3x + d) + Rn\frac{(n-1)x}{2} + Rn\frac{(n-1)(n-2)d}{6} \end{aligned} \tag{6.22}$$

and for *Case* 4, *four piles in the intermediate row,* formula (6.17) becomes

$$
\begin{aligned}
Pz &= R(n+4)z \\
&= R(6x+4d) + Rn\,\frac{(n-1)x}{2} + Rn\,\frac{(n-1)(n-2)d}{6}
\end{aligned}
\qquad (6.23)
$$

In the above cases, new general values of x and d may be determined as before, or the particular values for n for the case considered may be substituted in formulas (6.16), (6.21), (6.22), and (6.23). Similarly, the formula may be written for any possible grouping of piles.

3. *Eccentric about Two Axes.* In designing the spacing between rows, at right angles to the lines discussed above, the rows or groups may be treated as units and the spacing of rows or groups may be determined by formulas (6.14) and (6.15).

Graphical Method (Fig. 6.7*b*).[1i] Piles may be spaced for equal reactions, when the ground pressure is not uniform under the footing, by the following graphical method. Obtain the position of the resultant and the pressure intensities at the toe and heel of the footing and plot to scale. Extend *AB* and *DE* to intersect at *C*. With radius *AC*/2, describe semicircle *AFC*, and with radius *BC* and center *C* describe the arc *BF*. Erect a perpendicular cutting *AC* at *G*. Compute the number of piles required to carry the total vertical component P_V of the resultant thrust P_R and thus obtain the number of piles n. Then divide *AG* into n equal parts, and drop perpendiculars from *AG* to cut the semicircle. With *C* as a center, describe arcs from these points to cut the base at points 1, 2, 3, 4, 5, etc. Perpendiculars from 1, 2, 3, etc., cutting *DE*, will divide the pressure diagram into n equal parts, and the verticals through the centroids of each division will be the required center lines of the piles.

CHAPTER 7

STRUCTURAL DESIGN OF PILES

DIRECT LOAD AND BENDING

Forces Acting on Piles

Forces. Piles must be capable of resisting without damage (*a*) *crushing* under the permanent-design vertical load; (*b*) crushing caused by *impact force* during driving; (*c*) bending stresses occurring during *handling;* (*d*) *tension* from uplift forces or from rebound during driving; (*e*) bending stresses due to *horizontal forces;* (*f*) bending stresses due to *eccentric location* of pile with regard to applied load; (*g*) bending stresses due to *curvature* in the pile; and (*h*) column action for portions not receiving lateral support from the ground, but free-standing in air, water, or very liquid mud.

Piles must also provide sufficient surface area, in the case of friction piles, to transfer the load from the pile to the soil.

The load-carrying capacity of a pile may usually be better determined by the ability of the soil to support the loads from the pile than by the structural strength of the pile; or ability to withstand handling or driving stresses may govern.

Piles under Direct Load. Portions of piles unsupported laterally should be designed as columns in accordance with standard codes. Such design is unnecessary if adequate side support is available.

End conditions theoretically have a large effect on column capacity. It is general practice to estimate the point of full fixity of the embedded portions of piles, but end-restraint conditions at the top are often ignored.

Fixity at the bottom of piles free-standing in air or water is set by the New York City building laws at a point 5 ft below soil contact level in compact sand-clay soils or better materials, and 10 ft below in loose fine sand, medium compact sand-inorganic silt, stiff, firm, or medium soft clay, and loose saturated sand-clay soils.

The degree of fixity at the top is difficult to determine, depending as it does upon the type of connection, amount of load, relative rigidities and deflections of the floor beams or deck, and the pile. The effects from end conditions may be helpful or harmful.

The use of a reduced section in design computations should be considered, to allow for abrasion and decay.

Design of Piles for Direct Load and Bending

General Equation. The general equation for stress occurring in the free-standing portion of a pile subjected to direct load, eccentricity, and bending is as follows:

$$f = \frac{R_p}{A} + \frac{R_p(\Delta \pm e_1 \pm e_2)c}{I} + \frac{Mc}{I} \tag{7.1}$$

where R_p = load on any pile, in pounds;
Δ = deflection of pile curvature, in inches;
e_1 = eccentricity of applied load to top of pile, in inches;
e_2 = eccentricity of pile owing to deviation from plumbness, in inches;
c = distance from gravity axis to extreme fiber, in inches;
M = moment due to horizontal forces, in inch-pounds;
A = cross-sectional area, in square inches; and
I = moment of inertia, in inches^4.

Lateral Support and Buckling

The lateral support provided by practically any soil except the softest or most fluid has been generally found sufficient to prevent pile failure from buckling for the embedded portions. For unsupported portions of piles, both in water and aboveground, the piles should be designed as columns under direct loads, unless lateral forces are present, in which case the design should be made to resist these also.

For piles extending through water, mud, or soft clay incapable of supporting load, carrying the load in end bearing on rock or in friction in the lower portion, buckling should be checked. It has been shown theoretically by Bjerrum[147] that buckling needs to be considered only if

$$\frac{I}{A^3} \leq \frac{f^2_{yp\,\max}}{4cE} \tag{7.2}$$

where c = coefficient of lateral displacement (equals horizontal subgrade reaction times pile width; this varies from 100 to 600 psi for typical values of length and width of steel piles).

The right side of this equation is fairly constant in any particular soil. The left side depends only upon the shape and is smallest for compact sections. Using yield points and low values of c, it will generally be found that there is no danger of buckling in piles that remain straight.

A theory has been developed to cover buckling of piles having initial imperfections, but none for piles containing stresses imparted by curvature during driving.

Use of batter piles around a pile group in soft soil may approximately double the critical load values.

Buckling in piles in soil is generally of concern only for static loads, since the soil has not time to yield laterally to any great extent under an instantaneous driving blow or transient loading.

The following examples indicate how little lateral support suffices. Under a paper machine 1¼ in.-diameter rods were welded into lengths reaching through clay to bedrock 130 to 165 ft below and loaded to 13,600 psi. Jacking resistance was 5 tons, and maximum test load was 8 tons. In Gothenberg, pipes 2 and 3 in. in diameter and 130 ft long load-tested to the yield point of the steel.[168]

At the Potomac River Bridge at Ludlow's Ferry, Md., 14-in. 102-lb H piles 215 ft long driven through 90 ft of soft river mud to refusal at 190 ft carried for six weeks a test load of 200 tons on 114 ft of the pile in air, water, and soft mud with no visible tendency to buckle. At Houston, Tex., 12-in. 72-lb H piles 90 ft long driven through 60 ft of soft and 12 ft of hard clay with the tops exposed to sway of a pier were tested with 100 tons for a year with no signs of buckling. For the John Hancock Building in Boston, Mass., 14-in. H piles in 10 ft of medium clay, 81 ft of soft clay, and 10 ft of hardpan over rock were analyzed for stability, and found to be of conservative design, and capable of bowing even 1 ft with no possibility of buckling while still retaining the ability to carry a substantial load.[3w]

Eccentricity and Bending in Piles Acting as Columns

Effect of Curvature or Out-of-plumbness. No pile is likely to be entirely plumb or straight. Any curvature or out-of-plumbness causes bending stresses. An exception would be straight piles having opposite batters under a common cap, in which case they act as an A frame.

The possibility of deviations from alignment of long wood or H piles in the ground is usually ignored unless deviations during or after driving are seen. No inspections of solid piles can be made, and they are usually allowed to carry full design loads. In Oslo, pipes are welded to the sides of H piles and an inclinometer measures deviations in both directions; all piles showing a radius of curvature of less than 1,200 ft are rejected.[147] Slope indicators for hollow piles have been made which record bends greater than those permitting view of the bottom.

With long pipe or combination pipe and shell piles, bends are often seen and restrictions or rejections applied. Distortions may be bends of short or long radius or dog-leg bends. Many piles have been rejected in the past because a light lowered into the pile could not be seen from the top.

Tests conducted at a Detroit site by the Engineering Research Institute, University of Michigan, showed satisfactory results on piles with sweeping bends, and it was concluded that the lateral support afforded piles driven in the soil mass was quite sufficient to counteract the eccentricity developed by departure from vertical alignment. Tests made on long dog-legged piles in New York City resulted in acceptance by the Building Department.[125] Many load tests on bowed piles have been recorded, and it appears to be generally conceded that the ultimate structural capacity is equal to or greater than the working load times the required factor of safety. The ability of curved piles to carry significant loads satisfactorily, even if dog-legged or badly inclined, is due to the fact that soils provide substantial lateral support, usually in excess of that needed for stability. Methods of analysis for approximating the stress in dog-leg end-bearing piles are available.[125] If calculations show that such piles can carry all, or even part, of the design load safely, a great deal of money may be saved.

FIG. 7.1. Hemlock pile pulled showing break due to eccentricity. Broken 6 ft from tip which was reported 6 ft into rock.

Lateral Forces

It is important to give consideration to lateral as well as eccentric forces on piles, since stresses may mount rapidly from these causes when combined with stresses from direct axial loads. Wharf, pier, bridge, and bulkhead piles are often called upon to resist horizontal loads that will cause bending in piles unless adequate batter piles or bracing can be used.

Pile Load Limitations

Factors Limiting Loads. When making the choice of a pile type from among those that may be used, the number of piles and cost per pile of each type affect the decision. The number might be thought to be directly governed by the application of the driving and column-design formulas. However, this is not entirely the case, for limiting values for each type are generally used, being based on experience, recommendations of manufacturers and of technical associations, and building-code requirements. Some of the more common typical practices will be given, as indicative of what may be expected in the way of regulations or rules. However, often the strength requirements necessary to permit driving to the desired penetration, or the factor of safety desired, will result in heavier sections than those required by column designs for working loads, or by the practices discussed in this section.

Composite Piles. These should be limited to the maximum capacity allowed for the weakest member.

Pile Group Reduction. The Uniform Building Code of the International Conference of Building Officials contains a formula for reduction of values for single piles when driven in clusters, if they depend solely on friction.

Combined End Bearing and Friction. The Los Angeles Building Code prohibits assuming that frictional resistance and bearing resistance act simultaneously.

Friction. The Los Angeles City Building Code may prescribe arbitrary unit-friction values after performance of special foundation investigations as set forth in the code, although they permit for any cast-in-place pile an assumed frictional resistance equal to one-sixth of the bearing value of a foundation material at minimum depth, or 300 psf, whichever is the lesser.

Dynamic Load Tests. The Los Angeles Building Code limits the determination of pile capacity by driving formula to one containing the factors included in formulas (2.1*a*) and (2.1*b*), using a factor of safety of 4. Use of such a formula is permitted only in cases where redriving after 24 hr shows not over 25 per cent increase in resistance.

Static Load Tests. Several rules for interpreting load tests appear in Chap. 15.

Lateral Loads. The Los Angeles authorities may require load tests for lateral loads.

Factor of Safety. Some building codes establish the factor of safety. For instance, that set by the International Conference of Building Officials Uniform Building Code is 4, even for use with the more complete type of formula required by them.

Compression Value of Confined Rock and Concrete

Tests of H piles loaded beyond the elastic limit of the steel show that steel pile sections of lengths several times longer than the depth of section cannot be pushed farther after having been seated in rock by hammer blows and that the flanges or webs will buckle locally at about the yield point of the steel, without penetration.*

Confined rock in its natural beds requires many times the pressure to

* G. G. Greulich, in a paper presented before the AASHO meeting, 1941.

cause it to flow than that which an unconfined specimen block or cylinder will stand.[39] Lateral flow is prevented for the rock in place, whereas loose specimens fail by diagonal shear. For example, an area of Berea sandstone 1.25 in. square withstood 60,000 lb, which was 7.5 times the strength of 2-in. cubes. Baker's *Masonry Construction* showed that pressures of 50,000 to 75,000 psi were required to push a small piece of steel into sandstone. Devonian shale bedrock 120 to 175 ft below the surface, along the Cuyahoga River, indicated a crushing strength of 11,000 psi on 1.25- and 2-in.-square areas when supported laterally.

Confined concrete acts in a similar manner.[39] Laboratory tests of 3,500-lb concrete on 1.25-in.-square confined areas showed a value of 25,000 psi, and only 0.2-in. penetration at 60,000 psi. Disruption of adjacent surface areas did not occur until the value reached 85,600 psi. The true crushing strength of concrete when confined, as in rock sockets for Drilled-In Caissons or piles, appears to be eighteen to twenty times the test-cylinder strength.

Improved resistance to end bearing in a weathered rock stratum, far below the river bottom, that had stopped driving of pipe piles was obtained after mucking out by placing a 5-in. grout pipe with a concrete end plug. A 4-in. grout hole was drilled through the plug, and batches of grout forced into the rock. Repeated charges obtained 5- to 20-ft depths of treatment.[126]

Point Devices. H-pile tips may be reinforced by adding welded or riveted plates to reduce the pressure between the steel and rock to a 3,000- to 6,000-psi range, in comparison with small-rock-cube crushing strengths of 6,000 to 18,000 psi. Metal thicknesses should be made 2½ to 3 times the original for a height 2½ to 3 times the diameter. This also avoids local buckling if one corner hits rock first. Reinforcement is not generally used when rock is overlain by 6 to 8 ft. of hardpan, 8 to 10 ft of gravel, 10 to 12 ft of sand, or 12 to 15 ft of hard clay.

Oslo points are often used in Norway and Sweden at the tips of H and other metal piles bearing on rock that may be sloped.[147] The points are tempered-steel bars 3 or 4 in. in diameter, projecting several inches below the foot of the H. On an H pile, a slice is cut into the web and the bar welded in place. The end is hollow-ground so that the sharp edges can secure a hold on the rock. A chiseling procedure is followed, preferably using a hammer that can operate at reduced energy until a hole is established. These points avoid eccentricity from pile edge or corner bearing or from slipping of the pile foot.

Fin-type points[149] on pipe piles have aided penetration through obstacles and enabled good seating in firm rock to be formed, by giving a drill action. They consist of four or more thick rib plates cast or welded on the outside of a cone bottom.

Wood Piles

Design of Unsupported Lengths of Wood Piles as Columns under Direct Load. Extending above ground support, wood piles should be well braced. The portion below water level remains as an unsupported column, but the piles may be stiffened by cross bracing above water. Unless otherwise required by local building codes, the safe load for unsupported wood piles may be obtained by the following formulas:[2b]

$$R = Ap \qquad (\text{when } l/d \lesseqgtr 11) \qquad (7.3a)$$

$$R = Ap\left[1 - 0.33\left(\frac{l}{Kd}\right)^4\right] \qquad (\text{when } 11 < l/d < K) \qquad (7.3b)$$

$$R = \frac{0.274AE}{(l/d)^2} \qquad (\text{when } l/d \gtreqless K) \qquad (7.3c)$$

where R = safe load, in pounds;
p = allowable unit compression parallel to grain, in pounds per square inch;
l = unsupported height, in inches;
d = least dimension of square column having same area as a round pile at one-third point of unsupported height from small end, in inches;
A = area at one-third point of unsupported height from small end, in square inches; and
$K = 0.64\sqrt{(E/p)}$.

The above formulas include a reduction factor of 3, which may be increased if advisable.

The following formula for allowable fiber stress in wood piles acting as columns, and also in bending and compression, is given by the War Department,* based on an allowable working stress of 1,000 psi for continuously wet wood of the grade carried in depot stocks:

$$f = 1{,}000\left(1 - \frac{l}{60d_{\text{ave}}}\right) \qquad (7.4)$$

No piles should be used as a column in which the unsupported length is greater than fifty times the diameter.

Eccentricity in Wood Piles. The influence of crooks such as those common in wood piles probably does not cause as great a reduction in strength of a column as might be expected. Tests have shown that when a timber is subjected to combined compression and bending it develops a higher stress at both the elastic limit and maximum load than when subjected to compression only. This does not mean that crooks or

* *War Dept. Technical Manual* TM5-286, *Semipermanent Highway and Railway Trestle Bridges,* April, 1945.

eccentricity need not be restricted, but it may relieve some anxiety over the common bends.

Axial Load and Bending.[2b] *The Wood Handbook* gives formulas for wood columns subjected to axial loads and bending, which may be used for piles unless otherwise required by code.

Load Limitations. Wood piles are limited by the New York City building laws to a maximum working load of 20 tons on a pile having a 6-in. tip and 25 tons on a pile with an 8-in. tip.

The Uniform Building Code of the International Conference of Building Officials limits the load to 500 psi at mid-length. The Massachusetts Department of Public Safety limits the safe load on wood piles to 12 tons for spruce or Norway pine, and 15 tons for yellow pine or oak, except that piles driven to refusal shall be figured as columns with an area equal to the middle cross section.

The ability of the soil to support the load transmitted to it by the piles is usually the factor that determines the adequacy of the foundation. The permissible working stresses for piles of any material are seldom fully utilized. It makes no difference to the soil what material composes piles of equal dimensions. Failure to recognize this fact has resulted in the use of loads for wood piles much below those of concrete piles, even though the allowable unit stress of wood is above that of concrete. Design loads of 25 to 30 tons are now becoming common, and loads of 40 tons where soil conditions warrant.[2ci,2co]

Concrete-filled Pipe Piles

There are a number of methods for designing concrete-filled pipe piles, in cases where building codes do not govern. Four of the most commonly used methods will be given, using the following notation:

R = safe load on pile, in pounds.
A_c = area of concrete, in square inches.
A_s = area of steel in the pipe, in square inches; in some cases this is the net area of the pipe, in others, a reduction for possible rust has been made.
n = relation between moduli of elasticity of steel and concrete.
f'_c = crushing strength of concrete at 28 days, in pounds per square inch.
f_{syp} = yield point of steel in the pipe, in pounds per square inch.

The methods of design are as follows:

Proportionate Method. The assumption is made that the concrete and steel each carry a portion of the load, directly proportional to their areas and to their moduli of elasticity. However, recognizing that these two units work in more intimate relationship than is ordinarily the case in spirally reinforced concrete columns, owing to the fact that the concrete is confined by a steel shell, the normal loading of $0.255f'_c$ has been increased by 20 per cent. The fomula reads as follows:

$$R = (A_c + nA_s)(0.225f'_c \times 1.2) \tag{7.5}$$

where A_s is considered as the full area of the pipe.

Ultimate-load Method. This method takes into account the fact that we really do not know exactly how the load is distributed between the concrete and the pipe. However, a great deal of information on this matter has been obtained from tests carried out at the University of Illinois and Lehigh University on steel pipes filled with concrete. The ultimate strength of such members has been found to be at least 85 per cent of the crushing strength of the concrete after 28 days multiplied by the area of the concrete, plus the area of the steel mutliplied by the yield point. The working load is deduced from the utlimate load as given by dividing by a required factor of safety. The formula reads as follows:

$$R = \frac{0.85f'_cA_c + f_{syp}A_s}{FS} \tag{7.6}$$

where FS = factor of safety.

A_s is considered as the area of pipe remaining after deducting 1/16 in. for corrosion.

ACI Method. A third method frequently used is that laid down by the American Concrete Institute Building Code (ACI 318-56),[5ad] which has adapted the column formula to the peculiar condition of having a continuous spiral reinforcement in the shape of a pipe surrounding the concrete core. This formula is as follows:

$$R = 0.25f'_c \frac{1 - 0.000025l^2}{r_c^2} A_c + f'_rA_s \tag{7.7}$$

The value of f'_r shall be given by the following formula when the pile has a yield strength of at least 33,000 psi and an l/r_s ratio equal to or less than 120:

$$f'_r = 17{,}000 - \frac{0.485l^2}{r_s^2} \tag{7.8}$$

where R = safe load on axially loaded short piles, where unsupported length l is not greater than ten times the least lateral dimension d in pounds;

f'_r = allowable unit stress in metal pipe, in pounds per square inch;

l = unsupported length of pipe, in inches;

r_c = radius of gyration of concrete; and

r_s = radius of gyration of steel pipe.

A_s is considered as the area of pipe remaining after deducting $\frac{1}{16}$ in. for corrosion.

Chicago Building Code Method. The Chicago Building Code has established a formula as follows:

$$R = 0.25 f'_c A_c + 0.36 f_{syp} A_s \tag{7.9}$$

where A_s = area of pipe remaining after deducting $\frac{1}{8}$ in. for corrosion.

Conventional Precast or Poured-in-place Concrete Piles

Direct Load. *ACI Method.* The formulas for tied or spirally reinforced concrete columns contained in the ACI Building Code (ACI 318-56)[5ad,5b] may be applied to the unsupported lengths of conventional precast concrete or poured-in-place concrete piles. These formulas are as follows, the allowable axial load for a tied design being 80 per cent of that for spirally reinforced columns.

For tied columns,

$$R = 0.80 A_g (0.225 f'_c + f_s p_g) \tag{7.10a}$$

For spirally reinforced columns,

$$R = A_g (0.225 f'_c + f_s p_g) \tag{7.10b}$$

where A_g = over-all or gross area of column, in square inches;

f_s = nominal working stress in vertical column reinforcement, to be taken as 40 per cent of the minimum specification value of the yield point, namely, 16,000 psi for intermediate grade and 20,000 psi for hard grade or rail steel; and

p_g = ratio of effective cross-sectional area of vertical reinforcement to the gross area, A_g.

The maximum allowable load R' on axially loaded long columns having a length l greater than ten times the least lateral dimension d is given by formula (7.11), in which R is the allowable axial load on a short column as given by formula (7.10*a*) or (7.10*b*).

The maximum allowable load R' on axially loaded long columns having a length l greater than ten times the least lateral dimension d is given by the following formula:

$$R' = R\left(1.3 - 0.03\frac{l}{d}\right) \tag{7.11}$$

where R = allowable axial load on a short column as given above in formula (7.7).

The percentage of steel should be between 1 and 4 per cent, using $\frac{5}{8}$-in.-minimum-diameter longitudinal bars spaced clear from the face not less than $1\frac{1}{2}$ in. plus thickness of ties. For severe exposures, such as freezing or thawing, 3 in. of cover should be used.

The ratio p_g should be not less than 0.01 nor more than 0.04 for tied columns subject to axial loads, using at least four ⅝-in.-diameter bars, and increasing the 0.04 ratio to 0.08 in case of combined axial and bending stresses; for spirally reinforced columns the ratio p_g should not be less than 0.01 nor more than 0.08, using at least six ⅝-in.-diameter bars. Tests by the Research Board[21] have not indicated any particular value in

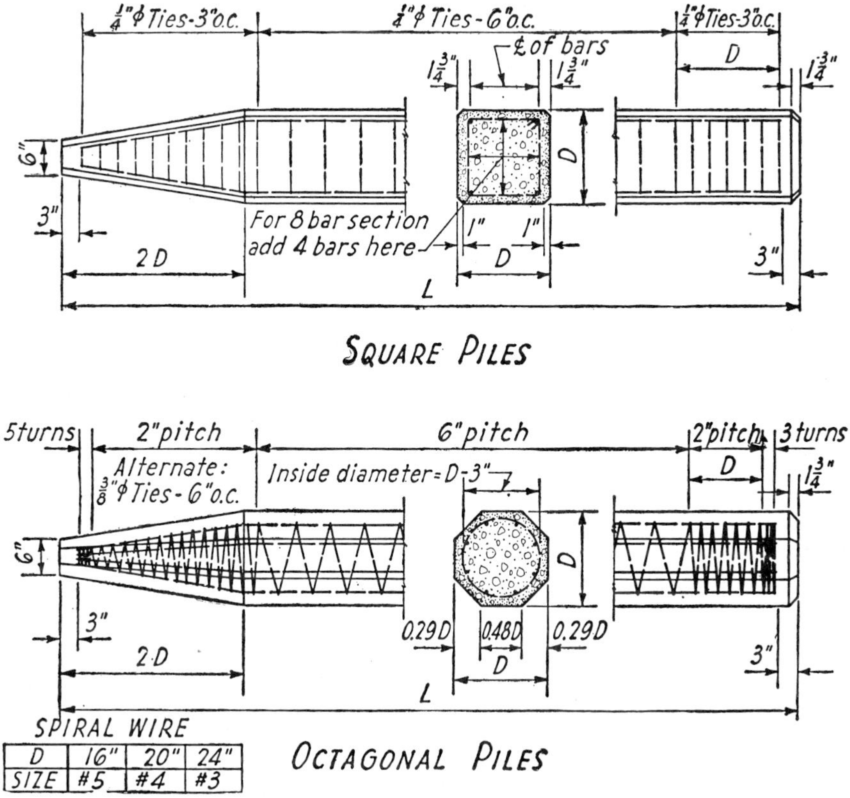

FIG. 7.2. Conventional precast concrete piles.

a covering thickness greater than 1 in., as far as strength during driving is concerned, but good practice for concrete buried underground or exposed to the weather should be followed in providing coverage for reinforcing bars in piles as well as in other structural member. For severe exposures, such as to alternate freezing and thawing of wet concrete, 3 in. should be used.

Lateral ties in tied columns should be at least ¼-in.-diameter, spaced apart not over 16 bar diameters, 48 tie diameters, or the least dimension of the column. The spiral reinforcement should not be less than the following:

$$p' = 0.45[(A_g/A_c) - 1]f'_c/f'_s \tag{7.12}$$

where f'_s = 40,000 psi for hot-rolled intermediate grade bars, 50,000 psi for hard grade bars, and 60,000 psi for cold-drawn wire.

Spirals should have a minimum diameter of ¼ in. for bars or No. 4 AS&W gage for wire, spaced apart not over 3 in. nor less than 1⅜ in. nor 1.5 times maximum aggregate diameter.

Precast piles 24 to 30 in. in diameter and up to 114 ft long have been cast. Large precast concrete piles are usually hollow, with a solid section near the top. Piles as small as 6 in. in diameter have been used.

Direct Load and Bending. The ACI Building Code[5ad] gives the following formulas for columns subjected to axial load and bending when the ratio of eccentricity to depth, e/t, is not over two-thirds in the directions of bending:

$$\frac{f_a}{F_a} + \frac{f_b}{F_b} < 1 \qquad \text{(for bending in one plane)} \tag{7.13}$$

$$\frac{f_a}{F_a} + \frac{f_{bx}}{F_b} + \frac{f_{by}}{F_b} < 1 \qquad \text{(for bending oblique to the principal axis)} \tag{7.14}$$

where e = eccentricity, in inches;
t = over-all depth of column section, in inches;
F_a = nominal allowable axial unit stress ($0.225f'_c + f_s p_g$ for spiral columns, and 0.8 of this value for tied columns);
F_b = allowable bending unit stress that would be permitted if bending stress only existed;
f_a = nominal axial unit stress (R/A_g);
f_b = bending unit stress (actual) = bending moment in inch-pounds divided by section modulus; and
f_{bx} and f_{by} = bending-moment components about x and y principal axes divided by the section modulus of the transformed section relative to the respective axes.

The ratio p_g should not be less than 0.01 nor more than 0.08 for tied or spirally reinforced columns subject to combined axial load and bending.

Formula (7.13) is applicable to both conventional and prestressed concrete piles.

In designing a concrete column subject to both axial load and bending, the preliminary section may be determined by use of an equivalent axial load given by the following formula:

$$R = N\left(1 + \frac{Be}{t}\right) \tag{7.15}$$

For trial computations, B may be taken from 3 to $3\frac{1}{2}$ for rectangular tied sections, the lower value being used with minimum reinforcement. For circular spiral sections, values of 5 to 6 for B may be used. If the load N has an eccentricity e greater than $\frac{2}{3}t$, the procedure set forth in the ACI Building Code may be followed.

When piles are subject to wind or earthquake loads in addition to combined axial loading and bending, the section need not be increased unless the allowable stress given by formula (7.15) is increased by more than one-third.

Handling Stresses. In foundation or bearing piles, the working load is generally direct compression, and the reinforcement is needed only for handling purposes, except where the piles act as columns or in bending.

In general, piles must be rolled over when removing them from the forms, picked up at the center or at two or more points with an equalizer, loaded, blocked in position for shipment to the site, unloaded, stored at the site, and picked up from a horizontal to a vertical position. The method of transportation, whether by train, truck, or boat, should be known when designing piles.

Reinforcement is often the major cost item and may be considerably reduced by providing proper handling devices to minimize handling stresses. A slight excess of reinforcement is sometimes desirable to permit more leeway in handling, stacking, and driving.

The pile should be designed, after selection of lifting points, as a beam or as a beam with cantilever ends, and lifting points should be marked on the pile. During handling, the sequence of operations may be such that several span combinations will occur, as shown in Fig. 7.3. The moment of inertia of square piles should be figured about both the normal and diagonal axes. Design charts for various sizes of concrete piles, of the types shown in Fig. 7.2, based on three methods of pickup shown in Fig. 7.3, are shown in Fig. 7.5.[5a] These pickup locations result in equal positive and negative moments. Design practice varies from 0 to 50, 75 or 100 per cent impact during handling of piles. If it is desired to design piles for impact, reduce the ordinates of the curves in Fig. 7.5 by $100/(100 + I)$, where I is the per cent impact.

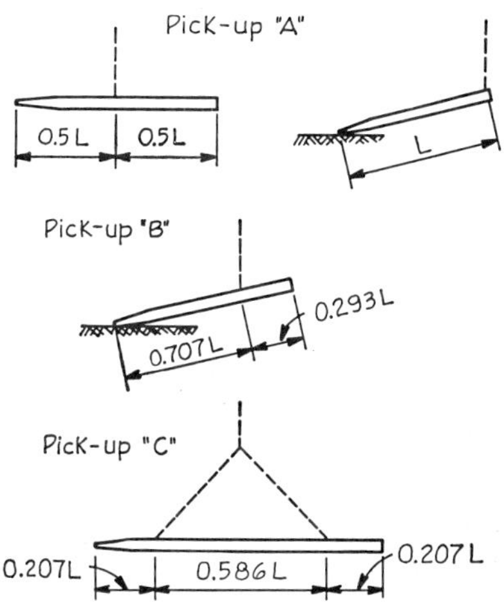

FIG. 7.3. Precast concrete pile pickups.

If the lengths of the piles are so great that more than two pickup points are desirable, Fig. 7.4[5a] shows span and moment coefficients that give equal reactions. The reinforcement required to resist these moments should be continued throughout the length of the pile, since the moments are nearly equal throughout. If the reactions are not equal, the bending moments for several arrangements of lifting points should be calculated until an arrangement is found that gives nearly the same positive and negative bending moments. It is common practice to disregard taper near tips in such calculations.

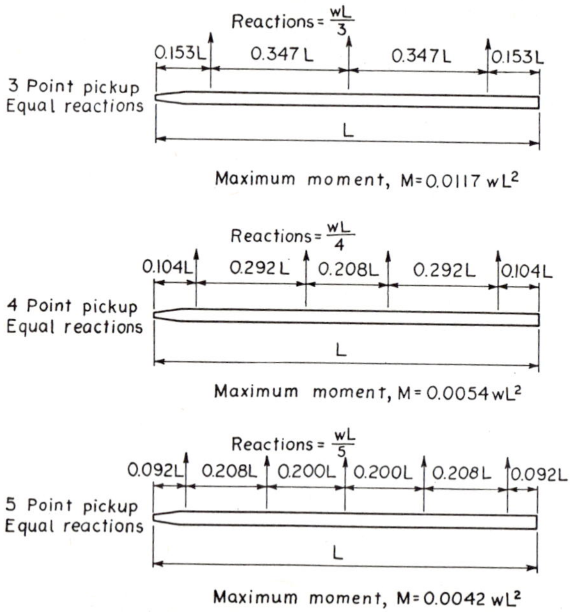

FIG. 7.4. Pickup locations and bending moments for equal reactions.

In designing a precast concrete pile, when determining the resisting moment, it is convenient to use the transformed section in which the steel area is considered equivalent to a concrete area n times larger. The resisting moment is $M = fI/c$, where f is the unit stress at the distance c from the neutral axis, and I is the moment of inertia of the transformed section. The value of I is the sum of the moments of inertia of the elements of the transformed pile section about its center of gravity. The concrete is assumed to have no tensile strength. The position of the neutral axis is first assumed and the transformed areas and the distances of their centers of gravity are calculated. If it is found that the center of gravity of the section coincides reasonably well with the assumed position of the neutral axis, the moment of inertia may be

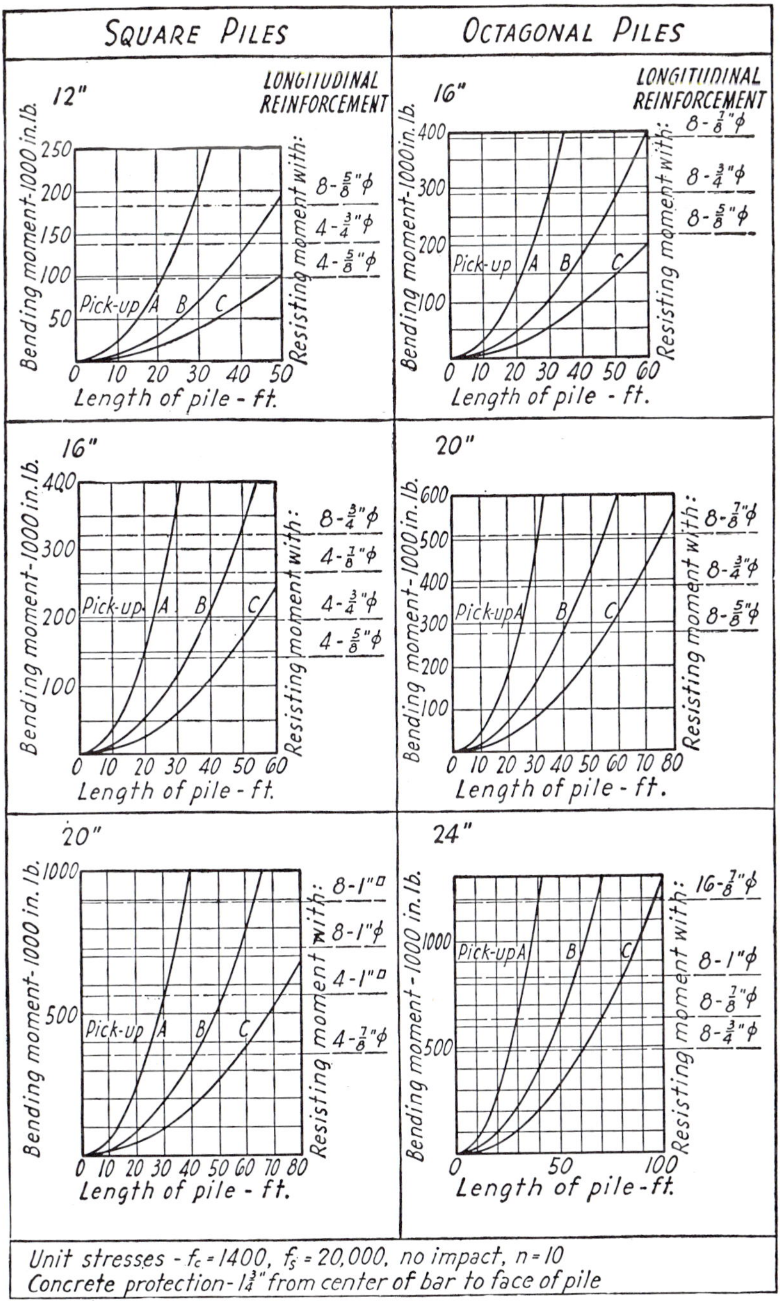

FIG. 7.5. Design charts for square and octagonal conventional precast concrete pile pickups.

calculated about that axis; otherwise a correction must be made. It is convenient first to calculate I_c, the moment of inertia about the center line of the pile section, and to determine I by the relation

$$I = I_c - A_t x^2 \tag{7.16}$$

where A_t = total area of transformed section, and
x = distance from neutral axis to center line.

Handling Arrangements. Where many long piles are used, a gantry crane with equalizer lifting device may be economical. For a few small piles, caterpillar or rail cranes are usually used. Piles under 25 ft long are usually designed to be picked up with a single line at any point. In Fig. 7.6[5a] are shown cable arrangements for multipoint pickups. Piles

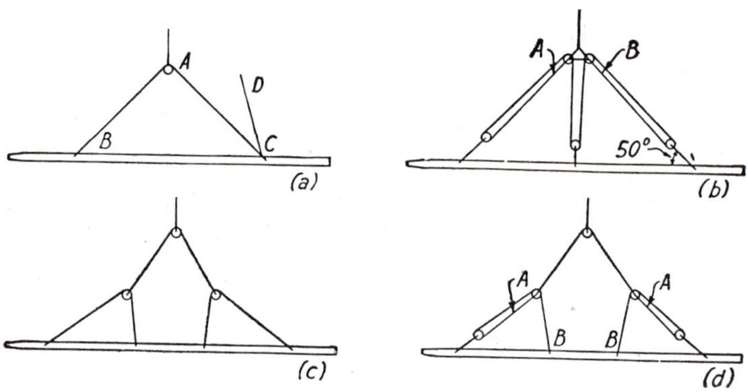

FIG. 7.6. Equalizer cable-lifting arrangements.

designed for a two-point support and lifted by cables require a sheave at point A so that the cable becomes an equalizer cable, giving equal reactions. Unless an equalizer is used, the pile must be lifted carefully so that the tension in the cable will be equal and the entire load not rest on one end. To raise the pile to the vertical, line CD is attached, and sheave A shifted toward C. A snubbing line is necessary to retain control. The three-point supporting arrangement acts on similar principles. The four-point arrangement in Fig. 7.6*c* is similar to the three-point arrangement, but the cables may be arranged as in Fig. 7.6*d* to increase the end reactions.

The possible arrangements for lifting piles by cables and sheaves are unlimited. To determine the forces, draw a force polygon, as shown in Fig. 7.7.[5a] From the given pickup points, draw to scale a trial arrangement of cables A, B, and C. The forces in these cables are equal. To construct a force polygon, draw line a parallel to A and lay off distance 1–2. Then draw b parallel to B and c parallel to C, the distances a, b, and *c* being equal. The direction and relative magnitude of *d* is the line 1–4. The magnitude of *e*, half the pile weight, is shown by line 4–5. Line 1–5 is drawn parallel to the pile. The magnitudes of *a*, *b*, *c*, *d*, and *e* are proportional to their lengths in the force polygon. The arrangement determined thus is satisfactory if bending moments are a minimum for the number of supports used, and if the height of the center sheave can be handled by the equipment.

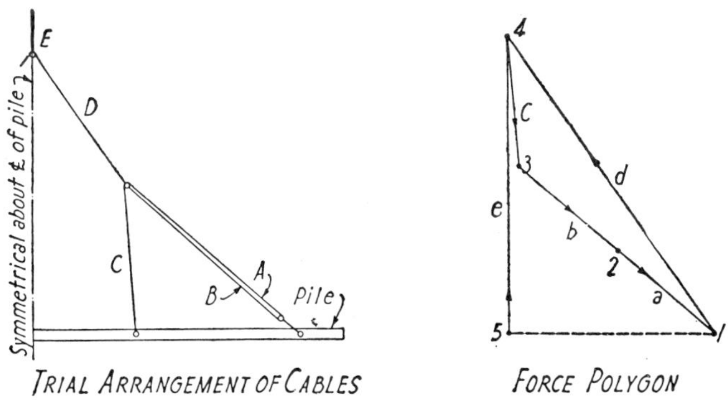

FIG. 7.7. Graphical stress determination in lifting cables.

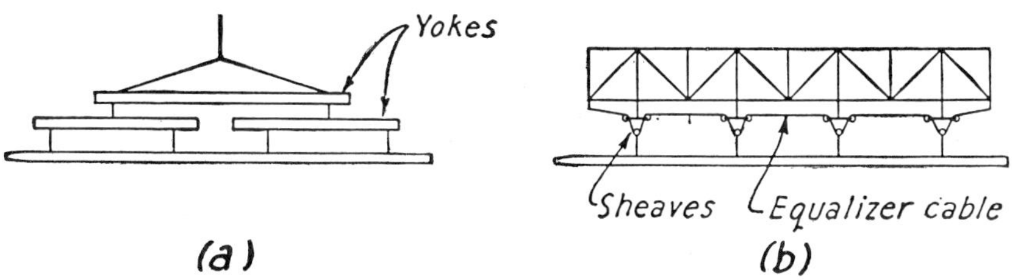

FIG. 7.8. Yokes or spreader beams.

Spreader beams or yokes as shown in Fig. 7.8*a*[5a] may be used instead of equalizer cables for very heavy piles. With them, the required crane-boom height is less and reactions may be made equal at all lifting points. With a three-point pickup, this is done by adjusting the leverage of the yokes. When a beam or truss is used, a continuous equalizer cable is needed as shown in Fig. 7.8*b*; otherwise the reactions depend upon the initial tensions in the cables and deflection in the beam or truss.

Lifting Holds. Usually piles may be picked up by cables and hooks looped around the piles. To protect the edges from damage and to prevent wear on the cable, short lengths of wood or other cushion should be used. Where long piles have been designed for definite pickups, hooks or bolts are often cast in place and cut off after the pile is in driving position. This also ensures that the pile will be handled right side up, where extra reinforcing bars have been placed on one side only. Lifting attachments are sometimes designed to be inserted when lifting the pile and removed when the pile is in storage or ready for driving. A typical hold is a 1½-in. I bolt or T bolt which screws into a nut and washer cast in the bottom of the pile. To form a hole, a wood pin is used when the pile is cast, turned from time to time while the concrete sets, and removed. The hole is kept plugged until the bolt is inserted.

Loading and Stacking. Where piles are to be loaded or stored in tiers, the blocking should be in vertical lines to avoid undue bending in a lower tier, as shown in Fig. 7.9.[5a]

FIG. 7.9. Misplaced blocking.

Load Limitations. The New York City building laws limit the allowable working stress in friction piles for controlled concrete to $0.25f_c'$ at the cross section located two-thirds of the length measured up from the tip, and for average concrete to smaller stated values. For end-bearing piles passing through loose saturated sand-clay soils, medium soft clay, or better, 75 per cent of the load is assumed as carried at the tips of piles over 40 ft long. For end-bearing piles less than 40 ft long the entire load is assumed as carried at the tip.

The Uniform Building Code of the International Conference of Building Officials limits the allowable working stress to $0.2f_c'$ on the average cross-sectional area.

The Los Angeles Building Code requires precast concrete piles to have a minimum tip diameter of 6 in. and minimum average diameter of 10 in., to be designed as reinforced-concrete short columns with not less than 2 per cent longitudinal reinforcing, and to have developed a strength of 3,000 psi prior to driving.

The Massachusetts Department of Public Safety states that the load on precast concrete piles driven through loose wet soil incapable of adequately resisting lateral bending shall not exceed 400 psi on the middle cross section plus 6,000 psi on the reinforcement, which shall be between 2 and 4 per cent and have 3 in. of cover and ¼-in. loops at 10-in. centers minimum. The average diameter is to be at least 11 in. and the tip 8 in. The length is not to exceed thirty times the average diameter in material having little lateral stiffness when driven to firm bearing, nor forty times in any case.

Splices in Conventional Precast Concrete Piles. An occasional disadvantage of precast concrete piles has been experienced when necessary driving resistance was not met at the expected depth and driving below grade was required. For conventional precast concrete piles with about 2 per cent of the area as bar reinforcement, substantial splices may be made by lapping or welding the bars; the pile can be driven after the extension has cured.

Pretensioned Precast Concrete Piles

The load-carrying capacity of the pile is more generally governed by the capacity of the soil to receive the load than by the structural strength of the pile. For determination of pile designs to carry the forces established from the soil characteristics, or where structural strength of the pile is the limiting load factor, the following methods of analysis may be used.

Direct Loads. Criteria for capacity of prestressed concrete piles in compression do not appear in current standards.*

Until standard criteria are available, it is suggested that designs for pretensioned piles be based upon the following assumptions:

1. The flexural modulus of elasticity of concrete (E_c), in pounds per square inch, may be assumed to be 1,800,000 plus 500 times the cylinder strength at the age considered. Actual values may vary as much as 25 per cent.[5ba] The value for pretensioned concrete piles acting as long columns carrying transient loads is often taken as 5,000,000; for sustained loads, over several years, the value should be reduced to 2,000,000 to allow for creep.

2. The modulus of elasticity for steel, in pounds per square inch, may be assumed to be 29,000,000 for cold-drawn wire, 27,000,000 for seven-wire strand, 25,000,000 for strands with more than seven wires, and 27,000,000 for alloy-steel bars.[5ba]

3. Prestress, after losses, should be based on a minimum value of 700 psi, with a maximum of $0.2f'_c$.

4. Pickup points for handling should be located based upon computed stresses, using an allowance of 50 per cent for impact, with a maximum tension of $7.5\sqrt{f'_c}$, in psi.

5. Shapes and sizes should be from 8 to 24 in., by 2-in. increments in both square and octagonal sections. Normally, use solid piles up through the 20-in. size, with circular voids in larger sizes. Holes should preferably extend the full pile length, because abrupt changes in cross section may be failure points; if cored holes must have closed ends, cone-shaped transitions with the cone length equal to at least twice the core diameter are suggested.[5ax] Precast concrete tips shaped as truncated cones, with dowels to insert into sleeves cast into the lower end of the pile, are sometimes used. Corners of square piles should be beveled ¾ to 1 in. or given a ¾-in. radius.

* At the time of this publication (1961) the Joint Committee of the AASHO and PCI is preparing a specification and standard plans for prestressed precast concrete piles.

6. Spiral reinforcement should be used in all piles. Arrangements used consist of five turns of No. 5 wire on 1-in. centers at each end, followed by 16 turns at 3-in. spacing, then by 6- to 9-in. spacing in the remaining central section.

7. Longitudinal reinforcement should be seven-wire, uncoated, stress-relieved, prestressed concrete strands, using a strand tension from the manufacturer's catalogue. Wire gages and numbers of strands should vary to suit the pile size and driving conditions.

Preston[5ax] proposes the following formulas for allowable stresses for piles acting as pin-ended columns (use 70 per cent of supported length for piles fixed at one end, or 50 per cent if fixed at both ends), when using concrete having an ultimate strength of 5,000 psi, initial prestress force of 800 psi, and final prestress force of 640 psi after losses:

For l/r 0 to 44:

f_{ca}, psi = 900 (usual regardless of amount of prestress in the pile, in the range of short columns, up to the point where capacity is determined by buckling and not by the yield point of the material, using a factor of safety of 4) (7.17)

For l/r 44 to 130:

$$f_{ca}, \text{ psi} = 585 + \left[\frac{130 - (l/r)}{86}\right] 315 \tag{7.18}$$

For l/r over 130:

$$f_{ca}, \text{ psi} = \frac{0.5\pi^2 E_c}{(l/r)^2} \tag{7.19}$$

where f_{ca} = allowable unit stress in concrete, based, respectively, on yield point for short columns, arbitrary values for intermediate lengths, and Euler formula with factor of safety of 2 for long columns.

Direct Load and Bending. For prestressed concrete subject to combined axial and bending, formula (7.13) may be applied, in which the value of available stress (f_b) should be taken as the ultimate compression stress (f_c') minus the prestress unit stress, times the reduction factor of 0.4 (as used for bridge members in compression, given in Sec. 207.3.2.*a* of the *Tentative Recommendations for Prestressed Concrete* of the ACI-ASCE Joint Committee 323[5ba]). With the allowable stresses F_a and F_b established, the total of the equation taken as 1.0, and either the nominal axial unit stress f_a or the bending unit stress f_b determined, formula (7.13) can be used to solve for the remaining stress term.

When the bending moment is large and the direct load small, stress on the tensile side may be the governing factor. In such cases, the ultimate strength in bending should be figured and provide a factor of safety not less than 2. The *Tentative Recommendations for Prestressed Concrete* of the ACI-ASCE Joint Committee 323[5ba] should be followed until the methods given therein are adopted or superseded and contain the following formula for ultimate bending moment as determined by the tensile side of the pile:

$$M_u = A_s f_{su} d \left(1 - \frac{k_2}{k_1 k_3} \frac{p f_{su}}{f'_c}\right) \tag{7.20}$$

where f_{su} = average stress in prestressing reinforcement at ultimate load, in pounds per square inch;

d = depth to centroid of force;

k_2 = ratio of distance between extreme compression fiber and center of compression to depth to neutral axis; and

k_1k_3 = ratio of average compressive concrete unit stress to cylinder strength, f'_c.

Tests have shown that the factor k_2/k_1k_3 may be taken as 0.6. Determination of f_{su} requires knowledge of the stress-strain characteristics of the steel, effective prestress, and crushing strain of the concrete. Assumptions must be made regarding the relation between steel and concrete strains, which will be different for bonded and unbonded construction. The following expressions for f_{su} may be used if the stress-strain properties of the steel are similar to those described in Section 304 of the *Tentative Recommendations* and if effective prestress after losses is not less than $0.5f'_s$:

$$f_{su} = f'_s \left(1 - 0.5 \frac{p f'_s}{f'_c}\right) \qquad \text{(for bonded members)} \tag{7.21}$$

$$f_{su} = f_{ce} + 15{,}000 \qquad \text{(for unbonded members)} \tag{7.22}$$

where f'_s = ultimate strength of prestressing steel, in pounds per square inch;

p = A_s/bd, ratio of prestressing steel; and

f_{ce} = effective steel prestress after losses, in pounds per square inch

Properties of typical prestressed concrete piles are shown in Tables V.7, V.8, and V.9. The allowable loads tabulated in Tables V.8 and V.9[12r] are maximum loads for average conditions. For adverse marine exposures, it is desirable to decrease the allowable tensile stresses and tabulated bending moments to increase the factor of safety against cracking. Where long unsupported lengths are used with combined direct loads and bending, it may be advisable to reduce allowable direct loads and moments to retain the desired factors, particularly if the pile deflection is appreciable. For ideal conditions, such as short foundation piles fully protected, a smaller factor of safety may be used and allowable loads increased. Tabular values can be modified to retain safety factors in keeping with conditions.

Examples of numerical solutions for various factors entering into the design of prestressed concrete piles appear in references 5*aw*, 5*ax*, and 12*r*.

For seating precast prestressed concrete piles in marl or soft rock, the Florida State Road Department uses steel H sections cast 6 ft into the concrete and projecting 6 ft. For 18-in. piles, 8-in. BP36 or built-up sections of ½-in. minimum plates or two 90-lb rails welded back to back are used; for 20-in. piles, 10-in. BP42 or built-up sections or rails are used. Typical spiral ties in the composite 6 ft consist of eight turns at 3-in. pitch, six turns at 6-in. pitch, and eight turns at 3-in. pitch, reading up from the foot of the concrete and using No. 5 gage wire.

Handling Prestressed Precast Concrete Piles. When lifting prestressed concrete piles, for equal positive and negative moments, they should be designed to be picked up at the points shown on Figs. 7.3 and 7.4. A single pickup point is located at $0.30L$ from the top, with a maximum pile length in feet of four times the width in inches; two-point pickups are located at $0.21L$ from each end, with a maximum pile length in feet of 5.5 times the width in inches, although without prestressing, three or four lifting points might be needed.

The maximum length of pile that can be picked up when using the above point locations can be determined for any prestressed concrete pile of a given size and moment capacity, or the moment capacity determined for any given length, by the following formulas:[5ax]

$$M_{neg} = \frac{w(0.21L)^2}{2} = 0.022wL^2 \qquad \text{(for 2-point pickup)} \qquad (7.23)$$

$$M_{pos} = \frac{w(0.58L)^2}{8} - \frac{w(0.21L)^2}{2} = 0.02wL^2 \qquad \text{(for 2-point pickup)} \qquad (7.24)$$

$$M_{neg} = \frac{w(0.30L)^2}{2} = 0.045wL^2 \qquad \text{(for 1-point pickup)} \qquad (7.25)$$

$$M_{pos} = \frac{wL^2(0.4L)^2}{8(0.7L)^2} = 0.041wL^2 \qquad \text{(for 1-point pickup)} \qquad (7.26)$$

Splices in Prestressed Concrete Piles. The small cables in prestressed piles represent only about 0.5 per cent of the area and are insufficient as nonstressed reinforcement for a suitable splice. If further driving is not required, a splice can be made by lapping the cables with bars and drilling in a few dowels, but if further driving is needed, the usual procedure has been to pull the pile and replace it with a longer one.[5bh]

Florok's Plasticement* is coming into use as a fast-setting, epoxy-type, inexpensive splicing compound for splicing sections of prestressed concrete piles; such joints have tested stronger than the concrete. Setting time is from 5 to 15 min, after which driving may be resumed. A steel

* Manufactured by the Chargar Corp., Hamden, Conn.

boot is fitted over the lower section, and the upper section set in, with a blocked clearance of ½ in. between sections. Dowel sleeves are cast in the head of the lower section, and dowels in the foot of the upper section. The hot cement is poured through side pockets in the boot, filling the space around the dowels in the sleeves and the ½-in. joint.

H Piles

Design for Direct Load. The unsupported portions of H piles should be designed as columns in the manner usual for structural-steel columns. The formulas given for columns by the American Institute of Steel Construction* are as follows:

For axially loaded columns with values of l/r greater than 120,

$$f_s = 17{,}000 - 0.485 \left(\frac{l}{r}\right)^2 \tag{7.27}$$

For axially loaded columns with values of l/r greater than 120,

$$f_s = \frac{18{,}000}{1 + (l^2/18{,}000r^2)} \times \left(1.6 - \frac{l}{200r}\right) \tag{7.28}$$

where f_s = allowable working stress, in pounds per square inch.

In general, the greatest efficiency is obtained by using the section having the largest over-all dimensions for a given weight, if adequate thickness is available for corrosion resistance. To secure sufficient contact with the soil, in order to develop a reasonably high capacity for the pile, and often to provide sufficient strength to withstand the driving stresses necessary to reach the desired depth of penetration and to obtain the desired factor of safety, it will be found that a pile with the necessary over-all dimensions will have considerably more cross-sectional area of metal than is required to carry the design or working loads when acting as a column. Present good practice limits the working load to two-thirds of the basic column value, or 10,000 psi, which provides allowance for such factors as damage during driving, corrosion, poor tip contact, and slight eccentricity due to out-of-plumbness or bowing. Since H piles are usually used in long lengths and stout sections are needed for handling and driving, the stress is usually below this figure except for short piles. Unit stresses due to vertical working loads usually range from 4,000 to 10,000 psi, a common rule of thumb allowing from 1 to 1¼ tons for each pound of section weight. The lower values apply to heavy and important structures whereas the higher values are for light manufacturing buildings, trestles, wharves, and piers. It is recommended that a maximum limit of 12,000 psi be set, in order to allow for possible corrosion

* *Steel Construction,* 5th ed., American Institute of Steel Construction, 1955, p. 284.

loss.[11a] Where combined bending and direct stresses occur, the limit can be raised somewhat. The New York City building laws set $\frac{3}{8}$ in. as the minimum thickness of metal and limit the compressive stress under working load to 9,000 psi, with protection against corrosion to be provided where required; local encasement may answer this problem.

Each nominal size of H bearing piles is furnished in both light and heavy sections. The lighter sections, which have a minimum uniform web and flange thickness of $\frac{7}{16}$ in., are usually used for driving through uniform, homogeneous materials under fresh water. The heavier sections, which have a uniform minimum web and flange thickness of $\frac{9}{16}$ in., are used where boulders, sunken timbers, cribbing, or riprap may be encountered, and in salt-water exposure. Sometimes H-column sections, which have webs thinner than flanges, are used for piles, but this should not be done without considering corrosion possibilities. The stubby, thick flanges of structural H sections may be desired when seating in or on rock, to avoid corner buckling.

Direct Load and Bending. There are two methods in use for specifying allowable working stress in a steel pile subjected to both axial stress and bending.

AISC Method. As required by the AISC, a satisfactory design is one in which

$$\frac{f_a \text{ actual}}{F_a \text{ allowable}} + \frac{f_b \text{ actual}}{F_b \text{ allowable}} < 1 \tag{7.29}$$

where f_a = axial unit stress (actual);
f_b = bending unit stress (actual), owing to both eccentricity and transverse forces;
F_a = axial unit stress that would be permitted if axial stress only existed; and
F_b = bending unit stress that would be permitted if bending stress only existed.

This method requires that members subject to both direct and bending stresses shall be so proportioned that the greatest combined stresses shall not exceed the allowed limits. For axial load, the allowed value is that given by the column formula, whereas for bending, the allowed value is determined by allowable beam stress in bending.

AREA Method. This method is much more severe. It requires that members subject to both axial and bending stresses shall be so proportioned that the combined fiber stress will not exceed the allowed axial stress.

Load Limitations. The United States Steel Co. recommends limiting the fiber stress in H-pile design to a maximum of 12,000 psi under working load, instead of the 15,000 psi used in structural design, to allow for corrosion.

Code Restrictions. New York City building laws limit lateral loads to 1,000 lb unless it has been demonstrated by tests that the pile will resist a lateral load of 200 per cent of the proposed working lateral load without lateral movement of more than ½ in., and resist the proposed working lateral load without a lateral movement of more than 3/16 in., at the ground surface.

The New York City building laws limit the fiber stress in H piles to 9,000 psi, provided that where injurious soil conditions exist the steel is protected.

FORCES ON MARINE STRUCTURES

Lateral Forces from Winds, Waves, and Ships

Wharves and piers are acted upon by a number of possible forces, either singly or in various combinations. Each of these forces will be discussed separately, but the judgments of the designer, owner, and ship operators should be combined to establish criteria suitable for the particular site, including physical features, observations and records of wave heights, meteorological and oceanographic conditions, and types and sizes of ships expected. Ships will not berth or remain at wharves or piers when winds and waves are too strong, so that lesser waves and winds may be used in conjunction with a vessel; maximum conditions for berthing should be agreed upon.

Wind, current, and wave forces on a berthing ship may be additive to lateral ship impact forces from the component of forward motion normal to the face of the structure, while the longitudinal motion of the ship remains unchanged. During strong winds or periods of rapid current, ships maneuver only with great caution on the windward side of a structure. Design to resist all possible ship impacts would generally be impracticable.

The velocity of approach normal to the wharf is affected by the type of construction. A rigid wall must be approached more carefully than a flexible fender, for fear of damaging the ship.

Several methods have been used to determine pressure from waves from recorded depth of water and wave dimension, observations, computation, or assumption. Various mathematical methods of wave-dimension determination show a wide spread of results and range from a simple formula with one variable to complex equations.

Ship Dimensions

Ship dimensions showing relationships among lengths, beams, dead weights, and shaft for various classes of vessels are given in reference 192.

Wind Forces

Wind forces on structures, wharves, dolphins, mooring platforms, and exposed presentments of moored ships acting upon them may be determined from the following Weather Bureau formula for wind on flat surfaces:

$$p_w = 0.003V_w^2 \tag{7.30}$$

where p_w = horizontal wind load on vertical surface normal to wind, in pounds per square foot; and
V_w = velocity of wind, in miles per hour.
The effect of the wind blowing from all possible directions may be determined from Duchemin's formula, as follows:

$$P_n = p_w\left(\frac{2\sin\theta}{1+\sin^2\theta}\right) \tag{7.31}$$

where P_n = horizontal wind load normal to plane vertical surface, in pounds per square foot; and
θ = horizontal angle between line parallel to direction of wind and plane vertical surface.
Designs for a pressure of 20 psf or the breaking strength of hawsers have been used.[103]

Trestles should be designed to resist lateral and longitudinal wind loads on the structure and on the side of a train.

Water Forces

Structures in water should be designed to resist horizontal forces from waves, tide drag, or water current acting on the structure and moored ships or on any obstructions that may lodge against them.

Wave Action. Information on waves is essential for design of marine structures. Preliminary information may be used to indicate whether waves are likely to be a governing factor in design. If of consequence, detailed wave studies can be made.

Available formulas for calculation of wave forces are not perfect but offer convenient design criteria when used intelligently.

Structures should be designed to resist the most forceful waves that may occur and also to resist such wave forces as may be permissible on moored ships that rest against the structure.

Prediction of wave heights, periods, duration and frequency or occurrence, and seasonal effects may be valuable when considering construction methods, particularly in exposed locations.[190]

Basis of Wave Determination. Wave heights, lengths, and periods can be determined if known or assumed values are taken for water depth,

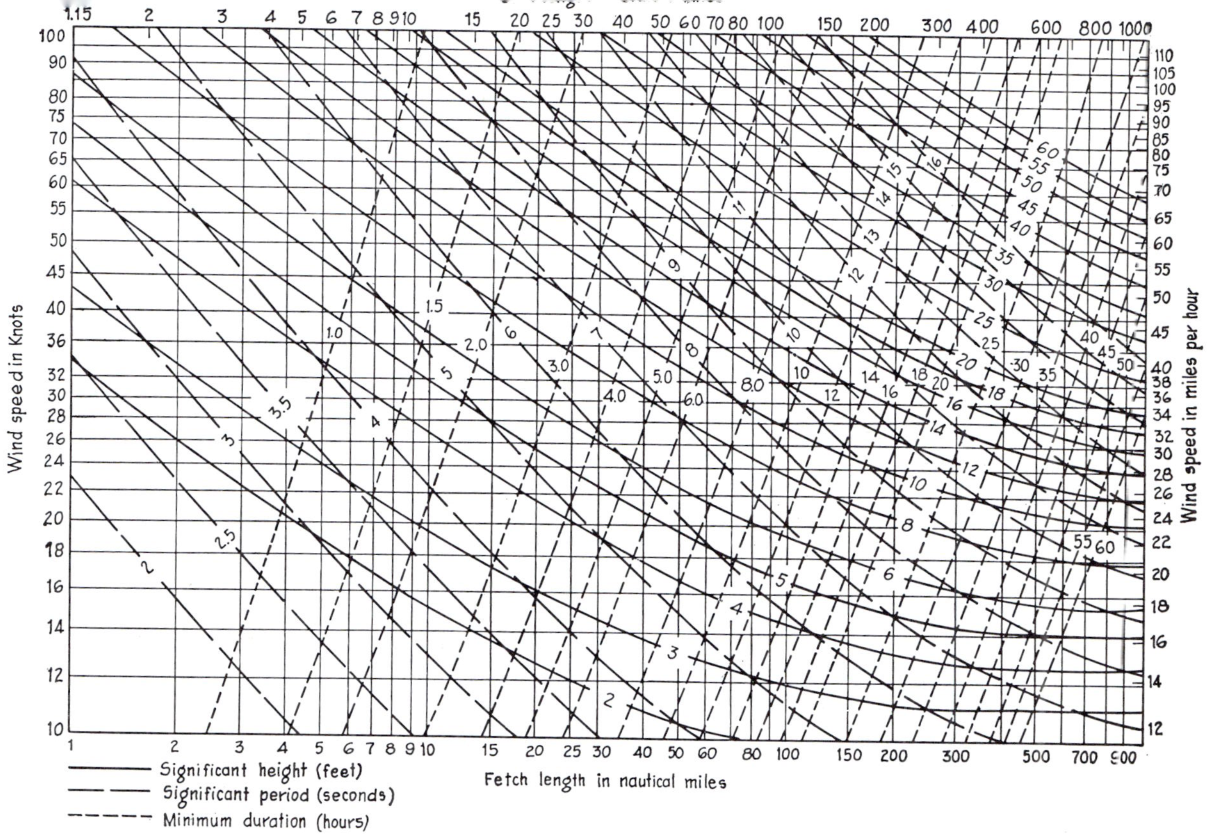

FIG. 7.10. Deep-water wave-forecasting curves as a function of wind speed, fetch length, and wind duration. (*From Bretschneider, Council on Wave Research*, 1958.)

slope of bottom, velocity and duration of wind, and length of fetch (unbroken distance over water the wind has blown).

Preliminary information may be obtained from visual observations, local recollection and records, and analogous situations adjusted to conditions. Other means of wave determination are hindcasting, forecasting, and climatic studies; for consideration of these methods references 179, 183, 190, and 195 may be consulted and services of meteorologists and oceanographers may be required. Sea and swell atlases are published by the U.S. Navy Hydrographic Office; such data are useful for indicating monthly variations and location differences. Methods are available for converting sea, swell, and wind data into frequencies of wave heights.[190]

Deep-water forecasting curves computed by the method of Sverdrup and Munk and modified by Bretschneider[195] are shown in Fig. 7.10. Deep water is considered as water deeper than one-half the wave length.

For forecasting wind waves in shallowing water, less information is available, but design methods are contained in references 128, 179, and 195. This condition assumes that the waves are generated in deep water but decay upon approach to the structure. The possibility should be investigated that such decayed waves might be higher than waves generated in the shallow water close to the structure. The effect of shoaling may be obtained from tables in references 179 and 181, in which the ratios of H/H_o' (wave height in shallow water to height deep-water wave would have been if H were unaffected at shallow-water depth d by refraction and friction on the bottom) to d/L_o appear: for values of d/L_o between 1.0 and 0.057, the ratio of H/H_o' is between 1.0 and 0.9 and may be ignored for engineering purposes; for values between 0.057 and 0.018, the H/H_o' ratio is between 1.0 and 1.25, and for values between 0.018 and 0.0001, between 1.25 and 4.5.

Before using the higher values of the ratio H/H_o', indicating shoaling waves much higher than the deep-water wave, the likelihood of the wave breaking should be investigated by such means as use of reference 179, Fig. 8. Similar results for wave decay due to shoaling may be obtained from reference 195, Fig. 8, in which the ratio H/H_o' is called the shoaling coefficient K_s. Additional loss from friction on the bottom may be obtained from reference 195, Fig. 7, but would appear to be of possible engineering significance only if T^2/d is greater than 0.5.

Forecasting curves for waves generated in shallow water of constant depth, by Bretschneider, appear in Fig. 7.11.

Wave Heights. A marine structure should be designed to withstand the effects of the highest predictable wave at that site, if economically feasible. Visual observations of storm waves are usually unreliable, and direct measurements costly, and observed waves are not necessarily the highest that may occur.

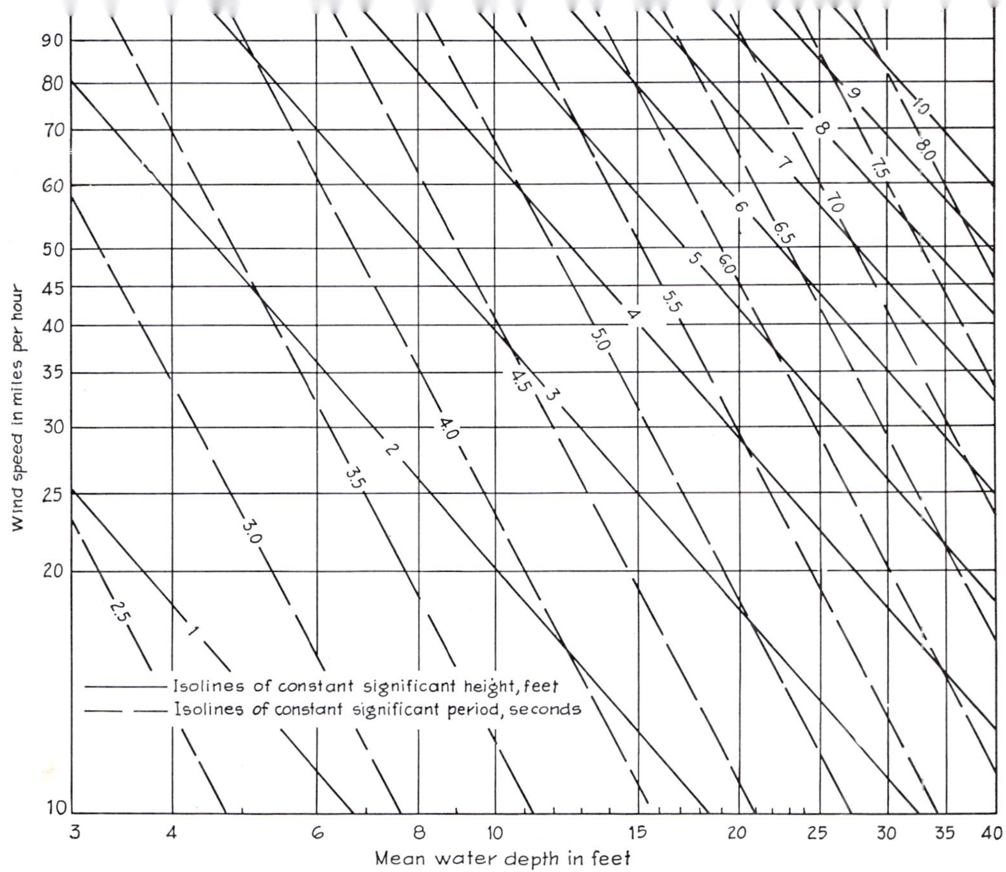

FIG. 7.11. Wave-forecasting relationships for shallow water of constant depth. (*From Bretschneider, Council on Wave Research,* 1958.)

Wind-wave characteristics depend upon wind velocity and duration, fetch length, and air-mass character.

Empirical pre-World War II formulas for obtaining wave heights have been used, which assume constant wind velocity, unlimited duration of wind, water deep enough so that the waves can be fully developed, and (except for the Molitor equations) either a sufficiently large velocity to generate the maximum possible wave for each fetch or a fetch long enough so that the given velocity will generate the largest possible wave, as follows:

$$H_{max} = 0.81 V_k \qquad \text{Cornish} \tag{7.32}$$

$$H_{max} = 0.742 V_k \qquad \text{Zimmerman} \tag{7.33}$$

$$H_{max} = 0.027 V_k^2 \qquad \text{Rossby and Montgomery} \tag{7.34}$$

$$H_{max} = 1.5\sqrt{F_n} \quad (\text{for } F_n > 39) \quad \text{Stevenson (used by Minikin)} \tag{7.35a}$$

$$= 1.5\sqrt{F_n} + 2.5 - \sqrt[4]{F_n} \quad (\text{for } F_n < 39) \quad \text{Stevenson} \tag{7.35b}$$

$$H_{max} = 0.196\sqrt{V_k F_n} \quad (\text{for } F_n > 39) \quad \text{Molitor} \tag{7.36a}$$

$$= 0.196\sqrt{V_k F_n} + 2.5 - 1.035\sqrt[4]{F_n} \quad (\text{for } F_n < 39) \quad \text{Molitor} \tag{7.36b}$$

$$H_{max} = 4.59\sqrt[4]{F_n} \qquad \text{Iribarren} \tag{7.37}$$

where H_{max} = maximum possible wave height, trough to crest, for conditions present, in feet;

V_k = wind velocity, in knots; and

F_n = fetch (the continuous length of water over which the wind blows in a continuous direction), in nautical miles (6076.1 ft).

The Stephenson and Molitor equations were most widely used before World War II. During and after the war great advances in oceanography were made. Sverdrup and Munk propounded the significant wave (the average period and height of the highest one-third). The Sverdrup-Munk method gives much higher waves than the others, partly because of inclusion of a duration factor. The Darbyshire formula was proposed, as follows, but later said by its author to give too low wave heights for high wind velocities over deep water:

$$H_{max} = 0.054 V_k^{1.5}(1 - e^{-0.23\sqrt{F_n}}) \tag{7.38a}$$

which can be solved by use of a table of Naperian exponential functions.

Later versions of the Darbyshire formula[202] for use in plotting maximum wave heights were proposed, in 1955, based on $H_{max} = 2H_{equiv}$, as follows:

$$H_{max} = 0.054 V_k^{1.5} \qquad \text{(coastal regions)} \tag{7.38b}$$

$$H_{max} = 0.076 V_k^2 \qquad \text{(deep water)} \tag{7.38c}$$

The methods most widely used since the war are the Darbyshire in England and the Sverdrup-Munk in the United States. Bretschneider has improved the Sverdrup-Munk method and shown graphical relationships among fetch, wind velocity, and deep-water wave heights in reference 195, Fig. 2. Use of this method is suggested. A graph of deep-water waves by Wilson (reference 195, Fig. 3) in 1955 gives somewhat different results.

Wave heights predicted by various formulas are shown in comparative graphical form in Fig. 7.12. This indicates the divergence of opinion and state of accuracy.

The significant wave height is defined as the average of the highest one-third of the waves and is generally reported because it discounts the lower waves. The maximum wave has been observed to be as high as 1.87 times the significant wave.[194] Darbyshire and Putz give the following relationships:

	Darbyshire	Putz
Mean wave height (average of all), H_{mean}	1.0	1.0
Equivalent wave height (height of sine wave of same energy as complex wave), H	$1.2H_{mean}$ (is ½ max wave height)	$1.2H_{mean}$
Significant wave, H_s	$1.6H_{mean}$	$1.57H_{mean}$
Maximum wave (largest of 100—Darbyshire; largest in 20 min—Putz), H_{max}	$1.5H_s$ $2.4H_{mean}$	$1.81H_s$

Significant wave heights and periods are of value to theorists for studying energies and durations, but are not considered here except to give their relationships to other heights and periods; note that some of the graphs and formulas are expressed in such terms. However, a given wind velocity will not produce the maximum wave of which it is capable unless the wind is of sufficient duration, as indicated in Fig. 7.10.

The maximum (design) still-water depth during a storm is the basis of wave computation and is given by

$$d = d_e + A_t + S_t \tag{7.39}$$

where d = design still-water depth, in feet;
d_o = mean low-water depth, in the absence of storm effects, in feet;
A_t = astronomical tide, in feet; and
S_t = storm tide, in feet.

Mean low-water depths appear on U.S. Navy Hydrographic Office navigation charts or may be obtained from soundings adjusted for tidal effects. Astronomical tides appear in U.S. Coast Guard and Geodetic Survey tables or may be observed. Storm tides are more difficult to predict.

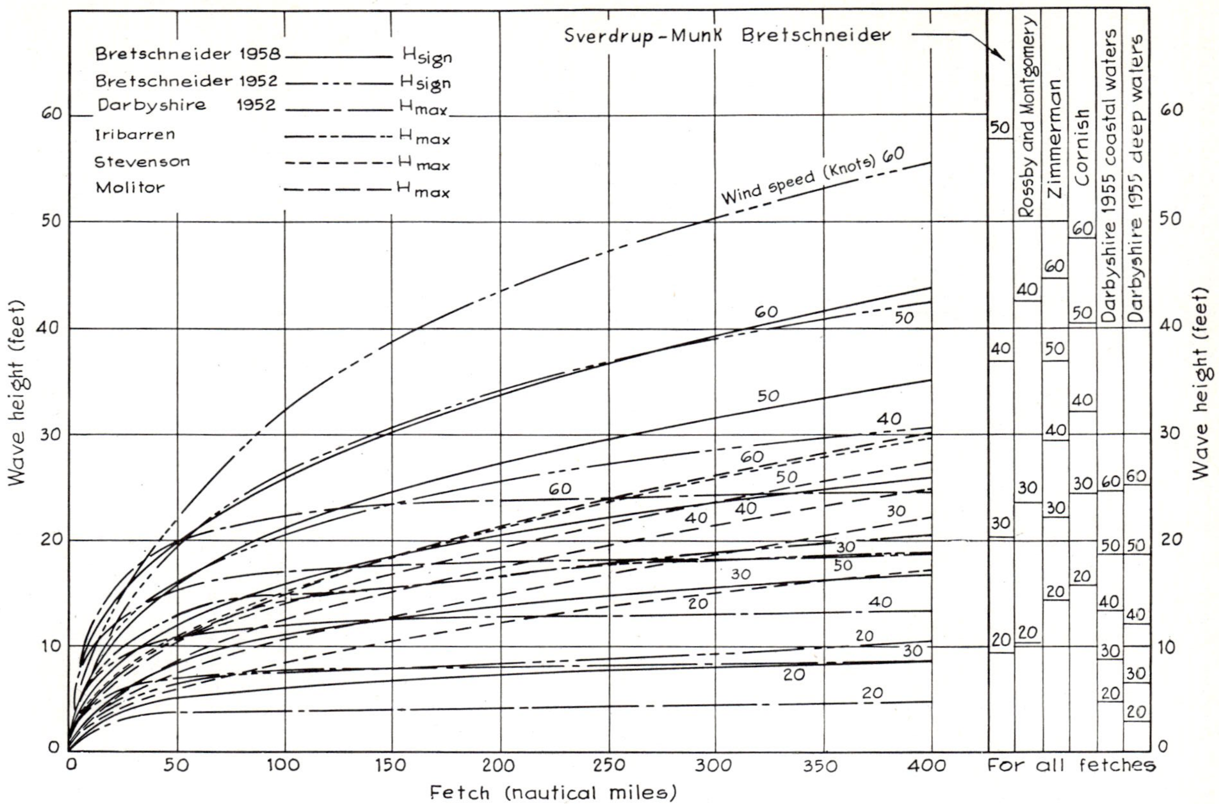

FIG. 7.12. Comparison of various formulas for wave height as a function of fetch and wind speed. (*After Gesler, Eaton, and Hall, Proc. 18th Intern. Navigation Congr.*, 1953.)

The maximum solitary wave height in a given depth of water is, for values of d/T^2 less than 0.08, as follows:

$$H_{max} = 0.78h, \text{ or } h = 1.28H_{max} \qquad \text{(in shallow water)} \qquad (7.40)$$

$$H = L/7 \qquad \text{(in deep water, predicted by Michell and Havelock)} \qquad (7.41)$$

h = distance of wave trough above the bottom, in feet; and
L = wave length, crest to crest, in feet; and
T = wave period, in seconds.

For values of d/T^2 greater than 0.08, the dependence of limiting wave height upon the period becomes important. A graph of relationships for all conditions appears in reference 128, Fig. 5.

Wave run-up is the vertical height to which a breaking wave will reach. A value of 1.5 times the wave height is most frequently used, although values of 0.9 to 2.0 have been suggested for individual designs.[191] The relative run-up of 1.5 should be adequate for steep design waves, with H_{max}/T^2 value of 0.8, although not adequate for waves less steep.

Refraction. For cases where underwater contours may affect the heights of waves by refraction, methods of study appear in reference 179.

Wave Pressures. Pressure exerted on a structure by nonbreaking waves is essentially hydrostatic. Broken and breaking waves exert additional pressures due to the dynamic effect of turbulent moving water and compression of trapped air pockets; these pressures may be much greater than hydrostatic. Structures located where waves may break should be designed to withstand much greater forces than those subjected to nonbreaking waves. Depths at which waves will break are shown in reference 179.

Nonbreaking waves may occur where the fetch is limited and in protected locations. However, most shore structures are so located that storm waves will break in the water depth at the structure. The most common method of determining pressures due to nonbreaking waves is that of Sainflou.* Various methods are compared in reference 187.

Sainflou-method pressures on a vertical wall from a clapotis wave are shown on Fig. 7.13. A clapotis wave is the standing wave caused by reflection of a wave train from a bulkhead, breakwater, or steep beach. If a wave of length L and height H hits a vertical wall, point A is the crest elevation on G, the trough. The mean level or orbit center of the standing wave is above the still-water level D, a distance h_o determined as follows:

$$h_o = \frac{\pi H^2}{L} \coth \frac{2\pi d}{L} \qquad (7.42)$$

* M. Sainflou, "Treatise on Vertical Breakwaters," Ann. des ponts et chaussées, vol. 4, 1928. (Translated by W. J. Yardoff, Corps of Engineers, U.S. Army.)

$DA = H + h_o$, while DG equals $H - h_o$. The still-water hydrostatic pressure wd at the base of the wall is plotted as CE. The triangle CDE is the still-water hydrostatic wall pressure. These relationships are shown in Fig. 7.13. As the clapotis moves up or down, the hydrostatic

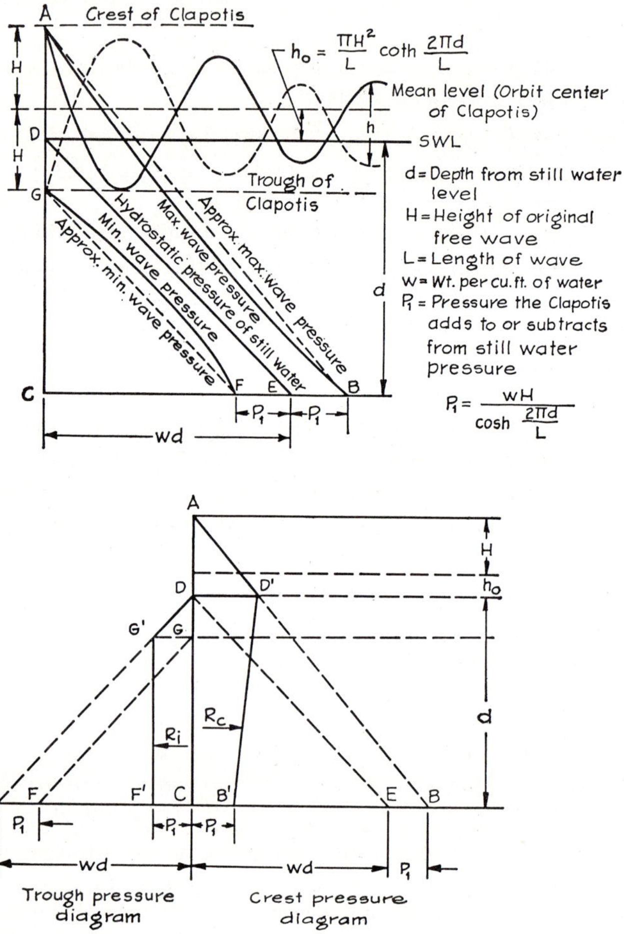

FIG. 7.13. Clapotis on vertical wall. (*From U.S. Army, Corps of Engineers, Beach Erosion Board.*)

pressure is increased or decreased by P_1. This pressure change is computed as follows:

$$P_1 = \frac{wH}{\cosh (2\pi d/L)} \tag{7.43}$$

These equations have been plotted graphically in Figs. 7.14*a* and 7.14*b*.

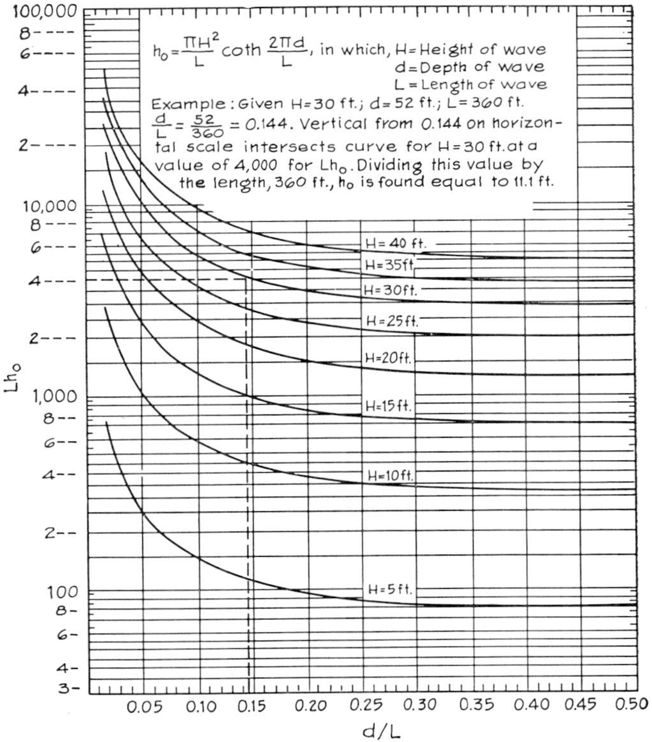

Fig. 7.14*a*. Determination of value of h_o in Sainflou's formula. (*From U.S. Army, Corps of Engineers, Beach Erosion Board.*)

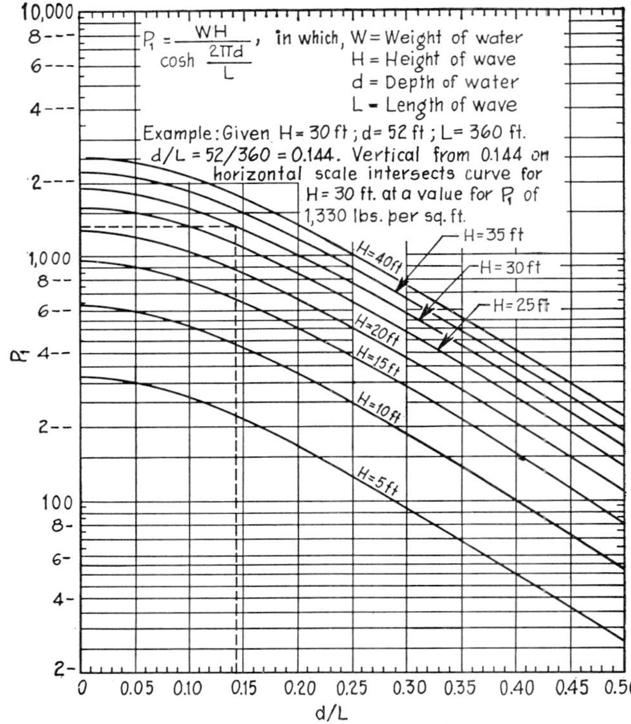

Fig. 7.14*b*. Determination of value of P_1 in Sainflou's formula. (*From U.S. Army, Corps of Engineers, Beach Erosion Board.*)

If still-water-level pressure also exists on the other face of the structure, hydrostatic pressures *CDE* cancel, leaving a clapotis pressure *ABED* or *AD′B′C* for pressure at crest condition and pressure at trough condition of *DEFG* or *DG′F′C*. If no hydrostatic pressure is present on the wall from the land side, the pressure triangle would be *ACB* when the crest is at *A*, as shown on Fig. 7.13.

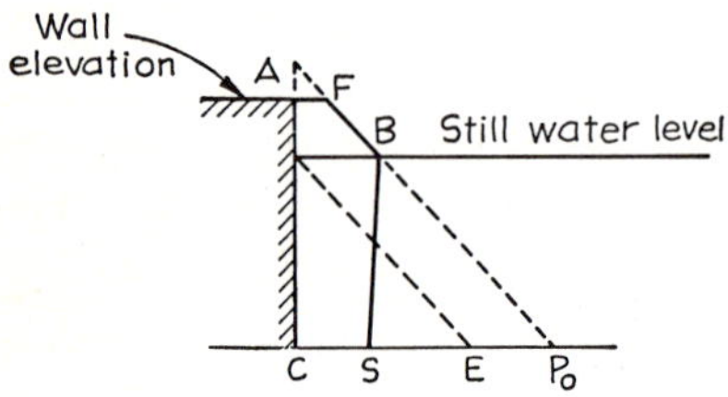

FIG. 7.15. Pressure on walls of low heights. (*From U.S. Army, Corps of Engineers, Beach Erosion Board.*)

If the wave overtops the structure, compute the pressure diagrams as though the wall were of equal height, then ignore the pressure above the structure, as shown on Fig. 7.15.

Tables of values for hyperbolic functions appear in reference 181.

A somewhat different method of obtaining the Sainflou pressure diagram is contained in reference 136, with formulas and graphs.

Theory and model tests indicate that wave pressures on a vertical wall do not vary appreciably with the angle of incidence.[179]

Waves breaking on a structure create pressures which may be large. The *Minikin-method* pressures probably represent the closest approach to actual pressures, although containing inconsistencies,[136,179] and explains failures otherwise unpredictable. The dynamic pressure is concentrated

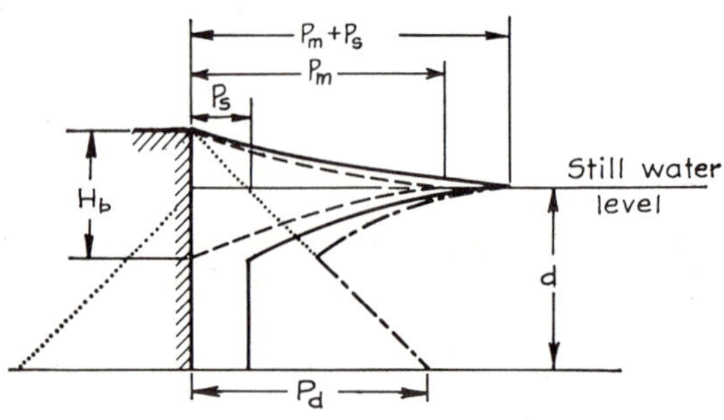

FIG. 7.16. Minikin wave-pressure diagram. (*From U.S. Army, Corps of Engineers, Beach Erosion Board.*)

at still-water level. The Beach Erosion Board[179] presents Fig. 7.16 and the following formula for application of the Minikin method:

$$P_m = \frac{101 H_b w}{L_D} \frac{d}{D} (D + d) \tag{7.44}$$

where P_m = maximum intensity of wave pressure, in pounds per square foot;
H_b = height of wave just breaking on structure, in feet;
w = unit weight of water, in pounds per cubic foot; and
D and L_D = deeper-water depth and wave length, respectively, in feet.

Approximate values of the deep-water wave length may be obtained from

$$L_o = \frac{g}{2\pi} T^2 = 5.12T^2 \tag{7.45}$$

where L_o = deep-water wave length, in feet.

With the ratio d/L_o, the wave length (L_d) at the structure at depth d may be obtained from tables in references 179 and 181.

Minikin,[136] does not show separate L_o and L_D symbols, but uses L and a coefficient m, which thus permits solution without the necessity of having tables or graphs at hand. This formula is as follows:

$$P_m = \frac{m\pi d}{LD} wHg \frac{D + d}{2} \tag{7.46}$$

where the value of the coefficient m is taken as 2, in which case formula (7.46) can be converted into formula (7.44).

Hydrostatic unit pressures on the seaward side at still-water level P_s and P_d at depth d are given by

$$P_s = \frac{wH_b}{2} \tag{7.47}$$

$$P_d = w\left(d + \frac{H_b}{2}\right) \tag{7.48}$$

The dynamic pressure occurs centered on the still-water level and falls away rapidly on parabolic curves to distances $H_b/2$ above and below. The hydrostatic and dynamic pressures on the seaward side are added. Hydrostatic pressure on the landward side should be subtracted if present. This construction is shown in Fig. 7.16. The dashed and dotted lines show dynamic and hydrostatic pressures, and the solid line the combined pressures, for the case when hydrostatic pressures are on each side of the wall.

The resultant wave force per linear foot of wall is determined from this figure and, for the case of hydrostatic pressure on each face, is as follows:

$$R = \frac{P_m H_b}{3} + P_s\left(d + \frac{H_b}{4}\right) \tag{7.49}$$

where R = total wave thrust, in pounds.

For the case of no hydrostatic pressure on the land side, the formula is

$$R = \frac{P_m H_b}{3} + \frac{P_d}{2}\left(d + \frac{H_b}{2}\right) \tag{7.50}$$

The Minikin method of obtaining pressures from breaking waves is recommended.

Molitor's solution for breaking-wave pressures, not shown here, appears in various publications. It is reported that pressures far in excess of those obtained by this method have been observed and that structures designed in this manner have failed.[179]

Some measurements of wave-pressure intensities have been made. Stroyer* reports values of 7 cwt per sq ft in comparatively quiet water up to 3.5 tons per sq ft maximum recorded. Minikin[136] reports 1.4 tons per sq ft from 6-ft rollers and 2 to 3.5 tons per sq ft from 20-ft high waves; Luiggi gives a peak pressure of 3 tons per sq ft from a wave 23 ft high; Rousilla and Petry measured a peak of 6 tons per sq ft from a wave 6 ft high. Tschebotarioff† states that pressures on the Great Lakes are seldom over 1 ton per sq ft.

Waves breaking seaward of a structure cause conditions for which methods of obtaining approximate solutions are shown in reference 179.

Wave Forces on Piles. Approaches to the design of piles against wave action appear in references 128, 131, 136, 168, 171, 178, 179, 182, 184, 185, 189, and 196. The approach in Morison[171] has been extensively verified.‡,[174] However, this method applied only where water depth was large compared with wave length and wave height small compared with wave length. Most deep-water waves meet these criteria, but in shallow water, only those of small steepness do so. Since the theory does not apply to shallow-water waves of large steepness, it is not applicable where piles are near the point of breaking waves.

The spacing of piles in a row lined up in the direction of wave travel has slight effect on wave characteristics where the spacing is greater than 2 pile diameters.[128]

Ship Responses to Wave Action. Surging, heaving, and pitching of a vessel occur in head-on seas; sway, heave, and rolling motions take place in beam seas. Mooring lines restrict the tendency of a ship to move under the influence of wave action.

It is not possible to treat the subject of ship responses to waves briefly. General principles can be mentioned to guide the designer in the more usual condition. Where solutions depend upon graphs or tables not shown, references are given to indicate where they may be found.

Approach of a Ship to a Berth under Wave Action. Berthing in moderate-head seas should not create a serious problem, but in a beam

* J. R. Stroyer, *Concrete Structures in Marine Work*, Knapp, Drewett & Sons, Ltd., 1937.

† G. P. Tschebotarioff, *Soil Mechanics, Foundations, and Earth Structures*, McGraw-Hill Book Company, Inc., 1951.

‡ Example in reference 171 has been repeated but revised in reference 179.

sea the impact on the structure may be large and kinetic energy from sway may have to be absorbed by the elasticity of the ship, fenders, and structure.

Methods of computing these forces are presented by Wilson.[192] Use of an approach speed of 1 fps may be inferred to provide a factor of safety inclusive of possible wave influence in normally protected ports. Forces are radically different on open structures on piles and on solid structures. An open-pile structure permits waves to pass without serious reflection, and ship movements will be those of freely drifting ships in progressive waves until impact. A solid wall will reflect waves, and a standing wave will result in radically different conditions.

Response of Moored Ships to Wave Actions. Mooring lines restrain ship movement caused by wave action. High stresses may occur in the lines. Experiments have been made with heaving, swaying, pitching, yawing, and rolling motions, which have received little attention compared with surge. Theoretical treatments have been largely confined to surge caused by standing waves or long-period seiches. Surge is treated in reference 192. Surge rarely breaks mooring lines and lets the ship go adrift.

Causes of surge may be complex, but generally the cause is the convergence of long-period waves from different directions; short-period waves seldom cause surge. Surge effect is also caused by general movement or sudden intensification of a storm which raises the water level.[178] Horizontal force from surge and swing is created when a moored vessel approaches the limit of drift permitted by hawsers. It is sometimes taken as about one-third of the wind force.

Sway is relatively more important than surge because the lateral impact can be much more damaging to a vessel, fenders, or structure.

Suction drag on a moored ship, due to close passage of another ship, may be large. A liner passing near a moored ship caused two 10-in. (circumference) manila mooring lines to snap.

Tidal Bores. Tidal bores create required mooring design characteristics which cannot all be met. To permit movements of low amplitude with tides, some cable slack is used. However, because of the obliqueness of the cables, the ship's movement is not only longitudinal and vertical, but transverse, and may result in abrupt lateral forces. This subject is discussed in reference 172.

Water-current Forces. The total water-current force on a piled structure with a moored vessel is given by the sum of such of the following items as are applicable in the conditions which may be concurrent:

$$F_t = F_{cw} + F_{cv} + F_d + F_{pd} \tag{7.51}$$

where F_t = total force on wharf due to current action on piles, structure, and vessel;

F_{cw} = total pressure of current on wharf piles and bracing, in pounds;

F_{cv} = total pressure of current on vessel and floating objects lodged against structure, in pounds;

F_d = friction drag from current flowing parallel to vessels, in pounds; and

F_{pd} = drag from propeller when current flows parallel to vessel, in pounds.

Current Pressure. Unit pressure from a steady-state current is given by $p_c = wh$, where the head which the velocity would produce is called the velocity head and is written $v_c^2/2g$. Therefore $p = wv_c^2/2g$. By canceling the value of weight of water per cubic foot (fresh 62.4, sea 64.0) and the practically equal value of 2g (64.4), the total current pressures may be expressed as follows:

$$F_{cw} = kB_w v_c^2 \tag{7.52}$$

$$F_{cv} = kB_v v_c^2 \tag{7.53}$$

where B_w = projected flat surface area of structure below water line, in square feet;

B_v = projected area of cross section of hull below water line for end or side presentments, in square feet;

k = form factor, for round piles 1.0, square piles and bracing 1.4, current normal to center line of vessel 0.8, and current parallel to vessel varying from 0.15 to 0.6 for fine lines and square ends, respectively; and

v_c = velocity of current, in feet per second.

Current velocity at the surface, when unaffected by wind, is usually about 0.85 of the mean velocity, which occurs at about 0.6 of the depth. Maximum velocity occurs between these points and may be assumed as 1.1 times the mean velocity. The velocity at the bottom is usually 0.75 to 0.85 of the mean. Formulas (7.52) and (7.53) utilize a constant average velocity; if varying velocities for different portions of the pile length are used, it would be necessary to use varying pressures. In case the current speed is not known, it may be assumed as 4 or 5 knots, since moorings are generally not located in swifter water.

*Friction Drag on Vessels.** Friction drag from a steady state of current parallel to the vessel, according to a Navy bulletin, is expressed thus:

$$F_d = \frac{S}{4}\left(\frac{V_k}{6}\right)^2 \tag{7.54}$$

* U.S. Navy, *Civ. Eng. Corps Bull.*, vol. 1, No. 2, pp. 29–35, January, 1947.

where S = wetted area of hull, in square feet; and

V_k = velocity of current, in knots (1 knot = 6,076.1 ft per hr).

Somewhat different results are given by the following formula, developed by Taylor* and appearing in reference 194:

$$F_d = 0.00657(V_k^2 + 0.641V_k)S \tag{7.55}$$

An additional 25 per cent should be taken on account of appendages, form factor, and degree of fouling.

Propeller Drag. This occurs when current flows parallel to the vessel and is expressed thus:

$$F_{pd} = 2.88A_pV_k^2 \tag{7.56}$$

where A_p = projected area of propellers, in square feet.

Ship Impact

The kinetic energy of a moving ship is absorbed by bending of fenders or structural members, deformation of floating fenders, compression of springs or rubber, movement of gravity fendering devices, elastic deformation of the ship, rolling of the ship, plan rotation of the ship, piling up and displacement of water between ship and wharf, deformation of the harbor bottom, damping from paying out of mooring lines, reversal of propellers, elastic deformation of ground, plastic deformation of ship, and damage to the structure.[137] The last two means should be avoided, but they increase the factor of safety against total collapse. The structure should not provide so much rigidity, or be so limber, that it will allow itself or the ship to be damaged.

In unbraced pile structures, the piles absorb energy in bending up to the limit of their capacity, and the ground will absorb some energy. In all structures, the soil provides the final resistance, and the structures can provide no greater value. If batter piles are used, they provide a horizontal component, and the vertical-component effect on vertical piles should be investigated. With braced structures, shock absorbers should be used.

A ship usually approaches a wharf at a slight angle with the face, and the curve between the straight side and bow will probably hit a fender system first. This reaction will be spread over a relatively small length of fender member and number of yielding supports, depending on their rigidities. Consensus is that spacing of fenders should be such that from three to ten units act. The engineer's experience and judgment usually decide. Fender spacing depends on kinetic energy, approach angle, shape and length of ship, individual fender capacity, and strength of the ship. Solution for the problem of beams on multiple elastic

* T. Thorpe and K. P. Farrell, "Fixed Mooring," *Inst. Nav. Arch.*, 1947.

supports is shown in modern volumes on structural design. The reaction will then be redistributed over a greater length of wharf and number of piles.

For a more complete treatment of the design of structures to resist ship impact, consult references 143, 144, 170 to 172, 175 to 177, and 192 to 194.

Kinetic Energy. The amount of kinetic energy from the impact of a moving ship normal to the face of a structure to be resisted by the structure is given by the following formula:

$$E_k = \frac{12(CW_s)v_s^2}{2g} \tag{7.57}$$

where E_k = kinetic energy acting on structure, in inch-tons (long).

C = energy-absorption coefficient of wharf, which may be taken as 0.05 to 1.00, depending upon the mass and elasticity constants of the ship and mooring construction. Coefficients of from 0.35 to as high as 0.91 are often used, although the higher values tend to apply to round-bottomed ships of slight draft; a good general guess by many designers is said to be 0.4.[144,175] Frequently taken as 0.5[136] for straight-sided ships and 1.0[137] for curve-sided ships. It would be nearly 0.9 to 1.0 in case of accidental direct, or nearly so, head-on ramming. For a straight-sided ship hitting a long continuous wharf a lateral blow squarely, the value may be 1.0, but a long length of wharf will resist; if the wharf deck is carried on widely spaced piles, the load will be concentrated upon them, and the value may be 0.4[139] to 0.7[140] if the bow of the ship strikes first, but only a short length of wharf will resist. The value of C may be calculated more closely, if desired, by differential equations.[138]

W_s = displacement or total weight of ship and contents, in long tons.

v_s = velocity of ship before impact (component normal to resisting plane), in feet per second.

The relationships of net effective ship weights and impact forces at different velocities for different deflections are shown in Fig. 7.17. The force to be dealt with is directly proportional to the net effective weight of the ship and the square of the normal component of its approach speed and inversely proportional to deflection.

Relationships of net effective ship weights and approach velocities to kinetic energies are shown in Fig. 7.18. The energy to be handled is directly proportional to the net effective weight of the ship (CW_s) and the square of the normal component of the approach speed.

When considering reported and recommended approach speeds normal

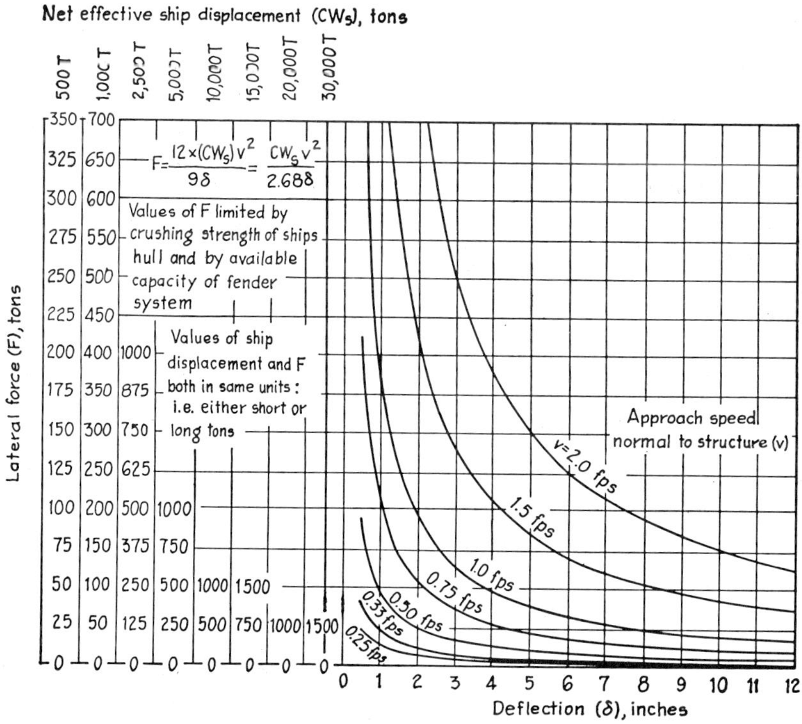

FIG. 7.17. Ship impact forces.

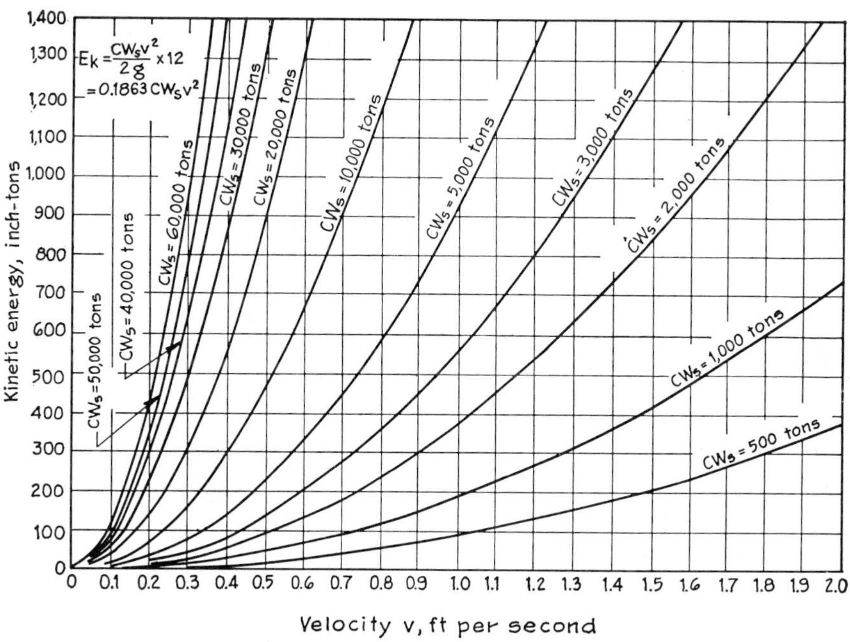

FIG. 7.18. Kinetic energies for various velocities of net effective weights of ships.

to a structure, bear in mind that these are based on calm-water approaches and that the kinetic energy is only that for a level approach.[192] A broadside wind can cause a ship to acquire additional momentum which may not be fully controlled, even by tugs. Tidal currents may also increase berthing speed. Very full discussion of these conditions appears in the *Proceedings of the Eighteenth International Navigation Congress.* Opinions prevailed that both wind and tidal current need not be considered to be simultaneous and that an approach speed, normal to the structure, of about 1.0 fps may be inferred to provide a factor of safety, inclusive of possible wave influence in normally protected ports.

Small ships usually approach wharves at considerably greater speeds than do large ones and therefore may exert comparable forces locally. Minikin[136] reports 50 cases, in two-thirds of which the normal component was from 0.21 to 0.33 fps; the remainder were up to 1.0 fps, with a single value of 2.08. Robertson[137] and Little[135] conclude that not over 0.5 fps is a suitable value for large vessels and 1.0 for small ones (less than 1,000 gross tons). Wharves for large tankers have been satisfactorily designed using a speed of 1 to 1.5 knots at 10 deg with the wharf face, giving a normal component of 0.28 fps. The criterion for ship impact used for the Creole Petroleum Corp. piers for ocean-going tankers in Venezuela was a 34,000-ton displacement fully loaded tanker approaching at a velocity having a transverse component of 0.27 fps and delivering a broadside blow compressing the spring fender 12 in. and deflecting the structure 1 in. The entire kinetic energy of the ship was assumed transformed into the work, deflecting a structure that was composed of long, unbraced, heavy steel box piles welded to a steel-framed deck.[103] For a 30,000-ton vessel at the New York Naval Shipyard, 0.4 fps was used.[175] The Port of New York Authority designed Hoboken Pier C for 25,000-ton (short) ships at an approach velocity of 1.0 fps and an energy-absorption coefficient of 0.4. Hopkins[141] states that an approach velocity of 0.3 fps is too low; and that 3 or 4 fps, sometimes used for accident forces against bridge piers, is too high for wharves.

The basis of design adopted for jetties at Thames Haven for oil tankers of 45,000 tons dead weight (60,000 tons displacement) was that the fendering should be capable of absorbing 2,800 inch-tons of kinetic energy, this being 25 per cent of the energy of the vessel when making an approach at a velocity of 1 fps normal to the face.[203]

Wilson[192] states that, in most ports, approach speeds normal to the structures are less than 0.5 fps, but with winds and currents, international opinion favors a design velocity of 1 fps. He finds that in the presence of waves having a trough-to-crest dimension of 3 ft in water 50 ft deep, such a design speed can be exceeded as a contribution from the waves alone.

Baker[139] suggests collision speeds normal to the face, depending upon the ship size and navigation conditions, as shown in Fig. 7.19; it is suggested that this graph be used as a check on the speed selected after consideration of each particular case.

Volse[144] summarizes common practices at the accident velocity-energy line as follows: 0.75 fps for 30,000-ton ships; 0.85 fps for 20,000-ton ships; 1.3 fps for 10,000-ton ships; and 1.8 fps for 5,000-ton ships. Note that these are accident velocities; also that the ships' weights are displacement tonnages not reduced by an energy-absorption coefficient (C).

Impact on pier corner construction and dolphins should not be taken at as low values of absorption coefficient as for piers and wharves

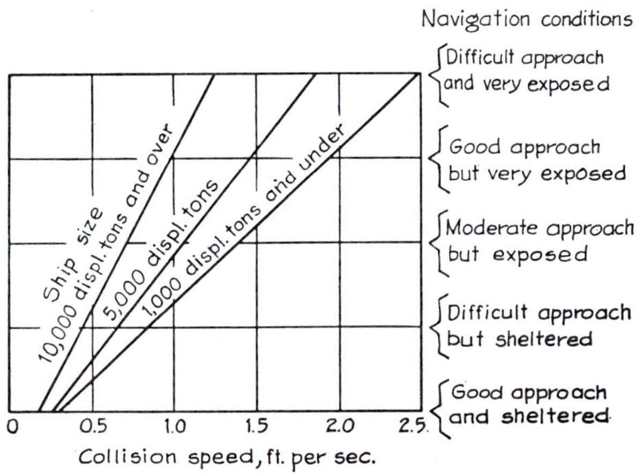

FIG. 7.19. Collision speeds for ships for fender design. (*From A. L. L. Baker, Proc.* 18*th Intern. Navigation Congr.*, 1953.)

generally. For turning dolphins the Creole Petroleum Corp. increased pier forces by 50 per cent. Risselada[176] reports successful designs of flexible-steel dolphins for 100,000-ton tankers in an enclosed dock approaching at 0.42 fps normal to the face and 135,000-ton tankers at 0.5 fps in a tidal basin. He previously reported design of steel dolphins for 22,500-ton ships at 0.67 fps.[177]

Impact against a Gravity Structure or Backfilled Wall. Assumption that ships will approach a solid, unfendered structure carefully, to avoid damage to the ships, may be disastrous to ships making bad approaches. Poor approaches may be unavoidable at times, because of wind, waves, current, or surge.

In theory, the proportion of energy taken by an unfendered pier to that taken by the ships is in a ratio of $W_s/(W_s + W_w)$. When the mass of the structure is small relative to the mass of the ships, the greater part of the energy is transmitted to the structure. With a heavy structure, most of the energy will be absorbed by the ship, and a fender system should be used to prevent damage to the vessel.[144] The mass of piled structure may be small, and although even a relatively small mass of the pier may reduce the kinetic energy somewhat, a large amount will remain to be taken by resistance of the structure to deformation.

When the mass of the pier is large compared with the mass of the ship, formula (7.57) and the first two parts of formula (7.59) should be replaced by the following formula for determining the kinetic energy to be absorbed by the fendering system:

$$E_k = \frac{12(CW_s)v_s^2}{2g} \frac{W_s}{W_s + W_w} \tag{7.58}$$

where W_w = weight of wharf structure, in long tons.

Absorption of Energy. Robertson[137] states that experience indicates that the impacts per fender from large and small ships do not differ much in magnitude. This is because small ships with curved and belted sides are usually more robust and approach with much less care and at higher speeds and most of the kinetic energy must be absorbed at a single impact point. Large straight-sided ships usually approach carefully and more slowly and either make contact over a long length or near an end, in which latter case the center of gravity is free to continue in motion and considerable energy is absorbed by other means than the fenders and structure. Using formula (7.57), kinetic energies from a 2,000-ton ship, using an energy-absorption coefficient of 1.0, approaching at a velocity of about 1.0 fps, and from a 20,000-ton ship, using an energy-absorption coefficient of ½, at a velocity of about ½ fps, are approximately equal and are on the order of 400 in.-tons. A value of 400 in.-tons appears to be a suitable energy for the design of an all-purpose wharf for usual impacts where the largest ship is of 20,000-ton displacement. This value would increase to 700 in.-tons for a large ship of 30,000 tons displacement and to 900 in.-tons for a large ship of 40,000 tons displacement. The above values do not apply to corner posts and dolphins.

Since impacts are proportional to the square of the velocity of the ship, small increases in approach speeds rapidly increase the kinetic energy of the vessel. It is impracticable to design fendering to resist all speeds and impacts, and an accident line should be established, suggested as three times the kinetic energy provided for at working stresses.

When adopting the average impact energy of about 400 in.-tons for design at allowable working stresses, 1,200 in.-tons is commonly taken as the accident impact-energy level. Common practice is to design fenders and piers so that if the velocity greatly exceeds the accident energy line, the ship will take up the excess energy in the form of hull damage. Though it is important to protect the ship, it is impracticable to design a pier that would collapse rather than cause hull damage to the ship.[137,144] Methods of computing ship damage appear in reference 200. The increase in velocity required to increase kinetic energy from a normal magnitude to an accident energy-line velocity may be read from Fig. 7.20. Thus for a usual kinetic energy of 400 in.-tons for a net tonnage of 10,000 CW_s, a velocity of 0.46 fps is required; for an increase to 1,200 in.-tons, a velocity of 0.8 fps would correspond.

Resisting elements which will be called upon to absorb energy, such as wharf piles, or constructions which are provided solely for this purpose, such as fender piles or shock absorbers, may be designed to resist lateral forces by equating the effective kinetic energy from the ship to the potential energy of the resisting members such as piles in bending or gravity fenders, as follows:

$$E_k = \frac{12(CW_s)v_s^2}{2g} = \frac{F\delta}{2} = E_a \tag{7.59}$$

where F = horizontal force remaining after deductions for dissipated energy, in long tons;

δ = deflection, in inches; and

E_a = minimum potential energy, or work for which means of absorption must be provided, in inch-tons (long).

Fender Piles. Relative potential energies (energy or work absorbed) of various fender-pile sizes and materials may be investigated in the manner shown in Fig. 7.20. The work absorbed may be obtained from the following formulas:

$$M_a = \frac{f_{ba}I}{c} \tag{7.60}$$

$$E_a = \frac{M_a^2 l_u}{6EI \times 2{,}240} \tag{7.61}$$

where M_a = allowable bending moment, in inch-pounds;

f_{ba} = allowable extreme unit fiber stress in bending, in pounds per square inch;

l_u = unsupported length of pile in bending, in inches;

E = Young's modulus of elasticity of pile material, in pounds per square inch;

I = moment of inertia of single pile, in inches^4; and

c = distance from neutral axis of pile to extreme fiber, in direction of bending, in inches.

By using the proportional limit or yield-point stresses in the preceding formulas, the accident energy line may be determined. By using a lower working stress to provide a factor of safety against the assumptions made, safe working energy is found. The energies are in the same ratios as the stresses.

A comparative list of worldwide woods for fender piles appears in reference 2*cm*.

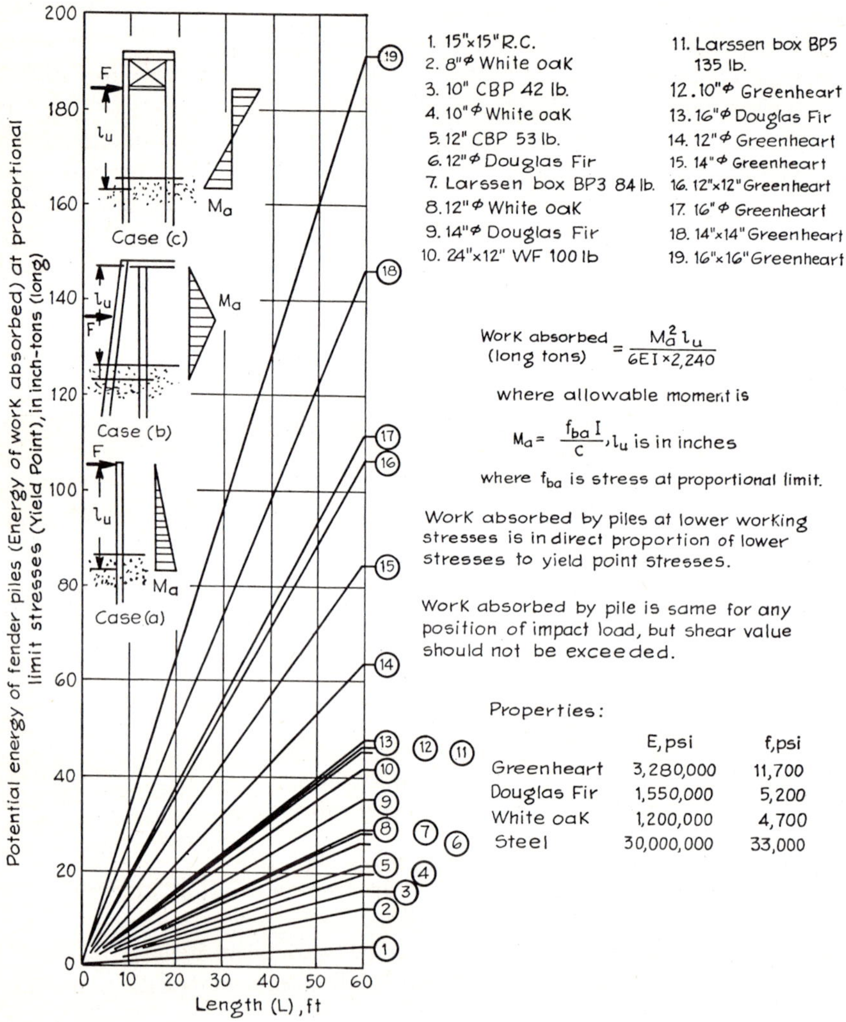

FIG. 7.20. Relative potential energies of fender piles at proportional limit stresses.

Working Stresses. The estimated speeds of approach are the assumed maximums for a ship not under proper control. The resulting impact stresses are not expected to be a usual occurrence, and higher stresses than the usual working stresses are permissible. Higher stresses result in greatly increased energy absorption, because the elastic energy in bending is proportional to the square of the fiber stress. If the working stress is doubled, energy absorption is quadrupled. Allowable stresses are sometimes taken as 50 per cent greater than normal.[141]

Wharves on Cantilever Piles. Piles hinged at the top, or unbraced sections of piles with fixed conditions at the top, may be designed to resist lateral forces.

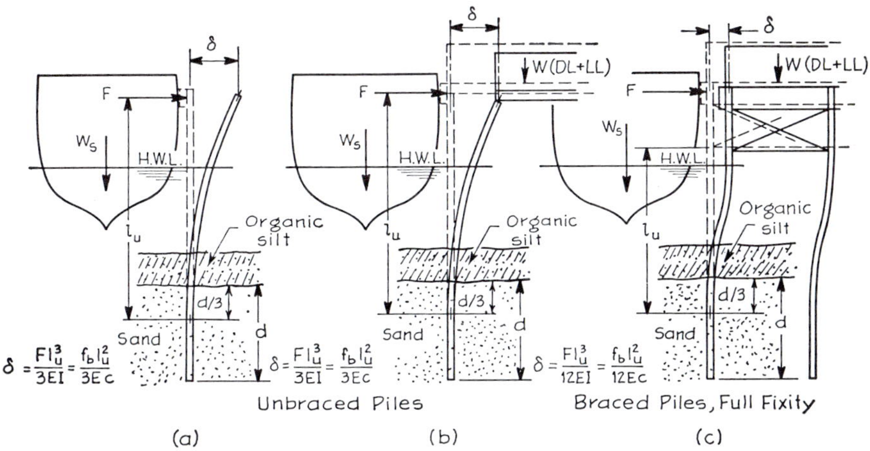

FIG. 7.21. Piles of equal unsupported lengths resisting lateral impact.

Where all piles have the same unsupported lengths (Fig. 7.21), the formulas for determining pile sizes are as follows:

$$\Sigma I = \frac{1.12CEc^2W_{sp}v_s^2}{f_{ba}^2 l_u} \tag{7.62}$$

$$f_{b(x,y)} = 1.06c_{x,y}v\sqrt{\frac{CEW_{sp}}{(EI)l_u}} \tag{7.63}$$

where I = moment of inertia of single pile, in inches⁴ (all assumed same).

ΣI = required sum of moments of inertia of individual piles about resisting axis, in inches^4. (Note that this amount must be provided in bents over which the impact is distributed. In a wharf, impact is often taken for half the ship length.)

W_{sp} = weight of moving object, in pounds.

$f_{b(x,y)}$ = fiber stress in bending about axes x-x and y-y, respectively, in pounds per square inch.

Pile sizes can be assumed, and vertical fiber stresses can be combined with fiber stresses from transverse impact and longitudinal forces. The longitudinal forces are limited by friction on the fenders, snubbing pulls on mooring lines, or catching of proud edges of lapped plates.

Where unsupported pile lengths vary (Fig. 7.22), the formulas for both free and top-fixed piles are as follows:

$$f_{1b} = 1.06cv_s \sqrt{\frac{CEW_{sp}}{N}} \left(\frac{K_1^2}{I\sqrt{\Sigma K^3}} \right) \tag{7.64}$$

$$f_{2b} = 1.06cv_s \sqrt{\frac{CEW_{sp}}{N}} \left(\frac{K_2^2}{I\sqrt{\Sigma K^3}} \right) \tag{7.65}$$

where $K_1 = I_1/l_1$, $K_2 = I_2/l_2$, etc.;

N = number of bents over which the impact is distributed (often taken in one-half the length of the ship); and

ΣK = sum of K values in a bent.

The above formulas indicate that stresses in the shortest pile are greatest, if all have the same I value, and that stresses are directly

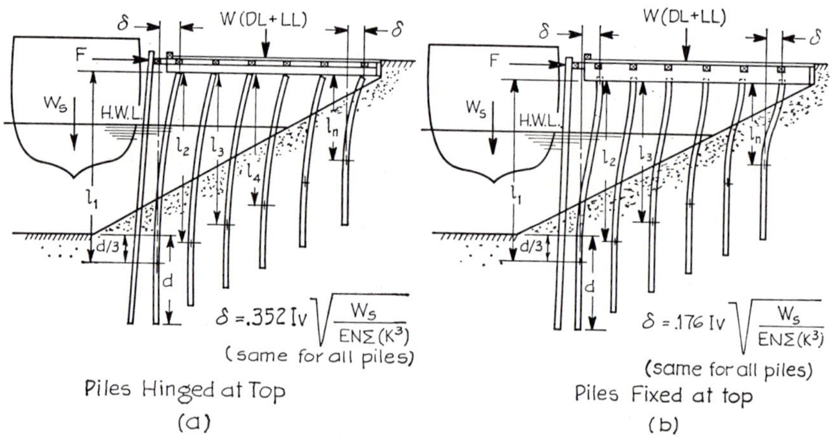

FIG. 7.22. Piles of unequal unsupported lengths resisting lateral impact.

proportional to the velocity of the ship and the square root of the weight of the ship.

Wharves on Cantilever Piles with Fendering Devices. The methods of design for wharves resisting all effective kinetic energy by bending of the piles have been given. Such designs might be used for relatively light service or investigations of existing structures. For wharves where only a small deflection or racking action is suitable, batter piles may be required together with springs or mechanical fendering systems, in which cases, for practical purposes, all effective kinetic energy may be considered as absorbed by these buffers. If, for any reason, batter piles are not wanted in wharves for ships docking with large kinetic energies and use of fendering is desired, the effective kinetic energy will be absorbed in both the wharf piles and the fendering, in proportion to their elastic abilities or potential energies. The simplest procedure is cut-

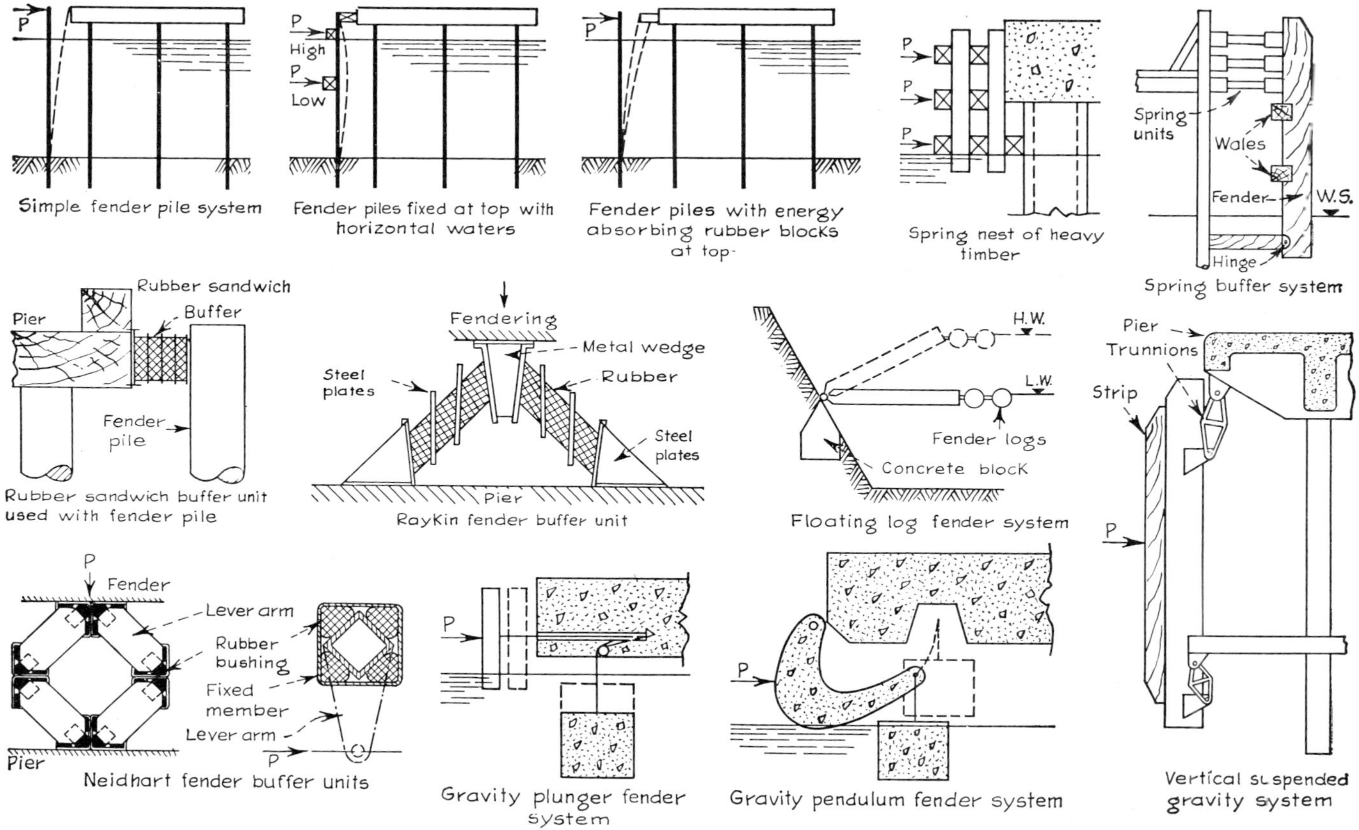

FIG. 7.23. Fendering devices. (*From L. A. Volse, Engineering News-Record.*)

and-try by the following steps: assume a pile type, spacing, and embedment, and from the capacity of the soil and pile in bending, determine the allowable horizontal force and the resulting deflection; using this force and distance, compute the potential energy of this part of the system and deduct it from the total effective kinetic energy; then provide potential energy in the fendering system to equal this remaining kinetic energy, determining the movement distance of the fender, which, when using the same force as necessary to deflect the piles, will give this result.

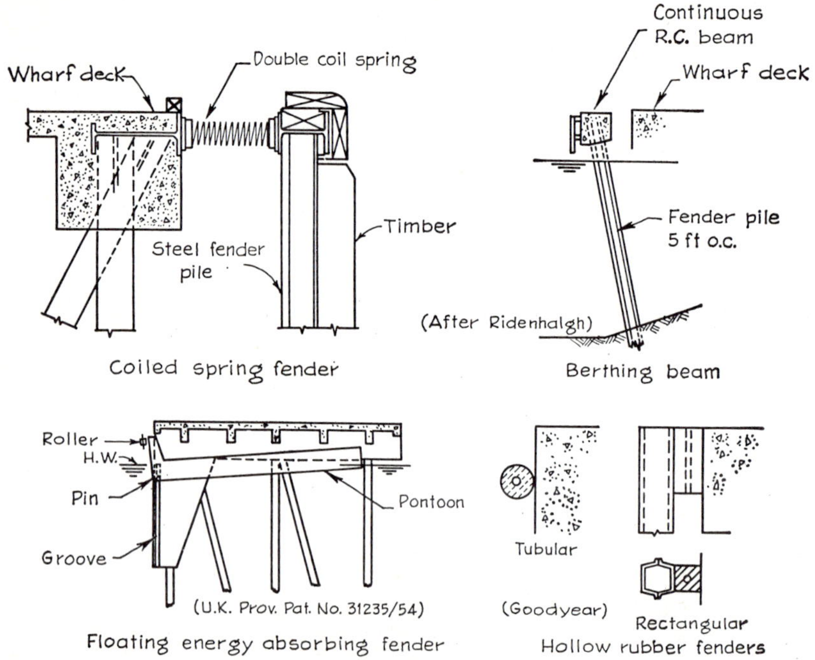

FIG. 7.24*a*. Other fendering devices.

Fendering Devices.[144] Berthing a ship without damage to pier or ship is a delicate operation, made more difficult by currents, wind, waves, seiches, tide changes, close quarters, differences in maneuverability and skill of crews, and by insufficient provisions for absorbing energy due to economies forced on the engineer by the wharf owners.

Means for absorbing energy are necessary, and ample fendering systems are relatively inexpensive insurance for the wharf and ship and may also cut down the berthing time and cost. More wharf operators are turning to the use of impact fenders, whose design, because of the large uncertain forces which must be resisted at reasonable cost, has forced engineers to a variety of ingenious and empirical solutions (Figs. 7.23 and 7.24). There is no such thing as a standard design. A study of practice has been made by the *Engineering News-Record*,[144] from which this section quotes extensively. Fender types are also described in reference 104. Even the most extravagant fender system cannot cushion the impact of a large ship when it comes in too fast, and probably both ship and wharf will be damaged. The problem of the designers is to provide a fendering system that will protect both ship

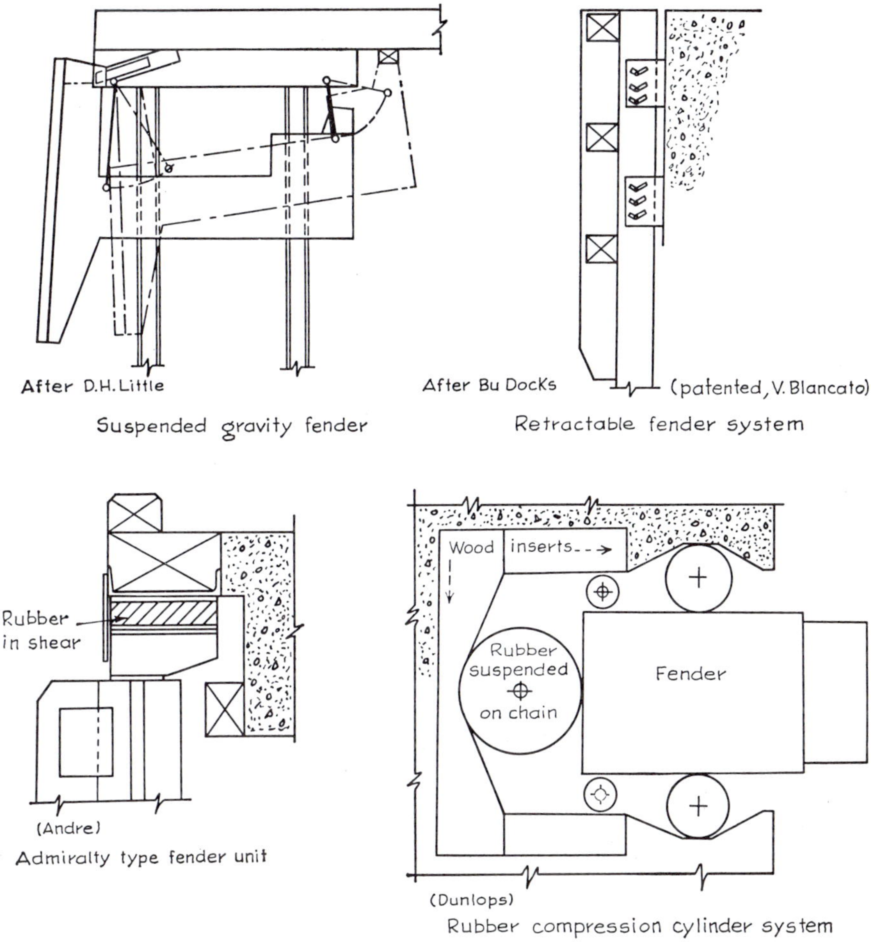

FIG. 7.24*b*. Other fendering devices.

and wharf as cheaply as possible but be durable and have low maintenance costs.

A wharf designed to be flexible will absorb much of the kinetic energy of the ship. If the wharf is rigid and immovable, little energy can be transferred to it and the fender system must absorb the shock to avoid serious damage. For a given kinetic energy, reaction from the fender is inversely proportional to available inward movement, so that fender deflection is economical for wharf design. Fenders range from old rubber tires to costly mechanical devices. The engineer should compare relative costs of wharves and fenders for different combinations to find the most economical solution.

With ships of over 10,000 tons displacement, rigid fendering unable to absorb energy can cause damage to the ship unless very low approach velocities are used. To be most effective, a system should be designed for ships of fairly uniform tonnage. For instance, a system designed as flexible for a 50,000-ton ship is quite rigid for smaller ships, and they might be damaged.

Pads or resiliency in wharf or fender piles may suffice for smaller ships, but springs, rubber fenders, cantilevered buffers,[135] or mechanical fendering devices may be needed for large ships.

Fenders can receive large longitudinal forces, which may be reduced by fender deflection under a glancing blow and reversal of propellers. Since mooring ropes are used to check longitudinal movement, the wharf should be designed to resist their pull; the breaking point may be used as a criterion. The pull on the hawsers need not be large; Minikin believes that a pull of 25 tons on each of two hawsers would be a reasonable assumption for a 10,000-ton ship. Ideal fenders would avoid longitudinal impact forces by receding or rolling and would provide no means of engaging ship projections. Generally, the fenders and longitudinal supports should be designed to resist longitudinal forces equal to 0.1 to 0.25 of the transverse forces applied at any point where collision may occur.[139]

Compression Springs. Springs may be helical single- or double-coil. Outside diameter, free height, solid height, wire diameter, total capacity, and steel grade should be specified. Material and workmanship may be specified to meet the Association of American Railroads Specifications M-114, *Springs, Carbon-steel, Helical,* and M-112, "*Steel Bars, Carbon, for Railway Springs.*" Spring design formulas are available in mechanical handbooks.*

Rubber Fenders. One type of rubber fender consists of hollow circular or square synthetic rubber tubing in sizes up to 15 in. round or 12 in. square.

Energies up to 22,000 ft-lb. per ft of length and loads up to 90 tons per ft can be resisted. Such fenders are hung horizontally and require a continuous wall or waler to react against. They can compress to about one-third of the original diameter. These are manufactured by the Goodyear Tire & Rubber Co., from whom tables of properties may be secured. Other types of rubber fenders utilize blocks in compression or rubber in shear.

* Such as L. S. Marks, Mechanical Engineers' Handbook, McGraw-Hill Book Company, Inc., 1958.

Raykin fender buffers (Fig. 7.23) consist of a V-shaped arrangement of series of rubber and steel plates. Properties can be secured from the manufacturer's literature.[128]

Mechanical Devices. Mechanical fendering devices are numerous and ingenious, with new ones frequently appearing. Gravity fenders are popular in Europe and are being increasingly used in the United States. Pneumatic and hydraulic plungers have also been used to some extent in Europe.

Ship Moorings

It is not practicable to construct dolphins and mooring platforms requiring piles in rock, shale, boulder, or coarse-gravel bottoms. Drilled-In Caissons have been used on rock bottoms.[211] Excessively long piles may be required in soft mud. Safety of design is of prime importance, because the life of the ship may depend upon it. Use of deadmen on shore or of dolphins or mooring platforms in as favorable conditions as possible is preferable. Designs should be based on loads from wind, surge and swing, current, waves, ice, snubbing of the ship's leeway, and drag from passing ships.

Dolphins. These are usually formed by a cluster of closely spaced piles pulled together at the top and bound with cable. Sometimes wood blocking and bolts are used. Because of their strong appearance and ability to withstand glancing blows, they are often assumed to be suitable for moorings, but they have little lateral strength to sustain steady forces. They are designed as the sum of single-pile values, neglecting friction between pile unless piles are connected to develop longitudinal shear and are embedded well in firm material. They usually must be designed for pull from several directions. H piles sometimes are used as batter piles with wood dolphins. Dolphin designs are shown in references 116, 136, and 143. A breasting dolphin for supertankers is described in reference 213.

Rigid Mooring Platforms. For mooring purposes, rigid mooring platforms are desirable. They may consist of a number of piles, all or part battered, with heavy concrete caps carrying bollards. Fender protection can be provided, and the batter piles set back so that they will not be hit by ship hulls. Mooring platforms may be used in up to 40 ft of water, if the bottom is able to hold piles from pulling out.

ICE ACTION

Ice Pressure[62,66,145]

Thrust from ice sheets depends upon thickness, maximum rate of air-temperature rise, degree of restraint of structure, and extent of exposure to solar radiation. Supplementary factors are drag of current, snow cover, skin friction from snow action, confined volume of ice, and fluctuation in surface level. Tests have shown the modulus of elasticity of ice to range from 250,000 to 1,400,000 psi, variation in time being the principal cause of difference; the modulus is greater at lower temperatures and decreases with applied load. Plastic action occurs under very small loads.

The crushing strength of ice has generally been accepted as 400 psi, but this depends upon rate of load application and temperature. Tests at the St. Lawrence Waterway showed values of 300, 693, and 811 psi at temperatures of 28, 14, and 2°F, respectively. Impact values from floating ice sheets may be larger than crushing values.

Buckling of ice sheets results from temperature rises and also from water-level variations. When the water level falls, tension cracks fill with water which freezes, and horizontal pressures are created when the level rises again. If open channels are formed in the ice in waterways, filling and freezing in cracks may cause ice to move out from shore and pull on any structure resisting this movement. Forces due to changes in level may be moderate in wide ice sheets, but for small extents such as 65 to 130 ft, forces may reach maximum values found for temperature changes.

Dams have been designed for ice pressures as large as 45,000 to 50,000 lb per lin ft.[179,204] Norway formerly used thrusts as high as 30,000 lb per lin ft on dams, but later used 3,350 lb per lin ft on dams with sloped upstream faces. Lofquist reports that investigations made by the Swedish State Power Board and calculations on the buckling of an ice sheet indicate that ice will buckle if as thick as 8 to 12 in. but rarely at 2 in. and that probable maximum pressure of from 20,000 to 27,000 lb per lin ft will result; Swedish practice has been to use from 3,350 to 20,000 lb per lin ft according to geography and judgment. The Hydro-Electric Power Commission of Ontario, Canada, has used a pressure of 10,000 lb per lin ft. Pressures of 14,000 to 20,000 lb per lin ft were measured at Eleven Mile Canyon Reservoir by the U.S. Bureau of Reclamation. The highest pressure recorded in field or laboratory is 23,000 lb per lin ft, and this pressure times a factor of safety of 5 was used for the Mackinac Bridge piers.[146]

Determination of ice thrusts is a complex problem, and further data and study are needed to reconcile variations observed. Meanwhile, graphs and formulas for the pressures appear in references 62 and 145 and may be used until further results are available. A list of other sources is contained in reference 145.

Structures subject to blows from floating ice should be capable of resisting from 10 to 12 tons per sq ft on the area exposed to the thickness of the ice.[179,204]

It is not customary to consider ice and wave forces as acting at the same time. To produce maximum thrust, a sheet of ice must have firm resistance at the opposite edge of the sheet. Pressures from large floating ice fields piled in large packs against obstructions and driven by wind and current may be great; isolated structures acting as obstructions should be given special attention.[204]

EARTHQUAKE ACTION

Earthquake Forces

The problems resulting from earthquake forces are complex, extensive, and incompletely solved, although much studied. A great deal has been published on the effects of lateral forces on buildings. Relatively little has appeared on design of pile foundations in earthquake regions. However, the actions of the structure, piles, and soil are interdependent. In some cases piles may be used to aid in stabilization.

For detailed consideration of foundation design against earthquakes, reference 114 and the *Proceedings of the World Conference on Earthquake Engineering* sponsored by the Earthquake Engineering Research Institute and University of California in 1956 may be studied; see also the *Proceedings of the Second World Conference on Earthquake Engineering* held in Tokyo in 1960.

No definite and simple design methods have been agreed upon. It is suggested that the foundation designer unfamiliar with this subject should obtain the services of a consulting engineering geologist or seismic expert.

Regional Occurrence. The world's belts of frequent, violent earthquakes are around the rim of the Pacific Ocean; from the Azores across the Alpine structures of the Mediterranean; across Asia to Burma and along the front of the Himalayas; and, in a minor degree, narrow belts in the Arctic, Atlantic, and Indian Oceans, rift zones internal in stable masses, and active areas marginal to continental stable masses. Major earthquakes of the world have been plotted on regional maps.* Locations of United States earthquakes appear in *United States Earthquakes,* published annually by the United States Coast and Geodetic Survey.

Factors in Earthquake Design. Strength, stiffness, mass, damping characteristics of the structure and the soil, natural periods of structure and ground, force, amplitude, direction and duration of impulses, and distances from faults and the epicenter of the quake are factors in design

* D. Gutenberg and C. F. Richter, *Seismicity of the Earth and Associated Phenomena,* Princeton University Press, 1949.

against earthquakes, as well as pile and soil properties and their elastic and nonreversible deformations.

Periods of Earthquakes. All structures have a natural period of vibration. This should be computed in order to avoid resonance with the range of damaging periods known to be experienced in earthquakes. The rapidly changing periods with amplitudes present in an earthquake are very effective against resonance effect, causing the stress limitation due to this factor to be important.*

The commonly observed periods fall between 0.1 and 0.7 sec at locations near the epicenter, and usually the largest acceleration occurs in waves having periods less than 0.5 sec, while periods increase with distance from the epicenter and the intensity of the shock decreases with distance.† In California, periods of 0.5 to 0.6 and of 1.0 to 1.1 sec are observed more frequently. Periods of around 1 sec occur infrequently as a rule at distances close to the epicenter, but probably in very strong shocks these periods may also play an important part at short distances.† Imamura and Crookes state the destructive range as 1 to 2 sec; Heck, perhaps allowing for minor damage, as 0.5 to 2.5 sec.‡ Analysis of earthquake records indicates that the effectiveness of the quake is inversely proportional (roughly) to the period, so that increasing the period usually means an increase in safety.* Within the range of probable maximum periods of acceleration, there is reported to be no evidence of resonance if the free period of vibration of the structure is kept above 2 or 2.5 sec.

The maximum deflection produced in pier structures having the equivalent of a flexible first-story construction by a given ground motion is not much affected by considerable changes in the period of free vibration of the structure, and the best measure of the earthquake-resisting qualities of this type of construction is simply the maximum horizontal deflection which can be sustained without exceeding the allowable stresses in the piles.§

Suyehiro states that a marshy alluvium, although having a greater amplitude than a firmer material, cannot exert more force on a structure than the soil will stand, and that the energy of elastic vibrations dissipates more quickly in the loosely fixed end conditions of foundations in such ground than in comparatively cohesive diluvial ground, which more

* M. A. Biot, "Analytical and Experimental Methods in Engineering Seismology," *Trans. ASCE, Paper* 2183, pp. 365-408, 1943.

† "Earthquake Investigations in California, 1934–1935," *U.S. Dept. of Commerce, Spec. Publ.* 201.

‡ S. I. Crookes, Jr., Structural Design of Earthquake-resistant Buildings," Leightons Ltd., Auckland, New Zealand, 1940, p. 20.

§ Norman B. Green, "Flexible 'First-Story' Construction for Earthquake Resistance," *Proc. ASCE,* vol. 60, No. 2, pp. 177–184, February, 1934.

than compensates for the greater motion of the soft layer.* He attributes much of the motion of a rigid building on soft soil to rocking caused by the yielding of the soil.

Imamura says that if the nature of the soil is hard and unyielding, the violence of the earth vibrations is comparatively little and damage to buildings due to unequal settlement seldom, if ever, occurs.† He further states that rock, hard clay mixed with gravel, and firm stratified gravel are superior soils, with dry sandy soil and hard loam next, and wet sandy soil, mud, and newly made ground the worst soils upon which to locate a building in earthquake regions. He remarks that in a severe earthquake, a change of pressure under the foundation often takes place, and that it seems that the pressure in some instances may become two or three times greater than its static value. It is also concluded by him that for deep soft soils, a pile foundation is the most suitable, using wood if the piles are entirely below the water line, otherwise concrete.

Ishimoto believes that the period increases with the depth of alluvium.

The commonly accepted conclusion is that soft soils having little cohesion are more dangerous than firm, compact soils, and although many examples confirm this belief, there are exceptions. For instance, in the 1923 Tokyo earthquake, the damage to brick buildings was greater on diluvial soil than on soft alluvium. Also, in Long Beach in 1933, damage to brick buildings on soft, recent alluvium with high ground water was less than to structures on older, firmer marine deposits with ground water not so near the surface.‡

There is a great deal of confusion regarding effects of earthquakes and proper design methods. Professor Martel believes that much of this results from uncritical acceptance of opinion as fact, from generalization, from dearth of quantitative measurements, and neglect of factors.‡ He feels, however, that trends appear, including the following: On hard soil, the greatest number of minor quakes appears to occur within a band of periods; on soft soil, the minor quakes appear to be spread over a wide range of periods; the fact that destructiveness has usually been greater on soft soil is probably caused by this characteristic period-frequency distribution and by the longer duration of the shaking; the occasional greater damage on hard ground seems to be caused by the peak on the period-frequency curve; short-period waves seem to become attenuated with depth, whereas longer period waves are only slightly reduced in intensity.

* K. Suyehiro, "Engineering Seismology," *Proc. ASCE*, Part 2, vol. 58, No. 4, May, 1932.

† A. Imamura, *Theoretical and Applied Seismology*, Maruzen Co., Tokyo, 1937.

‡ R. R. Martel, "Effect of Foundations on Earthquake Motion," *Civ. Eng.*, vol. 10, No. 1, pp. 7–10, January, 1940.

Natural Periods of Vibration of Unsupported Portions of Piles. The natural period of vibration of the structure should be computed in order to avoid resonance with the range of damaging periods known to be experienced in earthquakes. For a single pile, its period may be obtained by the following formula:

$$T = c_f \sqrt{\frac{W l_u^3}{EI}} \tag{7.66}$$

where T = period of vibration of the pile, in seconds;
W = vertical dead and live load on the pile, in pounds;
l_u = unsupported length of pile down to assumed point of fixity in ground, in inches; and
c_f = a constant depending on end conditions of the pile as a column (value between 0.18 and 0.09 for conditions shown in Fig. 7.25).

The value of $c_f = 0.18$ applies when full fixity is obtained at the bottom and no fixity occurs at the top of the pile. The value of $c_f = 0.09$

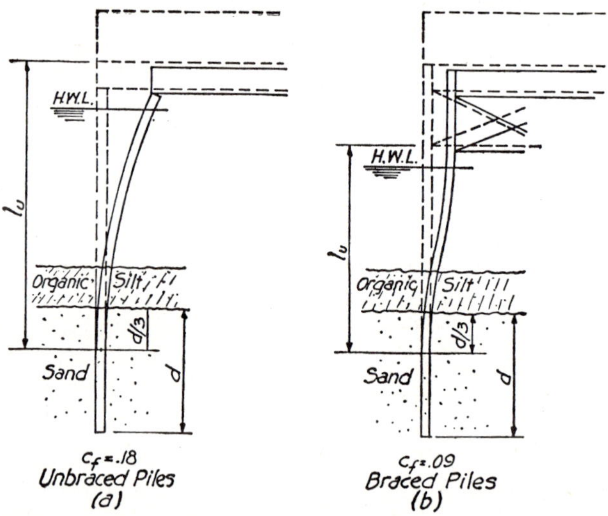

FIG. 7.25. Unsupported lengths of piles in earthquakes.

applies when both top and bottom of the pile are fully fixed. Practically, it is impossible to obtain full restraint, and likewise it is unlikely that the top of an unbraced pile is entirely unfixed in the deck of the structure; therefore the value of c_f to be selected should be judged accordingly, between the two values given for the limiting cases.

For computing the period of the entire unit or structure, in which the unsupported length of the various piles may vary, the following formula may be used:

$$T = c_f \sqrt{\frac{\Sigma(W)}{\Sigma(EI/l_u^3)}} \tag{7.67}$$

This formula may also be written, in order to avoid any possible misunderstanding, as

$$T = c_f \sqrt{\frac{(W_1 + W_2 + W_3 + \cdots + W_n)}{\left(\frac{EI_1}{l_{u1}^3}\right) + \left(\frac{EI_2}{l_{u2}^3}\right) + \left(\frac{EI_3}{l_{u3}^3}\right) + \cdots + \left(\frac{EI_n}{l_{un}^3}\right)}} \tag{7.68}$$

where the subscripts 1, 2, 3, . . . , n refer to the numbers of the piles.

The above formulas neglect the effects of the uniform weights per linear foot of the piles on the periods, but these should be slight, compared with the influence of the superstructure.

Lateral Earthquake Forces. A uniform seismic coefficient of the weight above, applied in all cases regardless of soil conditions, foundation type, and building construction and dimensions, is neither theoretically nor practically correct, although the basis of most codes. Shocks are actually choppy and irregular, causing superposition of impulses.

Lateral forces on the superstructure have sometimes been assumed to be transferred to the ground by shearing and bending resistance of the piles, and sometimes it has been assumed that lateral forces have been applied to the piles by ground movement while the inertia of the structure resists movement. Both methods of analysis should be used. Piles should be capable of deflecting with the ground, and the inertia forces from the building should be evaluated by permissible lateral resistance of the soils adjacent to the pile caps and grade beams.[132] This is particularly important for structures on piles through loose fills and marshland or bay muds, where shear waves of large amplitudes and slow velocities occur.

Footings should be well tied together with construction capable of transmitting, in tension or compression, forces equal to at least 10 per cent of the vertical loads. Piles and foundations and the soil may be subjected to a separation of several inches. Basement walls may provide thrust area which will reduce lateral forces on piles.

Batter piles are effective in resisting lateral loads applied above ground level; they are usually no more effective than vertical piles in resisting ground motion and may be subjected to bending forces from subsidence of soft surrounding upper strata.

Damping Action. The energy applied to a structure by an earthquake may be absorbed in the superstructure and substructure in elastic or plastic flexural deformations, shearing, or rocking motion. Any excess kinetic energy left over will result in damage. Shear and vertical components will act on the soil, and the damping action of the earth is important. Lateral displacement of piles may enlarge holes and permit more energy to be absorbed in pile bending, or some soils may close in immediately. Possibly piles may retain some eccentricity after an earthquake.

The piles must resist any shear that pressures of the foundations against the soil do not, and since the piles are stiffer, they probably should be designed to resist most or all of the shear and be capable of carrying it down into the ground until increased soil resistance equals the applied shear. Vertical loads from rocking action may cause temporary uplift, then pile settlement.

Building proportions should be such as to avoid harmonic action, or the forces may be increased by several times.

It seems probable that foundations without piles, if the ground can be compacted to carry raft or spread footings tied together, would have less rigid lateral support and provide more damping and absorption of energy and therefore possibly be preferable.

Unequal Vertical Reactions. Lateral earthquake force on a structure results in temporarily increased loads on piles under outer foundations.[132] If these foundations are made larger, their settlements under dead loads may be decreased and differential settlements increased relative to center foundations. Rocking action may tend to leave more load on center piles. Unequal settlements prestress portions of structures and leave them more susceptible to earthquakes.

Torsional Effects. Torsional effects may be present because of direction of the impulses or unsymmetrical shape and height of the structure.

Effects of Changing Soil Conditions. Damage may result from changes of ground-stress distribution or mechanical properties of soils during earthquakes. Saturated sands and sensitive clays may be dangerous under tremors. If a saturated sand is below critical density, an earthquake might cause the load to be carried temporarily by the pore water and result in liquefaction and soil flow. Such soils should be stabilized, or the piles supported in some other stratum. Buildings carried on piles supported in dense sand may not be damaged by temporary higher loads on the piles as long as a factor of safety remains, and allowable bearing pressures may be increased by 20 to 30 per cent. Although high pore pressure may be developed in clay from increased seismic loads on piles, the structure of the clay will not have time to readjust itself and little damage may be done to the foundations, and allowable pressures on clays may be increased by 25 to 100 per cent.

Japanese researches[133] indicate that the yield value of the soil diminishes with increasing accelerations of vibrations and that the amount depends upon moisture content, with the largest decrease near optimum moisture content. As the optimum moisture content of clay changes when vibrated, it is possible that a soil compacted near the optimum may not be the most resistant to earthquakes. It was observed that lowering of shearing strength was related to amount of damage and that the critical value of acceleration causing liquefaction of sandy soil coincided with accelerations of serious earthquakes. It is now the usual practice in Japan to place important structures on safer soils or to increase design provisions by such means as extending foundations to diluvial formations. Below a limiting value of acceleration, the mechanical properties of soils do not change for practical purposes. The increase of pressure from vibrations was found to be much larger near the surface of the ground than at lower depths. Because of frictional resistance under foundations, the soil just below behaves as part of the footing. When sand is saturated, the bearing capacity is reduced, because of buoyancy and the relative increase of the ratio of seismic force to gravitational force. The pore-water pressure increases because of vibrations, but the amount has not been determinable. Above a certain acceleration, liquefaction occurs and has resulted in much damage.

Unsupported Lengths of Piles. Supporting piles may be braced in their unsupported lengths to resist earthquake forces, or they may be designed by the flexible theory. The latter method may be forced on the designer for such structures as piers by the presence of deep water and soft bottom above firm support, making installation of bracing impracticable. There may be room above water level for bracing, thus reducing the unsupported height.

For piles of various unsupported lengths to firm support, carrying a unified structure, the horizontal force should be distributed to the various piles in proportion to their relative stiffness, as expressed by the ratio of I/l_u^3 for each pile to $\Sigma(I/l_u^3)$ for all piles in the unit, where I is the moment of inertia about any given axis and l_u is the unsupported length of the pile, as shown in Fig. 7.25. A reasonably deep penetration in good material is required to obtain fixity. For particularly slender piles, it is the practice of some designers to assume a maximum deflection for a pile, such as 2 in., if the computed deflection is larger than this amount, and to design for the corresponding moment.

Codes. In the United States, California has taken the lead in study of earthquakes. The Riley Act was enacted in California after the 1933 earthquake and was the first law in the United States to require a percentage of gravity as a lateral-force design factor. The Los Angeles Code introduced the concept of flexibility by varying the percentage of gravity in different stories, and the State Chamber of Commerce Code followed this principle in 1939. In 1948 the San Francisco Section of the ASCE and the Structural Engineers' Association of Northern California set up a Joint Committee on Lateral Forces to draft a model code. An objective was to state minimum design requirements to guard against major structural damage or loss of life in event of precedented lateral forces. For unusually unfavorable conditions, proximity to faults, unfavorable soil conditions, or importance of uninterrupted operation of a structure, the engineer should use his judgment in evaluating calculated risk. The Joint Committee presented a rational dynamic method for establishing base shear and the distribution of that shear on a structure, published in 1951.[114] San Francisco has adopted a code based upon this report.[134]

Determination of Base Shear and Overturning Moment. Earthquake phenomena are incompletely understood, and attempts to apply completely rigorous analyses are not warranted. Until further progress, judgment backed by available facts and engineering experience must be used for designs. The simplified recommendations of the California Joint Committee[114] are suggested for use in determining the forces affecting pile foundations, namely, base shear and overturning moment. Their procedure is summarized as follows:

1. Design for wind or earthquake, whichever is greater.
2. Assume wind is 15 psf below 60 ft and 20 psf above 60 ft for buildings and higher values as set forth for other structures.
3. Compute base shear due to earthquake forces from the following formulas:

$$V = CW \tag{7.69}$$

where V = total lateral force or shear at base;
C = a numerical coefficient; and
W = total weight of building dead load plus 50 per cent of design live load for storage floors and 25 per cent for other floors.

The coefficient C is obtained from

$$C = \frac{0.015}{T} \qquad (C \text{ shall not be less than 0.02 nor over 0.06}) \qquad (\textit{for buildings}) \tag{7.70a}$$

or

$$C = \frac{0.025}{T} \qquad (C \text{ shall not be less than 0.03 nor over 0.10}) \qquad (\textit{for other structures}) \tag{7.70b}$$

where T = fundamental period of vibration of building, in seconds, in direction considered.

In the absence of data to determine T, it may be assumed that

$$T = 0.05H/\sqrt{b} \qquad (\text{for buildings}) \tag{7.71}$$

where H = height of building, in feet; and
b = width of building in direction considered, in feet.

T shall be computed by recognized methods, assuming fixed base conditions, for structures other than buildings.

4. Provision for overturning moment should be made for earthquake forces on the top 10 stories or top 120 ft of the building, with constant moment below these points to foundations.

Wind and earthquake forces are not generally assumed to act simultaneously.

LATERAL FORCES FROM MECHANICAL CAUSES

Lateral Forces from Traffic

Traction and Braking Forces. Railway bridges are designed for a horizontal longitudinal force representing the effects of train traction or braking.

Nosing of Locomotives or Sway. Railway bridges are designed for a concentrated moving horizontal lateral force to withstand the effect of nosing of moving engines or for a uniform lateral force to withstand the effect of sway of engines and cars. Vertical effects of this lateral force are sometimes disregarded.

Centrifugal Force. Lateral forces from the effect of track curvature must be considered.

Longitudinal Expansion of Structure

Temperature may cause expansion or contraction of the deck, causing eccentricity from vertical loads as well as bending stresses. It is advisable to break a long pier into separate sections by expansion joints, to reduce effects of longitudinal expansion.

Pulling Piles into Position

Pulling pile tops may cause bending stresses and eccentricity under vertical loads.

Flow of Soil

Flow of soil may occur in soft banks and exert lateral force on piles.

Batter-pile Sag

Sag under the weight of pile, and possibly of fill, kelp, ice, or loads above, causes eccentric stress in compression batter piles.

Machinery Foundations

Reciprocating or rapidly rotating equipment may cause unbalanced horizontal forces which must be resisted by piles if the foundation is not on adequate soil. A heavy mass to absorb energy is desirable, but this will probably be below the horizontal impulse, and means must be provided to resist the remaining horizontal arm of the couple. Vertical piles will provide a balancing vertical couple. Batter piles will provide horizontal resistance. The elasticity of wood may be more satisfactory in absorbing vibrations than the relatively great rigidity of steel or concrete piles.[73] Prestressed spring caps fitted with screw jacks might be used on batter piles to maintain prestressing in case the piles move down after driving and loading, although the elasticity of wood piles is such that they have been used successfully without such caps.[73]

Compacted clay and sand and clay mixed soils are practically unaffected by vibrating machines, and settlements are about the same as from static loads. Graded sands and loosely compacted or saturated sand and clay mixtures are subject to extensive foundation settlement under vibration. Establishment of natural ground frequencies and of the weight of soil vibrating below the foundation, which is a factor in this value, is difficult. Effects of damping provisions must also be considered. Design of foundations under vibrating loads is at best semiempirical, and it is advisable to consider making provisions in the designed structure for easily altering the mass of foundation and damping arrangement. The effects of piles are difficult to evaluate; they can transmit vibration to depths that would be less affected without them; they also increase the weight of soil, thus decreasing the frequency of the system. Most recorded cases of objectionable frequencies due to resonance have been from machines operating at from 300 to 2,500 cycles per min.[159] Ranges of ground frequencies determined experimentally appear in Chap. 2. When the speed of the machine is greater than the natural frequency of the soil, little trouble is likely from momentarily passing through the critical range.

LATERAL RESISTANCE

Lateral Resistance of Piles

Methods of Obtaining Lateral Resistance. Since lateral forces cannot always be avoided, methods of resisting them must be considered. Cantilever action from a lateral force applied against the top of a vertical pile is not a very satisfactory method of resistance and generally should be avoided, where possible, in favor of batter piles, tiebacks, deadmen, or thrust surfaces.

The lateral resistance of a structure supported on vertical piles depends on the factors discussed below.

Stiffness of Piles. This depends on the material and size of the pile and conditions of fixity. If the heads of the piles are fixed in the structure, causing reversal of bending moments in the piles instead of only cantilever action, the resistance to lateral forces is increased, but the fiber stresses in the piles are greatly increased.

Stiffness of Structure. This has an effect on the degree of fixity of the pile heads, as well as on the distribution of lateral forces to the piles. Local increases in lateral forces may be resisted by a much larger portion of the structure if it is sufficiently stiff.

Resistance of Soil. This highly complicated subject depends on the resistance of the soil to compression, shear, and displacement, which vary not only with soil character, but with depth, and also with changes in conditions caused by pile driving, fluctuations in ground-water level, changes in unbalanced hydrostatic head, surcharge, and vibration. The shape of the lateral-loading curve applied to this soil of varying resistance depends on the degree of fixity at the head and in the ground. The shape of the loading curve and the soil resistance are interdependent, and a condition of elastic equilibrium must be reached. *Passive pressure between the foundations and the soil* may reduce the lateral forces which the piles are called upon to withstand, provided the soil is sufficiently firm and is not to be disturbed by possible removal, by changes in unbalanced hydrostatic head, or by vibration, if the structure is sufficiently stiff to distribute the load properly.

Character of the Loading. Repetition of loads, such as longitudinal traction and braking actions or lateral engine-nosing effect or sway of trains on trestles, impact of ships, impact of waves, forces of tides or currents, changes in water levels, wind, etc., may result in progressive bending of the piles and ultimate failure; it also may damage the structure supported by the piles. Some structures may move bodily and be unaffected, whereas others may be greatly damaged, depending on the character and use of the structure.

Grouping and Spacing of Piles. Large groups of piles subjected to large lateral loads do not seem to have so great an individual value per pile as do single piles. For small lateral loads, some tests have not shown any influence on the deflection, however.[35] The piles of a large group may grip enough soil to act as a diaphragm, so that the resistance of a vertical plane of earth bounding the group should be considered.

Classical Theories. The classical theories of earth pressures are not reliable for determining lateral resistance for bulkheads and even less for single piles. They assume mobilization of active and passive pressures, which do not occur, however, except at complete failure. Satisfactory methods of determining lateral resistances of single or groups of piles must be workable with small deflections. In some cases, the governing design criterion is the permissible lateral deflection; in other cases, it is the maximum load that the pile can take without overstress.

Approximate Calculations. Only approximate methods of investigation are available for stresses, since the three-dimensional elasticity problem of a load applied to an elastic medium by an embedded rod has not been solved theoretically. Two-dimensional solutions, treating a pile as a plane section of a bulkhead, have been made. Other solutions have assumed a constant soil modulus, which is not the true case, since it increases with depth in a given soil and decreases appreciably with increased load or deflection. It also varies with changes in the soil, the effects of pile driving on the soil, diameter, ground-water fluctuation, and with time. Soil disturbance or densification may affect resistance. Under loads of long duration, soil consolidation may occur as well as elastic deformation, thus increasing deflection. Methods of analysis proposed assume only elastic conditions. Furthermore, in practice, a pile often penetrates various strata, each having a different soil modulus.

A theory of subgrade reaction for design of vertical piles against lateral forces is presented in reference 208, with a range of coefficients for various types and conditions of soil.

Based on assumptions as to the conditions of restraint in the ground, approximate calculations have frequently been used. Such schemes usually assume a pivot point at some point in the ground. This may or may not be the case, depending on the relation between the rigidity of the pile and the probably varying elasticity of the soil. A point of contraflexure in the ground may develop, which will mean that something less than full fixity occurs at this point; the pile may act as a cantilever above a pivot point, with the portion below remaining vertical; or rigid piles in soft ground may have no pivot point and act more like cantilever beams fixed at the top and loaded with the earth load of the soil resistance.

For use in determining the safe lateral force applicable to foundation piles, such approximate calculations are not a safe guide since only a small lateral deflection or movement is generally permissible to avoid damage to the structure. In structures such as cantilever bulkheads of sheet piling or in some types of wharfs where readjustments can take place without damaging the structure, such approximations may be used.

Mathematical Solutions. One need is determination of the soil stress-strain characteristics; the other is determination of the pile deflection curve. The most versatile method is the difference-equation solution.[161,162,165] The accuracy depends largely upon the engineer's experience in assigning values to empirical terms. No definite comprehensive method has been established for field or laboratory determination of these values, although studies for particular conditions have been reported for transient loads.[162,164,166,167] A method has been suggested for using earth constants derived from tests on a single free-head pile to predict the action of a capped fixed-head group of piles.[162]

Available formulas, including definitions, tables, charts, and presentations and discussions of limitations, assumptions, and applicabilities, are too complex and voluminous to include here but may be studied in references 161 to 164. Broad conclusions as to effects of changes in the several variables have been discussed in reports of tests.[117,119]

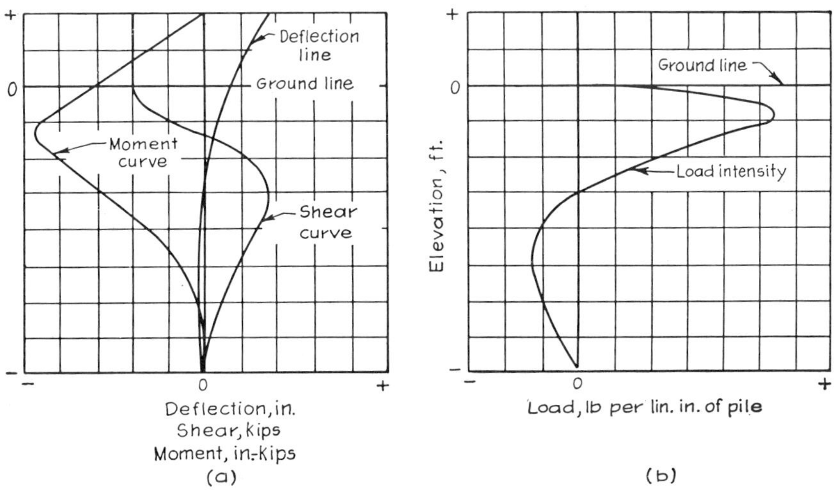

FIG. 7.26. Typical cantilever-pile characteristics.

No literature has been observed on any theoretical approach which would take into account the effect of irreversible soil deformations.

A typical shear, moment, and deflection diagram indicating the general characteristics of these functions is shown in Fig. 7.26*a*. A typical load diagram showing lateral soil pressures is shown in Fig. 7.26*b*.

Field Tests. Field tests of capacities of single piles and groups of piles are recommended as a means of determining safe working lateral resistances for piles under foundations, for projects of sufficient magnitude and where possible savings in batter pile or tieback costs would more than repay test costs. For small jobs, very conservative figures indeed are recommended, and a liberal use of other means of resisting lateral forces suggested.

A single pile, which moves laterally under a test thrust, does not provide the direct answer to most problems involving lateral resistance. Lateral movement is due to creep or plastic flow and may continue until objectionable deflections are reached. Results of tests on a single pile or small capped group should not be used without investigation of the soil structure to make sure failure will not occur along some plane of weakness. An example of such field tests appears in the description of work done for Lock and Dam No. 26 at Alton, Ill., by the U.S. Engineers.[35] At another lock site, movement of a number of inches occurred under a lateral load of only 3.5 tons per pile, the piles being 30-ft wood piles and the soil silty clay.

Based upon Bureau of Reclamation lateral load tests[163] on timber piles driven in various soil conditions, such as fine sand covered by several feet of organic silt, deep peat and silt beds over firm clay, and glacial till, the following broad conclusions for single piles were drawn: (*a*) Overdriving reduces the lateral resistance due to enlarged holes caused by bending of the pile during driving. (*b*) Increase in length does not improve lateral resistance if the pile is embedded enough to prevent movement in the lower portion. (*c*) Lateral resistance of a pile in poor material may be increased by lengthening. Increasing the size increases lateral resistance. (*d*) Ultimate driving resistance may not indicate lateral resistance. (*e*) Lateral resistance of a pile group in which free ends are rigidly tied together may be equal to (for deflection under ¼ in.) or greater than (for deflection ¼ to ½ in.) the combined resistance of each pile. (*f*) The strength and type of soil in the first 20 ft of depth have considerable effect on the lateral pile strength. (*g*) Lateral resistance may be improved by application of vertical load.

The following conclusions were reached by Navy engineers for a single steel box pile embedded 40 ft in uniform compacted sand fill above ground water and subjected to a 10,000-lb thrust:[164,165] (*a*) The stiffness factor (*EI*), width of pile, and depth of embedment have surprisingly small effects on the magnitude of the maximum moment and its location in the pile, for the pile tested. (*b*) The magnitude of the maximum moment is much less than usually supposed and as usually computed by assuming a point of fixity. (*c*) The depth below grade of the maximum moment is much less than is commonly supposed. (*d*) The magnitude of the peak earth pressure far exceeds the passive pressure as computed from classical earth-pressure theories. (*e*) As pile width increases, the maximum moment and its depth below ground surface decrease slightly and deflection also decreases. (*f*) With increased stiffness factor, deflection decreases, points of peak pressure and maximum moment shift slightly downward. For the sand soil tested, the point of maximum moment would be nearer the surface with a timber pile than with a concrete pile and maximum moment in the concrete pile would be greater. (*g*) The pile did not return to its original position after loadings. It was believed that the residual pressures and deflections were the result of movements of the sand as the pile deformed under thrusts.

Allowable Thrusts. In lieu of definite information, Table 7.1 gives suggested allowable thrusts for vertical piles, based on a safety factor of 3 applied to the load required for ¼-in. deflection.

Concrete piles should be reinforced to resist bending stresses caused by lateral thrusts.

TABLE 7.1. SUGGESTED SAFE ALLOWABLE LATERAL FORCES ON VERTICAL PILES, LB

	Medium sand	Fine sand	Medium clay
Free-end timber piles, 12-in.-diameter	1,500	1,500	1,500
Fixed-end timber piles, 12-in.-diameter	5,000	4,500	4,000
Free-end concrete piles, 16-in.-diameter	7,000	5,500	5,000
Fixed-end concrete piles, 16-in.-diameter	7,000	5,500	5,000

SOURCE: Reference 167.

CHAPTER 8

WOOD PILES

Required Characteristics

Manufacture. A good pile has the following characteristics: (*a*) freedom from sharp bends, large or loose knots, shakes, splits, and decay; (*b*) freedom from short or reverse bends and from crooks greater than one-half the diameter of the pile at the middle of the bend (in short bends it is good practice not to permit the distance from the center line of the pile to a line stretched from the centers above and below the bend to exceed 4 per cent of the length of the bend, or 2½ in.); (*c*) straight line between centers of butt and tip within the body of the pile; and (*d*) uniform taper from butt to tip.

Straight tree trunks, cut above the ground swell and with branches closely trimmed, are necessary. For important work or where reliance is placed upon skin friction, the piles should be peeled smooth, clean, and free from inner bark.

Defects. From the nature of trees, it is evident that some defects will be present. Experience has established permissible amounts and limitations. These appear in standard specifications in Appendix VI.

Sizes. For North American woods, piles 40 ft long or under are furnished in length mulitiples of 2 ft, and piles longer than 40 ft in multiples of 5 ft, with a 6-in. tolerance. Diameters are measured 3 ft from the butt and at the tip. For piles up to 40 ft long, butt diameters range from 12 to 20 in., and tips from 8 to 10 in. For piles 40 to 90 ft long, tips vary from 6 to 9 in. in diameter. For piles over 90 ft. long, tips are usually from 5 to 6 in. in diameter. Butt diameters must not exceed the clearance between pile-driver leads.

Storage. Wood piles stored for some time should be protected against decay and also against marine borers if in salt or brackish water. They should be free from bending strains while stored.

Common Woods Used for Piles

North America. *Southern pine* is a term covering longleaf, shortleaf, slash, loblolly, old field, rosemary, and other names. The heartwood resists decay fairly well. The character of the particular wood and not the species is the governing factor in suitability. Treatment is taken well. This wood is excellent for piles. *Red (Norway) pine* is suitable for creosoted foundation piles and is used to a considerable extent for this purpose in New England.

Douglas fir is a term covering many varieties of this western wood. The heartwood is moderately resistant to decay. It has a relatively thin ring of sapwood, and will not absorb as much preservative as woods with more sapwood. This wood makes excellent piles.

Bald cypress is recognized as tidewater red cypress, yellow cypress, and black, white, and gulf cypress. Suppliers have claimed variations in decay resistance according to geographical origin and color. Logs that floated high have been known as white and those that sank as black. It is maintained that red cypress grown near salt water is the most durable. The Forest Products Laboratory[2cn] observed that the outer heartwood of only a few per cent of samples was very resistant to decay and that color classification was not significant. The widespread belief that red cypress is the most durable seems likely to be correct, but its scarcity makes this point academic. Moreover, cypress sold as red may include dark brown, which is only moderately decay-resistant. The reputation of bald cypress for superior resistance to decay may rest on the wood from larger, old virgin trees. Second-growth timber has less resistance and less heartwood. There is no practical geographical distinction in decay resistance of present-day cypress. Generally, the indicated resistance would not justify recommending extensive use of the untreated wood in ground contact.

Cedar piles are resistant to decay but are soft and of low strength, and will not stand hard driving.

Maple piles are used in foundations or temporary structures.

Rock-elm piles are used in foundations or temporary structures and for fenders. They are tough and flexible.

Oak piles usually come from the South, Mississippi Valley, or Appalachian region. The wood is hard, strong, and heavy. *White oaks* are more decay-resistant than *red oaks.* Oaks should be seasoned before treatment, and even after seasoning, white oaks can be treated only in the sapwood. White oaks cannot be treated green, and red oaks are apt to be injured in the attempt. Air-seasoned red oaks usually take treatment well, although blackjack oaks and some other red oaks are difficult to treat. *Mixed oak,* often specified, allows any oak species, and permits inclusion of the less desirable ones. If piles are to be treated, red oak should be specified. The *white-oak group* contains the following oaks: white, chestnut or tanbark, rock, post or iron, burr or mossy, overcup, basket or cow, live, swamp post, and chinquapin or yellow oak. The *red-oak group* consists of red, black, blackjack or barn, scarlet, water, willow, pin, Spanish, turkey, and shingle or laurel oak. Oak piles are generally short.

Central America. *Greenheart* has an excellent record in temperate waters, but not so good in warm waters, although it is one of the best woods in such locations.

Angelique, manbarklak, and *foengo* have considerable resistance to borers. *Black kakaralli* and *purpleheart* are good woods to resist borers.

Tropics. *Palmetto* has low strength but some resistance to borers.

Mangrove and *palm* of some species have given fairly good borer resistance but are hard to drive and suitable for only light loads.

Asia. *Teak* and *eng* have high strength and durability.

India. *Sal, acle,* and *teak* have high strength. *Deodar, chir,* and *poon* have medium strength. *Jarul* and *white siris* are also used.

Australasia. *Ironbark, jarrah,* and *white* or *red gum* have medium strength and high durability.

Ironbark is practically immune to *Limnoria,* but little more resistant to *Teredo* than Oregon fir. *Turpentine wood* succumbs slowly to *Limnoria* and *Teredo* in sapwood, but heartwood is destroyed by *Limnoria. Totara* has considerable resistance to borers.

Europe. *Norway pine* has medium strength. *White deal* and *kail* have low strength. *Northern pine,* called Scotch fir, red deal or fir, Baltic fir or pine, or yellow deal or fir is elastic, strong, and durable. *Alder* lasts indefinitely when submerged; otherwise it should be dry. *Elm* is tough, and durable if dry or wholly submerged. *Dutch and cork-bark elm* are inferior. *Old English oak* is durable, strong and tough, and resists driving well but may not be straight in long lengths; it is heavy and expensive for ordinary driving, and corrodes iron or steel fastenings. It is superior to Austrian or Baltic oak. *Larch* is the toughest and most durable of the common woods, is free from knots and very durable, even when alternately wet and dry.

Characteristics of Woods. The characteristics of these woods are described later in this chapter, in Chap. 13, and in Table VI.

Wood for Fenders

A world-wide list of comparative properties of woods for fenders appears in reference 2*cm.*

Treatability of Various Woods

The usual method of classifying wood as "easy," "moderately difficult," and "difficult" to treat is relative and of limited use without experience. The following classification is based upon extent and duration of pressure required to impregnate to a given depth.[2bb] The weight of preservative is not mentioned, because it depends on the species and on the dimensions of the piece. Furthermore, the weight of preservative alone does not give a reliable estimate of protection against decay for a given amount may be more effective when it penetrates into the fiber than when it penetrates only into the vessels. Overlapping of classifications may occur because of variations in wood structure.

Treatment of the first two classes (impermeable or extremely resistant) is not practicable, but it appears that the first class is more suitable when impermeability is desired than is the second class.

Sapwood is usually easier to treat than heartwood. Sapwood permits piles to be treated deeply and uniformly. Heartwood resistance varies with species, often greatly as in Douglas fir and jarrah, but sapwood is more consistent. Permeable and resistant piles should not be mixed in the same charge, for great variations in absorption occur between heartwood and sapwood.

In Table 8.1, the categories are:

Impermeable. Cannot be penetrated to significant depth laterally or longitudinally, even under hard pressures. In practice, laterally not over $\frac{1}{25}$ in., longitudinally up to 1 in.

Extremely Resistant. Absorb very small amount even under heavy pressure treatments, and impracticable to inject sufficient preservative to protect against decay. In practice, laterally not over $\frac{1}{25}$ in., longitudinally in some vessels up to 12 in.

Very Resistant. Most resistant of woods practicable to treat, these require 4 hr or more treatment under 140 to 180 psi to obtain lateral penetration of $\frac{3}{4}$ in., the minimum for decay protection.

Resistant. Require 2 to 4 hr under pressure for lateral penetration of $\frac{3}{4}$ in.

Moderately Resistant. Can be treated fairly well under pressure, but not very well in open tank. Require 1 to 2 hr under pressure for lateral penetration of $\frac{3}{4}$ in.

Permeable or Absorbent. Can be treated easily under pressure, and also satisfactorily by open-tank process. Complete impregnation of 2 to 3 in. obtainable from short pressure treatment or by open-tank process.

Vessel-porous. Wood in which penetration is easily obtained along a large number of vessels by pressure, but fibers remain unpenetrated. As much as 12 lb of preservative may be absorbed, but so diffused that value is uncertain. Creosote may seep from the ends.

Comparative Values of Sapwood and Heartwood

There is practically no difference between the mechanical strengths of sapwood and heartwood, according to the U.S. Forest Products Labora-

TABLE 8.1. CLASSIFICATION OF WOODS IN ORDER OF PERMEABILITY

Reference numerals indicate that the wood is classified in more than one group because of variations in its permeability.

1. *Impermeable woods*		
Brush box	Jarrah[7]	Tallow wood
Camphor (east African)	Karri	Turpentine wood
Guarea		
2. *Extremely resistant woods*		
Cedar, African, pencil	Iroko	Oak, English
Gaboon	Keruing[4,7]	Padauk, Burma
Greenheart	Mahogany	Pyinkado
Idigbo	Nargusta	Teak
3. *Very resistant woods*		
Aspen, Canadian	Larch	Poplar[7]
Cedar, western red	Oak, American white	Spruce, Canadian
Chestnut, sweet	Oak, Tasmanian	Walnut
Douglas fir[5]	Obeche[4]	Willow
Gum, American red		
4. *Resistant woods*		
Balsa[7]	Maple, rock[5]	Spruce, English
Birch, Canadian yellow[5]	Muninga	Spruce, Sitka
Elm, rock	Obeche[3]	Whitewood, European
Keruing[2,7]		
5. *Moderately resistant woods*		
Ash	Maple, rock[4]	Pine, pitch
Birch, Canadian yellow[4]	Opepe	Pine, red
Celtis[7]	Padauk, Andaman	Pine, Scots (English)
Cottonwood	Pine, Corsican	Pine, western white
Douglas fir[3]	Pine, Honduras pitch	Pine, yellow
Fir, silver	Pine, jack	Redwood, European
Hemlock, western		
6. *Permeable woods*		
Alder	Coachwood	Odoko
Basswood	Hornbeam	Podo
Beech	Lime	Sycamore
Birch	Oak, American red	Tupelo gum
Chestnut, horse		
7. *Vessel-porous woods*		
Balsa[4]	Eng	Keruing[2,4]
Bombway, white	Gurjun	Poplar[3]
Celtis[5]	Jarrah[1]	

tory. From 1 to 3 in. of sapwood is desirable for piles to be treated. In coniferous or soft woods, the sapwood should not be restricted, as sapwood takes treatment better than heartwood.

Comparative Values of Timber Cut from Live and Dead Trees

Some specifications do not permit the use of timber cut from dead trees. Other than the weathered or charred exterior appearance, there is no known method of distinguishing the difference, and the U.S. Forest Products Laboratory considers that timber from fire-killed or insect-killed trees is just as good as that cut from live trees of similar quality, provided the wood has not been further injured by decay or insect attack.[2ai] If a tree stands on the stump too long after it is killed, the sapwood is likely to become decayed or badly infested by wood-boring insects, and in time the heartwood will be similarly affected. The same is true of logs cut from live trees and cared for improperly. The heartwood of a living tree is entirely dead, and in the sapwood only comparatively few cells are living, so that most of the wood cut from live trees is already dead. Specifications could well state that material showing evidence of decay or insect infestation exceeding a specified limit will not be accepted, instead of excluding dead trees entirely.

Effect of Partial Seasoning on Strength[2as]

The effect of partial seasoning on the strength of wood has been shown to be nonexistent on the fiber strength until the wood becomes drier than the fiber saturation point, which is the theoretical point at which no water remains in the cavities although the cell walls are still saturated (usually 23 to 30 per cent of the oven-dry weight, depending on the species of wood). Loss of moisture below the fiber saturation point is accompanied by increase in fiber strength. If drying were uniform in wood, strength values in air-seasoned or artificially seasoned wood could be predicted from moisture contents, but since generally moisture becomes low at the surface and increases toward the center, and moisture at the surface fluctuates with weather conditions about an equilibrium value (from 5 to 8 per cent in arid regions to 18 to 20 per cent in moist regions), average moisture content does not provide a rule by which strength changes can be predicted. Some woods in which partial seasoning has not reduced moisture below the fiber saturation point show increase in strength owing to partial seasoning, whereas others do not. Woods seasoned to an average moisture content below the fiber saturation point may or may not have a higher strength than would be expected from average moisture content.

Bark

Piles in which reliance is to be placed on skin friction for carrying downward or uplift loads should be barked before driving. If the bark is left in place, a decomposed slippery layer develops between bark and wood, which permits the bark to slip readily after some months' exposure to underground water conditions. This condition may apply both to bark embedded in soil and to bark gripped by foundations.

Preparation for Driving

Before driving, wood piles are trimmed. The butt is squared off and shaped to fit the pile cap, and the tip pointed or squared. Splices may be prepared or lagging attached. A pneumatic ax for pointing and heading wood piles has been developed, tripling the efficiency of hand axes and reducing accidents. A 4-in. frost blade is fitted in a pneumatic hammer.[2ch]

Butt Preparation. The butt should be cut square and chamfered to fit the pile cap if a recessed cap is used. If a cap is not used or if it is not recessed, it is still advisable to chamfer the butt, for the reduced area concentrates the blow on the central portion of the pile and reduces the tendency to split.

Pointing. It was formerly fairly standard practice to trim the tip to about a 4-in. diameter when driving through hard clay or coarse gravel. If damage to the point is expected, a shoe (Fig. 3.1) should be used. Pointed shoes have been extensively used, but a point may tend to cause the pile to drive out of line, and a square or recessed tip is preferable.

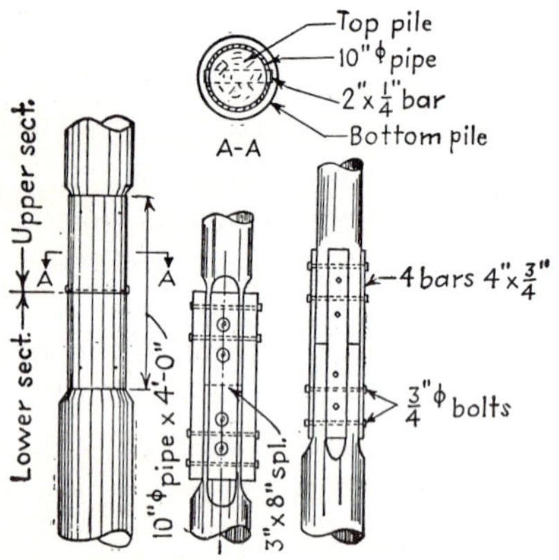

FIG. 8.1. Typical splices for wood piles.

Splices

Piles can be spliced (Fig. 8.1) if unavailability of long piles makes it necessary, or if long piles cannot be handled in the driver. Unless firm lateral support is present, splicing is poor practice, and splices should not be located in the middle portions of long unsupported piles unless they are made very long and strong. It is difficult to develop the bending strength of a pile in a splice, and most splices do not approach it. A strength of 50 per cent of that of the pile in bending is required by the New York City building laws. Splices which increase the diameter at the splice should be used only in water or very soft mud.

Uplift piles should not be spliced, unless some such strong detail as described below for a gunite-encased splice is used, together with bolts to transmit tension.

Composite piles are sometimes used to permit driving long untreated submerged lower sections of long piles, at savings in costs, with a splice connecting an upper treated-wood or concrete section.

Sleeve joints are made from 8- or 10-in.-diameter pipe, cut in 3- or 4-ft sections. The inside diameter of the pipe should be slightly less than the pile tip diameter, and the entering tips and butts should be trimmed to a tight fit. Wedges should not be used to fill out space. A crossbar keeps the sleeve in place during driving. Contact ends of the piles must be carefully squared.

Bolted splices may be made, using either timber or steel splice bars.

Gunite splices[85] in wood piles may be made by using steel pipe sections and encasing the splice in gunite as described in Chap. 13. Steel pipe sections of 12 in. diameter 4 ft long, covered by gunite encasement 10 ft long, are said to test stronger than the pile. Such splices are made in the yard and can provide wood piles from 80 to 130 ft long if desired for economy or quick delivery. Very long wood piles cost much more per foot delivered than shorter lengths and may also require a longer delivery period.

Gunited splices in wood piles, using 5-in.-thick concrete sections 6 ft long reinforced with twelve ¾-in. longitudinal bars in a cage of No. 10 4- by 6-in. wire mesh with five turns of ¼-in. spiral at the ends tested 25 per cent stronger than a simple wood pile, while 4-in.-thick concrete gave strength slightly under.[2bl] A drop-hammer energy of 40,000 ft-lb caused no damage.

Underwater splicing of old wood piles by grouting in a sleeve has been done.[2bz] Damaged piles were cut off at the mud line by a diver using an air saw. Metal sleeves 36 in. long were bolted to the bottom tapered 18 in. of sunken posts, a grout hose nipple attached to the sleeve at mid-height, and the extension floated to location, weighted, and lowered over the cut end of the old pile by the diver. The posts tops were wedged to the caps, and the voids inside the shell pumped full of grout.

Uplift Anchorage

The method of anchorage to wood piles for uplift requires care. Notches, about 1½ in. deep by 3 in. high, located far enough from the butt to avoid shearing the wood longitudinally, have been used, and may be saw or ax cut. Rings of ½-in.-square by 6-in.-long boat spikes, staggered, may be driven about 4 in. into the pile, leaving a projection of about 2 in. Such devices may be tested in the field by casting short lengths of piles, with the selected connection, into concrete walls and jacking apart. Such a notched key in a 10½-in.-diameter pile set in an 18-in. concrete wall showed no signs of distress under a push of 45,000 lb, the capacity of the jack. The distance from the center line of the notch to the end of the pile section was 18 in. This specimen was covered with bark and had been subject to tidal wetting and drying for 6 months, and a test on a similar slightly tapered specimen showed bark

TABLE 8.2. TESTS ON PILES EMBEDDED IN CONCRETE

The concrete mix was 1:2:4, plastic but not sloppy. Piles were of southern pine, rather inferior in grade, having 8 annual rings per inch in the outer 3 in. Piles were soaked in water several days before casting in concrete. Turned specimens to constant diameter; unturned specimens, having very small taper, peeled and roughly trimmed.

A. PILE HEAD BOND TESTS

Series	Age, days	Unit load at initial slip, psi			Remarks
		Max	Min	Av	
1	7	52	33	42	Turned
1	21	67	58	63	Turned
1	7	96	62	79	Unturned
1	21	63	49	56	Unturned
2[a]	21	73	62	65	Wedges
3A	21	73	57	66	2 railroad spikes[b]
3B	21	59	37[c]		4 railroad spikes[b]
3C	21	55	26[c]	57	6 railroad spikes[b]
3D	21	76	60[d]		8 railroad spikes[b]
7	21	83[e]	49	61	
8	21	28	21	25	Concrete placed under water

[a] On series 2, the bond resistance was approximately double after a 2-in. slip.
[b] Railroad spikes ⅝ by 6 in., driven 4 in. into piles.
[c] Poor pile (soft).
[d] Concrete cracked.
[e] Pile very close-grained.

B. ANCHORAGE TESTS

Series	Age, days	Load at initial slip, lb		Remarks
		Max	Min	
4A	7	15,370	12,000	Notched bar[a] driven before casting
4A	21	19,200	16,880	Notched bar[a] driven after concrete set
4B	21	26,120	13,000	Plain bar[a] driven after concrete set
5	21	17,430	16,000	Fox bolt[a] driven after concrete set

[a] Holes for bars and bolts drilled ⅛ in. less in diameter than bar or bolt.

slippage at 12,500 lb when embedded in a 24-in. wall, when pushing against the taper.

Bond and anchorage tests were run in a laboratory on a series of wood pile butts embedded in concrete, by the U.S. Bureau of Public Roads (Fig. 8.2).[2ae] Tests results are tabulated in Table 8.2.

Appreciable bond can be developed between wood piles and concrete, increasing with age of the concrete, and therefore the use of high-early strength or accelerated concrete may often be necessary. Casting under water seems to reduce the bond greatly, although the mix used in these tests was poorer than generally used in underwater placing. The bond on the end of the pile seems negligible. Bond did not materially increase by using railroad spikes, which tore through the wood. To test the efficacy of the spikes, it would be necessary to break the bond between

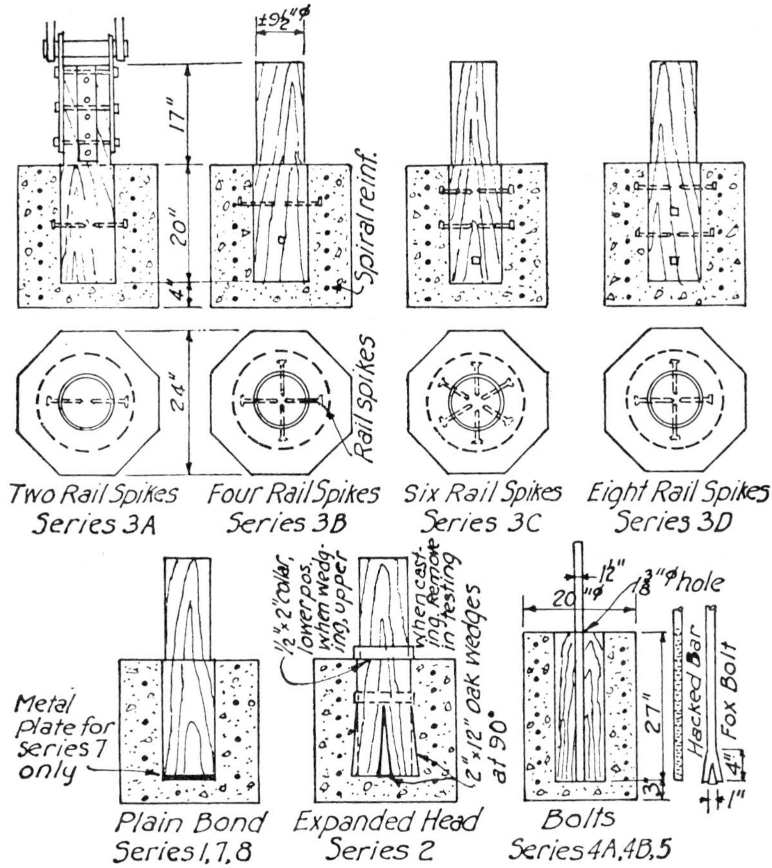

FIG. 8.2. Bond and anchorage tests.

wood and concrete, by coating or paper. Wedges did not have much effect on average bond values but gave very uniform results. Unturned specimens tended to give somewhat higher values than turned specimens but were less uniform in results, the slight taper and irregularities of the untrimmed specimens introducing unknown effects. It also appears that appreciable anchorage of steel bars may be obtained in the ends of wood piles. The bored holes were ⅛ in. less in diameter than the bar, and plain round bars appeared to develop as great an anchorage as either hacked bars or fox bolts, although results with fox bolts were slightly more consistent. All of the values in the table are taken at slip, and a suitable factor of safety should be selected.

An economical and effective uplift connection at the top of wood piles may be formed by the use of four 4-in. Teco claw plates, set in pairs on flattened side of the pile about 7 in. from the top to the first set and another 7 in. to the second set. A steel strap plate on each side of the pile, held by two ¾-in. bolts passing through the connectors, extends up into the concrete footing where it is anchored by bending or by horizontal bars through holes. Such a connection should develop an ultimate strength of about 12 tons.

It is especially advisable to use mechanical anchorage for uplift piles in structures such as intake screen wells that may be unwatered and permit large uplifts on piles.

Clumping

A common method of clumping wood piles is to take two or three turns with a heavy wire rope around the piles, dead-ending one end to the rig and placing the other around a power hoist drum, and pulling taut and holding while a galvanized strand rope is anchored to one of the piles and at least three turns are taken around the clump. Usually every third turn is pulled tight by an eccentric cable clamp attached to a line pulled by a winch. Each turn around the clump should be securely fastened to each pile with boat spikes. The end of the strand is cut off, the two ends twisted, bent over, and spiked down so that no ragged edges project. A safety line should be attached to the eccentric clamp to prevent kickbacks if the clamp should slip or the rope part. The line should be run through a snatch block so that if the strand breaks at the eccentric clamp, it can fly back only to the block and not to the winchman.

CHAPTER 9

CONCRETE AND PIPE PILES

Methods of Placing

In 1897 A. A. Raymond patented the Raymond pile system and was first to develop a practical, economical way of placing cast-in-place concrete piles. In 1903, R. J. Beale developed a method of driving a steel pipe, either plugged with a follower or with a cap, of filling it with concrete, and then withdrawing the pipe.

There are numerous types of concrete piles, both precast and cast-in-place. Methods employed in placing these types are driving, jetting, jacking, boring, ramming, pulling down, washing out, sand pumping, blowing out, coring, drilling, screwing, and using explosives. Long pipe piles have even been pushed down by weighting with water tanks loaded to provide the desired factor of safety over the design load, after driving as far as possible with a hammer.[55] The choice of the correct type and method for a given situation is not always simple. Soundings, borings, soil load tests, or test piles will usually save many times their cost by providing data for determining the correct type and most advantageous method. The resulting avoidance of delay may even be more important than the saving in foundation costs. The soil is the most vital part of the foundation, and a complete knowledge of it is essential to selection of the proper type of pile.

Concrete can be dropped successfully into dry watertight pile shells. Use a central vertical spout and measuring hopper so that voids can be detected. A plasticizing retarder may prevent arching. Uniform consistency is needed.

PRECAST CONCRETE PILES

Precast concrete piles (Fig. 7.2) are useful in carrying fairly heavy loads through soft material to firmer strata. They can be reinforced for bending and uplift. They can be used where decay would prevent the use of wood piles. Such piles are apt to become excessively expensive in lengths beyond 55 ft. Conventional precast piles 24 to 30 in. in diameter and up to 114 ft long have been cast. Piles as small as 6 in. in diameter and as short as 6 ft have been used.

Fig. 9.1. Steps in cutting off concrete piles. Sledge and chisel are used to remove outer cover of concrete, exposing reinforcing. Acetylene torch cuts off the bars. Line from the crane snaps off the head. Exposing reinforcing preparatory to splicing a large precast concrete pile. (*Courtesy of Portland Cement Association.*)

Precast concrete piles require time for curing, storage space, and equipment for handling. It may be difficult to predetermine lengths, which may involve expense in cutting off or building up.

This type of pile is usually constructed either square with constant cross section or taper, or octagonal, with or without a circular cored hole to save weight. Tips may be flat or pointed, and sometimes the lower few feet are tapered considerably. The publication *Concrete Piles* by the Portland Cement Association[5a] gives details of construction methods.

Solid-concrete circular piles of constant cross section, cast horizontally in round metal forms, have been used. Tips may be flat, or a taper may be obtained on the bottom few feet. Owing to the round shape, these piles can be handled readily in the yard. They require less concrete for equivalent strength than do other precast shapes.

Macco Spunpiles. These are centrifugally spun precast reinforced-concrete piles, first developed during a metal-pile shortage. They are from 15 to 50 ft long, usually with 16-in.-diameter shafts and 20-ft-long

Fig. 9.2. Casting yard for precast concrete piles for Chicago & North Western Railway. (*Courtesy of Portland Cement Association.*)

bottom sections tapering to 10 in. in diameter. They can also form the upper sections of composite piles. These piles are furnished by the Macco Corp.

Protected Precast Concrete Piles (Fig. 9.3)

Precast concrete piles have been cast inside of previously cast concrete shells coated with asphaltum (Conzelman piles)* or in bituminized concrete armor blocks impregnated under pressure,[5i,5k,5ac] located in the exposure zone. Concrete piles have also been given a pressure treatment of asphalt (Duocrete piles),*[,5l] and concrete piles having the upper section jacketed in concrete have been pressure-treated as a unit.[5l]

* *Eng. News-Record*, vol. 96, No. 18, pp. 732–736, May 6, 1926.

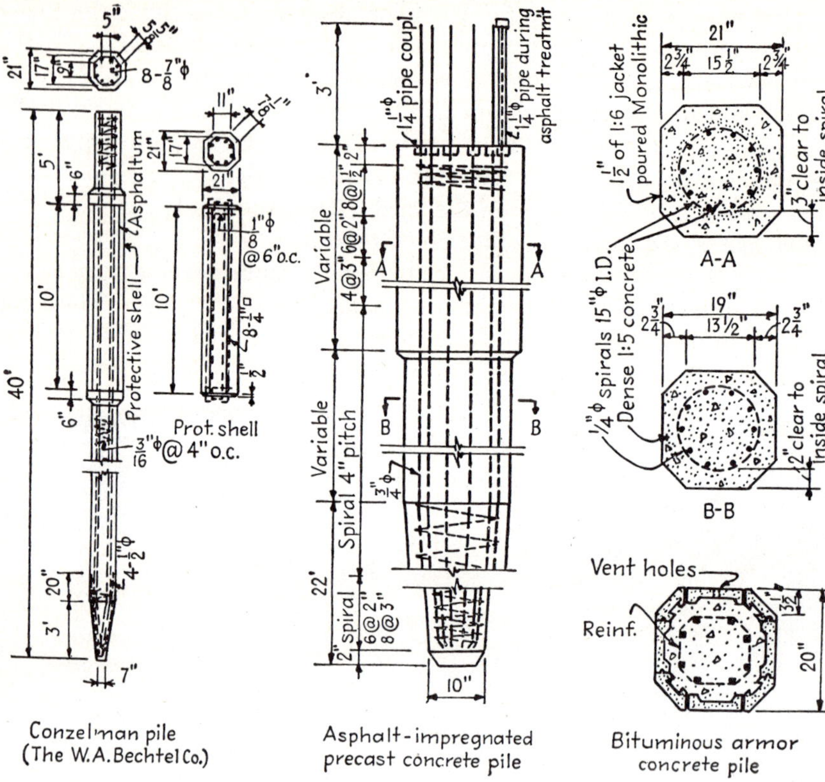

FIG. 9.3. Armored treated precast concrete piles.

Sectional Precast Concrete Piles

Hawcube Piles. These consist of precast reinforced-concrete pile sections, usually 5 to 10 ft long, and a base section with a tapered end. The sections are connected by inserting a grooved tenon on the bottom of an upper section into a socket on the top of a lower section, the socket containing grout. The tightness of fit and length of tenons are claimed to result in straight piles. The construction of the pile in sections makes it convenient in locations in which the predetermination of full-length precast concrete pile lengths might be difficult owing to variable elevations of suitable bearing strata. Handling problems are decreased. These piles are controlled by Holland & Hannen and Cubitts, Ltd., of Great Britain.

Precast Prestressed Concrete Piles

Long precast concrete piles have been constructed with prestressed reinforcing steel in Sweden since 1939. Solid and hollow prestressed piles were first driven in Great Britain in 1949. Prestressed concrete piles are lighter than conventional concrete piles and are more durable. Keeping the concrete in compression prevents cracking and spalling during lifting and recoil during driving.

Advantages of prestressed concrete piles include ability to withstand extremely hard driving far in excess of conventional types, virtual indestructability in sea water because of absence of cracks, greater column capacity, and reduction in handling costs because of lighter weight and fewer pickup points required. When a concrete pile is driven, it compresses, then recoils, with the momentum causing elongation, which makes cracks in conventional piles. An adequate prestress will keep the concrete always in compression. An initial prestress of 800 psi has been found generally to be enough for this purpose.[5bd]

Prestressed concrete piles have much larger moments of inertia than conventional piles of the same dimensions, the entire area contributing, since the concrete is all in compression; this is of value if the capacity depends upon the slenderness ratio.

Square piles of length 50 times the thickness can be handled with a single-point pickup and up to 60 times with two-point pickup. Piles 20-in. square and smaller are usually solid, but 24-in. piles usually have a 12-in.-diameter cored hole. Octagonal piles 20 in. in diameter by 132 ft long, with 11-in.-diameter cored holes, have been driven by a McKiernan-Terry No. S10 hammer through 30 ft of riprap and rock fill to carry 60-ton loads. Longitudinal stress-relieved wire strands with wire spirals are used.

Precast prestressed concrete batter piles as long as 143 ft have been used.

Elliptical precast prestressed concrete piles have been made and may be useful as batter piles or when lateral forces occur in one direction.[12r]

Descriptions of manufacturing steps are given in references 5*at*, 5*ay*, and 5*bg*.

Properties of pretensioned concrete piles are shown in Tables V.7, V.8, and V.9. Specifications appear in Appendix VI.

Sectional Prestressed Precast Concrete Piles

Raymond Cylinder Piles of Prestressed Concrete. During the past few years a new type of precast concrete pile has been developed which is particularly well adapted to soil conditions requiring very long piles of high carrying capacities. This unit often becomes both the subsoil pile and the aboveground column.

The pile is of hollow, cylindrical, centrifugally cast prestressed concrete made of a series of sections placed end to end and held together by cables of high-strength steel wires. The pile sections are manufactured by various processes, similar to the manner in which concrete pipe is commonly made. Each section is reinforced with a small amount of longitudinal and spiral steel to facilitate handling. Longitudinal holes for the prestressing wires are cast in the walls of the sections. The sections are placed end to end, with cable holes in alignment. A plastic

FIG. 9.4. Raymond cylinder pile, 36 in. o.d., has penetrated through 37 ft of water and 22 ft in bottom under its own weight with 170 ft remaining free-standing. Located 40 miles offshore for oil-production platform in Gulf of Mexico, for Magnolia Petroleum Co. (*Courtesy of Raymond Concrete Pile Co.*)

FIG. 9.5. End of Raymond prestressed hollow precast concrete pile, showing prestressed cables under tension. (*Courtesy of Raymond Concrete Pile Co.*)

joint compound completely seals the section joints. The strength of the hardened plastic both in tension and in compression is greater than that of the concrete itself. The number of cables can be varied, depending on design requirements. After the cables have been tensioned with jacks, the ends are anchored temporarily. Cement grout is pumped into the cable holes under pressure. After the grout is hardened the anchors are removed.

These piles weigh much less than solid piles and can be handled and driven more readily. Piles of 36 and 54 in. o.d. with wall thickness of 4 to 5 in. are most common. Both diameters, to lengths of over 175 ft, have been driven in a single piece with hammers delivering up to 66,000 ft-lb. Design loads have exceeded 300 tons per pile. Piles have been successfully loaded to 500 tons. An important feature of this pile is its exceptional strength in bending. Unbraced piles are in use supporting oil-well foundations in water depths exceeding 100 ft. In the high-level bridge, in Fig. 5.10 unbraced piles project over 70 ft above the mud line to the pile cap. The pile is controlled by Raymond International, Inc.

For properties see Table V.10.

Stent-Sykes Prestressed Concrete Piles. These consist of precast reinforced-concrete sections 5 ft long, hollow except for a diaphragm at the head of each section and the solid head and toe pieces. Each piece is socketed and mortared into the piece above. The units are assembled on rollers, prestressing cables threaded through corner ducts and hair-pinned through anchor loops in the toe piece, and the cables are jacked and anchored and the sockets grouted. Lengths of 70 ft can be handled. Provision is made for screwing rods into the anchors, for use if extensions are found necessary after driving. These piles are controlled by Stent Precast Concrete, Ltd. of Great Britain.

CAST-IN-PLACE CONCRETE PILES

Cased Driven-shell Concrete Piles

This is the most widely used type of cast-in-place concrete pile and is suitable in practically all ground conditions. The shell is driven into intimate contact with the surrounding soil and remains in place to maintain driving resistance and protect the concrete filling during the placing of other piles and during the critical setting period. This type of pile is easily cut off or extended to meet variations in shell lengths. When using this type, it is not necessary to predetermine pile lengths by driving test piles, which results in savings in time and money. One of its chief advantages is that it is subject to internal inspection after it is driven.

Solid-point Steel Pipe Piles (Fig. 9.7). This type of pile is economical where either light or heavy loads have to be carried to depths in excess of 30 ft and supported by end bearing or friction. Piles 10¾ in. in diameter, in lengths over 100 ft, carrying 65 to 75 tons, are not uncommon.

This type can be used to advantage for underpinning or for foundations to support loads in basements where such foundations were not provided when the building was constructed, since it may be formed from short sections requiring little headroom. Sections as short as 2 ft have been used, allowing work to be carried out in headroom of 5 ft or less.

Diameters range from 6 to 24 in. inclusive for seamless steel pipe, and 8 in. up to and including 36 in. for spiral welded pipe, although generally a closed-end pipe would not exceed 18 in. in diameter. Shell thicknesses range from $\frac{5}{16}$ to $\frac{5}{8}$ in. for seamless steel pipe and from $\frac{7}{64}$ to $\frac{5}{8}$ in. for spiral welded pipe. Pipe piles of less than 10 in. diameters are not allowed by some building codes.

FIG. 9.6. Driving 10-in. pipe piles in short lengths in existing building for lathe foundations, using No. 3A Union double-acting steam hammer. (*Courtesy of Union Iron Works.*)

Design of closed-end steel pipe piles, particularly of considerable length, is given detailed consideration in the building laws of the City of New York because of the prevalence in that locality of heavy load concentrations to be carried considerable distances through overlying soft material to rock. More liberal loads on pipe piles are allowed by the building codes of Cleveland and Chicago.

Method of Forming. (*a*) A cast-iron point is set on the ground and a steel pipe set on the point. The pipe is jacked, jetted, or driven into the ground. (*b*) A cast-steel drive sleeve is fitted into the top of the first section of pipe, and a second section attached thereby to the first section, or sections are welded together. Driving is continued, with as many sections being added as may be required to reach the final bearing strata. (*c*) The pile is cut off to the proper elevation for capping and filled with concrete. If the load is in excess of 1,000 psi on the total cross section, the pipe is left above cutoff elevation so that bond between cap concrete and steel pipe will deliver the load carried by the steel pipe to the cap; or a steel bearing plate is grouted on top of the pile.

Seamless, spiral-welded, or lap-welded pipe is used for piling. For hard or uncertain driving, or where obstructions may be encountered, the use of seamless or spiral-welded pipe is recommended. Pipe usually is specified to meet the requirements of the latest revision of the *ASTM*

Standard Specifications for Welded and Seamless Steel Pipe Piles (Serial Designation: A252) for sizes from $8\frac{5}{8}$ to 24 in. inclusive (Grades 1, 2, or 3); the *ASTM Standard Specifications for Electric-fusion (arc)-welded Steel Pipe* (Serial Designation: A139 for sizes 8 in. to but not including 30 in. for spiral-seam Grade B pipe, or Serial Designation: A134 for sizes 30 in. and over), except that hydrostatic testing is not required; the *ASTM Standard Specifications for Spiral-welded Steel or Iron Pipe* (Serial Designation: A211) for pipe and the *ASTM Standard Specification for Low and Intermediate Tensile Strength Carbon-steel Plates of Structural Quality* (Serial Designation: A283) (Grade B) for material, for the 6-, 8-, 10-, and 12-in.-o.d. pipe sizes only; or else in the case of seamless steel pipe, the applicable parts of the *American Petroleum Institute Specifications No. 5L for Line Pipe.* The yield points of pipes purchased under Grades 1, 2, and 3 of *ASTM Specification* A252 are 30,000, 35,000, and 45,000 psi, respectively. The minimum yield point for pipe purchased under *ASTM Specifications* A139 for Grade B pipe for sizes up to but not including 30 in. is 35,000 psi. The yield point of pipe purchased under *ASTM Specifications* A134 for sizes 30 in. and over is between 24,000 and 33,000 psi, depending on the grade of steel selected. The yield point of pipe purchased under *ASTM Specifications* A211 is 25,000 psi, and under *ASTM Specifications* A283 (Grade B) is 27,000 psi. The yield point of the *API* Grade C pipe is 45,000 psi. Most pipe purchased for piling use under the *API Specifications* is Grade C, although occasionally Grade B, having a yield point of 35,000 psi, is used.

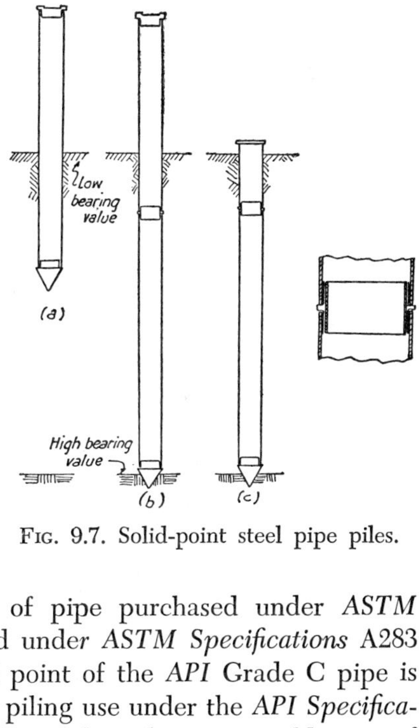

FIG. 9.7. Solid-point steel pipe piles.

The above specifications usually require a shop coating as specified by the purchaser and, if none is desired, this should be stated when purchasing. It is not customary to require cleaning and painting of pipe used for piling.

The piles should be driven with steel heads having a ring projecting 3 or 4 in. inside the pipe, with about $\frac{1}{4}$ in. clearance, and with wood cushion blocks.

Points or tips for closing the lower end are of hot pressed steel or flat steel plates, welded in place at the mill or on the job with equal success. Flat plates are less expensive.

Joints may be made between sections of pipe. Inside sleeves having a driving fit, with a flange extending between the pipe sections, are used by some engineers. By applying a small amount of bituminous cement or compound on the outside of the ring before driving, a watertight joint is

FIG. 9.8. Coupling joint in place in pipe pile. Second section about to be lowered into place, at Carbide & Carbon Co. plant, South Charleston, W.Va. (*Courtesy of National Tube Co. and Linde Air Products Co.*)

obtained. Other engineers weld the pipe, using straps to guide the sections and to provide more strength to the welded joint. If the two sections can be welded together before driving, a good way to secure alignment is to form a pyramid of three horizontal pipes, the top pipe being rotated for welding. Welding should be done by tacking opposite points first and then doing opposite sections, to avoid distortion of the alignment. Shielded electric welding is preferred. When the pipe wall is thinner than ¼ in., the welding should be done against a back-up ring*

* Such as Electric Type Split Chill Ring No. P-421, manufactured by Wedge Protectors, Inc., Cleveland, Ohio.

that enters the pipe loose, is not fastened to either, and merely acts to allow full fusion without the possibility of burning through. Another method is to obtain the backup ring from the pipe itself, by flame cutting a circle 2 in. wide from the end of the pipe, decreasing its circumference by removing a sufficient amount to allow the ring to slip inside the pipe end, and tack welding in position with about 1 in. projecting. With walls ¼ in. and heavier, a backing ring is not necessarily used. Sometimes quick-opening clamps are provided with three windows through which tack welding can be done prior to removing the clamp and welding the circular joint.

Properties are given in Tables V.5 and V.6.

A comprehensive treatment of pipe piles is contained in reference 6*f*.

Open-end Steel Pipe Piles (Fig. 9.9). This type of pile is used in the same way as the solid-point steel pipe pile for underpinning or for foundations in basements.

These piles are installed in less time than open piers, and are generally more economical than open piers more than 15 ft deep. In comparison with pneumatic caissons, the economies in cost and time are far more marked. Less underpinning of adjacent structures is required when using these piles than when installing caissons, and less loss of ground occurs. These piles may be driven tangent to adjoining walls, thereby eliminating cantilever-foundation expense.

The blown-out steel pipe is particularly useful where vibration and displacement due to the driving of any other type of pile would be liable to cause injury to adjacent structures. Blowing out may be done by air or water. Blowpipes are usually 2 to 2½ in. in diameter, but 3-in. pipes were used on a project having 18-in.-diameter piles 140 ft long.[6a] The pipes have gooseneck ends, which penetrate from a few inches to several feet into the material to be removed, the material being softened by water prior to blowing out, if necessary, in order to avoid blowing narrow holes through too stiff material. Pressures up to 100 psi are used, applied through quick-opening valves which explode the plug of soil above the gooseneck, shooting it high into the air. Bonnets may be used to deflect the ejected slurry into gutters or adjacent skips from which it may be trucked away.[82,91,66] Ample air receivers or pressure tanks and ample mains such as 6-in. diameter mains are desirable in order to avoid sudden pressure drops. For very long pipe piles, it may pay to blow out the upper portions of the piles with a shorter blowpipe, suspended possibly from a crane with a shorter boom, and then to complete the blowing out with a longer pipe hung from a longer boom. Unless the piles are short, quick-coupling joints in the blowpipe are worth while.

Where heavy loads are to be carried to rock, large piles of this type may be sunk to depths which could not be attained with any closed-end type. The blowing not only relieves end pressure against the pile, but by carrying the excavation slightly below the bottom of the pipe, side pressure and consequently friction may also be partially relieved. Piles of this type up to 36 in. in diameter and 340 ft long have been formed. The ranges of diameters and shell thicknesses obtainable in seamless-steel and spiral-welded pipes are as given under Solid-point Steel Pipe Piles.

In some cases, bedrock is overlaid by small boulders or strata of partially disintegrated rock. The open-end type of pile permits removal of this material and gives assurance that load is carried directly to bedrock.

Great depths may be reached by driving a lower section through a large upper section. On one project a 30-in.-diameter pipe was driven 80 ft and cleaned out, then a 24-in.-diameter pipe in one piece was driven through it open-ended to bedrock 200 ft from the surface.

When open-end pipe piles are used as friction piles, the last few feet of soil may be left in the pipe and compacted by a mandrel section fitted with a round plate on the bottom, which will also serve to clean the inside of the pipe.[82]

The design of open-end steel pipe piles, particularly of those of considerable length, is given much more detailed consideration and liberal treatment in the building laws of the City of New York than in most other codes, because of the prevalence in that locality of heavy load concentrations to be carried considerable distances through overlying soft material to rock.

For this type of pile, the New York City building laws allow a working load of 25 per cent of the 28-day strength (maximum 1,000 psi) on the contained concrete and 9,000 psi on the section of the steel at right angles to the load line, provided the pipe is at least ⅛ in. thick. There is no reduction for corrosion, but the steel must be protected where injurious soil conditions exist.

This type of pile may prove economical where (*a*) loads are large and concentrated; (*b*) rock can be reached within 60 ft (piles of this type have been driven well over 100 ft); (*c*) water conditions would require the use of air if caissons were sunk; (*d*) underpinning or inside foundations must be carried to rock through material which would seriously impede driving a closed-end pipe pile; (*e*) it is important to avoid jarring or soil displacement.

Method of Forming. (*a*) A section of steel pipe is driven, jetted, or jacked into the ground. (*b*) A drive sleeve is fitted to the upper end of the first section, a second section added, and driving continued. Additional sections are placed and driven until rock is reached. (The driving may be alternated with the cleaning shown in step *c*.) (*c*) The soil is removed from the inside of the pipe. This may be done by means of compressed air, water jets, coring out, or in larger pipes by the use of small orange-peel buckets. (*d*) The pipe is again driven to refusal in rock and any chips or loose material removed. (*e*) The pipe is filled with concrete and a bearing plate grouted on top.

Seamless, spiral-welded, or lap-welded pipe is used for piling. For hard or uncertain driving or where obstructions may be encountered or where the length is extreme, the use of seamless or spiral-welded pipe is recommended. The pipe is usually specified to meet the requirements of the standard specifications described under Solid-point Steel Pipe Piles.

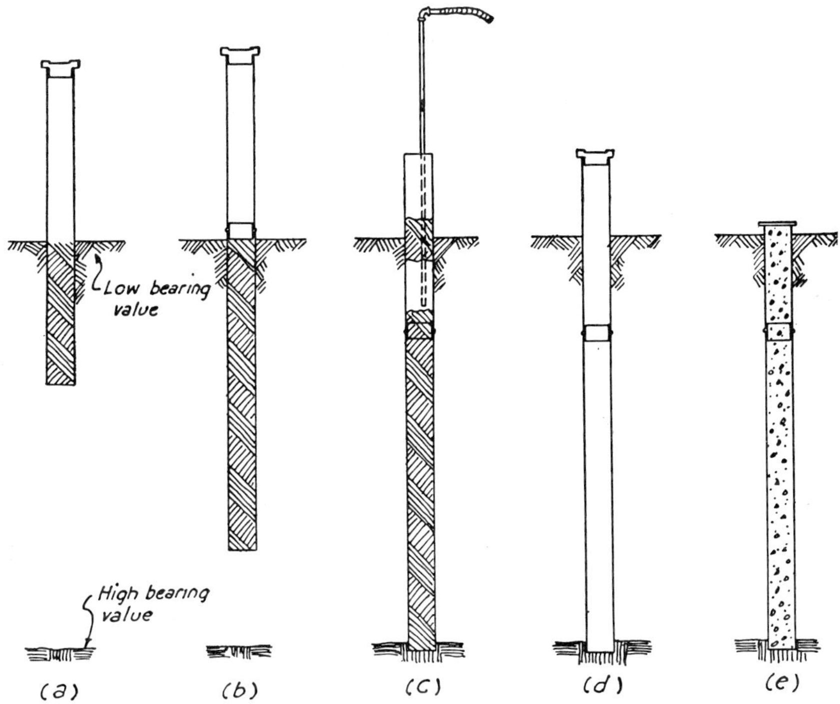

FIG. 9.9. Open-end steel pipe pile.

The piles should be driven with steel heads having a ring projecting 3 or 4 in. inside the pipe, with about ¼ in. clearance, and with wood cushion blocks.

Joints may be made as described under Solid-point Steel Pipe Piles. For exceptionally long piles, in order to reinforce the joints between sections of pipe, heavy cast-steel internal sleeves about three times the length of ordinary sleeves, with outside diameters greater than the inside diameter of the pipe and with tapers top and bottom, have been driven into place to provide rigid couplings.

Properties are given in Tables V.5 and V.6.

Swage Piles (Fig. 9.10). These piles are used to great advantage in some soils where the driving is very hard or where it is desired to leave watertight shells for some time before filling with concrete. A light steel shell is forced over a slightly conical precast concrete plug by driving with the hammer. No damage is done to the shell by considerable overdriving, and the plug seems able to withstand tremendous punishment. These piles can be driven with a crane in comparatively short lengths, thus reducing the cost of installation.

Method of Forming. (*a*) A thin steel pipe, usually 11 in. in diameter and ⅛ in. thick, is placed on a precast concrete plug, and a core which is not long enough to reach the plug is inserted. As the pipe is driven over the plug until the core reaches the plug, the pipe is swaged out by the taper of the plug, forming a watertight joint. (*b*) The pile is driven to a specified depth. The driving force is practically all exerted by the core on the plug, and the pipe is pulled down rather than driven although, since the shoulder on the core bears on the pipe, the stresses in the pipe should be approximately balanced. (*c*) The core is removed and the pipe left open until it is desired to fill it. (*d*) The pipe is filled with concrete.

This type of pile is furnished and driven by the Western Foundation Corp.

Union Metal Monotube Fluted Steel Shells. Fluted steel shells are suitable for a wide variety of soil conditions, from end-bearing to friction-load-carrying soils. The shells are of various gages from No. 3 to No. 11, providing rigidity while being of comparatively light weight to handle. They are tight against inflow of water.

The shells may be inspected after driving. The stiffness of the shell against crushing from driving adjacent piles is good. Owing to cold rolling, the yield point of the metal is high.

The shells may be driven with hammers of comparable size to those used for wood piles.

The pile shells are easy to cut off to proper elevation before filling to the proper level or after.

The shell, minus a corrosion allowance, may be considered as reinforcing, thus providing lateral strength as well as vertical. Corrosion-resistant alloys are available.

Method of Forming. (*a*) The shell is driven. (*b*) The interior of the shell is inspected. (*c*) The shell is filled with concrete and the excess shell cut off.

This type of pile is furnished by the Union Metal Mfg. Co., but must be driven by other companies. Properties are given in Table V.3.

Raymond Standard Concrete Piles (Fig. 9.11). The Raymond standard pile is used primarily as a friction pile, since its uniform heavy taper of 1 in. per 2½ ft usually results in shorter piles for equal driving resistance, or higher driving resistance for equal length, than piles of lesser or no taper. Whether this factor is useful or detrimental depends upon the character of the soil stratum providing the resistance to driving: it may have permanent load-carrying ability, or it may provide only temporary resistance to driving and have undue settlement characteristics under permanent load.

Shell sections are made of sheet steel varying fram 18- to 24-gage, with the heavier gages used for the lower sections, where greater strength is

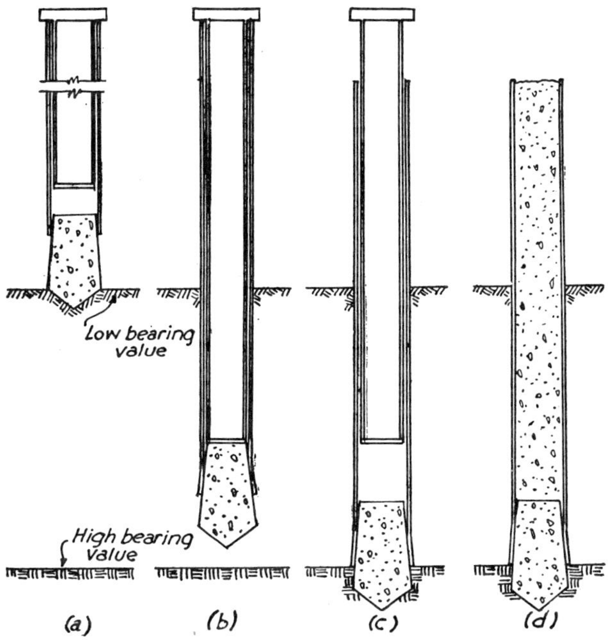

FIG. 9.10. Swage pile.

usually required. Selection of proper gages is a responsibility of the Raymond Concrete Pile Company. Shells are reinforced internally with spirally wound wire on 3-in. pitch. Normal tip diameter is 8 in., but a 10.8-in. tip is available. Maximum pile length is 37.5 ft when using an 8-in. tip.

Shells are driven with a mechanically expansible core or mandrel capable of withstanding high driving pressure. After the pile is driven, the mandrel is collapsed and withdrawn. The shell is inspected internally after driving and before placing concrete. Reinforcement is not required when the pile is cut off and capped at or below ground surface except where uplift or lateral forces may occur. Normal design loads range up to 50 tons per pile, although design loads as high as 80 tons have been used. See Table V.1 for details.

Method of Forming. (*a*) The shell is placed over a steel mandrel or core which extends to the bottom of the shell. (*b*) Mandrel and shells are driven to required resistance. (*c*) The mandrel is withdrawn, leaving the shells in place. (*d*) The shell is inspected internally. (*e*) The shell is concreted to cutoff elevation, and excess shell removed.

These piles are manufactured and installed by Raymond Concrete Pile Co., a division of Raymond International, Inc.

Raymond Step-taper Concrete Piles (Fig. 9.11). This type of mandrel-driven pile is used either as an end-bearing or friction pile and can be driven in any type of soil.

The shell sections are made of sheet steel in gages from 12 to 20, the heavier gages being used for the lower sections. The purpose of the shell is to maintain driving resistance and to protect and act as a form for the concrete filling. Step-taper shells are helically corrugated to provide strength against collapsing pressures. The Raymond Concrete Pile Co. assumes responsibility for providing proper gages to resist collapsing pressures. The usual section length is 8 ft, but 4, 12, 16, and 24 ft are available. Joints between sections are screw-connected, and the bottom shell is closed with a flat plate welded to the boot ring. Where serious pile heave is to be expected, a special slip-type joint for one of the lower shell sections of each pile can be provided. This permits the upper shell sections to heave with the ground without disturbing the seating of the pile point.

The shells are driven with a rigid internal steel mandrel or core which is stepped to conform with the shell sections used. A shoulder on the pile core bears on each shell collar ring and the boot ring. The heavy rugged core provides a high degree of penetration and efficiently transmits hammer energy to the bearing strata. The energy losses during driving due to elastic compression, vibration, and deformation of the pile are practically eliminated, although the inertia loss is greater; the heavy member can, however, sometimes be driven where the more elastic types of piles cannot.

The pile diameter increases in steps at the rate of 1 in. for each successive shell section. Normal point diameters used are 8, 9, 10, and 11 in., although 12, 13, and 14 in. are available. Maximum length available at the date of writing is about 120 ft, using an 8-in. tip. Longer all-shell piles could be installed provided the necessary heavy driving rig were available. Longer high-capacity piles can readily be installed using the Raymond pipe step-taper shell and pipe composite pile described under Composite Driven-shell and Pipe Piles.

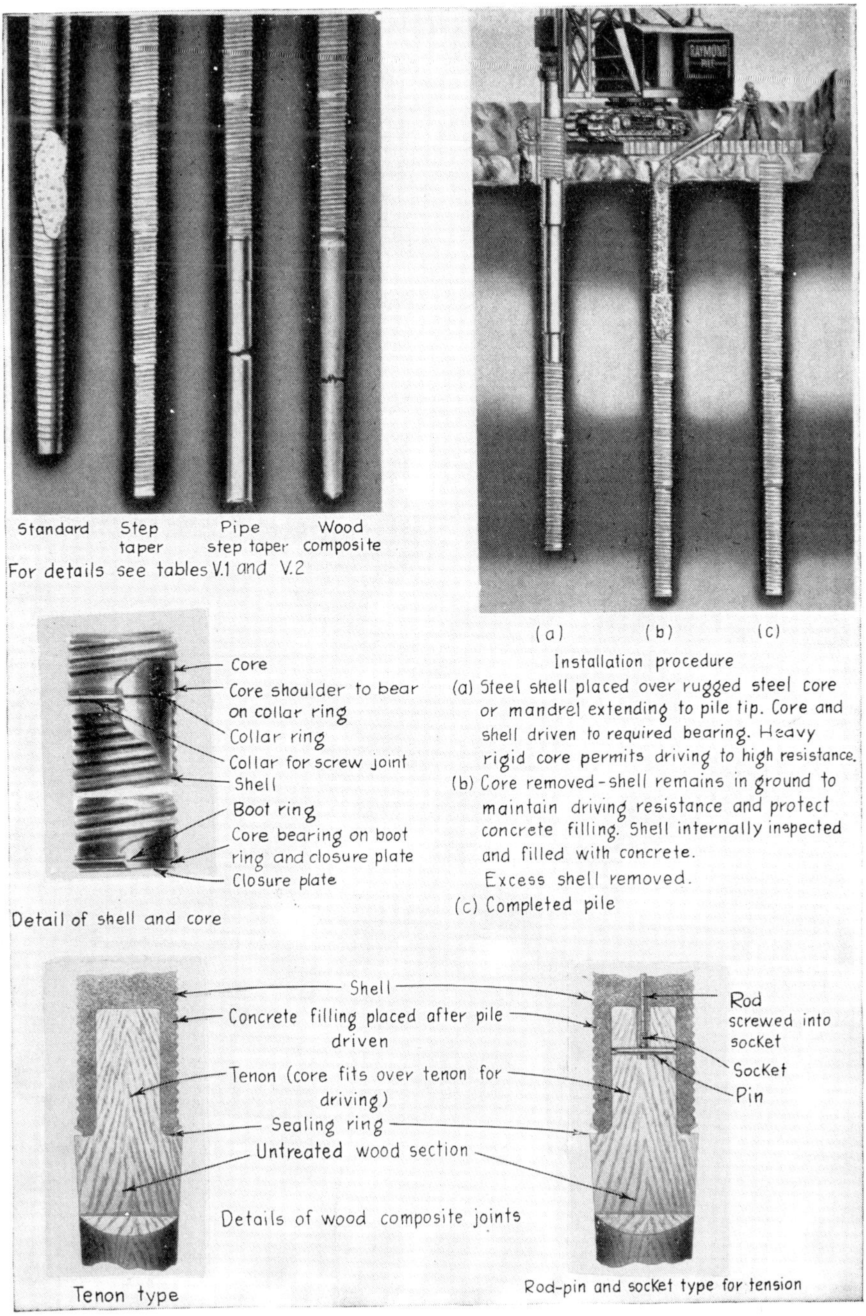

FIG. 9.11. Raymond concrete piles.

Internal reinforcement is not required for normal structure loading. Reinforcing can be installed for unusual conditions such as uplift, high lateral loads (where batter piles are not used), or unsupported lengths that will extend through air, water, or very fluid soil.

The step-taper pile has all the basic advantages of the driven-shell-type cast-in-place concrete pile, such as on-the-job length flexibility, internal inspection after being driven, and a steel shell left in place to maintain driving resistance and protect a fresh concrete filling. Tip elevations should be checked to detect possible accordioning. The pile is capable of being driven to very high resistance. Design loads up to 100 tons per pile have been used. See Table V.2 for details.

Method of Forming. The method of forming consists of the same steps as described under Raymond Standard Concrete Piles.

These piles are manufactured and installed by the Raymond Concrete Pile Co., a division of Raymond International, Inc.

Cementation Tapered Piles. These European piles are similar to Raymond standard piles and are controlled by The Cementation Co., Ltd.

Cobi Pneumatic Mandrel Piles.[5ae,5ag] This type of pile is used either in end-bearing or in friction-load-carrying soils of any kind. The piles are poured-in-place concrete and use thin-gage corrugated steel shells of uniform diameter such as Armco Hel-Cor; a Cobi mandrel is used in driving them. The mandrel is formed of four segments of curved checkered steel plate spaced by internal members about a small central expansible core. In its collapsed state the mandrel is ¾ in. smaller than the shell diameter and when inflated can expand to a slightly larger diameter than that of the shell, thus ensuring a tight grip. The shape of the mandrel conforms to the inside surface of the boot to form a solid driving point. The stiffness of the mandrel resists possible tendencies to curvature during driving.

The shells can be inspected before filling. Crushing of shells already placed, or filled, by driving adjacent piles should be watched. Reinforcement may be placed in the shells if needed.

For 12-in.-diameter shells, the approximate weight of the mandrel to be used in driving computations is 100 lb per lin ft. Standard mandrel lengths are 32, 45, and 60 ft.

Pipe extensions may be driven first, to obtain piles of long lengths, using special drive sleeve connectors.

Since the corrugated driven shell remains in contact with the earth at all times, full friction is always present.

Method of Forming. (*a*) The mandrel is attached to the hammer, is hoisted to the top of the leads and lowered in its deflated state into the shell. (*b*) The mandrel is expanded using nitrogen or air at a pressure of approximately 125 psi. (*c*) The mandrel and shell are driven as a unit to the desired depth. (*d*) The valve is opened and the mandrel collapsed and withdrawn from the shell. (*e*) The shell is filled with concrete.

The piles are controlled by the Pneumatic Pile Co. of New York.

Hercules Mandrel Piles. These are similar to Cobi piles, except that the straight-sided mandrel expands mechanically. They are driven by the Hercules Concrete Pile Co.

Caudill Piles.[*5bi] This is a method of pulling down light, thin-walled steel pipe by driving on the pile shoe through a heavy pipe mandrel. It avoids some of the energy losses that take place when driving on the pipe and saves expense by reducing the weight of the piles left in place.

Method of Installation. (*a*) A Caudill boot, consisting of a short circular collar with an interior enlargement at its top and slotted vertically, is welded onto a flat bottom plate of greater diameter than the thin pipe, then placed on the ground. (*b*) The thin pipe is set on the boot. (*c*) The mandrel, which is tapered at the bottom, is driven inside the collar which forces the enlargement at the top of the collar to expand against the thin pipe and deform it to act as a latch. (*d*) The mandrel is driven, pulling down the pipe. (*e*) The mandrel is pulled and the pipe filled with concrete.

This method has also been used with two sections of pipe, the upper being 12 in. OD and 0.125 in. thick, and the lower 11.269 in. OD and 0.1345 in. thick. A steel plate band fills the space between the pipes, which are lapped a few inches. The band is placed flush top with the top of the lower section to form a driving step to assist in placing the pile. The lower section of pipe is the same length as the mandrel below the shoulder, 30 ft.

This pile has been driven by the Guild Construction Company, Inc., using Armco foundation pipe.

BSP Base-driven Cased Piles (Fig. 9.12). These are concrete piles with permanent steel tubular casings. The casing is made from steel strip plate welded helically. Standard sizes are from 10 to 28 in. in diameter and from No. 11 gage (0.116 in.) to ⅜ in. thick. Other diameters and thicknesses are supplied on order. The length can easily be cut off or increased. A bottom shoe allows the concrete to be placed in the dry, and the casing can be inspected after driving. Driving is done on a plug of concrete in the bottom of the pile, and the shell is drawn down. The light shells are easily handled and can be driven with relatively light hammers. Properties are shown in Table V.6.

Method of Installation. (*a*) The first length of steel casing with shoe and concrete plug is set in the leaders. (*b*) The bottom length of casing

* Patent pending for Howard F. Caudill.

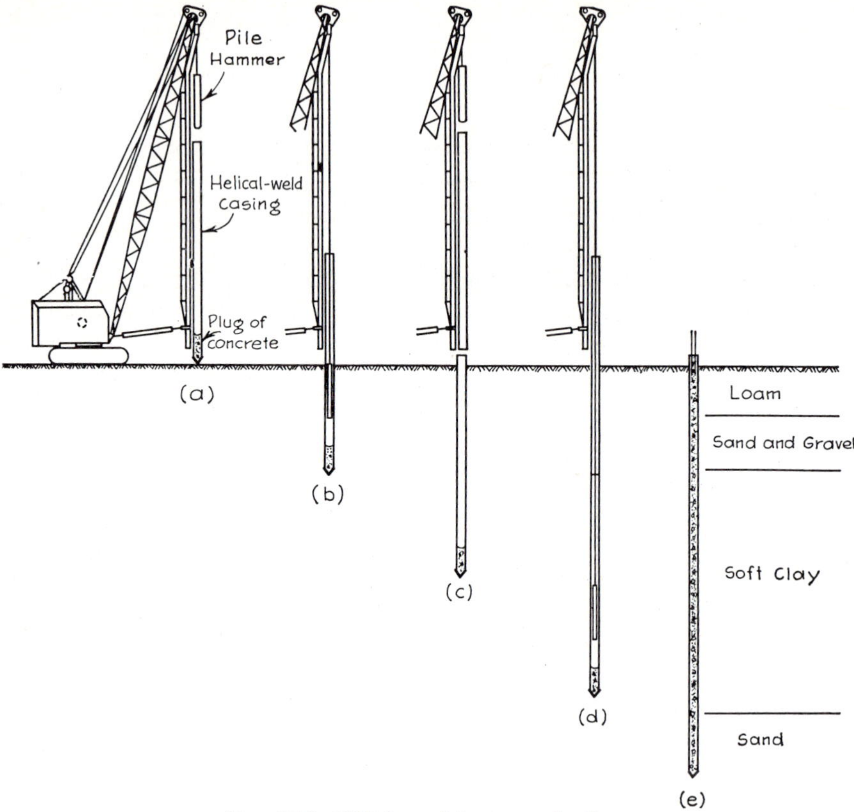

FIG. 9.12. BSP base-driven cased piles.

is driven by dropping a cylindrical ram of an internal drop hammer. (*c*) The top section of casing is placed in the leaders to butt-weld to the bottom section. (*d*) The complete casing is driven to the desired set. (*e*) The casing is filled with concrete.

These piles are installed by the British Steel Piling Co., Ltd.

Cased Dropped-in-shell Concrete Piles

This is a variation of the uncased cast-in-place concrete piles whereby a permanent light shell is placed inside the outer casing after it has been driven. The outer casing is then withdrawn. This variation is used when it is impractical to install properly uncased concrete piles or where an interior shell is required to protect the concrete filling. The annular space left when the outer casing is withdrawn should be filled to reestablish lateral support to the inner shell. If lateral support is not reestablished, column action should be considered.

Piles having a driven outer casing that is pulled after a shell has been set inside are not considered with favor by some engineers, because when the outer casing is pulled, the ground pressure is relieved and the carrying capacity that was indicated by driving is no longer the same.

Dropped-in-shell Concrete Piles (Fig. 9.13). The cased cast-in-place pile should be used where the soil is too soft and soupy to permit the use of an uncased pile or where the soil is difficult to compress and would deform an uncased pile.

Cased piles can be made in various diameters from 12 to 20 in.; the length limit is about 75 ft.

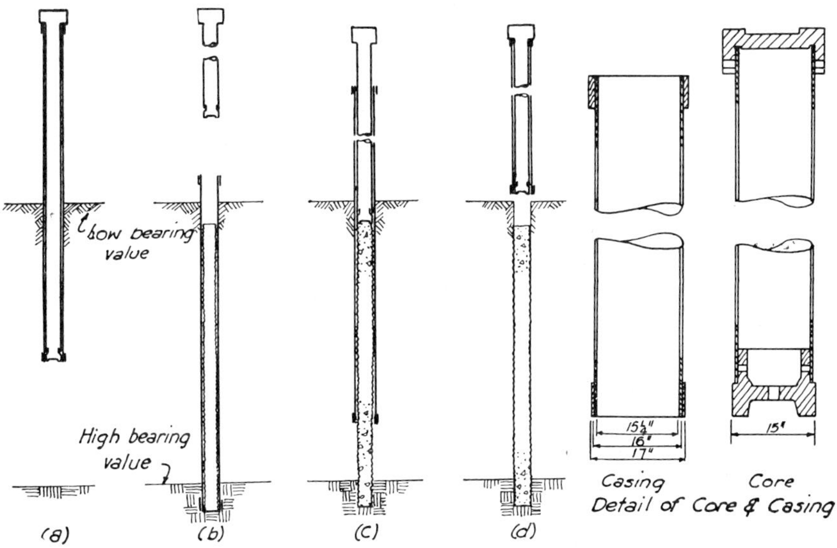

FIG. 9.13. Dropped-in-shell concrete pile.

Method of Forming. (*a*) The pile apparatus consists of a casing and a core. The bottom of the core is of such a size as to close the casing completely when the core is inserted in it. The core and casing together are driven into the ground until the required penetration is obtained. (*b*) The core is removed and a pile shell inserted in the casing. This shell serves the purpose of a container for the concrete prior to setting. It will be of light material and either corrugated or plain, riveted, or lock seam. (*c*) The core is replaced in the casing in contact with the top of the shell. The hammer is dogged so that the position of the core becomes fixed. The lower collar is drawn toward the upper collar. This causes the withdrawal of the casing, while the shell is so held that the newly formed pile cannot be disturbed by movement of the casing.

Dropped-in-shell Concrete Piles with Compressed Base Section (Fig. 9.15). This type of pile is adapted to use where upper strata of soil through which the pile must penetrate to reach bearing strata are too soft to permit the placing of concrete under such pressure as is required for the formation of the compressed-concrete pile shaft.

FIG. 9.14. Lifting corrugated shell for cased concrete pile, using standard MacArthur pile driver, at connection of Southern Railway, Cincinnati, Ohio. Note tilt rig suitable for driving either vertical or batter piles. (*Courtesy of MacArthur Concrete Pile Corp.*)

The lower section of high-compressed concrete develops the full value of the bearing strata.

Method of Forming. (*a*) The pile apparatus consists of a casing and a core. The bottom of the core is of a size to close the casing completely when the core is inserted in it. The core and casing are together driven into the ground until the required penetration is obtained. (*b*) A charge of concrete to form the lower section of the pile is deposited in the casing, and the core is replaced in the casing in contact with the concrete. (*c*) The casing is then withdrawn over the core, leaving the weight of the hammer and plunger on the concrete, in a manner which will prevent breaking apart of the concrete shaft and ensure forming a concrete shaft of a diameter at least equal to the outside diameter of the casing. A method which, it is claimed, precludes any possibility of lifting apart of the concrete column when removing the casing, consists of pulling the casing by drawing together the collar on the casing head and the crosshead over the hammer top; this results in a positive compression in the concrete column at all times, obviating breaking or arching of the

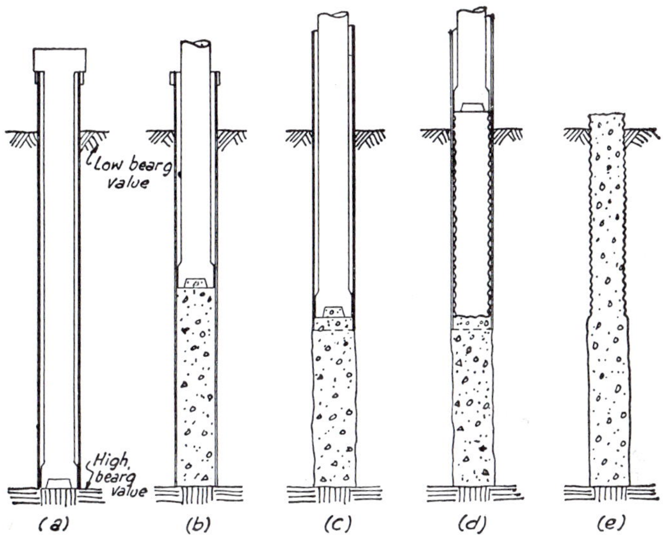

FIG. 9.15. Dropped-in-shell concrete pile with compressed base section.

concrete in the casing. This latter method is used by the Western Foundation Corp. (*d*) The core is removed from the casing and a corrugated shell is placed in the casing. The core is replaced in contact with the corrugated shell and is locked in this position. (*e*) The casing is withdrawn, leaving the corrugated shell in the ground. The corrugated shell is filled with concrete either before or after the withdrawal of the drive casing.

Dropped-in-shell Concrete Piles with Compressed Base Section, Reinforced Shaft. This pile is the same as the one just described, except that reinforcing has been added to the shell-enclosed section of the pile shaft. It is used when (*a*) the side support of the soil is so slight that the pile must be figured as a column, and (*b*) the pile must be designed to meet lateral forces arising from eccentric loading or from a possible displacement of soil.

Method of Forming. The method of forming is the same as for the pile just described, except that the corrugated shell contains a reinforcing cage consisting of spiral and vertical rods proportioned in accordance with anticipated loads. Means are supplied to assure that rods are maintained in their proper position and spacing within the shell.

Button-bottom Dropped-in-shell Concrete Piles (Fig. 9.18). These piles are used in locations where an increase in end-bearing area is desired. The button forms an enlarged hole in the soil during driving, which reduces side friction at least temporarily. Loading tests and redriving will show the amount of regain in friction for permanent loads.[5w] These piles have been driven through 16 ft of dumped rock fill with little delay.[5w] Lengths up to 76 ft may be driven and loads up to 50 tons carried. Longer lengths may be driven by adding a socketed top section of casing after the top of the original casing has reached ground level; in this case cables are attached to devices on top of the lower section and are pulled down into the ground; then they are used to lift the lower section with the upper one riding upon it.[5z] Lengths have been increased 25 ft in this way.

FIG. 9.16. Button-bottom pile base. (*Courtesy of Western Foundation Corp.*)

FIG. 9.17. Button-bottom pile ready to drive with No. 1 Vulcan single-acting steam hammer in suspended leads from caterpillar tractor. (*Courtesy of Western Foundation Corp.*)

Method of Forming. (*a*) A steel pipe, usually 14-in. in diameter with ½-in.-thick walls, with a reinforced base of cast steel, is set upon a concrete button which has a diameter about 1 in. larger than the pipe. (*b*) The pipe and button are driven to a specified elevation. (*c*) A corrugated shell is inserted in the casing, resting on the button. A steel plate with a bolt hole in it is welded on the bottom of the casing before placing it. This hole fits over a central bolt in the button bottom, and a

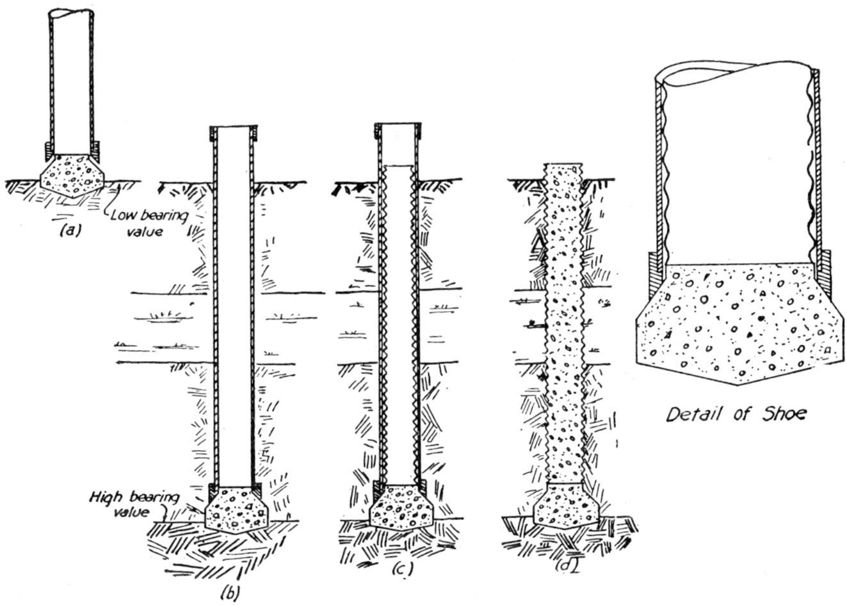

FIG. 9.18. Button-bottom dropped-in-shell concrete pile.

nut is held in a long socket wrench and placed to hold the casing from rising due to heave or flotation. (*d*) The casing is then withdrawn, leaving the button in place, and the shell is filled with concrete, using reinforcement if necessary. If uplift is severe, the shell can be partly or entirely filled with concrete before the casing is lifted.

This type of pile is furnished and driven by the Western Foundation Corp.

Dropped-in-shell Pedestaled Concrete Piles (Fig. 9.19). This type of pile is suitable where a soil, too soft to permit use of a compressed pile, overlies a stratum which may be developed into an adequate bearing stratum, provided the load may be spread somewhat. The pedestal serves to distribute the load over a larger area than would otherwise be the case.

This type may be useful on sloping rock surfaces, since the concrete may be driven out so that the pedestal grips the rock surface, thus avoiding eccentric stresses (that might be developed in a type of pile where one edge rests on the rock surface) and resulting in less possibility of slipping on steep slopes. If this type of pile is driven in an artesian stratum, a precast concrete wedge plug is inserted in the bottom of the casing before driving. This will keep out the water while the initial

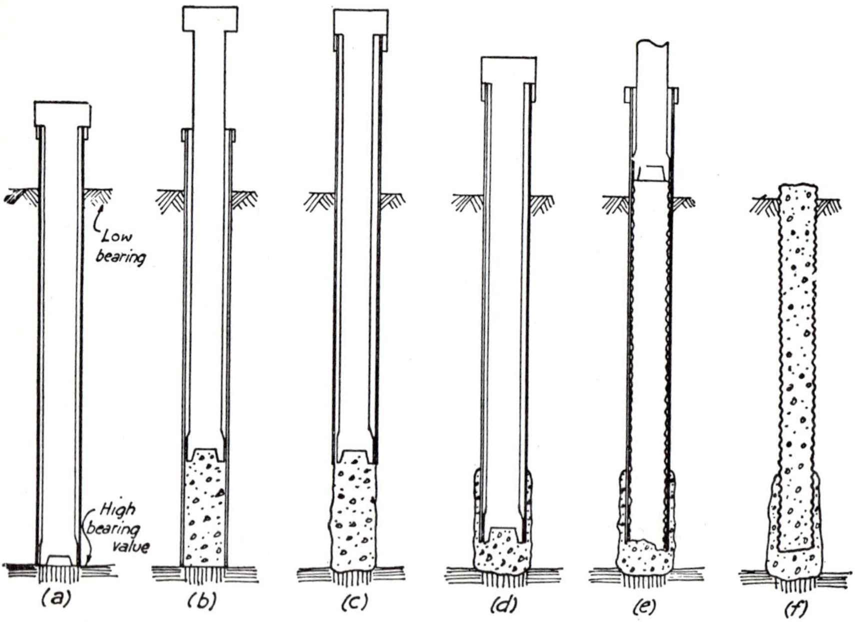

FIG. 9.19. Dropped-in-shell pedestaled concrete pile.

batch of concrete is placed, and will be driven out when the pedestal is formed. The pedestal concrete generally keeps the water out of the casing, but if it should not, an additional charge is placed and the casing redriven through it, or the pile may have to be redriven.

The dynamic resistance to driving, assumed to be a measure of the load capacity in cohesionless soils, is computed from the set obtained when redriving through the freshly deposited concrete of the pedestal.

It is impossible to "neck" the concrete in the shaft of this type of pile when lifting the casing, with a possible mud layer forming, as might happen occasionally with some uncased types, because the casing and core are redriven through the concrete of the pedestal and the shell placed down in the pedestal.

Method of Forming. (*a*) The pile apparatus consists of a casing and a core. The bottom of the core is of a size to close the casing completely when the core is inserted in it. The core and casing together are driven into the ground until required penetration is obtained. (*b*) The core is removed and a quantity of concrete is deposited in the casing. (*c*) The core is replaced in the casing in contact with the concrete and the casing drawn up over the core. (*d*) The core and casing together are again driven down through the previously deposited concrete. (*e*) The core is removed and a corrugated shell placed in the casing. The core is replaced in the casing in contact with the top of the shell. The hammer is dogged so that the position of the core becomes fixed. (*f*) The casing is withdrawn and the shell filled with concrete, or the shell may be filled before the casing is withdrawn.

Cased piles are made in diameters from 12 to 20 in., and the length limit is about 75 ft.

If occasional piles of this type are needed longer than the limiting length, a pipe section may be driven through the fresh pedestal concrete, which usually seals the junction around the bottom of the casing against water, or a composite projectile-type pile may be driven.

Dropped-in-shell Pedestaled Reinforced-concrete Piles. This type is the same as the one just described, with the addition of a cage of reinforcing which is placed in the shell before pouring concrete.

Uncased Concrete Piles

Cast-in-place uncased concrete piles may be used where it is certain that neither soil nor water will fall into the hole, or squeeze into and reduce the size of the hole left after withdrawing a driven mandrel or shell before pouring concrete, and where driving of adjacent piles will not damage the green concrete. Such piles need no storage space; they do not require cutting off excess lengths or building up short lengths, do not require special handling equipment, and are not subject to damage from handling. The concrete is not liable to damage from driving, except that caused by driving adjacent piles.

Some engineers have an aversion to the use of uncased concrete piles because of the possibility of piles being cut off or necked by ground pressure in the middle, the chance of soil falling into the hole, and uncertainty as to size or eccentricity of pedestal if used.

MacArthur Compressed-concrete Piles (Fig. 9.20). This type can be made in diameters from 14 to 24 in. and is satisfactory under any soil conditions which permit of the concrete being placed under pressure. If the soil is of a soupy nature so that it has little or no resistance to the flow of the concrete, a casing should be used. A flowing quicksand stratum contained by a drier stratum above it will give no trouble.

The length limit on piles of this type is usually 60 ft.

Method of Forming. (*a*) The pile apparatus consists of a casing and a core. The bottom of the core is of a size to completely close the casing when the core is inserted in it. The core and casing together are driven into the ground until the required penetration is obtained. (*b*) The core is removed and the casing filled with a concrete of coarse aggregate. (*c*) The core is replaced in the casing and in contact with the concrete. The casing is then withdrawn, leaving the weight of the hammer and plunger on the concrete.

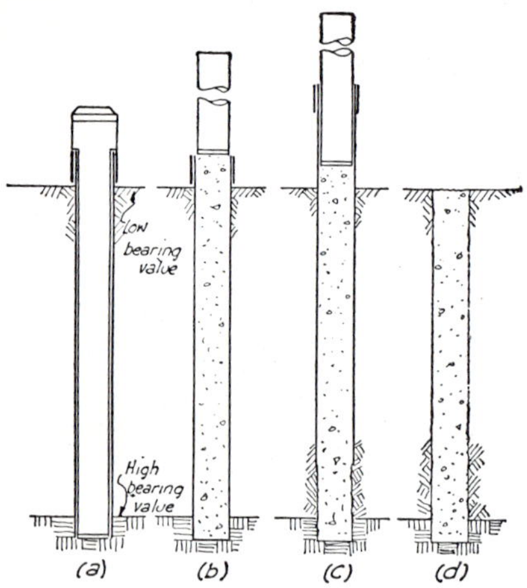

FIG. 9.20. Compressed concrete pile.

This type of pile, often referred to as a MacArthur straight-shaft pile, is frequently driven by the MacArthur Concrete Pile Corp. It is claimed that the weight of the hammer and core, approximately 7 tons, results in a solid dense shaft of concrete. A method that, it is claimed, precludes any possibility of lifting apart of the concrete column when removing the casing, consists of pulling the casing by drawing together the collar on the casing head and the crosshead over the hammer top; this results in a positive compression in the concrete column at all times, obviating breaking or arching of the concrete in the casing. This latter method is controlled by the Western Foundation Corp.

Compressed-concrete Piles with Mushroom Base (Fig. 9.21). This type of pile is of particular advantage where the bearing stratum, of limited thickness only, can be reached within economical depths. The mushroom-base pile gives the effect of a spread footing on this comparatively thin bearing stratum. This type of pile is of advantage where the bearing stratum is a rough or sharply inclined rock surface. The mushroom base ensures a firm grip on the rock. Very short piles of this type may be used to tighten up a surface stratum which is underlaid by poorer material.

The dynamic resistance to driving, assumed to be a measure of the load capacity in cohesionless soils, is computed from the set obtained when redriving through the freshly deposited concrete of the pedestal.

Method of Forming. (*a*) The pile apparatus consists of a casing and a core. The bottom of the core is of a size to close the casing completely when the core is inserted in it. The core and casing together are driven into the ground until the required penetration is obtained. (*b*) A charge of concrete is deposited in the casing. (*c*) The core is replaced in the casing in contact with the concrete and the casing drawn up to meet the core. (*d*) The core and casing are redriven through the deposited concrete. (*e*) The core is removed and the casing filled with

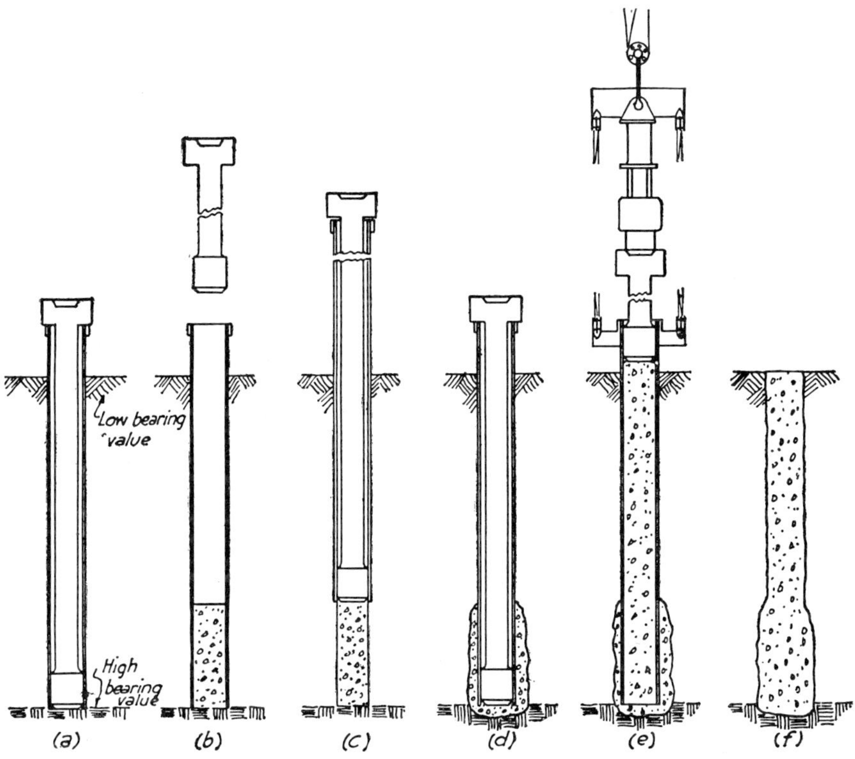

FIG. 9.21. Compressed concrete pile with mushroom base.

concrete. The core is replaced in contact with the concrete. (*f*) The casing is removed by drawing the casing over the core in a manner which will ensure a continuous concrete shaft by positive pressure.

The shape of the pedestal may be varied to meet soil conditions. In step *c*, instead of drawing the casing back until it is in contact with the core head, it may be drawn up only a few inches. The pedestal can then be formed by operating the hammer and forcing the concrete out, much as a bubble is blown. A pedestal of this type is usually preferable where the bearing is on rock or heavy gravel. The more elongated type is to be preferred where the bearing stratum is penetrated to some depth as in the case of a stiff-clay, a sand, or a loose-gravel bearing stratum.

MacArthur Compressed-concrete Pedestal Piles (Fig. 9.22). This pile is of particular advantage where the bearing stratum, of limited thickness only, can be reached within economical depths. The pedestal gives the effect of a spread footing on this comparatively thin bearing stratum.

This type of pile is also of advantage where the bearing stratum is a rough or sharply inclined rock surface. The pedestal ensures a firm grip

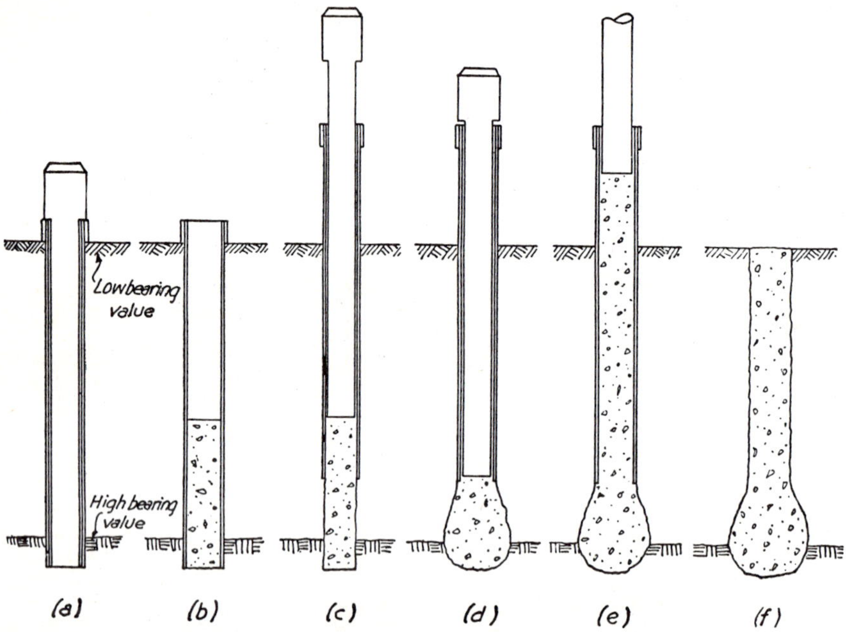

FIG. 9.22. Compressed concrete pedestal pile.

on the rock. Very short piles of this type may also be used to tighten up a surface stratum which is underlaid by poorer material.

Method of Forming. (*a*) The pile apparatus consists of a casing and a core, the lower end of the core being flush with the bottom of the casing, and the end made flat. A pressed-steel shoe is used where necessary to prevent the inflow of water in water-bearing soils or the entry of matter which might otherwise be forced into the casing through the back pressure of the soil when the core is removed. The core and casing are driven into the ground until the required penetration is obtained. (*b*) The core is removed and a charge of concrete is dropped to the bottom of the casing. (*c*) The core is replaced in the casing and the casing is pulled up 18 in. to 4 ft with the pressure of the core and hammer remaining on the concrete (approximately 7 tons). (*d*) The charge of concrete is rammed out. (*e*) The core is removed and the casing filled with concrete. The core is replaced in contact with the concrete. (*f*) The casing is steadily withdrawn while the concrete is under the pressure from the weight of the hammer and core.

FIG. 9.23. Three pedestal-type concrete piles, excavated, showing typical pedestal formations in difficult strata. Approximately 2,800 piles of this type driven for Boston & Maine Railroad. (*Courtesy of MacArthur Concrete Pile Co.*)

This type of pile, often referred to as a MacArthur pedestal pile, is frequently driven by the MacArthur Concrete Pile Corp. The largest diameter generally driven by them is 16 in. It is also driven by the Western Foundation Corp., the usual diameter being 14 in.

Simplex Concrete Piles (Fig. 9.24). This type of pile may be driven through soft or hard soils. The head of concrete forces it out against the soil as the pipe is withdrawn. However, the soil must be sufficiently firm to form a good mold for the green concrete after the casing is withdrawn, or else an inner casing of slightly smaller diameter than the interior of the shell must be inserted before pouring the concrete. The usual point is solid and is left in place, as shown; but sometimes, if the soil is firm enough to stand, an alligator point is used that is hinged to the shell and upon opening up is withdrawn with it. However, it would seem difficult to be certain that no soil would fall into the hole at the bottom. A reinforcing cage may be inserted. Occasionally piles are inserted through

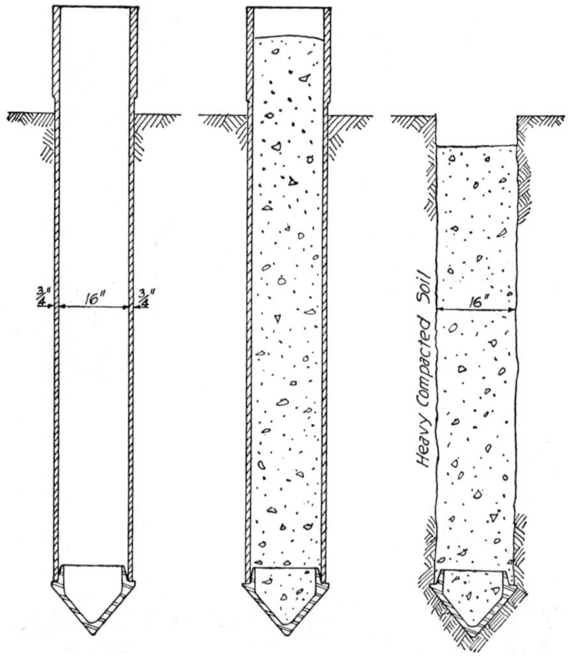

FIG. 9.24. Simplex pile.

the shell, instead of pouring concrete. The method illustrated is standard.

Working loads on British Simplex piles have been recommended by the manufacturer as 40 to 45 long tons for 16 in. diameter, and 55 to 60 long tons for 18 in. diameter, when a set of four blows with a 2-ton hammer falling 4 ft, or equivalent, produces a set of 5/8 in.

Tamped Simplex piles are similar to the standard type, but in addition have a patented loose sleeve fitted around the lower end of the driving tube and a heavy spring collar at the top. After concreting, the driving cap is placed on the tube and withdrawing bonds attached to the spring head commence to pull the pile. Every 2 in. or less, the hammer strikes the tube, thus tamping the concrete in the bottom sleeve.

Bulb Simplex piles may be formed with either the standard or tamped types. A long heavy mandrel working in the tube is used during withdrawal of the tube to compress the concrete and drive it out of the bottom to form a bulb. The shoe can remain in place. Where driving resistance is inadequate, the bulb may be enlarged to provide good support.

Simplex piles are in use both in Great Britain and the United States.

Express Piles. These piles are formed by driving an 18-in.-diameter steel tube to the required set. The lower end is covered by a solid point or sheet metal. A reinforcing cage is placed and the tube filled with concrete. As the concrete is filled in, a driving member and the hammer are used to tamp out the concrete below the tube. The diameter of the pile can be increased in this manner, and a bulb can be formed at the foot. Additional concrete is placed, the tube raised, and the operation repeated, thus producing a pile of varying diameter according to the strata.

These piles are controlled by The Cementation Co., Ltd.

Franki Piles (Fig. 9.25). These piles are of particular advantage where a bearing stratum, of limited thickness only, can be reached within economical depths. The mushroom base gives the effect of a spread footing on this comparatively thin bearing stratum. If used as friction piles, the enlargements aid in distributing the load to the soil. This type of pile is best suited to granular soils.

Franki piles are formed by ramming concrete in a heavy 20½ in.-o.d. removable pipe shell. After the enlarged base has been formed, a reinforcing cage may be installed if desired, having four to six vertical bars set in a 16-in. diameter with ¼-in.-diameter spiral at 6-in. spacing. The hammer goes inside the cage of reinforcement. The base often has a diameter of 3 ft or more.

Franki pile concrete shaft diameters vary from 22 to 24 in., according to the size of the tube used and compressibility of the soil. The piles may be driven on a batter up to 25 deg. The sponsors recommend working loads of 80 to 100 long tons for standard piles, and 45 to 65 tons for lightweight piles.

Method of Forming. (*a*) Set the tube on the ground and place a charge of dry concrete. (*b*) With a drop hammer, drive on the concrete. This forms a dense plug that penetrates the ground and drags the tube down with it. (*c*) When the tube has reached the desired depth, the tube is held in place by cables and the hammer applied to the concrete, forcing it down and outward. (*d*) The shaft is formed by introducing successive charges of concrete, ramming each in turn, while gradually withdrawing the casing.

This type of pile is used extensively in Europe, the British Isles, and Canada, and is installed by the Franki Compressed Pile Co., Ltd., and its representatives.

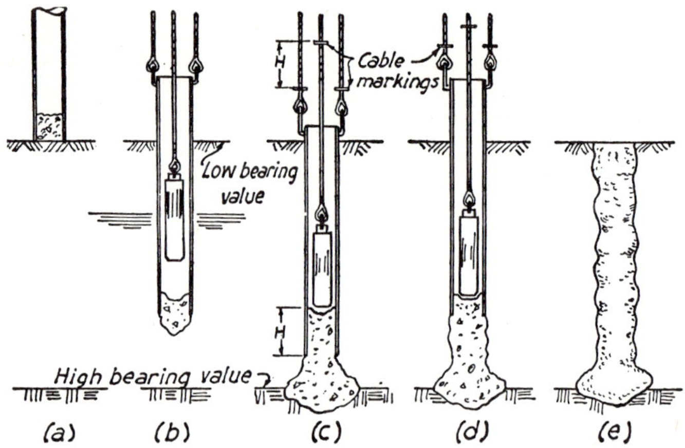

FIG. 9.25. Franki pile.

Vibro Piles (Fig. 9.26). Vibro piles are formed by driving a steel tube and shoe, filling with concrete, and extracting the tube, using upward extracting and downward tamping blows alternately, forcing the concrete against the soil to prevent necking of the concrete. A Vibro hammer with extracting links is used.

These piles are suitable when the condition of the ground is not so soft that it has little resistance to the flow of concrete. They are particularly suitable when ground of low bearing value but adequate consistency to resist the flow is encountered for a considerable depth. The base has an enlarged rim that diminishes the frictional resistance to driving. Reinforcing steel is used, consisting, for 13½-in. piles, of four ½- to ¾-in. vertical bars with $\frac{3}{16}$-in. stirrups at 6-in. pitch and, for 17-in. piles, of up to six vertical 1-in. bars with ¼-in. stirrups at 6-in. pitch. Piles of 19-in. diameter are also used. Piles with a finished diameter of 13½ in. are intended to carry loads up to 40 long tons, with 17-in. diameter up to 60 long tons, and with 19-in. diameter up to 75 long tons. These loads are based on a stress of 600 psi in the concrete. Greater loads can be carried at higher stresses if the ground provides adequate resistance.

Vibro tubes up to 85 ft long are available. If a sufficiently firm bearing stratum is not found at this depth, a second tube can be added by use of a muff coupling. Rakes of 1:3 with a 40-ft frame, 1:5 with a 50-ft frame, and 1:6 with a 64-ft frame are possible.

Method of Forming. (*a*) A steel tube with a cast-iron shoe of slightly larger dimension is driven to the required penetration. (*b*) The tube is filled with concrete and connected to the hammer by extracting links. The extraction of the tube and foundation of the pile are effected by the hammer, each upward blow being followed by a downward tamping blow. During the upward blow, the tube rises and concrete moves out under the rim. During the downward blow the friction of the tube on the concrete tamps it. (*c*) The finished concrete shaft presents a corrugated surface.

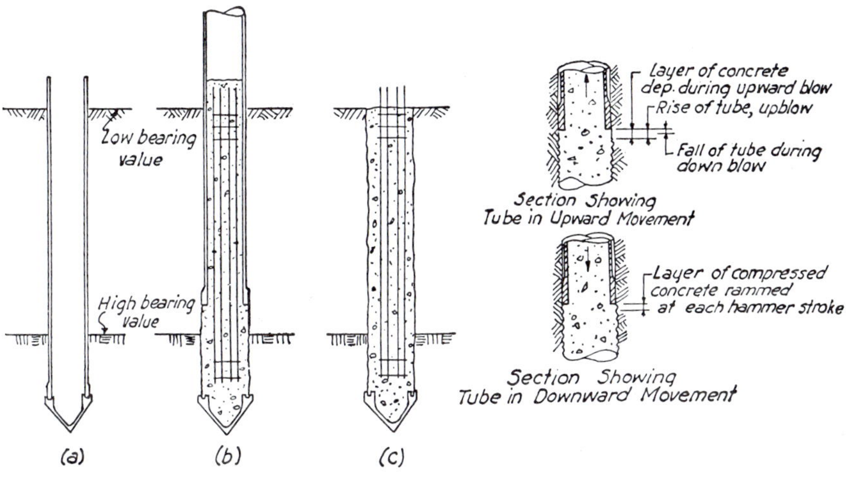

FIG. 9.26. Vibro pile.

These piles are used extensively in Europe and the British Isles and are driven by the British Steel Piling Co., Ltd.

Vibro Enlarged Piles. Vibro piles may be enlarged in the lower section or may have oversize shoes.

Vibro enlarged shaft piles (Fig. 9.27) are used where the pile is dependent on friction and does not reach a firm bearing stratum. Its bearing capacity may be increased by enlarging its diameter and therefore the surface area in contact with the soil. A larger tube may be used, or a second pile driven through the green concrete of the first, as described below. Where the soil is adequate to receive larger loads, enlarged piles are more economical than the standard.

Method of Forming. (*a*) A standard Vibro tube is driven to the required depth, and concrete is deposited up to ground level, without any reinforcement. (*b*) The tube is withdrawn prior to redriving. (*c*) The tube is fitted with a new shoe and redriven through the newly deposited concrete before it has begun to set. (*d*) A second pile, with suitable reinforcing cage, is then formed inside the first, resulting in doubling the cross-sectional area and obtaining 40 per cent more surface area. In extreme cases, the tube can be driven a third time to treble the cross-sectional area.

In some cases, if the ground will not cave and it is sufficient to expand only the lower part of the pile, the first concrete can be filled only to the desired part height before redriving through it.

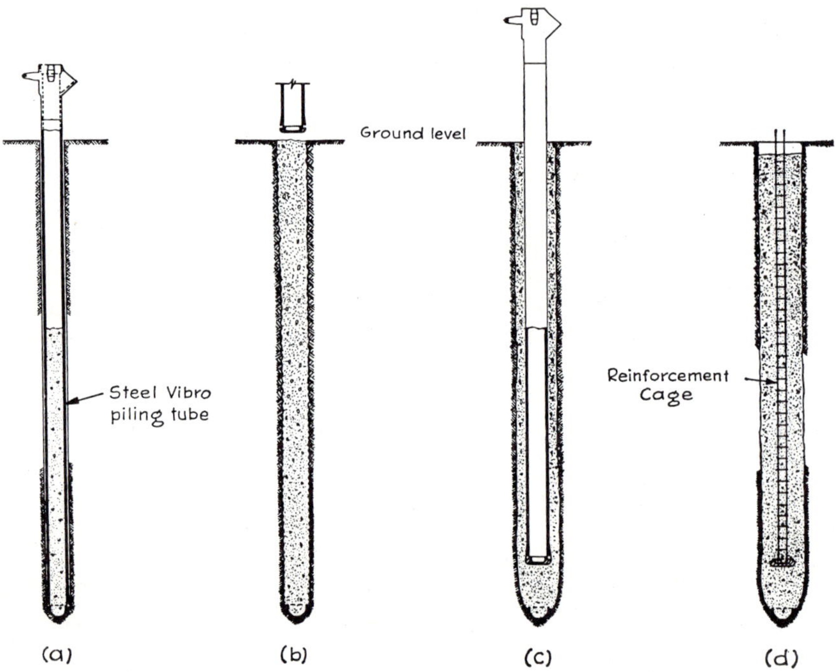

FIG. 9.27. Vibro enlarged shaft pile.

Vibro enlarged base piles (Fig. 9.28) are used where the piles depend mainly on point resistance. By increasing the end bearing, fewer piles are required and saving in cost is made. Oversize shoes up to 30 in. in diameter are available, which size provides more than double the area of an 18-in. shoe used on a standard Vibro pile.

Sectional Driven-shell Concrete Piles

Hercules Piles. These are pipe piles as placed by the Underpinning and Foundation Co., Inc. They may be either open-end or closed-end piles. Sections are joined by internal sleeve couplings, or in some cases by welding. The usual outside diameters are 10¾, 12¾, 14, 15, 16, 18, and 20 in., with shell thicknesses of ⅜, 7⁄16, and ½ in. When driven open-ended, they are usually blown out with a compressed-air jet. A chopping pipe, consisting of a pipe of slightly smaller diameter than the pile, with pointed teeth on the end, can be used for the simultaneous operations of loosening and removing the earth inside. Sometimes a series of conical buckets are lowered into submerged sand to raise the sand. Orange-peel buckets and other devices may also be used where suitable. Cast-steel points are used where it is desirable to obtain displacement piles or to drive through obstructions.

These piles may be driven with a hammer or, when used as underpinning, they may be jacked down by means of hydraulic jacks thrusting against the existing structure, a temporary framework, or an artificial load. When jacking, the gages show the jacking force and the factor of safety against working load. After jacking, the pile top is wedged against the footing, before or after all or part of the test load has been released, as desired for design and settlement purposes, and the exposed portion of the pile and the wedging steel are encased in concrete.

The most advantageous and economical conditions for uses of these piles are as discussed under Open-end Steel Pipe Piles.

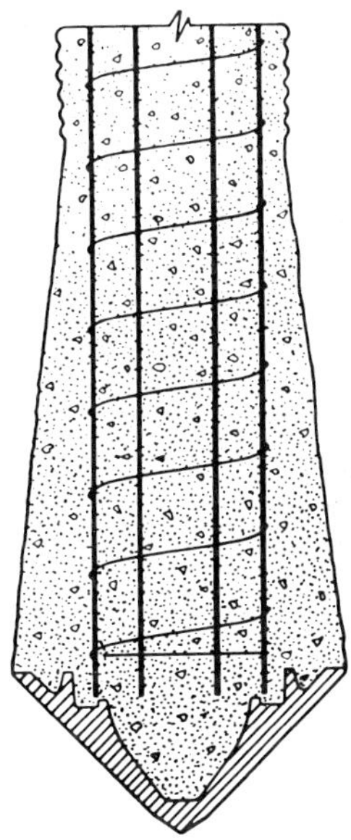

FIG. 9.28. Vibro enlarged base pile.

Tuba Piles. These are pipe piles driven by Spencer, White & Prentis, Inc. They are essentially open-end pipes driven to rock. The material through which the pile is driven fills the pipe and is removed to permit examination of the bottom and filling with concrete. The pipe consists of sections of lap-welded or seamless steel tubing. Stock lengths run from 18 to 22 ft. Sections are joined by tight-fitting internal sleeve couplings. The usual outside diameters are 10¾, 12¾, 14, 15, 16, and 18 in.

As the cylinders are considered to be free-standing columns deriving no resistance to settlement from the surrounding material, there is no minimum-spacing design requirement. The customary spacing for cylinders up to 18 in. diameter is 2 ft on centers. For any given loading, the most economical selection of cylinder sizes and most effective arrangement for supporting a billet or grillage should be studied.

The soil in the pipes is removed by hand tools or compressed air. After being driven, cleaned, and filled with concrete, the pipes are burned off and sometimes capped with steel plates or sections.

They are designed as composite piles, both steel and concrete carrying the load. An allowance of $\frac{1}{16}$ in. for corrosion should be made.

Tests are made by means of hydraulic jacks acting against existing structures, other cylinders, or platform loadings.

FIG. 9.29. Midget orange-peel bucket, only 12 in. in diameter, removing clay from inside of 16-in. Tuba pretest steel cylinders for underpinning footings in existing building through 37 ft of clay and 11 ft of sand. (*Courtesy of Spencer, White & Prentis, Inc.*)

A large number of these piles have been used in New York City, where bedrock may be from 50 to 100 ft, and sometimes 150 ft, below the surface.

The most advantageous and economical conditions for use of these piles are as discussed under Open-end Steel Pipe Piles.

Pretest Cylinders.[11t] The Pretest principle consists of jacking down steel cylinders in short lengths against the building reaction. The method was developed for underpinning, but has come into use to save time and money by its application during construction of the building. Short sections of concreted cylinders are set in pits and jacked down as the building load increases to provide sufficient reaction. This saves time otherwise required for driving piles and ensures that all cylinders have penetrated to the proper depth to take the required load.

The method is used to install Tuba piles, caissons, or Pretest cylinders in basements of old buildings during demolition to permit erection of the new building as soon as possible.

Method of Forming. (*a*) An excavation is made below the foundation. (*b*) The lowest section of pipe is placed and forced down by a hydraulic jack. (*c*) When the first section has been jacked almost flush with the bottom of the excavation, a second section is placed on the first and jacked down. (*d*) The operation is repeated until satisfactory penetration is obtained, as observed by the pressure gage on the jack. Common practice is to jack to an indicated resistance 50 per cent greater than the load to be carried. (*e*) The pipe is inspected and filled with concrete. The top section is wedged in place while the jacking pressure is still applied, so that no rebound can relieve the soil pressure. When piles are so close that the bulbs of pressure overlap, the group is tested.

Pretest cylinders are used extensively by Spencer, White & Prentis, Inc.

Raymond Sectional Pipe Piles. Methods and equipment to drive or jack closed or open-end sectional pipe piles in areas of limited headroom have been developed by the Raymond Concrete Pile Co. In addition to conventional methods for pile underpinning to stop settlement, Raymond methods include the driving of these piles through existing footings, which are then used as the pile cap, provided they are of adequate size and strength. This eliminates excavation around and under existing footings, temporary shoring and supports, and the necessity of load transfer to a new pile cap. It permits driving the piles close to the load center without the removal of the existing footings and with a minimum of interference. Where existing footings are not of adequate size and strength to be used as the pile cap, the same basic methods can be used, but special means are employed to reinforce the existing footing and transfer the load from the column to the piles. The Raymond method can also be adapted to situations where the footing is to be lowered.

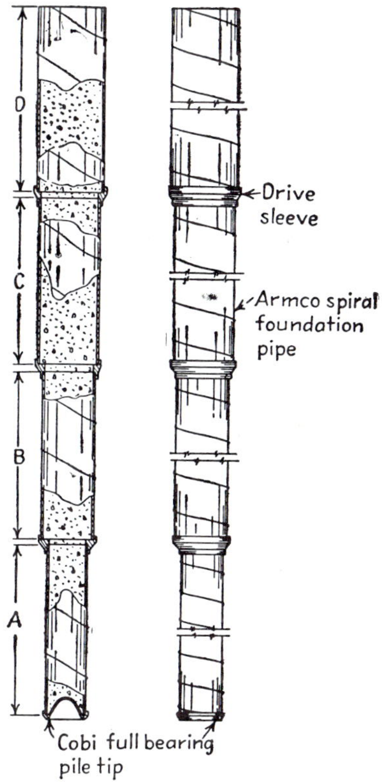

FIG. 9.30. Cuneiform steel pipe pile.

Cuneiform Piles (Fig. 9.30). This type of pile is made with lengths of Armco spiral pipe driven sectionally, using increasing diameter sections from tip to butt to form a tapered pile. Sections are joined by drive sleeves. Cobi pile tips are used. After driving, the pile is filled with concrete. These piles are suitable for end-bearing or friction-load-carrying conditions.

Shells may be of various gages and are light to handle. Piles may be driven in restricted headroom. They may be inspected before filling, and may be reinforced for bending or uplift if necessary. They may be driven with hammers of comparable size to those for wood piles, at a rapid rate because of the light weight. The shells allow for definite determination of length in the field, because of the sectional construction and the ease of cutting off.

These piles are a development of the Albert Pipe Supply Co., Inc., manufacturers of Cobi tips and drive sleeves.

Miga Piles. These piles are useful for underpinning existing structures. They can be installed in restricted headroom, close to existing walls, and in the vicinity of delicate equipment. They consist of a series of precast concrete sections 2 ft in length or less if required, jacked into the ground against the weight of the structure. They have a square reinforced cross section, with a 3-in.-diameter hole. The hole is used for inspection after placing, and for reinforcement. The length is theoretically unlimited; practically, a depth of about 40 ft has been attained.

Method of Forming. (*a*) An excavation is made below the foundation and a bearing plate placed to distribute pressure above. (*b*) The lowest section of pile, with a steel point, is placed below this plate and forced down by a hydraulic jack. (*c*) When the first section has been jacked almost flush with the bottom of the excavation, a second section is placed on the first and jacked down. Connection between sections is made by metal collars. (*d*) The operation is repeated until satisfactory penetration is obtained, as observed by the pressure gage on the jack. Common practice is to jack to a resistance 50 per cent greater than the load to be carried.

This type of pile is used extensively in the British Isles, Canada, and Europe, and is driven by the Franki Compressed Pile Co., Ltd., and its representatives.

Sectional Uncased Concrete Piles

Forum Piles (Fig. 9.31). These piles are useful when lack of headroom, congestion of the site, or necessity for freedom from vibration occur.

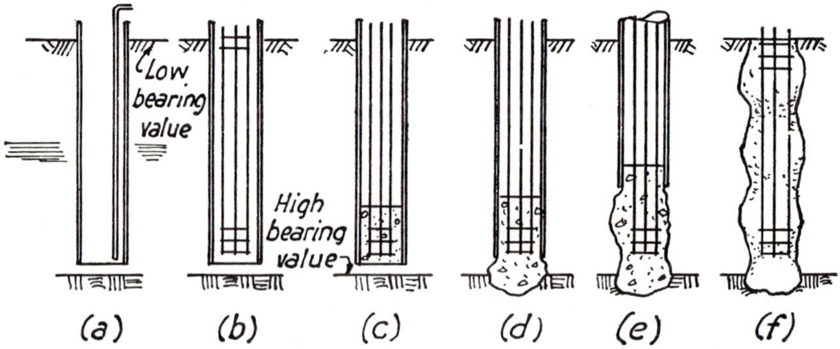

FIG. 9.31. Forum pile.

Method of Forming. (*a*) A tube consisting of sections 18 in. to 3 ft long is sunk in the ground. The sections have screw couplings. During the sinking process the earth is removed either by augers, sludge pumps, or mechanical diggers. (*b*) The water is pumped out, the bottom sealed, and reinforcement (if any) placed. (*c*) A charge of concrete is placed. (*d*) The concrete is rammed by a hammer, making an enlarged base. (*e*) The shaft is formed by introducing successive charges of concrete, ramming each in turn, while gradually withdrawing and uncoupling the sections of the tube. Underground water may be kept out by the use of compressed air while introducing the concrete through another connection.

This type of pile is used extensively in the British Isles, Canada, and Europe, and is driven by the Franki Compressed Pile Co., Ltd., and its representatives.

COMPOSITE PILES

Composite Precast Piles

Composite Precast Piles with H-section Tips.[4c] These have been formed from 20-in.-square reinforced-concrete piles with several feet of 15-in. 89-lb H piles with 1-in.-thick flanges projecting. Generally satisfactory penetration for heavy loads has been secured in Florida limestone having many pockets, crevices, potholes, and ledges, overlaid by sand or mud. The H extends 5 ft into the concrete, and close spacing of surrounding spiral reinforcement is used.

Composite Driven-shell and Wood Piles

This type of pile combines the economy of untreated wood with the permanence of concrete and eliminates the costly excavation sheeting and pumping required if the piles had to be cut off and capped below the permanent ground-water level. The timber section is usually untreated and driven below the permanent ground-water level. Pile loads are usually limited to the capacity of the wood section.

Raymond Wood Composite Piles (Fig. 9.11). This pile consists of a Raymond step-taper driven-shell cast-in-place concrete upper section and a wood pile lower section. The joint between the wood and concrete should generally be either the tenon type for normal loading or the tenon plus rod, pin, and socket for resisting uplift. In the latter type the rod is anchored in the tenon and extends through the concrete section to the pile butt. In both types the tenon is about 8 in. in diameter and 18 in. long. The wedge-ring type has little strength, although cheaper.

The step-taper shells used and the method of driving are described under Raymond Step-taper Concrete Piles. The shell just above the timber section is approximately 14 in. in diameter. A sealing ring is used between the bottom of the shell and the shoulder of the tenon to provide a watertight joint.

The pile can be driven in two operations: the wood first, followed by the step-taper shells, or in one piece with the wood and shell combined. In both cases a special drive head and/or core fits over the tenon and engages both the shoulder and the top of the tenon.

Cobi Wood Composite Piles. These are formed of a wood lower section and a constant-diameter thin-gage corrugated-steel shell, such as Armco Hel-Cor, for the upper section driven by the Cobi method and

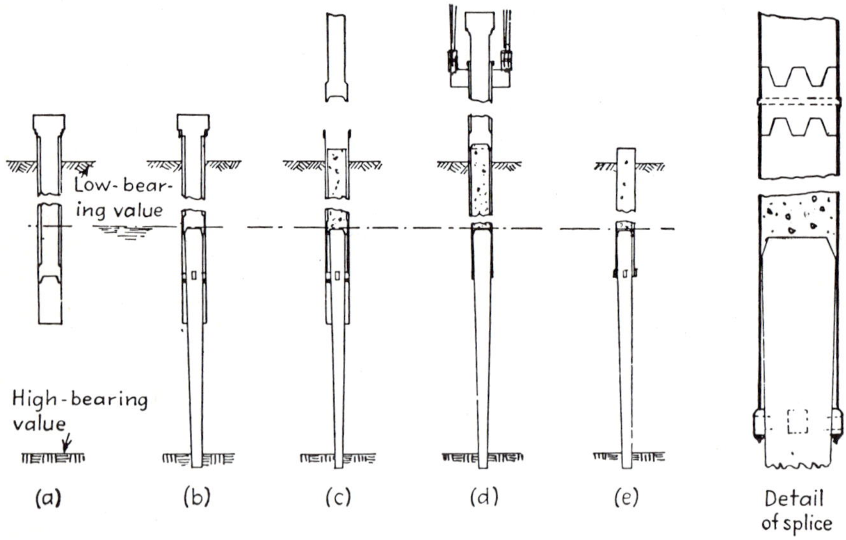

FIG. 9.32. Composite pile, waterproofed steel pipe and wood.

the upper section filled with concrete. These piles are similar in use and results to the Raymond composite piles described in the preceding section.

Composite Piles, Waterproofed Steel Pipe and Wood (Fig. 9.32). This type is a special adaptation by the Western Foundation Corp. of the standard cased concrete and wood composite pile, to meet a condition where the anticipated bending stresses were so great as to require that the splice and upper section of the pile have a greater resistance to lateral load than even the lower wood section.

Method of Forming. The method of forming is the same as in the standard cased composite piles, except that a heavy waterproofed steel pipe of sufficient diameter to enclose the head of the wood pile has been substituted for the light steel and reinforced cage.

Composite Dropped-in-shell and Wood Piles

Composite Piles, Dropped-in-shell and Wood (Fig. 9.33). This type of pile may be used to advantage under the same conditions as described under Composite Uncased-concrete and Wood Piles, in soil requiring casings. Piles of this type have been driven up to 180 ft long. Loads up to 35 tons may be carried. Bending and uplift may be taken in properly designed joints.

Method of Forming. (*a*) A core having a solid steel socket is fitted into a heavy steel casing. The two are driven together into the ground

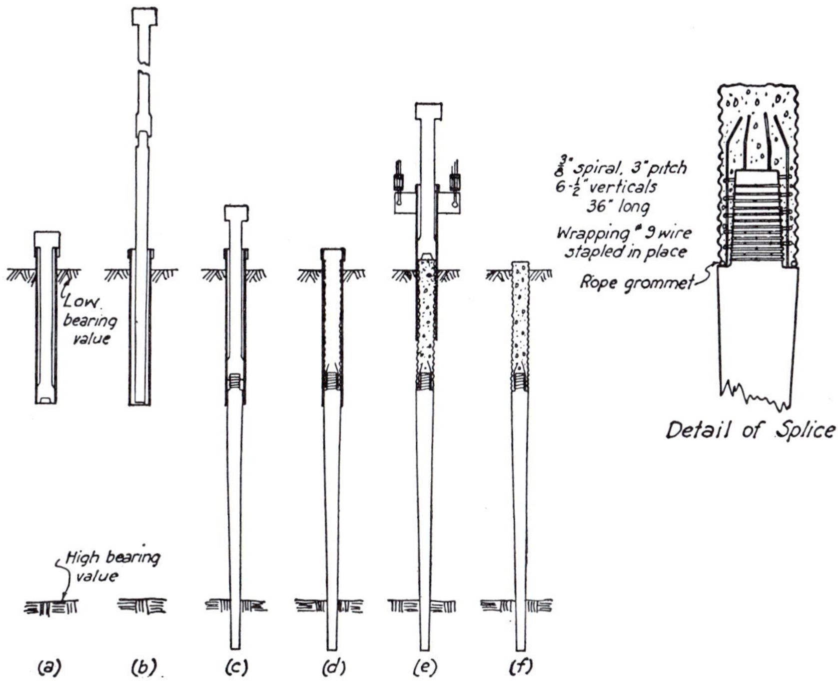

FIG. 9.33. Composite pile, dropped-in-shell and wood.

until the bottom of the casing is well below present or expected ground-water level. (*b*) The core is removed from the casing and a wood pile having a wire-wound tenon, as shown in the detail of splice, is inserted in the casing. (*c*) Using the core as a follower, the wood pile is driven through the casing until the head is below ground-water level and until the required resistance to driving is reached. (*d*) The core is removed and a corrugated metal shell containing a reinforcing cage (shown in the detail of splice) is placed through the casing so that it will surround the tenon of the wood pile. The cage may extend to the ground surface. Various locking devices are provided to prevent uplift of the shell from the wood sections. (*e*) The metal shell may be filled with concrete at this point or after withdrawal of the casing. The core is lowered into the casing until it contacts the metal shell. It is then fixed in this position so that the shell cannot rise with the casing, which is now pulled out of the ground. (*f*) The finished pile consists of a concrete section protected by a metal shell, a reinforced splice, and a wood pile.

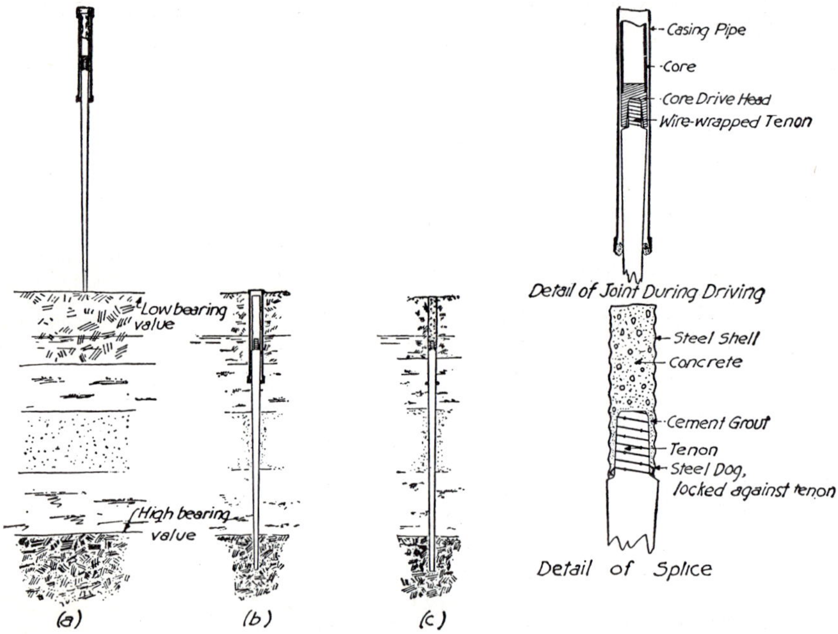

FIG. 9.34. Composite pile, waterproofed steel pipe and wood, cased-follow-down method.

If the diameter of the cased section is smaller than the butt diameter of the lower section, earth should be compacted around the cased section before driving adjacent piles, by jetting soil or by sluicing and tamping sand. If this is not done, there is often a tendency for adjacent driving to force the upper sections out of vertical.

Composite Piles, Waterproofed Steel Pipe and Wood, Cased—Follow-down Method (Fig. 9.34). This type may be used under the same conditions as described under Composite Dropped-in-shell and Wood Piles. The follow-down method may be used where the over-all length is not too great (about 70 ft), depending on job conditions.

Method of Forming. (*a*) The wooden section, provided with a wire-wrapped tenon, is slipped into the driving apparatus so as to form a unit. This driving apparatus consists of a steel casing with a bell bottom and an internal steel core 4 to 5 ft shorter than the casing. The core is recessed to take the tenon of the wooden section during driving. (*b*) The assembled unit is driven until the head of the tenon is below present or expected future water level and the required resistance to further penetration has been reached. A rope grommet is nailed to the wooden section and the bell bottom of the casing is forced over it to make a watertight seal. The core is withdrawn and a corrugated steel shell is lowered inside the casing of the apparatus to rest on the shoulder of the wooden section. This shell is provided at its lower end with steel dogs that engage the wire wrapping on the tenon, thus preventing separation of shell and wooden section even before the concrete has set. (*c*) A seal of grout is poured around the tenon, and the casing of the apparatus is withdrawn, leaving the shell to be filled with concrete later on, or the shell is filled and the casing withdrawn immediately, completing the pile at that time.

This type of pile is furnished and driven by the Western Foundation Corp.

Composite Uncased-concrete and Wood Piles (Fig. 9.35)

This type of pile may be used to advantage under the following conditions: (*a*) Where the permanent water head is not more than 60 ft below the ground surface, *i.e.*, 60 ft is about the length limit for the concrete section of this type of pile; (*b*) where the use of wood piles would require 10 ft or more of dry excavation (or as little as 4 ft of very wet excavation), which can be eliminated by the use of a composite pile; (*c*) where the over-all length of the pile is so great as to be economically impossible to obtain or handle in either straight concrete or wood piles. Cast-in-place concrete piles are usually limited to 70 ft, but this length can rest on top of wood piles of any length obtainable. Wood piles are expensive to buy and handle beyond 80 ft in length. Precast concrete piles have been made to lengths in excess of 100 ft, but they become excessively expensive in lengths beyond 55 ft.

Method of Forming. (*a*) The pile apparatus consists of a casing and a core. The core and casing are driven into the ground, well below present or expected ground-water level. (*b*) The core is withdrawn and a square timber placed in the casing. (*c*) The timber, guided by the casing, is driven down to necessary bearing. The core is withdrawn, a batch of concrete deposited in the casing on top of the timber, and the core replaced. (*d*) The casing is raised the necessary distance, with pressure of core and hammer remaining on the concrete (approximately 7 tons), and a pedestal joint formed around the top of the timber to make a connection between timber and concrete. (*e*) The core is withdrawn and sufficient concrete deposited to provide for voids and space occupied by the casing. (*f*) The casing is steadily withdrawn while the concrete is under the pressure from weight of hammer and core.

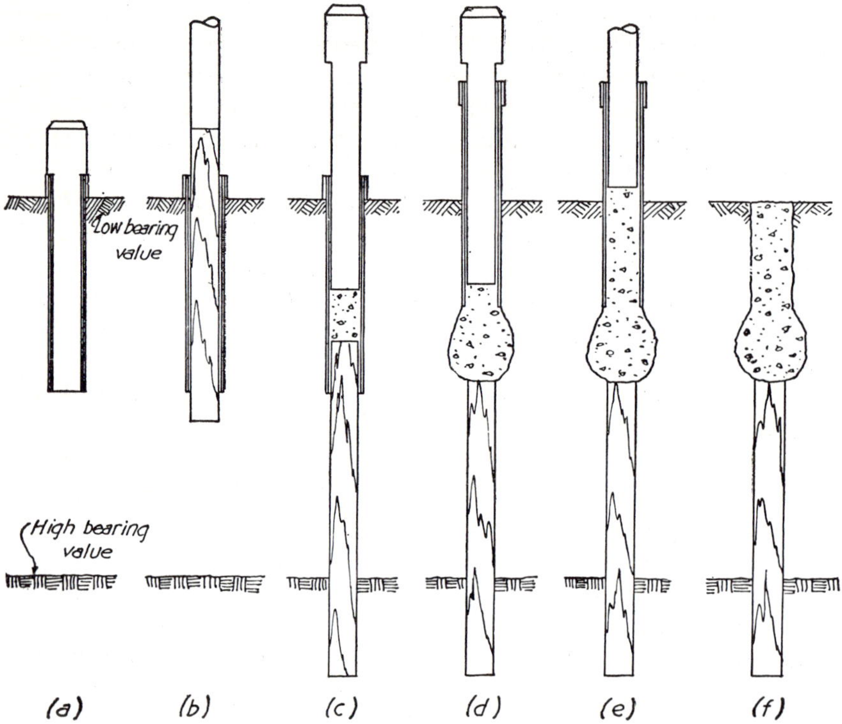

FIG. 9.35. Composite pile, uncased concrete and wood.

This type of pile, the MacArthur composite type, is often driven by the MacArthur Concrete Pile Corp.

Composite Driven-shell and Pipe Piles

Raymond Step-taper-shell and Pipe Piles. This pile consists of a Raymond step-taper-shell upper section and a closed-end steel pipe lower section. Lengths of shell and pipe sections can be varied as required.

The step-taper shells and method of driving are described under Raymond Step-taper Concrete Piles. The shell section just above the pipe is 11 or 12 in. in diameter when 10¾-in.-diameter pipe is used and 13 or 14 in. in diameter for 12¾-in. pipe. The pipe wall thickness used depends on driving conditions and length of pipe section required.

Hard driving, high required resistance, or long pipe sections require heavier pipe. Also, the wall thickness required would depend on whether or not the mandrel is extended through the pipe section to the boot plate. If the pipe is driven hollow, sufficiently heavy pipe must be used to transmit the hammer energy effectively without excessive losses due to elastic compression. If an internal mandrel is used full length, the driving stresses will be taken by the mandrel and a lighter-weight pipe can be used.

The joint between pipe and shell is a watertight slip-type joint. During driving, the step-tapered core extends at least 6 to 8 ft below the joint to assure alignment of the pipe and shell.

The pile can be driven in one piece—pipe and shell together—or in sections. If driven in sections, the pipe is first driven to whatever length desired. If the required length is such that two or more sections of pipe are necessary, the joints between pipe sections are butt-welded or pipe sleeves are used. After the pipe has been driven, the shell section is added and the pile driven to its required length. By driving the pile in two or more sections, shorter leaders and lighter driving equipment can be used. Thus long high-capacity piles can be driven beyond the normal length limit of the equipment to meet unforeseen length requirements or where only occasional extra-long piles are needed.

The pile is subject to internal inspection throughout its full length after being driven. The pipe and shell sections are filled monolithically with concrete.

This type of pile can be driven to high resistance to develop high load-carrying capacity.

These piles are manufactured and installed by the Raymond Concrete Pile Company.

Cobi Shell and Pipe Composite Piles. This composite pile consists of pipe lower section and a thin-gage constant-diameter corrugated steel shell, such as Armco Hel-Cor, for the upper section, driven by the Cobi method, and pipe and shell filled with concrete.

Composite Dropped-in-shell and Pipe Piles

Composite Piles, Projectile Type (Fig. 9.36). This type of composite pile may be of advantage where (*a*) the distance to a bearing stratum is too great to permit of the use of some types of cast-in-place concrete piles and where the permanent water level is too low to permit of the use of the wood-section composite; or (*b*) where the bearing strata are very hard (hardpan or rock) and a heavy load per pile could be developed by driving the projectile to a high refusal. Under these conditions it will usually be necessary to reinforce the section of the pile shaft above the projectile. Lengths up to and over 100 ft may be driven, and loads up to 75 tons carried.

Method of Forming. (*a*) A core having a solid steel socket is fitted into a heavy steel casing and the two are driven into the ground until the top of the casing is nearly at ground level. (*b*) The core is removed from the casing and a section of steel pipe, closed at the lower end with a plate or point, is placed in the casing. The upper end of the pipe is

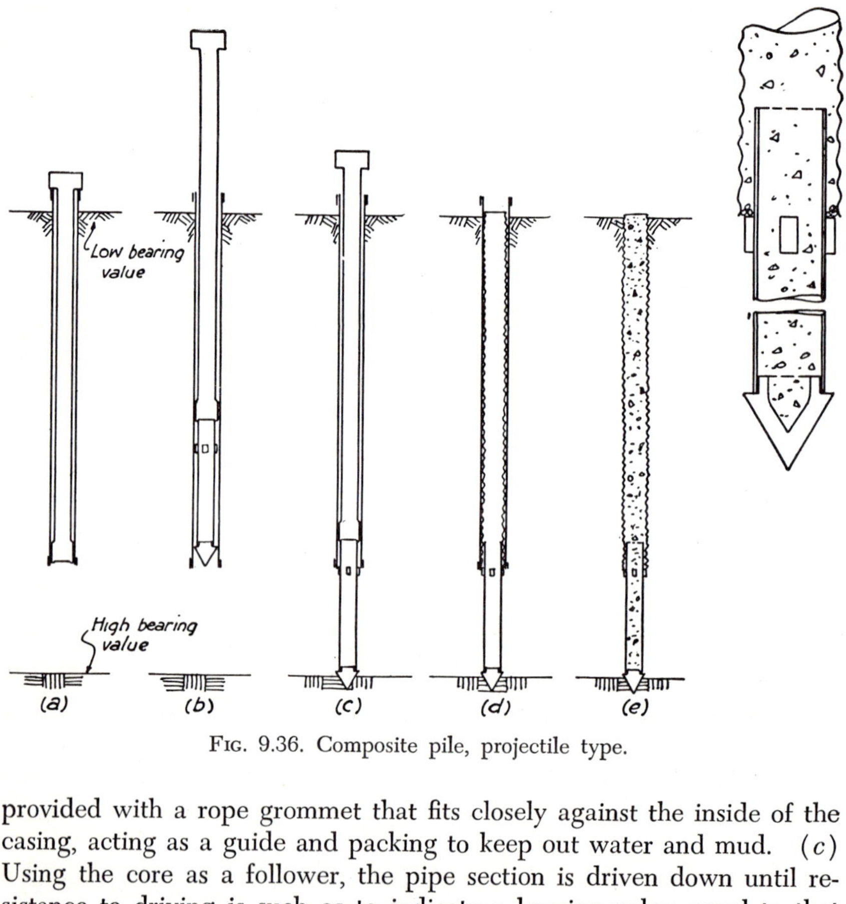

FIG. 9.36. Composite pile, projectile type.

provided with a rope grommet that fits closely against the inside of the casing, acting as a guide and packing to keep out water and mud. (*c*) Using the core as a follower, the pipe section is driven down until resistance to driving is such as to indicate a bearing value equal to that required. (*d*) The core is removed and a corrugated metal shell is lowered until it rests on the rope grommet. If reinforcement is called for, it is now placed and the shell is filled with concrete. The core is lowered into the casing until it contacts the metal shell, and fixed in this position so that the shell cannot rise with the casing. (*e*) The casing is withdrawn, leaving an all-concrete metal-encased pile in the ground.

A slip joint for use where heaving is encountered has been developed by the Western Foundation Corp.

A variation is the pipe composite pile with the pipe followed down by the driving casing. An H section can be used in a similar way.

Composite Driven-shell and H Piles

Raymond Step-taper and H Composite Piles. This type of pile is similar to the Raymond step-taper and pipe composite pile except that a steel H section is used in place of the steel pipe. A special joint is used between the concrete and steel H sections. This pile is not generally used since the pipe step-taper pile usually meets the driving and load-carrying requirements and can be inspected internally its full length.

Composite Dropped-in-shell and H Piles

Driven-core Composite Piles.[4b] Steel H sections are driven inside a corrugated shell that is then filled with concrete. This type is especially useful for heavy loads carried to rock or hardpan at moderate depths up to 50 ft as a possible alternate to open-end pipe piles, particularly where loads are too great for conventional precast or cast-in-place concrete piles but not heavy enough for Drilled-In Caissons.

The driving apparatus is a 19-in.-diameter casing and slightly shorter close-fitting core, driven together to refusal. A plug of dry concrete is introduced at the start to prevent water or soil from rising in the casing when the core is removed. The core is removed and a 16-in.-diameter open-end corrugated shell lowered into the drive casing. An 8-in. H section, centered by welded spacers, to carry the specified part of the load is placed in the shell and the shell filled with concrete. The H section is then driven, independent of the casing, to practical refusal to make a combination pile for very heavy loads. It can be driven through soft rock.

The New York City building laws allow this combination 1,000 psi on the concrete and 12,000 psi on the H, under 100 tons working load maximum except 200 tons if proved by tests.

This type is driven by the Western Foundation Corp.

PREEXCAVATED CONCRETE PILES

Preexcavated Driven-shell Concrete Piles

Cored Steel Pipe Piles (Fig. 9.37). This type of pipe pile is useful where difficulty is encountered in driving displacement piles at close centers through clayey soils.

Method of Forming. (*a*) A section of open-end steel pipe, containing an open-end steel pipe core with a trap arrangement, is driven into the ground. (*b*) At intervals the core is withdrawn and the contents dumped out by opening the trap at the bottom. The soil may be removed by means of compressed air, water jets, or jarring, if it does not fall out by gravity. (*c*) The core is withdrawn. (*d*) The shell is filled with concrete.

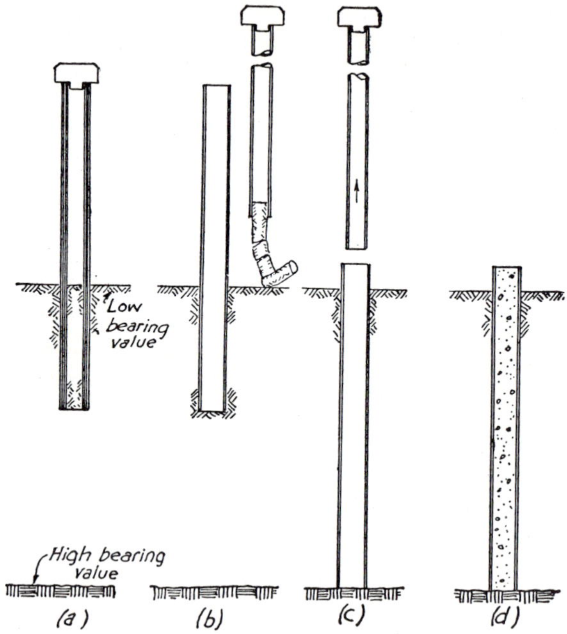

Fig. 9.37. Cored steel pipe pile.

If necessary, jetting may be done on the outside, to facilitate driving through sand and gravel layers. Composite piles may be formed, using the pipe section for the lower portion, and cased or uncased concrete sections of larger diameter for the upper section.

Raymond Preexcavated Piles. In addition to the routine methods of preexcavation such as augering and drive-coring, The Raymond Concrete Pile Co. offers a wet rotary preexcavation method for "jetting" soft to medium clays to eliminate or reduce ground disturbance and heave. The presence of soft to medium cohesive soils precludes normal jetting or other preexcavation methods.

The hole is predrilled under a combination of jetting and reaming to approximately the size and shape of the pile. The resulting slurry maintains the hole until the pile is installed. This method has been used successfully to depths of 100 ft. After the pile is placed in the preexcavated hole, it is driven to the required bearing.

Preexcavated Uncased Concrete Piles

Cored-out Concrete Piles (Fig. 9.38). This method of construction is particularly useful where it is necessary to carry the load to solid rock and where the soil conditions are such that the use of displacement piles is unsatisfactory. Such a soil is a clay that flows instead of compresses and that will cause the piles to lift off of the rock surface when

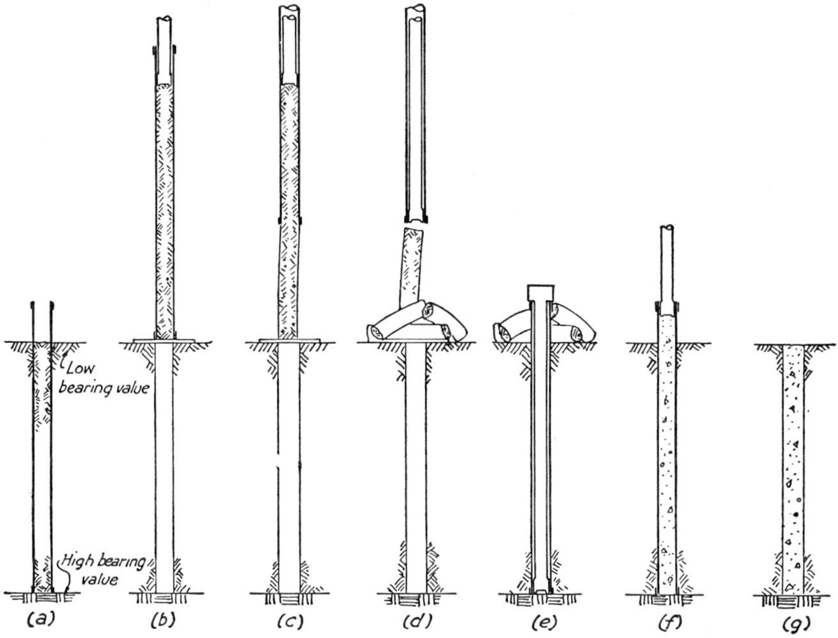

FIG. 9.38. Cored-out concrete pile.

additional piles are driven. By removing the core of material, the tendency to heave is, of course, entirely eliminated.

This type is also useful where the load per square foot is excessive so that it is difficult, if not impossible, to drive a sufficient number of piles to carry the load without crowding the piles; it is possible with this method to use caissons of a large enough diameter so that the diggers can work at the bottom of the excavation and make an enlarged footing and, through the addition of reinforcing, construct a caisson capable of carrying a full column load.

This method of construction can be used where there are strata of water-bearing material underlaid with clay, where ordinary caisson methods would require compressed air.

Method of Forming. (*a*) A steel casing with a cutting edge capable of cutting through the surface materials is driven to solid rock. (*b*) The casing is withdrawn and a platform of planks is placed across the hole. The casing with the cored-out material is placed on top of it and the mandrel is inserted. (*c, d*) The casing is drawn up over the mandrel exposing the core of material which is removed in sections of about 6 to 7 ft until the casing is entirely clear. (*e*) The casing and mandrel are redriven to the rock. (*f*) Concrete is deposited under pressure while the casing is withdrawn. The methods of withdrawal are as described for Compressed-concrete Piles.

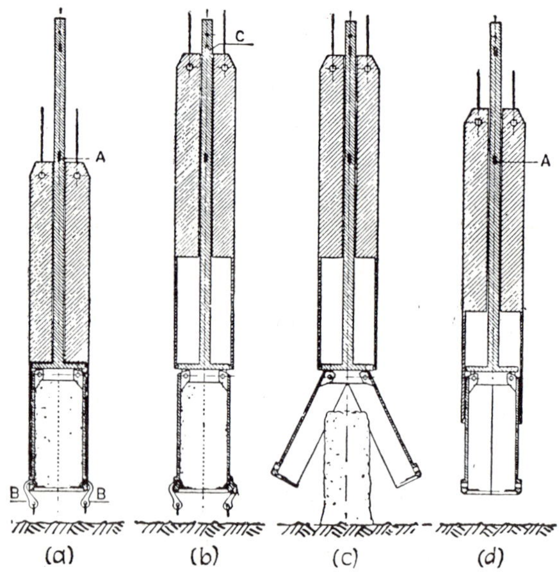

FIG. 9.39. Franki cored pile.

Franki Cored Piles (Fig. 9.39). These are used in clay where piles cannot be driven by compression, or where vibration or heaving might damage adjacent structures.

Method of Forming. (*a*) A steel casing in two semicircular sections is driven until full of soil. (*b*) The ram and extended circular section holding the jaws closed are raised to above ground. (*c*) The jaws are opened and the soil falls out. (*d*) The ram and extended circular section are lowered over the jaws to close them, and the assembly driven into the soil at the bottom of the hole.

Concrete Bored Piles. A hole is bored and lined where necessary with casing tubes with watertight joints. The tip is generally keyed into hard strata. Reinforcing steel is placed, and concrete rammed in small batches with a heavy drop hammer during extraction of the casing. A plug of tremie concrete may be used to seal the bottom for dewatering, or concrete may be placed through a bottom-dump bucket. Only light winches with small headroom are needed. The rake can be up to 25 deg. Standard diameters are 15, 17, 19, and 24 in., and average working load capacities 30 to 40, 40 to 50, 50 to 65, and 90 to 110 tons, respectively. These piles are installed by The Cementation Co., Ltd., of Great Britain.

Cementation Bored Piles. The boring is carried out as described for concrete bored piles. A reinforcement cage and a permanent injection tube are installed. Crushed stone or gravel is tamped in small quantities during extraction of the casing tube. A colloidal sand-cement grout is pumped in. For sandy or gravelly strata the grout can also be forced into the surrounding ground to form a cemented layer, greatly increasing the bearing power of the piles. These piles are installed by The Cementation Co., Ltd., of Great Britain.

SPECIAL TYPES OF CONCRETE PILES

Intrusion-Prepakt Piles.[5ak,5al,11ar] These are cast-in-place piles of various types for certain conditions. There are four types of I-P piles with high-strength fine-grain concrete including admixes and known as Intrusion mortar. Holes are augered, without shock or vibration, so that they can be used near adjacent buildings without causing damage. The installation is quiet, permitting work near hospitals and offices without disturbance. They can be placed on close centers without causing heaving. For interior or low-headroom locations, the mortar can be pumped from a distance. When work areas present unusual fire hazards, internal-combustion motors can be located at a safe distance.

Equipment ranges from mobile truck-mounted rigs to demountable units for indoor or congested locations. In such cases, an auger rig occupies the space of two or three cars; materials can be stocked in a less crowded location along with mixing and pumping units. Piles 40 ft long have been installed in 8-ft-headroom basements. Operations can be conducted in terrain where it would be difficult to move driving rigs. Settlement of existing buildings can be arrested by underpinning with these piles.

These I-P piles are installed by Intrusion-Prepakt, Inc.

I-P Cast-in-place Piles (Fig. 9.40*a*). A hole is drilled in self-supporting soils using a continuous-flight auger, and the hole is not cased. A grout pipe is inserted to fill the hole from the bottom up. The pumped grout floats out dirt and water. Since concrete is not dumped down the hole, soil is not knocked into the hole. Intrusion Aid in the mortar prevents water from diluting the mix and compensates for setting shrinkage. These piles are normally 12 and 16 in. in diameter and may be placed to 75-ft depths.

I-P Pakt-in-place Piles (Fig. 9.40*b*). One of the particular advantages of a pile without coarse aggregate is the convenience of placing in caving ground without the necessity of casing. These piles may be used in stable or unstable soils. They are of a pressurized cast-in-place design. They provide good friction contact with the soil for friction piles. Many projections extend out into the weaker spots in the soil. Installation is similar to that of I-P cast-in-place piles, except that flexible grout

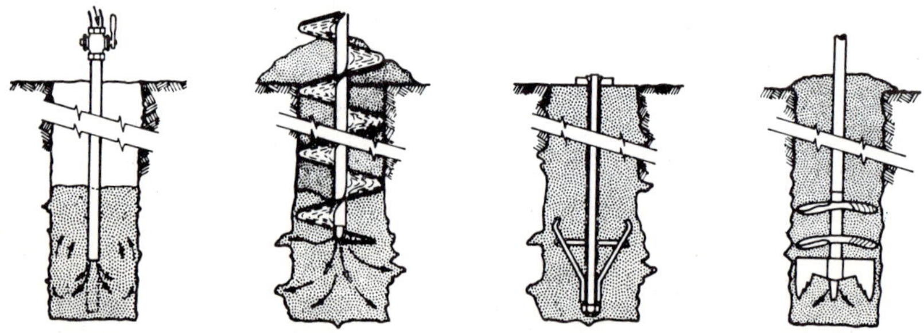

FIG. 9.40. Intrusion-Prepakt piles.

lines are connected to the top of the auger flight and Intrusion mortar is pumped through the auger shaft to fill the hole as the auger is withdrawn. The earth-filled auger acts as a pressure packer to force the fluid mortar deep into substrata fissures. The mortar pressure also consolidates the surrounding area. This type is also of 12- and 16-in. diameters and can be up to 75 ft long.

I-P Locked-in-place Piles (Fig. 9.40*c*). This type has the bearing advantages of the Pakt-in-place pile and also has much greater resistance to tension and bending than a conventional pile. They are especially adapted for use under footings of transmission towers, water towers, or antennae towers. Placing procedure is the same as for the I-P Pakt-in-place piles, after which a high-tensile-strength rod is set inside a pipe sleeve and the annular space filled with lubricant. Post-tensioning provides resistance to lateral loads and prevents cracking with exposure of the reinforcement to corrosion. Regulated post-tensioning provides bottom anchorage for uplift in excess of the applied posttension load.

When the pile is extended aboveground in a form or Sonotube, post-tensioning from pile tip to top of the column produces an integral pile-column with maximum stress resistance.

I-P Mixed-in-place Piles[5ar] (Fig. 9.40*d*). These were developed to overcome stabilization problems in fine-grained clay, silt, and sand. It is economical and effective in providing underground cutoffs, shoring, and bearing piles under many circumstances. The equipment has the same mobility as that used for the other types of I-P piles. The principal differences are in the auger and use of in-place soil as aggregate.

The soil is not removed. A combined drilling and mixing head injects Intrusion mortar into the disturbed soil on both down travel and withdrawal, through flexible hose attached to the hollow drive shaft. Reinforcement may be inserted before the mix hardens.

Strength varies with the soil type, but some adjustment can be made by altering the mortar composition and volume. For bearing piles, gravelly and sandy soils provide quality comparable with concrete with strengths up to 4,000 psi. These piles may be effectively placed in silts and clays, but strengths will be less with strengths of 500 to 1,000 psi. Soil may compose about 60 per cent of the pile. Good skin friction is obtained by the mortar intrusion under pressure into fissures or weaker zones.

These piles may be overlapped to serve as cutoff walls and can be carried out around obstructions. Used for coffeerdams or cutain walls around excavations, the need for shoring or backfilling is avoided, and the piles can serve as a foundation wall. Diameters up to 24 in. are available, and lengths can reach 60 ft.

Intrusion-Prepakt Cast-in-place Concrete Piles (with Coarse Aggregate).[5aa] These are the original type of Intrusion-Prepakt cast-in-place piles, which have been largely supplanted by the types of I-P piles described above. The later types have been found to be more effective and economical. The original type is formed by coring holes with an earth auger, inserting a ¾-in.-diameter injection pipe, funneling coarse aggregate of ⅜- to 2½-in. sizes into the hole, tamping, and grouting. Stovepipe may be inserted near the top of the hole, or where required, to keep soil from falling into the hole. In soft ground the hole may need full-length lining during augering. Sometimes the casing is required to remain until the concrete has set. The grout mix may be such as 3 bags of cement to 4 cu ft of sand and 1.3 cu ft of admixture.

Colcrete Piles. These are used to carry a load through poor soils to a bearing stratum. A pipe with precast concrete shoe is driven to desired set. Then the tube is filled with Colgrout (colloidal grout), well mixed and hydrated, and crushed stone is poured in. Reinforcement can be used for bending or uplift. This type was developed in Singapore and introduced into Great Britain.

The Colcrete method has been used for filling Drilled-In Caissons.[7k] The shell has been filled with coarse aggregate placed after excavation but without pumping out the water. A grout pipe, a sounding pipe, and the steel core are placed before the aggregate. Colgrout, pumped out of the bottom of the pipes, fills the voids in the aggregate from the bottom up, forcing the water to rise. The sounding pipe is slotted to admit grout from the outside, where its rise may be noted. Regular concrete sand, without admixtures, is mixed with cement to form a colloidal grout. Patented, double-drum mixers are made by Colcrete Structures.

Ridley Piles. These are formed by driving a heavy steel tube with a shoe, and partly filling it with cement grout. A precast concrete pile with a shoulder is then driven through the tube. The tube is withdrawn slowly, while pressure on the precast pile causes the shoulder on the concrete pile to force the wet concrete out laterally.

Franki Composite Piles (Fig. 9.41). These have enlarged bases driven in the same manner as for standard Franki piles, but with the

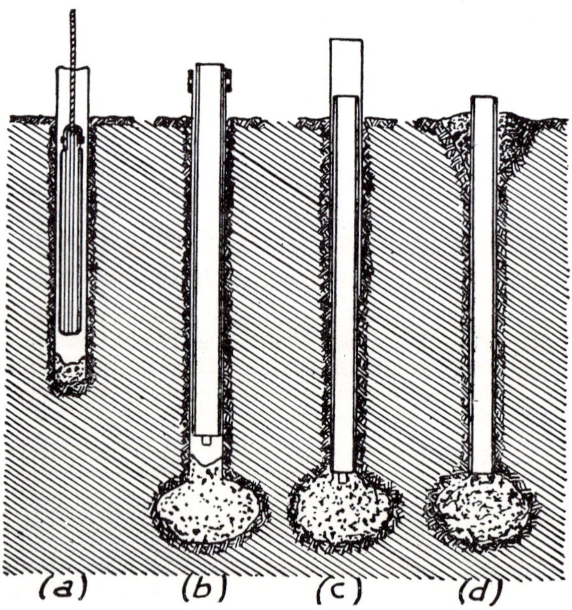

FIG. 9.41. Franki composite pile.

shaft of precast concrete. The head of the pile can be cut off at a point above ground, which is an advantage for bridge piers. They can be placed in water, and may have smaller diameters than possible with the standard Franki piles. The possibility of horizontal breaks during construction is avoided. In chemically active soils, dense or armored shafts may be used.

Method of Forming. (*a*) With a drop hammer, drive on the concrete, forming a dense plug that drags the tube down. (*b*) Hold the tube in place by cables, drive out an enlarged concrete pedestal, and lower a precast pile. (*c*) Set the precast pile in the pedestal concrete while green. (*d*) Withdraw the casing.

This type of pile is driven by the Franki Compressed Pile Co., Ltd., and their representatives.

West's Shell Piles (Fig. 9.42). This pile consists of precast reinforced-concrete shell sections 3 ft long, threaded on a steel mandrel for driving. By changing the number of sections, lengths can be driven to suit varying conditions without the need of splicing or cutting off. Shells can be driven from a higher level than cutoffs without wastage, by filling only to cutoff elevation. The units are light and easily transported to places of difficult access. Interiors may be inspected before concreting, and reinforcement can be set. Avoidance of possible damage to the green concrete shaft is possible because the concrete shell remains in place as part of the pile and is very strong against lateral forces from adjacent driving. Where corrosive water is present in the subsoil, the shells and shoe may be made with acid-resisting cement, thus forming a protective skin around the piles. Rigs are caterpillar-mounted or on wheels and require 18- to 57-ft headroom for leaders. Piles can be driven on a rake of 1:3.

The proportion of driving resistance exerted by friction and end resistance can be observed with this type of pile, thus providing valuable design data.

Piles are obtainable as follows:

External diam, in.	Core diam, in.	Shell length, ft.	Normal load range,* tons (long)
24	19	3	100–200
20–21	15	3	75–120
17½	12	3	50– 80
14½	10⅞	3	Up to 60

* Depending upon length and substrata.

Method of Forming. (*a*) Tubular concrete shell sections are threaded on a steel mandrel, a concrete shoe is set in place, and joints covered by steel bands are covered with bitumastic. (*b*) The whole of the tube, mandrel, and shoe is driven, new sections being added as needed, until the required set is obtained. Where friction is large, much of the force of the blow is taken by the shell; when tip resistance is large, most of the force will be carried to the shoe by the mandrel. A shoe with a

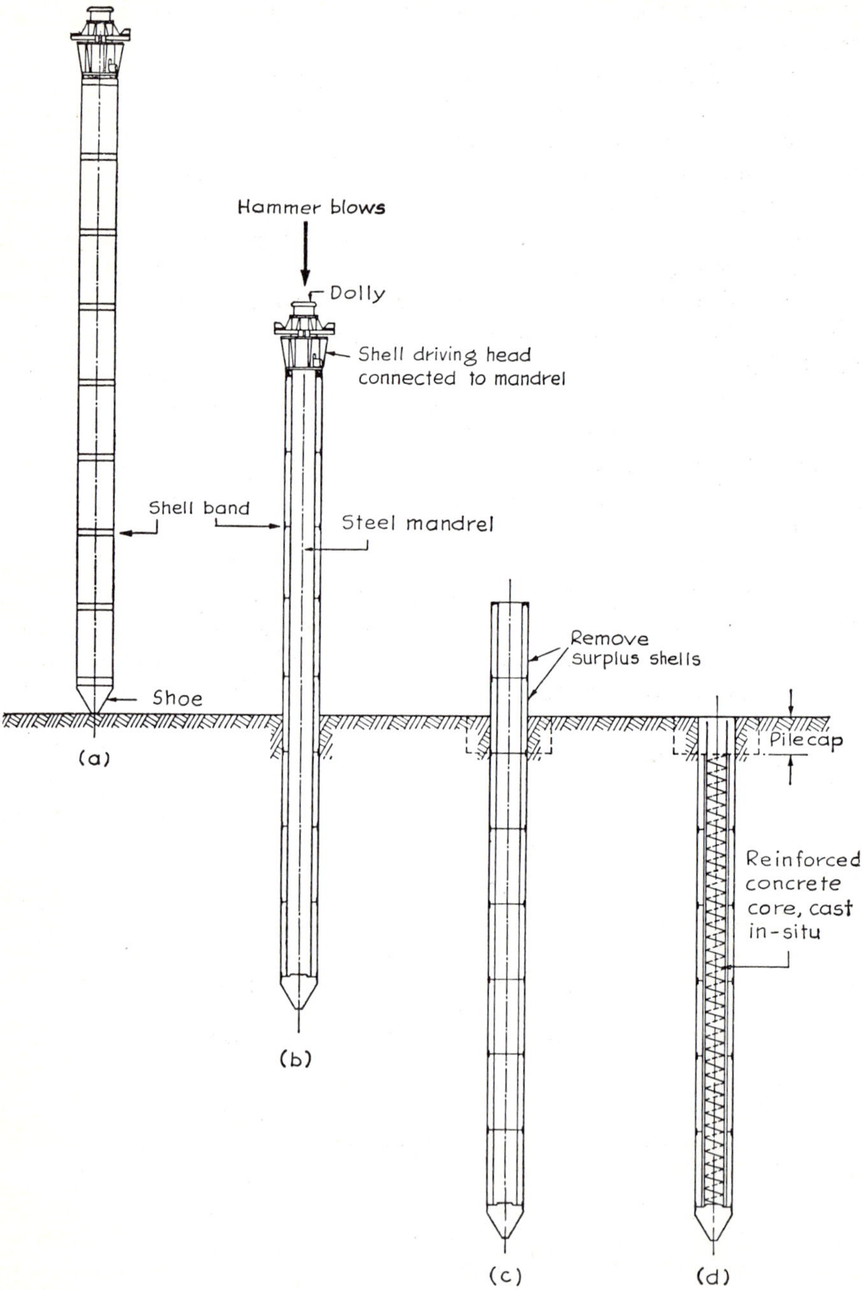

FIG. 9.42. West's shell pile.

steel disk and a steel point is available for penetrating a hard stratum. (*c*) The mandrel is withdrawn and the interior inspected. (*d*) A reinforcing cage is placed if required, and the core filled with concrete.

West's Extended Shell Piles. Piles up to 120 ft long have been placed by driving in two or three stages. The first section is driven in the normal manner, and the core is concreted to within 4 ft of the top of the driven shell. When the core has set, an 8-ft-long pipe or rolled section is embedded in quick-setting cement in the core. A mandrel is threaded over this member, and further shells placed. The solid shell section, with the empty shells above, is redriven to the required set. The process may be repeated once again.

These piles are controlled by West's Piling and Construction Co., Ltd., of Great Britain.

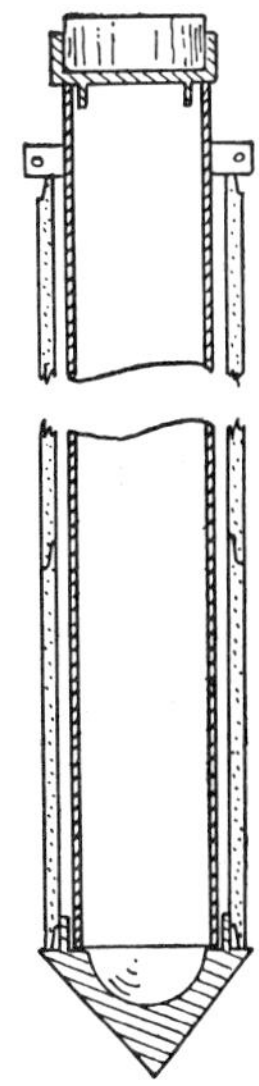

FIG. 9.43. Peerless pile.

Peerless Piles (Fig. 9.43). This pile is similar in general features to the West's shell pile described previously.

Combination Precast and Cast-in-place Concrete Piles. For depths too great for precast piles and where wood lower sections are unavailable, precast lower sections with projecting reinforcing rods are suggested, on which a cast-in-place upper section may be placed. A steel driving plate can be placed on top of the precast section and the rods projecting through holes in the plate can be welded to it. After driving the head of the precast section nearly to ground level, using a cap that will allow the rods to project through, a temporary casing pipe can be set on the driving plate of the lower section and driven down, filled with concrete, and withdrawn. If a cased upper section is needed in soft soil, the pipe may be left in place, or a corrugated metal shell may be used and the pipe withdrawn. The pipe or shell that is to serve as the permanent casing can be welded to the plate on the head of the lower section to obtain watertightness. Reinforcing cages may be inserted if required. It might be safer to drive on a mandrel in the outer pipe, to obtain a more central blow on the lower section and avoid possibility of spalling shoulders of the precast section. To obtain straightness of the two sections, a vertical pipe sleeve about 3 ft high can be welded on the plate cap of the lower section, using a close fit to the upper pipe. Out-of-straightness would cause eccentric loads in the pile, and also might result in driving on one edge of the lower section and thus cause breakage.

Prestcore Piles (Fig. 9.44). These piles are suitable where headroom is limited or vibration must be avoided. They are easily installed in waterlogged ground, and with slight modifications in the method, can be formed in 10 to 12 ft of water. The units are light and easily transported, and there are no waste in cutting off piles that are too long and no delay or expense for splicing. The simplicity of the equipment makes this system economical even if only a few piles are required, where the cost of transporting and erecting any ordinary type of equipment might be high. Diameters available are 14, 18, and 26 in., with corresponding safe loads up to 40, 60, and 150 tons recommended by the sponsors. Piles up to 90 ft long have been installed without difficulty. Vertical reinforcing rods are ½ to 1 in. in diameter, and four, five, or six are used.

Method of Forming. (*a*) A steel casing is sunk by well-boring methods. (*b*) A charge of concrete is placed in the bottom of the casing.

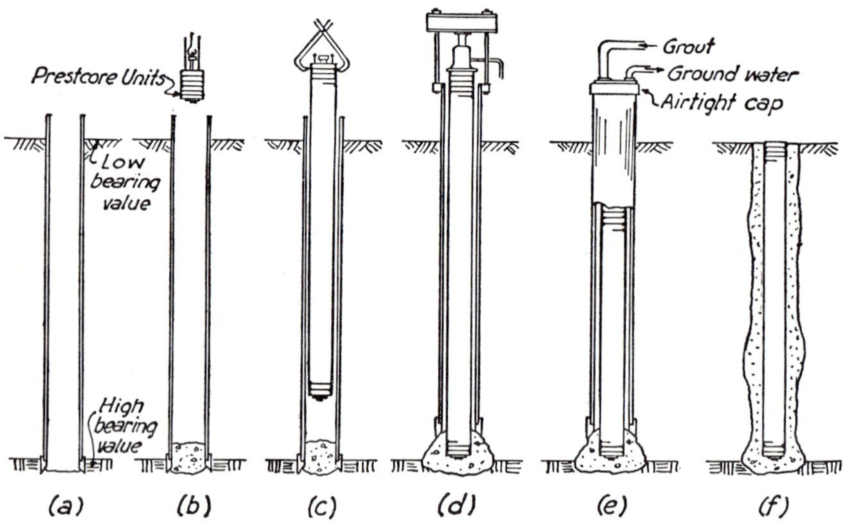

FIG. 9.44. Prestcore pile.

(*c*) A pile consisting of precast concrete sections assembled on a central steel pipe is lowered into the hole. (*d*) The casing is withdrawn by jacking it up against downward pressure on the concrete sections. (*e*) During withdrawal of the casing, the pile is grouted under pressure to expel subsoil water, so that the pile is enclosed in a thick skin of grout. Reinforcement rods may be provided if required. (*f*) An enlarged base may be provided on the finished pile if desired.

This type of pile is controlled by the British Steel Piling Co., Ltd.

Prestcore piled foundations in water (Fig. 9.45) may be constructed through water by bringing them up through permanent concrete cylinders. A precast tube is set on the river bed and sunk a few feet by internal excavation, leaving the top of the tube projecting above water. Water in the tube is not pumped out. The remaining steps are carried out as described for the standard pile until the desired depth is reached.

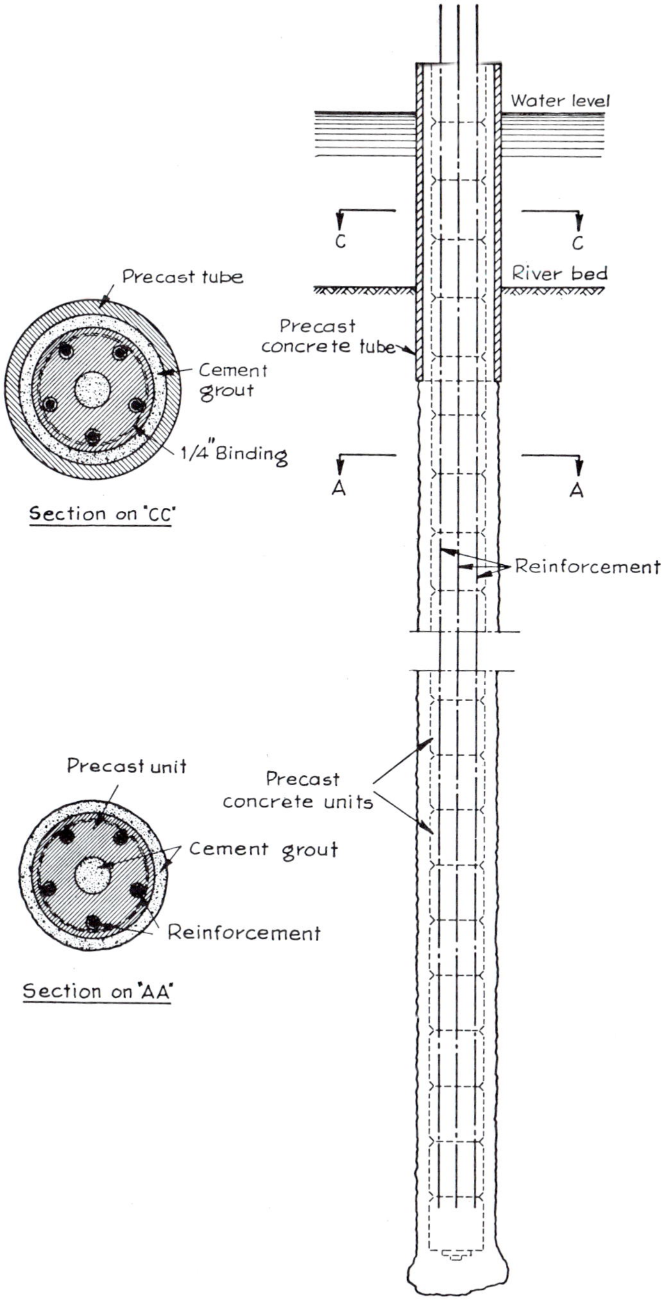

FIG. 9.45. Prestcore pile foundation in water.

The concrete tube remains as part of the pile, and the space between the tube and the core is fully grouted.

Bored Piles with Chemically Consolidated Ground Seal.[58] Where the bearing stratum is a sandy or gravelly material under hydrostatic pressure, bored piles must be filled by use of tremie, precast concrete segments on a rod, an air lock, or chemical solidification of the ground at the point. In this last method the soil may be injected with sodium silicate and calcium chloride, which form a gel filling the voids and creating an impermeable sandstone bulb several feet in diameter with compressive strength of over 700 psi. This process is usable if the soil grains are over 0.1 mm in effective size.

After sealing, the hole is cleaned and filled with concrete compacted by a heavy punner operated by a hand winch. The concrete is brought about up to ground-water level before the tube is raised and the seal with the injected soil broken. As concreting proceeds, the tube is raised short distances gradually and dropped sharply to compact soil and concrete.

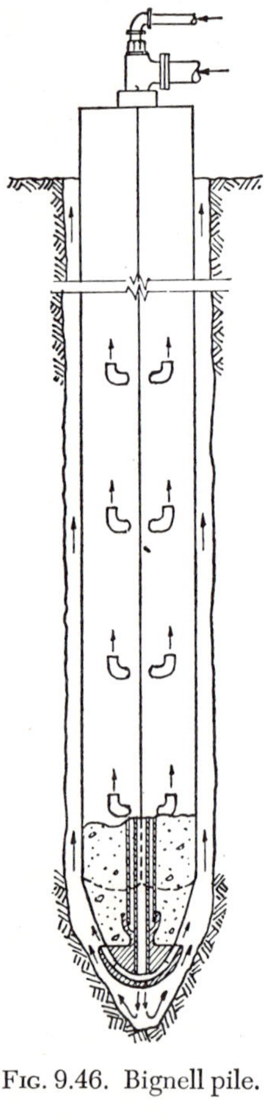

FIG. 9.46. Bignell pile.

Bignell Piles (Fig. 9.46). These precast concrete piles sink under their own weight, needing no hammer but merely a derrick. A 2-in.-diameter jet pipe extending to the tip is inside a 4-in.-diameter jet pipe that supplies numerous side jets which are turned up. The tip jet water softens the soil at the tip, while the side jet water lessens friction. Jetting pressure at the tip is about 200 to 300 psi, and at the sides is about 100 to 150 psi. This type is used considerably in Europe.

Screwcrete Piles. *Precast* Screwcrete piles are formed by installing a mandrel in a precast concrete pile, a screw having about one turn of considerably larger diameter than the pile being attached to the bottom of the mandrel. The screw is rotated to sink the pile. After the desired depth has been reached the mandrel is withdrawn and the hole filled with concrete. The pile diameters vary from 12 to 18 in. The screws may be up to 6 ft in diameter or more, the smaller ones being of metal and the larger ones of reinforced concrete if desired.

Cast-in-place Screwcrete piles are formed by attaching a metal or concrete casing to the screw, rotating the casing or not as desired, when the mandrel turns the screw. After the required depth has been reached, the mandrel is removed and the casing filled with concrete, reinforced if desired.

Diameters of casings are 19, 22, and 42 in., and loads reach to over 300 tons.

Screwcrete piles are installed by Braithwaite & Co., Ltd., of England.

SAND AND GRAVEL PILES

Sand Piles. These are formed by filling holes in the ground with sand. They are used for bearing purposes, the sand being much less yielding under load than the original soil, or for compaction of the ground. Compaction of permanent value cannot occur in fine-grained impermeable soils from which it is impossible to force the water content by quickly applied blows. In such impermeable soils, the increase in compaction is likely to be temporary and the permanent results uncertain. In soils which can be consolidated, satisfactory foundations may be obtained.

Holes may be drilled with an auger, or driven with a wood or closed-end pipe pile. The permanent use of sand instead of leaving the piles in place results in economy. The sand should be moist when placed, and should be tamped.

Sand piles intended for direct support of loads should not be used in earthquake regions or where there is danger of scour.

Sand piles were jetted for navy quay walls on the West Coast during World War II.[9b] Water depth was 40 ft; then came up to 10 ft of sandy loam, 25 ft of soft clay, and either firmer alluvial materials over rock or rock itself. The considerable vertical and horizontal forces required to hold the quay-wall pressures would have made the use of long vertical and batter piling very expensive and difficult. A jet pipe was lowered into the soft clay while the hydraulic dredge pumped water that cut into the clay beneath the pipe, washing down a hole vertically to hard material. The dredge cutter was then lowered into a bank of sand so that sand was delivered through the jet pipe, which could be drawn upward while the hole previously cut was being filled with sand. This operation was repeated at 5-ft intervals so that the sand piles occupied about one-third of the area of the foundation. It was felt that the pinnacles of clay remaining were not detrimental to the support of the foundation, and slip-circle analyses showed a favorable factor of safety. Timber crib structures were placed on the sand-piled foundation.

Sand piles have been used to stabilize and drain soft fill. In the case of a ship-construction basin, the walls of which were formed by cellular steel cofferdams filled with dredged soft marl, such horizontal pressures developed that 25 to 40 sand piles were used in each cell of 24 by

40 ft.[9a] They were made by driving a 12-in.-diameter pipe with a closed plate end which dropped off when the pipe was withdrawn. The pipe was filled with 15 per cent of sand and 85 per cent of gravel thoroughly mixed, and then the pipe was capped and 20 to 30 psi of air applied to force out the filling when the pipe was withdrawn. In the center of the middle pile in each cell, which was the first driven, a 4-in.-diameter pipe with a slotted bottom was placed before filling, to serve for bailing or pumping. The later piles squeezed the water from the marl and it was removed by the 4-in. pipe. Permanent horizontal drainpipes were driven into the central sand piles near the bottoms to drain the fill into the basins.

Gravel Piles. These are similar to sand piles and may be formed by driving a steel tube which is pulled out slowly as gravel is tamped in by a hammer.*

Franki Soil-compression Piles.[10x] These piles have been used principally for soil consolidation. They are formed in the same way as Franki concrete piles, except that a graded granular fill, without cement, is used. They can be used in any soil in which pore volume can be reduced, and since no cement is used, high ground-water level has no adverse effect. A more complete description is contained in Chap. 14.

* Wilhelm Loos, "Comparative Studies of the Effectiveness of Different Methods for Compacting Cohesionless Soils," *Proc. Intern. Conf. Soil Mech. and Foundation Eng.*, Cambridge, Mass., vol. 3, 1936, p. 179.

CHAPTER 10

CAISSON-TYPE PILES AND CAISSONS

Caisson Piles

There are several types of foundations which may be called caisson piles, since they partake of features of both caissons and piles.

Drilled-In Caissons (Fig. 10.1). This type is a composite, restrained, fixed-end column terminating in a rock socket in which the load is delivered by direct bearing and bond to the bedrock. Because of these features, it is possible to support heavy loads with a small column cross section, and to reach considerable depth. Numerous installations from 100 to 200 ft long have been made.[7a,7b,7c,7d,7g,11z] Depths can usually be reached at less cost and with greater speed than for pneumatic caissons. It is possible to pass through large boulders that would stop driven piles. The diameters are small compared with caissons which have no steel cores. The use of H or built-up cores is possible owing to the removal of the core load by bond into the concrete and thence into the rock in bond, thus avoiding the otherwise large intensity in bearing on the rock. If the load is too small to require a core of full height, the load carried by the steel shell may be transferred to a stub core projecting the necessary bond distance from the rock socket. Since the soil is removed and not displaced, these caissons can be put down close to existing structures. Shells are usually of 24 or 30 in. in diameter, and are provided with a high-carbon steel cutting edge. Steel H-core weights may range from light to several hundred pounds per foot. Accuracy in alignments and bowing has been excellent. The depth of rock socket depends on the bearing and shear values of the rock, using a reasonable factor of safety.

The depth of socket required to transfer the load into the rock may be figured by the following formula recommended by the Drilled-In Caisson Corp.:

$$\text{Socket depth, in inches} = \frac{R - 0.35f_c'A_c}{0.05f_c'C_s} \qquad (10.1)$$

where C_s = circumference of interior of caisson shell, in inches.

On work designed and built to date, a bond stress of 200 psi on the periphery of the rock wall socket has been used. The shear value of the

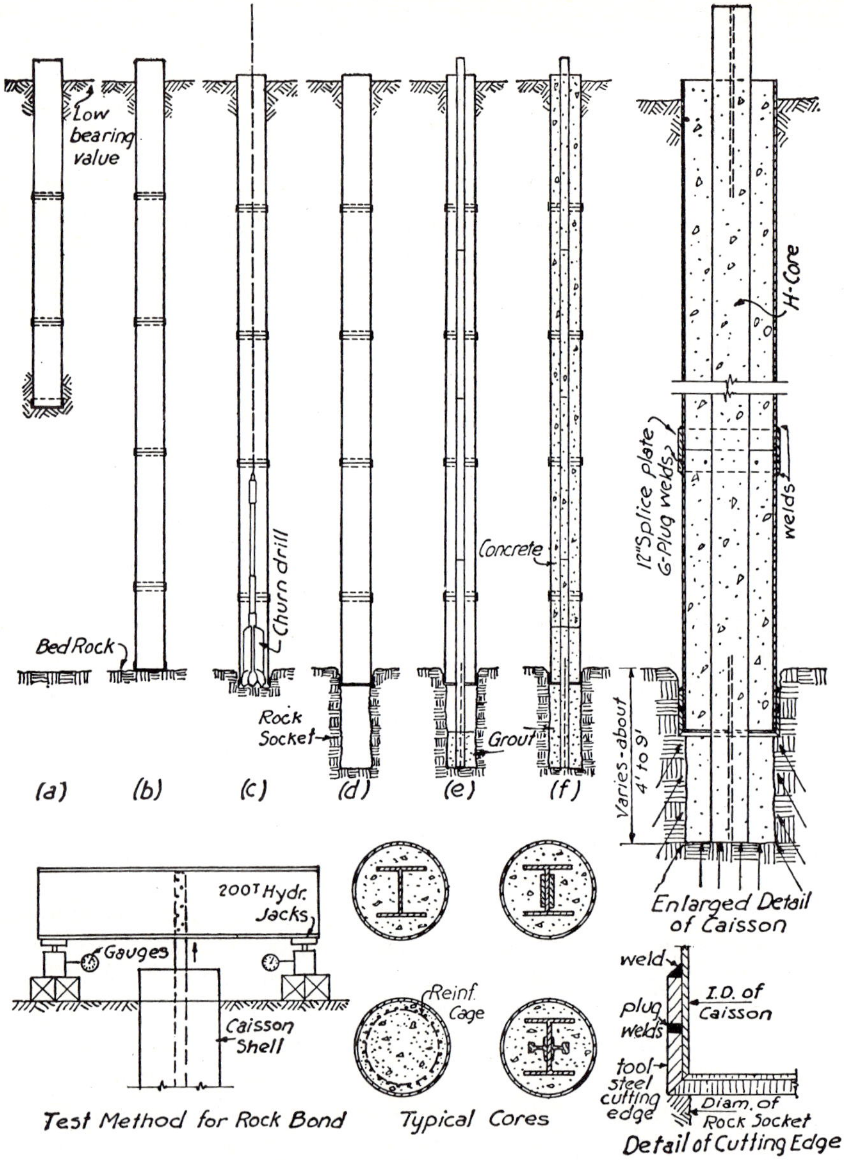

FIG. 10.1. Drilled-In Caisson.

rock may be tested readily, by grouting the H core in the rock socket but not filling the remainder of the caisson with concrete, and jacking.

The steel shell is counted as carrying its part of the load, although it is usually advisable to deduct a corrosion allowance of $\frac{1}{16}$ in. Concrete of high strength is generally used.

Method of Forming. (*a*) For maintaining alignment, a heavy timber template is often buried 1 or 2 ft below ground. A portion of the shell is driven (usually to a depth of 60 to 70 ft), jetting if necessary. Material from inside the pipe is cleaned out by jetting, spudding, sand pumping, blowing with air jets, or coring, and sections of shell added and welded as needed, using butt joints and exterior 12-in. welded plate sleeves. Sandy silts and muds are easily removed by jetting and sand pumping. When hard layers occur, a drill bit several inches less in diameter than the caisson is worked up and down, and the material jetted or bailed out. Small boulders are easily overcome by introducing a churn drill and spudding for a short time. Heavy boulders have been drilled through by the same means, and a charge of dynamite used to shatter the obstruction into pieces smaller than 8 in., which can then be removed by the plunger bucket. Clay may be removed by coring out with a short sectional cylinder, and blowing out the plug on the ground. (*b*) The shell is driven to bedrock, and the clay is churned up by churn drills and removed to rock. (*c*) The churn drill is drilled 1 or 2 ft into rock and the shell is driven down from 6 to 24 in. to seat in rock and seal off any inflow of water. (*d*) Churn drilling is continued to full depth of the rock socket, which is then cleaned out, washed, and inspected by a man lowered into the socket. In case natural gas leaks in, air may be forced in, and a miner's lamp or an oxygen mask may be used. (*e*) A few feet of concrete grout are placed in the bottom of the rock socket and the core set in it, or else the grout is placed after the core is set. If it is necessary to finish the caisson in the wet, enough grout is placed in the bottom under full water head, by means of a bottom-dump bucket, to fill the socket and extend about 3 ft up the shell, after which the steel core is inserted and the grout allowed to set for several days. Then the caisson is pumped out. (*f*) Grout is carried up to a point several feet above the rock socket into the shell, after which the shell is filled with concrete.

Loads of the magnitude of 1,500 tons are readily carried by this method. A TV camera has been successfully used to inspect rock sockets.

The Colcrete method has been found advantageous for filling the shells containing water and avoiding tremie concrete and pumping out.[7k]

This type of caisson pile has been installed by Spencer, White & Prentis, the Western Foundation Corp., and the Raymond Concrete Pile Co.

Rotinoff Caissons. These are formed by driving a hollow concrete pile of large diameter with a specially constructed heavy hammer. The upper strata are shot out by means of compressed air which is blown in at the bottom end of the hollow shaft; through the last few yards of compact sand the pile is driven without any shooting, thus forming a consolidated bulb under the toe similar to that formed under a solid pile, in order to obtain more bearing capacity than that given by nondriven caissons. On top of the caisson, for the driving operation, is placed a heavy steel apparatus containing a protecting cap, hammer, and exhaust tube for soil. The hammer is composed of several rings arranged around the exhaust tube and supported by compressed-air jacks with a maximum stroke of 0.35 m. Compressing and exhausting the air moves the hammer up and down about twenty times a minute. The hammer weight is suited to the conditions of the soil, and driving starts with a small weight, which is gradually increased. Very large test loads have been carried.

These caissons are controlled by West's Piling Construction Co., Ltd., of Great Britain.

Gow Caisson Piles (Fig. 10.2). These piles are useful in spreading heavy loads on the selected bearing stratum. This spreading out will

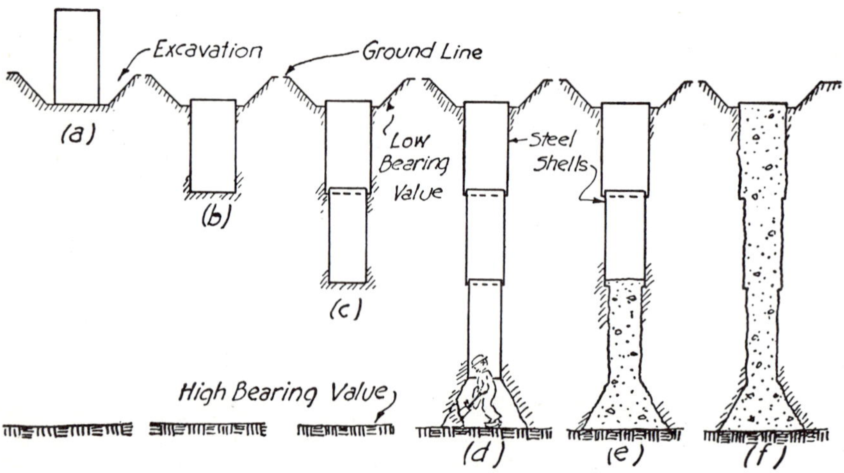

FIG. 10.2. Gow caisson pile.

also permit a load to be carried on a fairly thin firm stratum above a poorer stratum to obtain the value of its load-distributing ability, provided the settlements to be expected from the underlying softer stratum are within permissible limits. Since the casings are installed in short sections, they can be put down in limited headroom and may be used for underpinning. They can only be used when the bottom stratum is sufficiently firm and free from ground water to permit a man to excavate the bell.

Method of Forming. (*a*) A shallow pit is excavated. (*b*) The top cylindrical casing is driven down in this pit, and the soil inside excavated. (*c*) A second cylinder about 2 in. smaller in diameter than the first one is placed inside it and driven down, and the soil inside is excavated. This process is repeated until the caisson reaches full depth, the lowest cylinder being of the specified shaft diameter. The lowest cylinder is driven into the final bearing stratum to seal off. (*d*) Where necessary for increased bearing area, soil below the bottom cylinder is belled out by hand. (*e*) The lowest section is filled with concrete, and the lowest cylinder is withdrawn through the upper ones. (*f*) This process is repeated until the shaft has been completed and all cylinders withdrawn.

The excavation of the shaft can be done mechanically, by a rotary excavating machine similar to a large posthole digger, in clay below water level. If used in sand above ground-water level, water must be added. It can be used only in soils that will stand for fairly long periods, but it can go to considerable depths. This equipment is called a Gow rotary excavator and has sometimes been known as the Hunt machine.

No heavy equipment is required to install Gow piles, and the casings are salvaged for reuse. Casings are in 8- to 12-ft sections, and diameters vary to suit the load, the largest being more nearly caissons than piles.[7j]

These caissons are installed by the Gow Division of the Raymond Concrete Pile Co.

Western Caisson Piles (Fig. 10.3). These are caissons placed by pile-driving methods, having the characteristics and advantages of both types

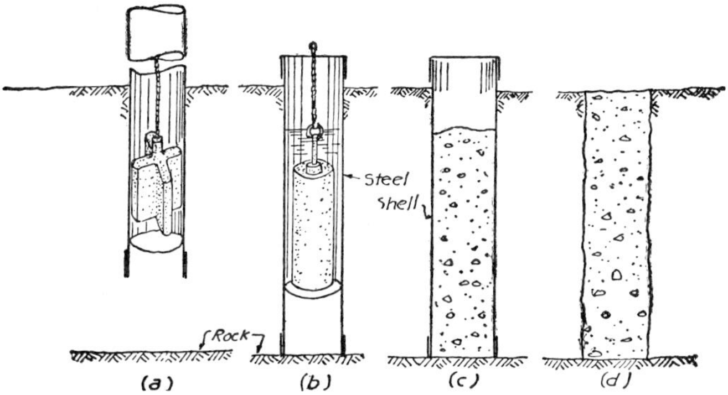

FIG. 10.3. Western caisson pile.

of foundations, namely, great supporting power, relatively large diameter, and opportunity to inspect bearing conditions at the base. Diameters range from 24 to 36 in. Loads up to 400 tons may be carried because the full value of the concrete in compression can be utilized. Since the pipe is removed, economy may result.

Method of Forming. (*a*) A heavy steel caisson pipe is sunk to rock, boulders being broken by a churn drill. (*b*) Materials are removed by sand pumps, bailers, etc., until bedrock is exposed. (*c*) When cleaned, the pipe is filled with concrete, either in the dry or by tremie. (*d*) The pipe is withdrawn for reuse.

The caissons are installed by the Western Foundation Corp.

Drilled Piles. Drilled piles are 16 to 24 in. in diameter. Load may be carried by friction and bearing, but principally by whichever is firmest. Lengths are unlimited but usually range from 40 to 50 ft up to 100 ft. They can be battered from 10 to 45 deg. Loads of from 50 to 150 tons are carried.

Drilled piles are formed by drilling holes and filling them with concrete. Advantages of a drilled foundation are no soil displacement or heave; rapid installation; reduced noise problem; no vibration; visual inspection of bottom of hole; penetration of ground and fill that cannot be driven through; and minimization of surface water problems. Drilled foundations are useful for underpinning. They may be more economical than long piles or than spread footings.

Drilling. Cohesive soils are readily drilled with augers or bucket-type drills. Granular, noncohesive materials are drilled by (*a*) augers or bucket-type drills with the aid of a binder or cementing element, such as water (capillary action), bentonite grout, or chemicals (sodium silicate), or (*b*) rotary drills, using drilling mud or water and either washing out the cuttings by circulation or by inserting a steel casing, which drops by its own weight, then removing the drilling fluid. Bentonite may be added to the drilling fluid so that it can be removed by auger.

Rock and Boulder Penetration. Rock may be penetrated, and boulders or old concrete passed by blasting, use of core barrels, or rotary drilling with roller rock bits in dry shafts where cuttings may be removed by air or in wet shafts where cuttings may be removed by fluids.

Belling. After drilling a caisson shaft, a belling bucket can mechanically enlarge the bottom of the shaft. When the bucket hits the shaft bottom, the kelly shaft weight forces cutting arms to open out of the bucket sides. As the bucket is rotated, a bell is shaped and the bucket filled with soil to empty. Upon lifting the bucket, the arms are retracted to permit bucket removal and hold in the soil. For large bells or under conditions of running soil, a tight-fitting casing may be inserted to the depth of top of the bell, then steel rods or oak laths driven as lagging spaced 6 to 12 in. on centers. As hand excavation proceeds, short wallboards and packing are inserted behind the lagging.

Under dry conditions, belling in clay presents no complications, but lagging may be required in sands.

Under wet conditions, belling may be done by (*a*) lowering the water table by deep wells and pumping; (*b*) consolidating the ground surrounding the bell by chemical injection; or (*c*) dewatering by sumping and pumping, using lagging and supports where necessary.

Heaving at the bottom of the excavation should be prevented by interior pressure or dewatering to a greater depth than the bottom of the excavation; otherwise a serious problem results.

Casing. Steel shells, designed for specific problems, are used to prevent sloughing of shaft walls that can reduce the pile friction capacity. Also, casing used when necessary prevents walls of soft, cohesive material from closing in and running silts or sands from entering the shaft. Resulting loss of ground may cause a real settlement.

Steel shaft casings are used where needed to shut off water, keep unstable ground out of the hole, or to protect diggers and inspectors. Casings are removed before concreting or during concreting in unstable ground. By placing a head of concrete in the clean bottom of the shaft, then raising the casing nearly to the top of the concrete, then repeating, soil and water may be kept out of the shaft and firm bearing secured with an unbroken shaft.

Reinforcing. Caissons or piles may be reinforced by cages inserted in the shafts before concreting.

Inspection and Preparation of Bottom. Test holes are usually drilled 3 to 10 ft into exposed bedrock at the bottoms of excavations to determine its character and the absence of seams or crevices. Shear dowels may be set into sloping rock, or steps may be cut.

Inspectors should check plumbness of shaft, depth, and bell dimensions. Written reports should be made of these factors and of foundation conditions, water conditions, and unusual circumstances.

Several leading foundation companies in this field have developed, through research and experience and use of new metals, specialized equipment and methods to meet a variety of conditions previously requiring the more expensive methods of pneumatic or Chicago caissons or piles. Solutions to ever-changing problems will continue to be developed. Some firms may not have the equipment or experience to handle all conditions. Among the leaders in installing drilled foundations are Case Foundation Co., Bell Bottom Foundation Co., George F. Casey Co., and McKinney Drilling Co.

Presscrete Piles (Fig. 10.4). These piles can be installed in very limited space, even under existing structures and within basements where only low headroom is available, since no cumbersome equipment is needed. Installation of these piles is noiseless, and also is done without vibration or disturbance of the ground adjacent to existing structures.

These piles are formed by sinking a sectional steel casing in the ground by ordinary simple shallow well-sinking methods, or by jetting, jacking, or other means. This casing, which is gradually withdrawn by counterpressure as the concreting work progresses, is closed at the top by an airtight pressure head having the necessary connections for the air and concrete pipes and pressure gages. The concrete is injected through the feed line to the bottom of the casing after the ground water has been expelled by air pressure. The casing is gradually withdrawn as concreting progresses. The pressure causes the concrete to expand into the annular free space with such force that irregular bulges are formed wherever the pressure exceeds the ground resistance. This provides a uniform contact pressure with the soil and increases the bearing capacity.

Piles up to 30 in. diameter and 100 ft long have been used. There is no length limit. They may be installed on a batter or horizontally.

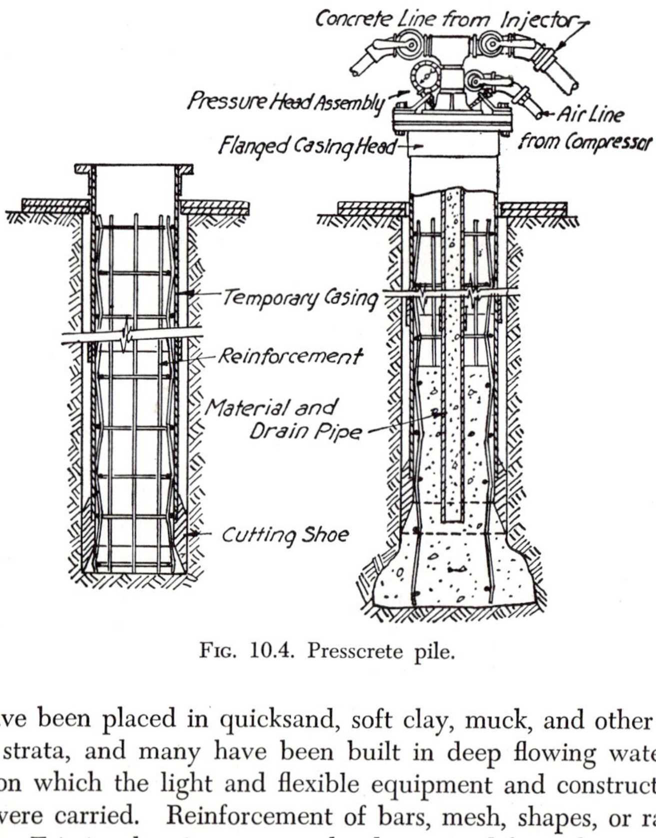

FIG. 10.4. Presscrete pile.

Piles have been placed in quicksand, soft clay, muck, and other difficult ground strata, and many have been built in deep flowing waters from barges on which the light and flexible equipment and construction materials were carried. Reinforcement of bars, mesh, shapes, or rails may be used. Frictional resistance may be determined from the air pressure used.

These piles are controlled by The Presscrete Co., Inc., of New York.

Pneumatic Caisson Piles. These are formed by lowering 5-ft-long sections of 16-in.- or 18.5-in.-diameter steel tubes screwed together as the pipe descends and sunk into the ground by their own weight in a hole dug by a cutter in clay and a bailer in sand. A cage of reinforcing bars is set, and concrete placed through an airlock and subjected to air pressure at varying intervals while the shell is slowly raised in 5-ft lengths by winch. An enlarged grout base can be created before concreting, for heavy base loads. The air pressure can balance any hydrostatic pressure and consolidates the concrete. When waterlogged ground is encountered, the shell is kept free of water and the concrete placed in the dry by the use of the airlock. Vibration during construction is avoided, and piles may be installed in 6-ft headroom.

This method is used by Piling and Construction Co., Ltd., of England.

Patent Pressure Piles. This type of pile can be installed where the headroom is limited, and close to adjacent structures. It is often used for underpinning.

These piles are formed by sinking heavy 12-in.-diameter tubes in the ground, practically by their own weight, while excavating the soil by well-boring methods. An enlargement is often made in the soil at the foot. Reinforcement is then placed. Ground water is driven out by closing the top and applying compressed air. Cement grout is admitted through the cap and used to form the pedestal; then concrete is introduced and forced into the soil while the tube is gradually raised by the air pressure. The bottom of the tube is kept below the top of the concrete at all times, and the air pressure is reduced as the tube ascends, to avoid enlarging the concrete shaft near the top. Collars of concrete are forced out wherever softer strata are encountered.

It is possible to install this pile on a batter. For installation under water, air locks are used.

These piles are controlled by the Pressure Piling Co., Ltd., of England.

Caissons

Drilled Caissons. These are from 18 in. to 10 ft 6 in. in diameter and carry load by end bearing. Rock socketing may be used. Mechanical belling at the bottom to 12-ft diameter, or hand belling to 30-ft diameter, may be used to obtain large bearing capacities on a firm stratum. Bell diameters should be consistent with shaft diameters. Lengths up to 150 ft are practicable. Loads are limited only by the capacity of the ground and are large. Economies result from elimination of pile caps.

The descriptions given under Drilled Piles apply equally to drilled caissons, including drilling, rock and boulder penetration, belling, casing, reinforcing, and inspection; the list of firms installing drilled caissons also is the same.

Caissons can be placed in alternate locations, and intermediate caissons driven before the concrete is too hard, to form a solid foundation wall or cutoff (Fig. 10.6).

Fig. 10.5. Drilling caisson for Mutual Trust Insurance Co. Building, Chicago, Ill. First 25 ft ringed and lagged to 9-ft diameter (Chicago method) for cutoff elevation prior to basement excavation. Caisson shaft 8 ft in diameter, 80 ft deep to bearing on hardpan, belled to 12-ft diameter mechanically and to 22-ft diameter by hand. Installation time was 4 hr for lagged section, 5 hr for shaft, and 8 hr for belling. *Architect-engineers:* Perkins and Will. *Caisson contractor:* Case Foundation Co. (*Courtesy of Case Foundation Co.*)

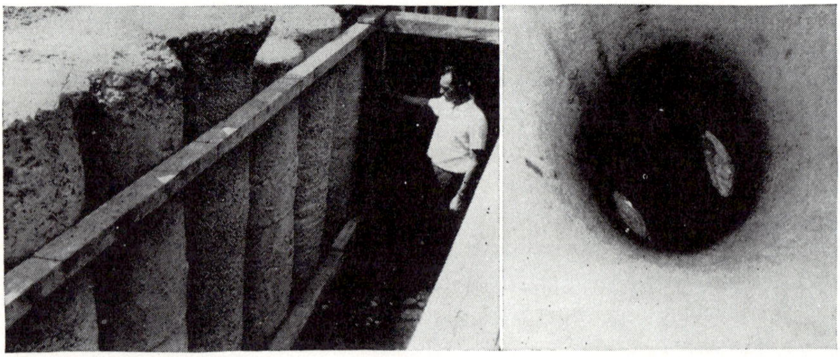

Fig. 10.6. Foundation wall formed by drilled caissons. Completed wall in left view. Alternate caissons placed first, followed by overlapping caissons between, which drilled out recent concrete of first caissons to form interlocks. (*Courtesy of Case Foundation Co.*)

Benoto Caissons.[8e] The Benoto caisson-boring machine was developed in France a number of years ago and has been used in Europe, Asia, South America, and Australia and is now in use in the United States. It can sink shafts from 20 to 60 in. in diameter and 350 ft deep and is self-powered and self-propelled by a diesel-powered hydraulic unit. Casings are steel shells which are forced into the ground by semirotary motions and thrusts. The first section has a cutting edge, to which sections are added as the casing goes down. Digging is done by a hammer grab weighing about 3,000 lb, which falls freely, thus cutting into, or disintegrating through impact, any kind of material from soft to hard rock, whether dry or submerged. Following impact of the hammer grab, the blades close and the muck is hoisted out. Firm anchorage can be obtained in hard base strata, and socketing will give high bearing capacity in rock. The bottom of the hole is inspected and core-sampled if desired. The casing is filled with concrete, and the same rams that delivered the vertical thrust move the shell up and down a short distance to corrugate the concrete surfaces and force concrete into close contact with the ground to improve friction. The semirotary movement is also used during pulling of the casing, to reduce friction between the shell and concrete. These actions and weight of fresh concrete prevent introduction of foreign materials into the shaft. Caissons can be driven on batters up to 15 deg. Caissons can be placed in alternate locations, and intermediate caissons driven so that the edges overlap, to form a solid cutoff or foundation wall, if done before the first concrete has set too hard. Installation time is rapid, and high loads may be carried. The rig walks about on a system of rail wheels and skids worked by rams and can operate on soft ground. The rig is fitted on a trailer for movement on roads and in city streets. It is sold by Benoto, Inc., Chicago, Ill. These caissons are installed by Spencer, White & Prentis.

Hollow Prefabricated Caissons.[7i] These consist of steel plate covered by spirally reinforced poured concrete, filled with water kept from causing corrosion by a tight cap; they are sunk into place by heavy weights. These piles were developed for use in the highly corrosive, *Teredo*-infested warm waters of Lake Maracaibo, Venezuela. Lengths up to 185 ft have been used in 100 ft of water. Diameters are from 4.17 to 5.50 ft. They may be installed in 1½ hr.

These caissons have the following advantages: (*a*) They may be produced by quantity production. Use of short sections permits welding into desired lengths. (*b*) They may be installed by conventional construction equipment. (*c*) Safe load-carrying capacity is known from applied load. (*d*) All parts in contact with water are corrosion-resistant and borerproof. (*e*) Time working over water is reduced to a minimum by assembly on land. Operations may be suspended without damage in case of sudden storms. (*f*) Rapid completion without delay for testing, other than driving a pipe test pile.

Method of Forming. (*a*) Two half cylinders of ¼-in. checkered floor plate are rolled and welded to form a 15-ft section. (*b*) A spiral cage of ½-in. wire spaced 4 in. on centers is placed midway between shell and outer face line, with the shell standing on end. (*c*) Two halves of a steel form are bolted, and a 4½-in. concrete shell poured. (*d*) Sections are placed end to end on a skid rack and projecting steel plates are welded. A steel form is placed around the joint and concrete is poured, using low-shrinkage cement. (*e*) A point section is welded in place. (*f*) A steel plate is welded across the open end to make it watertight, and the caisson is rolled into the water and floated into place. (*g*) The caisson is raised by a stiff-leg derrick and boom, after cutting a hole in the closure plate and filling with a hose as soon as the caisson is vertical. (*h*) Four pairs of 50-ton concrete blocks are set on the caisson, sinking it to safe bearing. (*i*) The caissons are cut off and tied together with prefabricated superstructure rigid-frame girders. (*j*) The caissons are sealed off by a welded plate and concrete cap fill, to prevent interior corrosion.

Chicago Caissons. These were developed for use in that city, where a deep bed of clay overlies hardpan, gravel, or rock. The upper 6 to 12 ft of clay is stiff; below, it is softer but still sufficiently stiff to permit excavation for short distances without caving. The caissons are formed by excavating about 4 ft, sheeting the hole with vertical tongue-and-groove lagging 2 or 3 in. thick and 4 to 6 in. wide, beveled to a circle, and bracing the lagging with two or three segmental steel hoops of bar, angle, or T section. As soon as one section is completed, the operation is repeated, butting the lagging with that above. An improved type of brace consists of four sections of T bars held by a spider having a hollow central hub and radiating screw jacks. The clay is hoisted by buckets or elevators. The lagging and bracing are generally left in place when concrete is placed. When the caisson rests on hardpan it is usually belled out to decrease the unit soil bearing value. The usual range of sizes is from 3 to 12 ft in diameter.

Continued exposure of the shaft may cause the clay to swell and crush the caisson so that heavy jacking frames may be necessary.[8a]

Marsh gas, which is explosive and poisonous, has sometimes been encountered.[8a] It is odorless and tasteless but can be detected by bubbles appearing through puddles in the bottom, or by the flicker of a Wolf-type safety lamp which detects small concentrations of the gas. It can be forced out by ventilation.

Concrete lagging, built in T sections with the flanges keyed together and the stems turned in against steel rings, has been used successfully in Chicago.[8b] The flanges are $5\frac{1}{2}$ in. wide by $1\frac{1}{2}$ in. thick, and the 2-in.-deep stems are $1\frac{1}{2}$ in. wide at the throat and 2 in. wide at the end. Haydite concrete is used, reinforcing the flanges with welded fabric and the stems with single $\frac{3}{8}$-in. bars bent at the ends. The T's are poured in a flat position with stems down, and vibrated. It is claimed that this concrete lagging decreases the diameter of the caisson by twice the thickness of wood lagging, since wood lagging which is usually left in place cannot be counted on for bearing value. It is claimed that this results in less excavation, concrete, and ring steel; there are also fewer pieces to handle.

Sheeted Caissons. With *vertical sheeting*, these are similar to Chicago caissons, except that the sheeting is continuous and is driven down either in advance of excavation, if possible, or as excavation proceeds, with bracing in the form of timbering or steel rings being placed every few feet. Driving may be by hand or by pile-driving hammer. If the sheeting is driven down as excavation proceeds, the soil must be sufficiently stiff not to flow into the bottom. Excavation generally is by hand, the material being hoisted by buckets, but in some larger caissons clamshells can be used. These caissons may be either round or square.

Where lack of vertical clearance prevents driving sheet, it is necessary to use *horizontal or box sheeting*. Use of wood sheeting which is left in place may cause serious settlement of surrounding structures, and the use of precast concrete sheeting may be advisable. The cost has been found to be reasonable and only slightly above that of the most economical wood sheeting.[8c] Lightweight concrete may be preferred, since the planks may be handled manually. Hand loops facilitate handling and also identify the reinforced face. It is customary to use louvers or shuttering construction to permit the space back of the sheets to be tamped solid, in order to equalize pressures on all sides.

Pneumatic Caissons. These consist of shells sunk into the ground by hand excavation, the water and earth being held back from filling the shell by compressed air. They are then filled with concrete. Pressures of up to 48 psi can be used. All workmen and materials enter and leave through an air lock. This is a slow, complicated, dangerous, and expensive operation, but in some cases it is the only way in which a caisson can be constructed. In locations where the rather bulky equipment for other methods cannot be used, pneumatic caissons may be required.

In 1820, Lord Cochrane patented a system of air-caisson construction that contained most features of pneumatic caissons. In 1839, Triger designed a lock-and-air chamber. Sooysmith largely adapted and perfected the method in the United States. Until 1900, most air caissons used timber, although sometimes cast iron or steel. In 1906, the first reinforced-concrete air caisson was developed by The Foundation Co. under D. E. Moran and J. W. Doty.

Prior to the more recent inventions of various drilled caissons, the pneumatic method was the only way of sinking caissons through soft, quick, or water-bearing strata. Present methods call for an air lock well above possible high water, attached to the top of the casing. It is necessary for an air lock to be as economical as possible in the amount of air used for locking in and out, and also to require the least possible time for decompressing, equalizing, and opening and closing doors. There are four types of locks—man, material or excavating, concreting or combinations of all three, and medical or hospital locks. Material or excavating locks are used to pass out excavated material and bring in tools and supplies in buckets. The use of buckets is too slow for concreting and requires a man at the bottom to dump the buckets; consequently concrete locks handling 1 cu yd or more at a time are used, generally as attachments to the material locks. One type of concrete attachment can be operated entirely from the outside, and another requires a man inside the lock to operate the door of the attachment. Most states have air or caisson laws with detailed requirements as to use, equipment, hours, care of men, etc. Such laws require a medical or hospital lock for treatment of men afflicted with caisson disease or "the bends." Shafts may be without ladders, or have ladders on the inside or in recesses in the shell walls.

Liner-plate Caissons. Curved segments of corrugated flanged steel tunnel liner plates, such as made by Armco Drainage & Metal Products, Inc., or steel liner plates, such as made by The Commercial Shearing & Stamping Co., are used for circular caissons.

CHAPTER 11

H PILES AND OTHER METAL PILES

Metal piles have been used since 1838, in the form of cast-iron pipes or solid wrought-iron shafts with disks or screw flanges penetrating only short distances. Use of structural steel I-beam and built-up piles originated before 1900 in highway bridges in Nebraska and adjoining states, to avoid failure due to ice floes and deep scouring. After 1908, rolled H sections were introduced into the United States by Bethlehem Steel Company and superseded the older types.*

H piles. Steel H bearing piles are suitable where it is desired to penetrate to rock or through hard material with the least effort and time; because of their smaller soil displacement, these piles are sometimes the only ones that can be driven to the desired penetration without recourse to jetting, coring, or other similar operations. They have in numerous installations been driven many feet in sand and gravel to depths below the scour line. Penetrations of 10 to 12 ft in hardpan are common, and 22 ft has been secured in cemented sand and gravel. Where close spacing is necessary, their small soil displacement is of importance. H piles are often used in reconstruction of bridges where they can be driven through existing construction in small spaces. They are also useful for driving close to existing structures, since they cause little displacement of the soil. Where large lateral forces may be exerted or where earthquake forces may occur, the large bending strength is of great value. It is sometimes possible to reduce the weight of the lower section in firm material, using a heavier section above where great unsupported length occurs. H piles require less space for shipping and storing than wood, pipe, or precast concrete piles. They do not require special slings or special care in handling unless they are long, in which case they should be supported with the web vertical, or attachments made at several points. Unspliced lengths of 127 ft 6 in. and spliced lengths of 304 ft have been driven.

* E. L. Durkee and E. L. McIntosh, "Structural Steel Bearing Piles—Their Use and Capacity," *J. Boston Soc. Civil Engrs.* vol. 24, No. 2, April, 1937.

Steel H piles are also used extensively in retaining walls, bulkheads, etc. In trench cuts, they are driven on 5- to 8-ft centers, prior to any excavation, and planks are dropped in the grooves formed by the flanges as excavation proceeds. This construction is useful when driving a sheet-piling cofferdam or trench and encountering some horizontal obstruction such as a pipe across the excavation; by driving an H pile on each side of the obstruction, planks can be dropped in the grooves of the piles below and above the obstruction. H piles are readily withdrawn for reuse.

Handling holes may be furnished at one or both ends by the mill, if desired.

A comprehensive study of H piles appears in reference 3*af*.

Properties of H piles are shown in Table V.4.

Pile Caps (Fig. 11.1). Caps built of flat plates with stiffeners can be furnished, and many other types may be used. The sizes should be such as to limit the bearing against the concrete to 0.25 of the ultimate strength of the concrete, unless otherwise governed by code requirements. However, research tests have been made by the State of Ohio[3t] that it is believed present conclusive evidence that where the top of a steel H pile is embedded in a concrete footing or cap, if the pile itself is of adequate section and the concrete member is of adequate size and arrangement and properly reinforced for pile reactions, there need be no concern regarding the strength of the connection for compressive force, and it is unnecessary to provide a bearing plate or other auxiliary bearing device at the top of the pile.

Splices (Fig. 11.2). Splices may be riveted, bolted, or welded. Splice connections are furnished by the mill if desired. Welded splices may be welded plate or bar splices, or butt-welded splices. Butt-welded splices require scarfing of one section, with the use of small backing plates if desired, and are coming into considerable favor. Splice material should be kept on the inner faces, if it is desirable to avoid forcing a hole in the ground larger than the pile, which would lose frictional and lateral support at least temporarily and might result in bending, eccentric stresses, or failure. If piles are restrained throughout their full length in firm soil, or if the pile is one of a cluster or the splice grade is staggered with regard to adjacent piles, and if driving is moderate, a splice consisting of a full butt weld or single web and flange plates welded on in the field suffices. Where splices are all at approximately the same grade and where hard driving occurs, even though the piles are restrained laterally for their full lengths, the splices should be designed to develop one-third of the moment value of the section. Splices in long piles not braced laterally, particularly if the piles form part of a trestle bent, should be designed to develop the full moment value of the

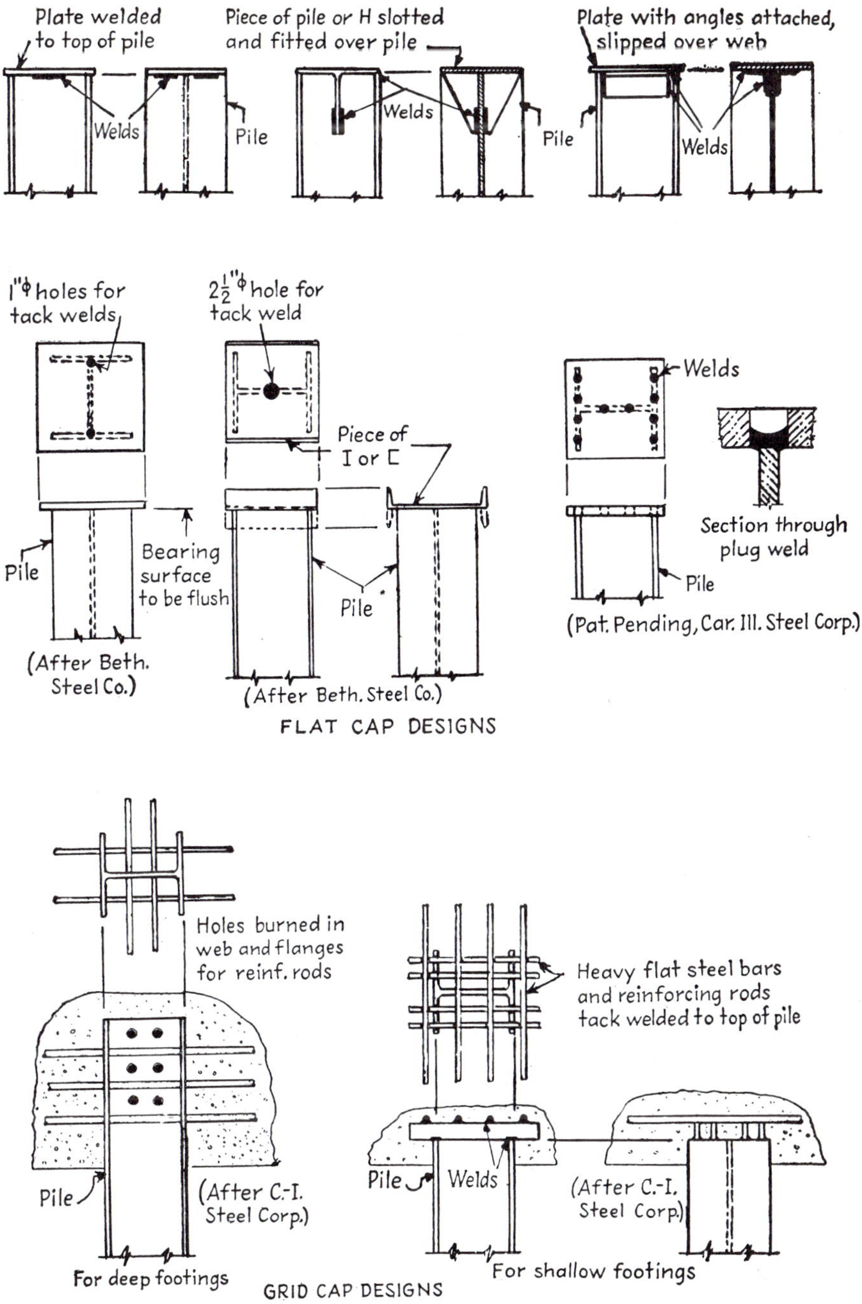

FIG. 11.1. Typical caps for H piles.

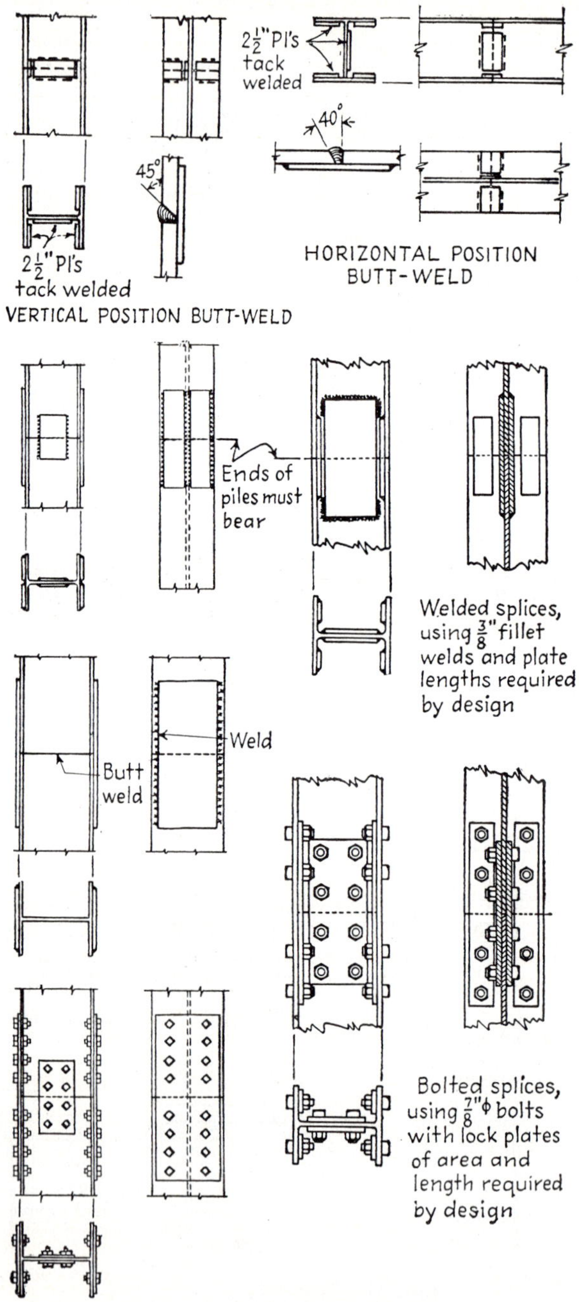

FIG. 11.2. Typical H-pile splices.

section. For bolted splices, the ends of the pile should be milled for bearing. For a table of typical welded splices, see Table V.11.

Followers (Fig. 4.19). Followers consisting of a section of H pile with down-standing plates to hold the alignment may be used. For follower designs, see Table V.12.

Points (Fig. 11.3). Points may be reinforced by adding welded or riveted plates to bring the pressure between the gross area of the steel and the rock down to a range of 3,000 to 6,000 psi. This is in comparison with crushing strengths of 6,000 to 18,000 psi found for small cubes of rock. It is advisable to build up the thickness to 2½ to 3 times the original, for a height of 2½ to 3 times the diameter. Although this reduces the pressure on the rock, the principal object accomplished is that of distributing the load into the shaft in case only one corner rests on rock, avoiding local buckling. In such cases, the welds or rivets should be proportioned to develop stresses of 10,000 psi in the added plates to ensure against angles and plates tearing loose during driving. Reinforcement is not generally used when 6 to 8 ft of hardpan, 8 to 10 ft of gravel, 10 to 12 ft of sand, or 12 to 15 ft of hard clay overlie the rock. Cast-steel points are sometimes valuable for piles which are to be pulled and redriven several times but are of no advantage for permanent piles. It is not recommended that points be chamfered or sharpened, unless driving to bearing on sloping surfaces of rock so hard that the piles cannot be driven into it, as experience shows that blunt flat ends drive straighter and penetrate farther into soft rock.

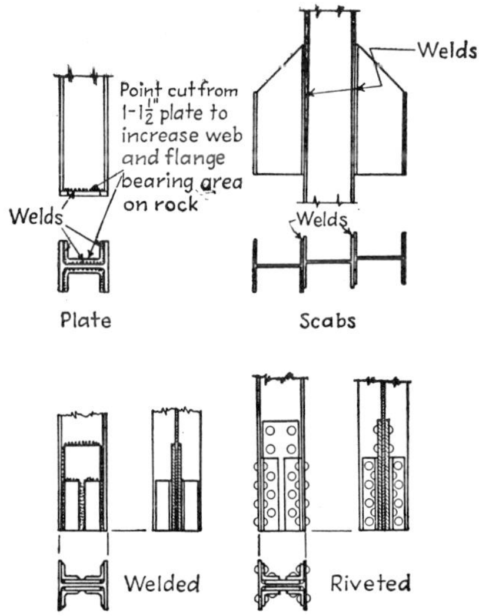

Fig. 11.3. Typical built-up points for H piles.

Fig. 11.4. Short H sections attached to H pile to increase bearing capacity in hard stratum. (*Courtesy of Bethlehem Steel Co.*)

Devices to Increase Bearing Capacity. For H piles that do not reach hard bearing, devices to increase bearing capacity are usually of little value, and additional length of plain section is often a less costly way of obtaining the same capacity. Sometimes short sections of H piling have been fastened to each flange of the pile at the bottom in order to provide greater bearing area in hard soils, but this method is not recommended for bearing on rock because of possible unevenness of the rock surface. Recent patented tapered enlargements of wood, concrete, or steel have been found of value; they form a spear point 10 to 25 ft long, enlarging up to three times the pile diameter. When piles extend through soft materials into firmer friction-load-carrying strata, such points in the firm material may reduce the pile cost appreciably.

I-beam Piles. Where additional lateral strength in one direction is needed, I beams may be used for piles. On the banks of the lower Mississippi, 30-in. beams were driven as wharf piles through water and silt and a long distance into a dense hard stratum 65 ft below mean water level, to resist thrust from a sliding bank.

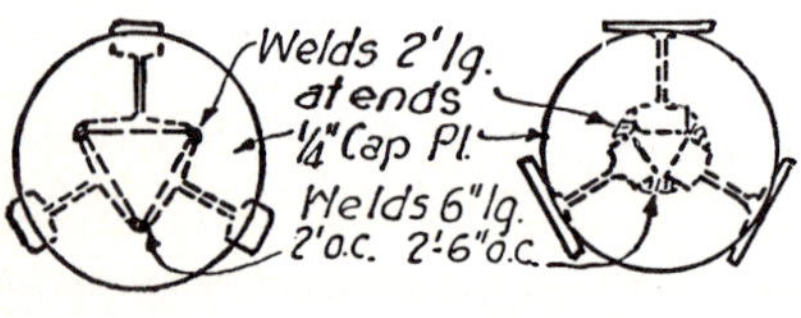

FIG. 11.5. Rail piles.

Rail Piles (Fig. 11.5).*,[3aa] Old rails have been used extensively by the Southern Pacific Co. for piles, from designs developed by them. Usual lengths are 30 or 39 ft, but have been welded to 90 ft. They are stiff for pickup in 90-ft lengths and may provide more lateral resistance and displace obstructions with less deviations than H piles. They can possibly be obtained more quickly in emergencies than some other pile types.

Beam-and-sheeting Box Piles (Figs. 11.7*a* and 11.7*b*). Great lateral strength can be obtained with box piles composed of deep beams and steel sheeting. The pile in Fig. 11.7*a* has successfully supported a wharf on the lower Mississippi, where deep water, silt, and sliding banks are present. The design shown in Fig. 11.7*b* was developed as individual supports for suction pipe lines extending out into the same river. The long direction was intended to resist sliding thrust from the bank, while the narrow direction provided great strength to resist a possible downstream component of bank sliding and the force of the current. Both types of piles were at least 100 ft long. Concrete fill was tremied after blowing out down to sand and gravel at a depth of about 70 ft.

Larssen Box Piles (Fig. 11.7*c*). These are formed by welding together two sections of Larssen steel sheet piling at intervals along the interlocks. Properties are given in Table V.13. They can be driven into soft rock and perform the same functions as a pipe pile. They

* *Eng. News-Record,* vol. 123, No. 11, p. 358, Sept. 14, 1939.

usually are not filled with concrete but can be cleaned out and filled to any desired depth for strength and protection of the interior against corrosion. Shoes can be furnished if desired. This piling is furnished to meet *British Standard Specification No.* 15 for steel, including 0.25 to 0.35 per cent copper content if desired, or it may be of Atlantes high-tensile-strength rust-resisting steel. The sheeting comes from the mill coated with an acid-free tar coating. The sections may be spliced by

FIG. 11.6. Larssen steel box piles being driven inside cofferdam to form new foundation for shipway extension for Harland & Wolff, Ltd., Belfast, Northern Ireland. *Consulting engineers:* Hobart & Heron. *Contractor:* McLaughlin & Harvey, Ltd. (*Courtesy of The British Steel Piling Co., Ltd.*)

welding. They are sometimes used as horizontal struts to brace cofferdams.

Larssen piling is rolled in England by the Cargo Fleet Iron Works of the South Durham Steel & Iron Co., Ltd. It has been imported into the United States.

Dorman Long Box Piles. These are formed from any structural sections and plates rolled by Dorman Long & Co., Ltd., to suit the requirements of the purchaser. The general types consist of a pair of channels welded toe to toe, or a pair of channels with toes facing each other but spread so that the box is completed by a plate welded to the flanges on each side.

Algoma Box Piles (Fig. 11.7*d*). These Canadian box piles are formed by welding together sections of Algoma steel sheet piling in various shapes (properties are given in Table V.14). They can be driven in soft rock and perform the same functions as pipe piles or caisson piles. They can be filled with concrete to any desired depth, to protect the interior against corrosion or to give added bearing value.

Normal practice is to weld the interlock bar using two or three 6-in. stitch welds on both sides and at each end of the piling, but the amount of welding depends upon pile length and driving conditions.

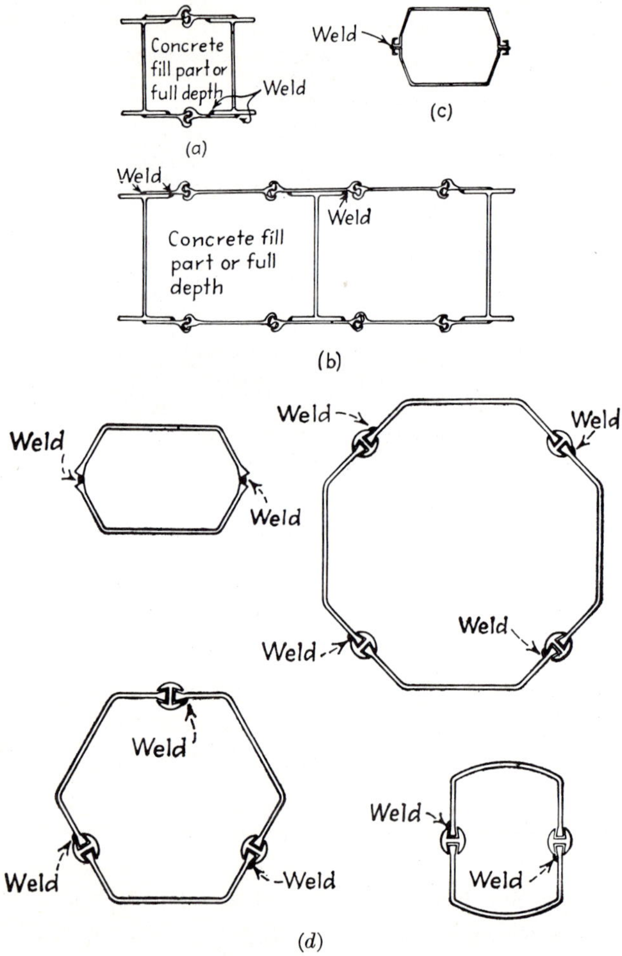

FIG. 11.7. Box piles.

Rendhex Piles (Fig. 11.8*a*). These are hexagonal box piles (properties given in Table V.15), having constant cross section, with shoes if desired, of lengths up to 120 ft without joints. They are rolled in England by the Cargo Fleet Iron Works of the South Durham Steel & Iron Co., Ltd.

Frodingham Octagonal Box Piles (Fig. 11.8*b*). These octagonal box piles (properties given in Table V.16) are formed by welding together two special rolled sections. Quite uniform radii of gyration and uniform strength in all directions are obtained. They are particularly suitable for marine structures, and bracing may be added to any flat face. They are generally driven open-ended, but may have a welded flat plate shoe or a conical loose or welded shoe.

Iron Piles. Cast-iron and wrought-iron piles were used to quite an extent some years ago. Some were filled with concrete. The shells were usually at least 1 in. thick. The chief advantage of these piles was their great durability, service records of 50 years or longer being noted, which is remarkably long for piles in sea water. Below water they are usually found still to be in good condition. Rusting occurred above, in spite of painting, but many are still serviceable. The cost today would be rather high. The bracing generally has rusted much faster than the piles. Among their disadvantages are high initial cost, the need for driving very carefully in order to have the bolt holes for fittings come at the right grade, the tendency to break at bracing points under impact of ships, and the relatively short lengths which can be used.

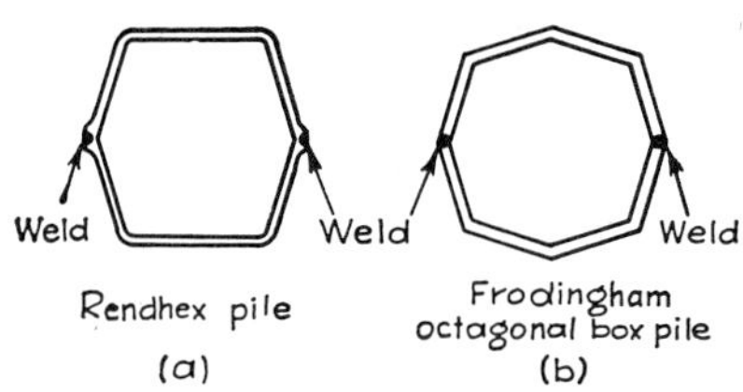

FIG. 11.8. Rendhex and Frodingham box piles.

There are three general forms of these piles: ordinary flanged pipe, disk piles, and screw piles. In the flanged type, a blank flange with a small orifice is attached to the tip and sinking is accomplished by water pressure. Truncated cones with a 2½-in. central orifice and four ½-in. openings halfway up the cone have been successful.

Disk Piles. Hollow cast-iron pipes with a plate or casting of enlarged size at the bottom are not often used now. They may be economical for light work on sandy silt or sand. For a 9-in. pipe, a 36-in. disk might be suitable. Shafts of 12-in. diameter with 72-in. disks have been used. These piles are usually jetted, and precautions are needed to keep them plumb and in location.

Screw Piles.[3r,3x] These are much superior to disk piles, causing less disturbance. They fell into disfavor after a period of use for such purposes as anchoring buoys, for lighthouse foundations, and for small bridges. The supporting power can be considerable and the pulling power is large since the weight of a cone of earth must be lifted. They can be put down without disturbing adjacent structures. They generally consist of an iron shaft from 3 to 10 in. in diameter, having at the foot one or two turns of cast-iron screw of a diameter from 1½ to 5 ft. They were screwed into the soil by men working capstan bars, or by horses or motive power. The screw penetrates most soils without great difficulty and will push aside boulders that are not too large. Dry sand presents the most difficulty. A water jet is sometimes used.

Later designs are hollow and open-ended so that the soil can be jetted and ground broken up if screwing becomes too hard. In Great Britain in late years, there has been a development of this type, screws of 7 ft 6 in. with cylinders 3 ft in diameter having been used successfully in deep mud. Steel instead of cast iron has been used to save weight. For heavy loads in poor soil up to three turns of the screw have been used. Four-foot blades have penetrated sand and clay at 10 ft per hr and hard chalk at 4 ft per hr, using a 5-in. pitch of blade. The ultimate bearing pressure is usually based on Rankine's formula for passive resistance:

$$R_u = wLA_h\left(\frac{1 + \sin\phi}{1 - \sin\phi}\right)^2 + \sum(\pi DL_sF_s) - W_p \qquad (11.1)$$

where w = weight of soil, in pounds per cubic foot;
A_h = area of helix, in square feet;
D = diameter of shaft, in feet;
L = depth to helix, in feet;
L_s = depths of various strata, in feet; and
F_s = skin friction per square foot on shaft, for various strata.

A factor of safety must then be applied. The thickness of metal at the root of the helix, and the pitch must be determined.[1k]

Braithwaite-type screw piles,[3r] comprising a 42-in.-diameter corrugated steel shell ⅛ in. thick to act as a form for the concrete, attached to a metal helical tip having an 8-ft diameter and steel-sheathed concrete-filled conical point, were installed in 20- to 65-ft lengths 15 ft into hard sandy bottom at Alexandria during World War II to carry 250-ton pier loads at 15 ft on centers. A removable rotating pipe of 1-in. thickness was inserted though the shell and screwed into the base during installation. The electrically operated rotating head was carried on a movable gantry, each pile requiring 1 hr for installation.

CHAPTER 12

SHEET PILING

Types of Sheet Piling

Sheet piling may be *wood, concrete,* or *steel.* For many uses, any of these are suitable, whereas for other purposes or in certain circumstances, one particular type may be advantageous from the standpoint of cost, ease of installation, availability, salvage, corrosion, decay, ability to withstand driving, lateral strength, number of wales required, ease of making connections, etc.

Uses of Sheet Piling

Sheet piling is used for a great variety of purposes and this versatile material is invaluable in construction operations. Many ingenious uses have been made of it, and new designs will always be worked out in the future to suit some particular condition. Among the types of service for which it is used are the following.

Cofferdams. Wood sheeting has long been used for cofferdams. Steel sheeting has greatly extended the use of cofferdams by reason of the ease and certainty with which it may be driven, the positiveness of its interlock, its watertightness without puddling, and the ability to pull and reuse it repeatedly. There are firms who furnish steel sheeting on a rental basis.

Straight, arch-web, and Z-shape types of steel sheeting are used for cofferdams, depending upon depth of excavation, soil, type of water levels, and number of wales and struts desirable from the standpoints of economy and noninterference with excavating or other construction operations.

Many cofferdams are of square or rectangular shape, with braced single walls. Double walls, with the space between the walls filled with puddled soil, have often been used with wood sheeting.

The unbraced self-supporting cellular cofferdam is the most efficient solution for certain types of large deep foundations. Ring-braced circular cofferdams have been found to be economical.

Building Foundations. Sheeting for building foundation or pier excavations is usually either wood or steel, and may be pulled or left in place to serve as forms. Sometimes it is lined with light tongue-and-groove wood sheathing or building paper,[3j] or greased to allow pulling after serving as forms, and the joints of steel sheeting have occasionally been greased or graphited with this end in view. Concrete sheeting has been used for lagging Chicago caissons and left in place to serve as part of the bearing area.[8b]

Bulkheads. All classes of materials have been used for bulkheads. If wood is used, it should be treated, unless it is for temporary service only. Such walls may be anchored back to deadmen or to pile bents containing batter piles, or braced with batter piles or struts.

Shoring. Shoring of adjacent structures may be done by driving sheeting close to their foundations prior to excavating. Since it is inadvisable to pull such sheeting, it serves as the forms for the concrete wall of the new structure. Batter piles, struts, or wales can remain in place in the wall. The wall design should take this into account.

Load-carrying Piers. The use of steel sheet piling to form both pile foundations and piers for bridges and wharves is increasing, often saving cofferdams, excavation, and superstructure concrete work. The piers are sometimes made of only a few pieces of sheet piling and corners, and filled with concrete to either part or all of the depth. The piers may be blown out or cleaned by any of the methods used for pipe piles. Larger piers may be filled with gravel and capped with concrete mats, if not filled with concrete.

Sea Walls and Groins. Sheet piling offers a good solution to the protection of low areas and shore property, and the building up of beaches. Wood has long been used for th_ purpose, but should have a 10- to 12-lb per cu ft creosote treatment for inland and fresh-water structures, and a maximum practicable retention by the full-cell process for salt-water locations exposed to marine-borer attack.

Steel sheet piling is very strong and rugged for this type of work. Sometimes the projecting portions are encased in concrete, and if the horizontal forces are large this top may be designed as a reinforced-concrete counterfort wall. Abrasion is sometimes reduced by using creosoted timber plank sheathing.

Concrete sheet piling has been used successfully on the Mississippi Sound, but designs heavy enough for more exposed localities have not yet been developed, although it seems likely that they could be, using dry vibrated dense concrete with 3 to 4 in. of cover over the reinforcement.

Permeable groins may be made by driving occasional piles below the general grade, spaced to permit passage of water and sand.

Small Dams and Cutoff Walls. Sheet piling is sometimes used for small dams but more often as cutoff walls under dams. The flow of water often occurring under stream beds in arid regions thus can be cut off and brought to the surface to form reservoirs.

Steel sheet piling is frequently used as the diaphragm for large earth dams, to prevent trickles from developing into permanent channels. The flexibility of steel sheeting makes its use advantageous, since it will accommodate itself to the settlements and adjustments of the dam.

Steel sheet piling is also used as cutoff walls under concrete dams to prevent underscour and, for this purpose, may be either single line, or double line with cross walls to form cells to retain the material upon which the dam rests.

Steel sheet piling is often used as a curtain or cutoff wall under earth or concrete levees.

Trench Sheathing. Wood and steel piling are both used for this purpose, the choice depending upon cost, availability, depth, magnitude of the pressures, etc.

Docks and Wharf Walls. Steel sheet piling is coming into common use in dock and wharf designs. It is used as the face wall, or as a wall at some distance back of the face. Wales transmit the earth thrusts to tie rods that extend well back from the wall.

Another type involves the use of heavy master piles spaced at regular intervals, connected by steel sheet piling driven in arcs (Fig. 12.1).

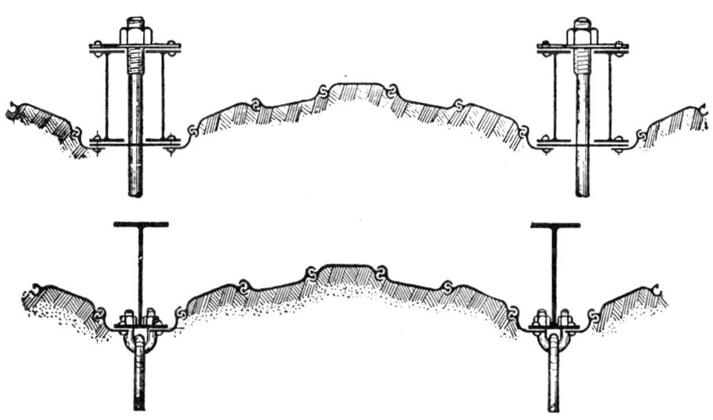

FIG. 12.1. Typical arc buckstay sheet-piling walls using master piles. (*United States Steel Co. designs.*)

The master piles are held by tie rods. Ocean-going vessels require large depths of water close to the wharf, and great pressures develop on the walls. In the arc master-pile type, the large section modulus needed in bending can better be developed by deep master piles than by the much shallower individual sheeting members if used in straight-line sheeting, provided sheeting with adequate interlock strength in tension is selected. Only straight-web sheeting is suitable for use in arcs, since the arch webs would flatten and twist the interlocks. The arc sheeting need not be driven as deep as the master piles but only far enough to hold the fill. Use of relieving platforms may result in lowered pressures.

Intake Caissons. Steel sheeting, driven in circular form with circular steel or concrete* wales, has successfully been used to form caissons for water-intake supplies. Excavation and the driving of holding-down piles for a tremie mat can take place through water. After this, the bottom can be sealed by a tremie mat, and wales placed as the caisson is unwatered.

Design of Sheet Piling

Classical earth-pressure theories do not apply to all types of soils, moisture contents, and yielding conditions of walls, and before designing the wales and span of sheeting, modern textbooks on the subject should be consulted. This is very important, since, in certain cases, the pressure on the upper part of the wall has been found to be several times the pressure obtained by the older theories.

There may be varying conditions of loading caused by possible changes in dry and wet or submerged conditions.

WOOD SHEET PILING

General Features (Fig. 12.2)

If only an earth bank is to be supported, a single or double row of planks will serve as wood sheeting. If watertightness is required or pressures are large, Wakefield, Martinez, or tongue-and-groove sheeting may be used. The foot is usually beveled to force the sheet against the one previously driven. The tongue should always lead, in order to prevent clogging a groove with wedged stones. The top corners should be cut back 3 to 4 in. at 45 deg in order to avoid brooming. Triple sheeting seems to resist driving better than single sticks, as defects cannot extend through the entire sheeting, and warping is minimized. Wood sheeting has been laid in panels on the ground and swung into position in movable jigs ready for driving the individual pieces.[110]

Types of Wood Sheeting

Wakefield Sheeting. This consists of three planks, 2, 3, or 4 in. thick, nailed together, with the middle plank offset to form a tongue and

* W. S. Colby, "Design and Construction of a Circulating Water Intake," *ACI Proc.*, vol. 46, No. 7, pp. 497–508, March, 1950.

groove. The planks are usually fastened together using pairs of staggered bolts at 6-ft centers, with spikes at 18-in. centers between, the bolts being ½ in. in diameter for planks up to 2½ in. thick, and ⅝ in. for thicker ones, and the spikes being driven half from each face. The planks are sometimes spiked together with nails ½ in. longer than the total thickness of the three pieces and the ends clinched, or else boat spikes may be used. Half the nails should be driven from each side, using 1-ft spacing for 3 to 6 ft at the ends and 3 to 4 ft in between. The ends are beveled to aid driving and prevent brooming. All planks

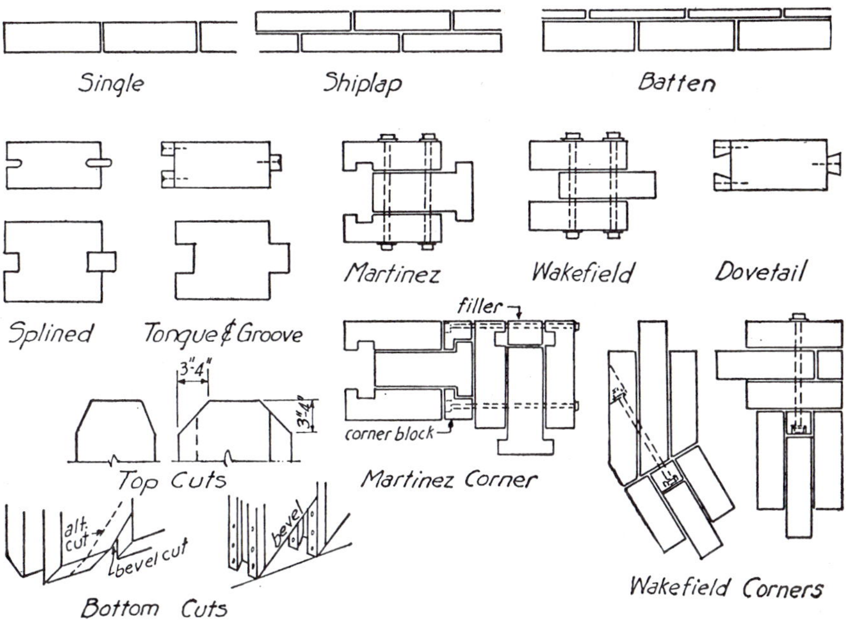

FIG. 12.2. Wood sheeting—types and details.

should be dressed four sides. The patents for Wakefield sheeting have expired.

Martinez Sheeting. This is an improved form of wood sheeting which has been used extensively for levee work along the Mississippi River. It has performed successfully where other types have failed. The planks are nailed together, as in Wakefield sheeting.

Splined Sheeting. This consists of single-thickness planks, with grooves on each edge into which separate splines are inserted. The splines can be of tough hardwood.

Tongue-and-groove Sheeting. This consists of a single thickness of planks, with a tongue cut in one edge of the timber and a groove in the other.

Dovetail Sheeting. This is formed by nailing beveled strips to the edge of the plank to form dovetailed tongue-and-groove joints. This sheeting in 4-in. thickness is standard with the Southern Pacific Co. It is made up to 12 or 15 in. thick. The tongues can be of tough hardwood.

Bending Strength

When designing wood sheeting against bending, the question arises as to whether the section modulus can be considered for the full combined section, or whether the nailing is inadequate to develop the longitudinal shears thus involved. This can be tested by loading several sections of sheeting on supports spaced at the wale spacing. Deflections can be measured and moments of inertia and section moduli computed. Test results seem to indicate that the section modulus of Wakefield sheeting cannot be depended upon to be any considerable amount in excess of the section moduli of the individual planks.

STEEL SHEET PILING

American Sections

Specifications. Manufacturers' standard specifications[3v] for steel sheet piling and the *ASTM Standard Specification for Steel Sheet Piling* (Serial Designation: A328) should be used.

Regular manufacturing practices of the American Iron and Steel Institute[3v] state rules in use for surface finish, minor repairs on conditioning, cleanliness, identification, test reports, inspection, loading, lifting, invoicing, certification, quality, chemical- and tensile-property limits, mechanical-test limits, cutting, and tolerances.

Cutting to length may be done by sawing or gas-cutting at the mill's option, unless specified.

Tolerance in mill length is plus 5 in., minus 0 in.

Standard practice is to provide handling holes or single or double extracting holes in one end of straight or arch web sections, and for Z wet sections in both ends, without extra charge. Holes will be on the center lines of webs and, if specified, may vary from 4 to 22 in. from the ends.

Special Manufacturing Practices.[3v] Certain requirements, when specified, are considered special or restrictive, such as specified interlock tests; nonstandard tolerances; nonstandard handling or extracting holes; painting or other rust deterrent; subjection to inspection after descaling by purchaser; handling of sheet piling for surface inspection independently by purchaser; refusal to permit minor conditioning repairs; and requirements for additional or different markings, other than indicated under regular manufacturing practices.

Copper-bearing steel (0.20 per cent minimum copper) can be obtained at slight additional cost, for cases where greater resistance to corrosion is needed than is obtainable from ordinary steel.

Profiles and Properties. Elements of sections of steel sheet piling are given in manufacturers' catalogues. Sections are grouped into three general classes: large interlocks, small interlocks, and Z sections.

Large interlocks are illustrated by sections MP-117, MP-101, and MP-102, which interlock together and are intended for use where high interlock strength is required, hard driving is encountered, or for repeated reuse.

Small interlocks are illustrated by sections MP-110, MP-112, MP-113, MP-115, and MP-116 which interlock together, sections MP-112 and MP-113 being proportioned for maximum interlock strength in tension for cellular and master-pile arc construction; these two sections are also used for corners and fabricated connections for other members of the group. Sections such as MP-110, MP-115, MP-116, MZ-38, and MZ-32 have interlocks with metal so distributed as to provide high strength in beam action.

Z piling has the highest beam strength for its weight and, although harder to roll in the mill and to drive straight, its great strength makes it invaluable in very heavy work.

Interlock Strength. This is the value of the interlock in tension per linear inch of sheeting. Although important in master-pile arc construction, it is of little importance in ordinary construction using wales, or in circular cofferdams or caissons where the action at the joints is compression rather than tension.

The values of interlock strength usually given are yield-point values, and it is recommended that half these amounts be used for working stresses. Two-third of these values have been used at times, with occasional difficulty.

Driving in the reversed position, that is, with the arches, instead of being staggered as in the normal position, all on the same side of the piling wall, results in only about one-half the normal interlock strength being developed. In cases where interlock strength is not important and where concrete seals or walls are poured directly against the sheeting, this results in a saving of concrete. It also permits pouring foundation walls closer to property lines, if the outsides of the webs all lie on the property line. Because sections AP-8 and MP-117 are rolled with one interlock reversed, the interlocks are in the normal position when driven either with the arches lined up on the same side of the wall or with the arches on opposite sides of the wall. Full interlock strength is thus obtained with either arrangement.

Lateral Strength. Values of the various sections are tabulated as section moduli of a single piece. The use of the interlocked-section modulus computed from the combined interlocked reversed section is not recommended unless slippage is eliminated longitudinally by welding or by other means. Because the interlocks of deep standard Z piling are located where the longitudinal shear is zero, they need never be welded together, and the section modulus of a single uninterlocked pile is the same as when interlocked with the adjoining pile.

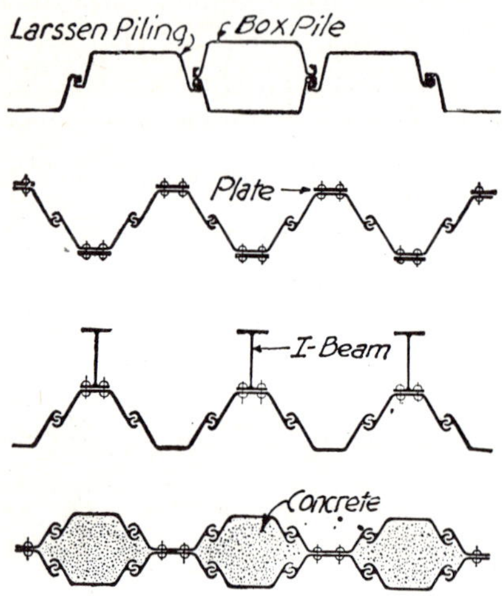

FIG. 12.3. Typical methods of reinforcing steel sheet piling.

The shallow Z sections WZ-22 and WZ-27 are used in locations where arch-web, straight-web, or deep sheet piles are generally used, and provide greater strength in bending for the same weight.

Increased section modulus may be obtained by attaching steel beams or flat plates to the flat webs, or by forming double sheeting so that concrete cores may be obtained (Fig. 12.3). These methods will permit the use of higher walls or surcharges.

Canadian Sections

Algoma Sections. These are made in Canada.[1ae] The ultimate interlock tension strength ranges from 6,000 lb per lin in. in the lighter interlock bars to 10,000 lb per lin in. in the heaviest bars, for sections A, B, and C. It is 16,000 lb per lin in. for S sections and 12,000 lb per lin in. for FA sections.

Box piles (Fig. 11.7*d*), consisting of several Algoma sheet piles, may be formed. Properties are given in Table V.14. For further information, see Chap. 11.

European Sections

Larssen Sections. This German steel sheet piling has been used extensively in Europe and to some extent in the United States. It is rolled in England by the Cargo Fleet Iron Works of the South Durham Steel & Iron Co., Ltd., and sold through the British Steel Piling Co., Ltd. Its uses, layouts, and driving are covered in *The B.S.P. Pocket Book.*[11c]

Standard specifications specify Siemens-Martin open-hearth steel conforming to *British Standard Specification No.* 15. Copper-bearing steel as above, but containing 0.25 to 0.35 per cent copper to increase rust resistance, may be obtained. Rust-resisting steel, of Atlantes high-tensile-strength specification, may also be obtained.

Tar is furnished on piles used for permanent work, and should be renewed on the exposed parts from time to time.

Lateral strengths of the various sections are tabulated both as the section moduli of a single piece and of a double-thickness wall. The section modulus of a line of Larssen sheet piling is claimed by the manufacturers to be that obtained by considering the interlock as fully effective.

Box piles (Fig. 11.7*c*), consisting of two ordinary Larssen sheet piles welded together at intervals along the interlocks, may be formed. Properties are given in Table V.13. They can be used individually, or at intervals in a wall to support occasional vertical loads. For further information see Chap. 11.

Imported Sections. Other European sections, such as Belval,[11ao] Rombas,[11ar] Peine,[11as] and Differdange[11am] may be imported into the United States.

Frodingham Sections. These are British sections.

Angular Deflections in Interlocks

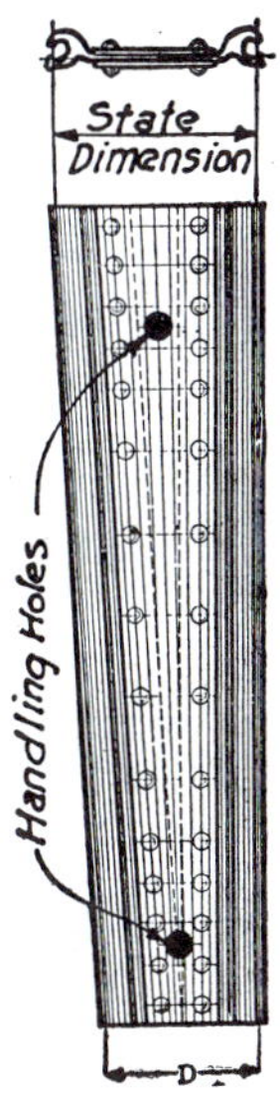

FIG. 12.4. Typical taper sheet pile.

Steel sheeting sections, except Z piling, can be driven to form an angle between adjacent pieces. This angle is 10 deg for American sections, except for such sections as SP-9 and I-21, for which it is 15 deg. The very tight interlock of Z piling precludes forming an angle between adjacent sections. If it is necessary to drive steel sheeting to a smaller radius than permitted by the above angles, it may be obtained with bends of 10 or 20 deg made in the web. Cylinders as small as 2 ft by 8½ in. diameter can be driven with SP-9 piling having a 20-deg web bend. Half of the piles must be bent with the fingers out and half with fingers in, except in the case of section AP-8 that has the fingers at one end reversed.

For Algoma sections, the standard interlock bars allow a swing between sheets up to 12 deg, permitting construction of a cell with a radius of approximately 7 ft. A much greater deflection, and smaller radius, can be obtained from using oversized interlock bars.

Taper Sheet Piles

Taper sheet piles (Fig. 12.4) can be furnished to form the closure pieces often required owing to the fact that toes of piles frequently tend to lead when driving. They are made by splitting a pile and cutting out a triangular-shaped piece, then riveting or welding the edge pieces together in the tapered form desired.

Splices

Splices may be obtained with the piles. Details of splices for American sections are shown in Manufacturers' Catalogues.

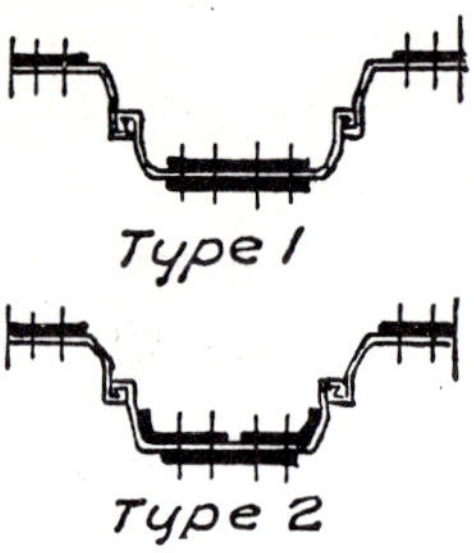

FIG. 12.5. Larssen pile fishplates.

Splices for Larssen piling may be formed by the use of fishplates, as shown in Fig. 12.5. The first type is standard and is formed of flat plates developing almost the full strength of the section. The small plates and half the large plates of the first type can be provided with square holes and placed on the outside of the piling, to ensure that the bolts do not turn when nuts are unscrewed from inside a cofferdam. The second type is a more rigid and watertight connection, but more costly. Corners and junctions may also be spliced.

Followers

Followers (Fig. 12.6) are made of sections of sheet piling with downward projecting plates on each side of the pile, shaped to fit the pile, and

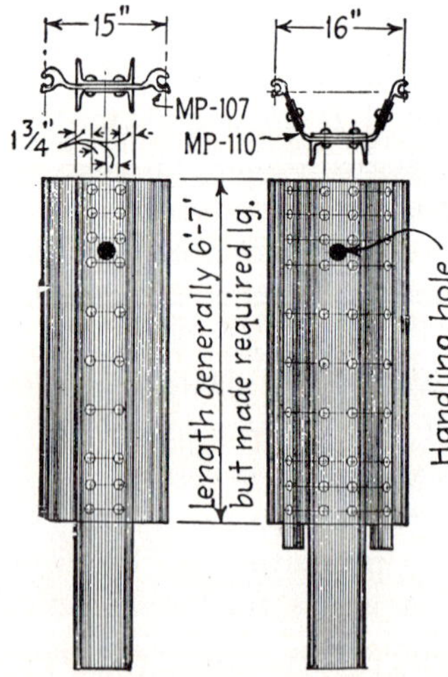

FIG. 12.6. Typical followers for steel sheet piling.

are used for underwater driving. Since hammers operate at reduced efficiency under water, followers are often used in these circumstances.

Caps

Caps (Fig. 12.7) for steel sheeting walls may be flat steel plates, steel channels or I beams, reinforced concrete, or wood. Space between

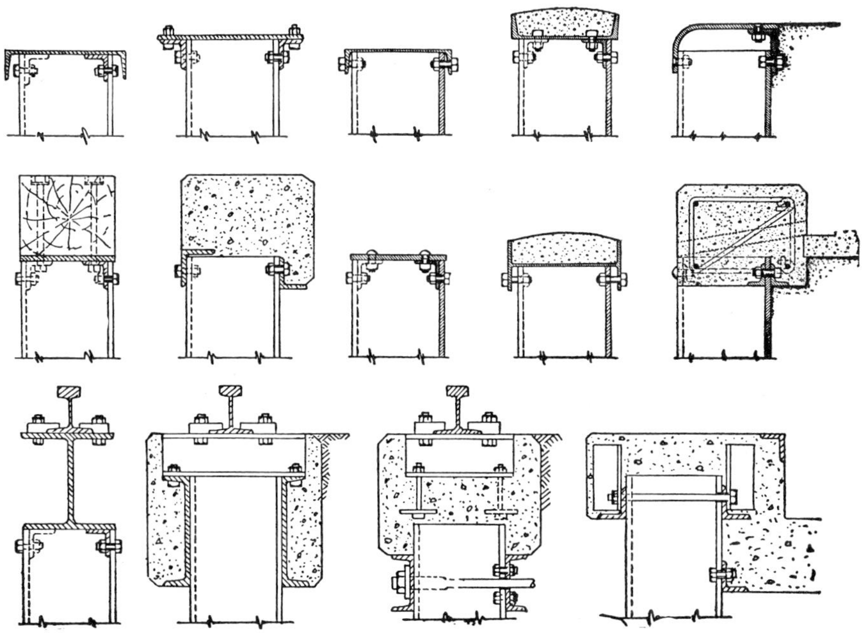

FIG. 12.7. Typical caps for sheet piling.

upturned flanges should be filled with concrete to prevent corrosion. Wood should be treated.

Value of Interlock in Section Modulus

There has been controversy as to whether the section modulus of steel-sheet-piling walls, when the sections were driven in the alternating positions, should be computed upon the basis of section moduli of individual pieces or upon the section modulus of the combined pieces using the total over-all depth of the two piles. Locating the neutral axis at the interlock means that the interlock must transmit longitudinal shear. Its ability to do this depends upon the tightness of the joint, driving conditions, tension between pieces, and amount of dirt in the joint. No doubt some shear is generally transmitted. There has, however, been severe criticism of some American designs for assuming use of a large proportion of the combined section modulus. However, in American sections there is a good deal more clearance and flexibility in the joints than in European or Canadian sections. With the latter types, it is frequently assumed that full section modulus of the combined sections is developed. The manufacturers of such sections claim that no failures have occurred in designs based upon this assumption, and that pressures computed would have caused the yield point of the metal to be exceeded had not the interlock been effective. Tests in which deflections and section moduli of loaded combined sections of European piles have been measured have been reported to show that even with quite free oiled interlocks, about 60 per cent of the combined section modulus was developed. Sand might be expected to materially increase this figure, possibly to 100 per cent. However, even with Canadian or European sheeting, when driving through mud it would appear advisable to use less than the full combined section modulus unless interlocks are welded together. The usual American method of entirely discounting the value of the interlocks may be slightly too conservative, whereas the European method of allowing full value for it may be too optimistic. When dealing with sheeting with interlocks in the shear web, it is advisable to weld the joint in a pair of sheets to ensure that the full section modulus is developed, although the entire length need not be welded, but possibly only 50 per cent.

Methods of Driving

There are two methods of driving steel sheet piling. One calls for driving a single pile or a pair of piles at a time; the other for assembling all piling in wall form first, including the closure piece, and driving continuously along the line.

In the first method the driving leaders should be stable and vertical at all times, with the hammer centered over the neutral axis of the pile, and it has been found that even long lengths of steel sheet piling can be driven accurately by this method without the use of taper piles. Driving one section at a time is generally confined to work in which a stable and level foundation can be created for the pile-driving equipment.

The second method, of assembling the sheeting in wall form, is often economical and can be used in deep cutoff walls. Success depends upon the accuracy with which the piling is set with regard to plumbness of both axes, and the hammer should be held rigid for best results. If swinging leads are used, the leads and the hammers should be held rigidly in the same vertical plane. Vibration in the hammer or piling will result in inaccuracy and creeping of the tops of the piles.

Sheet piles are often driven in pairs, as this saves time, makes guiding easier, and uses less power. The power required to drive a pair of piles is stated to be not over 1½ times the power needed to drive a single pile.[11c] Larssen sections may be obtained from the mill interlocked in pairs. Z piles should be driven in pairs, to prevent driving out of line.

The best practice is to drive the ball end leading, to prevent soil from becoming trapped in the interlock and tending to force it open during driving.

Provision of suitable guide walings is essential to obtain a well-driven cofferdam or line of sheeting, and the avoidance of later difficulties will well repay their cost. Distance between walings should be about ⅛ in. more than the back-to-back distance of the sheeting. Each pile should be wedged off at the walings by a wood block in the trough of the pile. Sometimes walings can be built in the form of a trestle that can be moved along.

Creep can be prevented but must be corrected as soon as it appears, otherwise it goes beyond control. The simplest method of driving short piles in soft ground is to drive to full depth singly or in pairs, but this should not be done with long piles or in hard ground. To prevent creep in such cases, the piles can be driven in panels by driving a pair carefully to part depth, then also setting a dozen piles or pairs in the walings; after this the last pile or pair in the panel is driven part way, the intervening piles driven to full depth, and the last and first piles fully driven, except that one-third of the last pile should project to guide the first pile of the next panel.[11c]

Long piles should be driven in panels of two stages. If the piles have begun to lean over, they should be pulled back, using one of the driven piles as anchorage. To avoid interlock damage, it is safer to pull on the pile prior to the last, although this is not so effective. Driving of center on the pile also aids in straightening. As a last resort, taper piles can be used.

Drawing down may occur in soft ground, or if unusual friction is present in the interlock because of leaning over. This can be counteracted by bolting piles to a stiff waling. It is hard to jack up a pile drawn down, and it is usually cheaper to weld on an additional length. Sometimes a rivet in the leading-edge interlock will aid by preventing the soil from entering, thus making driving the next pile easier. Such a rivet may also aid in breaking up stones.

Methods of Extraction

Sheet piling may be extracted by using extractors or double-acting hammers reversed. A steady pull may be obtained from a several-part line from a cable drum. Bond may sometimes be broken by driving down the adjacent pile before applying the extractor. Piling has been extracted in the Soviet Union by vibration.

CORRUGATED STEEL SHEET PILING

Description

Corrugated steel sheet piling consists of 8- to 12-gage steel sheets rolled with ⅝- to 2⅝-in.-deep corrugations. Widths vary from 6 to 32 in. according to the make. Some types have interlocking joints. Available makes are Armco, Caine Corr-Plate, Foster, and British Steel Piling Co.

Uses

Corrugated steel sheet piling is light, easy to handle, watertight, salvageable, and economical. It is driven by hand or by light hammers and is used for trenches, dams, erosion and river control, and foundation excavations.

PRECAST CONCRETE SHEET PILING

Description

Concrete sheet piles are precast piles of square or rectangular cross section, driven side by side to form a continuous wall (Fig. 12.8). To keep the piles in line, some form of interlock is needed, such as tongue-and-groove joints. The tongue may extend the full length of the pile or only for the portion below the water line, the groove above being grouted to ensure watertightness. T-shaped concrete sections have sometimes been used in Great Britain. Precast T sections have been widely used on the Continent under the name Coignet-Ravier, and are anchored into the backfill by reinforced concrete ties anchored to the webs. Steel interlocks are sometimes cast into precast concrete sheeting.

The foot is usually beveled on one side so that it will be forced against the adjacent pile and maintain contact during driving. The bevel slope in Fig. 12.8*a* is better than that shown in Fig. 12.8*b* because more pressure is developed to hold the piles together. The bevel is shown on the groove side in Fig. 12.8*a* and on the tongue side in Fig. 12.8*b*. This latter location is better because the pile is driven with the groove side fitting over the tongue of the pile already in place, which prevents clogging. Since the tongue in Fig. 12.8*a* is only on the lower portion of the pile, the bevel is placed on the groove side. In this case it is not possible to fit the groove over the tongue or the pile in place, since the top of the tongue is below ground level.

Prestressed Concrete Sheet Piles.[5bb] Prestressed concrete sheet piles have practically superseded the conventional type in Florida. In bulkheads, the face wall is capped by a continuous concrete member and tied back with a system of precast prestressed tie beams to an anchorage system of concrete deadmen or piles. As a practical rule of thumb, face piles usually penetrate in good material a distance of 0.6 of the height above grade, or more in poor material.

Florida standard sizes are 30 in. wide by 6 to 12 in. thick, with a groove full length on one side. On the other edge there is a groove

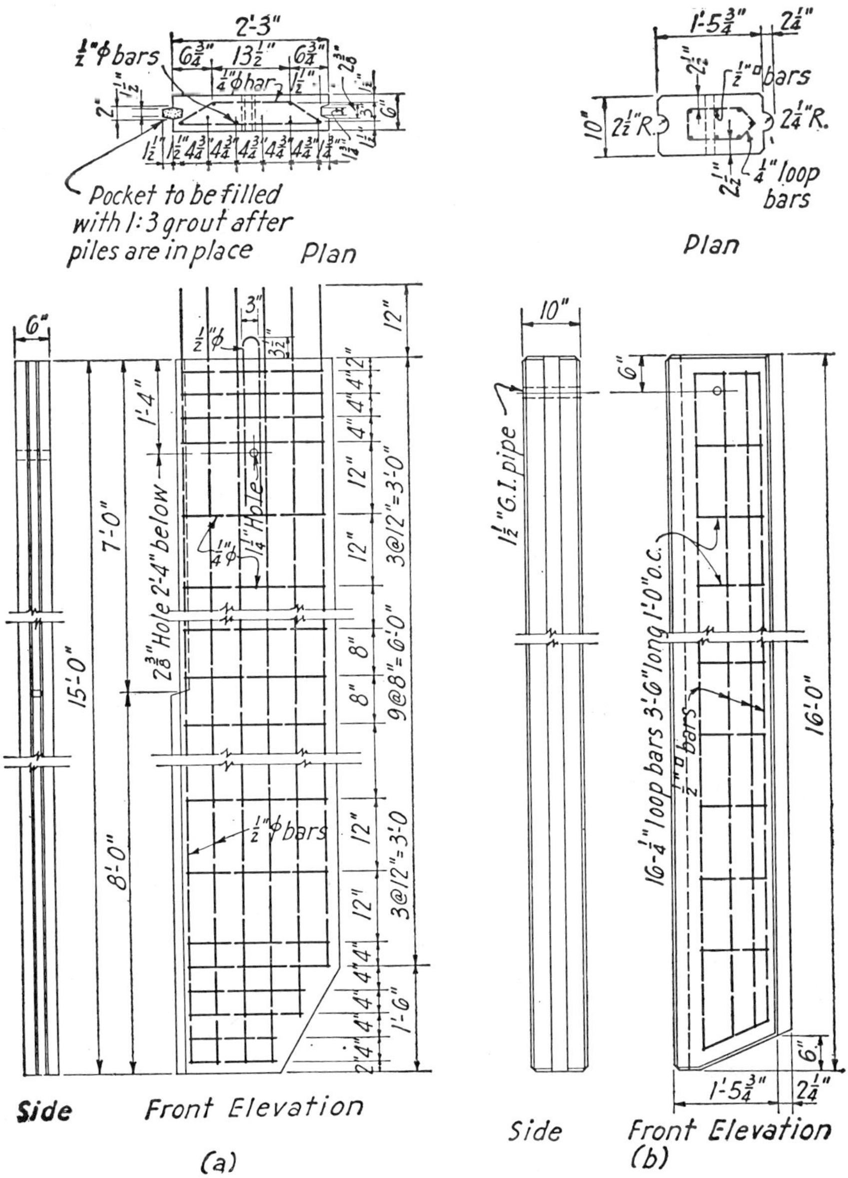

FIG. 12.8. Conventional precast concrete sheet piling.

from the top to 3 or 4 ft below the water line and a tongue below. The upper grooves in pairs are grouted.

The pickup point is offset 4 in. forward from the pile axis so that the tip will incline toward the previous pile. Alignment is controlled by templates. A minimum concrete stressing of 700 psi has permitted safe handling and placing. Reinforcement is made the same in both faces so that the pile can be handled from either end or side. It is impractical to set up prestressing arrangements for special pieces at corners. Sheet piles 28 in. by 4 ft were found to be competitive with steel walls for Pensacola harbor pier walls. Sheet piles in bulkheads can also serve as bearing piles; this was done at the Chesapeake & Ohio Railway facilities at Newport News, Va., using 24-in.-square tongue-and-groove

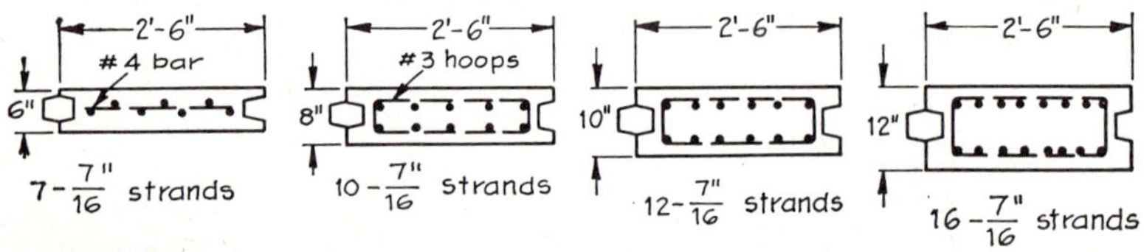

FIG. 12.9. Sections of typical prestressed concrete sheet piles. (*From W. E. Dean, Florida State Highway Department.*)

members 63 ft long, with core holes and 40 prestressing strands ⅜ in. in diameter. Broad sheet piles, if thick enough may have a pair of core holes.

Properties appear in Table V.9.

Driving

Sheet piles must be driven in good alignment and particular care should be given to be sure the first pile is accurately driven. In some soils a timber frame should be provided to keep the piles in line. Such a frame may consist of two pairs of wales above the ground, one set slightly above the other, braced by studs and diagonal braces, with stakes set at the feet of the braces.

At changes in alignment, gradual curves may be accommodated by adjusting the framing without change in pile shape. For sharp changes in alignment, special pieces must be cast.

Since many marine structures, in particular, are built on sandy soils, jetting is extensively used in driving sheet piles. Jet pipes are cast in the piles, or external pipes may be used. In soils composed of fine sand and clay, it is sometimes possible to wash a hole with the jet below the foot of the pile deep enough to place the pile in its final location. This hole tends to fill up quickly so that speed is required to line the pile correctly and fit it to the adjacent pile.

Joints

Most sheet-pile walls must be watertight, and joints are usually grouted after the driving is finished. The groove is first flushed out by a water jet from a pipe long enough to reach the bottom of the pile. A cement grout composed of one part cement and two parts sand is then deposited by means of a small sheet-metal pipe used as a tremie. The tremie pipe is lowered to the bottom of the hole and then filled with grout. As the tremie is withdrawn and the joint filled with grout, the water is forced out at the top. If the earth fill back of the wall is to be drained, occasional joints may be left without grouting for their lower portion. An easy way to grout to secure tight joints is to put a thin polyethylene tube over the grout pipe and let the grout force the tube tight against the concrete.[5bb]

To provide for expansion and contraction, joints having flexible fillers may be provided at intervals of 25 to 50 ft, or it may be more convenient to cast a special unit which is solid below ground and split above, the slit being filled with flexible joint filler. A cap is generally placed on the sheet-pile wall, and flexible joints should also be continued up through the cap.

Design

Sheet piles should be adequately reinforced for bending moments, taking account of conditions of earth and hydrostatic pressures and points of support. Details and descriptions of several varieties of concrete sheet-pile retaining walls are shown in the Portland Cement Association's booklet *Concrete Piles*.[5a]

BORED-PILE COFFERDAMS AND SHEETING

Bored concrete piles have been used to form cofferdams. The piles are placed prior to excavation, and later may be incorporated in the finished structure. They may be installed where headroom is low, and where vibration from driving would be harmful. This method has been used in Great Britain by the Cementation Company, Ltd., and also in the United States.

Drilled caissons have been used to form cutoff walls, drilling alternate caissons, then the intermediates a day later, cutting overlaps. Rotary caisson drilling methods are used (Fig. 10.6). Intrusion-Prepakt mixed-in-place piles have been used to form cutoff walls.[5al]

Dual-purpose concrete piles have saved foundation costs. Holes 20 in. in diameter have been drilled on 3-ft centers and trimmed to 29-in.-wide flat faces having keyed ears by blades cutting soil to drop into a bucket. Reinforcing cages are then placed and these soldier piles concreted. An 8-in. concrete wall is later poured, after excavation of the building area, to key with the soldier piles and make a combined basement wall and load-carrying wall.[5am]

PRECAST HOLLOW PRESTRESSED CONCRETE PILE BREAKWATERS

Cylindrical precast hollow prestressed concrete piles have been driven to form breakwaters. Piles are set several inches apart, and treated

FIG. 12.10. Cylindrical hollow prestressed concrete breakwater wall in Louisiana and Mississippi. Sand-filled 54-in.-o.d. piles spaced 5 ft on centers, with concrete cap and 10- by 10-in. creosoted closure timbers extending below mud line. (*Courtesy of Raymond Concrete Pile Co.*)

wood closure pieces have been placed to below the mud line. Concrete fillers may be used. Concrete capping is placed after the piles are filled with sand.

CHAPTER 13

DETERIORATION AND PRESERVATION OF PILES

The following factors should be considered when selecting pile material: (*a*) required length of life, (*b*) character of structure, (*c*) availability of materials, (*d*) type of loading, (*e*) factors causing deterioration, (*f*) amount and ease of maintenance, (*g*) estimated costs of types of piles, taking into consideration initial cost, life expectancy, and cost of maintenance, and (*h*) available funds.

The principal factors which cause piles to deteriorate are (1) *decay,* (2) *insect attack,* (3) *marine-borer attack,* (4) *mechanical wear,* (5) *fire,* and (6) *corrosion.* In the case of foundation piles buried in the ground, only the first, second, and last factors need to be considered. In the case of piles supporting water-front structures, all factors must be considered.

WOOD PILES

Cause of Decay in Wood Piles

Fungi. All decay in wood piles is caused by the growth of fungus, a form of plant life which, by deriving its food from the wood, breaks down the cellular structure. Fungi must have moisture, air, favorable temperature, and food in order to exist. By depriving the fungi of any of these elements, decay can be prevented. For example, if wood can be kept dry, or if it can be kept continuously submerged at a low temperature, or rendered unsuitable for food by poisoning, it will not decay. Untreated wood is serviceable only if conditions are not and will not become favorable for the growth of fungi. The fungi develop within the wood and at maturity produce fruit bodies on the surface, which are the most familiar parts. These appear only at an advanced stage of decay. They are in the form of mushrooms, shelves, or encrustations which shed clouds of dust-fine spores that are scattered by air and, at such time as favorable conditions for growth occur, infect any wood upon which they fall. Each spore starts a tubular growth which penetrates the wood tissues, feeding upon and destroying them.

Spores of fungi may lie dormant in wood for years and spring to life when changed conditions favor growth. For instance, when the water table is lowered, untreated piles may start to decay.

Microfungi and other microorganisms and bacterial organisms may cause fungal infection and decay of the soft-rot type in marine piling most often noted between high- and low-water levels, where the wood is alternately wet and damp, but not wet or dry for long enough periods to stop spore growth. The depth of attack is small, but the wood is more susceptible to abrasion, after which the process is repeated. Even untreated durable hardwood fenders can have surface soft rot.

Air. Air to provide free oxygen is required by wood-destroying fungi and, if the air supply is cut off by submersion or is poisoned by gases, the fungi will finally die. The air supply in soil becomes deficient a few feet below ground surface and at depths of more than 5 or 6 ft the rate of decay is usually very slow, especially in dense, compact soils, with decay extending deepest in light, sandy, or gravelly soils.[2aq]

FIG. 13.1. Wild mallard duck nesting on top of rotting piling of heavily traveled Wisconsin Avenue Bridge, Milwaukee, Wis. (*Photo by Harrison Nowell, Courtesy of Life.*)

Moisture. Moisture in moderate amount is required by wood-destroying fungi for growth, and wood in an air-dry condition is not attacked. Decay may develop when the moisture content of the wood is somewhat below fiber-saturation point (theoretical point at which no water remains in the cavities although the cell walls are still saturated, usually 23 to 30 per cent of oven-dry weight, depending on species of wood), and increases rapidly with more moisture, until insufficient air limits growth.

Temperatures between 65 and 95°F provide conditions for optimum growth of fungi; between 104 and 115°F growth ceases. Higher temperatures, however, are needed to kill fungi. The temperature required depends on the duration of heat and the humidity. All fungi become inactive as temperature approaches freezing and remain dormant until warmer conditions return.[2ap]

Rate of Decay. This depends upon the kind of fungus, character of the wood, and degree of exposure to moisture, air, and warmth. Decay may be so rapid that the wood is destroyed in a few months, or may be so slow that decay will scarcely be noticed. In dry, cold, or high-altitude climates, decay is generally near the ground line, but in the Mississippi Valley and Southern states, decay may also frequently be found on the tops and side of piles.[2aq]

Reduced Strength. Strength of wood is greatly reduced as rot develops from all decay-producing fungi. White rot, most prevalent in soft woods, has little effect on strength in early stages, but even the incipient decay due to brown rot has a serious effect. Brashness, or brittleness, is a characteristic effect of incipient rot, although it may also be caused by other factors.

Standing timber may contain decay, particularly in heartwood. This decay may develop in piles at some later date when conditions become favorable, even though no new source of infection is present. Felled timber has opportunity to become infected, and trees cut in early summer are more apt to contain spores than those cut in winter.

Decay Resistance. The natural decay resistance of all common native species of wood in the United States and Canada lies in the heartwood and, when untreated, the sapwood of all has low resistance and short life. The resistance of heartwood is greatly affected by the species of fungus, conditions of exposure, and character of the wood. For piles, no reliance should be placed upon permanent resistance of untreated native North American woods to decay. The relative resistance of heartwoods of these species is, however, given in the *Wood Handbook*,[2b] and elsewhere,[2am,2aq] if of interest for temporary work.

No species of wood is entirely immune from attack of some species of fungi if conditions are favorable for growth, but there is much difference in susceptibility.[2ap]

Service records showing good results are not generally a reliable indication upon which to base the use of untreated piles in conditions favorable to development of fungi, since wherever the environment is suitable, decay is apt to start, and spores may even be present in the new wood to be used. Also, the life processes in fungous growth are so complex that minor variations in humidity and temperature of soil and air, texture and composition of soil, and precipitation may greatly affect progress of decay. The principles outlined herein should be used only as a guide. In locations where no fungal growth can occur, any species of wood will last indefinitely, in so far as decay is concerned, and the choice of material should be governed by other factors.

Preservative treatments give most protection near the pile surface, and if abrasions, cuts, or bored holes occur, fungi spores may enter and start decay when conditions become favorable.

Identification. Identification of fungi species is generally unnecessary, although sometimes it has been of value in tracing the source of infection. Rots are often classified as white or brown rots, according to the color of advanced stages of decay. Some white rots attack only the lignin and not the cellulose of the cells of the wood, while others attack the cells. In brown rots, the cellulose is destroyed and the wood can finally be powdered in the fingers. Some fungi infect only sapwood of living trees, others dead softwood, sapwood, or heartwood. Still others are limited in range by dependence on a certain class of hosts.[2ap]

During the invasion stages, the wood is not definitely disintegrated and the appearance is normal, except possibly for some discoloration which may be hard to differentiate from stains. In later stages of decay, the color, texture, and strength of the wood are affected, and it becomes punky, stringy, spongy, crumbly, pitted, or ring-shaked. Narrow irregular colored lines, mostly black or brown, may appear, generally with white rot, and may be of value in diagnosing incipient decay.

Stains are due to fungi or chemical changes during seasoning, and usually occur in sapwood. Staining fungi propagate by spores from minute fruit bodies and develop under the same conditions as wood-destroying fungi. They do not feed on the cell walls, and general opinion is that stained wood is as strong mechanically as unstained,[2ap] but since the fact that it is stained indicates that conditions favorable for growth of fungi exist or existed, the occurrence of stain should cast suspicion on the wood, and it should be noted that the deeper discoloration of the stain may mask incipient decay. Staining fungi appear, in their early stages, much like wood-destroying fungi.

Molds may develop, similar to those on damp bread or cheese. These are cottony surface growths, ranging in color from white or other light shades to black, whereas the threads of wood-destroying fungi are usually compacted into strands or fan-shaped patches. The mold threads entering the wood do not bore into or dissolve the wood fibers, as do those of the young wood-destroying fungi. Molds rarely discolor wood except superficially, and may be brushed off. There is the possibility that moldy wood may be decayed, since the conditions favoring growth of molds also favor decay-producing fungi.

Ground-water Level

Records of sea, lake, or river water-level fluctuations are often available, or observations can be made. In the ground, information as to water levels is less plentiful except in certain localities where records are kept. Below a certain grade known as the water table, all ground is saturated, and such water is known as ground water. The water table fluctuates with rain and drought, and slightly with barometric pressure, temperature, and vegetation. The term "water table" is somewhat misleading, for the top surface of the ground water is not necessarily level, but approximately follows the irregularities of the surface of the ground. Ground water also flows in the same direction as the surface water, but at a very much slower rate. Thus it may be seen that a ground-water observation, possibly from a boring record, at a given spot at a certain season will usually not give all information needed for selecting pile cutoff grades over a large site, or possibly even a small one. It should not be assumed that the water table is at least as high as the water level in adjacent bodies of water—for instance, ground water in the Williamsburg section of Brooklyn, N.Y., was found, on a recent reconstruction project close to the bank of the East River, to have been drawn down over the years to a point far below the water level in the river, which had evidently formed a practically impervious channel lining.

When the Campanile of St. Mark's in Venice fell in 1902 because of structural defects, the submerged wood piles driven in 900 A.D. were found to be in such good condition that they were used to support the reconstructed tower.

All of the houses in Amsterdam are built on piles. The upper stratum of natural soil is loam and loose sand, upon which no permanent structure can be erected without driving piles 14 to 60 ft long. The Royal Palace in Amsterdam was built in 1648 on 13,659 wood piles driven 70 ft in the ground. The piles have been preserved by the continued high level of the ground water and water in the canals. On the other hand, the wood piles under a metropolitan power station and under a noted public library in the eastern United States were exposed and found to have dry rot due to the drawing down of the ground-water levels over the years. In the latter case, where ground water was drawn down after the construction of large intercepting sewers, upper parts of wood piles in the district showed signs of decay. It was necessary to dam the sewers to raise the ground-water level. In another instance where a new building was constructed recently, provision was installed in the foundations for introducing water into the ground should it ever be found necessary. The possibilities of ground-water control should be borne in mind when considering new or investigating old wood-pile projects.

Modern industry and air conditioning are two enormous users of water that affect ground-water levels far more than anything has done in the past, including scant rain and excessive runoff. In the Rubbertown district of Louisville, Ky., the first two years of synthetic-rubber production lowered the water table by 21 ft. In 1883, an industrial plant in Cincinnati had flowing wells, but by 1938 they were pumping 90 ft. In Indianapolis, the water table has fallen 15 ft in recent years. New York State now has a law regulating ground-water levels, and other states are accumulating data to enable them to establish controls. Artificial recharging of the ground water by returning the water after use is established practice in some areas of the western part of the United States, and is beginning to be done in the eastern part.[28]

Untreated wood piles are successfully used under structures where the full length of the pile is always below ground-water level. Composite piles, consisting of wood below the water line, and some other material, usually concrete, above this point, are often used with economy to avoid treatment of the parts permanently saturated.

Water Level

Water levels in the ocean are relatively regular but may vary more in estuaries and rivers because of dams, locks, irrigation uses, reclamation projects, droughts, flood-protection works, floods, etc.

It is necessary to watch the degree of salinity caused by encroaching sea water at periods of low river flow.

While *Teredo* infestation may be stopped by the exposure of piles above mud flats owing to receding waters, *Limnoria* can usually still work if the mud is wet. Such conditions will permit inspection of portions of the piles, however, and possibly the undertaking of some repair work, although piles at the deeper ends of wharfs will probably still be submerged with no reduction in severity of attack at those points.

Insect Attack on Wood Piles

Termites. One of the forms of life most destructive to wood piling is the termite or white ant. Their damage may often be mistaken for decay. They live in the wood and therefore are rarely seen. There are two types of termites in the United States: the first burrows through the soil and attacks the wood from the earth, with which it must maintain contact to retain life; the second flies to and attacks the wood directly. The second type does not require much moisture and can survive on less than the 12 per cent in air-dried wood. There are other still more destructive types in tropical regions.

The chief food of termites is cellulose in the wood. There is no known species of wood entirely resistant to termite attack, although it has been reported[3k] that ironbark, swamp mahogany, titree, turpentine wood, and cypress pine, if perfectly sound, are not attacked, and the heartwood of certain trees is very resistant. Among the more resistant woods are redwood, incense cedar, western red cedar, and species of

(*a*) (*b*) (*c*)

FIG. 13.2. Adult termites. (*a*) Winged sexual, or reproductive. (*b*) Worker, sterile. (*c*) Soldier, sterile. (*Courtesy of U.S. Bureau of Entomology.*)

FIG. 13.3. Termite honeycomb attack revealed by removal of outer surface. (*Courtesy of U.S. Bureau of Entomology.*)

junipers in the United States; cypress pine and camphorwood in the Orient; teak and sal in India; turpentine wood, jarrah, and red gum in Australia. Longleaf pine from Texas, cut from butt logs having a large resin content and locally called fatwood or lightwood, appears to be resistant to termite attack, although ordinary pine wood is very susceptible. Since very few termite-resistant woods occur in the world, it is probably more practicable, in the United States at least, to treat with preservative woods readily obtainable commercially.

Cracks are incipient points of trouble, for the termites gain entrance, enlarge, bring clay, and promote the destructive action of damp and fungi.

Beetles. There are a number of beetles which are responsible for damage to piles and bulkheads above high water. There are four distinct stages, namely, the egg; the larva, also called worm or grub; the pupa, or transforming stage; and the adult, or beetle stage. The beetles, particularly larvae, also cause much damage by carrying along the mycelia of destructive fungi, thus spreading decay.

Bark beetles confine their activities to the bark and burrow between it and the sapwood, often scoring the latter. The adults are short, cylindrical, reddish-brown to black, from 1/16 to 1/4 in. long. They push out fine brownish-white sawdustlike particles, or frass, and lay eggs in the tunnels. The larvae are tiny, whitish, cylindrical, slightly curved, and legless. They mine in all directions and loosen the bark.

Ambrosia beetles, or pinhole borers, bore through the bark into sapwood and sometimes heartwood. They closely resemble bark beetles, but their work differs considerably. The beetles bore into the wood and extend galleries in all directions, each making a hole about the size of a pinhead. They push out white sawdustlike particles which either fall out loosely in piles or come out in stringlike masses. The galleries are round, free from frass, and are often stained black. The damage by these borers is caused almost entirely by the adult beetles, as the larvae stay in the original galleries or cells until mature. These beetles work in wood only if the moisture content is above the fiber saturation point, which is approximately 50 per cent. Serious damage to piles above tide level has been reported from these beetles (see Chap. 16).

Wood-borer beetles include *powder-post beetles, roundheaded borers,* and *flatheaded borers. Powder-post beetles* are short cylindrical reddish-brown to black hard-shelled insects from 1/8 to 1/2 in. long. In the eastern United States the only species causing much damage is the red-headed shot-hole borer, about 1/4 in. long. The adult bores the wood, making a cylindrical tunnel around the wood just under the surface. Eggs are laid in cells. Loose whitish dustlike borings may be found on the bark below the entrance hole. The larva has a curved form. This beetle is exceedingly destructive in both adult and larval stages. The larvae extend to the pith, completely destroying the wood and often reducing it to powder. In larger pieces, the attack is mostly in sapwood, which may nearly all be destroyed. The larval attack is entirely in the interior and cannot be detected without cutting. The borings are similar to those of adult *Ambrosia* beetles but are tightly packed behind the larvae.

Roundheaded borers vary in general appearance and are from ¼ to 1½ in. long. Eggs are laid on the bark surface, on the sapwood, in crevices under the bark, or in slits or pits gnawed by the borers. The larvae are entirely responsible for the damage. They are elongated fleshy yellowish-white grubs, tapering slightly toward the tail. Some bore into the inner bark or outer sapwood; others make large oval mines deeper into the sapwood and heartwood. The grubs can completely riddle the wood in a few months. The galleries may be ½ in. wide and 2 ft long. Large quantities of frass may be pushed outside. Grubs of some species pack the frass tightly behind them. The composition varies from fine, white powdery material to coarse brownish particles or shreds of wood fiber.

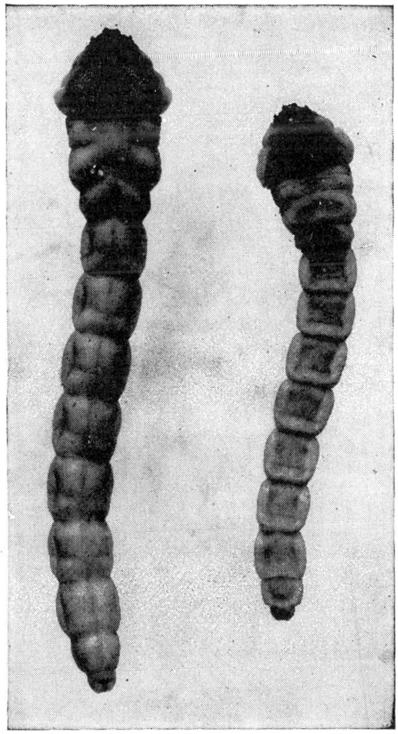

FIG. 13.4. Grub of the flatheaded borer *Chalcophora angulicollis*. (*Courtesy of U.S. Bureau of Entomology.*)

Flatheaded borers are more uniform in appearance than are roundheaded borers. They are slightly flattened metallic-colored boat-shaped beetles from ¼ to 1¼ in. long. The eggs are laid singly or in a mass

(*a*) (*b*)

FIG. 13.5. *Nacerda*. (*a*) Larva. (*b*) Mature beetle. (*Courtesy of Dr. William F. Clapp.*)

on the bark or in crevices. The larva is a slender flattened grub having conspicuously widened segments next to the head. The young borer mines in the inner bark or wood, making a flattened oval more or less tortuous wormhole which widens into a resting cell. This connects to the outer surface by a short oval exit hole through which the new beetle emerges. The mines may extend through the entire piece and are filled with tightly packed frass.

Nacerdes melanura larvae, commonly known as the European *wharf borer*, are among the most important beetle borers, and have done much damage to bulkheads above high-water line in Nova Scotia, Maine, New York, and Massachusetts. They are fond of damp partly rotted pile tops. The larvae are white, about 1 in. long, with enlarged heads. The adult is a slim elongated yellowish-brown beetle about ⅜ to ½ in. long.

FIG. 13.6. Damage from *Nacerda*, building foundation, Berlin, N.H. (*Courtesy of U.S. Bureau of Entomology.*)

FIG. 13.7. Bulkhead damaged by European wharf borer.

Oedemeridae larvae have caused damage, sometimes serious, to marine piles above high water in New York Harbor. Elsewhere they have been found in creosoted timber and driftwood.*

Saferda condida and Chrysobathrid femorata beetle grubs, the borers which are such a pest in apple culture, have been reported as damaging untreated wood piles.[2u] They appear to like a little dampness and to develop rapidly at temperatures of from 60 to 70°F, whereas at freezing temperatures they remain practically dormant. Destruction of timber may occur in from 1½ to 3 years.

* *AREA Bull.* 352, December, 1932.

Prevention of beetle attack,[2bc] once the wood is infested, depends largely upon the seasonal history and habits of the insects involved. Seasonal cutting is important. Seasoning beyond certain limits will create unfavorable conditions for *Ambrosia* beetle attack. If it is desired to retain the bark, a repellent chemical may be used. *Ambrosia* beetle damage may be greatly reduced by spraying with one part of dichlorodiphenyl oxide and three parts of kerosene. Certain powder-post and roundheaded borers cannot be controlled by seasonal cutting and seasoning methods, and creosoting is necessary.

Marine-borer Attack on Wood Piles[2o,2k,2cj]

Research. Marine-borer research in the United States is based to a great extent on the investigations of the Marine Piling Committee of the National Research Council,[2n] and in the British Empire on the Reports of the Committee of the Institution of Civil Engineers.[3k,2cj] Testing grounds are also being operated by them in the tropics, using hundreds of test timbers including treated woods from Europe, California, Oregon, South and Central America, and Australia. The published report of the San Francisco Bay Marine Piling Committee is a valuable treatise on all phases of the marine-borer problem.[2o] Investigations of the problem in the North Atlantic harbors have been conducted by the New England Marine-piling Investigation Committee. The report of the Marine Borer Research Committee, New York Harbor, covers that region excellently.[2bg] The work of Dr. William F. Clapp and A. P. Richards on marine biological research at The William F. Clapp Laboratory at Duxbury, Mass., has greatly clarified the problem and thus advanced solutions which depend upon accurate knowledge of the conditions, involving the characteristics, life cycles, habits, favorable environments, and distribution of the numerous borers.[2k,2l,2m] These researches have involved the setting and checking of test blocks and recording of water conditions, as well as observations of existing structures, and have proved invaluable. Many test boards are located under important wharves and prove inexpensive insurance against sudden damaging attacks. Cooperation of owners of existing and proposed structures with the researchers should result in mutual advantage and is strongly recommended.

Proceedings of the Annual Marine Borer Conference held at Wrightsville Beach, N.C., through the cooperation of the Seahorse Institute and Marine Biological Deterioration Conferences, discuss latest progress. A general review of borers and an annotated bibliography appear in reference 2*cc*.

Classification. The more important organisms included in a broad classification of marine borers are the following: in the subfamily *Mol-*

lusca, the genera *Teredo, Bankia, Lyrodus, Martesia, Hiata, Xylophaga, Pholas, Lithophaga* (syn. *Lithodomus*), *Zirphae,* and *Petricola;* in the subfamily *Crustacea,* the genera *Limnoria, Chelura,* and *Sphaeroma.* Several hundred species have been described in these genera. Because methods of protection which are of value for one species are often valueless for another, proper classification is essential before selecting the protective treatment. The problem is one for the taxonomist.

Mollusca. These are bivalves, as are oysters and clams.

Teredo has a wormlike slimy gray body with the two valves of the shell on the head used for boring. It has two unequal siphons, like a forked tail, projecting from the burrow. A pair of plumose or paddle-

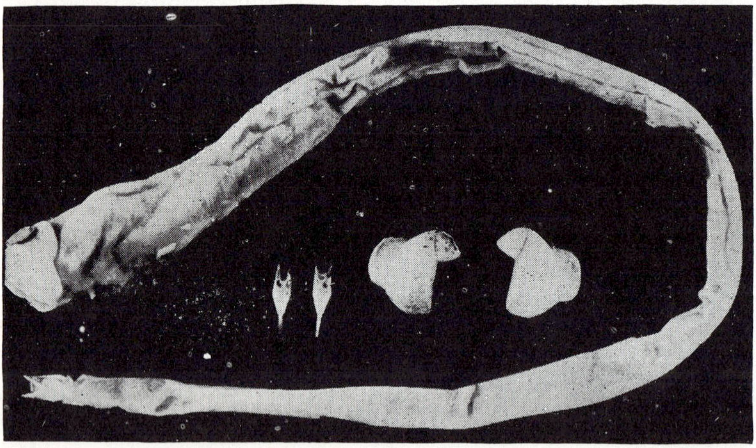

FIG. 13.8. *Teredo navalis*—animal, pallet, and shell. (*Courtesy of Dr. William F. Clapp.*)

shaped pallets near the siphons serve as plugs to close the burrow against intruders or undesirable elements in the water. Members of this genus vary from ⅜ in. in diameter and 6 in. in length to 1 in. in diameter and 4 to 6 ft in length. In Pacific Islands, specimens 3 in. in diameter and a number of feet long have been found.

Bankia resembles *Teredo,* but generally is larger, sometimes growing to 1 in. diameter and 3 to 4 ft long. It leaves two long plumose pallets projecting behind the siphons.

Xylophaga dorsalis resembles *Teredo* in shell shape but has no pallets.

Lyrodus is also a shipworm.

Pholadidae remain enclosed in their clamlike stout shells which are roughened for drilling. They gain entrance by opening and closing their shells, rapidly abrading the wood and making the small entrance

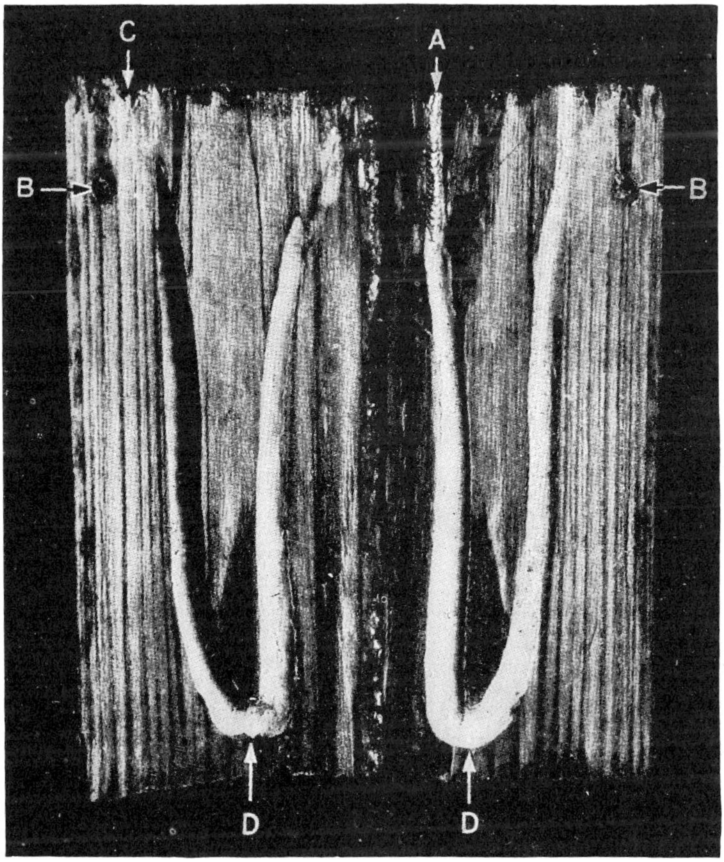

FIG. 13.9. *Bankia setacea,* 12 in. long, June, 1921, to October, 1922, Coos Bay, Ore.

FIG. 13.10. *Pholadidae penita,* Los Angeles, Calif. (*Courtesy of Dr. William F. Clapp.*)

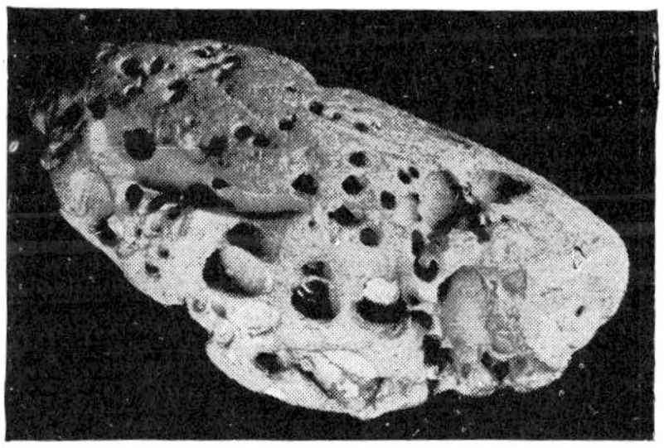

FIG. 13.11. *Zirphae* in rock. Nantucket, Mass. (*Courtesy of Dr. William F. Clapp.*)

needed when young. Some *Pholadidae* burrow in mortar and concrete, although they are stopped by rock aggregate.

Martesia resembles a clam, with its body entirely enclosed within the shells. It grows to 2 in. length and ¾ in. diameter. It is highly destructive. The wood disintegrates rapidly in a heavy attack, and may reduce the pile diameter by 4 in. in a year.

Hiata resembles *Martesia*, but is a little smaller and has a circular opening in the anterior end, whereas *Martesia striata* shows only a straight seam.

Lithophaga is a bivalve, has no structure by which it can scrape or bore, but secretes a substance that has a solvent action on rock. It may be found in emptied *Bankia* tubes, and also has been found drilling into wood.

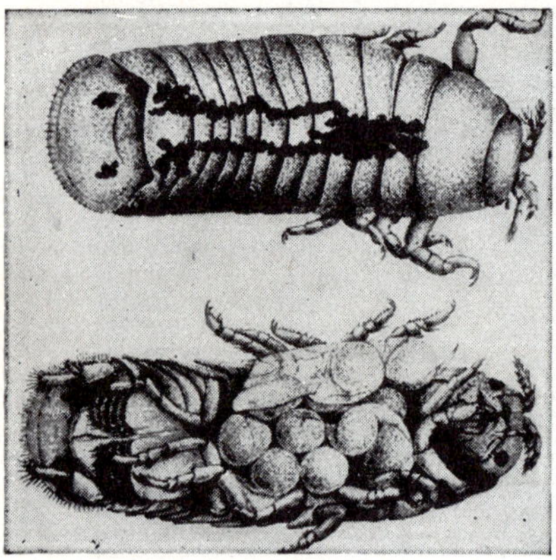

Fig. 13.12. *Limnoria lignorum.* (*Courtesy of University of California, Publications in Zoology.*)

Modiolus is a bivalve which is harmless to wood piles but may be found in emptied *Bankia* tubes.

Crustacea. This subfamily of borers is related to crabs and lobsters. Some of the principal ones are described below.

Limnoria lignorum is the common gribble or sea louse, the most widely distributed species of crustacean. It has somewhat the appearance of a wood louse, being from ⅛ to ¼ in. long, and one-third to one-half as broad. It is slipper-shaped, with horny boring mandibles, and has two sets of antennae and seven sets of legs with sharp hooked claws. It can roll itself up into a ball and can swim, crawl, and jump. In spite of its shape it is sometimes known as the "surface worm." The several species of *Limnoria* do not vary greatly in external form.

Chelura is allied to sand hoppers and is slightly larger than *Limnoria.* The joints in the body, the antennae, and the legs are heavily feathered with long hairs. In the male, the terminal segments of the last pair of abdominal legs are developed as two long, smooth, stout club-shaped structures nearly half as long as the body; these are used to block the burrow. It swims and jumps. A long, curved, pointed spike projects from the middle posterior of the back of *Chelura terebrans. Chelura insulae* have longer antennae, enormously large front claws, and no dorsal spike. *Chelura* may be distinguished from *Limnoria* by its pinkish tinge, and by becoming red in the sun after ½ hr.

FIG. 13.13. *Chelura terebrans.* Male, dorsal and lateral views (top); female, dorsal and ventral views (bottom). (*Courtesy of University of California, Publications in Zoology.*)

FIG. 13.14. *Chelura insulae.* Male (top), female (bottom). (*Courtesy of University of California, Publications in Zoology.*)

Sphaeroma and Exosphaeroma are sometimes known as "pill bugs" from their habit of rolling up in a ball when disturbed. *Sphaeroma pentodon* is similar to *Limnoria* but larger, being up to ½ in. long and ¼ in. broad. It rolls up into a ball about ¼ in. in diameter. It is much broader in proportion to length and more oval than *Limnoria.* It is dark olive to slightly reddish-brown, often mottled with lighter dull-yellowish areas in the middle of the back.

Worms. All creatures in *Teredo* burrows are not *Teredo. Pterochates* are worms that scavenge dead *Teredo* but do no damage themselves.

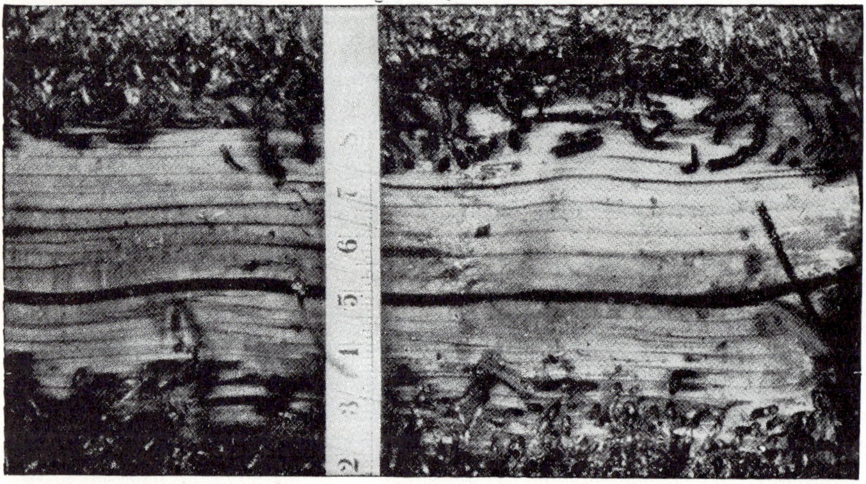

FIG. 13.15. *Teredo navalis* penetrated green Douglas fir pile 4 in. Driven August, 1920, pulled December, 1920. Mare Island Navy Yard, San Francisco, Calif. (*U.S. Navy photo.*)

Methods of Attack. The methods of attack by these two subfamilies are different. The molluscan group drills a tiny hole when young and grows inside it, destroying the wood during growth; the crustacean group destroys the surface of the wood. Molluscan attacks can only be detected by cutting the wood, or most careful surface inspection; crustacean attacks are visible on the surface. Molluscan attacks are capable of being much severer than crustacean attacks. X rays have been used to observe *Teredo* attack.[2cl]

Teredo larvae are ready to attach themselves to wood in 36 hr, and if they have not found a suitable home in the next 36 hr they become incapable of entering the wood and die.[2br] The chances that an adult *Teredo* will enter wood are very remote.

Burrows. *Teredo navalis* enters horizontally in minute holes which are as small as 0.008 in. when started but may reach a 0.03-in. diameter later. The hole usually turns down immediately, expanding after 1 or 2 in. to full diameter. Underwater wood is covered with a slimy coating of microscopic algae and other unicellular organisms, and the closest inspection with a hand lens is necessary to detect the entrances. There is nothing to indicate on casual observation the extent of damage inside the pile. The entrance can be fully closed by the creature's pallets. The burrows are lined with a thin wall of nacre that scales off readily. The burrows may turn to avoid crossing each other, or twist abruptly; or the borer may partially withdraw and start a new route. Minor deviations are noted, but not to the same degree as for *Bankia*. The burrows have no cross ridges or middle partitions, as for *Teredo norvegica*. Attack at some locations is heaviest at the mud line and at other locations it is confined to the between-tide area. The zone of attack is not related to the barnacle line, which may be anywhere from 6 in. below mean high water for *Chthalmus stellatus* to 100 fathoms below mean low water. *Teredo* burrows have been found to extend

FIG. 13.16. Damage inflicted by *Teredo* upon wood piles along Pacific Coast. (*Courtesy of Raymond Concrete Pile Co.*)

FIG. 13.17. Teredine attack in 13 months on untreated Douglas fir pile, mud line to 10 ft above, in 25 ft of water, at Redondo Beach, Calif. (*Courtesy of Stone & Webster Engineering Corp.*)

quite a few feet above the high-water line, the borer sucking up such water as it needed.

Bankia setacea burrows enter the pile about pinhole size at right angles and turn obliquely, generally down, and enlarge quickly to ¼ in.

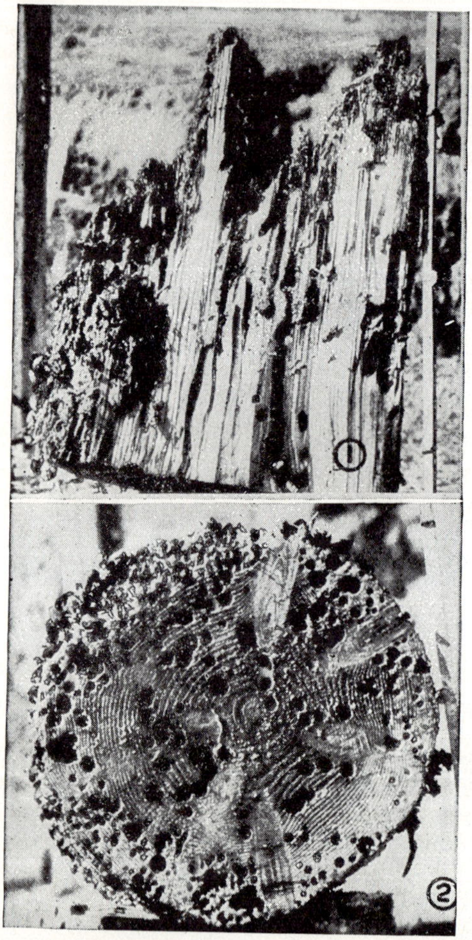

FIG. 13.18. Note large tubes of *Bankia* and smaller ones of *Terredo navalis*, with erosion by *Limnoria* at breaking point. *Sphaeroma* also present. Section cut 1 ft above break. Dolphin pile driven December, 1917, removed July, 1920. Municipal pier, Richmond, Calif. (*Courtesy of University of California, Publications in Zoology.*)

diameter within 2 in. of the surface, and to ⅜ or ½ in. within 4 in. of the surface. *Bankia setacea* does not restrict itself to sapwood and often bores obliquely in before turning to run with the grain, so that surface samples may not reveal the extent of infection. Burrows as large as ⅞ in. diameter and over 30 in. long have been found in San Francisco Bay.[20] The nacreous-lined burrows penetrate deeper than those of *Teredo* in light infections, and in heavy infections all go toward the center. The burrows show minor deviations from the straight or curved

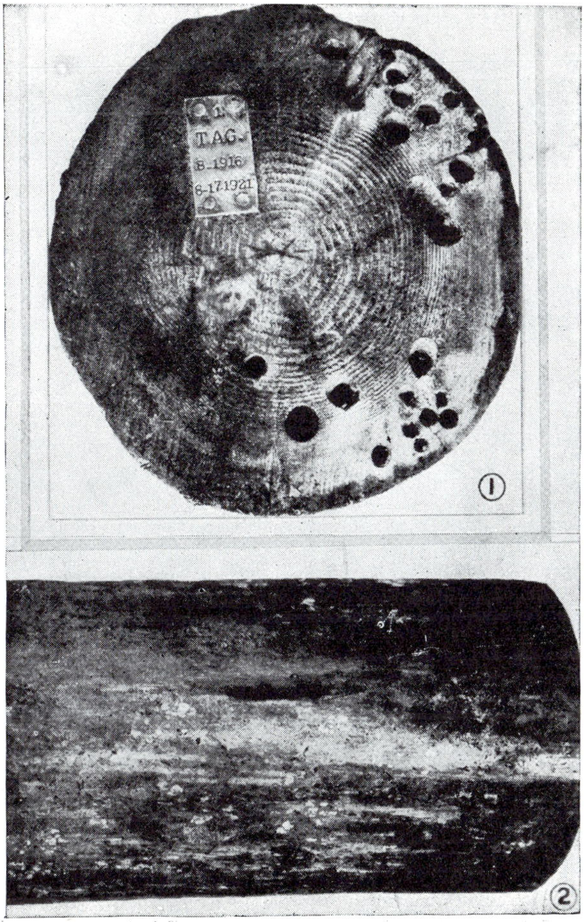

FIG. 13.19. *Bankia* burrows. Note thin treatment through which borers penetrated into interior (upper view). Surface of the same pile showing entrance holes of *Bankia* (lower view). (*Courtesy of University of California, Publications in Zoology.*)

routes of *Teredo*. The burrows of *Teredo* are more superficial in light infections. Attack may be concentrated at the mud line, or just above mean low water. The latter is true at Martha's Vineyard and at many locations in the tropics. In general, *Bankia setacea* makes larger burrows and bores faster than *Teredo,* but farther apart so that the wood is not so fully destroyed. However, its attack may equal that of the worst *Teredo.*

Xylophaga dorsalis makes a shallow burrow 1½ in. deep but does not line its burrow with nacre.

Martesia striata and *Hiata* make tiny round entrances that rapidly enlarge to good-sized circular holes inside the pile. The burrows usu-

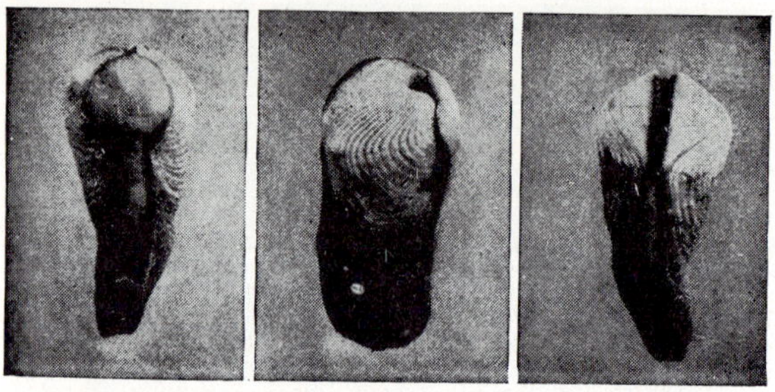

FIG. 13.20. *Martesia striata* from Pearl Harbor.

FIG. 13.21 *Martesia striata* in wood pile. Burrows with calcareous lining are *Teredo.* Cavite, Philippine Islands. (*Courtesy of University of California, Publications in Zoology.*)

ally reach in a year to about the length of the borer, or 2 in., with a ¾- to 1-in. diameter. The depth is apparently limited by the desire of the borer to keep his siphon ends in the entrance. Rapid disintegration of the wood allows the borer to burrow deeper.

Pholadidae openings, although slightly larger than those of *Teredinadae,* are still small and hard to find. Their burrows increase in size with depth and growth, and are not lined. *Pholadidae* are a serious problem, as they make cavities of up to 1½ in. deep in the hardest woods.

Limnoria destroys piles by gnawing interlacing branching burrows into the surface, as many as 200 to 300 to the square inch. The shallow burrows are 0.05 to 0.025 in. in diameter, seldom over ¾ in. long, and follow the softer rings. Thus the surface ⅛ to ¾ in. is left a mass of thin walls between burrows that breaks away and exposes a new surface to attack. *Limnoria* can burrow into soft woods such as pine and spruce to a depth of 1 in. a year and can reduce a pile diameter 2 in. a year in severe attacks. The wood presents a spongy appearance that forms a hairlike growth over piles. *Limnoria* works at all levels from mud line to high water, but is usually most active below low-water line,

FIG. 13.22. *Limnoria* attacking wood. (*Courtesy of Dr. William F. Clapp.*)

sometimes at the mud line even at 70-ft depths, and occasionally between tide levels. Attack results in a narrow-waisted effect on the pile, which is eaten away until it breaks. *Limnoria* avoids hard knots, which are left projecting. They will attack piles in other than the preferred zone, if that becomes unavailable. If fill is placed around piles up to the low-water line, which may have been the top limit noted for damage, it is to be expected that attack will occur above that point. The small size of this borer is deceptive when compared with its destructive power. Samples coated even lightly by fouling organisms can be quite free from attack.[2cd]

Chelura burrows appear much the same as those of *Limnoria*, and it now seems to have been proved that *Chelura* drives *Limnoria* out of the original tunnels and occupies them. Thus *Chelura* is blamed for destruction caused by *Limnoria*.

Sphaeroma pentodon makes round openings up to ½ in. diameter, of characteristic size, separated location, and appearance. The tunnels

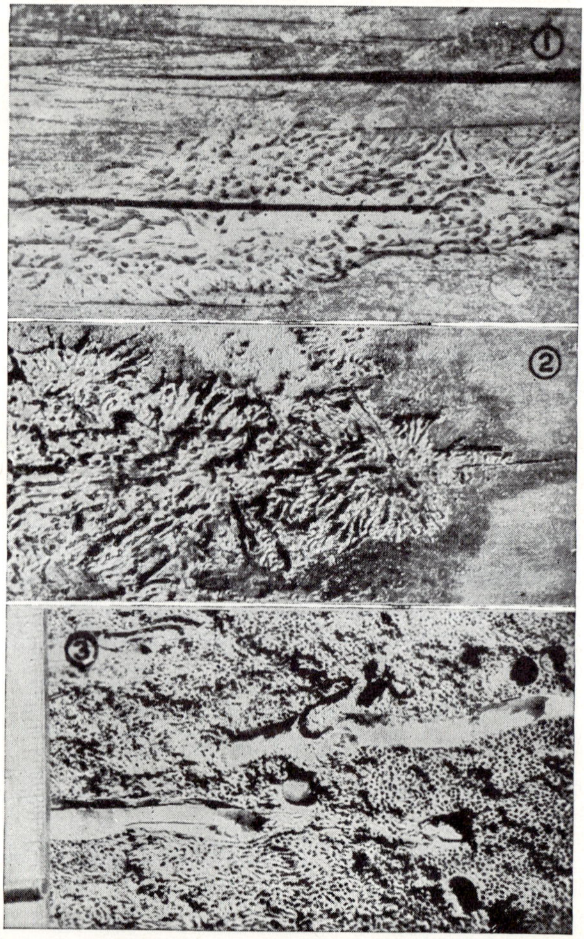

FIG. 13.23. Typical stages of *Limnoria* attack. (1) Initial attack of *Limnoria*, untreated Oregon fir pile exposed 1 month, July, 1920, Municipal quay, Oakland, Calif., estuary. (2) Deeper penetration by *Limnoria*, Southern Pacific Co., Oakland, Calif., mole. (3) *Limnoria* attack uncovered *Bankia* tubes in untreated Oregon fir pile for fender at Oakland, Calif., municipal quay. (*Courtesy of University of California, Publications in Zoology.*)

enter horizontally, then turn quite abruptly and run with the grain in the softer layers, but they do not expand as do those of the molluscan borers. These borers work principally between tide levels, although they are found at all depths. They produce a pitted appearance between tide levels, showing large, dark open burrows, with sometimes occasional channels in the surface, especially in deeper submerged parts. Their burrows are harder to detect below low water where they may be confused with attack by other borers.

Distribution. No place in the world seems to be really safe from attack by marine borers. Some waters and seasons or cycles of years are less conducive to reproduction and attack by borers but conditions change and damage may be done before it is suspected.

Some species predominate in certain waters and other species are rare. Waters may experience a shift of borer population. There is not much to indicate that any species is by geographical conditions excluded from any waters. About all that can be told definitely of species range is to name the bore populations as they have been found.

Eight species of boring *Mollusca* have been found along the North Atlantic Coast of the United States. All are capable of destroying wood, and some have been found drilling in soft rock, poor concrete, insulation, and rope. The *Pholadidae, Petricolidae,* and *Lithophagidae* have been reported as rock or concrete borers.

Probably the most destructive of these molluscan borers in northern waters is the common shipworm, *Teredo navalis,* which has been reported in salt-water harbors in Europe from North Cape to Italy and the Black Sea, and on the Atlantic and Pacific coasts of North America. Other species of *Teredo* have been found in the Canal Zone, Caribbean Sea, and Pacific Ocean. *Teredo parksi* is dominant at Pearl Harbor and other Pacific islands. Other teredine borers in northern regions are *Teredo dilatata, norvegica, thompsoni,* and *megotara. Teredo dilatata* is more damaging than *navalis* but less common. Teredine borers have been found in North American harbors only in salt or brackish waters, although they occur in fresh water in India, Australasia, and parts of South America.

Nausitora, a large shipworm, is very damaging in low salinities in tidal reaches of some Australian rivers, in which *Teredo* were unable to exist.

Bankia gouldi, another and larger type of shipworm, found from New Jersey to the Gulf of Mexico, is even more destructive than *Teredo navalis.* It seems reasonable to expect that the same factors which have caused the sudden invasion and survival of *Teredo navalis* in destructive numbers north of Cape Cod might result in a similar extension of *Bankia gouldi. Bankia setacea,* or Northwest shipworm, also known as the giant, or plumed, pileworm, is one of the most destructive shipworms and has a wide distribution on the Pacific Coast and in Alaska. It has been very destructive on the northern Pacific Coast, attacking creosoted piles through knotholes and damaged spots in the creosoting. In Seattle, an interesting case is on record where piles extending 50 ft through water were inspected by a diver, who hit with a hammer the *Bankia setacea,* which must leave a portion of their bodies projecting from the hole to act as a siphon for food; the damage was thus stopped for a time at least. This method gives merely temporary protection, for new arrivals take the place of the others, and destruction proceeds at the usual rate.

FIG. 13.24. *Bankia setacea* attack after 5 months. Mole at Oakland, Calif. (*Courtesy of University of California, Publications in Zoology.*)

Pholadidae are widespread and can bore quite hard rock. They have not, however, done a great deal of damage to sound concrete containing hard rock aggregate. Examination of concrete-jacketed piles in Los Angeles Harbor in 1922 revealed well over half more or less attacked, and one-fifth badly damaged.[20] However, the concrete was generally cement mortar poured in forms around the piles after driving. It would be considered poor by present-day standards although some was quite hard. Shipworms have penetrated through *Pholas* holes in casings in Alaska, attacking the wood. Most damage was caused by the edible *Pholadidae penita,* known as the *rock clam.* Other species found were *Platyodon cancellata* in very poor mortar, and *Petricola carditoides,* known as the nestler, which were probably inhabiting burrows made by other species.

Martesia striata has been active in Hawaii, the Philippines, Gulf of Mexico, Florida, Cuba, and Puerto Rico. Although it does not work as fast as the shipworm, it is able to destroy a pile in two years.

Hiata has been active in Florida.

Xylophaga dorsalis belongs to another type of borer. It has caused severe damage in Europe but to date has appeared only in the deeper waters of harbors on the New England coast.

Several kinds of boring *Crustacea* are destructive in the North Atlantic region, among these being *Limnoria lignorum* and *Chelura terebrans.*

Limnoria may be found in all salt or brackish waters of any temperature, whether polluted or clean. It abounds in English waters and up to the Arctic Circle, and has been found in Alaskan waters as far north as Kodiak Island, in Puget Sound,[2n] the Black Sea, the Baltic Sea, South Africa, New Zealand, Australia, Japan, the Falkland Islands, etc.

Chelura terebrans is reported as a wood destroyer in Norway, the Black Sea, the Cape of Good Hope, and New Zealand, and in recent years has infested the Atlantic coast of North America from Florida to Labrador, driving out the original *Limnoria* in some instances. *Chelura insulae* occurs in Samoa and Hawaii.

In northern Europe, Australia, and other countries where it is plentiful, including harbors of the islands in the Pacific, it was formerly considered that *Chelura* was much more destructive than *Limnoria,* and that if it continued to spread and increase it would become a major problem to owners of marine properties on the East Coast of the United States. It seems to have been proved that *Chelura* has been greatly overrated as a wood destroyer. Both *Limnoria* and *Chelura* work so rapidly that few other forms of marine life can obtain or maintain a hold on the same surface.

Sphaeroma is most common in tropical waters, from Florida to Brazil, and in South Africa, India, Ceylon, New Zealand, and Australia, but it appears occasionally in temperate waters.[2az] This species is very destructive and can adapt itself to low salinity. *Sphaeroma pentodon* is known to occur on the Pacific Coast to Alaska. *Sphaeroma destructor* requires brackish to almost fresh water, and has made severe attacks in the St. Johns River in Florida. *Sphaeroma terebrans* is very destructive in Queensland.

Infestation. Although the distribution of marine borers has become fairly well known, at least in certain regions during recent years, it is constantly changing, and frequently new infestations occur when conditions are favorable. The previous absence of borers in a location is no assurance that protective measures may be omitted with safety.[*,2m] In Newfoundland, Nova Scotia, and New Brunswick, during 1937, the attack of *Limnoria* on test boards increased two to ten times over previous

FIG. 13.25. *Chelura insulae* after 8 months, at Tutuila, Samoa. Ends of a few *Teredo* burrows exposed. One *Limnoria andrewsi* in lower-left-hand corner. (*Courtesy of University of California, Publications in Zoology.*)

years. In 200 locations in New England waters, *Limnoria* showed an almost unbelievable increase though serious attacks on piers in Boston Harbor had occurred years earlier.

Teredo navalis has been considered the most destructive borer in New England water, prior to 1936, but its activities were almost entirely south of Cape Cod. In 1937 a great increase in teredine activity was noted in southern New England waters, and *Teredo* appeared in destructive numbers in harbors from northern New England to Newfoundland which were known to have been previously immune. In addition, *Teredo tryoni* and *Teredo dilatata,* never previously recorded in New

* For examples of attack on structures in previously immune waters, see Chap. 16.

England harbors, caused considerable damage. Both *Teredo* and *Limnoria* increased to large numbers along the North Atlantic Coast in 1939, 1940, and 1941, typically indicated by records at New York, Portland, Maine, and Liverpool, N.S. In the 1940 peak there were one hundred times as many borers of both species as a few years earlier. *Limnoria* were still decreasing rapidly in 1945. *Teredo* nearly disappeared in 1943 but by 1945 were rising rapidly.[2bg]

Infested shipping or driftwood may import borers into locations previously free. Changes in food supply and natural environment and cyclical changes also may create conditions favorable to attack. Removal of old untreated timber marine structures and prevention of use of harbors as dumping grounds for wood waste are two measures advocated to reduce the borer population, which can spread with astounding rapidity in the presence of an adequate supply of wood coupled with other favorable environmental factors.

Salinity, temperature, current action, depth of water, pollution, hydrogen-ion values, dissolved oxygen, and sulphureted hydrogen all have great bearing on the presence, or possible presence, of borers. A variety of combinations of such conditions may occur in any one harbor, and wide variations in severity and nature of borer attack may thus occur at relatively short distances apart.

Studies by the Clapp Laboratories, based on the inspection of test boards and records from hundreds of locations throughout the world, now indicate that several factors which were at one time considered as controlling the variations in marine growth in reality influence this growth only indirectly through their effect on food supply for the borers. It is true that favorable conditions with respect to these several factors are necessary for continued borer activity, but they do not, in themselves, explain the periodic variations in marine growth.[2bg,2bh]

FIG. 13.26. Pile after *Limnoria* attack at Portland, Me. Original diameter indicated by projecting knots. (*Courtesy of Ralph H. Mann, American Wood-Preservers' Association.*)

Salinity. This is the most important single factor in marine-borer life. Most borers require salt or brackish water for continued life and reproduction. Two effects are noted owing to changes in salinity—retarded activity and death. Salinity may vary at different depths, and a large number of adults and larvae may be found near the bottoms of piles in some cases.

Salinity ranges from 30 to 35 parts of salt per 1,000 in the ocean to negligible amounts in fresh river water, and the danger point is reached at a salinity of about 15 parts over a continued period. *Limnoria* and *Bankia* seem to require 20 parts and all borers are found active at this figure.[2o] *Limnoria* found a salinity of 6.5 parts fatal in 24 hr, were sluggish and unable to swim after 2 weeks in 10.7 parts, and were retarded in activity by 12, 14, and 16 parts but continued to bore.[2o] In locations where freshets and high precipitation occur colonies of *Limnoria* may be exterminated. The Danish Engineering Association reports that the limiting salinity in Scandinavian harbors is 16 parts.

Teredo activity appears to decrease in salinities of less than 9 parts per 1,000. This species has the ability to plug the entrance to its hole with its pallets, retaining salt water inside and keeping out undesirable fresh water, thus greatly deferring the ill effects of fresher water. Experiments have indicated 5 parts per 1,000 as lethal generally, although some borers functioned in 4 parts for a time. Six weeks of salinity below 4 parts seems required to be fatal. *Teredo* is unusually resistant to low or fluctuating salinities. However, in a few of the warmer rivers of the Orient, such as at Rangoon, it is said that *Teredo* is found even in fresh water, and that some *Teredinidae* require fresh or almost fresh water. *Teredo norvegica* and *Teredo thompsoni* can rarely exist in water which is even slightly brackish.

Bankia setacea is much less resistant than *Teredo* to low and changing salinities. It is limited in San Francisco Bay to areas where the average salinity is about 25 and the minimal for the year 10 parts per 1,000. *Bankia gouldi* is understood to thrive in a salinity as low as 7 parts per 1,000. *Bankia* is the most active borer at Rangoon, which is 25 miles upriver from the sea and has a salinity one-third of that of sea water. *Nausitora* flourishes in low salinity in some Australian waters, and is adversely affected when it rises. Species occur in tropical fresh-water rivers.[2az]

Temperature. Water temperature was formerly considered one of the most important factors in borer activity. Studies by the Clapp Laboratories now indicate that, while favorable water temperatures are important, other vital factors must be considered also.[2bg] New England *Teredinadae* were found to be only slightly affected by the considerable variations in temperature during the seasons of activity in spring, summer, and fall. The water may be cold enough to kill the embryos in Nova Scotian waters during the short breeding season of one or two weeks. Temperature effects are greatly reduced once borers have entered wood.[2bh]

Current Action. Tests have shown that *Teredo* will not attack when the current velocity is over 1.4 knots, nor *Limnoria* when over 1.8 knots.[2br]

Effect of Light. *Teredo* larvae have been found to be much more active in the dark, and to show reduced activity in the presence of light.[2br]

Breeding Seasons. Borers have definite breeding seasons, but these vary among species so that if several species are present, attack may be practically continuous. The breeding season is usually associated with temperature. Reduced salinities may retard reproductive activities.

Limnoria has been reported to breed only when the water temperature is over 40°F in waters having a large seasonal change, but is also found above the Arctic Circle where waters are always colder than this.[2o] When conditions of temperature and salinity are favorable, *Limnoria,* of which the adults are free to move from place to place, appears to be able to breed throughout the year.[2o]

In the case of *Teredo* and *Bankia,* the destruction of wood occurs in the period following the breeding season, when the borers grow to maturity.

Bankia setacea starts breeding in San Francisco Bay in February or March and ceases by July. Temperature appears to govern, the borer favoring cold water. The breeding season starts at the beginning of the year in Alaska.[2o]

Teredo navalis breeds in late summer and early fall.

It has now been determined, chiefly by the use of test boards, that in northern waters the period of attack from marine borers is limited to little more than 2 months, although once inside the surface the borers continue to eat unless the water temperature falls to just above freezing, when they lie dormant.[2l]

Teredo navalis, at least, has shown marked cyclic intensity at 10-, 30-, and 70-year periods.

Pollution. Excessive pollution from mill wastes, oil, and catch-basin overflow usually is a barrier to most borers. Cases are known where removal of habor pollution in recent years has been followed by a large infestation of marine borers. In observations of marine-borer activity published prior to 1937 little distinction was made between the various types of pollution, but the nature of the pollution must be considered in order to prophesy with any degree of accuracy what changes in marine life may be expected should pollution be removed.[2m] Industrial wastes and catch-basin overflows usually appear to be far more effective barriers to borer activity than is sewage. It is reported that sewage pollution cannot be expected to afford protection from borers except actually at the mouths of large sewers.[2o] Limnoria appears to have no scruples regarding sewage.[2af] Invasion of certain harbors by borers thriving in the cleaner waters outside may be expected as a result of sewage-disposal and industrial-pollution programs, and owners of marine structures should be prepared to adopt protective measures before great damage has occurred.

No measurable change has been found in marine life in certain harbors where purification programs have been undertaken. Such programs may be limited to one or more types of treatment, and the discharge still contains industrial wastes, or additional chlorine. In an instance where all treated sewage effluent was carried outside the harbor, there was a sharp increase in borers.[2bg]

Annual variations in borer activity cannot be related directly to changes in pollution. Such variations seem to occur whether the water is polluted or clean.[2bg]

Effect of Hydrogen-ion Concentration and Dissolved Gases. The amounts of oxygen and hydrogen sulphide present in the water affect marine borers, the former being necessary for respiration and more than a tiny amount of the latter being fatal. If the pH value is above 7, the water is alkaline; below 7, it is acid. An increase in the pH value represents a decrease in hydrogen ions. Normal sea water has a pH value of 7.5 to 8.5. Any marked change in these figures is apt to be fatal to borers. A hydrogen-ion concentration lower than 7 on the acid side is usually an indication of heavy industrial waste or other heavy pollution.[2bh] In San Francisco Bay, the maximum amount of hydrogen sulphide was 0.42 cc per liter; the average, 0.13 cc. These amounts are considered small.[2o] High oxygen content is encouraged on warm sunny days. High oxygen and low hydrogen sulphide contents are favorable for marine borers. These factors, however, although of importance, are minor compared to the effects of temperature, salinity, and sources of infestation, except where pollution is present.

Silt. Heavy silt is a factor of great importance inhibiting *Limnoria* activity. Also it is said that *Teredo* cannot continue to thrive in muddy or turbid waters.

Variations in Food Supply. Much has been published to show that borers burrow to obtain food. Studies by the Clapp Laboratories appear to indicate that the borers feed upon microorganisms in the water, as do clams and oysters.[2bg] Borers which burrow in rock or concrete cannot obtain food from the rock or concrete. Published records show that oysters are poor in years when their food supply is limited during the growing period, and conversely. In dry seasons, the food supply decreases, and oysters are poorer. Heavy rainfalls carry down large supplies of food. It is now thought that these same factors affect marine borers. Records indicate that borer activity in some harbors varies at more or less regular intervals, which may be associated with the food supply.[2bg] Observations by Dr. Walton Smith suggest that adult shipworms make use of the wood in which they dwell to maintain their carbohydrate content.[2br]

Investigations along these lines are presenting a promising field for study. If borer activity could be predicted, this would be of economic value.

Resistance and Geographical Distribution of Various Woods. Although it has been claimed that certain woods are immune to attack by marine borers, such as greenheart in Europe and America, billian in the China Seas, teak in India, turpentine wood in New South Wales, black cypress and titree in Queensland, and spotted gum in Tasmania, experiments have not confirmed these statements. The only woods showing great resistance are a few tropical ones. Several woods which have fairly good records in cool waters do not give nearly as good service in the tropics. Jarrah, greenheart, and blue gum last much longer in English waters than in the tropics.

Wood having a good record in one place may not do so well in another. Test results in one location should be checked against local experiences when possible. Age, density, sapwood proportion, time and method of cutting, seasoning, etc., undoubtedly have their effects on the ability of the wood to withstand borer attack.

The availability of woods preferred by borers may cause incorrect inferences to be drawn as to immunity of other species. If both oak and pine untreated piles are used in a wharf, it may be found that the oak piles are left untouched by the borers, which will concentrate their attack on the soft wood. It might appear that oak is immune, but let the pine piles be removed and it will be found that the oak is severely attacked.

Silica inclusions in the wood appear to explain the resisting qualities of manbarklak, in which the percentage runs up to 1.5 per cent. The minimum useful amount seems to be 0.5 per cent and, since the amount is quite variable, a limit should probably be specified.[2ce] Of 814 Indonesian woods examined, 181 had silica inclusions. Kolaka was one rich in silica.

Poisonous substances contained in the wood appear to contribute to the resistance of a few woods. Greenheart contains an alkaloidal poison, and the ironwoods from Indonesia also contain poisons.

The worms do not grow so freely or so large in hard woods, which consequently generally have a slower rate of destruction.

As far as the United States is concerned, it does not generally appear economical to import the more resistant woods, with the possible exception of greenheart, unless home-grown treated woods become more expensive. In Caribbean regions, where only 8 to 10 years of life is generally expected from treated soft woods, study may show use of some native woods to be economical.

The names of the tropical woods used are the most common or scientific ones, but many of them pass under a variety of local names, which are given in the *Report of the Committee on Marine Piling Investigation*[2n] or may be ascertained locally.

Africa. The resources of the forests are little developed. Kussia, kaku, and potrodum from Ghana (known respectively as opepe, ekki, and erun in Nigeria), Rhuan palm from Gambia and Sierra Leone, Azobe from the Cameroons, and several woods from Kenya have shown considerable resistance.

Asia. India has teak, acle, and sal, which are extensively used for marine construction. Indonesia has a number of woods with fair powers of resistance, such as billian, ballow, rassak, camphor, and tampinnis. The Celebes have kajoe lara and kolaka, and Sumatra kajoe malas. Burma has thande, teak, and pyinkado. The Philippine Islands have several woods which offer resistance; mancono is the first but it is expensive;[2n] others are liusin, aranga, dungon, malabayabas, and pagatpat. Although these are not immune locally, they might give long life in temperate waters. In general, the forests are hard to develop.

Australasia. Australia has many eucalypts, more resistant to borers than pine or oak. Western Australia has jarrah, blackbutt, and karri. Eastern Australia has turpentine wood, ironbark, stringybark (red), Murray red gum, and tallowwood. The virtue of turpentine wood seems to lie in the bark, which must not be removed or damaged. Queensland has cypress pine, satinay or satiné, and swamp mahogany. Tasmania has blue gum, Huon pine, and stringybark. New Zealand has totara. Some other resistant Australasian woods are becoming scarce.

Europe. The Baltic and White Sea ports export pines, spruces, and firs, much used in England. These are soft woods, but they are reasonably good in some localities when creosoted. Southern and Central Europe have extensive coniferous forests also.

North America. Large quantities of southern yellow pine are used, creosoting being necessary for borer resistance. Much Douglas fir from the western region is used, particularly for long piles, with creosoting necessary for borer resistance. Other pines and spruces also need protection. From California come red or gray gum, manna gum, sugar gum, and blue gum. Cottonwood, while poor structurally and subject to rapid decay, has given good resistance in Alaska, though not in the northern West Coast states. Palmetto, palm, and mangrove have been used in the South, and although not immune, these are generally not badly attacked. They must be jetted entirely, since the soft core in the hard shell of the wood cannot stand driving with a hammer, although a light tap is generally safe to assist sinking. They are suitable only for light loads and must be cut off below low water to prevent rapid decay. Subject to these restrictions, and possibly attack from *Martesia* if present, these woods should generally give quite good service, much longer than pine.[2n]

South America and the Caribbean. South American resources are not developed. The best known wood used and exported for piles is greenheart from British and Dutch Guiana, which has records in temperate waters varying from fairly to very satisfactory, although its record in tropical waters is not always good. Mora from British Guiana, Honduras, and Trinidad resists rot and insects well but does not resist borers. Bulletwood or balata is said to be good, if obtainable. Angelique, manbarklak, anoura, foengo, and spensi hodoe from Dutch Guiana, alcornoque from Panama, and black kakaralli from British Guiana have shown up well in regard to resistance.[2ao]

Elimination of Damage from Marine Borers. This is a difficult problem. Persons unfamiliar with it often draw too hasty conclusions. Lack of sufficient service data and conclusions drawn from tests in locations where borer attack is relatively mild, such as in cooler waters, lead to unsafe decisions. Great danger lies in basing conclusions on short-time tests, for even under severe conditions several years may be needed to show the effectiveness of the proposed protection. Unless the owner is willing to take a chance of early failure, he should never accept a method of preservation which has not had sufficient widespread use to prove its effectiveness under conditions comparable to those expected. Protection against marine borers becomes more difficult, in general, as warmer or tropical waters are approached, and is a much more difficult problem than protection against decay and insects. Measures found satisfactory for treating inland piles in water or inland piles buried in the ground or exposed to air and moisture may be practically worthless against marine borers.

The method of protection selected will depend upon the local conditions, type of construction of the property to be protected, expected economic life of the structure, severity of service, costs, etc.

The two general methods of eliminating or reducing attack are (*a*) *poisoning the wood by treatment,* and (*b*) *armoring.*

The consensus has been that marine borers attack wood to obtain cellulose for food. It may be that *Teredo* burrows for protection and can live on marine microorganisms siphoned through its body by tubes projecting from the burrow. If cellulose is required as food, poisoning the wood can be considered as a remedy. Some materials considered as toxic have little effect on the health or digestion of *Teredinidae.* Some species can plug the burrow entrances so tightly with their pallets for several weeks that poisoning the water has little or no effect. Some molluscan borers are not able to do this and may be temporarily exterminated in this manner. No natural enemies of borers are known for certain. Since the food of borers is wood, the introduction of a new plentiful supply of untreated wood in a locality may be the means of causing a large increase in the number of borers.

Treatment of wood may not always prevent destruction of wood by *Limnoria punctata, Pholadidae, Martesia,* and *Sphaeroma,* which have sometimes been found boring freely in heavily creosoted piles,[2at,2au,2cc,2cd,2ct] in which case some type of encasement must be used if the wood is to be protected. If *Limnoria, Martesia,* or *Sphaeroma* are present in the water, together with a borer which is known always to be poisoned by a sufficiently treated wood, the other borers may pass through the outer treated portions of wood destroyed or bored by the *Limnoria, Martesia,* or *Sphaeroma,* and thus attack the inner untreated portions. *Bankia setacea* has been occasionally found at work inside creosoted shells but not traversing them.[2ab] *Martesia* has penetrated in a few months wood sheathing with an underlayer of tarred felt, and it seems probable that metal sheathing or concrete are required to stop it. *Limnoria* is particularly adept at finding and entering cracks, spots damaged by dogs or cant hooks, or boltholes in creosoted piles, and will reduce such piles to mere shells. It is sometimes found in such piles actually at work in the creosoted zone itself.[2af]

In order to reduce checking of the heads of wood piles during driving, a minimum butt diameter of 14 in. is advisable so that the head can be shouldered, thus providing fibers outside of those actually struck to furnish support. Checking provides points where decay and *Limnoria* can start attack.

The foregoing is only sufficient to introduce the difficulties of the subject. The services of an expert taxonomist should be sought when performing work in infested waters, to predict whether or not danger is likely to occur in the future, and to determine what method of protective action will be most effective.

Mechanical Protection of Wood Piles

Fill. Fill placed around piles damaged by borers will stop further damage, and may be used if there is still sufficient strength in the piles to carry their loads, and if there are suitable means of holding the fill.

Riprap seems to provide protection, probably because the oxygen content of the water is reduced.

Barnacles sometimes provide considerable protection.

Bark. Bark has been advocated as a protection against borers, but to be of value it must be unbroken. *Limnoria,* although preferring softer wood, will enter through exposed knots. This requires great care in driving and handling, and the covering of all knots and abrasions with sheet metal or other protective coating is necessary. Bark protection is suitable only for temporary structures.

Charring and Tarring. This ancient method has only temporary value.

Carbo-Teredo Process. This consists of applying a surface coating of Vaseline and heating with a blowtorch, which results in fairly even charring to a depth of ¼ to ½ in., and reimpregnation of knots and ends combined with hammering to consolidate the char at these points. This patented process has been used in Australia and strong claims are made for it. Cuts and damaged spots may be repaired.

Armor for Wood Piles. The great difficulties with armor are to keep it firmly attached to the pile and intact. Constant attacks usually occur from abrasion, unequal expansion, rupture, corrosion of fastenings, wave action, shellfish and borer attacks, etc. The following methods are among those most commonly used at the present time; listed also are some methods that are not now used extensively or that have been abandoned, which are included to provide background to the subject: individual pile casings consisting of steel, iron, vitrified pipe, concrete cylinder,[2y,2ac] gunite[2ab,2ah] or shotcrete,[2z,2aa,2ab] mortar, creosoted battens, copper, zinc, corrosion-resistant aluminum, or Muntz metal sheathing, fabric sheathing, etc.

Methods proved satisfactory and economical in one location may prove otherwise elsewhere. The subject is one which is receiving constant study. Borers are not generally active below the ground line or under encasement in good condition. Scour or dredging may change the location of the mud line when providing protection against borers to this point.

Tile or cast-iron casings should be filled after placing, with sand topped with mortar or with lean concrete using fine aggregate, since this makes them much more resistant to impact. If sand is used it also affords a means of detecting a break below water by observing whether or not the sand is still in place at the top. Tile is brittle and should not be used where exposed to wave action or current, since it is readily broken by drift, but it is efficient so long as it is unbroken. Cast-iron casings are expensive but are sometimes justified owing to long expected life.

Steel and iron sheathings are seldom used now and may have sufficient length of life to justify their use, particularly if they are of light-gage metal.

Metal sheathing requires that the piles be prepared before driving. Knots and projections should be removed and the surface made as smooth as possible, and then an even coating of asphalt-saturated felt or burlap applied. The sheet metal is tightly nailed over this. *Copper sheathing* is subject to abrasion and theft. *Zinc sheathing*, while undoubtedly prolonging the life of the pile, has not proved to be economical. *Muntz, or "yellow," metal* must be carefully checked to be sure that the alloy, consisting of about 60 per cent copper and 40 per cent zinc, is homogeneous, or else electrolytic action may be set up, as seems possible in view of conflicting reports of service attributed to variations in composition. Nails must be of the same composition as the metal. Corrosion-resistant aluminum is a later development.* Cupronickel (90 per cent copper, 10 per cent nickel) has shown great resistance in sea water; nails must have the same composition.[2ch]

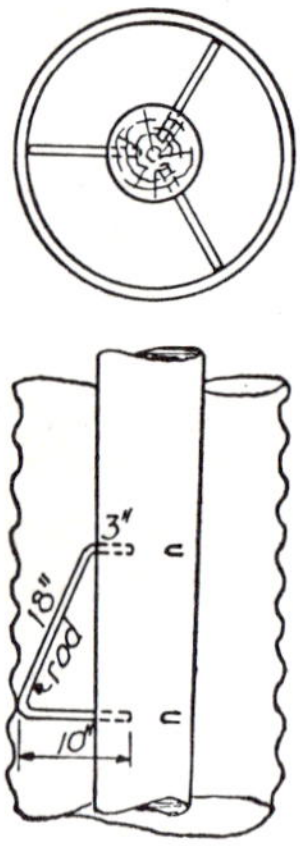

FIG. 13.27. Corrugated steel shell and concrete armor.

Fabric sheathing abrades easily during driving or in storms violent enough to wash gravel.

Untreated wood battens bolted to creosoted piles often permit borers to attack the treated wood. Battens may be easily damaged.

Scupper nailing with iron or steel as a method of protection has merit and can be used where labor is cheap. Copper nails have not been found to give much protection,[2at,2au] and the following remarks are based on iron or steel nails. Only the portion of the pile exposed to attack needs this protection. Nails such as 3d or 4d spaced ¼ to ½ in. apart have been used, as have 1-in. nails having ⅜- to ½-in.-diameter heads driven with ½-in. wire mesh as a template. Protection is given both by the nailheads and the rust incrustation formed. Borers appear to have an aversion to the iron compounds formed in the wood. Scupper nailing gives better protection against *Limnoria* than *Teredo,* since *Limnoria* burrows only about ½ in., so that if a nailhead is missing the damage is local, whereas if *Teredo* finds such a spot it penetrates deeply and destroys the pile. This method is still used to some extent, particularly in Denmark and Germany.

Corrugated steel shells (Fig. 13.27),[2x] 24 in. in diameter, 14 gage, have been slipped over the tops of piles after the driver has been backed away. They are then sunk into the mud bottoms for a distance of about 2 ft by the weight of the hammer or by a light tap. A corrugated section is used as a follower, with wood cushions and fittings top and bottom. The casings, which may be of any desired length, extend

* Patents held by W. Horace Williams Co., New Orleans, La.

several feet above high-water level. They are guided in place by rod brackets, after which they are filled with tremie concrete, without reinforcing.

Coatings. Present-day heavy creosoted treatments, together with restriction on the presence of cuts and framing connections and on the use of cant hooks and dogs, etc., result in an average pile life probably twice as great as previously. An intensive search has been made for an armored coating which would both keep out borers and withstand abrasion from rafting and moving objects.

Coatings in great variety have been tried, consisting in general of some kind of heavy paint of a bituminous, asphaltic, or similar nature, combined with burlap, felt, roofing paper, wire mesh, wood battens, or other material.[2af] The object of the combined materials is to strengthen the paint coat and make it less liable to be torn off or broken. Variations and changes have been made as conditions changed, results were observed, and opinions formed. These coatings are applied only to the portions of the pile requiring protection.

Concrete Encasements. These have been used successfully both to protect new wood piles and to repair damaged piles. Rock borers have been known to open holes in 1½-in. concrete encasement, permitting *Teredo* to enter the wood in a few instances.[2n] Typical examples of these methods are described.

Hay Process. This is a method of restoring damaged wood piling to the strength of new piling by means of a concrete encasement.[2ab] It consists of a patented form so designed as to allow the placing of a concrete encasement to any desired depth, even down to and into the ground, and has been successful in 30 ft of water. The concrete is mixed and placed in the watertight form suspended above water and lowered to the required depth, where it is left until the concrete has set. The forms are then removed. Should damage from marine borers or abrasion be confined to the portions near or above the water level, there is no need of carrying the encasement to any great extent below the damaged parts. This method is controlled by the Cement-Gun Co., Inc.

Presscrete Pile Encasing. This is a method for reconstructing, protecting, strengthening, and fireproofing weakened, deteriorated, burned, or ruptured wood, steel, or concrete piling to restore the piles to the original or increased strength, and avoid the cost of replacements.[2ax] Concrete of great density is applied by air pressure in metal forms from which the entrapped water has been ejected by compressed air. The encasement is usually from 2 to 4 in. thick and is reinforced with wire mesh or other steel. The usual mix is 1 part cement to 1.5 parts of sand and 2.5 to 3 parts of coarse ⅝-in. aggregate, silicified to resist injurious chemical influences and injected into the forms under a pressure of 5,000 to 8,000 psf. The concrete is injected into the form bottom, forced upward, and compacted within the form space. The pressure head is equipped with controls for the air and concrete. The forms remain until the concrete has hardened sufficiently to permit safe removal. Only such portions need be encased as require strengthening or protection, but the process may be carried below the mud line if necessary. No heavy driving equipment is needed, and work can progress with little interference to the use of the structure. Under warehouses, piers, and bridges, the compressed air and concrete are conveyed to each pile in flexible hose or pipe lines from shore or floating equipment, while the form work is done from small boats wherever the tidal headroom permits. This method is controlled by The Presscrete Co., Inc.

Precast Concrete Jackets. Jackets may be single piece or sectional. The Koetitz patent covered a special design of precast sectional jackets. By driving the piles butt end down, the casings were slipped over the ends after driving. Precast jackets could have round, octagonal, or square exteriors, but were usually round inside, allowing a space to be pumped out and grouted. The minimum shell thickness was 2½ to 3 in. If the wood pile was driven at the time of jacketing, if the bottom was soft, and if the pile penetration could be predetermined, lugs might be bolted to the pile before driving to support the bottom of the jacket. This type was first used in San Francisco in 1908 and has given excellent service, being affected only by those factors influencing the life of good concrete in sea water.[20] Care was needed in handling, to avoid developing cracks which allow the reinforcement to rust. The cost was relatively high, and the economic life had to be such as to justify use of this type.

A large navy pier was constructed in 1942,[2ac] only the batter piles of which were of untreated wood, driven butts down. Over these were slipped square precast concrete jackets 38 to 52 ft long, with 4-in.-thick reinforced-concrete sidewalls, designed to penetrate 15 to 18 ft below the mud line. In order to slip encasements over the tops of driven piles, the piles were driven butt end down. This increased the resistance to driving and also resulted in higher unit stresses in the pile. A rubber gasket, placed where the bottom of the jacket would set, was driven with each pile to seal the bottom sufficiently to permit grouting the bottom 5 ft of the jacket to the pile. The grouting space was filled with rich concrete. For protection in the tidal range, the jackets were coated with asphalt mopped hot over a primer consisting of 70 per cent distillate and 30 per cent roofing asphalt.

Gunite or Shotcrete (Fig. 13.28). Concrete, shot on before or after driving, has been successfully applied to wood piles.[85] A large installation by the Port of Tacoma in 1922[2ab] had a coating shot on before driving. The coating was 2 in. thick and from 28 to 51 ft long, extending from 2 ft below the top to a point 5 ft below mud line, reinforced with 2- by 2-in. No. 12 wire mesh. This installation was found, in 1945, to present no evidence of cracking, spalling, or detrimental effects from sea water, and is reported to have appeared good for an indefinite period. The concrete was beveled into a dap in the wood at the bottom of the encasement to present a smooth surface.

In 1942 and 1943, during World War II, because they were unable to secure treated piles of sufficient length in time, the U.S. Engineers drove a large number of untreated wood piles for wharves. The piles were from 100 to 125 ft long, with butts 14 to 18 in. and tips 5 to 9 in. Shotcrete coatings ranging from 15 to 48 ft long were added to encase the upper portions and extend a few feet into the mud to ensure protection

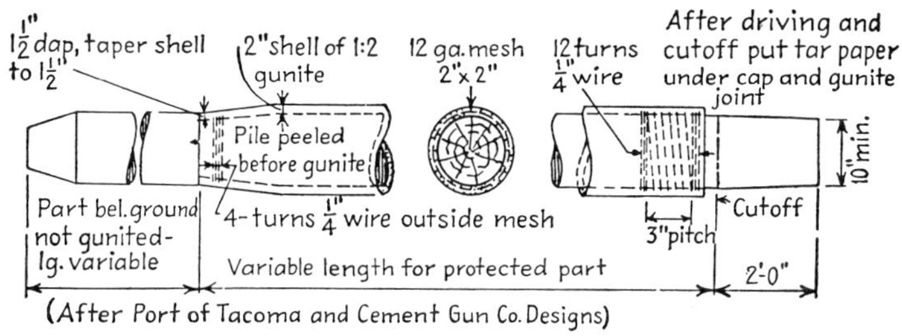

FIG. 13.28. Typical gunite encasement for wood piles.

against borers.[2z,2aa] The coating was 1½ in. thick, reinforced with 2- by 2-in. No. 12 or 14 wire mesh, with the top and bottom 6 in. having additional wrappings of six courses of No. 9 wire each. It was found advisable to wet the piles before coating, to keep the shotcrete as dry as possible, and to lap joints in the mesh 6 in. In order to prevent any slipping of the coating, it was found necessary to notch the piles 1 in. deep with ax cuts having a 4- by 4-in. area, spaced 18 in. apart in three or four rows. It was found that the piles withstood much harder driving than creosoted piles and could be driven much deeper without damage. They could also be driven deeper than precast concrete piles. As a test, over 5,000 blows on a pile driven to practical refusal, with a hammer having a rated energy of 15,000 ft-lb, failed to cause any damage other than cracking in the top 2 ft of coating. It was found that the increased stiffness of the coated piles added considerably to the rigidity of the wharf and dolphins, and that the piles could be handled and driven without damage after a 3-day curing period when using standard portland cement. This construction also had the advantage of fireproofing the piles. The costs indicated a very favorable result when compared with precast concrete, precast concrete-jacketed untreated wood, and treated wood piles.

In Fig. 13.29 are shown details of shotcrete jacketing of creosoted wood piles installed in 1945.[2ay] A 1:3 cement-sand mix 1½ in. thick was used, reinforced at the center with wire mesh held in place by metal clips driven into the pile. The clips had shoulders to place the steel ¾ in. from the pile and recesses to hold the mesh. A hammer blow

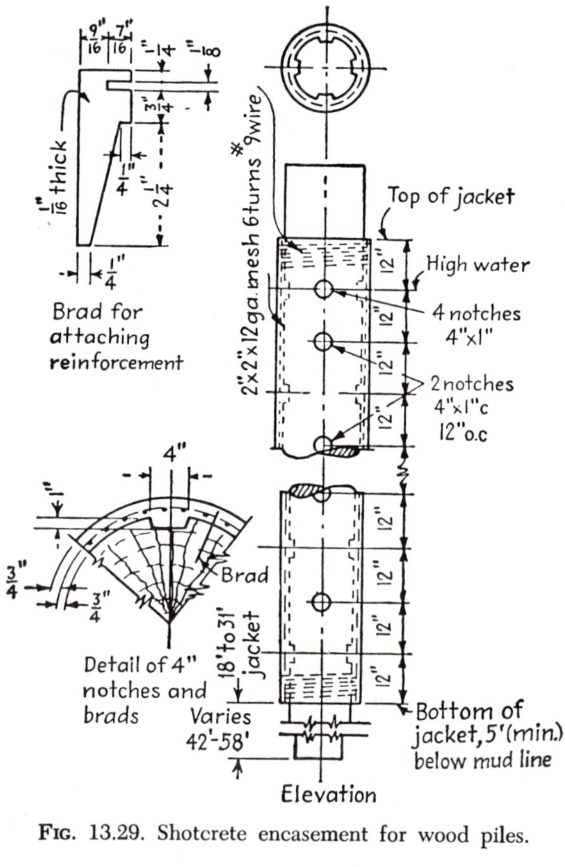

FIG. 13.29. Shotcrete encasement for wood piles.

crimped the clip to secure the mesh. Protection was applied from 2 ft above high water to 5 ft below the mud line. The piles were wrapped with burlap and kept wet 4 days, then dampproof coated as a curing aid. Adequate bond was found without surface cleaning of the treated wood, but 4-in.-diameter holes 1 in. deep were drilled at 1-ft alternate centers to provide mechanical bond as a safety measure. The jacketed piles were not damaged during handling or driving.

Many thousands of gunite-encased wood piles have been driven successfully under very hard driving. When properly made, minor failures in the jackets have occurred in less than 1 per cent of the piles, and then always in the top few feet where the damage may be seen and repaired.[85] This is a better record than might be expected from concrete or creosoted piles. Wood piles gunited before driving cost about half as much as precast concrete piles or wood piles gunited after driving.[85]

Greater protection than from either creosoting or encasement alone should be obtainable from the use of both. This was tried in 1944 on a Southern Railway coal pier reconstruction in Charleston, S.C., and the results should be instructive. The original pine piles driven in 1914 to

FIG. 13.30. Piles wrapped with wire mesh ready for application of shotcrete. Coal pier, Southern Railway Co., Charleston, S.C. (*Courtesy of Concrete for Railways, Portland Cement Association.*)

1915 had a 20-lb creosote treatment and cast-in-place concrete jackets extending from 2 ft below low water to 1 ft above high tide. *Limnoria* attack was noted in 1924. Attack by *Bankia gouldi* was observed in the spring of 1943 in the deep-water sections, and by fall 30 per cent were destroyed below low water. The reconstruction resulted.*

Concrete encasements in the marine-borer range will not prevent decay from entering the top of the pile or attacking the upper portion above the encasement.

Sectional Gunite Encasements. These provide a method of repairing piles damaged by decay and marine borers, consisting of guniting circular concrete jackets in 5-ft lifts above the water, and lowering this jacket 5 ft at a time until it is extended from high-water line to 5 ft

* *Proc. AWPA,* vol. 45, p. 299, 1949.

below the mud line. The piles are first cleaned of marine growths; then 1- by 2-in. vertical spacer battens are nailed on the pile. The battens are tilted out slightly at the top. Asphalt felt around the battens makes a form. Hooks are set in the bottom section, with the inner ends curved into guides to hold the shell equidistant from the pile. After the first section has set and is lowered, guniting on the next 5 ft is continued, using the same battens with fresh felt. A water jet is used to sink the shell into the mud. Mud forms a bottom seal for the grout tremied by hose into the annular space. Sway bracing is removed during the operation, then replaced.[2bi] This method has been used by the California

FIG. 13.31. Shotcrete-jacketed piles in foreground. Old piles in background, to be replaced, previously encased with concrete while in place, but jackets did not go below the mud line, so complete protection was not provided and piles were destroyed by *Teredo*. Coal pier, Southern Railway Co., Charleston, S.C. (*Courtesy of Concrete for Railways, Portland Cement Association.*)

Division of Highways.[2bj] The method is understood to have been developed during World War II both by Ben C. Gerwick, Inc., and by the Case Construction Co., the latter being succeeded by the Johnson Western Co.

Pressure-jacketed Concrete Encasements. These can be applied to piles before driving, by the use of pneumatic pressure in airtight metal forms equipped with pressure heads for air and concrete connections.[2ax] Steel mesh or other reinforcing may be used. The concrete may be silicified to provide resistance to chemical action. This method is controlled by the Presscrete Co., Inc.

Intrusion-Prepakt Encasement Method.[11ak] This is used for the repair of wood piles. For that section of the pile from the desired depth below mud line to such level as will not be subject to severe drying, Intrusion

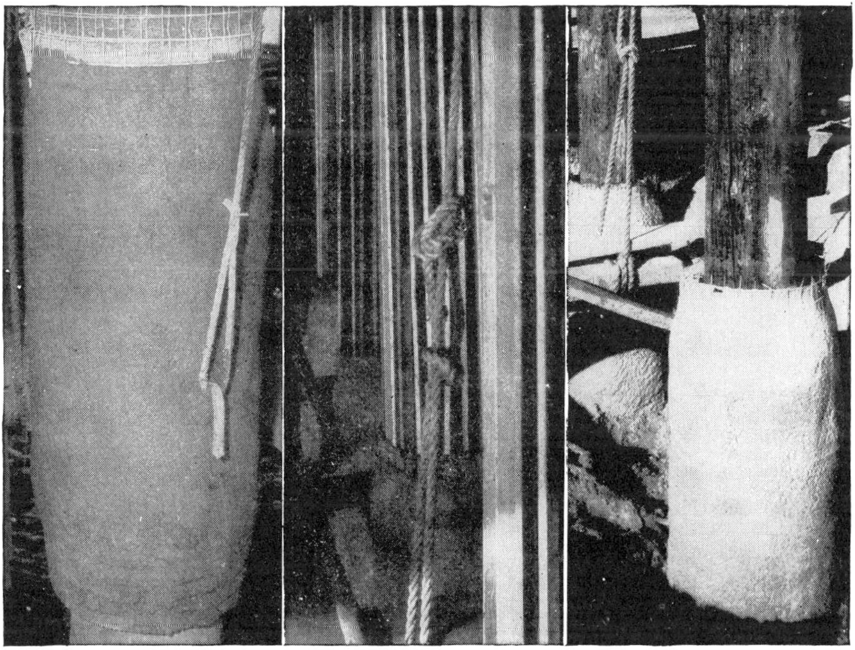

(*a*) (*b*) (*c*)

FIG. 13.32. Steps in Johnson Western pile-encasement method. (*a*) Bottom section gunited ready to lower. (*b*) Tilted battens ready to receive next course of asphalt felt and reinforcing mesh. (*c*) Top section of completed shell ready for tremie grouting by hose. (*Courtesy of Johnson Western Co.*)

FIG. 13.33. Tops of finished encasements by Johnson Western method showing treatments of pile, pairs, and groups. (*Courtesy of Johnson Western Co.*)

mortar with reinforcement is used. From the drying level upward, reinforcement and graded aggregate are first placed in the forms, then solidified with Intrusion mortar. Stovepipe forms may be used under water and left in place. Jetting can be used to lower the forms and to clean them out. Temporary spacers are needed to hold them properly away from the pile. If the cone method of excavation is used, mortar should be pumped into the bottom of the cone around the pile, and the form pushed into it. When forms are jetted, a seal is placed in the bottom of the form. After seals harden, forms are flushed. Above water, stovepipe or reusable split-steel shells are employed. Spacers on reinforcement cages are wire springs or concrete blocks. Thickness of Intrusion mortar or Prepakt concrete is usually 3 in. The increase in strength of such piles over wood piles is estimated at about 30 per cent, and fewer piles may be required or some bracing eliminated by the repair work.

Prepakt concrete has less than half the drying shrinkage of standard concrete, and strengths up to 6,500 psi at 90 days. This concrete, and also Intrusion mortar, has no setting shrinkage, is unaffected by salt water during curing, and is highly impervious.

Intrusion mortar contains sand, cement, Alfesil, and Intrusion Aid. Alfesil is a fine siliceous material that combines with the lime to form strong, impermeable compounds. Intrusion Aid inhibits early stiffening, neutralizes setting shrinkage, and produces the same effects as air entraining agents. This method is controlled by Intrusion-Prepakt, Inc.

Dri-Por Method. This method of repair uses a watertight box around a damaged pile to permit removal of poor concrete and inspection and placing of new dense encasing concrete in the dry. New reinforcing may be welded to old. No tremie or grouting is required. This process is controlled by Masonry Resurfacing and Construction Co., Inc.

Chemical Protection of Wood Piles

Electrolysis. Electrolysis of the water around the pile to free chlorine was proposed, but experiments seemed to indicate that water currents removed the chlorine and that *Teredo* merely ceased feeding while the concentration was high, so that frequent repetition of treatment would be needed to kill the swarm of new larvae coming with each tide.[20]

Reduction of Salinity. Barriers for the purpose of keeping salt water out of estuaries, rivers, and portions of bays have been suggested as a means of creating conditions in which borers cannot live. Fresh water was pumped into the sea at Hartlepool, England, to dilute the water sufficiently to kill borers at a log storage.

Toxic Treatment of Water. The possibility of using slowly soluble toxics near piling, by attaching tubes or cloths to the piles, has been suggested.[2n] There are no doubt a number of toxics that would kill both larvae and adult borers of both types in the immediate vicinity of infested structures; but they would also kill all other marine life there, as well as in other regions washed by the same tides, currents, and winds. The possible distribution might have to be checked by observing the spread of nontoxic solutions or dyes before proceeding. The treatment would have to be repeated every 2 to 4 weeks during the breeding season. Legal complications might ensue.

Dynamiting Borers

Dynamiting in the water has been suggested as a means to kill borers. It also kills all other marine life and may be contrary to public regulations. It would only kill borers in the immediate vicinity, and fresh larvae would arrive constantly so that the process would require frequent repetition. This method was used successfully at the San Francisco Naval Shipyard during World War II to preserve wood piles for 3 months until gunite encasements could be installed to a point 5 ft below the mud line. The explosive comes in strips and can be lowered vertically into the water or placed in horizontal positions between pile bents under water.

Preservative Treatment of Wood Piles

Creosoting. The common method of poisoning the food supply of the fungi, marine borers, and termites in wood piles is by creosote treatment, generally by pressure processes in preference to immersions in hot (200°F) creosote in open tanks. The first to propose creosoting was Franz Moll, who took out a patent in 1836. Practical adoption of the process followed John Bethell's patent in 1838. Piles treated in this manner, when completely buried in the soil, should not decay regardless of the ground-water level. This does not mean, however, that creosoted piles will have an indefinite life, for this may not be the case.

Life of Creosoted Piles Aboveground. The life of pressure-treated poles aboveground has been estimated by the Forest Products Laboratories of Canada to average 35 years,[2ar] and the results confirmed by study of telephone poles over a long period of years.[2j] Poles having more than the usual 8 lb pole treatment should have longer life.

Life of Creosoted Piles Buried in Ground. According to the above studies,[2j] buried creosoted piles will retain the preservative treatment practically permanently if properly pressure-treated. The Forest Products Laboratories of Canada, from a study of records, estimate the conservative life of properly pressure-treated creosoted piles embedded in the ground under masonry caps at 100 to 150 years.[2ar]

The American Wood Preservers Institute has published records of many creosoted piles projecting above the water table, extending back to 1890, that are still in excellent condition.[2ca,2ci]

Marine-piling Life. This is lengthened considerably by creosote treatments, but attacks progress, though at a slower rate. For instance, in Pacific island and Caribbean waters an average 8- to 15-year life is all that is expected from creosoted piles. Longer periods may generally be expected in colder waters, where the activities of marine borers are less violent. In San Pedro Harbor, the life is placed at 20 to 30 years, and in San Francisco Bay at from 15 to 25 years.[2a] Experiments at Gulfport, Miss., indicated that the life in years was roughly equal to the number of pounds retention of creosote per cubic foot.[2at,2au] On the New York and New England coasts life expectancy is 50 years.

Many reports of longer life in various localities are available. Such long life may or may not be obtained in future work, and in analyzing the economics of a new project, it is often advisable to adopt a conservative attitude. The definition of pile life also depends upon whether it is considered as the start of damage, damage to only a few piles, or failure. In cases where damage occurs to some of the piles and they can be renewed, the life of the structure, or economic pile life, is greater than where renewal is impossible or where encasements must be used. Damaged spots permit rapid entrance of borers and may account for some reports of short life. Variations in composition and quality of preservative affect the life greatly.

Creosote treatments protect the outer surface of the pile, the depths of penetration ranging from 1 to 1½ in. for Douglas fir up to 3 to 4 in. in southern pine piles. For example, in 1927 it was reported that 80 per cent of all holes in the treated shells of creosoted piles through which borers had gained entrance and started attack were caused by dogging the piles without plugging the hole.[2o]

Borers generally damage the pile first at the point where the impregnation is thinnest, or where damage has occurred. It is important to secure uniformity in the treatment. The use of cant hooks, rafting dogs, dogs on shoes, chain slings which may crush the fibers, and framing connections or boltholes should be prevented on the portion of the pile between the mud line and high water. A point of even minute damage may permit entrance of the borer, which can then work within the pile without difficulty.

Floating Collars. Floating collars for applying a creosote coating have been extensively used to provide protection against marine borers in Australia, where attacks seem to be concentrated near the low-water mark.[2az,*] Attack is not confined to this zone, however, in harbors of

* "Pier Piling Creosoted by a Floating Collar," *Eng. News-Record,* vol. 146, No. 17, p. 54, Apr. 26, 1951.

the Atlantic and Gulf coasts of the United States, where much damage may occur at the mud line. The method was developed for use in preserving existing piles not too seriously damaged to still support their loads. It is especially suitable for wharf piles where attack is from crustacean and not teredine borers, and where attack is mostly between tide marks. The collar is a round hinged galvanized-iron casing, supported on floats, with the top about 18 in. above the water line. The depth of the cylinder may be as much as 8 ft. In treating piles more than 3 or 4 ft below low water, a fixed one-piece envelope of oiled fabric can be passed down into the water and clamped together as it enters. Any toxic nonsoluble liquid lighter than water may be poured in the small space between the cylinder and the pile, gradually displacing the water. No oil escapes. This oil travels up and down the pile with the tide and collar. Twenty-four-hour treatment has been found to kill all crustacean borers. The treatment is repeated annually for a few years. The method is apparently effective, of low cost, and simple to use. It permits inspection of the piles and does not add weight.

Salts. Another class of preservative is toxic water-borne salts. They are highly effective against fungi, especially for sawn timbers moderately exposed. They are not best suited for piles. The initial cost is less than for adequate creosote absorptions, but life may be shorter. Many arsenic, copper, and other highly toxic salts in soluble and insoluble form have been used experimentally but have failed to give protection to marine piles.[2at] Suggestions have been made that salts insoluble in water and soluble in the digestive juices of higher forms of life be used, but these have not been effective, possibly because the digestive apparatus of the borers does not possess chemical substances to make the salts toxic. Paint may be applied after drying.

Oil-borne Preservatives. Toxic ingredients can be added to petroleum solvents in accordance with the standards of the American Wood-Preservers' Association.

Pentachlorophenol.[*] This is toxic to fungi, termites, and wood-boring insects. It is stable to dilute soil acids and alkalies. It is not appreciably soluble in water and does not leach readily. Approximately 5 per cent by weight of pentachlorophenol is combined with a petroleum solvent. The solution is impregnated into the wood by standard vacuum and pressure processes so as to permeate all of the sapwood and as much of the heartwood as practicable. It can be used for foundation piles in contact with the ground, with recommended minimum retentions per cubic foot of 10 lb in arid regions, 12 lb where high moisture exists, and 16 lb for tropical service. It is not recommended for marine piles.

[*] Ira Hatfield, "Pentachlorophenol Comes of Age," *Proc. AWPA*, vol. 45, pp. 84–89, 1949.

Creosote Treatments for Wood Piles

For lasting preservation of wood piles, the most effective method of treatment is by a pressure process of which there are a number, all of which employ the same general principle but differ as to details of application. The piles to be treated are loaded on train cars which are run into a large steel cylinder. After the cylinder door is closed, preservative is admitted and pressure applied until the required absorption has been obtained. The two principal types of pressure treatment are the full cell (Bethell) and the empty cell.

Full-cell (Bethell) Process. A preliminary vacuum is applied to remove as much air as is practicable from the wood cells. The preservative is then admitted without air. After the cylinder is filled with preservative, pressure is applied until the specified absorption is obtained. A final vacuum is commonly applied immediately after the cylinder has been emptied of preservative, to free the charge from dripping preservative.

When the timber is given a preliminary steaming and vacuum treatment, in the case of green wood which must be treated without waiting for it to air-season, it is first steamed and then a vacuum is applied. Although this does not remove enough water from the wood so that it may be considered as seasoned, it does remove enough so that the wood is more penetrable to preservatives. After this preliminary preparation, the preservative is admitted. In case the charge has received a preliminary treatment by the Boulton, or boiling-under-vacuum, process, the unfilled space at the top of the cylinder is filled with preservative and pressure is applied as soon as this conditioning process has been completed.

It is impossible to remove all of the air from the wood cells regardless of the method of treatment, and therefore even under the most favorable conditions there is some unfilled air space in the cell cavities of the treated wood after use of the full-cell process.

Empty-cell Processes. The empty-cell process consists of forcing preservative into the wood cells when they are filled with air. When the preservative pressure is released, the confined air, which is under pressure in the wood cells, drives out part of the preservative, leaving a lower net absorption than is obtained by the full-cell process. Good treatment depends very largely on the depth of penetration, which is in a general way proportional to the gross absorption.

Two empty-cell treatments, the Lowry and the Rueping, are in common use. The cells of the wood are necessarily partly filled with preservative as in the empty-cell process. The difference between the full-cell and empty-cell processes lies in the degree to which the cells are left filled after treatment.

Lowry Process. In this process, which is also known as the empty-cell process without initial air, the preservative is admitted to the cylinder at atmospheric pressure. When the cylinder is filled, pressure is applied and the preservative forced into the wood against the air originally in the cells. After the specified absorption has been obtained, pressure is released, and the air under compression in the wood forces out part of the preservative. This makes it possible, with a limited net retention, to inject a greater amount of preservative into the wood and obtain deeper penetration than with the full-cell process.

Rueping Process. The main difference between the Lowry and Rueping empty-cell processes is that the latter employs air pressure above atmospheric. In the Rueping process, air is forced into the treating cylinder before the preservative is admitted. The air pressure is maintained while the cylinder is filled with preservative, thus leaving the wood cells more or less impregnated with air under pressure. This process is also known as the empty-cell process with initial air.

Boulton Process. This is a process for seasoning or conditioning timber by boiling in creosote under vacuum. The treating cylinder is filled with hot preservative oil so that all wood is covered. The oil is kept heated while a vacuum is applied. The oil serves to keep the wood hot, while the vacuum lowers the boiling point of the water in the wood and causes part of it to evaporate.

When an empty-cell treatment is specified, the cylinder is emptied of preservative after the conditioning period, and air at atmospheric pressure or higher is admitted as desired. The preservative treatment is then applied as for air-seasoned material.

In treating by the full-cell process, the cylinder is filled after the conditioning is completed and the pressure is applied at once. Some preservative is absorbed during the conditioning period.

The Boulton process has long been used for Douglas fir and is now coming into use for other woods, such as unseasoned red oak, which check severely by the steam-and-vacuum treatment but which check but little with the Boulton process. Green beech and southern yellow pine have also been treated by the Boulton process.

Specifications for Creosote Treatment for Wood Piles

Piles are usually treated to purchasers' specifications, which sometimes refer to general specifications developed by various associations, and sometimes embody the purchaser's own ideas which he considers important. Often the specification fails to protect the purchaser and may result in harmful procedures or increased cost. The specification should cover species, dimensions, proportions of heart- and sapwood, degree of seasoning at time of treatment, method of seasoning or conditioning for treatment, kind of preservative, absorption, penetration, and details of the treating process. The specification should avoid fixing the details of the treating operation. Provision should be made against the use of temperatures, pressures, and treating periods that are likely to damage the wood. It should also define the methods that will be used in judging the product. The general specifications of the American Society of Civil Engineers, of the American Wood-Preservers' Association, and of other associations are generally a good starting point, and may be limited or modified if necessary to meet particular requirements (see Appendix VI).

There has been at times, such as during and following World War II, difficulty in specifying and obtaining first-class creosoted piles for marine work, owing to the quality of preservatives available. Probably the first requirement is to purchase piles from a reliable producer. Second, the preservative should meet the specifications and these should call for heavy retentions as discussed later. In addition to complying with the main body of text in the AWPA specifications, the preservative should meet the specific gravities of fractions as given in the footnotes. The addition of any petroleum greatly impairs the effectiveness of a creosote treatment in marine installations. It frequently helps to request the supplying of piles and treatment at a fixed price, and to pay for the oil on the basis of the quantity actually used. When purchasing structures under lump sum contract, the owner can buy the treated piles, to avoid purchase by the contractor in the cheapest market. Certificates should be obtained as to grade or class of material, grade of preservative, and final retention in pounds per cubic foot.[2a] It is desirable to have piles branded with the producer's name at 5 and 10 ft from the butt, costing a few cents each.

To obtain adequate protection in all waters where borers are or may become a problem, A. P. Richards of the William F. Clapp Laboratories[2g] sets forth several more severe requirements than those contained in standard specifications AWPA P2 and Federal Specifications TT-W-556c, as follows:[2cg] "Do not permit use of distillate creosote or low residue distillate in place of creosote-coal tar solutions, tests having shown better results with 70/30 solutions where *Limnoria* are involved; increase the specific gravities of fractions specified as 1.025 to 1.030, and 1.085 to 1.100; use not less than 20 lb treatment for southern yellow pine, and 16 lb for Douglas fir; require the brandings 5 to 10 ft from the butt to show the year of treatment, preservative used, and retention, as well as the name of the testing company, and to use the 'assay' method for determining retention. The assay method is reported to show retention within 1 lb plus or minus if twenty 3-in. borings are extracted from each charge of piling. Determination of weight of treatment is very difficult for an inspector by other methods, and the less preservative used, the larger savings in material and freight costs for the seller. Since the service life of the piling is increased with better treatment, it is suggested that purchaser pay for added retentions, the cost being slight in comparison with the added advantages, and all bids being on a competitive basis."

Avoidance of Injurious Treatment

In the steaming and vacuum process of conditioning green wood, there appears to be no reason why pressures above 20 lb gage (about 259°F) should be employed. No known advantages gained by higher pressures offset the greater danger of the higher temperatures damaging the wood. The steaming period should be kept as short as possible, and depends on the diameter of the pile. Since 50 to 60 per cent of the total water removed by this process is taken out in the first hour, and 70 to 80 per cent by the end of 2 hr, and only 90 to 95 per cent at the end of 4 hr, it appears that there is no good reason for continuing the treatment longer than 2 hr. It is probable that very little is gained by attempting to heat the wood to higher than 200°F at 3 in. from the surface. In larger sizes, this may not even be obtainable without damage. If specifications require that heating continue until the wood is thoroughly sterilized, this may increase the steaming period, and it should be recognized that the risk of damage is increased. This process is now largely confined to green southern yellow pine, and may be employed to advantage under certain limited conditions, when skillfully done, but otherwise it may be dangerous to the strength of the wood and very uncertain in its effect. For a full discussion of this subject see the *Manual on Preservative Treatment of Wood by Pressure*,[2c] which gives charts from which safe limiting values of time and temperature may be selected.

In the Boulton process of conditioning green wood, time and temperature are again the important factors to control in avoiding damage. Specifications of the American Wood-Preservers' Association permit a maximum oil temperature of 220°F during the boiling-under-vacuum period for Douglas fir piles and 200°F for sawed timber. In the light of present information, these values should not be exceeded. The same specifications provide that the boiling under vacuum shall continue until the amount of water collected is $\frac{1}{10}$ lb per cu ft of wood per hour. Variable results will be obtained from wood of different sizes, ages, and growth, and it is not established that it is necessary to carry boiling to this limit. Probably in some cases boiling can be discontinued sooner. No maximum time limit is ordinarily specified, but reliance is placed on the fact that the operator is not likely to continue treatment longer than necessary.

Treating pressures specified should be maximum allowable pressures, not required pressures. The operator should be allowed to use any lower pressure which will give the required absorption and penetration and avoid damage. Some of the pressures used in treating plants have been found to be too high if applied for a considerable period or when used with preservative temperatures which are more favorable for treatment. With preservative temperatures of 195 to 200°F, gage pressures of 100 to 125 psi are generally as high as should be used in the full-cell treatment of low-density species such as spruces and true firs, and other species which show tendency to checking and collapse during treatment. In the empty-cell treatment of such woods, the maximum preservative pressure will depend somewhat on the initial air pressure, but it should always be less than the amount equal to the pressures recommended for a full-cell treatment plus the initial air pressure.

Selection of Treating Process for Wood Piles

The relative merits of the full-cell and empty-cell processes are often not clearly understood. Basically, the effectiveness of treatment depends on the preservative, absorption, and depth of penetration, and not upon the treating process except in so far as the process used may affect these factors.

The names "full-cell" and "empty-cell" are confusing, since the "full-cell" process does not leave the cells completely filled with preservative, nor does the "empty-cell" process leave them empty. The principal difference between these processes is in the relative amount of air retained in the cells at the time the preservative is injected. The difference in the results obtained is that the full-cell process gives a greater concentration of preservative in the treated portion than does the empty-cell process. Higher gross absorptions and therefore deeper penetrations are normally obtained by the empty-cell process unless very high net retentions are specified. In selecting the treatment, the object should be to obtain the maximum penetration practicable with the absorption specified. This important consideration has been overlooked frequently in specifications. Some have required full-cell treatment of wood largely sapwood, and with this have specified that all sapwood be penetrated, while at the same time requiring net retentions that will not permit complete sapwood penetration by the full-cell method. An empty-cell treatment should be employed whenever the penetration can be improved, and the operator should not be restricted to the use of the full-cell process when a limited absorption is specified. Full-cell treatment should be specified only when the maximum possible net absorption is desired. The operator should, however, be allowed to use the full-cell process when, because of a limited amount of sapwood, resistance of the material, or for other reasons, the specified net absorption cannot be obtained by the empty-cell treatment.

The merits of steam curing are the subject of considerable controversy. In some species of wood, steaming makes good penetration of preservative possible, and sufficient steaming will ensure sterilization of the wood in which fungus infection may not be reached by the preservative. Operators opposed to steaming contend that all piles should be air-seasoned for treatment and that infected piles should be culled and not treated. It would be desirable if both recommendations could be met, but this is not always practicable. Often it is necessary to use green piling and, in some places, seasoning conditions are unfavorable or uncertain. Steaming is now used only rarely with seasoned piling, and seldom for any wood other than southern pine. Seasoned wood can generally be heated to better advantage by using a sufficiently high preservative temperature.

Specifications should not attempt to fix all the details of treatment. When treating green piles, variations in the wood will require different steaming or boiling-under-vacuum periods, depending on the diameter of the piles, moisture content, proportions of sap- and heartwood, etc. A certain degree of latitude should be permitted in treating pressures and periods, since the same conditions do not give equally good results with all woods. Maximum limits on steaming pressures, periods, and other operations should be set instead.

Gross absorption as well as net retention is sometimes specified in an attempt to ensure deep penetrations, but since the net retention may vary widely for any given gross absorption, such specifications are impractical and often impossible to meet. If sapwood is thin or if the wood has a high moisture content when treated, it may be impossible to secure the required gross absorption. It is sufficient to specify the net retention and whatever penetration may reasonably be specified.

The amount of sapwood should not be limited by specifications since this merely makes the wood more difficult to obtain, adds to the cost, and may add to the difficulties of treatment. Preservatives make sapwood as durable as heartwood, and permit better absorption of preservative. The strength of sapwood is as high as that of heartwood.

Long vacuum periods are not needed, except in the boiling-under-vacuum process. Preliminary vacuum for the full-cell treatment of air-seasoned materials needs to be extended little, if any, beyond the point of reading the maximum vacuum. In order to allow time for the preservative to drip from the piles, final vacuums after full-cell or empty-cell treatments sometimes need to be held longer than a preliminary vacuum.

The vacuum after steaming should not be held longer than necessary to obtain a practicable moisture reduction. The most effective periods are not definitely determined, but often too long periods are used to accomplish the best results.

Absorption of Preservative Required

Wood Piles. Most piles treated under pressure are southern yellow pine and Douglas fir. Although a full-cell treatment will give a greater concentration of preservative in the treated portion, an empty-cell treatment is preferable where there is any question about complete sapwood penetration with the absorption specified. Where particularly heavy concentrations of the preservative are needed, absorptions should be specified that will ensure good sapwood penetration with the full-cell treatment. This can best be obtained by specifying full-cell treatment to refusal, and specifying the minimum rather than the maximum absorption. Even a 16-lb full-cell treatment will often fail to give complete sapwood penetration when the sapwood is deep.

Specifications for most species used for fresh-water and land piles generally require absorptions of 10 lb per cu ft for Douglas fir, oak, and red (Norway) pine and 12 lb for southern pine, in accordance with American Wood-Preservers' specifications. Both creosote and coal-tar treatments are used.

The *Standard for Creosoted-wood Foundation Piles C*12 (1951) of the American Wood-Preservers' Association calls for treating foundation piles (defined as entirely embedded in the ground and capped with masonry) with Grade 1 creosote to a final retention of not less than 12 lb per cu ft, using the full-cell process only when the stipulated retention of preservative is greater than can be obtained by an empty-cell process.

Experience has shown the importance, from the standpoint of economy, of treating with heavy absorptions piling exposed under severe conditions such as occur in at least the more temperate and warmer parts of coastal regions. This not only provides a reserve supply of creosote to compensate for loss over a longer time but also ensures deeper and more uniform penetration. Piles to be used in salt water where borers are active should be treated by the full-cell process to virtual refusal. The American Wood-Preservers' Association specifies a minimum absorption of 20 lb of creosote or creosote-coal-tar solution per cubic foot in southern pine. They also recommend 12 to 16 lb creosote treatments by the full-cell process for Douglas fir piles according to the severity of attack expected. Their recommendation is creosote or creosote–coal-tar treatments to refusal by the full-cell process for oak piles. Observations have shown that borer attack started where penetration was thin, and these spots are undoubtedly where the effects of loss of creosote were first felt most severely. Once borers have obtained access to the interior of the pile, destruction takes place rapidly. The effect of *Limnoria* attack on the creosoted surface is to honeycomb it and greatly hasten loss of creosote and facilitate the entrance of shipworms.

The highest absorptions can be obtained only in piles which have been thoroughly air-seasoned.

Wood Sheet Piling. For wood sheet piling used in beach and shore protection in fresh water or inland structures, a 10- to 12-lb creosote treatment is generally adequate. When exposed to salt water and marine-borer attack, the maximum practicable absorption of creosote is advisable and treatment by the full-cell process is recommended, with final retentions of 20 lb per cu ft for pine piles and 16 lb for Douglas fir piles.

Penetration of Preservative Required for Wood Piles

Timber species found in the United States can almost all be treated successfully with coal-tar creosote by standard impregnation methods. In coniferous or soft woods, the amount of sapwood should not be restricted, as sapwood takes treatment more readily than heartwood. Hardwood piles to be treated should be limited to species which can be readily treated, such as the red oaks. There is practically no difference between the mechanical strengths of sapwood and heartwood, and from 1 to 3 in. of sapwood is preferable for treatment.

The minimum average and the minimum acceptable penetration should be specified whenever practicable, taking care not to establish requirements impossible to meet. Specifications of the American Wood-Preservers' Association set forth required minimum penetration in terms of inches depth and percentages of sapwood that should be treated. Depending upon the required treatment, such values range from 3.5 to 4 in. and 90 to 100 per cent for southern pine to ¾ in. for Douglas fir. Some general specifications require that all of the sapwood and as much as practicable of the heartwood be penetrated. Complete sapwood penetration is not always strictly obtainable. Even a 16-lb full-cell treatment will often fail to give complete sapwood penetration when the sapwood is deep. Merely to specify penetration as much of the heartwood as practicable is not specific enough, however, and the best method is to specify minimum and average penetrations of the heartwood faces. All holes made in treated wood for inspection should be plugged with thoroughly treated plugs.

Effect of Treatment on Strength[2b,2c,2e,2f,2g,2h,2i]

Coal-tar creosote, water-gas tar, wood-tar creosote, creosote-tar mixtures, and creosote-petroleum mixtures have no effect on the strength of the piles. Two to five per cent solutions of zinc chloride have no important effect on the strength. However, frequently green wood must be treated without waiting for it to air-season, in which case it is customary to apply an artificial heat treatment so that it can be penetrated by the preservative. The principal method used is first to steam the piles and then to apply a vacuum. If steaming or boiling under vacuum is used to season green wood to the degree necessary to permit the desired retention of preservative, the strength may be seriously reduced if extreme temperatures or heating periods are employed. The duration of conditioning and temperature should be kept as low as possible, using a steam pressure of not over 20 psi (or approximately 259°F) for steam conditioning. Higher pressure (or temperature) is much more likely to damage the wood. No advantage is known to occur from using higher values. The average amount of water removed from southern pine piles by the steaming-and-vacuum process is not over 5 to 6 lb per cu ft, most of which comes from the sapwood. Since about 70 to 80 per cent of the moisture in the sapwood is removed in the first 2 hr, there is little need to carry the vacuum much longer than that.

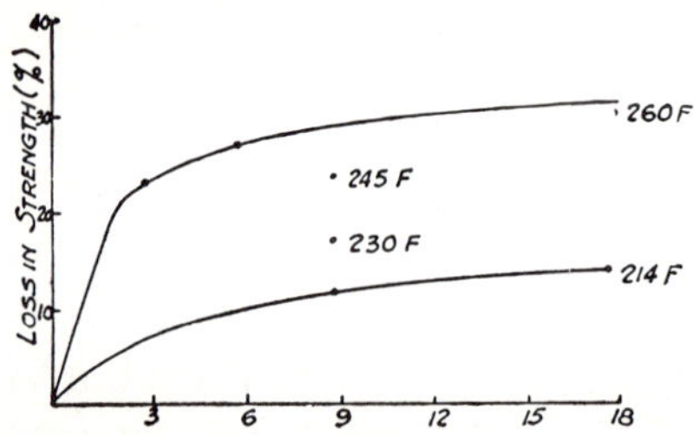

FIG. 13.34. Effect of different steaming conditions on the modulus of rupture of small clear specimens of unseasoned shortleaf and slash pine sapwood.

The percentage of reduction in strength caused by the steaming-vacuum process is generally more severe in other woods than in southern pine, with which the process is generally used. However, the reduction in strength by the process of steaming-and-vacuum depends upon many factors, such as the kind of wood, initial moisture content, density, rate of growth, etc., and, although information on the subject is rather scarce, it is suggested, as a rough practical guide, that the maximum fiber-stress values given for green wood in Table VI be reduced by approximately one-third in cases where such conditioning is required. The value of the modulus of elasticity is also decreased by steaming, but little information on the amount is available.

To indicate the general type of relationship between steaming and loss of strength in wood, observe the graphs in Fig. 13.34. Although these curves represent the results of test on shortleaf and slash pine, they indicate the mode of action.

If piles are submerged in creosote which is heated above the boiling point of water, under atmospheric pressure only, the steam will escape and have the same effect as seasoning the wood. However, rapid seasoning is apt to weaken wood, and it is suggested that the fiber-stress values for green wood treated in this manner also be reduced by one-third, as above.

Strength reduction caused by the boiling-under-vacuum process appears to be considerably less than that from steaming, or boiling in creosote at atmospheric pressure. Tests on various sizes of unseasoned Douglas fir timbers, approximating pile sizes, made by the U.S. Forest Service and by wood-preserving firms, in which the maximum temperatures were not over 200°F and the duration from 20 to 30 hr, showed average strength reductions of different series ranging from none to 17.5 per cent as compared with untreated control specimens, with individual values having from 69 to 114 per cent of the value of wood not conditioned.[2g]

Application of the above conditioning treatments to several series of tests on seasoned Douglas fir timbers, approximately of pile sizes, having from 15 to 16 per cent water content, made by the U.S. Forest Service, indicated approximate average reductions in strength as follows (individual minimum and maximum values in parentheses expressed as percentage of strength of unconditioned wood): boiling in creosote at atmospheric pressure—one-third (42 to 127); steaming—one-eighth (40 to 131); boiling under vacuum—one-tenth (64 to 130).[2g] The wide individual variations above and below the averages should be noted.

Creosoted piles have been found to become brittle and to disintegrate, presumably because the above precautions were not observed.

Piles are often wet in service, but because of the increased strength from seasoning, air-seasoned piles will stand harder driving than green piles. This is particularly true of treated piles, for tests have shown that, although the strength of green wood may be considerably reduced by treatment, after thorough seasoning treated wood may be but little weaker than untreated.[2b]

Treatment of Field Cuts

An ideal creosoted-wood structure would have no field cuts, holes, or damaged spots. This is difficult to achieve fully in practice, but every effort should be made to approach it as nearly as possible, since such points may permit start of decay or borer damage at an early period and thus cause the waste of a considerable part of the investment for treatment.

It is most desirable for extending the life of a structure in borer-infested waters that *no bracing extend below the high-water line,* for borers attack through the boltholes and ends of the braces more readily. Design of the structure may make this requirement difficult to meet, but consideration should be given to all possible methods of avoiding this condition.

There are two causes for the failure of creosoted piles from decay, if the timber was sound when treated. One is poor penetration and the other is mishandling in the field.[2ag] The former need not exist if standard specifications for treatment, such as those of the American Wood-Preservers' Association (Appendix VI) or American Society of Civil Engineers, are followed. The latter is the problem of the engineer and field forces.

Preservative treatments generally give the most protection near the surface of the pile. The penetration of creosote is apt to be variable, and attack most readily occurs where such penetration is least. If this outer shell is abraded, punctured, or cut, or if holes are bored, fungi spores may enter and start decay, provided the other conditions described as necessary for growth are present; or marine borers may find entrance. Some borers locate these spots quickly and obtain entrance to the interior of the pile where they progress rapidly, leaving the creosoted outer fibers as a mere shell.

FIG. 13.35. Bolt hole in creosoted fender pile, enlarged by *Limnoria*. (*Courtesy of Dr. William F. Clapp.*)

Cant hooks, peaveys, lifting hooks, and *rafting dogs* should not be applied to the portions of treated piles which may fall in the locations liable to attack. *Nails* or *spikes* for scaffolding should not be driven into treated piles. Creosoted plugs should be driven into such holes as do occur.

Framing and *boring* should be done prior to treatment in so far as possible. Treated piles should not be cut for bracing, but treated filler blocks should be used if necessary to fill out spaces. Boltholes drilled in the field should be filled by pouring hot creosote in through a bent funnel, or preferably by pressure equipment.* Cuts should be painted thoroughly with two coats of hot creosote. Some foremen make a practice of dipping all bolts in sealing compound just before driving them in place, in order to close any possible crevices around the bolt. The upper heads of all drift pins and vertical bolts, particularly if countersunk, should be given a liberal coating of sealing compound after they are in place. Instances have been recorded where field treatments with excellent re-

* A special device for pressure treatment of boltholes, manufactured and sold by Greenlee Bros. & Co., Rockford, Ill., is very useful for this type of work and is widely used. The straight-shank type is intended for treating holes in piles without removing the sway braces and is made for ⅝-, 11⁄16-, ¾-, and 13⁄16-in. holes. The universal type has tapered shanks accommodating all hole sizes from ½ to 1 in. Both types consist of a pump which is screwed into one end of the hole, and a plug for screwing into the other end. A pressure of 120 psi is exerted.

sults have been given to cuts and boltholes in bridges 10 to 20 years old, where such treatment was neglected at the time of construction. Bolts were removed long enough to sterilize the holes with a pressure device. It is reported that considerable amounts of preservative can be injected in this manner.

The American Wood-Preservers' Association *Standard Instructions for the Care of Pressure-treated Wood after Treatment* (Appendix VI) should be followed or incorporated in specifications for treated-wood-pile jobs.

Protection of Wood Piles below Ground Line

Much damage to the outer creosoted shell formed in wood piles during treatment may result from driving through heavy gravel or small sharp boulders, and if this damage is in a zone where protection is desired against decay, insect, or marine-borer attack, the piles should be placed by jetting or in auger-drilled or churn-drilled holes, or should be driven through a pipe casing which is withdrawn later. Severe splintering or gouging may occur otherwise, as can readily be observed by pulling piles. It would not be wise to neglect this source of short life or failure while taking careful precautions to prevent damage by mechanical handling methods.

Butt Protection of Wood Piles

Exposed Butts. If the wood in a pile butt is allowed to dry out and check, water, ice, dirt, vegetable growths, and fungi can enter. Decaying cutoffs are unsightly and may lead to unnecessary maintenance expense, whereas adequate protection immediately after construction is inexpensive and ensures years of freedom from maintenance or replacement expense. Both exposed cutoffs and cutoffs capped by timbers should be protected. If no field cutoff is required for an exposed butt of a pressure-treated pile, sometimes no decay will occur, but no dependence can be placed on this and all exposed butts, whether cut in the field or not, should be protected.

Experienced engineers generally require that cutoffs receive two or three coats of hot preservative, followed by one or two applications of sealing compound.[2ag] This compound can be made on the job by mixing creosote with coal-tar pitch to a suitable plastic consistency; sometimes it is mixed in advance and heated for use as required. For some Gulf Coast bulkhead and dolphin piles, no other protection than sealing compound, renewed every 2 or 3 years, is used. Exposed heads of piles are sometimes domed before sealing.

Permanently exposed butts of creosoted piles for such structures as fenders, dolphins, and ferry slips should be given two heavy coats of creosote oil upon being cut off, and two coats at least yearly, to provide increased life, according to the recommendations of the *ASCE Manuals of Engineering Practice No.* 27.[1h]

Creosote penetrates wood from eight to fifteen times as far along the grain as tangentially, which makes possible effective butt protection for piles cut off after treatment. As much creosote should be applied as the wood will absorb, instead of following the usual practice of giving two or three coats of hot preservative and allowing each to sink in before application of the next. Great improvement is secured by impounding the hot creosote for some time on the cut surface.[2av] The Illinois Central System has tapped metal rings lightly into the cutoffs, the diameters being slightly less than those of the piles and the heights being 1 to 1½ in., and filled this cup with creosote, allowing it to stand until absorbed. An equally effective method was to fasten a strip of roofing felt around the top of the pile with wire, to form the cup. When the rings or strips were removed, the cutoffs were given a heavy coating of a mixture of creosote and pitch and the cap timber placed while the mixture was still warm. Wet clay or earth dams have been used with general satisfaction, but when the sapwood is thin, felt or metal outer bands are best. The Southern Railway uses standard nested sets of 12-gage metal rings 4 in. high with inside beveled bottom edges, ranging from 11 to 16 in. in diameter.

FIG. 13.36. The first step toward proper field treatment of cutoffs is to saturate cut surface with hot creosote and apply a heavy coat of coal-tar pitch. (*Courtesy of American Wood-Preservers' Association.*)

On inland structures, customary practice is to place a waterproof cover over the sealing compound and fasten it to the sides of the piles with lead-headed or galvanized roofing nails. No nails should be driven in the top. The most common covers have been copper, galvanized iron, lead, heavy building paper, or at least three layers of canvas which has been saturated with tar or sealing compound. Four layers of Fiberglas set in pitch have been used.

Lead caps on oak piles have been observed to corrode, owing to the action of the acetic acid in the wood. Charring the tops of oak piles before lead capping has been done in Europe with some success. Copper and lead caps have often been stolen.

Capped Butts. Butts of creosoted piles to be covered by capping timbers should be thoroughly painted with two heavy coats of hot creosote, followed by a heavy coat of coal-tar pitch, after which a sheet of heavy roofing paper or a metallic cap should be placed before setting the timber cap, according to the recommendations of the *ASCE Manuals of Engineering Practice No. 27.*[1h]

Butts of Foundation Piles. Butts of piles entirely embedded in the ground and capped with masonry should be brushed liberally with two coats of hot creosote, followed by the application of a coat of coal-tar pitch, after cutoffs have been made to final elevation, according to the *Standard for Creosoted-wood Foundation Piles* C12 of the American Wood-Preservers' Association. There should be sufficient interval between applications to permit absorption of each coat before the succeeding one is applied.

FIG. 13.37. Three plies of pitch-soaked canvas coated with sealing compound waterproof these pile tops and thus provide complete protection. (*Courtesy of American Wood-Preservers' Association.*)

FIG. 13.38. Piles thus protected can be depended upon for long life in the completed structure. Here the waterproofing fabric was brought down and wired around the pile. (*Courtesy of American Wood-Preservers' Association.*)

Treatment of Old Cutoffs. Treatments of cutoffs, sometimes 10 to 20 years old, have been made successfully, where initial protection was omitted, by cleaning out the decayed heartwood, thoroughly swabbing the cavity with hot creosote, filling with sealing compound, and covering.

Preliminary Cutoffs. Preliminary cutoffs, made prior to final cutoffs, should be given at least one application of hot creosote oil.

Water-line Protection of Wood Piles

Additional protection is sometimes required on wood piles at the zone of fluctuating water levels to protect untreated wood against alternate wetting and drying and the pile against abrasive action. Soft rot due to marine microfungi may facilitate wear from repeated abrasion by continually attacking freshly exposed surfaces. Among the methods used are treated wood battens, gunite or shotcrete,[2z,2aa,2ab] concrete encasement,[2y,2ac] and metal sheathing.

Corrosion in Metal Fastenings in Wood Piles

Corrosion of some metals in some species of timber piles is a common occurrence. Iron fastenings in oak deteriorate rapidly, particularly where there is an excess of moisture and air. The natural acids in the wood, acetic in the case of oak, cause corrosion. In oak, the iron salts from this corrosion then react with the tannins in the wood to form the characteristic bluish stain. The causative agents for some other woods are not known at present. Where conditions are favorable for such action, metals other than iron should be used.

STEEL PILES

Corrosion of Steel Piles

Nature of Corrosion. Corrosion is generally considered to be the result of a difference in potential between two points in a conductor exposed to an electrolyte, with material moving from the anode, which in this case is the surface corroding, to the cathode, or noncorroding material. Corrosion cells are formed on piles between areas of different aeration. The steel surface plentifully supplied with oxygen is cathodic vith respect to the portions of the pile less accessible to oxygen.

Pitting. Many reports of rates of corrosion refer to depth of scattered pits, and while these are important to pipelines, they do less to impair the structural strength of piling. The pitting rate may be several times the loss of weight rate.

It has been claimed at times that marine organisms cause pitting or corrosion of metal piles, but this does not appear to be borne out by the evidence, except for some reports of pitting caused by shellfish adherence in tropical waters.[3f,3g]

Experience As a Guide. Experience with similar piling in the locality should be studied as the best guide to the needs and efficacies of various measures. Laboratory tests and field investigations of steel piles in various soils and waters have been carried out over a period of years, and their results should be studied.

Corrosion of Steel Piles in Soil. The rate of corrosion of steel piles in soil varies greatly with the texture and composition of soil, depth of embedment, and moisture content. In coarse-textured soils, due to the circulation of air, corrosion may approach the condition that will occur in the atmosphere. In heavy clay, the deficiency of oxygen results in conditions approaching those of submerged corrosion. Deeply situated sands under dense strata may have little oxygen, particularly during periods of low ground water.

In general, steel piles projecting into the air from ground may corrode near the ground line and for a short distance below, because of the aerated condition of the top strata and presence of organic surface material and water. Coal-tar paints or concrete encasements extending short distances above and below the ground line, at least 2 ft each way, are efficacious. Atmospheric corrosion above this point is combated by painting, as for any exposed steel.

Tests in Norway on steel piles embedded in coastal areas indicated corrosion due to electrolytic action well below the line of free air circulation, attributed to ion exchange in the presence of certain salts.

Corrosive Soils. Swamps, peat bogs, and alkali spots are considered corrosive soils. If corrosive materials are present in the soils from coal storage, acids, new cinder fill, caustic alkali fills or soils, industrial or mine wastes, etc., corrosion may occur. There are several tests that will indicate whether or not soils are potentially corrosive. The possibility of avoiding these locations should be considered; consideration should also be given to the type of pile and to methods of protection.

Corrosion from Anaerobic Bacteria. Bacteria of the anaerobic type may cause corrosion of iron and steel underground and under water.[3c,3d] They have been found in many soils and at great depths. The principal groups are sulphate-reducing and iron-consuming. The most common sulphate-reducing bacteria are the *Sporovibrio desulfuricans,* which require oxygen but derive it from the reduction of compounds such as sulphates, sulphites, thiosulphates, or organic substances rather than obtaining it from free air. Moisture and sulphur are essentials to them. Hydrogen sulphite, which attacks iron severely, is liberated, and hydrogen is removed from the cathodic areas of the metal with the formation of ferrous or iron sulphide. The *desulfuricans* group is thought to be active only in nearly neutral materials nearly free from oxygen. Presence of this group is indicated by abnormal amounts of iron sulphide in the corrosion products; this can easily be detected by the black stain of sulphide that appears in the subsoil and by the odor of H_2S when dilute HCl is added to the products of corrosion. The most common iron-consuming bacteria are the *Crenothrix, Leptothrix, Spirophyllum,* and capsulated form *Cocobacilli,* and although these do not actually consume the iron as food, they require ferrous iron in solution for growth, exuding it as red ferric hydroxide or a similar insoluble ferric compound. Occasionally soils have been classified as noncorrosive, as a result of the common mineral analyses of both soil and water, and yet corrosion has occurred.

Bacteriological examinations of the soil and products of corrosion have revealed the presence of bacteria which have, at least, contributed to the corrosion, owing to their production of chemicals which facilitate electrochemical action or attack the metal. Sulphate-reducing bacteria are most active in poorly aerated swamps where pH of the soil water is about neutral and there is enough organic matter and soluble sulphates to support them.

Corrosion from Aerobic Bacteria. Aerobic bacteria may also cause or accelerate corrosion.[3e] The *Thiobacillus* is of this type, sustaining itself from the oxidation of sulphur compounds. Bacteria of the soil flora oxidize sulphur, if present, forming a strong solution of sulphuric acid that will, in turn, combine with any basic material present. These bacteria develop in a drained or open soil, if sulphur is present. The necessary sulphur is frequently obtained from the natural decomposition of organic matter, and sometimes from iron pyrites. They do not multiply in water-saturated soils but will survive for considerable periods in such conditions.

Experiences with Corrosion in Ground. In New York City, pipe piles that have been in the ground for over 25 years have been exposed and cleaned of the protective coating of rust, and in no case has any pile shown a loss of more than $\frac{1}{64}$ in. of metal.[6b] The closely adjacent soil has been permeated and made water resistant by ferrous oxide.

Steel sheeting in soil for a sewer at Newark, N.J., in contact with concrete on one side and soil on the other, showed blue-black mill scale after 18 years.

A long 12-in. 65-lb H pile driven in a swamp at the Bonnet Carré Floodway in Louisiana showed no measurable corrosion after 17 years, although two strata were found corrosive in the presence of oxygen by a National Bureau of Standards test. Soil ranged from blue sandy clay down through sands, humus with sand, and clays.[3ad]

Corrosion of Steel Piles in Fresh Water. Steel piles generally corrode little in fresh water. If the water is polluted corrosively, piles may be painted with coal-tar paint before driving. Such a coating is desirable at the water surface where rusting is most active, although generally not severe.

Experiences with Corrosion in Fresh Water. The first known permanent installation of steel sheet piling in the United States was driven in 1901 for the Randolph Street Bridge in Chicago, Ill. Thirty years later it was found to be in excellent condition below the water line and poor above. A number of other sheet-piling installations in fresh water have been found to be in excellent to very good condition after 20 years, including cases along the Calumet River in Hammond, Ind., where, in spite of the high acid content of the river, paint marks made at the rolling mill were still visible.[11a]

In fresh water, steel sheet piling pulled from the Black Rock Ship Canal near Buffalo, N.Y., in 1931 after 19 years service in water polluted by sewage and industrial waste showed a very thin rust scale but practically no corrosion where in contact with earth, whereas the surface exposed to water showed more corrosion, with some pitting of ⅓₂ in., the loss at 7 ft below water level being 2½ per cent, and 3½ per cent where alternately wet and dry.

Ordinary open-hearth 8-in. 17.3-lb steel beams driven as piles for the Lake Francis River Bridge in Arkansas in 1913, in easily drained sandy loam having a neutral reaction and about the same salts as found in the Mississippi River, were removed in 1934. A few shallow pits and a maximum loss of 4 per cent were found at the ground level, with the exposed portions free from pitting. It was estimated that an effective safe life of 50 to 100 years would have been obtained.

In only very few of the old county bridges in Nebraska having steel pile foundations were the piles protected in any way except by coats of red-lead paint. On a great many such piles examined, it was estimated that the decrease in section had not been more than 1 per cent in 20 years. Exceptions were found in the Salt Creek Valley, which is saline to a marked degree, where several old steel piles showed an estimated loss of 2 to 2½ per cent. Little if any deterioration took place at a greater depth than 18 in. below the stream bed or at ground water. When the steel was not protected by concrete, the corrosion at an elevation of 1 ft or so above the normal water line was not as great as in the average steel superstructure.[3h]

Steel sheet piling, Carnegie section M-101, 12 in. wide, 35 lb per ft, driven in 1912 for the Tenth Street Bridge over the Monongahela River in Pittsburgh, was exposed in 1931 when rebuilding the bridge. Little rusting was found. The pool level was about 1½ ft below the top of the piling, and practically all loss of weight was confined to the top 3½ ft. The loss in the top section was 14.9 per cent, none in the middle, and 0.4 per cent in the bottom. The piling was 24 ft long, with the bottom 10 ft in hard mud and gravel, the next 7½ ft in soft mud or ooze, and the top 6½ ft standing in water with 1½ ft above the pool level. The Monongahela is contaminated by drainage from mines, and only during high water does it become neutral or slightly alkaline. Most of the year it contains sulphuric acid, up to over 100 ppm. This piling lost approximately 1 per cent of weight annually in the upper portion, under these conditions. Protection of the top 3½ ft would have prevented practically all loss of weight in the pile.

Corrosion of Steel Piles in Sea Water. Corrosion cells are formed on piles between areas of different aeration. Certain zones of marine piling are subject to much more intense corrosion than others; typical relative intensities are shown in Fig. 13.39. Heavy corrosion may occur in the splash zone just above high-water level, because of the severe electrochemical corrosion cells set up on the piling, which cause currents to discharge from some areas on the pile and collect on the other areas.[3ah] Severest corrosion may occur in piles extending well above high water, around a foot from the top of the pile, and may be up to four times as great as elsewhere.[3ak] Caps may be needed.

In the zones between tides, corrosion is often severe. The area just below MLW is anodic when coupled with the area just being submerged

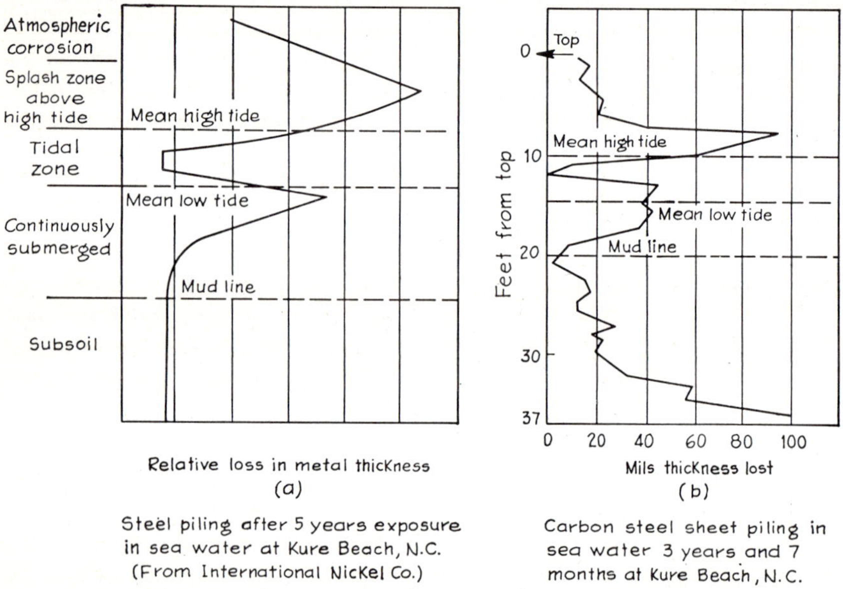

FIG. 13.39. Corrosion profiles of steel piling in sea water. (*a*) Steel piling after 5 years' exposure in sea water at Kure Beach, N.C. (*From International Nickel Co.*) (*b*) Carbon steel piling in sea water 3 years and 7 months at Kure Beach, N.C.

by the rising tide, and corrosion is intense because of the short lines of current flow and low resistivity of the sea water acting as an electrolyte. Pollution of the water by silt, oil, or grease that floats on the surface and coats the metal often gives considerable protection. In warm regions, the ratio of corrosion above low water to that below will be high because of accelerated growths providing protection below low water and increased attack on the unprotected section above; the ratio may be 2 to 1 in temperate zones and up to 4 to 1 in warmer climates.[3ah]

Another location for corrosion cells occurs on piles between the unaerated anodic area below the mud line and the cathodic area in aerated water. This corrosion intensity is not usually as great as those described above, because of longer current flow lines and higher resistivity of the mud.[3ar]

The difference between amounts of corrosion of steel in sea water due to variations in salinity, temperature, and marine growths is less than might be expected. High-water temperatures accelerate corrosion, but also accelerate protective coverings by marine growths.[3ak]

Around the world, the spread of a number of observations of corrosion has been reported as between 0.001 and 0.0077 in. per year, with an average of 0.0043 in. per year.[3ah] Loss of weight may be assumed as directly proportional to time.

Corrosion by sea water increases with the velocity of the water for areas in contact, if the steel is not covered by microorganisms or slime. Such organisms are apt to attack only if the velocity falls at some time below 3 fps, but if attached, they remain during higher velocities. Films of slime can form in higher velocities. If no growths or films are formed, or if formed in very high velocities which may remove them, corrosion may increase with velocity from such a value as 0.005 in. per year in still water to 0.02 in. per year at 5 fps, 0.03 in. per year at 10 fps, and very slightly more thereafter.[3ak]

The shape of piles should be considered. Corrosion on H piles is apt to proceed more rapidly on the edges of the flanges, because of differences in oxygen concentrations there and on flat surfaces. This may cause loss of strength disproportionate to the loss of metal. Round piles avoid such galvanic current concentrations and may corrode less, as well as more uniformly.

Experiences with Corrosion in Sea Water. The Sea-Action Committee of the Institution of Civil Engineers reports (Table 13.1) the following corrosion losses and depths of pitting in sea water (including for comparative purposes one set in fresh water) on ½-in.-thick plates of medium carbon-steel specimens having a low sulphur and phosphorus content, the analysis being carbon, 0.345 per cent; silicon, 0.20 per cent; sulphur, 0.025 per cent; phosphorus, 0.027 per cent; manganese, 0.715 per cent; copper, 0.076 per cent.

At Colombo, the aerial corrosion was especially severe, some specimens having swelled from the original ½ to 1½ in. and almost fallen to pieces. The aerial corrosion at Halifax was relatively slight, consisting of a fine scaly rust, whereas at Plymouth, the bars were covered with a laminated scale of blackish rust swelling up sometimes to 1½ in. total bar thickness. The half-tide specimens at Auckland and Colombo were covered with fauna and flora, including barnacles, but some mill scale still adhered at Auckland. The half-tide specimens at Halifax were fairly free from fauna and flora; they were blistered with heavy rust scales and waisted, especially near the bottom. At Plymouth, the half-tide bars had a few barnacles and were covered with red rust, swelling to a total bar thickness of ¾ in. The sea-water submerged bars, except those at Halifax, were covered with fauna and flora, and at Colombo and Auckland this was underlain by a layer of soft black oxide. The Halifax submerged bars were heavily corroded, and the sea-water bars at Plymouth were covered with flora and large rust nodules. The fresh-water submerged bars at Plymouth had no organic growths, but large scales and nodules of rust projected 1 in. from the surface, the quiet water not providing the mechanical agitation occurring in the sea water.

TABLE 13.1. CORROSION EXPERIENCED IN SEA WATER[3k]

Location of specimen	Average depth of corrosion on one face, in.[a]					
	With scale on			Descaled		
	5 yr	10 yr	15 yr[b]	5 yr	10 yr	15 yr[b]
Halifax, N.S.:						
In air	0.004	0.009	0.017 (*0.024*)	0.003	0.008	0.015 (.....)
Alternately wet and dry	0.014	0.016	0.031 (*0.196*)	0.009	0.013	0.049 (*0.154*)
Submerged	0.019	0.046	0.068 (*0.057*)	0.022	0.044	0.072 (*0.095*)
Auckland, N.Z.:						
In air	0.009	0.020	0.035 (*0.051*)	0.007	0.013	0.024 (*0.109*)
Alternately wet and dry	0.006	0.011	0.009 (*0.036*)	0.011	0.015	0.014 (*0.027*)
Submerged	0.017	0.029	0.056 (*0.194*)	0.023	0.040	0.061 (*0.054*)
Plymouth, Eng.:						
In air	0.032	0.064	0.096 (*0.102*)	0.010	0.038	0.022 (*0.041*)
Alternately wet and dry	0.012	0.023	0.036 (*0.094*)	0.023	0.040	0.063 (*0.057*)
Submerged (sea water)	0.019	0.031	0.034 (*0.223*)	0.021	0.037	0.043 (*0.070*)
Colombo, Ceylon:						
In air	0.079	0.115	0.155 (*46%*)[c]	0.088	0.146	0.190 (*70%*)[c]
Alternately wet and dry	0.066	0.091	0.132 (*P*)[d]	0.064	0.098	0.119 (*P*)[d]
Submerged	0.024	0.043	0.068 (*0.298*)	0.029	0.051	0.067 (*0.059*)
Plymouth, Eng., submerged (fresh water)	0.011	0.018	0.025 (*0.104*)	0.018	0.023	0.029 (*0.037*)

[a] Computed from total loss of weight.
[b] Figures in italics and parenthesis are deepest pits on 15-year bars.
[c] Thinning of bars.
[d] *P* denotes perforated.

Submerged bars lost much more weight than did bars exposed at half tide, except at Colombo, thus showing that the popular idea that corrosion is always most severe at half-tide level is not always true.

Mill scale, so long as it adhered, provided considerable protection against corrosion, but where it cracked or flaked, serious local pitting occurred. Aerial corrosion, except at Colombo where the bars were subjected to spray, caused less loss of weight on descaled bars, and the depth of pitting was less. The half-tide tests showed slightly greater mean losses of weight for the descaled bars, but the bars were less deeply pitted. Some oil was present in the water at Halifax, forming a deposit on the half-tide bars. Complete immersion showed pronounced differences in uncleaned and cleaned bars, the cleaned bars being quite uniformly corroded whereas the uncleaned bars showed severe localized corrosion where the mill scale had become detached. Thus, although the cleaned bars lost more in weight, this advantage was more than countered by the less serious pitting. In fresh-water immersion, the cleaned bars also lost more weight, but the pitting was generally less serious. It appeared advantageous to remove mill scale under all the conditions of the tests.

No specimens containing the 0.20 per cent copper content commonly used in the United States for copper-bearing steel were tested, but a series of bars containing 0.635 per cent copper was tested with the scale on. This copper content reduced the corrosion at Colombo by about one-third for the submerged and alternately wet and dry bars but had no effect on the aerial bars; at Plymouth no effect was observed other than some improvement at 5 and 15 years on aerial bars, except that a reduction in corrosion of about one-third on bars submerged in fresh water was noted; at Auckland corrosion generally was reduced from one-third to one-half; at Halifax the only effect was a reduction of about one-third in the corrosion of the aerial bars.

From the report it appears that the average rate of corrosion in British waters may be considered as 0.003 in. per year in sea water, and 0.002 in. per year in fresh water. Fabricated steel sheet piling in Guantanamo Bay, Cuba, showed the portions below water line in perfect condition after 32 years of service, whereas the portions above were practically destroyed. Fabricated Freistedt-type piling, used temporarily twelve times between 1907 and 1912 when it was installed at Key West, was found in 1936 to be in such good condition below the low-water line that the mill branding was clearly visible. Plate and sheet piling removed from Glenwood Landing, N.Y., in 1929 from the apparent zone of worst corrosion about 1 to 1½ ft below high water showed loss of $\frac{3}{32}$ in. in 21 years and $\frac{3}{64}$ in. in 11 years. These rates are about twice those found at Plymouth, England.

The investigations of the Sea-Action Committee disclosed that some tropical sea waters contain a type of shellfish which, adhering to unprotected steel, causes abnormal pitting, but that this is prevented by encasing the steel in concrete. From Panama, this corrosion has been reported to eat completely through the steel in 3 to 5 years. Because of this, gunite encasements were used on H piles for a long wharf.[3f,3g]

In salt water, 22-year-old steel sheet piling driven at Jacksonville, Fla., was found in fair condition at the water line and excellent condition 10 ft above; at Key West, Fla., after 19 years in a ferry slip, steel sheet piling was found in excellent condition except poor where exposed to salt spray and wave action; and at Key West, Florida, after 20 years in a railroad abutment, the condition was excellent below low water, serviceable above low water where backed by earth on one side, and entirely corroded where exposed to spray on both sides.[11a]

Cylindrical steel piles installed 31 years at Dry Tortugas in the Gulf of Mexico, with mean low water 36 in. from the top and a tide of 1.5 ft, showed, when examined in 1933, the following total loss of section: 3 in. from top, $\frac{1}{4}$ in.; 12 in. from top, $\frac{7}{32}$ in.; 24 to 36 in. from top, $\frac{1}{16}$ in.; 51 in. from top, $\frac{1}{32}$ in.

In 1933, an inspection of the 7-year-old Oceanside, Calif., pier H piles, which had been painted with asphalt before driving, required removal of the barnacles with a chisel between tide levels. Here, in many cases, mill scale was found, and there was not over 1 per cent corrosion loss. However, from 3 ft above tide level to the deck, the paint had disintegrated and the steel corroded because of action of the salt air.

Electrolysis causing corrosion and pitting of H piles at the mud line was reported in one instance in sea water, where 2 ft of organic mud overlaid shale into which the piles were driven. It was thought that a battery was formed by the nascent oxygen in the sea water above, which was an electrolyte, and decaying organic matter in the mud, with the steel pile acting as a conductor.

The mean rate of corrosion of steel sheet piling on the Atlantic Coast of the United States, based on a total of weighted averages, was found by the Beach Erosion Board to be about 0.008 in. per year.[3aq] Average rates of loss for harbor bulkheads were 0.0049 in. per year from 2 to 8 ft above MHW to about 0.0025 in. per year at MHW and mean tide level and to 0.0035 at MLW. The rates for beach bulkheads were one-half or less of these figures. For groins and jetties, losses were from 0.010 in. per year between 2 to 8 ft above MHW to one-half of this value at MHW and over double from MHW to MLW. Between MLW and the ground line, the average was 0.0035 in. per year, being highest near MLW. Soil cover on one face of harbor bulkheads reduced the rate from 0.0075 in. per year to one-third of this value; for beach bulkheads the rate was reduced from 0.027 to 0.010 in. per year; but for groins and jetties the rate was the same, 0.020 in. per year, whether one or both faces were covered. Soil cover on both faces reduced the loss on beach bulkheads to 0.0017 in. per year and on groins and jetties to 0.0026 in. per year. Partial or temporary covers gave intermediate values. The average losses for harbor bulkheads subject to heavy, moderate, and light spray were 0.0083, 0.0041, and 0.0024 in. per year; for heavy spray on beach bulkheads, groins, and jetties it was 0.016 in.

per year. Structures painted even only once showed losses reduced by one-half.

Corrosion of Steel Piles in Brackish Water. No exact prediction can be made as to rates of corrosion in brackish waters as compared with sea water of full salinity because they vary widely and are often associated with sewage or industrial pollution. Factors may occur, such as lower dissolved oxygen content of highly polluted waters. Some tests have shown that rates of corrosion of continuously immersed steel were of the same magnitude as for clean sea water.

Experiences with Corrosion in Brackish Water. Corrosion in Lake Maracaibo, Venezuela, where the water is brackish due to connection with the sea, is especially severe and extends all the way down to the mud line. Steel in this lake sometimes corrodes entirely through in a few years.

Protection of Steel Piling

Effect of Pile on Rate of Corrosion. Allowance for corrosion can be made when selecting the thickness of metal. In the cases of some types of piles in which the interior of a pipe or shell is filled with concrete, no allowance is made for the supporting capacity of the steel, which is used merely as a form for the concrete, whereas in others the steel shell is assumed to work with the concrete in carrying load. In such cases, an allowance for corrosion should be made. A common figure for this purpose is $\frac{1}{16}$ in. Building codes often specify the amount of thickness of shell not to be considered as permanent load-carrying metal.

Round piles resist corrosion better than H piles, the absence of edges preventing galvanic current concentrations.

Silicon and *manganese* contents may influence corrosion adversely, but the effect is small.

Sulphur is harmful, and the content should be low.

Cast iron has an enviable record against corrosion.

Wrought iron has good corrosion resistance.

Copper-bearing steel probably adds to corrosion resistance, although conflicting reports are made.

Copper-bearing Steel. Copper-bearing steel containing a minimum of 0.20 per cent copper seems to provide more resistance against atmospheric corrosion than plain carbon steel. No added resistance seems to be obtained when it is immersed in fresh or salt water or within the tidal range of sea water.[2h] It should not be used in contact with plain steel or wrought iron, especially in sea water, for local electrolytic action may result.

Mill Scale. Mill scale is highly resistant to corrosion and tends to protect the pile unless in contact with the base metal in the presence of an electrolyte, in which case the rate is accelerated. Mill scale is efficient protection if intact, but if discontinuous, corrosion starts beneath it and peels the scale, carrying with it any surface coating. This results in formation of deeper pits and greater loss of weight than with descaled steel. Scale may be removed by weathering and scraping, pickling, wire-brushing, flame-cleaning, or preferably by sandblasting.

Protection by Products of Corrosion. It is known that the rate of corrosion slows up after formation of a rust coating, since this acts as a protection itself. Under certain conditions, these products of corrosion also permeate the ground for a short distance, thus helping to protect the steel. Steel piles driven in sand become coated with an impervious insoluble coating of ferrosilicate, which acts as a protective encasement.

Low Unit Fiber Stress. Low unit stress under working loads in many steel piles automatically provides a considerable factor of safety. It has been a common practice to use steel piles without any added protection when they are entirely under water or in ground, on the basis that the supply of oxygen is limited under these conditions, but, as previously discussed, there may be corrosion even well under ground. Unprotected piles have also often been used above water or ground, but contradictory results have been obtained. Rivets should be avoided, if possible, in unprotected steel piles in zones of corrosion, since they are often incipient points of damage. Methods of protection in whatever may be considered the danger zone under the particular conditions obtaining are coming into more frequent use. Among protective methods are surface coating, local increase of section, and encasement.

Surface Coatings. Conditions of exposure and corrosion rates vary widely. Numerous methods of protection against the severe splash-zone corrosion are being tested constantly. Oil paints may serve for only a few months; ordinary coal-tar paints for a longer time, although unsatisfactory in the sun. Pneumatically applied concrete coatings and concrete caps are effective but expensive. Other coatings studied include grease, sheet neoprene, flame-sprayed zinc, Monel jacketing, zinc-rich paints, hot-sprayed mastic, and plastic adhesive tapes.[3ah] Other possible coatings are phenolic mastic, Saran resin, and vinyl resin with aluminum or mastic. Until some other coatings prove superior, sandblasting followed by two coats of cold-applied bitumastic or by one coat of hot-applied bitumastic enamel might be used.

Coatings should be extended below 2 ft below MLW if necessary, as in the presence of certain rarely found marine growths which are said to exude acid that causes pitting, certain types of marshes, and waters badly polluted with sewage, industrial wastes or coal. Such conditions may indicate the desirability of coating the steel full length, or at least to the mud line.

Soft and fine materials below the mud line will not usually damage enamel coatings enough to destroy their value. If backfill material is very rocky, use of selected backfill may be made. When driving will be in coarse and hard materials, consideration should be given to excavating before driving or encasing in concrete instead of enamel. Coated surfaces should not be driven in contact with the ground unless it can be shown that they will not be damaged thereby.

Abrasion caused by flowing sand in water near the mud line may be severe on hard coatings, but less so on resilient materials.

Contact surfaces on the interlocks of steel sheeting should be primed but not coated with heavier materials if they interfere with driving.

Driving templates should be equipped with rollers if necessary to avoid scraping off the protective coating during driving.

Preparation of Surfaces. For steel waterfront structures, all surfaces above 2 ft below MLW, except the rods and fittings, should be flame-cleaned and wire-brushed, or preferably sandblasted, then primed or coated as soon as practicable. Steel to be encased in concrete should be coated at least 6 in. into the concrete, except where the concrete extends down to MLW and coatings are omitted below this level. Surfaces to be welded may be given a very thin coat of primer. Welds should be primed. Steel may be cleaned and primed in the shop or at the site. Surfaces primed before delivery should be reprimed at the site. Where primer is not recommended under cold coal-tar base coatings, a light shop coat of the coating may be applied to reduce field cleaning expense.

Bituminous Paints. This term applies only to *thin-film, cold-applied cutbacks,* which sometimes have small amounts of filler. The principal bituminous materials used in paints are *asphalt* or *coal tar.* This paint may be brushed or sprayed to a dry-film thickness of 1.5 to 3 mils per coat. The minimum total dry thickness should be 5 mils. These cutbacks have no rust-inhibitive properties and act only as mechanical barriers. Films are thin unless such fillers as calcium carbonate, slate, mica, silica, clay, or fibrous inorganic material like asbestos are included. Unfilled cutbacks are only suitable for temporary use in noncorrosive soils. Under severe exposures a rust-inhibitive primer should be used; this is easy under asphalts, which many paints can withstand. The solvents in coal-tar paints are too powerfully destructive for conventional primers.

Coal-tar base coatings generally should be applied to bare metal. Wash primer may be used for temporary protection of blast-cleaned steel until the coal-tar base coating can be applied; the chromate may exert some inhibiting effect, but otherwise the paint should be applied on bare metal.

Coal-tar paints are used for the same types of coatings as *asphalt paints.* They are usually coal-tar pitch in solution and can be filled or unfilled.

Hot-applied enamel or hot-applied coal-tar pitch (no filler) is used over a cutback coal-tar primer. The coal tar has better water resistance than asphalt but poorer weather and acid resistance. *Hot coal-tar paint* is better on buried objects. Thick mastic-type coal-tar coatings are available.

Coal tar was found by the British Sea-Action Committee to give good results and to give much better service than iron oxide or red-lead paints.[3k] They also found that one or two coats of red-lead primer were beneficial. Coal tar is good for underground use where appearance does not matter; it tends to alligator in the sun.

Epoxy-coal-tar paint is formed from epoxy resin and coal-tar pitch and is cold-applied. The epoxy has two components that cure throughout the thick film by polymerization instead of by evaporation of solvent. The color is black. It resists alligatoring, oils, abrasion, salt water, acids, and alkalies. Two coats should be minimum. Coverage is 160 sq ft per gallon per coat, and a thickness of 14 to 16 mils dry is obtained from two or three brush coats or four to six spray coats. No primer is needed. Steel should be sandblasted. This paint has been available for too short a time to have long experience records, but it shows good promise. *Poxitar* is the name under which it is made by the Inertol Company, Inc.

Cold-applied Filled Bituminous Coatings. These are coal-tar or asphalt solutions that may be applied in several consistencies. *Medium films* are of high consistency and are designed to be brushed or sprayed to a dry thickness of 5 mils per coat minimum or 10 mils maximum. Total dry film should have a thickness of 12 mils. *Thick films* are of very high consistency and are to be brushed or sprayed to a dry thickness of 10 mils per coat minimum or 25 mils maximum; the total dry-film thickness to be 50 mils minimum. *Extra-heavy films* are very thickly applied by brushing, spraying, or troweling. Minimum dry thickness should be 35 mils per coat, and the total dry thickness as selected.

Bitumastic No. 50, made by the Koppers Co., Inc., is a cold-applied, heavy-duty, self-priming, plasticized coal-tar pitch used in severe service. Films of at least $\frac{1}{16}$ in. may be obtained by a spreading rate of 55 to 70 sq ft per gallon. It resists brine and is useful on buried surfaces. On piles and marine structures the surfaces should be cleaned to bare metal before application of two coats. If in sunlight, a top coat of *Bituplastic No. 28* should be added. No. 50 withstands temperatures of —10 to 160°F; No. 28 from —50 to 200°F. For use below —10°F both coats should be No. 28. Good life should be expected from cold-applied No. 50 coating applied in two coats to a total thickness of $\frac{1}{32}$ to $\frac{3}{64}$ in.

Bitumastic Enamel No. 70B, made by the Koppers Co., Inc., is a hot-applied coal-tar base filled enamel that has given good service under severe conditions. It may be applied in a single coat $\frac{3}{32}$ in. thick, plus or minus $\frac{1}{32}$ in.

Bituminous Emulsions. These have an asphalt or coal-tar resin. A primer is required on steel, and it may be bituminous paint or a pigmented primer with rust-inhibitive properties which is known to be compatible. Very thick coatings may be built by successive applications. Bituminous emulsions have value as top coats over coal-tar paints, coal-tar enamels, or other bituminous paints to prevent alligatoring and cracking in sunlight.

Several British authorities, in describing world-wide projects, have stated that best life should be secured from well-cleaned steel, preferably sandblasted, cold-primed with coal-tar base or asphaltic-base bituminous primer, covered by a coal-tar base or asphaltic-base bituminous enamel containing a filler, such as slate powder or talc, applied hot in a thickness of $\frac{3}{32}$ to $\frac{1}{8}$ in.[3ai,3aj,3ap]

A United States authority suggests the following treatment, until some other coating proves superior: sandblasting, wash primer, anticorrosive coating, coal-tar base paint applied cold, and top coats of coal-tar emulsion.*

Coatings and Cathodic Protection. Cathodic protection is not effective in the splash zone and only partially in the tidal zone. Below MLT, corrosion can be practically avoided by a current of 3 ma per sq ft after an initial greater current has caused formation of a calcareous deposit. To obtain long life from the entire pile length, a protective coating should be applied extending from above the splash zone to as close as possible to MLT. Failure of bare piles should first occur in the splash zone. If piles are painted on only the continuously submerged portions, worse corrosion would be expected in the tidal zone because the painted areas would not provide galvanic protection to the tidal zone, and corrosion would be severe at breaks in the protective coating. If piles are painted on only the portion above MLT, which is cathodic, sacrificial corrosion would be mostly avoided and failure eventually would occur below MLT; also, a large degree of protection would be given at breaks in the coating in the tidal zone, except for slight pitting at breaks, and without seriously affecting lower portions, because of the larger ratios of areas below to the areas of the coating breaks. It therefore appears that painting above MLT, coupled with cathodic protec-

* Letter from L. A. Palmer, Engineering Consultant, Navy Department, U.S. Bureau of Yards and Docks.

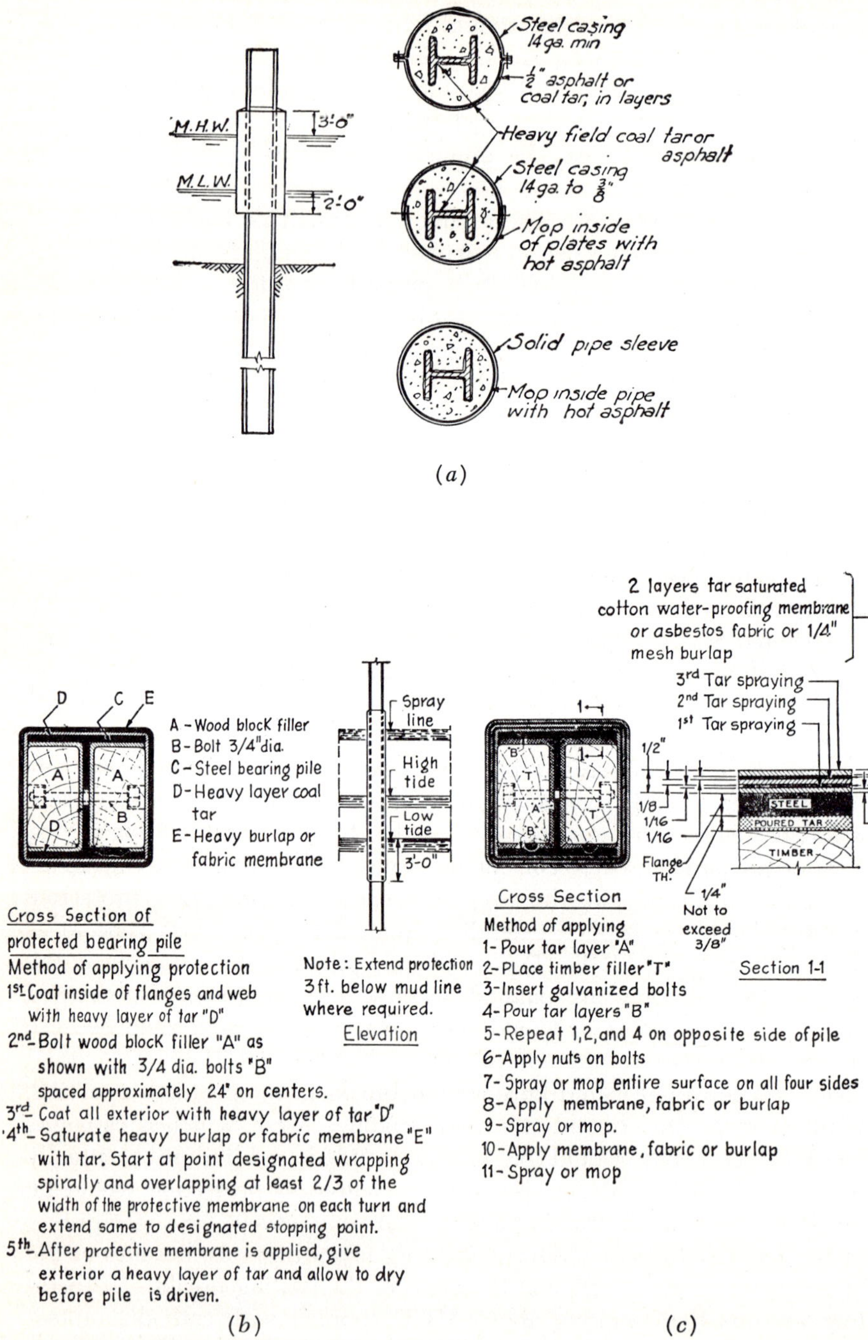

FIG. 13.40. Typical details of protection against abrasion and corrosion. (*Courtesy of United States Steel Co.*)

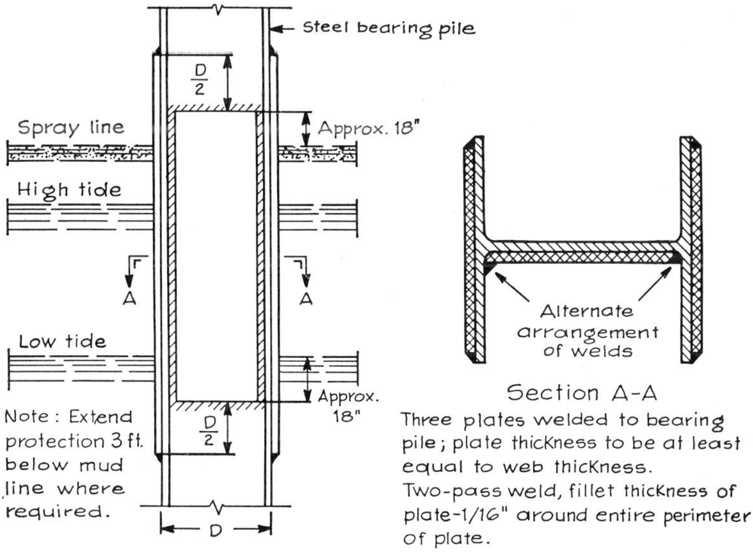

Steel Plate Protection For Steel Bearing Piles in Tide Water

(*d*)

FIG. 13.40. (*Continued*)

tion, should provide practically full protection to continuously submerged surfaces.[3as]

Organic Growths.[3k,3av] Organic growths may be expected to increase corrosion by any destructive action which they exert on the coating, although by their tenacity they provide a certain degree of mechanical protection. The predominant effect is one of local conditions. Corrosion may be set up where one organism is smothered by a faster-growing variety, which produces hydrogen sulphide. Where the water velocity does not exceed 4 mph, it is possible for thick layers of growth to develop which may decrease the corrosion rate. Barnacles materially reduce the life of bituminous coatings and may cause them to peel in the tidal zone, but they have little effect on harder coatings. Marine organisms develop at a greater rate with a rise in temperature of the water, and the breeding season is longer in southern waters.

Fuel Oil. Oil is becoming increasingly common in harbors, and, as it rises and falls with the tide, the sticky deposit collects mud and slime which provides considerable mechanical protection.[3k] Fuel oils are likely to exert a deleterious effect on paints containing vegetable oils.

Thickening of Metal (Fig. 13.40*d*). Added plates ¼ in. thick, welded to H piles, are claimed to form a satisfactory protection with an estimated added life of 25 years.[3a] These plates are usually welded before driving. Such protection is rugged, not unduly expensive, and convenient to handle and drive. The protection might well be confined to the zone of corrosion.

Encasement. Encasement of steel piles is often necessary at the zone of maximum corrosion, which is usually between the high- and low-water marks but may be in the splash zone. Such encasement should begin at least 2 ft below low water and extend above high water (Fig. 13.40). Above this point, painting may be practicable.

Poured Concrete. Poured concrete encasements (Fig. 13.41) are most common. When used for protection in the zone of changing water levels, the encasement should extend at least 2 ft below mean low water and 3 ft above mean high water, and higher if required by spray or wave action. When used through water, the encasement should extend at least 4 ft into the bottom. Encasements are often used for protection several feet above and below ground line, or up to the pile caps, in stream beds normally dry. Encasements have sometimes been used on H piles under footings, the concrete jacket extending down 2 to 4 ft from the underside of the footing, in corrosive soils, in the hope that the oxygen necessary for corrosion would not be available at that point. Encasements are very necessary in ground or streams carrying mine wastes or other corrosive materials. When the encasement can be poured in the dry and in the open, the usual square wooden forms with chamfered corners, or octagonal forms, may be used. A circular sheet-metal or pipe form is convenient for use in the ground or under water, may be left in place, and may be streamlined if desired.

Concrete jackets should be reinforced with some such reinforcement as ¼-in.-diameter spirals and four ½-in.-diameter vertical corner bars, or a welded fabric such as 4 by 4 in. of No. 8 wires with eight vertical rods, with the reinforcing set out at least 1 in. from the column steel, and with at least 3 in. of cover over the bars for use in sea water or underground, and 2 in. for use in fresh water.

One method of installing a protective steel form, later to be filled with concrete, consists of sliding down a circular steel form, jacking, driving, and jetting as necessary, having a tight-fitting closure at the bottom.[3a] The form is then filled with concrete. If steel forms are left in place, that much additional protection is obtained and, in locations where abrasion from sand and gravel or ice floes is expected, the forms should be left in place.

Where it is desired to protect the upper end of a steel pile by a concrete encasement without resorting to prior excavation, the following method has been suggested as convenient by Greulich (Fig. 13.42).[3a] The steps are (*a*) Shell and mandrel are driven to the required depth; (*b*) the mandrel is pulled, the centering cap placed, and the pile driven

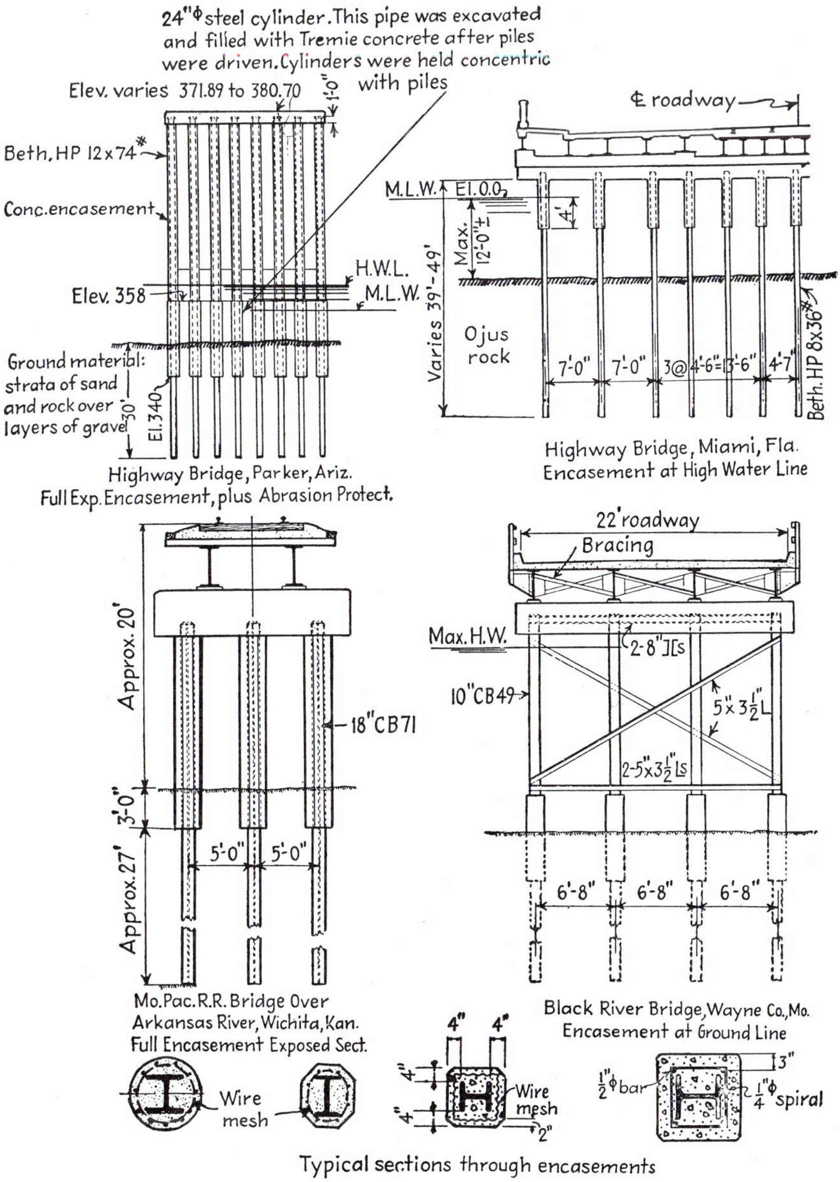

FIG. 13.41. Typical concrete encasements for steel H piles.

to grade; (*c*) the centering cap is removed, and the shell filled with concrete; and (*d*) the shell is pulled as soon as concrete is poured. This method is very useful when the upper part of the pile extends through deleterious fill such as unleached cinders or material saturated with corrosive industrial wastes. It is also useful where piles pass through water polluted with mine wastes or sewage, or where the mud is of a damaging nature.

Where the bottom of the encasement is below ground water in permeable soil, a seal is desirable before placing the concrete in the encasement form. The method shown in Fig. 13.43 accomplishes this purpose

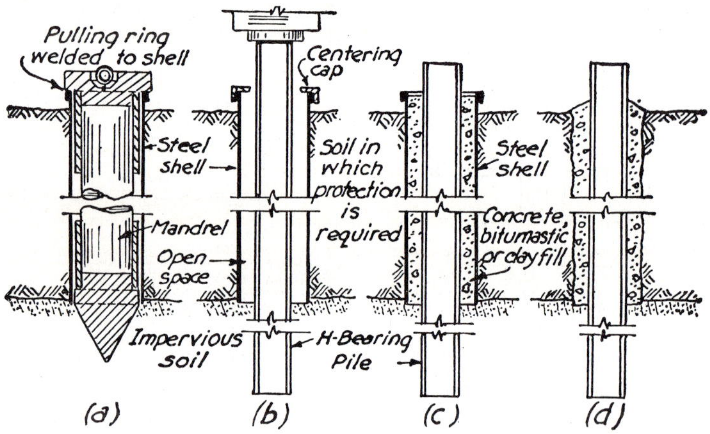

FIG. 13.42. Method of installing encasement for H piles without excavation.

and is installed after the piles are driven, by a separate crawler-crane rig with light hammer and pulling reeving of cables.[38] A steel casing and a steel core 12 in. shorter than the casing with slotted base to fit the H pile are required. The method of installation is as follows: (*a*) Place stiff coarse concrete in a shallow excavation around the pile and set the casing and core. (*b*) Drive casing and core, with the concrete packing against the bottom of the core, to form a hard seal against mud and water. The underside of the bottom plate is fitted with angles which force the concrete against the pile surface during driving, thus scouring it clean. (*c*) Remove the core and fill the casing around the pile with soft concrete. (*d*) Reenter the core in the casing on top of the concrete fill, and pull the casing while exerting a downward pressure on the concrete caused by the method of reeving cables. (*e*) Remove the core. On good days 30 to 40 piles a day were coated for 8 ft, using a crew of nine, on the first project using this method. This method has been used by the Western Foundation Corp. and Raymond Concrete Pile Co.

Gunite (Fig. 13.44). Gunite encasements have been used on steel piling in tropical waters where marine growths may eat through steel in a short time.[3f,3g] It has been found feasible to handle and drive steel piles gunited prior to driving. Bars are welded to the piles, and preformed wire mesh tack-welded to the bars. A thickness of 1½ to 2½ in.

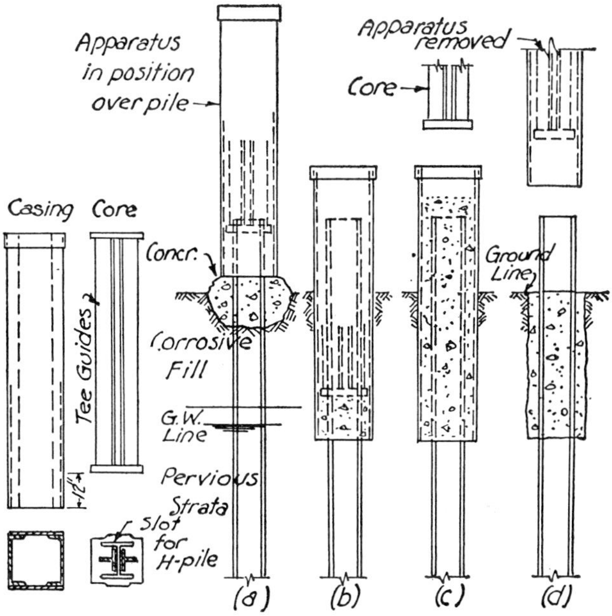

Fig. 13.43. Method of installing encasement for H piles below ground water.

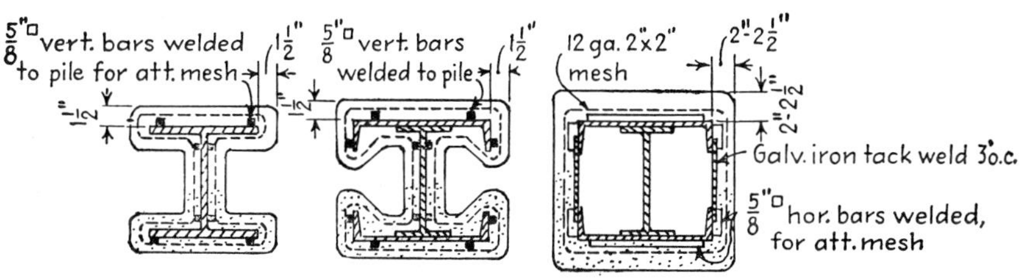

Fig. 13.44. Typical gunite encasements for steel pile sections.

of gunite is required. To avoid the expense of following the contour of the section, sheet metal can be welded across the spaces between flanges and the web space left open except for a few feet of solid concrete at the ends. Sandblasting may be necessary prior to guniting. Coal tar has also been applied to the steel prior to guniting.

If desired, protective encasements generally may be used on piles that require lagging, since the encasement usually extends down to a point a few feet below the mud line, while the lagging can be on the necessary portions of the embedded length. In one case, 14-in. H piles up to 120 ft long were driven with gunited jackets placed on the upper portions before driving, and two 10- by 12-in. timbers bolted through the webs in the lower portions for use in a soft bottom in tropical waters.[3f,3g]

A gunited H pile wrapped with mesh spaced ¾ in. from the pile by short sections of reinforcing bars, following the contour of the pile, and keyed by 1½- by 1½- by ⅛-in. lug angles in lieu of the notches cut in gunited wood piles, was cured 7 days; it was then driven to refusal and hit 1,500 blows with a No. 1 Vulcan hammer without any sign of cracking.[85] Single-length gunited H piles have been driven 160 ft long.[85]

Shotcreted H piles, 115 ft long and driven on a batter, were used in salt water.[3ag] Two inches of gunite was applied in two layers, after sandblasting. A cage of 2 by 2 in., No. 12 galvanized mesh was held out by ¼-in. rods crimped at 16-in. centers to provide ¾-in. clearance from the steel. Cracking at the upper ends under driving was rectified by easy patching at first, and later greatly reduced by locating shear bars on the upper 15 ft.

Gunite or shotcrete encasement will provide protection against electrolysis.[85]

Membrane Encasement. Patented methods of membrane protection[11a] in the tidal range consist of coating the webs and inside faces of H pile flanges with heavy tar layers; bolting timber fillers to square out the pile, using galvanized bolts countersunk in the timbers; and coating the exterior with tar. One method then uses a spiral wrapping of burlap or fabric membrane and an exterior final coating of tar. Another method wraps the tarred pile and timbers with two layers of tar-saturated cotton waterproofing membrane, asbestos felt, or burlap, over layers of sprayed coal tar, with a final sprayed coating of coal tar.

Neutralization of Salt Water. For temporary protection against corrosion in the Gulf of Mexico for H piles set in pipe casings for offshore oil-drilling platforms, the entrapped salt water has been neutralized with caustic soda to a pH of 10 and sealed with heavy oil.[97] For proven wells permanent concrete encasement and cathodic protection are later installed.

Cathodic Protection. Cathodic protection has been increasingly used since the early 1930's as an alternative or supplement to protective coatings. The driving force in electrochemical corrosion is the difference in potential between the anodic areas where the current leaves the steel to go into the electrolyte and the cathodic areas where it returns to the metal. Steel in contact with highly aerated sea water acts as a cathode, and steel in contact with less aerated sea water or mud acts as the anode. Metal can be kept from corroding if sufficiently large countercurrents can be impressed to neutralize corroding currents and leave the entire structure cathodic to its surroundings at all times. Cathodic protection is effective where the structure is in contact with a good electrolytic conductor such as sea water or most soils.

Corrosion of steel piles below low water was formerly a serious economic problem. Cathodic currents can give full protection to this region. It is hard to paint piles near low-tide level, but this region will also be protected cathodically. Full protection will extend up to about two-thirds of the height of the zone between low- and high-water levels, because of formation of calcium deposits. Without cathodic protection, it has been common practice to increase the thickness of metal at considerable cost, but this has become unnecessary. Cathodic protection will not be effective against scour. Otherwise, steel structures should last almost indefinitely, with practically no maintenance costs outside of occasional inspections to enable small adjustments to be made or anodes replaced. This protection can be obtained at a cost of a few cents a square foot per year.[3ah,3am,3an] All known major H-pile installations in New York Harbor have, or will have, cathodic protection. Many organizations, such as the large oil companies, use it extensively.[3ah]

At a cathodically protected metal surface, molecules of hydrogen gas form, and this film of gas protects the metal. Since hydrogen gas can be removed mechanically or by dissolved oxygen in the water, more electric current will be needed for structures in rapidly moving aerated water than in quiet water.

Higher electric-current densities are also required, even if no oxygen is available, in the presence of sulphate-reducing bacteria.

Following a brief general discussion of features of cathodic protection, several installations will be described. The variety of situations and systems possible is nearly endless, and these examples are given only to convey an idea of the type of arrangements and magnitude of installations. They are not presented as designs to be copied, for each case will require its own design. Further descriptions of actual installations appear in the references shown and in bibliographical items to which they refer. General treatments of the subject are contained in some of the references in this section.[3aj,3am] Textbooks and a corrosion handbook[3ak] treat the subject in more detail. The services of a corrosion engineer or electrical engineer versed in this subject should be utilized.

Information on corrosional prevention is contained in many articles in *Corrosion,* the monthly publication of the National Association of Corrosion Engineers, and reprints of articles are available; most articles contain bibliographies. Reviews of current articles in other publications and of new books on corrosion are given, as well as abstracts of foreign publications. Technical reports of their committees contain extensive information on various aspects and may be purchased. Their *Bibliographic Surveys* for years are available and contain abstracts.

Methods of Cathodic Protection. Cathodic protection may be obtained by using a rectifier to convert alternating to direct current and impress it through graphite, scrap iron, or steel anodes, or it may be secured from galvanic currents generated from replaceable, sacrificial anodes of magnesium, aluminum, or zinc. Each method has merits and drawbacks under certain circumstances. Choice of the most efficient and economical method should generally be made by a corrosion engineer or electrical engineer familiar with the subject. Factors to be considered include distances to other structures which may affect the arrangement or efficiency of the system; electrical resistivity of the surrounding media; practicable means of supporting the anodes where they will be effective, yet protected from damage by waves, ice, shipping, or water velocity; and ease of access and availability of trained personnel for maintenance.

Impressed Current. When using impressed current from rectifiers, piles should be electrically connected if the structure does not provide a connection. If there is no bracing, they can be connected through floor reinforcing bars or by cables, rods, or straps. Electrical conductivity may not occur in steel sheet piling, and several inches of weld is recommended at each interlock.[3ap] The amounts of current applied can be varied and controlled at a rectifier. To obtain quick initial protection, rectifiers usually are designed to produce current at a high initial rate, which means that they are oversized thereafter, thus adding to the cost.[2ah]

During initial treatment with high-density current, calcareous deposits may form under rust or barnacles and permit these layers to be removed by wave action. Cathodic protective-maintenance current does not prevent growth of marine organisms on coated surfaces. The calcareous coatings will protect the cathodic surface film of hydrogen from oxygen, enabling the current to flow to more remote metal areas. This provides considerable protection for many months after maintenance current is stopped or fails. Protective potentials can be reestablished at the previous maintenance-current density without application of a high current density.[3an]

While large amounts of current can be obtained from a few rectifiers, the wiring system required to distribute these large concentrations of power uniformly over the structure to numerous anodes at piers, wharves, or bulkheads may be complex and expensive, needing strong anode mounts and splices and cables free from disrupting leaks. If water movements are allowed to flex the cables, fatigue may soon cause failure.

Graphite anodes buried on land usually are surrounded by coke breeze. This absorbs the oxygen liberated at the anode and which would attack it. The coke actually becomes the anode, and the graphite becomes the connection to the breeze. This can permit the graphite to have potentially unlimited life if the volume of coke is relatively large. Typical sizes used have been 3-in.-diameter anodes 30 to 60 in. long surrounded by coke breeze 12 to 14 in. in diameter and 10 to 20 ft deep.[3ao]

Selenium-type rectifiers may require replacement in about 10 years; silicon or germanium types may last longer.

Galvanic Protection. Galvanic anode installations are widely used to protect steel in sea water. The anodes may be supported by rigid attachments to the piles and do not require connecting wires. An anode may be required for each pile. Maintenance costs are low and consist of replacing anodes after they have been sacrificed. The life of an anode depends upon its size and shape. Installations may be designed to last for a short or long period of years. Data are plentiful on the use of high-purity zinc anodes in sea water. Because of the high activity of magnesium, it is better suited for anodes in fresh water than in sea water. Galvanic anodes will provide quick initial protection, after which the rate of sacrifice decreases to about one-half of the initial rate.

Galvanic protection systems for piers, wharves, and jetties in various parts of the world have been of such sizes that 500,000 to 1,000,000 lb of zinc anodes has been used for a single installation, with total current outputs up to 3,000 amp.[3ah]

For oil wharves, where inflammable or explosive materials are handled, rectifier installations should be avoided because they have concentrations of current and voltages that may spark. Zinc anodes in a galvanic protection system produce only a fraction of a volt and practically eliminate the possibility of sparking.[3ah]

Galvanic protection for steel sheet piling may or may not require several inches of electrical bonding of interlocks by welds, depending upon details.[3ah]

Cathodic Protection and Applied Coatings. Coatings have not been found to be entirely effective when used alone, since they may contain small holes at which corrosion would start and force the coating away. Cathodic protection can prevent this and preserve the coating and also prevent corrosion from starting if abrasions or shellfish break the coating at any later time. Some engineers have considered it desirable to coat the entire lengths of piles even when driving into coral rock, because the current flows to any base metal and the coating would reduce the consumption of current.[3ap] Others consider it uneconomic to paint below the low-water line.[3ah] A comparative estimate can be made of the cost of painting below low water against the increased current requirements of bare metal. It is probable that the best coating obtainable should be applied to cathodically protected steel piling in sea water, because the amount of current required is proportional to the area of bare metal.

Cathodic Protection of Offshore Drilling Platforms. Life of offshore platforms may be required to be 35 to 50 years. Typical corrosion rates of unprotected steel in sea water are 0 to 10 mils per year submerged, 25 to 40 mils per year in the splash zone, and 3 to 5 mils per year in the atmospheric zone.[3at] Corrosion under water is caused by stray currents, reaction between the steel and chemicals in the water, and galvanic action at welds. Galvanic action in the splash zone results from slight potential differences in the steel and in oxygen concentrations. Mechanical damage increases corrosion. In the atmospheric zone, corrosion is due to reaction between steel and oxygen in the presence of spray and condensation of humidity.

The cost of steel replacements offshore may be triple that onshore, so that economics of protection should be studied.[3at] Cathodic protection is the best means of preventing corrosion under water. Because of low resistivity of sea water, bare steel can readily be polarized by an initial current density of 5 to 10 ma per sq ft, later reduced to 3 ma, by using galvanic anodes.

In the splash zone, considerable expenditure is warranted. Methods of protection have included thickening of metal, wrapping with tape, glass-fiber plastic material, rubber, and Monel sheathing. In the atmospheric zone, less costly treatments can be used because of easier renewal, although frequent maintenance is too costly. Tubular members receive best cathodic protection and are easiest to sheathe or coat. For maximum protection by a coating, one oil company has found vinyl to offer the most promise and to have necessary chemical resistance and ability to be renewed by recoating.[3at] A five-coat vinyl system over a sandblasted surface comprised an acid primer to prevent early rusting and improve adhesion, a "build" coat containing flake mica, a "mastic" coat containing flake mica and fibrous asbestos to increase the film and abrasion resistance, and a thin-film finish coat to serve as a binder and provide a smooth surface of the desired color. A later promising development is epoxy-coal-tar coating.

Experiences with Cathodic Protection in Soil. Sheet-piling walls for railway cuts at El Paso, Tex., were given cathodic protection after piling-to-soil potential readings were taken inside and outside.[3ab] Because of restricted space, a galvanic magnesium-anode system yielding sufficient current density could not be used. A rectifier system was provided. Ground beds were buried at various depths, using a number of 4- by

80-in. graphite anodes set vertically in coke-breeze backfill in 12-in.-diameter drilled holes. Anodes were spaced at 10-ft centers, set 5 ft from the outer faces of walls, and at 20-ft centers, set 15 ft from the inner wall faces where more space was available. A 24-year life of anodes was estimated.

Experiences with Cathodic Protection in Sea Water. *Galvanic Systems.* For the Creole Petroleum Corp. piers in Venezuela, study showed galvanic anodes more economical.[103] Magnesium was used as the galvanic anode because of its high voltage differential in combination with steel, and its high electrical storage capacity, amounting in actual use to 500 amp-hr per lb. The current requirement for protection of clean steel in sea water is over 10 milliamperes (ma) per sq ft of structure. However, currents of this order will in about 3 months reduce the potential of the steel below the corrosion point. The current required for protecting steel piles in soil is approximately 2 ma per sq ft of bare steel. Two or three 50-lb magnesium anodes about 20 ft apart were hung from the structural-steel deck by neoprene-jacketed copper wires connected electrically and mechanically to the steel which, because of its welded connections, formed one huge cathode. Monel metal was used for the supporting eyebolt and thimble to resist corrosion. Anodes are to be replaced yearly. All pipe lines from pier to ship or shore have insulating gaskets.

Offshore drilling platforms in the Gulf of Mexico were protected by sufficient permanent magnesium anodes to give a steady current of about 2.5 ma per sq ft. These anodes were supplemented initially by cored magnesium ribbon, sacrificed during the first few days, to provide a high initial current density of 100 to 200 ma per sq ft, to ensure rapid deposition of calcium and magnesium salts from the sea water. The permanent anodes then suffice, with current intensity regulated by a resistance at the head of each lead to an anode. The totally submerged steel area of one platform was 16,320 sq ft, and the unprotected piling potential was —0.68 volt. Twenty-eight magnesium-alloy anodes of 59 lb each were used, which provided protective potentials of at least —0.90 volt and give protection for 2 to 3 years.[3am,3an]

At Newcastle, England, steel piling under a 900-ft-long quay had corroded to about one-third of its original thickness of 0.42 in. during 22 years. One hundred magnesium anodes, each weighing 200 lb, were hung from bracing at 9 ft on centers in two staggered rows, each being connected to the steel piling. It was expected that corrosion would cease, while the piles would receive a deposit of magnesium hydroxide. The estimated cost was much less than that for concrete encasement and surface protection.[3aj]

Impressed Current Systems. For the Port of Long Beach, Calif., wharves constructed on Monotube piles in front of steel sheet piling employed two types of cathodic protection on different sections, one using a steel anode with rectifier, and the other galvanic anodes. Each section was 250 ft long, on 130 Monotube piles. Effective protection was about equal. Initial costs of the rectifier system were greater, but maintenance and operating costs were less. The anode for the rectifier system consisted of three superimposed layers of steel rail 30 ft long laid at 3 ft on centers on the harbor bottom 75 ft in front of the wharf face, connected to a 500,000-cir-mil cable; the piles were bonded together in the concrete deck. The anodes for the galvanic system consisted of two 51-lb anodes strung vertically, suspended in the water, for each bent of four piles, locating the strings staggered; bents were 10 ft apart.[3aj]

Integration of cathodic protection using rectifiers instead of galvanic anodes in designs for three piers in New York Harbor has been described.[3ar] All have steel H piles driven to rock. Alarms were provided for use when any circuit fell below 50 amp. Control piles give a check on effectiveness of the protection. Before driving, portions of piles to remain above the mud line were cleaned and coated with a coal-tar preparation. Piles were bonded by welds of reinforcing bars. Capacities of 5 ma per sq ft in the water and 1 ma per sq ft in the mud were assumed. Graphite anodes of different types or Duriron were used, depending upon location in water, in firm fill in the 8 ft below MLW, or in coke breeze where fill was above MLW.

CONCRETE PILES

Deterioration of Concrete Piles

Deterioration of Concrete Piles Aboveground. Piles aboveground are subject to the usual weathering of exposed concrete and air-borne destructive elements. In damp seacoast regions moisture may penetrate the concrete and cause spalling and rusting of reinforcement. Where conditions favor corrosion, reinforced-concrete piles are threatened.

Deterioration of Concrete Piles in Earth. Plain or reinforced-concrete piles in earth are generally considered permanent. Ground water containing destructive alkalies, acids, or salts may cause damage, more severe in sandy soils which permit rapid leaching and provide more air than in clays where the retarded movement may not be important.

Sulphate salts are found in very small amounts in most clays and ground water, and sites may be classified as having low, moderate-severity, or high risk.[5aj] Dense, rich portland cement precast concrete mixes should be safe against low or moderate concentrations of calcium sulphate salts (gypsum) if principally gypsum, such as up to 1,000 ppm of sulphur trioxide in ground water or 0.5 per cent in clay, unless buried over a long time; for cast-in-place concrete, sulphate-resisting cement has been suggested; if the salts consist of magnesium and sodium sulphate, the use of any cast-in-place piles is not recommended. For any sulphate contents exceeding the above limits, it would be well to avoid concrete.[5aj] Conditions may vary widely in short distances and with fluctuating moisture conditions.

Destructive chemicals in ground water may come from manufacturing-plant wastes, leaching from coal storage or cinder fill (particularly soft coal or recent cinder fill), leaching of sodium or magnesium sulphates from the ground, organic acids from decay of vegetable matter, or leaky sewers. Old cinder fills are not necessarily detrimental and may be tested. Waters of some streams and lakes in the western United States are very destructive because of alkalies, as are some sea-water salts.

Effects of chemical attack on concrete piles by a variety of substances are contained in reference 5*ap*; also in Appendix I of the report of the Joint Committee[5b] together with recommended protective treatments. Most active on concrete are nitric, sulphuric, sulphurous and hydrochloric acids, sulphates, acid sulphate, and nitrate of ammonia. Slight attack occurs from phenol, creosote, chlorides, sulphide ores, molasses, silage, milk, linseed oil, lactic acid, tannic acid, olive oil, coconut oil, and cottonseed oil. Suggested treatments for all except the acids are applications of fluosilicate, sodium silicate (water glass), vitrified brick or tile laid in litharge cement and rubber, bituminous applications, etc. Spar varnish helps in the cases of phenol and olive oil.

Deterioration of Concrete Piles in Sea Water. This is due to both mechanical and chemical actions.[5l]

Abrasive Action. This occurs from ice, debris, wind, waves, and spray and causes serious disintegration in even the best quality of concrete. The lime carbonate surface deposit resulting from carbonation of the lime probably attributable to the carbonic acid content of sea water, that provides some protection against chemical action, is removed by abrasive action.

Excessive Stresses from Horizontal Forces. These may occur from blows of storm waves; waves that break just before striking cause much more stress than do reflected waves. Blows from ships approaching to moor, or from surges of moored ships, especially in winds or currents, cause bending and can cause localized cracks which may permit water to enter, resulting in corrosion of the reinforcing steel and in spalling.

Mechanical Action. This may result if freezing water in the pores causes progressive disintegration and finally exposes the reinforcing. Concrete is porous to some extent, no matter how well made, and water penetrates to depths in proportion to the porosity; in sea water, as it penetrates farther it may become less saline and is more easily frozen, so that ice may form and expand inside the concrete while water outside remains liquid.[5as] Nonprestressed precast concrete piles generally have a number of crazing and hair cracks, and fine horizontal cracks, caused by shrinkage, differences of temperature between interior and exterior surfaces, handling, tension, and shear. Below low water the cracks tend to close themselves sufficiently to eliminate rusting of reinforcement, the cracks often being filled with clay at the surface and with carbonate crystals behind the surface. It has been concluded that small cracks below tide are not detrimental to the life and usefulness of the pile. Alternate wetting and drying is not limited to the tidal range, but spray, waves, and capillary action carry the sea water to considerable heights, where evaporation leaves crystalline deposits in the surface interstices with resulting disintegration from the mechanical effects of expansive action.

Prestressed concrete is less subject to deterioration than ordinary cast concrete.

Chemical Decomposition. Chemical decomposition of concrete in sea water is promoted by the presence of the cracks described above. It will ultimately expose the reinforcing steel to rusting in air or oxygen-bearing water and since rust occupies twelve to thirteen times the volume of steel, a slow irresistible action occurs which completes by expansive forces the destruction of the concrete. The chemical action during hydration essentially results in the decomposition of higher silicates into lower silicates and calcium hydroxide. The calcium hydroxide crystals dissolve slowly in water; this is followed by decomposition of the clinker grains liberating new quantities of calcium hydroxide, and eventually breakdown of the cement is complete. The free lime in the concrete also reacts with magnesium sulphate in sea water, forming calcium sulphate that occupies more space than the original calcium hydrate, thus causing a swelling. These attacks continue until the steel is exposed.

Deterioration begins shortly after driving and proceeds at a slow rate for possibly 7 or 8 years, then more rapidly. Some untreated concrete piling in Los Angeles Harbor lost one-third of its section in 11 years with more rapid disintegration thereafter. An inspection of practically all reinforced-concrete structures on the Pacific Coast from Vancouver to San Diego showed deterioration in all structures over 15 years old, some of which was quite serious.[85,86] On the other hand, reinforced-concrete pile jackets of superior workmanship and only 1 in. cover over the reinforcement showed no deterioration by rusting of the steel after 36 years in San Francisco Harbor.[5x]

Borer Attack. Neat cement or poor concrete may be rapidly destroyed in tropical or semitropical waters by rock borers such as *Pholads.* Good sand or gravel seems necessary to resist them. Damage appears to have been done only in poor concrete, reported from such widely scattered locations as Los Angeles,[20] Panama, and Plymouth, England.

Electrolysis in Concrete.[5y] The effects of passage of electric current through concrete have been investigated by the U.S. Bureau of Standards. It has no effect on crushing strength except in the close vicinity of the electrodes. When the reinforcement is anodic, insoluble products of corrosion are formed on its surface and, if in sufficient volume, may cause cracking of the concrete.

When the reinforcement is made cathodic, electrolysis has no ill effects if there are no appreciable amounts of alkali salts in the concrete. If there are, such as from use of sea water in mixing concrete or from salt added to lower the freezing point of green concrete, electrolysis leads to formation of alkali hydroxides at the cathode, which attack and soften the concrete around the reinforcement and impair or destroy bond.

Passage of direct current through concrete increases electrical resistance, owing largely to the formation of calcium carbonate near the cathode. If alkali salts or chlorides are present in concrete in appreciable quantities, formation of calcium carbonate is inhibited and increased resistance does not occur.

The effects all depend on application of current of relatively high current density. No damage seems to occur from a steady direct current of less than 15 volts. Therefore there is little likelihood of damage from leakage currents in practice unless leakage is abnormally high.

Damage during Handling and Driving of Concrete Piles

Handling Concrete Piles. Handling of precast concrete piles must be done in the manner for which the piles were designed, to avoid cantilever or beam action which will cause overstresses. Portions overhanging trucks or cars may vibrate and fail, especially if subject to impact from bouncing. Since it is practically impossible to prevent minute cracking of conventional precast concrete piles during manufacture, curing, handling, and driving, particular attention should be given to protection of portions of piles from a few feet below water line to the tops. Abrasions should be avoided, particularly for piles in sea water or alkali soils.

Driving Concrete Piles. *Early driving* before sufficient time for curing is a common cause of damage to precast concrete piles.

Impact during driving may cause damage to precast concrete piles, but this is usually remedied by casting the piles long enough to permit cutting off the top portion. Banding reduces this damage, as discussed under that heading. No great increase in impact strength has been noted for mixes richer than 1:1½:3.[21] Prestressing avoids most damage.

Crushing of concrete at some point in the length of the pile may occur in cased or uncased concrete piles or in concrete piles poured in thin steel shells if the pile centers are too close. A center-to-center distance of 5 ft would be desirable as a remedy for this situation, but this is not always practicable, especially if the piles are grouped in footings. The spacing should be kept as wide as possible and not closer than 3 ft. Careful inspection of shells should be made to detect the probable damage point. When using uncased piles there is no way to detect such damage except by excavating a few, or by coring a continuous concrete sample, so no chances should be taken on too close spacing. In general, it is desirable to drive the center piles in a group first in order to give the soil a chance to move slightly.

Manufacture of Precast Concrete Piles

Sound Ingredients. These appear necessary. There is evidence of at least one prominent failure of concrete piles exposed to sea water which can partly be attributed to lack of durability of part of the aggregate[5x] (Chap. 16, *Case* 33).

Reactive Aggregate-cement Combinations. These may occur.[5x] If salt-laden moisture can penetrate the concrete and build up large amounts of salts, the concrete may disintegrate through action of the sulphates on components of the cement and aggregates. Expansive reactions have been found between reactive minerals in local aggregates and high-alkali cements.

Cement Composition. Composition can be selected to improve resistance to sulphate attack by use of Type II (moderate sulphate-resistant, 8 per cent C_3A) and Type V (sulphate-resistant, 5 per cent C_3A) cement.[5x] Type IV has characteristics similar to Type V. There is evidence that concentration of sulphates such as found in alkali soils and sea-water exposures through alternate wetting and drying may cause serious disintegration in a short time in the presence of cements with substantially greater than 8 per cent C_3A, particularly where the concrete is permeable. Type II is recommended for use in sea water.

Waterproofing Dipping.[5ab] Dipping in a 6 per cent bath of sodium silicate was done to precast piles of rich mix made with a cement low in C_3A, for driving in Baghdad through the bricks of the ancient city. The use of gypsum mortar had made the sulphate content of the ground water so high that it would weaken unprotected concrete.

Air Entrainment. Use of a suitable air-entraining agent reduces the absorptive power of concrete and improves resistance to sulphate attack in the same manner as does any other means of increasing imperviousness, such as mix design.[5x] Air entrainment of about 5 per cent has been recommended for use in sea water. Use of 10 per cent reduces resistance to abrasion. Addition of up to 2 per cent of calcium chloride to air-entrained cement to give high early set will not cause loss of beneficial effects of the air.[5as]

Integral Waterproofing Compounds. These are intended to make more workable mixes and promote impermeability. Varying degrees of success have been obtained with lime, clay, pozzuolanas, tars, asphalts, soaps, emulsions, diatomaceous earths, barium salts, and waterproofing cements.[5ac]

Pozzuolan Cement. Pozzuolan cement originally referred to materials basically of volcanic origin containing compounds of silicon, aluminum, and other elements. The term now includes certain shales, trass, and blast-furnace slag that are silicious. These materials are mixed and ground with calcined limestone, but are not clinkered and reground as for portland cement. Pozzuolanas have been used in some marine structures in sufficient quantity to react with the free line of hydration of the portland cement to form a rigid gel of monocalcium silicate that is not soluble in water and helps watertightness.[5as]

At least one brand of portland-pozzuolan cement has shown considerable resistance to sulphate attack as well as to cement-aggregate reaction, regardless of the C_3A or alkali content of the cement clinker.[5x]

Amount of Cover. Two inches should be the minimum. The Bureau of Yards and Docks requires 3 in. from 3 ft below low water to above the splash zone. The mix should be well worked in around the bars.

For prestressed concrete which is centrifugally cast, the cover can be reduced to 1½ in. even under severe salt-water exposure. This process produces a very dense concrete with an absorption rate less than half that of concrete produced by ordinary methods, even though prestressed.

Heads. Heads of precast piles should be smooth and flat, and cast normal to the axis, to avoid eccentric stress and local failure.

Prestressing. Prestressing the reinforcement may reduce cracking.

Water-cement Ratio. This should be kept as low as possible, because the use of wet mixes is a major cause of failure.

Curing. Curing by a continuous moisture blanket from sprinklers or by burlap regularly dampened for at least 14 days is of great assistance in developing the potential concrete strength. A membrane type of cure (such as Hunt Process "Clear") may be used.

Storing. Great damage may occur in storing unless proper care is taken regarding locations, spacing, and levels of supports.

Conclusions. Observations and tests indicate that the following procedures are advisable in the manufacture of precast concrete piles and should give high resistance to alkali attack: (*a*) Use sound cement and aggregate; (*b*) use a nonreactive cement-aggregate mix; (*c*) use a low water-cement ratio and not over 2 or 3 in. slump and at least six bags of cement per cubic yard, and up to seven bags where exposure is severe or a higher water-cement ratio is used (U.S. bags of 94 lb); (*d*) avoid segregation and honeycombing; (*e*) use Type II or Type IV or V cement in alkali soils and sea water; (*f*) use not less than 2 in. cover over reinforcement; (*g*) use an air-entraining agent to aid impermeability.

Methods of Protection of Concrete in Sea Water

Records of service of concrete piles in sea water are spotty. The conclusions given above regarding manufacture should, when followed, result generally in long life. However, in view of past performances, many attempts at securing economic protective methods have been made.

Painting. Painting with thin coats of paint or preservative has been reported by many engineers to be effective only temporarily. It seems practically impossible to secure adequate bond for any length of time.

Asphalt Impregnation. Successful methods were developed by the Los Angeles Harbor Department and used by them and by private industry in many installations prior to World War II.[5l] Inspections in 1942 of piles in service 20 years showed no signs of disintegration. It was estimated that such piles would have a life of 75 years or more and that higher initial cost would result in lower final cost. Prices following World War II appeared to make the expense of this treatment considerable.[85,86]

Finely emulsified asphalt, known as *Hydropel,* made by the American Bitumuls Co., was used in precast concrete piles for a wharf in California in 1948. This was added to the mixing water, and it was expected that the asphalt would fill the interstices between the crystallized particles of cement. Laboratory tests showed greatly increased durability, and field experience showed the piles tougher and less brittle. Results of this treatment should be watched.

Concrete Armor. This has been used to protect the portions of concrete piles above low water, and is described in Chap. 9.

Shotcrete Encasement. Shotcrete has been applied to precast concrete piles.[85] This may be done before driving. The practically impervious jacket protects the pile from disintegration in sea water, sewage, and industrial waste.

Shotcrete over tar and felt has been used to protect concrete piles, from below MLW to above MHW, with a thickness of reinforced shotcrete of 1½ in. Inspection of two wharves in New York, 14 years old, showed serious damage or complete destruction.

Wrought-iron Armor. This was used to repair and protect existing precast concrete piles at the water line for the Bass River Bridge at Yarmouth, Mass., about 1948. The piles had spalled to the reinforcing steel, and two channel-shaped ¼-in. wrought-iron plates 6 ft long were bolted together and filled with concrete around the pile to form a jacket.

Creosoted Wood Jackets.[5an,5ao] Deterioration of precast concrete piles in tide waters around New York City has been prevented for periods up to 33 years by pressure-creosoted wood jackets in the tide zone, with concrete and wood still sound when inspected. Unprotected piles were mostly damaged. Creosoted jackets were used for new viaducts in Long Island in 1952 and 1956. Jackets are of 3-in. pine, or two layers of 2-in. pine, with 20-lb coal-tar creosote or creosote–coal-tar solution, with jacketing used as permanent forms extending from 2 ft below low water to the underside of the cap. Galvanized-wrought-iron straps and bolts are used. Damaged piles have been repaired similarly, concreting under water from the bottom up while withdrawing the grout pipe and vibrating the forms.

General References

Quite comprehensive information on deterioration and protection of concrete piles, with extensive bibliographies, appears in references 5*aq*, 5*ar*, and 5*as*.

ECONOMICS, INSPECTION, AND REPAIRS

Economic Life of Piles

Untreated wood piles should last indefinitely in conditions not permitting fungal growth or attacks by insects or borers. Adequately creosoted wood piles should last from 8 to 10 years in tropical sea waters to many years in fresh waters, above ground-water line, or in air; they may not resist certain marine borers, and any breaks in the creosoted shell permit entrance of fungi or borers and thus cause loss of investments on the structure and the treatment as well. Encasements in damage zones may be economically valuable on both untreated and treated piles.

H piles have shown good records of long life with generally slow rates of corrosion, except in danger zones previously described. Encasements for such portions, or cathodic protection, should be relatively inexpensive and permit obtaining the service implied in selection of this type of pile.

Renewal or replacement of concrete piles is expensive, and unless a long life is anticipated, their selection over wood piles in the United States might be unlikely. If they can be assured of a conservative life of 40 to 75 years, which may be from two to four times as long as may be expected from wood, their higher initial cost will usually be justified, particularly when taking into account replacement costs of wood piles, removal of part or all of the superstructure, redriving through riprap, added fire risk and insurance owing to wood piles, and inconveniences of having the structure out of service.

Selection of piles and protective measures should be consistent with the expected life of the structure. One should not be deterred by initial costs for this purpose, for proper design justifies the expense through maintenance savings.

Inspection during Life of Structure

Piles once in service are usually forgotten. Safeguarding of life and of the investment in the supports and supported structure sometimes makes periodic inspections or check of conditions advisable. Damage found before it progresses too far can usually be repaired at small expense compared with that involved in allowing it to progress.

Ground-water-level records may indicate dangerous conditions for untreated wood land piles.

Ordinary inspection of marine piling can be made only above low water. Experience may show such inspection to be sufficient, particularly on small, inexpensive structures. Where expensive and important structures are involved, inspection should be carried down to the mud line by pulling a pile or by diving. Experience may show uniformity of conditions so that only a percentage of piles need be checked.

The effect of borers on wood piles should be watched for carefully, for once the pile has been penetrated, other causes of deterioration also can proceed more rapidly. The presence and action of marine borers may also be observed by the use of test blocks which can be withdrawn, but this method does not give comparable results with other types of piles. The use of test blocks is invaluable where the presence of *Teredo* or *Bankia* is suspected, because their entrance holes are minute and hard to observe, and are often concealed by marine growths. The degree of attack can be determined only by cutting the wood. It is possible to suspend test piles, typical in every way except length, which only penetrate the mud a short distance so that they may be lifted for inspection. Because of the effects of decrease in salinity on the life of some borers and the fact that fresh-water inflows may change the salinity, seasonal changes should be watched. Variations of salinity with depth make it necessary to be careful not to draw general conclusions from the results at any one depth. Test boards should preferably be installed just before the breeding season, and care should be taken not to draw false conclusions of immunity from lack of attack on fresh boards installed after this period.

Instructions for the preparation of test boards and determination of salinity content of the water are given in the final report of the San

Francisco Bay Marine Piling Committee.[20] It is suggested that test-board designs made by the William F. Clapp Laboratory be used.

Damage above low water to concrete and steel piles may be seen and measured. Conditions under water can be explored to some extent by a diver, but pulling a pile may be necessary.

Systematic inspections and salinity tests should be made and records kept, which will serve as a basis for future engineering decisions during maintenance or for replacement purposes. Such data are also useful for collaboration with other owners nearby or with public authorities. Photographs may form an excellent supplement to a written description.

Repairing Damaged Piles

Repairs are usually well worth while and often prolong the life of the structure considerably. In creosoted piles, small holes made by cant hooks, rafting dogs, etc., can be plugged with creosoted plugs, and larger holes can be covered with tar-saturated felt and copper securely nailed, or they can be filled with mortar. Concrete encasements can be formed and poured, or gunite collars or shells used on either wood or concrete piles. Any marine growths interfering with the patch should first be removed. Floating collars can be used for application of a creosote coating. Fill may be placed around the piles to prevent further attack from borers, provided conditions favorable for decay are not established.

Ravages of borers were confined to the upper parts of wood piles in No. 9 Pier, Victoria Dock, and the Melbourne Harbour Board spliced new sections over 30 ft long onto 600 stumps. Divers with air saws cut off the riddled sections at the mud line. New sections were turned to fit the stumps, cast sleeves placed, and the assembly secured by driving with a pile hammer.

Borer damage to the upper parts of wood piles under the Fisher Flouring Mills in Seattle Harbor was repaired by replacing the portions of the piles above the mud line or at a point slightly below. A diver excavated at the mud line, determined the soundness, and cut off the piles. If it then was found that the cut should be made at a lower point, this was done. The diver then drove a dowel and set a measured length of creosoted pile over the dowel. A reinforcing cage and corrugated steel culvert pipe for a form were placed and grout pipes set in the form. Gravel was placed in the form and then grouted.

CHAPTER 14

SOIL STRENGTHENING

Soil Strengthening as Applied to Pile Foundations

The two solutions to the problem of foundation design are, (*a*) to provide a foundation that will not cause failure of the soil, and (*b*) to strengthen the soil. The first method is the usual one in the United States and Great Britain. The second method has been used successfully, particularly on the continent of Europe. Soil strengthening may meet particular needs and probably will be developed for greater uses.

Soil strengthening is considered here only so far as such procedures might be applicable to pile foundations. The expression means changing the stress-strain characteristics of the soil by solidification either with or without change in the soil mass.

Selection of Method

Strengthening of soft material may be done by filling voids with insoluble material, by drainage and consolidation, by changing the chemical structure, by solidifying void water, or by reducing the volume of voids.

Injection grouting has been widely used. The most common method for filling voids is by grouting with a slurry of water and portland cement. Fine sand, clay, or similar materials may be added to the grout or used alone. Other injection methods include bitumens or various chemicals either singly or in combination. Theories of permeability appear in references 10*ar* and 10*as*. An entire chapter on the mechanics of chemical grouting appears in reference 10*ax*.

Improved properties of certain soils may also be obtained by compaction. This may be done by driving displacement piles, sand-pile drainage, explosives, vibroflotation, or electroosmosis. Freezing provides strength and impermeability as long as suitable temperatures are maintained.

There is no single method of grouting that can be used successfully regardless of the nature of the porous soil. Methods and materials best suited for the particular problem should be chosen.

A review of available information on grouting made by the Waterways Experiment Station at Vicksburg, Miss., appears in reference 10*ar*. Detailed descriptions of various methods and their actions also appear in reference 10*as*.

In case of settlements involving structures already built, methods must be used which do not change the soil mass but in which voids are filled with new materials rather than with rearranged soil grains.

In case of grouting soil to stabilize piles against vibrating loads, it is necessary to stop vibration until the solidifying medium has attained sufficient strength to hold the ground firm against the forces. Test samples may be made for such control purposes if times of hardening are not known. Weighting the surface may be necessary, in the case of grouting with high pressures near ground level.

Coarse-grained Soil. Coarse-grained materials can be concreted by cement grouting, made into artificial sandstone by chemical injection, or made nonpervious to water by bituminous grouting. Void water may also be temporarily solidified by freezing. Consolidation by vibration, rammed piles, or explosives is possible. Although sandy and gravelly soils generally represent a satisfactory site, in some cases it is necessary to bring about additional consolidation.

Fine-grained Soil. Fine-grained materials forming cohesive soils are composed of particles of small or colloidal size, each of which is covered with a film of adsorbed water which grips it tightly and is not free water. Normal drainage methods are ineffective. Great and long-continued pressure would be required to eject this water through the pore spaces, and other more practicable means of stiffening these soils are required. The adsorbed film carries a negative charge, and can be caused to detach itself from the particles and flow along an electropotential gradient, also carrying free water with it, to well points, as described under Electroosmosis. Electrolysis appears to provide a method of stiffening some clays. Both electroosmosis and electrolysis could probably be used at the same time.[10w]

Uses of Soil Strengthening

Strengthening may be of use in the following cases: (*a*) *creation of conditions which will permit use of less elaborate foundation designs;* (*b*) *stopping horizontal vibrations* in a structure, caused by moving machinery, if the upper strata around piles are so soft that movement is occurring in the piles; (*c*) *preconsolidating strata subject to vibrations* from machinery, so that objectionable settlements will not take place from this cause; (*d*) *removing considerable inequalities in soil conditions* over a site; (*e*) *sealing bottoms of caissons or pile holes;* (*f*) *prevention of lateral movement* by forming a wall; (*g*) *stopping or reducing settle-*

ment, by solidifying a mat around and below the pile tips, or by forming an enclosing wall; (*h*) *providing a means for supporting additional load on existing piles,* by solidifying a mat around and below pile tips or by forming an enclosing wall, if the piles and strata in which they are embedded are not capable of permitting transfer of the combined load from one to the other, but provided the underlying strata are capable of sustaining the total load. The last two uses, if a solidified mat is formed, actually transform friction piles into end-bearing piles.

Grouting

Grouting of granular soil with silt, cement grout, chemicals, or emulsified bituminous-cement slurry has been practiced fairly widely in Europe. Application of these processes has been growing in America.

Desirable Properties of Grout.[10av] No materials are available which have all of the desirable properties, such as: (*a*) The grout must be able to modify the soil properties as desired, usually strength increase and permeability decrease. Temporary improvement may suffice during construction, but permanent effects are usually sought. (*b*) The grout must be placeable in adequate quantities. This requires a liquid solution. Materials should be mixable on the job with water, to reduce transportation costs. (*c*) The grout must be able to penetrate material of the existing grain size and density. (*d*) Time of reaction should be controllable. (*e*) Grout should have low viscosity to obtain best penetration and to ease pumping. Viscosity should remain unchanged until the stabilizing reaction occurs. (*f*) Grout should be relatively insensitive to impurities in the soil or water. (*g*) The process should be irreversible. (*h*) Improvement of the soil properties must not decrease with time. (*i*) Chemicals should be noncorrosive to equipment. (*j*) Chemicals should be nontoxic and nonexplosive. (*k*) Materials and methods should be inexpensive enough to justify grouting instead of other possible means of obtaining desired results.

Silt Injection Method.[10aq] This method can be used for consolidating loess or other porous soils and consists of pumping silt slurry into the moist soil. To be successful, the soil must have enough porosity to serve as a filter, allowing moisture in the soil and the added water to be forced out of the area being consolidated; the slurry being pumped in must be sufficiently porous to allow the water to move under pressure through the slurry when the walls of the filter begin to hold the solids of the slurry; and the slurry must have sufficient fines to lubricate the mixture sufficiently so that it can be pumped in place. If the soil will not pass the water out under pressure, the mass will remain liquid, and if the injection work continues under these conditions, there will be a failure.

Cement Grouting. Cement grout may consist of mixtures of cement, clay, and sand; cement and clay; or pure cement; more recently colloidal grouts have been used. When the grain size is such that cement, even in colloidal grout, is not practicable, a stabilized clay may be used, and for fine sands use of chemicals may be necessary. Cement-base grouts can be transformed into colloidal mixtures by mixing, by addition of a dispersing agent, or by raising the temperature.

Cement grouting is the best known and most commonly used method of solidifying soils.[10a,10b,10y] It is applicable only in granular soils. Injection of cement suspensions is possible only at a pore-size diameter larger than 0.1 mm. The required coefficient of permeability to grout sands successfully is about 1,000 by 10^{-4} cm per sec.[10ar,10as] Because of the size of cement grains, it is impossible to force grout into fine sand and, according to the theory of filters, a granular material with grains ten times larger than those in suspension will prevent passage of most of the suspended grains. The median grain size of cement may be approximately 0.1 mm, with small percentages several times larger or smaller; therefore, if the effective grain size of compact sand is about 1.4 mm or smaller, of loose sand about 0.5 mm, or if the material is a mixture of gravel and fine sand, passage of grout would probably be blocked. Increased pressure might permit formation of a local grout bulb around the grout pipe, but the cement would not go far. If the sand contains layers of coarser materials, the grout will follow these and fill voids therein before attempting to penetrate finer portions of the strata. In order to introduce the grout into a bed of coarse-grained sand or gravel, 3-in.-diameter holes are usually drilled, 1-in.-diameter grout pipes inserted, and the space between grout pipe and casing filled with coarse sand. The casing is then withdrawn. After a short period of grouting, it will probably be necessary to reestablish connection between the grout pipe and the soil, which may be done by firing small charges of explosive to split open the pipe and cemented shell. Grout has also been placed through grout pipes in which perforations have been shot by a special gun after driving.[10d] The spacing of grout holes must be determined by results. The first holes may be spaced 10 to 20 ft apart, depending on the pressure used, and several sets of intermediate holes used as needed. Holes should be drilled, not driven, to avoid solidifying the shell of soil surrounding the hole. It is best to start grouting at the bottom of the hole and work up. Small holes and high velocity of grout are advantageous against plugging and sealing.

Chemical Effects of Ground Water on Grouting.[10ar] The type and concentration of chemicals harmful to cement or concrete depend on several physical and chemical properties of the cement. Sulphates of sodium, calcium, magnesium, and potassium are detrimental and, if over 0.5 per cent, may destroy mass concrete in 10 years. A concentration of 0.2 per cent may destroy lean or thin sections. Also harmful are sulphides that may be oxidized to sulphates. Organic acids and dissolved carbon dioxide are harmful. Chlorides have little effect unless concentrated, but vary the setting period. Waters with pH values outside the 6.0 to 8.0 range may be considered harmless.

François Cementation Process. This process was introduced into the United States some years ago.* It consists of high-pressure injections of liquid cement mixture into cavities, fissures, pores, or cracks so that they are filled. Pressures up to 3,000 psi are used. Fissures are rendered more tractable to the reception of grout by the lubricating action of a patented chemical process which, it is claimed, seals the finest hair cracks. In one instance, a 16-ft stratum of clay and decomposed rock soil over rock compacted 15 per cent by this method.[10i] A 2-in. grout pipe was inserted 4 ft at 7½-ft centers to form a bulb at each spot, and before the grout set holes were drilled through to form second bulbs. Before the second bulbs set, a smaller pipe was driven through the 4-in. pipe to rock, withdrawn 18 in., cored with an auger, and the hole tamped and grouted. Actually, the grout followed cleavage planes in part of the soil which was fill. The soil, which had previously settled under a load of 2,500 psf, now carried the load without settlement.

Prepakt Intrusion Method.† This consists of stabilizing a foundation by drilling holes at intervals and injecting an intrusion mixture consisting of portland cement, a powdered, finely divided mineral filler called Alfesil, an intrusion agent, water, and, in some cases, fine sand. Features of the process are characteristics of the mixture, fittings and connections, sequence of solidification, and determination of adequate penetration.

The Alfesil tends to prevent agglomeration of the cement grains and thus increases their effectiveness, and in the hardened mixture it combines with the liberated lime to form insoluble calcium silicates. Intrusion Aid is a patented admixture which acts as a protective colloid to inhibit early stiffening of the grout and facilitate pumping; to completely neutralize setting shrinkage; and in conjunction with Alfesil, to hold the solids in suspension until the grout has set to assure complete filling of voids. Small amounts of calcium chloride may be used to hasten the set.

Holes are drilled to the various depths required. A smooth slurry is made and pumped in such a manner as to expel air and water ahead of the mixture. Pressure is maintained long enough to ensure any ad-

* By the Dravo Contracting Co., *Eng. News-Record,* vol. 99, No. 25, p. 1023, Dec. 22, 1927.

† Intrusion-Prepakt, Inc., Chicago, Ill., Cleveland, Ohio, and Toronto, Ont.

ditional penetration possible after first refusal. The intrusion mixture passes quickly through sand which cannot be penetrated by cement grout of equal consistency, and has far less volumetric shrinkage.

During floods, it is not unusual for mud or gravel to be washed from under piers on piles. If the amount of material removed is considerable, the lateral support of the piles may be removed and permit weaving of the structure. In one case, holes were drilled through the footings, mud and sand flushed out, and the space filled with intrusion mixture, which also spread around the pier and stabilized an area greater than the original.[10j] The holes were extended farther downward and intruded to form a solid wall to prevent future scour. Still farther down, the existing layer of coarse sand and gravel was flushed out and solidified by intrusion. In stabilizing piers of other bridges, in some cases the opening beneath the pier has been so large that it was filled with riprap or prepacked aggregate in order to reduce the amount of intrusion mixture required.

The method has been used for encasing groups of H piles in concrete to stiffen them and eliminate bracing.[10aa]

Bituminous Grouting. *Shellperm Process.** The slurring used consists of an emulsified asphaltic bitumen, practically as thin as water, mixed with cement grout.[10e] This treatment can be used only in rather porous soils containing only a few grains finer than 1.0 mm. The soil can be solidified to develop a compressive strength of about 10 tons per sq ft and a shearing strength of 50 psi. The time of setting can be controlled.

Bituminous Emulsions. Bituminous emulsions may be used for soils possessing a coefficient of permeability of about 1.0 by 10^{-4} cm per sec.[10as] Such injections are used mainly to decrease permeability; they impart little or no cohesion to the soil. A coagulator is added just before injection, to cause the emulsion to flocculate and fill pore spaces. Viscosity of the emulsion is much less than that of cement grout and but little more than that of water. This process can be used with fine sands from 0.1 mm to coarse sands up to about 2.0 mm.[10v] This method is not used for very coarse sands or gravel, owing to the possibility of ground-water flow removing the emulsion before flocculation, and because the cost exceeds that for cement grouting.[10v]

Chemical Grouting. The principle of chemical grouting consists of filling the voids with two or more chemical solutions which react and harden upon contact. Sodium silicate is usually used, with calcium chloride, aluminum sulphate, or any heavy metal salt. A silica gel is formed, which is sticky and semisolid, and gradually hardens upon losing its water content, binding the sand grains together in a solid, permanent

* Developed and patented by the Royal Dutch Shell Group.

mass similar to sandstone and insoluble in fresh or salt water. The chemicals penetrate nearly as freely as water and cannot leach out. The materials are either injected one after the other, or both together while using some means of delaying the reaction.

Chemical grouting is of value particularly because of its greater ability to penetrate finer grained soils.[10a,10b,10h] It is applicable in soils having an effective grain size down to about 0.1 mm,[10v] and a permeability between 0.0001 and 0.1 cm per sec for a liquid of the viscosity of water at 60°F. Waterlogged soils of this type may be grouted. Clays, dust-fine sands, sand-silt and sand-clay mixtures, or mud cannot be grouted by chemical penetration, since their permeabilities may range from 1 by 10^{-6} to 1 by 10^{-8} cm per sec.

In material of too fine grain size or in soils containing lenses of coarser materials, when using two separate injecting liquids, the second may not be forced into good contact with the first, so that only local hardenings are obtained. Use of two solutions can result in high soil strength, but a solid deposit tends to form around the injection pipes and impede grouting. Since the radius of action is very small, a large number of boreholes may be required. Use of a mixture of two solutions having a predetermined definite time of gelation allows a greater soil mass to be impregnated and reduces the number of grout holes.

Analysis of the chemical nature of the soil is necessary. With knowledge of the permeability and pressure to be used, spacing of grout pipes can be determined. Pipes should be driven, not jetted, to avoid loss of ground. Driving is done with air hammers, hydraulic pipe pushers, or drop hammers. Pipes may be driven on any angle. Air compressors are used for the chemical pumps. Pipes may be withdrawn from solidified soil by a jack. By injecting through the lowest pipes first and proceeding upward, an impenetrable base is formed which confines the later material and ensures filling all voids. The withdrawn pipes should be tested under water pressure to be sure that perforations are not clogged.[10h] Water spraying out of the perforations indicates that the pipe is clean and that the soil zone treated by it has been evenly saturated with chemical. In order to be sure that no sections have been missed, test pipes may be driven and further injections attempted.

Single fluid injections are naturally cheaper than those requiring two operations, but while they decrease permeability, they are said not to increase the shearing strength of the soil markedly.[10v]

Chemical grouting is used in one type of bored pile to solidify permeable strata under hydrostatic pressure at the foot of the tube, so that concrete may be placed in the dry hole, after which the tube is withdrawn.[58]

Costs of chemical treatment are comparable to those of cement grouting, and vary with accessibility, depth of grouting, and other factors, but only in extremely difficult cases do they exceed those of reinforced concrete.

Joosten and François Methods. The Joosten method* uses sodium silicate (water glass) or hydrofluosilicic acid and calcium chloride.† The François method uses sodium silicate and aluminum sulphate. In both of these methods, the silicate is injected through a grout pipe resembling a well point. The pipe is driven into the soil, and the salt solution is injected as the pipe is raised. Grout pipes of 1-in. diameter with the lower 2.5 ft perforated, are spaced approximately 2.5 ft apart. A pressure of about 200 psi is used. In clean fine sands having an effective grain size of about 0.1 mm, results are very successful; artificial sandstone having a compressive strength of 20 tons per sq ft or more is obtained. Very successful results were obtained in submerged sand having a grain size from No. 42 to No. 80 U.S. sieves.[10h] Laboratory tests on Chicago, Florida, Atlantic and Pacific Coasts, and London sands, solidified by the Joosten method, have shown compressive strengths from 450 to 900 psi.[10m,10l] In London "ballast," containing 24 per cent silt and clay, a strength of 16 tons per sq ft was developed in 30 min. Tensile strength of 157 psi was tested on treated Thames "ballast." Permeability of treated Chicago sand was less than that of Lake Superior sandstone.[10n]

Sand to be grouted may be dry or under high water pressure, as high as 1,400 psi having been reported[10o] in coarse-sand values.

Later methods used ammonia or carbon dioxide to retard the time of set.

Gayard Method. This method is based on the same principle as the Joosten but calls for the use of some salts, such as sodium or potassium bicarbonates in the proportion of 5 to 30 per cent of the weight of silicate; or 3 to 15 per cent each of sodium or potassium carbonate, with 3 to 15 per cent of sodium chloride and 0.3 to 1.0 per cent of sodium or potassium hypochlorites.

Rodio Method.[10as] One solution is used, having a sodium silicate base and generally sodium aluminate as a reagent. Sands with a permeability of 1 to 10 by 10^{-4} cm per sec may be treated. Time of solidification may be regulated. This method may be used in conjunction with cement injection. The gel is not attacked by corrosive ground waters and is highly impervious and stable.

Langer Method. The Langer method, a later development, reduces the pH content of the silicate to a determined amount, permitting control

* For practical improvements by C. M. Riedel and Lawrence Scully of the Chemical Soil Solidification Co., see reference.[10h] This method was introduced into the United States in 1939, and is patented.

† Chemicals furnished by Philadelphia Quartz Co. and Solvay Sales Corp.

of the time of set from ½ min to several hours, thus making it possible to mix the chemicals before injection, predict the proper proportions, save chemicals, eliminate danger of silica-gel formation about the pipe, and obtain better uniform penetration and solidification of the soil.

The injection material is a clear colloidal silicate solution, the stability of which has been decreased by means of an acid in a way that, by the addition of at least one soluble heavy metal salt, gel may be formed after a predictable time. Since the liquid is clear, a large radius of action may be obtained in the soil.[10as] The gel is insoluble in water and is not sensitive to chemical solutions that are detrimental to concrete.

KLM Method.[10v] This method uses simultaneous injection of silicate solution and two admixtures or reagents. Proportions can be adjusted, after analysis of the acidity or alkalinity of the soil and its permeability, to cause setting in from 30 sec to 6 hr. This promotes homogeneity and avoids pockets. Less time is required for the single injection than for two injection processes.

This process is suitable for materials having a permeability of not more than 0.1 and not less than 0.0001 cm per min. If the permeability is over 0.1 cm per min, cement grout is said to be preferable; if less than 0.0001 cm per min, no grouting treatment will give good results.

Strengths up to 1,200 psi are obtainable in gravel and 500 to 700 psi in very fine sands, with possibly higher values in thoroughly compacted deposits. Strengths are obtained gradually over a period of weeks, which often is not objectionable and permits excavation to be made soon in treated areas if necessary. Impermeability occurs rapidly. The chemicals used have no bad effects on concrete.

Calcium Acrylate Grouting.[10ar] This process was developed for military use as a surface treatment and later studied for suitability for use in injection methods. A water solution of calcium acrylate to which has been added a catalyst (ammonium persulphate) and an activator (sodium thiosulphate) causes a polymerization reaction. The concentration of catalyst and activator, called the redox system, controls the rate. The low viscosity favors uniform injection into all but the finest sands. Kaolin may be used as an admixture to lower the cost of the materials.

AM-9 Chemical Grout.*[10ap,10at,10av] This is a mixture of two organic chemicals, acrylamide and N,N′-methylene bisacrylamide, in proportions that form very stiff gels from dilute aqueous solutions. Treatments known as AM-955 and CJ-1 are synonymous with AM-9. AM-9 is an extremely water-soluble powder. Up to 50 per cent by weight can be dissolved in water fairly rapidly, and solutions up to 30 per cent are very stable. Viscosity approaches that of water. Above 4 per cent con-

* Trademark of American Cyanamid Co.

centration, gels with constant reproducible properties will always form. Most field work has used concentrations of 7 to 15 per cent.

Gelation, or polymerization, is caused by adding small quantities of other chemicals which catalyze the AM-9 solution. Many catalyst systems will cause gelation, such as ammonium persulphate (AP) and β-dimethylaminopropionitrite (DMAPN). Continuing research may produce other cheaper and more effective catalysts. When the concentration of catalysts reaches a certain value, gelation is rapid. Time before gelation may be controlled is from a few seconds to several hours, depending upon the amount of catalysts. Gels form best in soils saturated or nearly so, but can form in dry soils if high catalyst concentrations are used. AM-9 pumping operations are generally very rapid compared with cement grouting, which helps to balance the higher cost of materials to a considerable degree.

Wherever the natural soil is such that sufficient water is present to cause construction problems, it is usually possible to pump in enough chemicals in a short time to effect stabilization. In addition to being effective in sands, this process has worked in coarse and fine silts, even with some clay soil in them.

Once the gel is formed, it appears permanent, and no solvents have been found at ordinary temperatures. No strength change or change in gel under nondehydrating conditions has been apparent with age in the several years since the process has been known. With long exposure to air, the water in the gel can evaporate, causing shrinkage. As the gel shrinks, capillary pressures cause increase in intergranular pressures, resulting in increased soil strength and shrinkage in some soils. Completely dried soil strength may approximate that of cement mortar. The wet strength is generally of more interest; very loose soil becomes rubbery and elastic; very dense soil becomes relatively brittle with unconfined compression failures similar to those of triaxial tests under high lateral pressures. In their loosest laboratory state, unconfined strength can be 15 to 20 psi. Wet strength of 100 psi may be obtained in well-graded fine sands, or 50 psi in coarse sands, using a 7 per cent solution. For a given concentration of AM-9, strength increases with decreasing grain size and increases with increasing spread of grain size. Unconfined compressive strength varies directly with the concentration of AM-9. Compression increases with load, and virtually all settlement occurs at once as in natural sands.

There do not appear to be any chemicals normally present in soils which would prevent gelation. Low pH values may increase the required induction period, but this may be overcome by adjusting the amount of catalyst.

For all practical purposes, the stabilized soil is impermeable if it is not allowed to dry out. Soil stabilized by this method will exhibit plastic flow, but will not consolidate; that is, water will not be forced out by loading. Alternate freezing and thawing will cause deterioration.

AM-9 has some toxic properties. Repeated contact may temporarily affect the central nervous system. Use of a mask when handling the dry powder and of rubber gloves and coveralls when working with the solution are effective safeguards. It is believed that the gel is not toxic except that it may contain small amounts of unreacted acrylamide AM-9. Only unreacted AM-9 can leach out of the gel, and any amounts would be small. AP is an oxidizing agent and may ruin clothing and cause skin irritation and corrosion of steel. However, the treatment is usually performed by specialists, who know how to handle the materials. Each application should be analyzed if potable or recreational water is nearby.

This system is useful in cutting off water flows and installing some types of caissons in water-bearing soils. Injections at the feet of pipe piles or caissons can form pedestal bulbs to seal the bottoms and increase end bearing.

Precatalyzed AM-9 can be mixed with soil, clay, or portland cement to form a slurry that can be placed before gelation.

AM-9 is the trademark name for this process, which was developed in 1951 by the American Cyanamid Co. Injections have been performed by the Chemjet Corp., Franki of Canada, Ltd., Boyles Bros., Selby Drilling Co., Halliburton Oil Well Cementing Co., St. Joseph Lead Co., and others. (Rights acquired by Chemical Soil Solidification Co.)

Chrome-lignin Process. Use of this process appears promising. The lignin, which forms the major part of the grout, is a waste product from paper mills and is stable, easy to handle, and of low cost. Lignin acts as a dispersing agent. The dichromates used in this process are hazardous to handle in the field.

Terra Firma is a chrome-lignin grout made by the Concrete Chemicals Co., Cleveland, Ohio, a subsidiary of Intrusion-Prepakt, Inc. It is a soluble powder which is mixed with water and can penetrate anywhere that water can. Sands may be treated to become as impermeable as clay or lean concrete. The gel is a dark-brown rubbery substance, nonreversible, soluble in water and organic solvents, and substantially impermeable to water in normal dilution. Gelling time can be controlled from a few minutes to several hours. Dilutions of 1:4 or 1:5 are normal. One hundred pounds is sufficient to treat from 20 to 40 cu ft of soil having 30 per cent voids. This treatment is well suited for consolidating soil too firm to accept cement grout. Under certain conditions, it is used as a priming or lubricating agent, and it is injected into the ground to facilitate and improve final consolidation with intrusion cement grouts.

Compaction of Soils

Compaction by Displacement Piles. Loose cohesionless soils may be compacted by driving displacement piles. In soils having a density below critical, the displacement volume of the piles usually more than compensates for decrease in soil volume owing to vibration from driving. The method should not be used where damage might be done to adjacent structures.

Compaction by Sand-pile Drainage. Water may be withdrawn from soils by pumping from sand piles that act as wells. Sand drains have been placed successfully by jetting a double-shelled 19-in.-diameter pipe, 62 ft long, having a ring of water jets at the foot of the annular space and fewer air jets on the outside. As the tube is withdrawn, sand from a hopper fills the hole.[10r] In another method, an 18-in.-diameter jet tube with a perforated plate on the bottom is driven as a mandrel inside a casing, then pulled and sand placed, after which the casing is pulled.[10r] The plugged-mandrel method has been used considerably; it consists of driving a closed-end mandrel in a casing and pulling it, then filling the casing with sand and pulling the casing. Compressed air frequently is used to expel the sand as the casing is raised.

Special equipment for installing sand drains is made by the McKiernan-Terry Corp.[12l] Stiff lightweight compression leads guide the pipe and pile hammer and resist the pull exerted when withdrawing the pile. The pipe is from 12 to 24 in. in diameter and is closed with a hinged plate or conical point. Lengths of piles vary from 10 to 100 ft. Coarse sand is dumped into a sand skip, which is hoisted up to empty into the pipe through a trap door. This door is then closed, air pressure applied, and the pipe withdrawn. Sand columns are usually placed at 6- to 20-ft centers, depending upon the time for consolidation and character of soil. A sand blanket is usually placed on the ground first, to support the driving equipment and to provide weight for consolidation. Silt has been consolidated 20 ft without mud waves. The sand blanket becomes part of the fill necessary to replace the displaced water.

A multiple pile driver for installing sand piles more rapidly and economically might be advantageous.* Compaction takes place both while the casings are being driven and when additional sand is placed in the casings and tamped as they are withdrawn. Existing types of pile drivers, driving one pile at a time, are too slow for compacting large areas of fills to the usual 7- to 15-ft depths required for airport sites.

* "Soil Compaction for Airports," *Eng. News-Record,* vol. 136, No. 12, pp. 422–426, March 21, 1946.

Franki Process.[10x] This method of compacting sand is to drive a steel tube of approximately 19 in. diameter with a special pile driver and heavy ram having a long drop falling within the tube. The blows of the ram strike a plug formed of sand, gravel, and broken stone at the base of the tube, which is open at both ends. The plug expands under the blows of the ram and pulls the tube down by friction. After the desired depth has been reached, the plug is forced out by holding the tube while tamping, and sand and broken stone are fed into the tube and tamped while the tube is withdrawn gradually. This forms a dense pile of frictional material but with no binding content, which compacts the sand during placing. Soil compression is produced during ramming by vibration, but principally by sidewise displacement of the soil. Internal stress in the soil may reach the value of passive pressure, and may partly remain afterwards. For greatest effectiveness, piles should be driven first at the corners of a square, then at the center, and then at intermediate points on each side. To obtain best results, the water level should be as high as possible, to reduce apparent cohesion of the soil, even if artificial raising by a supplementary water supply is necessary. Since large lateral pressures are developed during driving, special attention must be paid to foundation walls or sheeting nearby. These piles were first proposed for this purpose by the Franki Compressed Pile Co., Ltd., in 1935.

Compaction by Explosives. This was demonstrated to increase by a large amount the degree of compaction in moderate to great depths of submerged fine and medium sands, including those having an appreciable silt content.[*,10ay,10az] The average degree of compaction at the Franklin Falls dam site in New Hampshire was increased from 26 to 58 per cent over a large area, with surface settlement of 2 ft and compaction distributed over 20-ft depth, using five coverages.

Size, depth, and lateral distribution of the explosive are based on the spheroidal shattering effect around each charge. The charge should fracture the ground but not form a crater. The charge forms a gas cavity in the saturated soil, and the shock and vibratory waves rearrange the grains more compactly, with the surplus pore water causing liquefaction. The gas is then squeezed out to the surface, with no cavities remaining. In a few minutes, boils and water geysers break out, as in quicksand, and last up to 30 min. Repetition further settles the mass, but in decreasing amounts.

Observations of the results of exploding charges in submerged strata provide a method of testing the degree of compaction of submerged cohesionless strata in advance of construction operations.

* A. K. B. Lyman, "Compaction of Soils by Explosives," *Civ. Eng.*, vol. 10, No. 4, pp. 205–207, April, 1940.

Vibroflotation. This is a method of compacting granular soils by a vibrator that sinks into the soil. It was invented by Steuerman and developed by the firm of Johann Keller in Germany and was used to compact sands there prior to and during World War II.

Originally, 16-in. cased wells were bored and a vibrator with nozzles lowered, after which sand was filled in and jetted as the device was raised. This scheme was too costly, and the method described below succeeded it.

Keller Vibratory Ram Pressure Process.[10x] This method of compacting by vibration sinks a vibrator consisting of a spindle with eccentric weights driven by a 40- to 50-hp motor. The amplitude is ¼ in. and the frequency 3,000 rpm. Water is supplied through a pipe of a 4 to 6 in. diameter, ending in nozzles at and directly above the base of the vibrator, at a rate of about 2½ gps in fine sand and 30 to 40 in coarse sand. The bottom nozzle is first opened, and the jetting effect is so great that the desired depth is usually reached in a few minutes without vibration. The lower jet is then closed, the jets above opened, and the vibrator started. Densification is indicated by increased power consumption. The vibrator is then raised slowly. Success is independent of water level.

FIG. 14.1. Vibroflotation tool. (*Courtesy of Merritt-Chapman & Scott Corp.*)

Vibroflotation Process.[10y,10ac,11aq] This is the name by which the above method, with improvements, has been known and promoted in the United States since World War II.* The vibrator is sunk to the desired depth of consolidation by vibration and injection of a stream of water from the bottom of the vibrator. The upward-directed pressure of this water largely eliminates friction between the granules, the soil becomes temporarily quick, and the vibrator sinks rapidly with little energy required. During lowering, a compaction of the surrounding material occurs and a crater several feet in diameter is formed.

* By the inventor, in conjunction with Parsons, Brinckerhoff, Hall, and Macdonald as consulting engineers, and Merritt-Chapman & Scott Corp. as contractors. This process is now used by the Vibroflotation Foundation Co.

When the vibrator reaches the predetermined depth, the water feed is changed to the top, so that the granules are acted upon by a stream directed downward. At the same time, sand is dumped into the crater to replace lost soil volume. After compaction at the lowest position, the device is retracted step by step, resulting in a cylindrical compacted column of 8 to 10 ft diameter.

Fine granular material, such as loose hydraulic fills, may be compacted. The method will cause settlement and is useful only where this is not harmful at the time.

Vibroflotation was used to underpin existing foundations for a hangar at Harmon Air Force Base in Newfoundland, using 650 penetrations to depths of 20 to 30 ft to consolidate 25,000 cu yd of loose sand fill.

Electroosmosis

If direct current passes between two electrodes in moist soil, water moves from the positive electrode (anode) to the negative electrode (cathode). Well points work in permeable soils, and may also be made to work in silts and clays by this means, using the well points as cathodes and driving small pipes for anodes.[10s,10t] It has been found that (*a*) when large currents are employed, considerable heating occurs at the anode, which assists the water movement; (*b*) there is a linear relation between amount of water expelled and clay content; (*c*) the weight of water expelled is proportional to coulombs of electricity; (*d*) the quantity of water expelled is greatest for sandy soils, and least for heavy clays; and (*e*) drying out occurs mainly at and near the anodes. The moisture content at the cathode is not changed much, if the expelled water is drained.

The method was used in Germany and Norway during World War II with success, to prevent slips in cuttings and excavation.[10u] In general, well points consisted of 10-in. auger holes, each containing a 4-in. soil pipe surrounded by a graded sand-gravel filter, with a central pumping pipe, spaced about 30 ft apart in each row, using two or three rows. Anodes were 1- or 1½ in. gas pipes driven in an intermediate row. Voltages used were from 90 in ordinary silts to 30 in salty soil. Amperages required for each point ran from 30 to 15.

Water yields per well point were increased in one case from 0.5 to 80 cu yd per 24-hr period, and soil moisture content reduced from a range of 20 to 24 per cent to a range of 14 to 17 per cent, so that material which was unstable in banks stood on a 1:1 slope. In another case, yields increased from a range of ¼ to 11 gal per hr to a range of 2.5 to 105 gal per hr, and slippage of sheet piling was stopped.

Silting of cathodes limited the life of a well point to about 3 weeks. Corrosion of anodes in salty ground required some replacement.

A deep excavation for an extension to the Joppa power station was made, with a high bank of soft wet clay which started to move in. Electroosmosis reduced the water content of the clay from 25.8 to 19.2 per cent. Unconfined compression tests increased from 0.19 to 1.87 tons per sq ft, and vane tests showed an increase in shear resistance from 21 to over 100 psi. Anodes and cathodes up to 50 ft long alternated on 5-ft centers each way. Anodes were 1½ in. standard black pipes with 1-in. reinforcing rods driven down inside them. Cathodes were 1½-in. well points. More than 300 electrodes were used. Applied current was 90 volts dc.[10al]

A discussion of the action of electroosmosis is contained in references 10*ad*, 10*aj*, 10*al*, 10*am*, 10*an*, 10*ao*, 10*av*, 10*aw*, and 10*ax*. Installations have been made by well-pointing contractors.

Tests[10aw] indicate that the dehydration in clay produced by the current is not permanent but is reversible and that the mechanical properties of the dehydrated clay run the risk of being inferior to those of natural clay. The treated clay is more compressible and is subject to greater and more rapid swelling under a reduced load. Use of alternating current results in hydration, with the clay more compressible and subject to greater swelling.

This process might be used for cuttings, tunnels, shafts in bad ground, runway stabilization, and ground hardener under tracks and roads.[10s] It could be used to increase pile capacities if the pile is sheathed in metal at the foot, although permanency is not known. The method might be used to reduce friction between a pile or caisson and the soil, by using these as cathodes, thus creating a lubricating action to aid in sinking them.

Presumably the application of this method is limited to pore diameters of approximately 0.1 to 1.0 mm in diameter. The process can be used to accelerate settlements in fine-grained soils. The effect of preconsolidation can be obtained. Loading speeds up the action. Prospects of developments in these processes are promising.[10ah]

Electrolysis

Electrolysis is a development for hardening soft and plastic clay soils, which shows promise and may become commercially practicable.[10c,10p] Investigations and large-scale experiments on pile foundations were carried out in Germany for a number of years before World War II with very satisfactory technical results.

The method is based on precipitation of insoluble metal salts by electrolysis. Aluminum anodes and copper cathodes are used, with direct current passing until the soil is hardened. If aluminum is used at both electrodes, the soil is consolidated at both. Moisture content of the clay is considerably reduced,* and remains unchanged, even under water.[10p] Compressibility is reduced to a small percentage of the original. The angle of internal friction is increased* up to 30 deg, and more in heavy clays.[10p] Precipitation of insoluble aluminum salts around the piles increases the bearing value of friction piles to possibly double the untreated value. The consolidation is permanent and irreversible.[10p] The process might be used to induce rapid consolidation before erection of a structure.

This treatment is fairly expensive, reaching $10 per cubic yard in Germany at the time reported. Consolidation of solid layers of clay requires from 22 to 220 kwhr per cu yd plus costs of aluminum and labor, which have been estimated at from 65 to 6.5 per cent of the figures for the cost of electrical energy given above. Pile foundations were found to require from 1 to 2 kwhr per lin ft on pile, while doubling the bearing capacity of the pile.

A clay with 80 per cent water content was hardened so that a ¼-in. square rod failed to penetrate under 20 lb load. Immersion did not soften the clay sample. The method was used for bridge piles in Königsberg, Germany, where the piles passed through a 33-ft swamp layer into a very fine powdery sand having 1 per cent clay. Some piles were covered with a 30-gage aluminum sheet for a height of 25 ft and subjected to the current, while others were not. Piles through which the current passed showed one-quarter of the settlement of the other piles.

Freezing

Freezing of soils is sometimes used to stabilize soil temporarily or during the existence of an adjacent excavation. Freezing may prevent settlements, but settlements of several inches have been observed after the treatment stopped, perhaps from thawing of ice lenses formed around the tubes. Possibly the use of lower temperatures and quicker freezing might avoid these, if such settlements are considered serious.[10v]

The process can be used in any water-bearing soil of any permeability or grain size. It may be used with silts under high hydrostatic pressure.[10v] The general method consists of sinking 4- to 6-in.-diameter pipes, with closed bottoms, spaced about 3 ft apart, inside of which are 2-in. open-end pipes. Brine at about —4°F is circulated down one pipe and up the other. This permits a solid wall to be frozen.

The length of time of stabilization depends upon the continuance of the freezing operation and might presumably be of indefinite duration provided it were economical.

* L. Casagrande, *Die Bautechnik,* vol. 15, 1937, and vol. 17, 1939, and W. Bernatzik, *Final Report,* Second Congress of Int. Assoc. for Bridge and Struct. Engrs., Berlin, 1936.

Double-walled circulation pipes 2 and 4 in. in diameter were driven 60 ft into the ground, and the earth temperature was lowered to —20°C for 8 months, to stabilize a lens of fine wet sand under a skyscraper carried on piles, and to prevent settlement caused by sand flowing into the excavation for an adjacent structure, after chemical- and cement-grouting attempts had failed.[10f,10g] (See Chap. 16, *Case* 20.)

Freezing was used in Russia to form a permanent barrier against sliding at a large hydroelectric development. It was felt that there would be sufficient time to repair possible failures of the refrigeration equipment without danger, since it was estimated that it would take 3 months for the barrier to thaw. Thin frozen walls would thaw much more rapidly, possibly in 2 days.

The sizes of compressor, condenser, cooler, and pumps can be determined by calculating the number of Btu required to freeze the ground solid,[10ab] as follows: (*a*) Number of Btu to cool ground from average temperature to 32°F is weight of water in voids to be frozen times temperature drop, plus weight of soil to be frozen times temperature drop times specific heat of soil (approximately 0.2). (*b*) Number of Btu for ice formation is weight of water to be frozen times 144 (latent heat per pound of ice). (*c*) Number of Btu for further cooling of above volume of ground from 32°F to average between 0°F at coldest point and 32°F is the weight of ice times 0.5 (specific heat of ice) times drop from 32°F, plus weight of ground times drop from 32°F times specific heat of soil. (*d*) Number of Btu for cooling ground from boundary of frozen soil for an assumed distance of several feet is the additional weight of water times half the difference between the original temperature and 32°F, plus the additional weight of soil times half the difference between the original temperature and 32°F times the specific heat of soil. The total number of Btu required is the sum of the above plus the Btu for atmospheric and insulated piping losses.

Freezing has been found to be a structurally safe method.

The effects of the formation of ice on various types of soils should be considered, standard works on soil mechanics should be consulted, and the services of a soil-mechanics expert should be retained. The effects of soil expansion, disturbance of soil structure, and heaving may or may not be detrimental to existing structures, depending upon the soils, moisture conditions, type of foundation, and type of structure.

The cost of this method is high, because of the plant required, and because of the time required for freezing, which may be several months. Freezing may be done at great depths, and in such cases the method may become economical.

Dry ice has been used successfully for forming ice cofferdams below the frozen surfaces of rivers.[10ag] Dry ice is solid carbon dioxide, having a temperature of about −110°F. It comes in 9- or 10-in. cubes, weighing 20 to 25 lb, that are placed in the trench 6 in. apart and covered with canvas and straw. The dry ice vaporizes and absorbs from adjacent materials the heat necessary to raise its temperature to that of the surroundings, and also an amount of heat equal to the latent heats of both fusion and evaporation. One pound of dry ice will absorb about 260 Btu in reaching 32°F. Care must be taken to prevent asphyxiation of personnel. The process can be used on small jobs and requires only hand tools.

Poetsch Process.[10v] This process uses ammonia as a refrigerant and circulates chilled calcium chloride brine. A compressive strength of 150 psi is obtained.

Dehottay Process.[10x] This process uses single tubes as refrigerant tubes in which direct vaporization of carbon dioxide occurs. Although freezing has often been used in Europe for construction of shafts in unstable soils, it has seldom been employed in building foundations, since cheaper methods were available for this purpose. In Germany during World War II, efforts were made to simplify freezing processes so that they could compete with steel sheet pile walls and save steel. Previous methods required one circuit containing a freezing material, such as carbon dioxide, and another circuit containing a saline solution. The Dehottay process was successfully developed, and would be useful especially where rammed foundation solidification methods are not possible owing to lack of headroom or need for avoidance of vibration.

Thermal Treatment. Two methods of permanent stabilization of loess and similar soils have been used in the Soviet Union. Loess is porous and semipermeable; in dry state it has very high bearing capacity, but becomes soft and highly compressible if moist.[10ak] These methods have also been used for underpinning. The cost is said to have been less than for chemical grouting.

The *thermal method* blows hot air at 1100 to 1470°F under pressure into sealed boreholes, causing permanent soil changes giving complete resistance to liquefaction; absence of differential compressibility; increase of several times in cohesion, compression, and shear strengths; and cessation of tendency to settle when the ground is wet.

The *thermal-chemical method* uses solid, liquid, or gaseous fuels in the boreholes, where combustion takes place. Temperatures range from 570 to 2000°F at a pressure of 0.5 atmosphere. Treatment in one 4- to 8-in.-diameter borehole for 5 to 10 days resulted in a consolidated zone 33 ft deep and 5 to 8 ft in diameter.

CHAPTER 15

PILE LOAD TESTS

Purpose of Pile Load Tests

What is the purpose of making a pile load test? In answering this question one must bear in mind the fact that a satisfactory pile foundation must have a sufficient ultimate bearing capacity (*i.e.*, the soil shear strength must be adequate) and that settlement under working load must be tolerable. A pile load test can give information only about the ultimate bearing capacity.

FIG. 15.1. Pile load test by cantilevered load. Applying 90-ton vertical reaction on test pile. Variable reaction obtained by reloading after moving carriage. Cyclic loading obtained by allowing pile to rebound to rest after each unloading.

In cohesive soils, settlement is a function of the length of time of load application. Since the time in a load test is generally so short relative to the time required to approach full settlement, the load test tells nothing about the settlement behavior of a single pile, without even considering the difference in settlement behavior of a group of piles as compared with a single pile.

In cohesionless soils, the load test will show the settlement behavior of a single pile but tells little of the settlement behavior of the group, although the settlement of the group will be more than that of the single pile.

Basically, therefore, a pile load test can determine only the ultimate bearing capacity and not the settlement characteristics of the pile group. Settlement computations are a separate matter and the subject of soil-mechanics calculations. It is impossible to evaluate tests unless adequate boring records present a complete picture of the underground at or close to the test pile.

FIG. 15.2. Pivot and uplift points of cantilevered load test on pile (prototype of test in previous photograph). Wheeled trolley suspending carriage carrying load of reinforcing bars on track visible on top of beam, giving varying load by changing location. Jarrah pile at Adelaide, South Australia. (*Courtesy of Hurren, Langman & James, Consulting Engineers, Adelaide, South Australia.*)

Methods of Making Pile Load Tests[9]

There have been innumerable arrangements of apparatus developed for making loading or pulling tests on piles. Only a few can be illustrated here, but these will serve to indicate the permissible flexibility in design, and the ingenuity which may be applied to make tests with greatest economy of time and use of available equipment.

Test loads may be applied (*a*) by direct load from a platform on which heavy weights are placed (Figs. 15.3*a* and 15.3*c*); (*b*) by direct load from a platform on which water tanks are placed, to be filled as desired; (*c*) by jacking against a loaded platform (Fig. 15.3*d*); (*d*) by jacking against an existing structure; (*e*) by jacking against previously driven piles (Figs. 15.3*b*, 15.3g, and 15.3*e*); and (*f*) by application of load by means of a cantilever arm (Figs. 15.3*f*, 15.3*g*, 15.3*h*, and 15.3*i*), thus reducing the amount of load needed.

Direct load may consist of pig iron, earth, sand bags, precast concrete blocks, water tanks, etc. Cement bags have been used, but one must be quite sure of the climate to risk this. The procurement of sufficient fixed load, such as pig iron, rails, etc., is sometimes difficult, and the removal of such full load for repeated loadings and releases on the same pile, usually desirable, is nearly impracticable. Water tanks may be arranged for draining and refilling fairly readily. There is a danger from improperly piled fixed loads. Corner supports should be placed close under loaded platforms to catch the load should tilting occur because of shifting of the load or yielding of the soil. When loading, wedges or jacks should be used. These should not be removed until the load has been placed and balanced, after which they should be lowered slightly. If considerable equal settlement under direct load is anticipated, the safety supports can be changed in elevation to stay quite close under the platform at all times. Jacking against fixed load on platforms is preferable to resting the load on the pile. The platforms always remain resting on cribbing, and less danger occurs.

Jacking is usually done with hydraulic jacks or jacks employing a gas under pressure. Another advantage of jacking is that the loads can be applied and released quickly and at will, permitting quick determination of the net settlement of the pile, or movement in the soil, after rebound has occurred.

The *ASTM Method of Test for Load-settlement Relationship for Individual Piles under Vertical Axial Load* (Serial Designation: D1143) appears in Appendix VI.

Pulling Tests

Pulling tests also may be made by a wide variety of arrangements. A typical platform loading, using another pile as a fulcrum, is shown in Fig. 15.7*d*.[5a]

A jacking load is shown in Fig. 15.7*a*. A crane pull on one end of a cantilever is shown in Fig. 15.7*b*, with a means of observing the reaction, from which the pull on the test pile may be computed by statics. A light load on a long cantilever, using another pile as a fulcrum reaction, is shown in Fig. 15.7*c*.

Plastic and Elastic Deformations

Movement at the pile head is caused by elastic deformation of pile and soil and plastic deformation of soil. The last causes undue settlement of structures and must be guarded against. This is the value which is the significant one to be obtained from load tests, and not primarily the total downward movement of the head of the pile under the test load. In Fig. 15.8 is shown the split of total movement into elastic and plastic. The curve of plastic deformation is the most significant, and this is the one from which the working load and factor of safety should be set.

Cyclic loading provides a means of determining whether the test loading is being carried by the stratum selected for this purpose, or if the

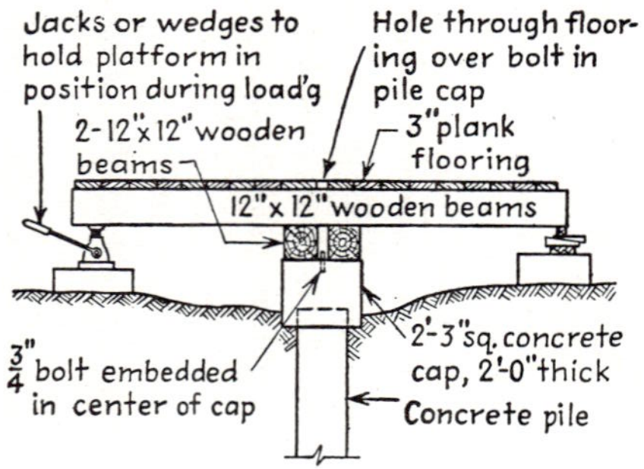

(a) Platform load test

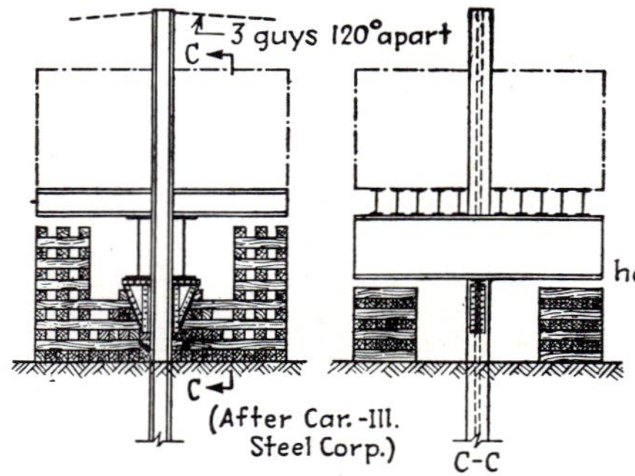

(c) Platform load test

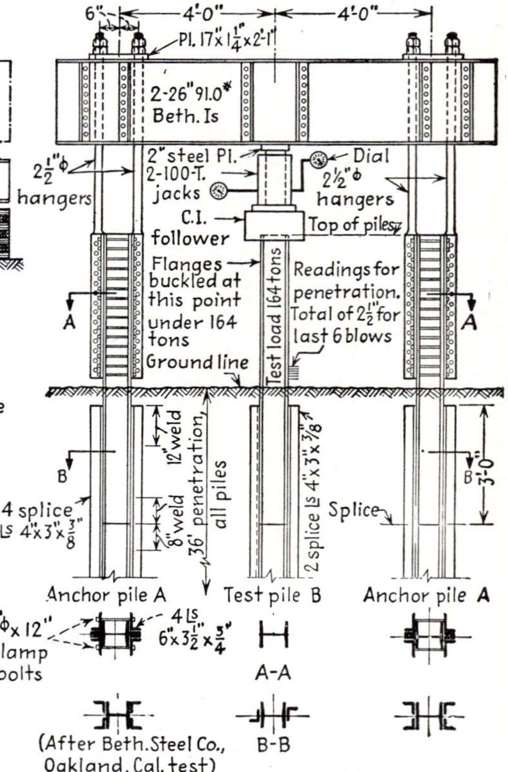

(e) Jacking load test against piles

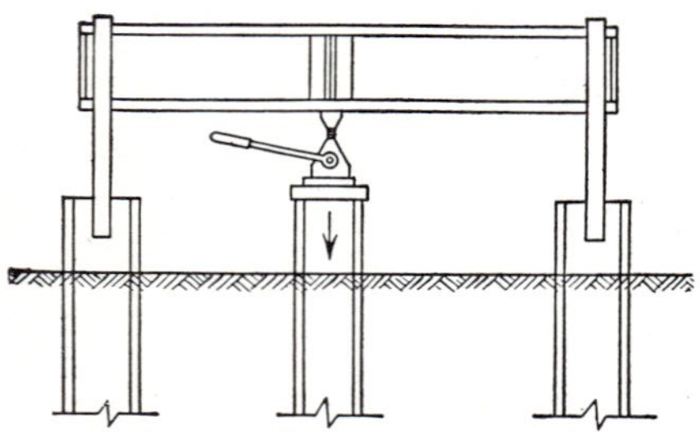
(b) Jacking load test against piles

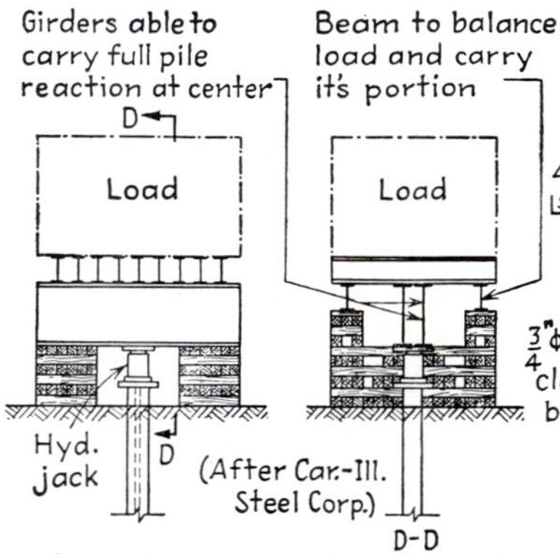

(d) Jacking load test against platform

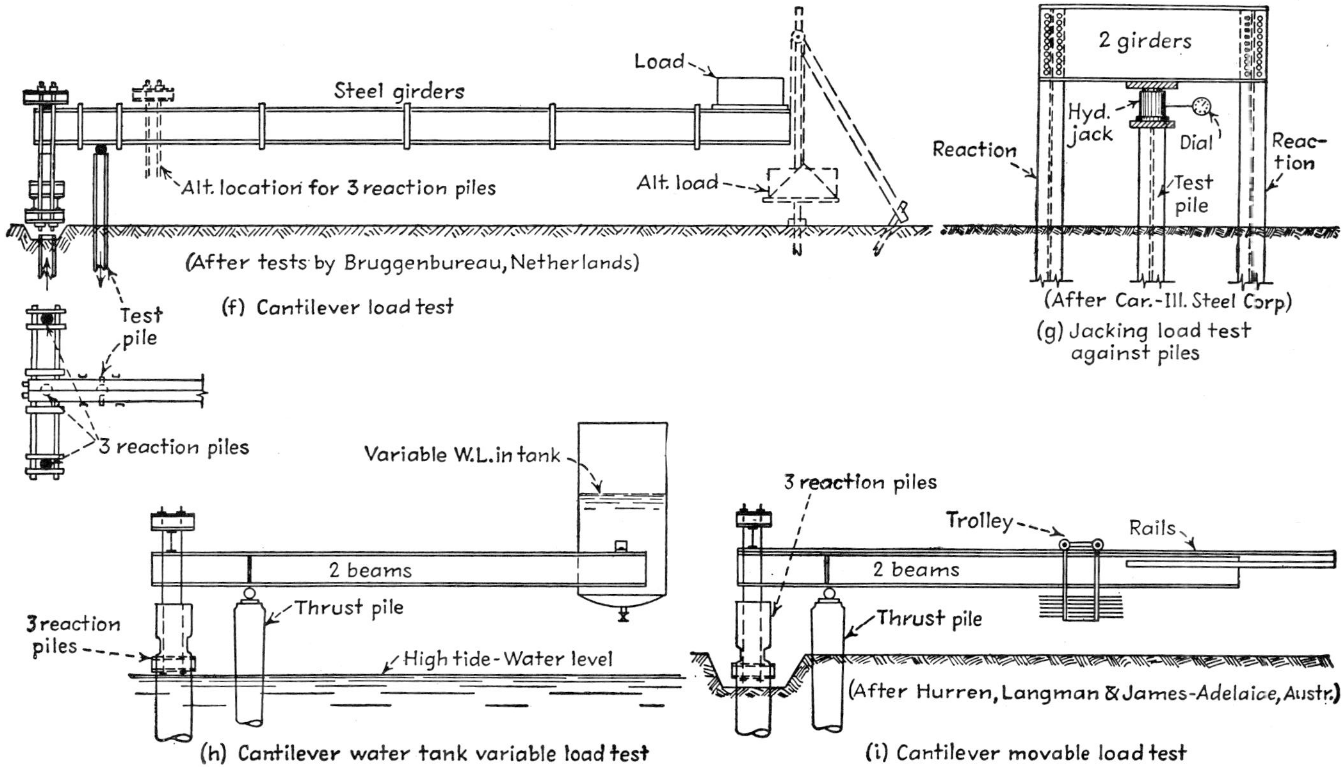

Fig. 15.3. Typical test-loading arrangements.

Fig. 15.4. Pile load test on wood pile, jacking against dead load of pig iron supported on pile butts driven at same time as test pile to avoid cribbing on too soft soil. For Geneva Iron Works, Columbia Steel Co., Provo, Utah. *Consulting Engineer:* R. V. Labarre. (*Courtesy of W. F. Swiger.*)

Fig. 15.5. Testing 8-in. 32-lb H pile with hydraulic jack. Note dial at left. Oakland, Calif. (*Courtesy of Bethlehem Steel Co.*)

FIG. 15.6. Cantilever load test on 10¾- by ¼-in. pipe pile 128 ft long, loaded to 157.5 tons using 100-ton hydraulic jack. (Frame at right not part of test; used in future test to right.) (*Courtesy of Stone & Webster Engineering Corp.*)

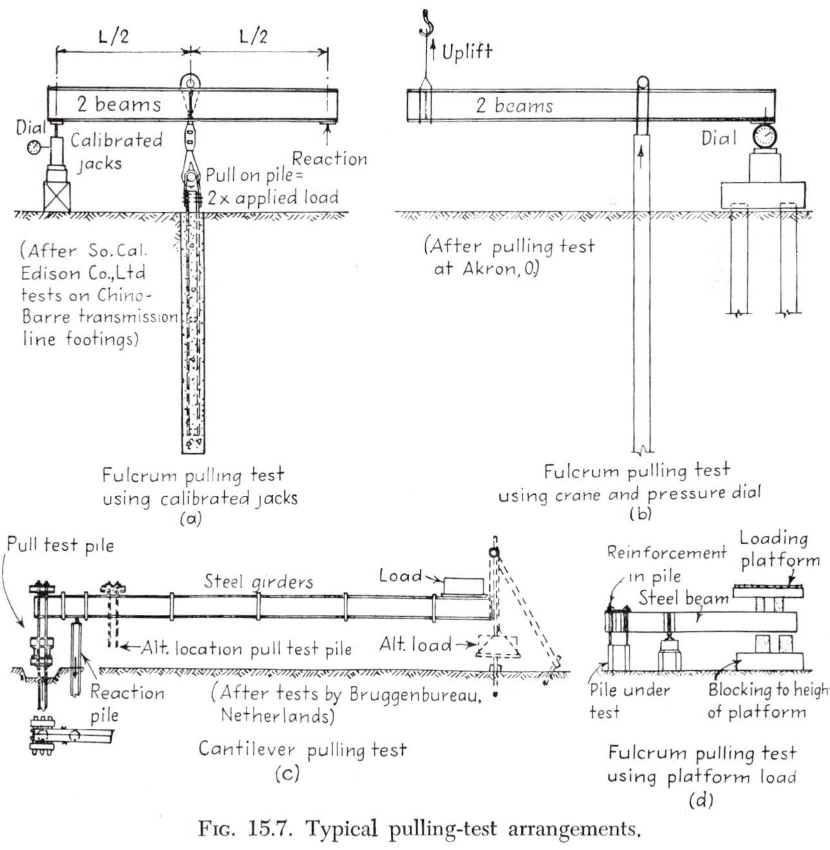

FIG. 15.7. Typical pulling-test arrangements.

center of resistance is at a higher point. If the distribution of load to the various strata is not determined by any of the other methods discussed in this chapter, this procedure will be of assistance. By removing the load from the pile several times during the process of adding loads by increments, and observing and plotting the rebounds as shown in Fig. 15.8, the curve of plastic deformation can be obtained. The

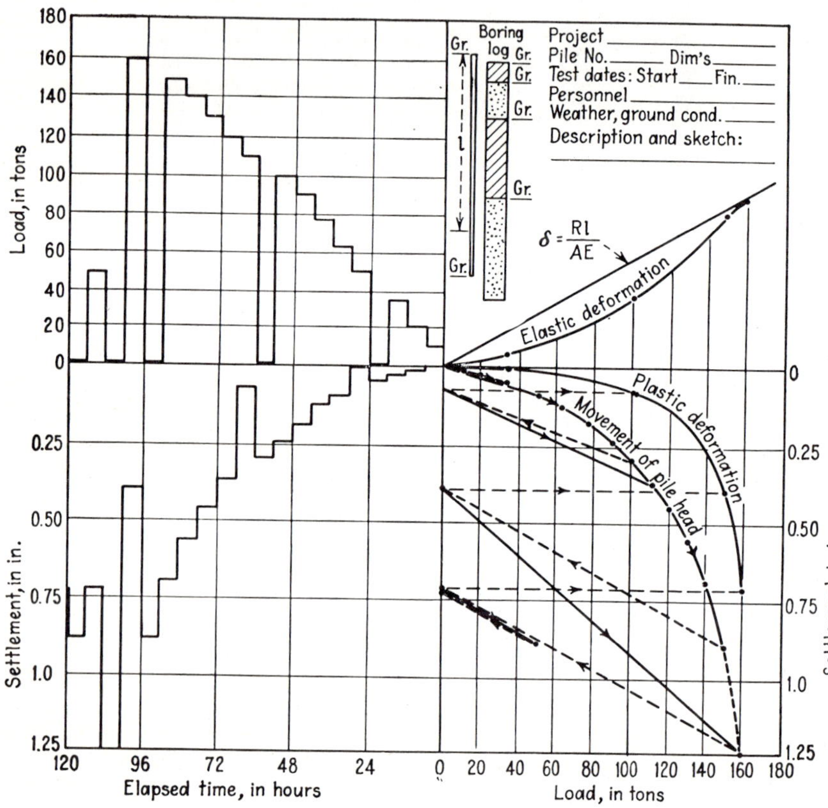

FIG. 15.8. Test-loading diagram showing relationship between loading, settlement, and time.

points where the rebound lines intersect the settlement axis are projected horizontally to points under the respective loads removed, and these points are then connected by a smooth curve. This is the plastic-deformation curve which should be studied for determination of ultimate pile capacity. By subtracting the values of this curve from those of the curve of total settlement, the elastic-deformation curve may be plotted and compared with the theoretical elastic-deformation line of the pile. By observing the differences between the actual elastic-deformation curve and the theoretical elastic-deformation line, the gradual lowering of the center of resistance with additional load may be observed and determination made as to whether or not the test load is actually reaching the stratum which it is desired to test.

Distribution of Load to Soil

Since piles should be designed to perform a definite function and not be driven by guesswork, it follows that in order to perform that function, which is to transfer the load to certain strata only, the distribution of load from the pile to the various strata should be known or closely approximated.

By comparing the curve of elastic deformation described above with the theoretical elastic-deformation line if the loads were carried at a center of resistance located at the desired penetration, it may be observed whether the loads are reaching this point or are being supported at some points resulting in a higher center of resistance. If the actual elastic curve has smaller ordinates than the theoretical at any points, this indicates that the upper undesirable strata, not selected for permanent load-carrying purposes, are supporting some of the load, at least temporarily. With increasing test load, the ordinates of the actual elastic curve should eventually increase to coincide with the theoretical elastic line.

The theoretical elastic deformation in the case of an end-bearing pile unrestrained by friction can be computed from the formula $\delta = Rl/AE$ and will be a straight line, and for other conditions, the theoretical elastic deformation of the pile can be computed by assuming the location of the center of resistance and considering l as the distance down to this point. By substituting for δ in the formula the elastic deformations read from the curve obtained from the cyclic loading tests, values of l may be obtained which show the location of the center of resistance to that load.

By driving test piles to different depths in a load-carrying stratum underlying poor materials, or by redriving the original test pile deeper and retesting and then considering the difference in test values as the carrying capacity of the difference in lengths, a square-foot unit friction value is obtained which may then be applied to the surface area of the entire length of embedment in the good stratum. The effects of the undesirable upper strata, and of tip bearing, being common to both test piles, are canceled. By this principle, the friction values of any of the upper crusts or strata also may be determined, for use in considering load-test or pile-driving results.

If it is possible to drive and blow out a pipe through the upper strata, friction on the lower stratum only may be obtained. Sometimes a closing shoe on the pipe can be used, and driven out by the pile when it goes through the casing.

The following method of determining the load distribution from the pile to the soil requires no special equipment other than dial gages.[23] A closed-end tube extends nearly the full length of the pile. It is provided with anchors to secure it into the concrete of a concrete pile or it may be welded to an H pile. Shelves are provided at several predetermined points in the inside of the pipe, on which will rest the bottoms of steel rods. The bottom of a rod moves with the point in the pile at the level of the shelf on which it rests, and the rod is free to move in the pipe. A dial gage at the top measures the movement of each rod. The gages are supported independently from the piles, and should read in thousandths of an inch. The load remaining in the pile at any point may then be determined, assuming proper values of E, by use of the formula $R = \delta AE/l$.

Carlson strain meters below ground level and Whittemore strain gages above were used successfully at Sepulveda Dam to obtain axial-stress readings at different elevations.*

The relative motion of the soil and piles may be determined by measuring, in addition, the soil movements. This can be done by driving ¾-in. pipes in the soil to the desired elevations.[23] A ¼-in. rod within the pipe rests in the bottom of a metal cup, loosely fitted over the bottom of the pipe. Heavy oil is poured into the pipe to prevent the entrance of soil and water and to keep the rod well lubricated. The pipe is then withdrawn 6 in., leaving the cup embedded in the soil. The rod moves freely within the pipe, and the vertical movement of the soil in which the cup is embedded is measured by a dial gage in contact with the top end of the rod.

Rules for Determination of Working Load from Tests

The ultimate pile capacity is defined in most soil-mechanics literature as the load beyond which the pile will begin to break into the ground, or expressed mathematically, when $\Delta s/\Delta p$ approaches ∞, that is, when the tangent to the load-settlement curve becomes vertical, with the settlements plotted as ordinates. In pile-engineering practice, however, it is customary to select a point beyond which the rate of change increases markedly or the settlement increment excessively in comparison with the load increment, to denote this as failure, and then apply a factor of safety. Although such a method involves the consideration of magnitude of settlement, separate settlement computations under the pile,

* *Bearing Pile Investigations for Sepulveda Dam,* U.S. Engineer Office, Los Angeles, Calif., April, 1940

footing group, and entire structure should be made in accordance with established theory, as shown in standard textbooks on soil mechanics.

Pile-load-test increments should be allowed to come to rest for 24 or 48 hr. If additional loads are applied before coming to rest, this results in a load-settlement graph which distorts the shape of the curve and may cause a larger safe load to be selected. This is particularly important because the false distortion of the curves may fall in the working load ranges. Money is ill spent on tests unless full obtainable information is secured.

Many arbitrary or empirical rules have been used or are contained in codes to serve as criteria for determining the allowable working load from load-test results. Some take into account plastic and elastic deformations; others do not. In some test graphs, the point of failure on either the gross settlement or plastic settlement is so evident that rules are unnecessary. In other cases, the curve changes slope so gradually that picking the failure point is difficult and the application of one or all of these rules, representing the judgment and experience of many engineers, may be of assistance. A few such rules are as follows:

1. The test load shall be twice the contemplated design load and shall be maintained constant for at least 24 hr and until settlement or rebound does not exceed 0.22 in. in 24 hr. The design load shall not exceed one-half the maximum applied load provided the load-settlement curve shows no signs of failure and the permanent settlement of the top of the pile, after completion of the test, does not exceed ½ in. (Boston Building Code).
2. Observe the point at which, no settlement having occurred for 24 hr, the *total* settlement including elastic deformation of the pile is not over 0.01 in. per ton of test load, and divide by a factor of safety of 2 (Department of Public Works, State of California).
3. The safe allowable load shall be considered as 50 per cent of that load which, after a continuous application for 48 hr, produces a permanent settlement not greater than ¼ in. measured at the top of the pile. This maximum settlement shall not be increased by continuous application of the test load for 60 hr or longer (AASHO).
4. Observe the point at which the plastic curve breaks sharply, and divide by a factor of safety of 1.5.
5. Tests shall be made with 200 per cent of the proposed load, and considered unsatisfactory if, after standing 24 hr, the total *net* settlement after rebound is more than 0.01 in. per ton of total test load (building laws of the City of New York).
6. Observe the point at which the *gross* settlement begins to exceed 0.03 in. per ton of additional load, and divide by a factor of safety of 2

for static loads or 3 for vibratory loads (W. H. Rabe, Design Engineer, Bureau of Bridges, State of Ohio).

7. Draw tangent lines to the general slopes of the upper and lower portions of the curve, observe the load at their intersection, and divide by a factor of safety of 1.5 or 2.

8. Observe the point at which the slope of the curve of *gross* settlement is four times the slope of the graph of *elastic* deformation of the pile, and divide by a suitable factor of safety.

9. The allowable axial load on an isolated pile shall not exceed: (*a*) 50 per cent of the yield point under test load. The yield point shall be defined as that point at which an increase of load produces a disproportionate increase in settlement; or (*b*) one-half of the load which causes a net settlement, after deducting rebound, of 0.01 in. per ton of test load, which has been applied for a period of at least 24 hr; or (*c*) one-half of that load under which, during a 40-hr period of continuous load application, no additional settlement takes place (optional rules of International Conference of Building Officials Uniform Building Code).

10. Take two-thirds of the maximum test load in a case where settlement is not excessive and where load and settlement were proportionate and the curve remained a straight line. Where the test load was carried to failure, take two-thirds of the greatest load at which settlement was not excessive and at which loads and settlements were proportionate (United States Steel Co.).

11. With several consistent tests over the area of the structure, take from one-half to two-thirds of the failure load, considered as somewhere in the vicinity of the break in the curve showing increased settlement per unit of load added (Bethlehem Steel Co.).

12. The safe allowable load shall be considered as 50 per cent of that load which, after a 48-hr application, causes a permanent settlement of not more than ¼ in. (New York State Department of Public Works).

13. One-half of the test load shall be allowed for the carrying load, if the test shows no settlement for 24 hr and the total settlement does not exceed 0.01 in. multiplied by the test load in tons (Chicago Building Code).

14. Observe the load at which is produced an increase in settlement disproportionate to the increase in load, and apply a factor of safety of 2 (Los Angeles Building Code).

15. Observe the load carried without exceeding a total permanent settlement of ¼ in. in 48 hr and divide by a factor of safety of 2 (Louisiana Department of Highways).

16. For important permanent structures, take the safe load on well-driven timber and concrete piles, with a final set of, say, ten blows to 1 in., at one-half to two-thirds of the test load which produces a final set-

tlement gradually of ½ in. after a period of 10 days' rest. For well-placed undriven concrete piles, tested to twice their estimated bearing capacity, the safe bearing load has been taken in practice at one-half the test load which gives a settlement of ⅜ in. after a period of rest of 10 days (W. Simpson, "Foundations," Constable & Co., Ltd., London, 1928).

17. Observe the point at which the *gross* settlement begins to exceed 0.05 in. per ton of additional load, or at which the *plastic* settlement begins to exceed 0.03 in. per ton of additional load, and divide by a factor of safety of 2 for static loads or 3 for vibratory loads (Dr. R. L. Nordlund, Raymond Concrete Pile Company).

Various impressions of the steepness of the load-settlement curves may be obtained by varying the scale of the graph; therefore a finite limit of the change-of-load to change-of-settlement ratio is desirable. Rules 6 and 17 appear to be the most rational and justifiable, although rule 6 may be too conservative.

CHAPTER 16

FAILURES OF PILE FOUNDATIONS

The remark is sometimes made that no one ever heard of a pile foundation failing. This impression is far from actuality. Scores of pile-foundation failures have occurred in every country in the world, and there are many published accounts of these failures. Some are abstracted in this chapter to show that expensive pile failures do occur, and to illustrate typical causes of failures, means of prevention, and possible remedies.

Causes and Prevention of Failures

A study of the causes of pile-foundation failures will generally indicate the means of prevention that should be taken. Failures are due to many causes, or combinations of causes, the most common of which are as follows:

Lack of adequate borings causes many failures. Borings are always necessary. In some cases, it is sufficient merely to determine the elevation of rock, whereas in others, the character of the overburden may be sufficiently judged from wash-water samples of wash borings. Sometimes dry samples from wash borings will provide all necessary information, but in other cases, the exact character of the soil becomes of vital interest and undisturbed samples are necessary.

Inaccurate classification of soils is fatal on many projects, often as the result of failure to employ a soil-mechanics expert to select the proper type of borings and to supervise making them.

Use of a dynamic driving formula for bearing resistance in predominantly cohesive soils results in many settlement failures. Such soils are not suitable for the use of a dynamic formula. Many such failures may be avoided by a check using the static formula for friction and end bearing.

Use of inadequate dynamic formulas often gives false results.

Misinterpretation of test-load data often occurs, in consequence of assuming (*a*) that the load test on a single pile gives results applicable to a group or a building; (*b*) that long-term settlements can be predicted from a short test loading; and (*c*) that the strata being tested are those which will finally have to carry the load.

Soft underlying strata below the pile tips are frequent causes of settlement.

Damage to an uncased pile in dense ground may occur, and if suspected, a pile may be excavated, or casings which can be inspected after driving adjacent piles may be used either for all piles or for a few as a test.

Movement of earth into an open pile hole in soft ground may be prevented by the use of cased or precast piles.

Collapse of thin shells of cased piles may be prevented by more careful driving or by use of a lighter hammer and jetting. In an instance where pressures developed by fine-grained water-bearing sand, estimated at not less than 4,000 psi, caused thin shells to collapse when withdrawing the mandrel, the use of heavier gage metal and welding circular beads around the corrugations in the lower 8 ft reinforced the boots so that they did not collapse.[94]

Settlement due to heaved piles which fall back into their original positions under load may be avoided by checking for heaving and redriving when it has occurred.

Overloading due to added weight from settling fill may cause failures. Such fill is either that through which the piles are driven or further fill that may be anticipated. Such overloading may be taken into account in designing the piles.

Compensating excavation for a basement occasionally can be made to balance added building load and avoid the load of a stratum over a soft material through which the piles must penetrate.

Buckling of piles may occur as a result of inadequate lateral support, removal of side support, increased load, or overdriving.

Breakage of wood piles may readily occur as a result of overdriving, particularly with heavy hammers and small tips.

Lateral forces, either static, intermittent, or vibrating, require adequate provision in the design.

Lack of batter piles may permit vibration in structures, caused by moving machinery. Grouting or chemical stabilization of porous soil to stabilize a mass around loose or vibrating piles sometimes may be used.

Flowing of a stratum carrying pile load, caused by adjacent excavations, may be prevented if the exact character of the soil strata is investigated prior to construction operations.

Sliding of piles owing to sloughing banks may sometimes be prevented by stopping dredging, or by washing down silt deposits.

Wrong choice of pile type is responsible not only for some physical failures but for uneconomic installations.

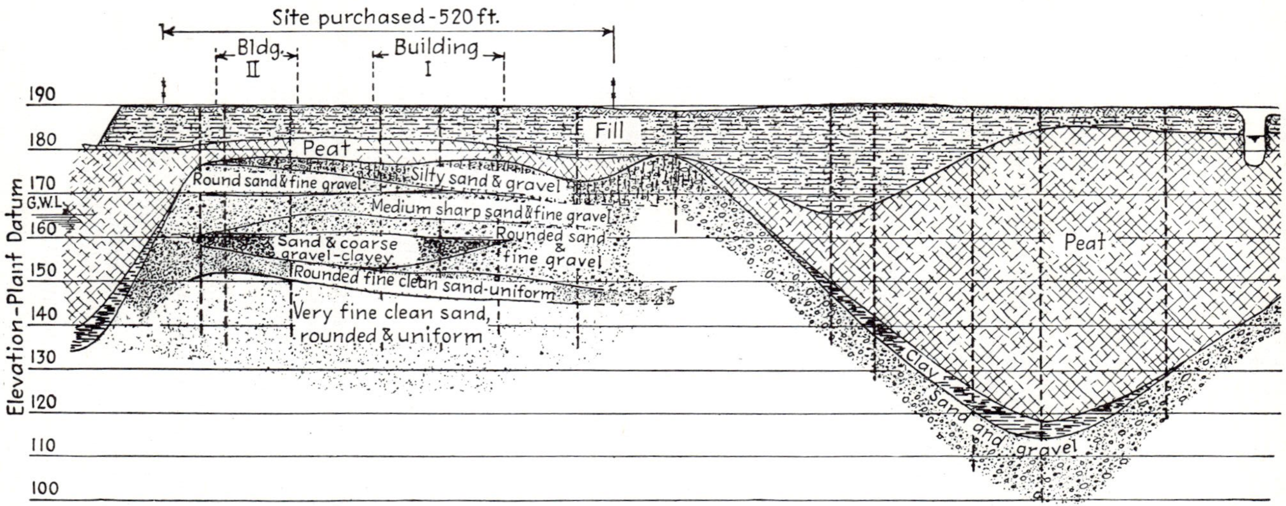

FIG. 16.1. Subsurface section through site in Ohio purchased before making borings, showing fortunate location of buildings over firm underground where short piles sufficed. Deep peat pockets bordered site.

Tension failures may be prevented in concrete piles by the use of reinforcement and also by prestressing.

Incomplete filling of cast-in-place concrete piles can be detected by proper inspection.

Piles bowed or out of plumb beyond acceptable tolerances should be rejected.

Decay in wood piles may be prevented by keeping the cutoff below the lowest possible ground water, by maintaining an artificial ground-water level, by preservative treatments, or by the use of a composite section.

FIG. 16.2. Group of overdriven and decayed piles exposed in connection with underpinning slab and turbine support for Portland Electric Power Co., Portland, Ore. (*Courtesy of Raymond Concrete Pile Co.*)

Insect and marine-borer attack in wood piles may often be prevented by preservative treatments or encasements.

Disintegration of concrete piles may be greatly reduced by proper attention to the mix and location of reinforcement; also possibly by the use of armor or asphalt treatment, and also by using prestress.

Abrasive action may require the use of protective coverings or guards.

Improper driving and testing methods can result in falsely high resistances being indicated, and in inadequate lengths of piles, with resulting large settlements. Some of these practices have been followed through lack of understanding of their effects or of the seriousness of the results, while a few, on rare occasions, have been used unscrupulously. Some of these circumstances are not widely known, possibly because of the lack of contact which sometimes exists between office engineers, who may not be informed about latest pile-driving theories, and field engineers and pile-driving foremen who may not always be aware of the serious effects which these conditions may have upon the desired results. If the engineer is aware of these conditions, he may guard against them in specifications or otherwise. Some of these procedures are given.

Use of light hammer for the particular work to be done may result in inability to reach the necessary depths or in stopping driving too soon owing to the appearance of small sets. Since a considerable proportion of the energy of even a proper-sized hammer is lost before it can be effective for actual work on the soil, if a hammer in which a very small percentage of energy is net is used, it becomes very sensitive, and small

FIG. 16.3. Safety of a heavy structure depended upon these piles. (*Courtesy of Dr. William F. Clapp.*)

obstructions or tightening up of the soil will appear as unduly high resistances.

Cushioning material is thrown into the head of the driving block in some types of equipment, and when such blocks of soft wood are added, they temporarily absorb a considerable portion of the net energy available for driving. Set readings should not be taken just at this time. With undersized hammers or small sets, the effect is particularly marked.

Speed of double-acting hammers is a vitally important factor to observe when recording sets, and although it has not been common practice to note it on pile inspectors' reports, this should always be done. Reductions in speed caused by reduced steam pressure reduce the net energy enormously and may shorten the pile a number of feet and greatly increase probability of settlement. For instance, if 30 per cent of the gross energy at full speed is available as net, and if the gross falls off one-sixth, the net may be reduced by 50 per cent.

Stroke of single-acting hammers is equally important, and for the same reason given above, since shortening of a few inches in stroke may make a large change in net energy. Almost any single-acting hammer can be, and often is, slowed down by the operator, reducing the stroke several inches. Sometimes the stroke has been shortened by changing the ports, which is almost never noticed since very few specifications call for measuring the stroke. Unscrupulous resetting of the slide bar has been known to occur, so that a large back pressure develops, thus greatly reducing the set. This was done by a foreman who established driving records on a job having several competing rigs.

Stoppage of driving for a few minutes may show an increased setup, or "freeze," which often is large. Such a setup is often only temporary. The longer the stoppage, the greater the setup may be. When stopping to splice a pile, capacities have been computed upon the redriving set, although the pile might be much shorter than a continuously driven pile. Usually if the suspension of driving occurs while the tip is in homogeneous soil, the sets and indicated driving resistances fall back to the previous rate within 2 or 3 ft after resumption of driving. If the setup is permanent, upon redriving, a load test should confirm the indicated value of the pile. Comparable values can be obtained only if driving conditions are constant. Specifications are written and field judgments as to reaching firm strata are developed on this basis. Determination of the actual pile capacity from a load test allows correction with driving requirements. One rule of driving conditions, whatever its basis, should be applied to all similar piles on the project. An alternative is to not apply the set-resistance criterion to stopped piles, but to drive them to the same tip elevation as adjacent piles that were not stopped. All stops should be reported on the pile inspector's reports, and if numerous, the engineers should scrutinize the driving procedures in an attempt to reduce them.

Sequence of driving affects the soil-consolidation pattern and depths at which the various piles reach specified set. An unsymmetrical arrangement of tip elevations may be undesirable.

Linking of a hydraulic jack has been unscrupulously done so that, after initial jacking sufficient to show evident crushing at the pile head or blocking, the rest of the indicated value was internal pressure in the jack.

Remedies for Failures

Occasionally damage can be caught in time so that moderate repairs may be undertaken. Examples of this would be gunite encasement of partially decayed or abraded pile sections, cutting off and replacing damaged top sections, removal of part of the load, underpinning or driving of supplementary piles if inadequacy of the number of piles is the fault, provision of resistance to horizontal forces before failure, and raising of ground-water level.

In the case reports that follow, where the remedy for failures is given as "underpinning," this has usually been expensive and often dangerous and ingenious work, usually bringing about far greater ultimate cost than would the provision of ample foundations during the original construction. The subject of underpinning is a complicated and interesting one, but it is outside the scope of this volume, except as some of the piles described are used in underpinning work. (Consult references 46 and 217 for details.) The above failures are physical failures from inadequate piles.

Economic Failures

There is another class of failures, not designated as such, however; these are economic failures, or actually failures in good engineering design, and result in driving too expensive a foundation. These failures may be nearly as costly as failures from inadequate structural designs, although the presence of the extra cost may never be brought prominently to attention as in the case of physical failures. Examples of this economic failure are hard to find in print because they are not made the subject of engineering articles. Several of the examples of test piles in Appendix IV indicate some of the causes of such failures in economy.

Actual Cases of Pile-foundation Failures

The following actual cases illustrate briefly some of the causes of pile-foundation failures.

Failures Primarily Due to Lack of Borings or to Acceptance of Hearsay Evidence. *Case* 1. The effects of reliance upon hearsay and of short individual test piles are illustrated. For a large mill on a lake shore in Canada, no borings were made. Other mills adjacent to water fronts in other parts of Canada were understood to have been built satisfactorily without piles. But as a precaution it was decided to drive 6,000 50-ft untreated wood piles. Short load tests on individual short piles were made and considered satisfactory. After 3,000 piles were driven, some being spliced twice to give 150 ft driven lengths, and many small resistances were noted, a question was raised as to the satisfaction which the foundation might be expected to give.

A report was authorized, and borings were made. These revealed that under a surface bed of sand varying in thickness from 0 to 25 ft there was a bed of highly plastic clay of the consistency of soft cold cream, varying from 50 to 100 ft in thickness, under which was a bed of silty fine sand under considerable hydrostatic pressure extending to sloping rock. The clay was valueless for support but use of the silty sand was considered possible for load-supporting purposes provided piles of sufficient length to penetrate into it were driven, using composite piles with concrete upper sections to resist decay above ground-water level.

Further investigations of stability of the sloping ground adjacent to the lake shore led to the belief that there was a reasonable possibility of the weight of the mill causing upheaval of the lake bottom in front of the mill, resulting in a landslide into the lake.

The driven piles were abandoned. A new site nearby was used, where short pedestal piles could be driven to rock.

Lack of borings resulted in additional construction cost, because part of the foundations and structural steel had already been erected upon the original site, and in the loss of a year's profits by setting back the construction program. *Cause of failure:* Lack of borings. In addition to the huge settlements which would have occurred, untreated sections of piles above the water table would have decayed. Reliance was placed on short load tests on individual short piles. Furthermore, danger of a landslide was present, which could be revealed only by adequate site borings. *Prevention:* Study of boring results to determine proper pile lengths, and use of decay-resistant material above the water table. *Remedy:* Abandonment of site and wood piles in favor of short concrete pedestal piles to rock in a location where no landslide danger was present.

Case 2. An armory was built in Minneapolis on a swamp site. It had been reclaimed by filling with sand and gravel, so that at the start of construction the strata were 15 to 20 ft of sand and gravel fill, 6 in. to 41 ft of mud, then sand and gravel, with ground water 1.5 ft below cutoffs. Wood piles 30 to 40 ft long were driven, those at the southeast corner passing through the mud into the base sand and gravel, whereas at the northwest corner about 10 ft of mud remained below the pile tips. The southeast corner did not settle, the southwest corner settled 3 in., and the northwest corner 37.5 in. After 8 years, the building was condemned. *Cause of failure:* Piles stopping in mud under part of building because no test borings were taken. *Prevention:* Adequate borings. Even the use of a pile formula should detect a case like this. *Remedy:* Underpinning, if undertaken in time. (Miller,[25] *Eng. News-Record,* vol. 82, pp. 1607–1609, 1919.)

Case 3. Borings supplied by the previous owners of the site for the 12-story Westinghouse Building in Philadelphia, Pa., showed 8 to 10 ft of loose fill underlaid by uniform clay and sand to rock at 45 ft. The concrete-pile contractors drove test rods to determine pile lengths and, as a result of penetrations considered satisfactory, tapered concrete piles 20 to 30 ft long were driven. Before the building was completed, settlement occurred. Further investigation by core borings and an open caisson disclosed that the soil was 27 ft of loose fill, 4 ft of peat, 1 ft of gravel, 15 ft of silted peat, 8 ft of silt, and 2 ft of sand and gravel on hard mica schist. The strata were not uniform, and the peat lay mostly under the east side of the building where 4 in. of settlement had occurred by the time steel was erected. Differential settlements of nearly 1 ft took place in 11 years. Elaborate and ingenious underpinning to rock by means of concrete-filled pipe piles was required. *Cause of failure:* Acceptance of inadequate borings, poor classification of soils, and reliance upon resistance of rods as a means of determining pile lengths. *Prevention:* Proper borings and longer piles. *Remedy:* Underpinning to rock. (Miller[25]; *Eng. News-Record,* vol. 91, pp. 192–193, 1923.)

Case 4. At the Brooklyn, N.Y., Navy Yard, during the rush construction period in 1917, an 11-story warehouse was constructed on piles 10 to 34 ft long in a fill over an old marsh. Serious settlements occurred during erection and continued after completion. Later investigation showed 1 to 7 ft of peat 34 ft below the surface, subjected to a load of 2½ tons per sq ft. Tests on samples gave 35 per cent compression for this load. It was necessary to underpin part of the foundation with pipe piles. *Cause of failure:* Peat bed below pile points. *Prevention:* Borings would have shown the need for longer piles. *Remedy:* Underpinning. (Miller[25]; *Public Works of the Navy, Bull.* 35, pp. 32–33.)

Case 5. To bridge a navigable 80-ft-wide river in Westbrook, Conn., through a marsh, U-shaped concrete abutments were constructed on 90 12-in. average-diameter wood piles 40 to 53 ft. long, intended to carry 7½ tons load each. No preliminary borings were made. When 6 ft of roadway fill had been placed, a severe forward movement of one abutment occurred. The fill was removed and relieving platforms on piles built. The bridge was erected and gravel fill replaced, causing a movement of 16 in. The fills were removed and the abutments returned to their former positions. Approach spans were built on new piers. When the new road fill reached the approach piers, the abutments again moved together, buckling the bottom chord of the truss. The road fill was again removed and the abutments moved back, releasing the bow in the truss. Heavy rock fill was started in front of the abutments but discontinued for fear of breaking the batter piles. Borings were then taken showing bedrock covered by strata of hard sand and gravel, and clay. The marsh material above was saturated, with a 4- to 5-deg angle of repose. The piles had fixity at their lower ends but were free to move through this top material. The road fill exerted a pressure on the mud, forcing the abutments forward. Two steel trusses were placed under the river bed between the abutments, set as inverted arches in order to maintain the navigable character of the river. This construction has succeeded in preventing movement of the abutments. *Cause of failure:* Design inadequate owing to lack of borings. Weight of new fill on marshy soil caused horizontal pressure on abutments, which piles could not resist as cantilevers fixed at bottom. *Prevention:* Use of a type of approach construction that would not cause thrust from the marsh soil. *Remedy:* Strutting abutments apart. (C. L. Nord, "A Case of Bridge Abutment Movement," *Proc. Intern. Conf. Soil Mech. and Foundation Eng.*, Cambridge, Mass., vol. 1, pp. 93–100, 1936.)

Failures Primarily Due to Inadequate Borings, Disregard of Boring Results, or Inadequate Pile Lengths. *Case* 6. For the Naval Aircraft Building in Philadelphia borings showed silt to a depth of 35 to 55 ft below mean low water, with sand and gravel below sloping from a high at one corner to a low at the opposite one. The building was carried on clusters of 30-ft-long poured-in-place tapered cased concrete piles driven with a mandrel. The reason cited for not using longer piles was that longer piles of this type were not then available. Insufficient driving resistance was noted. Resistance increased considerably after rest, and since this had been adopted as the solution on many projects under conditions considered apparently similar, construction was continued. Additional piles of the same length were also driven. An early load test on one pier showed a 3-in. settlement in 45 days under full building load. In 6 months, the settlements varied from 6 to 16 in., corresponding to the thickness of the silt and the typical dish-shaped settlement depression, and 3 years later, when underpinning was finally undertaken, the settlements varied from 12 to 24 in., with cracking of the floors and walls. *Cause of failure:* Too short piles with compressible silt below the points, reliance on resistance readings after setup of piles, and the assumption that differential settlements of a large loaded area can be predicted from a load test on one pier. The increase in number of piles decreased somewhat the too high friction value in the silt, but did not improve the bearing power of the poor material below the tips. *Prevention:* Use of piles of sufficient length to reach firm material. *Remedy:* Underpinning and repairs to structure. (E. D. Graffin, Lt. (CEC), U.S.N., "Sinking Building Underpinned by Unusual Procedure," *Eng. News-Record*, vol. 98, No. 24, pp. 988–989, June 16, 1927; "Soil Reactions in Relation to Foundations on Piles," *Trans. ASCE*, vol. 103, pp. 1211–1212, 1938; *Public Works of the Navy, Bull.* 34, pp. 88–122.)

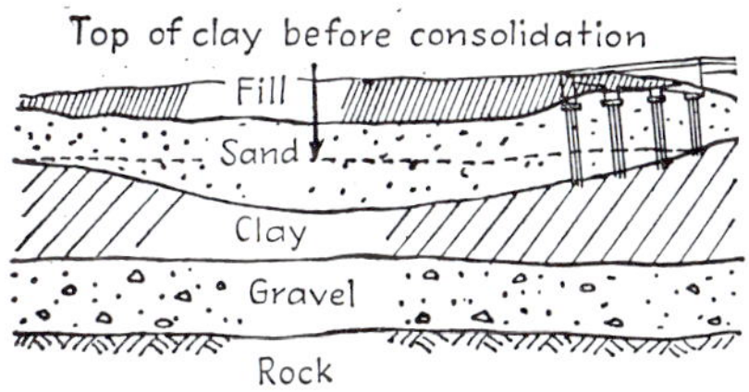

FIG. 16.4. Subsurface section showing later effects on granite abutment supported over deep clay bed, due to placing approach fill.

Case 7 (Fig. 16.4). The 14th Street Bridge abutment in Washington, D.C., was supported on 40-ft friction piles driven to a 50-ton bearing by the *Engineering News* formula. Borings taken prior to construction showed that under the base of the abutment was a layer of sandy clay extending nearly to the pile tips. Below was a deeper bed of softer clay.

A year after the U-shaped abutment was built, the landward end had settled 1.14 ft and the river end 0.11 ft. As the 17-ft-deep long approach fill was placed during the ensuing year, the added weight caused consolidation of the soft clay. Since the abutment rested near the edge of the characteristic dish-shaped depression formed, it tilted backward. Total settlement at back and front were estimated as 2 ft and 10 in., respectively. *Cause of failure:* Inadequate pile lengths. *Prevention:* Use of piles of sufficient length, with capacity to carry clinging load from the upper sand stratum when the clay settled, in addition to the abutment load. *Remedy:* Underpinning. ("New D.C. Bridge Abutment Settles," *Eng. News-Record,* vol. 143, No. 17, p. 30, Oct. 27, 1949.)

Case 8. Ten years after building a bascule bridge on wood piles in Toledo, Ohio, the abutments had moved together 0.9 ft, and had a vertical settlement of 1½ in. The piles were found to be only 8 to 12 ft long, whereas they should have penetrated to a much greater depth to transmit the load properly. *Cause of failure:* Inadequate pile lengths and lack of lateral support. *Prevention:* Use of piles of sufficient length, with proper design for lateral support. *Remedy:* Extensive repairs involving addition of abutment piles of sufficient length, piled bracing struts between abutments, and structural repairs and alterations to superstructure. (O. J. Pilkey, "Arresting Abutment Shifting on a Bascule Bridge," *Eng. News-Record,* vol. 108. No. 20, pp. 725–726, May 10, 1932.)

Case 9. The Lethbridge Viaduct, 312 ft high and 5,327 ft long, in Alberta, Canada, rested on concrete piles about 10 ft long, under the 20-ft-deep footings, driven into clay which was under the whole site. These piles stopped about 20 ft above shale rock.

Settlement of 2½ in. was noted in 4 years. *Cause of failure:* Stopping piles in clay short of rock. *Prevention:* Use of piles to rock. *Remedy:* Underpinning by caissons to rock 50 ft below surfaces. (F. W. Alexander, "Maintenance of Substructure of the Lethbridge Viaduct," *Eng. J.,* vol. 17, p. 523, 1934.)

Case 10. An oil tank of 52.5-ft diameter causing a load of 1 ton per sq ft was constructed on wood piles 35 ft long driven into sand and gravel. Undoubtedly the penetration was hard and the driving formula gave satisfactory results. However, the piles extended nearly through the sand and gravel, the tips being a very short distance above the top of a mud stratum varying from zero thickness at one edge of the tank to about 30 ft at the other. Settlement reached 1 ft in 4 years, with no signs of a decrease in rate. *Cause of failure:* Consolidation of soft material, and reliance upon a pile load test and surface-bearing test. *Prevention:* Obtaining adequate borings, consideration of need of piles, and study of consolidation characteristics of underlying strata. *Remedy:* Not stated. ("Progress Report of Special Committee on Earths and Foundations," *Proc. ASCE*, vol. 53, No. 5, pp. 777–820, May, 1933.)

Case 11. The Dunwoody Industrial Institute in Minneapolis started settling soon after construction. In 15 years, the settlement became so bad that major reinforcement of the foundations was necessary to save the building. Piles were composite concrete and wood 70 ft long, with the wood section 30 to 35 ft long. In some places, over a foot of wood blocking was found between pile caps and bottom of foundation concrete. Some interior footings had settled 2 ft. For underpinning, 14-in. H piles were driven in 37-ft lengths, using milled splices, to an average depth 40 ft greater than the composite piles. *Cause of failure:* Too short piles. *Prevention:* Use of longer piles. *Remedy:* Extensive underpinning. (Walter H. Wheeler, "Building Settlement Checked Successfully," *Eng. News-Record*, vol. 123, pp. 531–533, Oct. 26, 1939.)

Case 12. For a Navy building at Washington, D.C., tapered concrete piles, too short to reach an adequate substratum, were driven through a recent fill over an old marsh. A settlement of 4 in. was observed shortly. The greatest settlement did not occur at the heaviest loaded piles, but in the area of greatest depth of fill where the settling of the fill itself carried the piles and structure down with it. *Cause of failure:* Too short piles, failing to reach adequate bearing material, aggravated by consolidation of marsh under weight of new fill. *Prevention:* Use of piles long enough to reach hard stratum, and provision of sufficient pile capacity to withstand the dragging-down action of the fill on the piles, as well as to support the building load. *Remedy:* Underpinning to firm material. (Miller[25]; *Public Works of the Navy, Bull.* 33, p. 43.)

Case 13. A 175-ft chimney was built during World War I at the Navy Yard in Philadelphia, Pa., on 64 straight-shaft uncased concrete cast-in-place pedestal piles 36 ft long. Borings showed silt to a depth of about 50 ft. below mean low water with sand and gravel below. The penetration of the piles was steady during driving, and indicated safe loads of 130 tons each by the *Engineering News* formula. A few months after completion, the angle of lean was 1:9. *Cause of failure:* Too short piles, with reliance placed on indicated driving resistance, without due account of the compressible nature of the silt, and use of the *Engineering News* formula with this heavy type of pile having small sets which indicated falsely high bearing resistances. *Prevention:* Use of piles long enough to reach the sand and gravel below the compressible silt. *Remedy:* Underpinning to firm material by means of a piled ring foundation. (Miller[25]; *Public Works of the Navy, Bull.* 34, pp. 88–122.)

Case 14. An apartment house in Berlin, Germany, was built on cast-in-place concrete piles which, on one side of the building, extended through fill, marshy soil, sand and shells, clayey ground, and into silty sand in which the tips rested. On the other side of the building the piles were longer and penetrated into a stratum of coarse gravel. The short piles settled. Chemical grouting was used to solidify a 6-ft-thick mat 60 ft long in the silty sand around and below the tips of the short piles to increase their bearing value, although the pile tips were 60 ft below ground-water level. *Cause of failure:* Some of the piles underlaid by silty material. *Prevention:* Use of longer piles. *Remedy:* Chemical stabilization of the soil. (Lewin.[10a])

Failure Primarily Due to Inadequate Load Tests. *Case* 15. The power plant of the Continental Motors Co. at Muskegon, Mich., started settling immediately after completion. This was a large modern steel-framed power station with brick walls and 175-ft-high superimposed stacks.

Logging and lumber operations had been conducted on the site, leaving up to 18 ft of sawdust and wood refuse as shown by borings, over which made ground had been formed from dredged sand and silt.

From tests on piles driven at 8-ft centers it was concluded that the safe pile load was 30 tons, although only 16 tons were actually used in the design. Two thousand piles were driven on spacings varying from 2 to 3⅔ ft. Piles were planned to be 40 ft long, but the piles reached the assumed resistance at lesser depths than anticipated, and considerable lengths were cut off.

Settlement soon reached 15¼ in. at the greatest point and 3$\frac{1}{16}$ in. at the least point, with no indication of decreasing rate.

New borings showed fill of various sorts underlaid by sawmill refuse to a total depth of 45 ft. Below this was sand and clay for 30 ft more, then hard clay.

Pretest 16-in. open-end steel cylinders 75 ft long were jacked down, blown out, and filled with concrete. *Cause of failure:* Reliance upon test loads on single piles for value of piles in groups. Also dependence upon indicated driving resistance, without consideration of lengths required to reach firmer strata below the pile tips. It is also likely that the recent sand fill gripped the upper portions of the piles and added much load. *Prevention:* Study of adequate borings, and use of piles of sufficient length to reach a stratum having adequate bearing capacity for the intensity of load from the structure and the old and new overburden, and inclusion of load from the upper new sand fill in the load each pile must carry. *Remedy:* Underpinning. (Harry Spillman, "Cylinder Underpinning Checks Sinking Foundation," *Eng. News-Record,* vol. 108, No. 15, pp. 544–545, Apr. 14, 1932.)

Failures Partly Attributable to Group Action of Piles. *Case* 16. Ten high circular concrete grain elevators were constructed on a 30-in. concrete mat 59 by 144 ft for the Chicago & North Western Railway at Green Bay, Wis.

The original timber elevators burned down leaving 1,280 wood piles 20 ft long, in good condition. These were tested under 20-ton loads for 8 days and showed no settlement, so they were considered safe for a 10-ton design load. In addition, 560 50-ft wood piles were driven.

Settlement started immediately, and after 19 years amounted to 2½ ft on the land side and 4 ft 2 in. on the river side. Borings revealed 37 ft of sand above 50 ft of clay to rock. *Cause of failure:* Consolidation of soft material below the pile tips, and reliance upon short-time loading tests of individual piles above deep clay stratum. *Prevention:* Obtaining adequate borings, and use of piles to rock. *Remedy:* Underpinning to rock with Chicago caissons, including removal of interfering piles. ("Caisson Underpinning Checks Settlement of Ten Tall, 20-year-old Concrete Grain Elevators on Waterfront," *Construction Methods,* vol. 28, No. 2, pp. 78–80, 148, 150, 152, 154, February, 1946.)

Case 17. A Pacific Coast factory 1,000 ft long was built on piles 60 to 80 ft long, using loads of 10 to 15 tons per pile. At a depth of 100 ft was a clay stratum. Before selecting the design, behavior of nearby structures was checked. A railroad bridge carried on narrow piers supported by 50-ft-long piles had not settled appreciably in 40 years, and it was concluded that even longer piles loaded with a conservative loading per pile would be satisfactory. The load from the factory averaged 2 tons per sq ft on the clay stratum. However, the load from the narrow bridge piers was negligible on the area of clay over which it was spread at the depth of the clay. Factory settlements were soon over 1 ft,* causing damage particularly detrimental to the machinery. *Cause of failure:* Consolidation of clay stratum located considerably below pile tips. It was erroneously assumed that since a narrow load on piles above this stratum had not settled, a large extent of loaded area would not settle. *Prevention:* Investigation of the character of the underlying strata and of the shape of the distribution of the load from the piles into the soil, by the principles of soil mechanics. *Remedy:* Not stated. ("Notes of Proceedings, Committee on Bearing Value of Pile Foundations, Waterways Division," *ASCE, Paper No. H*-18, *Proc. Intern. Conf. Soil Mech. and Foundation Eng.,* Cambridge, Mass., vol. 3, p. 150, 1936.)

Case 18. For a grain elevator in Portland, Ore., test piles were driven through a very soft soil to a resistance indicated as 25 tons by the *Engineering News* formula. One pile carried 40 tons satisfactorily.

* Over 28 in. in 18 years.

Piles were driven at 2½-ft. centers, and a 3-ft-thick mat poured. Marked settlement was observed during construction and, when the load reached 9 tons per pile on part of the area, subsidence varied from 9 in. to 2 ft. *Cause of failure:* Assumption that each pile in a group will carry the same load as an individual test pile. Also, short-time test on the bulb of pressure surrounding a single pile has no relation to long-time action on the deep bulb under the entire structure. *Prevention:* Study of group action, and of distribution of load from the entire group to the surrounding and underlying strata by the principles of soil mechanics. (At the time of this failure, the principles of soil mechanics were not well understood, but the example is illustrative of much present-day practice, nevertheless.) *Remedy:* Not stated. ("Foundation Troubles of Portland City Grain Elevator," *Eng. News-Record,* vol. 84, pp. 176–179, 1920; Miller.[25])

Failures Primarily Due to Flow of Unconfined Fine Soil. *Case* 19. Piers for a bascule span at Bridgeport, Conn., rested on 25-ft piles driven into material classified from wash borings as fine sand. The load on each pile was 27 tons. The piers tipped in opposite directions, breaking the tie and causing an increase in the previous small settlement of over 1 ft. Movement downstream also occurred. *Cause of failure:* Piles driven in fine unconfined submerged sand. *Prevention:* Obtaining soil samples by methods which would bring up the finest or colloidal material to determine the true nature of the strata, and driving the piles into the gravel bed or to the rock below the sand for the piers in question, as was done for some of the other piers. (D. P. Krynine and C. C. Nord, "A Case of Settlement of a Bridge Pier," *Proc. Intern. Conf. Soil Mech. and Foundation Eng.,* Cambridge, Mass., vol. 1, pp. 100–103, 1936.)

Case 20. A 24-story office building in São Paulo, Brazil, leaned 2 ft out of plumb in two directions, toward the street and a new adjacent excavation. This new excavation started movement in a lens of fine wet sand surrounding a group of concrete piles that supported the corner of the building. After attempts at cement grouting and injection of an aluminum salt proved futile, the soil was frozen over an 8-month period, lowering the earth temperature to —20°C, using the double-walled circulation pipes of 2- and 4-in. diameters driven 60 ft into firm ground. Concrete piers were then installed through the basement floor. The building was jacked up from these piers. The cost was 50 per cent of the original cost of the building. *Cause of failure:* Removal of side support of stratum of fine wet confined sand. *Prevention:* Selection of proper precautions for adjacent excavation. *Remedy:* Underpinning and jacking the settled building. (*Eng. News-Record,* vol. 131, No. 17, p. 623, Oct. 21, 1943; *Dowflakes.*[10g])

Failures Due to Lateral Forces Including Vibration. *Case* 21. For a compressor plant in Mississippi, a 200-kw multicylinder gas-engine-driven electric generator unit, operated at a speed of 412 rpm, was set on a separate concrete foundation resting on six vertical wood piles about 20 ft long, in accordance with local practice. The soil of the swampy area was alluvial sand and clay of poor bearing value. Vibration was felt all over the 10-acre site. *Cause of failure:* No provision was made for resisting the unbalance horizontal forces from the equipment. *Prevention:* The use of batter piles to resist the lateral forces. *Remedy:* Not stated. (H. R. Marsh, "The Importance of Soil Mechanics in Design of Compressor Foundation," *Natl. Petroleum News,* pp. R544–R549, Nov. 3, 1943.)

Case 22. For a compressor plant in Mississippi, horizontal 1,000-hp compressor units were installed. Preliminary investigation in summer showed that this plant was located over a deep sand bed. Individual foundations were set upon vertical wood piles. The sand was so firm that it was necessary to use pile shoes and bands during the difficult driving. In the fall when the plant was ready for operation, ground water had risen considerably, and horizontal movement was so great that it was unsafe to operate the plant because of stresses in the piping. The building movement was so great that building framing bolts were sheared. *Cause of failure:* Loss of horizontal resistance in the sand when submerged. *Prevention:* Use of batter piles. *Remedy:* Excavations were made under the compressor foundations, steel diagonal bracing installed on the piles, and the cavities refilled with pumped concrete. (H. R. Marsh, "The Importance of Soil Mechanics Design of Compressor Foundations," *Natl. Petroleum News,* pp. R544–R549, Nov. 3, 1943.)

Case 23. Vibrations set up by the compressors of a gas station caused serious trouble in the pile foundations of machines and building. The vibrations were practically stopped by grouting the 50 ft of sand through which the piles were driven. Application of grout under pressure was through solid-wall pipes, perforated after driving by a gun that shot bullets through the pipe at desired locations. Neat cement grout was supplied at 150 psi. *Cause of failure:* Lack of stability against horizontal vibrating forces. *Prevention:* Use of batter piles or stiffer type of foundation support if sand was denser than critical density, otherwise solidification of the sand. *Remedy:* Grouting the soil mass around the piles. ("Grouting Checks Foundation Vibration," *Eng. News-Record,* vol. 129, No. 17, pp. 95–96, Oct. 22, 1942.)

Failures Due to Negative Friction. *Case* 24. The Jurgens Margarine Co. erected an oil mill containing heavy equipment at Zwyndrecht, Netherlands. The site was land made of hydraulic sand fill. The soil consisted of 15 ft of sand fill, 43 ft of peat and clay, 17 ft of fine sand, then coarse sand and gravel. All soils were saturated. Test piles were driven to a resistance of 50 tons per pile. Working load was taken at 5 tons per pile. Creosoted wood piles 65 ft long were used.

Pile points stopped in the fine sand about 10 ft above the coarse sand and gravel. In four years the oil-hardening building had settled 27.6 in. near the center, threatening collapse. Maximum settlements did not occur under the heaviest loaded piles, which carried 18 tons each, although the average was about 3 tons. Piles under an outside extension with almost no load settled similarly. *Cause of failure:* Negative friction owing to settlement of the hydraulic fill which added an estimated 15 tons per pile and overloaded the pile points in the fine sand. The dynamic pile-driving formula used was not an adequate guide, since the friction which resisted the blow was in the upper strata which were settling. *Prevention:* Study of the strata by the principles of soil mechanics and consideration of static-friction values in the lower strata would have shown the need of longer piles to firm material. *Remedy:* Underpinning to firm material under the fine sand. (Miller[25]; *Engineering*, vol. 17, pp. 174–176, 1924.)

Case 25. Wood piles for a water-front structure were driven through rocky fill into firmer soils. Driving resistances were considered ample according to dynamic formulas, but the forces imposed on the piles by subsidence of the fill and compression of soft bay deposits was so great that a pile was pulled down from the concrete foundation of the structure it was intended to support, so that the head of the pile was several inches below the concrete caps.

A later boring showed 40 ft of sandy loam and rocky-fragment fill, 28 ft of soft bay mud, 17 ft of firm clayey soil, then decomposed rock against which the tip rested. The potential downward load of the gripping fill was estimated to be 175 tons, and that of the bay mud to be 12 tons. Some portion of these loads probably fractured the pile, permitting the upper part to slide by the lower part resting on the rock. *Cause of failure:* Additional load from fill and compressible stratum, and reliance upon driving formula. *Prevention:* Consideration of loads from settling strata, and use of pile type and size of adequate strength to resist frictional downward pulls. *Remedy:* Not stated. (W. W. Moore, "Experiences with Predetermining Pile Lengths," *Proc. ASCE*, vol. 73, No. 9, pp. 1341–1358, November, 1947.)

Case 26. A serious case arose from the placing of a heavy fill surrounding a concrete stadium which was provided with sufficient concrete piles to carry safely the loads from the structure, but where the piles were unable to carry the added load arising from the settlement of the soil compressed by the much greater load from a fill placed outside of, and around, the structure. *Cause of failure:* Addition of load from fill above compressible stratum. *Prevention:* Provision for this load, omission of fill, or provision of means of keeping it from piles. *Remedy:* Not stated. (G. A. McKay, *Trans. ASCE*, vol. 103, p. 1220, 1938.)

Case 27. A steel mill was constructed by supporting the column and crane loads on pile footings and the floor slab on ground, in accordance with a common practice in mills and foundries. Original ground consisted of a deep bed of plastic clap underlaid by hardpan and rock. H piles over 100 ft long were driven; then 15 ft of slag fill was placed to bring the site to grade for the floor slab. Loads as great as 17 ft high of armor plate were placed on the slab.

Within a year settlements of 1 ft had occurred and the H piles jackknifed. The slag had cemented to form a slab that also cemented itself strongly to the piles, putting on each pile an estimated total load of 350 tons. *Cause of failure:* Additional load on piles owing to cementing action of slag fill in itself and to piles. *Prevention:* Avoidance of use of fill, or use of free pipe sleeves around piles through fill, are among possible methods which could have been considered. *Remedy:* Not determined.

Case 28. Batter piles were driven to resist outward movement of a quay wall constructed in 30 ft of silt-laden water. Sheet piles for retaining the fill below and to the rear of the relieving platform were driven 40 ft back of the face of the wall, thus exposing the batter piles under the relieving platform to the accumulation of silt deposit. Frequent dredging left banks of soft mud under the wall held against rapid sloughing down into the stream by the piles. This added load from skin friction on the batter piles produced settlement of the piles and, instead of resisting the outward movement of the wall, they pulled it forward, so that the wall moved out several feet. *Cause of failure:* Drag from mud sloughing caused by dredging. *Prevention:* Consideration of this factor in design, or adoption of different type of construction. *Remedy:* Not stated. (G. A. McKay, *Trans. ASCE*, vol. 103, pp. 1220–1221, 1938).

Failure Due to Ice Uplift on Piles. *Case* 29. Uplift action by ice on 12-in. diam wood piles wrecked a 1-year-old bridge over a river behind a dam. The piles were driven by a 940-lb ram falling 20 ft, to a set of 0.5 in. The water level changed 2 or 3 ft several times during the first 3 months of winter. The piles held and the ice broke away at a short distance around them, until it was 15 to 16 in. thick, when the piles started lifting. When the ice thickness increased several more inches, all piles lifted. *Cause of failure:* Force exerted by ice adhesion during change in water level, while live load was generally not acting. *Prevention:* Deeper embedment in soil, avoidance of taper on piles, painting with tar to absorb the sun's heat, breaking up of the ice, etc. *Remedy:*

Construction of new bridge. (J. W. James, *Proc. Inst. Civil Engrs.* (*London*), vol. 41, Part 2, pp. 191–201, Paper 1374, 1875.)

Failures Due to Poor Construction. *Case* 30. Fracture and necking of uncased poured-in-place concrete piles was found in most of the 173 piles extending through 5 ft of hard clay, 9 ft of soft clay, then 6 ft of silt. Piles were designed for 50-ton loads, spaced 3 ft 6 in. on centers in groups, but with no pile driven within 6 ft of a pile that had been driven that same day.

A 14-in. inside-diameter tube with a grommeted shoe was driven and filled with concrete. Withdrawal was accomplished by fastening a pile line to the winch on the pile rig, the 15,000 ft-lb hammer alternately driving the tube up and driving it down part of the distance raised on the upstroke. The downstrokes were intended to tamp the concrete and force it out to the outside diameter of the tube. Corrugation spacings of up to 2.4 in. were measured, compared with the required 0.5 in. for each cycle. No reinforcing steel was used.

Excavation for footings revealed waisting in the soft clay stratum in varying amounts down to 8 in. diameter. Some waists were fractured. Clay had been forced through horizontal, oblique, and vertical cracks. *Cause of failure:* Driving of adjoining piles caused side pressure in confined soft clay stratum and raising of upper hard clay stratum, the grip of which lifted the upper portions of the piles, with resultant waisting, fracturing, and forcing clay into fractures. *Prevention:* Preferably use of cased or precast concrete piles, or other types not poured in place. *Remedy:* Abandonment of poured-in-place uncased piles and driving of 75- to 100-ton rail piles formed of three 60- or 70-lb rail sections welded together. (Muntz.[3aa])

Case 31. Under a gasholder, 188 cased concrete piles 25 ft long were driven, with embedment of 17 ft. The mat on top of the piles was flush with yard fill around the holder. The piles had corrugated steel shells, 14½ in. outside diameter, filled with concrete, with the upper 15 ft reinforced with four ⅝-in. bars and a spiral on a 2-in. pitch and 10-in. diameter. Removal of portions of the steel shells exposed sections which were never filled with concrete, probably because of the use of a concrete mix which was poorly proportioned for the materials, and too dry when placed, which, combined with the restrictions imposed by the 10-in. diameter spirals, prevented the concrete from completely filling some spaces between the spiral and the shell, and also induced arching of the concrete with resulting voids in the core of the spiral. Casings were removed to below ground level, visible voids filled with gunite, and a 2-in. gunite encasement shot on a mesh having 3-in. spacing. *Cause of failure:* Too dry a mix for workability. *Prevention:* Better inspection of mix and placing. *Remedy:* Gunite repairs and encasement.

Case 32. On a large precast-concrete-pile job in San Francisco, corner cracking occurred over main longitudinal bars, necessitating extensive repairs. These piles were cast in a horizontal position. Three of their sides were thus poured against forms; their top sides were troweled. Cracks occurred on the troweled side. *Cause of failure:* "Water gain" (a fine film of water on the underside of the top longitudinal bars, which water film later left a void space wherein rust could form and progress) appears to be the explanation of failure in this case. *Prevention:* Care in casting the piles. *Remedy:* Expensive repair work after piles were driven. (H. M. Hadley, "Concrete in Sea Water: A Revised Viewpoint Needed," *Trans. ASCE,* vol. 107, pp. 384–385, 1942.)

Case 33. Many concrete piles under the Ford Motor Company's wharf in Long Beach, Calif., soon developed serious disintegration between high-water level and the mud line.

Patching and jacketing in the damaged zone was unsatisfactory, and 600 new H piles and 83 new pipe piles were driven for all concrete piles exposed to sea water, irrespective of condition.

The H piles were driven through steel jackets suspended under the deck and filled with concrete after driving the piles, to obtain protection in the danger zone. *Cause of failure:* Possibly the use of unsound fine aggregate, containing variable percentages of kaolinized feldspar, was the primary cause of failure, permitting the secondary disintegration effects to proceed; however, with the normal testing procedures used on aggregates in this region, it seems unlikely that material decomposed to an extent sufficient to cause failure would be used for this purpose. A possible cause of failure was the formation of silica gel and swelling, owing to reaction between the high-alkali-content cements and silicious matter in the aggregates. *Prevention:* Investigation of the aggregates proposed. *Remedy:* Installation of new piles under the existing structure. (H. M. Hadley, "Concrete in Sea Water: A Revised Viewpoint Needed," with discussions by J. W. B. Blackman and H. E. Squire, *Trans. ASCE,* vol. 107, pp. 345–394, 1942; "Wharf Gets 600 New Piles Through Old Deck," *Eng. News-Record,* vol. 123, No. 23, pp. 750–752, Dec. 7, 1939; Stanton.[5x])

Failures Due to Decay of Untreated Wood Piles above Ground Water. *Case* 34. A grain elevator on the bank of the Chicago River was found leaning toward the river, supposedly because of a movement of the earth and foundation. The top 6 in. of the untreated piles had rotted. It was necessary to expose the entire foundation top, cut off the tops of the piles, and underpin. Insufficient allowance was made for the range of rise and fall of the water level. *Cause of failure:* Untreated wood-pile cutoff above ground-water level. *Prevention:* Slightly deeper concrete foundations to keep cutoff of wood piles below ground-water level, use of composite piles, piles other than wood, or use of creosoted piles. *Remedy:* Cutting off decayed portions of wood piles and underpinning structure. (F. R. Judd, "Creosoted Piles, Good as New after 13 to 21 Years' Use," *Eng. News-Record,* vol. 108, No. 10, p. 353, March 10, 1932.)

Case 35. A timber trestle on untreated wood piles was built across swampy ground near the shore of Lake Washington near Seattle. When it was desired to provide a heavier deck on the piles, they were found to be badly decayed at ground level but in perfect preservation a foot or two below, where permanent saturation existed.

The heads of the piles were cut off and encased by concrete piers. *Cause of failure:* Use of untreated wood piles above permanent water level. *Prevention:* Use of composite sections or treated piling. *Remedy:* Cutting off decayed portions above water level and replacing with concrete piers. (T. D. Hunt, "Concrete Structure Built on Old Timber Piling," *Eng. News-Record,* vol. 109, No. 7, Aug. 18, 1932.)

Failures Due to Decay Caused by Lowered Ground Water. *Case* 36. A large power station in Brooklyn was originally constructed on a concrete mat on untreated wood piles. A few years ago, it was decided to construct a modern station, and the superstructure was demolished. Extensive decay had occurred owing to lowering of the ground-water level over the years in this vicinity. It was necessary to remove the mat and drive new piles. *Cause of failure:* Lowering of ground-water table. *Prevention:* Consideration of possible changes in ground-water level, and use of treated wood, concrete, steel, or composite piles. *Remedy:* Underpinning or removal.

Case 37. Buildings in the Brooklyn, N.Y., Navy Yard began to settle owing to decay of untreated wood piles from lowering of the ground-water level caused by a greatly increased rate of pumping for industrial and other uses in the western end of Long Island. In some areas, ground water was below tide level, but mud cover on the shore prevented inflow of salt water. *Cause of failure:* Lowering of ground-water table. *Prevention:* Consideration of possible changes in ground-water level, and use of type of pile resistant to decay. *Remedy:* Not stated, but generally underpinning in similar cases, unless economic considerations indicate demolition. ("Pile Foundations and Pile Structures," *ASCE Manuals of Engineering Practice No.* 27.)

Case 38. About 1929, the Boston Public Library, built in the Back Bay reclaimed basin, showed cracks. The tops of some piles were completely gone and others were badly decayed, so that, for about 40 per cent of the area, underpinning and cutting of pile heads and replacing with concrete was necessary. The piles were most affected nearest a deep sewer laid in 1912. In 1885 ground water in the district was at elevation 7.7. A building ordinance was passed requiring piles to be cut off at elevation 5 or lower. By 1936, about 80 per cent of the area was covered by buildings and paving, and considerable quantities of ground water were pumped from sumps in subways and buildings. Ground-water levels had receded several feet. Cracks in sewers may have occurred during the 1925 earthquake, or owing to the general settlement of the area. Damming of the sewer near the library resulted in raising ground water rapidly. This test dam has been left in place by the city as a means of ground-water control. Observation wells were installed in the district so that, in the future, owners will have definite records available. *Cause of failure:* Decay of untreated wood piles owing to lowering of ground-water level. *Prevention:* Use of treated piles, lower cutoffs of wood piles, composite piles, piles other than wood, or a means of maintaining ground-water level. *Remedy:* Underpinning, cutting off and replacing damaged piles, and raising ground-water level. (B. F. Snow, "Tracing Loss of Ground Water," *Eng. News-Record,* vol. 117, pp. 1–6, July 2, 1936, and "Drainage in Reverse at Copley Square," *Eng. News-Record,* vol. 154, No. 17, Apr. 28, 1955, p. 47.)

Case 39. Cases in New York City illustrate the effect of the change of ground-water level upon wood-pile foundations. A deep fresh-water pond fed by springs formerly existed at the site of the Tombs prison and surrounding buildings. During subway construction the tunnel passed through the pond site and water was pumped. The pumps were located about 25 ft below the curb, and operated for a year.

Nearby, a seven-story brick warehouse showed evidence of settlement. The piles had been cut off about 10 ft below curb level, an elevation which was then below mean ground-water level. Subsequently the ground-water level was lowered from 5 to 10 ft in this locality by heavy pumping and drainage, chiefly in connection with the subway. Within 2 years thereafter, the foundations of this building settled about 2 in. Tops of piles were found 8 ft above ground-water level, and were no longer protected by saturation, so that decay had commenced.

It was believed that with the cessation of pumping for the subway, the lowering of the ground-water level would not only cease but might be reversed, so that in time it would rise at least part way to its former level. It was, therefore, considered that the safety of the building would be sufficiently ensured by safeguarding that portion of the foundation between the bottom of the concrete footings and the present ground-water level, below which the piles were durable. It was determined to cut the piles down 5 ft and extend the footings down. This was done at great expense.

An instance in which the lowered ground-water level did not return occurred at the Cambridge Hall Building on 33rd Street. This was supported on wood piles and, during construction of the cross-town tunnels, settlement of the building took place. After completion of the tunnel, it was thought that the stream which had been flowing into the tunnel the year before would back up and submerge the piles; but this did not occur, the stream having been permanently diverted from its old location in some unknown manner, with the result that the ground-water level never returned to its former grade. It was necessary to underpin the building to rock. *Cause of failure:* Decay of untreated wood piles, caused by permanent lowering of the ground-water level subsequent to selection of pile cutoff grade. *Prevention:* Setting untreated wood-pile cutoffs below

FIG. 16.5. Decay in wood piles following lowering ground-water level 14 ft below cutoff. Piles so badly rotted they could be pulled apart with fingers. (*Courtesy of Raymond Concrete Pile Co.*)

possible future lowered ground-water levels which may result in urban areas, use of treated wood piles, composite piles, or steel or concrete piles. *Remedy:* Underpinning. (*Cement Age*, April, 1910.)

Case 40 (Fig. 16.5). Wood piles exposed at the corner of Bethune and Washington Streets in New York City in 1924 were originally cut off at elevation 101, the water level at that time. The water level when piles were exposed was 87. For 3.5 ft below the original cutoff, the piles were so badly rotted that they could be pulled apart with the fingers. *Cause of failure:* Decay caused by permanent lowering of ground-water level. *Prevention:* Use of nondecaying material to sufficient depth to allow for future lowered water levels. *Remedy:* Demolition or underpinning. (*Raymond Concrete Piles*, Raymond Concrete Pile Co., 1926.)

Failure Due to Decay Caused by Lack of Butt Protection on Treated Wood Piles. *Case* 41. A bulkhead at Lynn, Mass., supported by several thousand southern-yellow-pine piles, was given a 12-lb creosote treatment with Grade 1 creosote. In 10 years, half of the piles were so badly decayed that extensive repairs were necessary. Destruction proceeded from the top down to the low-water mark, a distance of 10 ft, averaging 1 ft per year. In a similar nearby untreated structure, similar damage occurred in 5 years. *Cause of failure:* Lack of butt protection. *Prevention:* Use of adequate butt protection. *Remedy:* Replacement. ("Report on Destruction of Marine Organisms and Possible Ways of Prevention," *American Railway Engineering Association.*)

Failures Due to Insect Attack. *Case* 42. In 1930, settlement of a building built in St. Paul, Minn., in 1886 as four stories high, with two stories added in 1912, and carried on wood piles of 14- to 16-in. butt diameter and 40 to 50 ft long, revealed serious pile damage from insects.

The ground was a swamp bottom of clay, with partial fill. The piles were cut off 5 in. above existing ground surface, and masonry footings were built of large stones having the joints filled dry with spalls.

When the two stories were added 2- to 3-in. settlement had occurred at spots, attributed to drawing down of ground water by sewers. No immediate change was noted upon adding the two stories, but gradual settlement soon reached additional 3 to 5 in. at some columns.

Exposing the foundations revealed that many piles had disappeared for several feet down, leaving uncaved holes in the clay. The clay, for a few inches surrounding the hole, was dry, but farther in it was plastic and soft. The clay under many footings was also soft and plastic and seemed, in the opinion of the engineers, damp enough to have prevented rot had the tops been sealed. This fact and the further fact that so little remained of the piles indicated that they had been eaten out by borers. Beetles were found to be under the wood floor on the ground, and these seemed to be the cause of damage. The apparent method of operation was to attack the pile 1 to 2 ft down from the head, eat it away completely for some length and, working from the outside toward the center, leaving the dirt of the worms with the little chips in strings. The beetles apparently laid their eggs on the exterior surface, which was just damp enough to suit the borers. Starting 2 ft from the top of the pile, which seemed to be a little too dry for nice chewing, they cut off the moisture that had been brought to the top by capillary action, so that the extreme top rotted into brown punk, which later dropped down into the holes.

The wood borers hatched from the larvae of *Saferda condida* and *Chrysobathrid femorata* beetles, deposited on the surface of the pile 2 to 3 ft down. The borers are most active at 60 to 70°F, and practically dormant at freezing temperatures. The borer seems to work for 1½ to 3 years before changing into a beetle.

An expensive lawsuit developed between the lessor and the lessee, hinging on whether damage was due to decay or borers. *Cause of failure:* Destruction of untreated wood-pile heads in moist soil by beetle grubs. *Prevention:* Treatment of wood, or keeping cutoff below groundwater level and sealing the heads of the piles. *Remedy:* It was suggested that the damaged parts be burned out, thus also drying the adjacent clay to serve as a form, then placing reinforcement and concrete. (If as little water as this were present, possible decay might have occurred, and an underpinning operation might ultimately have been required in any event.) (Turner.[2u])

Case 43. Puget Sound Navy Yard Drydock No. 1 was constructed in 1896. In 1930, after marked evidences of decay, exploratory tunnels were driven. Tops of piles above mean tide level, as well as the inside of the dock lining, had been seriously attacked by rot and by *Ambrosia* beetles. It was necessary to remove lining and blast and remove the concrete behind the altars for 6 ft up from the floor, to enable reconstruction to be made. *Cause of failure:* Extension of wood piles above mean tide level. *Prevention:* Keeping cutoffs of wood piles below mean tide level. *Remedy:* Expensive reconstruction. (C. E. Dickeman, "Drydock Reconstruction at Puget Sound Navy Yard," *Eng. News-Record,* vol. 107, No. 27, pp. 1044–1045, Dec. 31, 1931.)

FIG. 16.6. Effect of *Limnoria* attack on Ocean Avenue Bridge over Sheepshead Bay to Manhattan Beach, N.Y., October, 1943. (*Courtesy of Ralph H. Mann, American Wood-Preservers' Association.*)

Failures Due to Marine-borer Attack on Untreated Wood Piles. *Case* 44 (Fig. 16.6). The Ocean Avenue Bridge over Sheepshead Bay to Manhattan Beach, N.Y., was opened in 1917. The piles were untreated southern yellow pine. Collapse of a section occurred in 1943 owing to *Limnoria.* Only a few small *Teredo* were found, near the mud line. It is believed that the untreated piles had been replaced at least once previously. Pine piles with 20-lb creosote treatment were used in reconstruction after this collapse. *Cause of failure: Limnoria* attack on untreated pine piles. *Prevention:* Treatment of piles. *Remedy:* Replacement with treated piles. (Letter from Ralph H. Mann, Engineer, Service Bureau, American Wood-Preservers' Association.)

Case 45 (Fig. 16.7). An untreated pile pier, 2 miles north of Flagler Beach, Fla., collapsed in 6 months from marine-borer attack. In contrast, the Flagler Beach Municipal Pier, built in 1928 on 25- to 44-ft long southern yellow pine piles given a full-cell process with 20 lb of creosote treatment, is still free from decay and marine-borer attack (Fig. 16.8). *Cause of failure:* Use of untreated wood piles. *Prevention:* Adequate creosote treatment. *Remedy:* None was used, only ruins of a few bents remain. (Ralph H. Mann, "Old Creosoted Piers in Borer Infested

FIG. 16.7. Remains of untreated pile pier located 2 miles north of Flagler Beach, Fla., which collapsed as a result of marine borer attack 6 months after installation. (*Courtesy of Ralph H. Mann, American Wood-Preservers' Association.*)

FIG. 16.8. Flagler Beach Municipal Pier, after 20 years, is built entirely of creosoted piles and timbers, with the exception of the deck planks, which are cypress. (*Courtesy of Ralph H. Mann, American Wood-Preservers' Association.*)

Waters," *Wood Preserving News*, vol. 26, No. 1, pp. 1–3, 12, January, 1948.)

Case 46. The Alaska Central Railway constructed a wharf at Seward, Alaska, in 1904. It was supposed that there were no marine borers in the bay and the wharf was built of unprotected native spruce. After about 18 months of service, the wharf failed because the piles were honeycombed by shipworms. The wharf was replaced. A similar failure occurred in Uyak Bay, Alaska. *Cause of failure:* Inadequate investigation as to presence of marine borers and lack of provision against them. *Prevention:* Use of construction resistant to borers. *Remedy:* Replacement of structures. (*Report of the Marine Piling Investigations of the Natl. Research Council,* 1924.)

Case 47. Greenheart piles at Swanage Pier were seriously damaged by *Teredo* and *Limnoria,* principally between high- and low-water levels. They were cut and replaced by reinforced-concrete sections. *Cause of failure:* Marine-borer attack on one of the most resistant types of untreated wood. *Prevention:* Use of fully resistant construction. *Remedy:* Cutting away and replacing damaged portions with borer-resistant construction. (Du-Plat Taylor, *The Structural Engineer, May,* 1929.)

Marine-borer Failures Due to Changed Conditions. *Case* 48. Commonwealth Pier No. 5 in Boston Harbor was started in 1897 by building a granite sea wall around the site and filling the interior. The wall was carried on spruce piles, using stone-chip filling around and in front of the piles, with a facing of large stone. This wall was surrounded by a 50-ft platform supported by untreated oak piles, driven with the bark on, and with heads covered with coal tar.

In 1925, marine borers were becoming active in Boston Harbor, and 55 per cent of the oak piles were attacked. In 1934, 92 per cent of the piles were affected, with 10 per cent half or more destroyed. Piles upon which the original bark remained were in good condition. The major part of the damage was done by *Limnoria.*

The riprap and silt in each area were excavated to as great a depth as feasible, and a steel cassion 54 in. in diameter was driven 5 ft minimum into the subsoil. The material in the caisson was then excavated to a depth of 15 ft below the anticipated level of the new pile tops. Four H piles were driven in each caisson to bedrock. Two feet of sand and gravel were placed in the bottom and the caisson was concreted. The steel piles had a maximum length of 137 ft. *Cause of failure:* Attack by marine borers on untreated wood piles where bark was damaged, in a location free from borers when the piles were driven. *Prevention:* Use of pile material resistant to marine-borer attack, in anticipation of possible future borers. *Remedy:* Reconstruction. (C. M. Spofford, "Pier Reconstruction in Boston Harbor—Reconstructing a Wood-pile Pier," *Civ. Eng.,* vol. 7, No. 12, pp. 843–844, December, 1937.)

Case 49 (Fig. 16.9). The wood-wharf substructure of the Army Base at Boston, Mass., built in 1919, contained 28,000 southern-pine piles 30 to 65 ft long, driven with the bark on. The piles were cut at about high-tide level and braced together between high- and low-water levels.

Marine borer activity was just noticeable in 1922. By 1932, approximately 30 per cent of the original pile cross section had been eaten away, as well as practically all of low-water and diagonal bracing and some of the longitudinal bracing near the high-tide level. Attacks extended from the mud line to some distances above low water. Piles on which the bark remained were not subject to attack.

A steel-sheet-piling bulkhead was driven around the entire base to retain sand fill placed around the wood piles and defective wood piles having a diameter of less than 4 in. at the mud line were replaced either by cutting at the mud line and capping with a new pile, drilling holes in

FIG. 16.9. *Limnoria* and *Chelura* damage, Boston Harbor. (*Courtesy of Dr William F. Clapp.*)

the deck and driving twin piles on each side, or encasing the old pile in a concrete shell. Pile encasement was done by use of a steel form 14 in. in diameter and 4 ft long, with a sheet-metal collar at the bottom and provision for withdrawing after the concrete had set. *Cause of failure:* Attack by marine borers on untreated wood piles where bark was damaged, and on wood bracing, in a location free from borers when the piles were driven. *Prevention:* Use of materials resistant to borers and sea water, in anticipation of possible future borer attack. *Remedy:* Reconstruction. (R. L. Miller, "Pier Reconstruction in Boston Harbor—Repairing Substructure of the Army Base," *Civ. Eng.*, vol. 7, No. 12, pp. 844–845, December, 1937.)

Case 50 (Figs. 16.10, 16.11, 16.12). In San Francisco Bay, sporadic but not serious attacks by *Teredo* were known to have occurred as far back as 1870, but it was thought that fresh-water inflow from rivers would prevent severe injury owing to the decreased salinity of the water.

In 1917, *Teredo* caused serious damage to untreated piles, chiefly Douglas fir, at the Mare Island Navy Yard, and progressed so rapidly that, by 1919, parts of the water-front structure began to fall, whole docks being affected and many millions of dollars worth of destruction occurring. The American Wood-Preservers' Association and the Forest

FIG. 16.10. Several loaded freight cars plunged into bay, Oleum, Calif., October, 1919. One of first docks which failed in San Francisco Bay attacks by marine borers. (*Courtesy of Dr. William F. Clapp.*)

FIG. 16.11. Municipal wharf and house collapsed from *Teredo* attacks in San Francisco Bay, Benecia, Calif., October, 1920. (*Courtesy of Dr. William F. Clapp.*)

Products Laboratory combined in a survey that resulted in full study of the causes and methods of prevention possible. It was found that the salinity was unfavorable for *Teredo* only in certain months. One conclusion reached was that all great ports are subject to repeated invasions by borers from other localities. *Cause of failure:* Introduction of borers, probably brought by foreign shipping. *Prevention:* Adoption of a policy of restricting uncontrolled use of untreated wood, supplemented, if possible, by a policy of removal of unused infested structures and prevention of use of the harbor as a dumping ground for waste wood; otherwise the use of protective casings or piles other than wood. *Remedy:* Reconstruction and installation of protective appliances. (Hill and Kofoid.[20])

Case 51 (Fig. 16.13). An unexpected influx of marine borers into waters previously considered free of them, combined with decay, at a pier built at Playland, on Long Island Sound, Rye, N.Y., in 1929 on untreated piles, caused the cost of repairs in 15 years to exceed the original cost. Untreated piles were used since other untreated piles in surrounding waters had been in service some years without signs of attack. Practically all piles have been replaced. Early replacements were made with southern pine having 12- and 16-lb Grade 1 coal-tar creosote treatments. In 1944, a retention of 20 lb was specified. Special care is now being paid to protection of pile cutoffs, field cuts, and boltholes, in accordance with the American Wood-Preservers' Association *Instructions*

FIG. 16.12. Untreated ferry-slip fender piles destroyed by *Teredo,* Carquinez Strait, San Francisco Bay. (*Courtesy of Southern Pacific Co.*)

FIG. 16.13. *Teredo* attack on pile in Playland Pier at Rye, N.Y., on Long Island Sound. Damaged pile shown at left, and creosoted-replacement-pile cutoff at right. (*Courtesy of Ralph H. Mann, American Wood-Preservers' Association.*)

for Field Treatment of Creosoted Timber and Piles. Pressure treatments are used for boltholes. *Cause of failure:* Destruction of piles by marine borers in water previously thought uninfested, and by decay. *Prevention:* Use of creosoted piles, with adequate butt and bolthole protection. *Remedy:* Replacement of all piles, and use of good butt and bolthole treatments. ("Repairs to Untreated Timber Pier in 15 Years Exceed Original Cost," *Wood Preserving*, vol. 22, No. 8, pp. 83–87, August, 1944.)

Case 52. The pier head of the Old Southern Pier in London, England, was built on untreated Memel fir piles, 12 to 14 in. square, driven 10 ft into sand or clay, sheathed to sea bottom with copper coated with pitch and tar. Twelve months after completion it was found at neap

FIG. 16.14. Collapsed bridge over Manasquan River near Brielle, N.J., resulting from *Teredo* damage to untreated wood piles exposed by scour after dredging channel. (*Photo by Frank Beardsley, Asbury Park Press, Asbury Park, N.J.*)

tides that nearly all piles were damaged at the base by *Teredo* and *Limnoria*, and in a few years more, all piles were completely eaten through and it was necessary to replace the structure, using cast-iron piles. *Cause of failure:* Local denudation of coast line after the piles were driven, so that the copper no longer reached the sea bottom. *Prevention:* Use of treated piles, sheathing driven into sea bottom, or piles impervious to borers. *Remedy:* Replacement of entire structure. (*Proc. Inst. Civil Engrs.* (*London*), vol. 9, 23.)

Case 53 (Fig. 16.14). Collapse, in 1946, of two spans of a bridge constructed in 1924 over the Manasquan River near Brielle, N.J., was blamed on *Teredo* destruction of untreated wood piles. The mile-long bridge crossed the wide tidal channel of the river, and consisted of a number of short spans carried on low concrete piers resting on wood piles driven to sand. One of these piers collapsed, dropping the ends of two spans several feet. In falling, one of the spans wedged the lift span. Parts of the bridge which did not fall were cracked and pronounced unsafe. It was expected that repairs might require a year. The gas main serving shore towns was severed.

Since construction, the navigable channel was dredged to 20 ft, and this deepening was believed to have changed the flow under the bridge, resulting in scour that exposed the piles of the collapsed pier, making them subject to *Teredo* attack. *Cause of failure:* Scour caused by dredging exposed untreated wood piling to marine-borer attack. *Prevention:* Use of treated, composite, concrete, or metal piling. *Remedy:* Removal and reconstruction of damaged spans, using resistant piles. ("Marine Borers Blamed for Bridge Collapse," *Eng. News-Record,* vol. 137, No. 9, p. 206, Aug. 29, 1946; and *Philadelphia Evening Bulletin,* Aug. 21, 1946.)

Case 54 (Fig. 16.15). Untreated wood piles were used for a pier for the Howard Company in Oakland, Calif. Following satisfactory service, further similar piles were used. Then the discharge of effluent from a gas plant, which had been emptied into the water nearby, was ordered by public authorities to cease, and it soon became necessary to replace the piles with treated ones, because of borer attack. The effluent had formed a temporary coating on the untreated piles. *Cause of failure:* Use of unprotected wood piles in sea water, failing after removal of temporary pollution. *Precaution:* Study of conditions and use of protection. *Remedy:* Replacement of piles. (Hill and Kofoid.[20])

FIG. 16.15. Untreated Douglas fir piles "hourglassed" by *Limnoria.* Good piles shown are replacements. Oakland, Calif., estuary. (*Courtesy of Moore Shipbuilding Co.*)

Case 55. The owners of a paper mill in Newfoundland employed scientists to determine the absence of borers before constructing a large wharf. The waters were correctly pronounced safe, and piles were driven. Two years later, a tug tying up at the wharf pulled the end away. It was found that lumber, not the piles, had come from a harbor in Nova Scotia where *Teredo* abounded. The *Teredo* multiplied rapidly and infested the piles. *Cause of failure:* Infestation of *Teredo* from imported infected lumber. *Prevention:* Inspection of imported wood by the owner, or use of treated piles if the possibility of imported infection existed from other sources over which this owner had no control. *Rem-*

edy: Expensive encasement or replacement of piles with treated piles. (*Science Illustrated*, vol. 2, No. 7, pp. 42–45, 64, July, 1947.)

Failure Due to Marine-borer Attack on Temporary Wood Piles. *Case* 56 (Fig. 16.16). An unusually severe attack by *Teredo* on the piles of a temporary trestle at a long jetty at Fire Island Inlet on Long Island caused the collapse of three bents of the trestle in 1940, 98 days after building. The piles were secondhand spruce that had been in service in the polluted waters of Long Island Sound near North Beach since 1924 without attack, but diver inspection revealed that new white-pine and mixed-oak piles also had been attacked badly. The worst attacks

Fig. 16.16. Bent of a temporary trestle at Fire Island Inlet on the south shore of Long Island after only 98 days' exposure to *Teredo* attack. Infestation was known, but untreated wood was used for a 6-month job. (*Courtesy of Ralph H. Mann, American Wood-Preservers' Association.*)

occurred 150 ft offshore where the water is 6 ft deep at low tide. Attacks extended from mud line to low water, being most severe 2 to 4 ft below mean low water. The rapidity of attack was notable. New oak replacement piles were driven, some with bark on and brush-treated with creosote, others wrapped with felt paper and galvanized iron in the zone of heaviest attack. *Cause of failure: Teredo* attack on untreated piles in unpolluted water. *Prevention:* Treatment or encasement of piles. *Remedy:* Driving new protected piles. (*Eng. News-Record,* vol. 124, No. 5, p. 163, Feb. 1, 1940; *Science Illustrated,* vol. 2, No. 7, pp. 42–45, 64, July, 1947.)

Failure Due to Marine-borer Attack on Creosoted Wood Piles. *Case* 57. The wharf at the Charleston, S.C., Lighthouse Depot was built in 1916 with longleaf-yellow-pine piles treated with 18 lb of creosote oil per cubic foot. Gradual but increasing attacks by *Limnoria* resulted, by

1932, in need for replacement or reinforcement. Piles were encased in concrete for 1 ft above high water to 3 ft below the usual line. The lengths of encasements varied from 7 to 28 ft. This was done by the Hay process, using 4-in.-thick concrete, which also filled all holes and damaged sections with concrete. *Cause of failure:* Lack of protection against types of borers immune to creosote. *Prevention:* Use of type of pile not subject to attack by borers immune to creosote. *Remedy:* *Encasement* of piles. (Cement-Gun Co., Inc.)

Case 58. Lillian Bridge over Perdido Bay between Alabama and Florida was built in 1916, as a toll bridge. The bay averages 8 to 12 ft in depth, with a bottom of very soft mud as much as 22 ft thick, underlaid by strata of sand and mud. In 1926, purchase for toll-free operation was proposed. Engineering advice was against purchase.

In 1929, the bridge was closed after four crashes demonstrated the great danger of continued use. When the bridge was taken over, the principal defects evident were above water level, and rotted piling was cut off at low water and replaced with braced columns. No marine-borer damage was noted above low-water level. Frequent floods of fresh water from the Perdido River kept salinity down. One accident disclosed that one of the piles which had been sawed off at low water and capped had broken off at the mud line 10 ft below tide level. Although other piles were practically perfect at tide level, they were practically eaten through at the mud line, with but little damage even 3 ft above. A large portion of the damage was caused in 1928 by *Bankia gouldi*, which were later killed by fresh water. *Sphaeroma destructor* entered these holes and continued the destruction. The few *Bankia* which entered the wood in 1929 did not thrive, but more were found in healthy condition in 1931.

When designing the new bridge, the great length of piles required gave rise to the consideration that the cost of concrete piles would be prohibitive, and heavily creosoted wood piles were selected. They were encased in a 4-in. jacket of mesh-reinforced concrete inside a sheet-metal form, painted on both sides with tar, extending from 3 ft below the mud line to 2 ft above tide level. The jackets were used to protect the piles from *Sphaeroma destructor*, to which creosoted piling is not immune. Also, the presence of *Limnoria* in nearby Pensacola Bay led to a fear that, in a future drought, high salinity might occur and permit this organism, which is not stopped from attacking by creosote, to thrive. *Cause of failure:* Destruction of piles at mud line by marine borers. *Prevention:* Encasement of piles from just below mud line to just above water level. *Remedy:* New piles and, in this case, a new bridge. (W. E. Wheat, "Jacketed Creosoted Piles Repel Marine Borers," *Eng. News-Record*, vol. 108, No. 6, pp. 208–209, Feb. 11, 1932.)

Failure Due to Effect of Treatment. *Case* 59. At Lake Charles, Louisiana, 1,500 creosoted wood piles were driven for a wharf for the Dock Board, without suspicion that anything was wrong with the piles. Dredging operations in front of the wharf caused some piles to fall over. When these were pulled it was found that from 15 to 25 ft of the pile lengths had the wood fibers so completely separated that they looked like old, worn hemp rope. It was found that 80 per cent of the piles pulled were in this condition. The Port Commissioners rejected the structure.

The specifications for treatment were the same that had been used in many other places without report of damage of this nature and extent. The contractors brought suit for payment, and during the trial it was disclosed that in the latter 2 hr of the steaming period, which had been at 30 psi, compressed air had been introduced into the cylinder to such an extent that the combined pressure of steam and air was from 70 to 90 psi. This appeared to be the only variation from practice during treating or driving. The Port Commission claimed that this introduction of compressed air was a violation of the specifications, and the company which treated the piles claimed that such treatment was not prohibited by the specifications. The United States District Court held that there had been a breach of the specifications, and handed down a decision for the Dock Board. The Court did not rule as to the cause of failure, however, and on this point expert witnesses disagreed. *Cause of failure:* Possibly breach of specifications during steaming treatment. *Prevention:* Possibly strict adherence to limitations on pressure, and avoidance of excessive time of steaming. *Remedy:* Removal and rebuilding of entire construction. ("Court Decides Suit over Piles in Favor of Dock Board," *Eng. News-Record,* vol. 109, No. 7, p. 209, Aug. 18, 1932.)

Failure Due to Chemical Attack on Wood Piles. *Case* 60. Oil tanks 60 ft in diameter and 50 ft high of the Sun Oil Co. plant at Marcus Hook, Pa., were supported on wood piles that were almost severed by ground water containing a high percentage of sulphuric acid filtered out from a nearby large acid pool.

The ground was a mixture of cinder fill, top soil, loam, silt, and 80-mesh sand. A ring foundation of soil was solidified under each tank wall by the Joosten method, then pierlike areas were solidified to support the bottom slabs and interior column loads. *Cause of failure:* Chemical attack on wood piles. *Prevention:* Avoidance of soil contamination. *Remedy:* Soil solidification. (C. M. Riedel, Chemical Soil Solidification Co.)

TABLES

GROUP I

TEMPORARY COMPRESSION FIGURES

TABLE I. TEMPORARY COMPRESSION ALLOWANCE C_1 FOR PILE HEAD AND CAP[a]

Material to which blow is applied	Easy driving, $p_1 = 500$ psi on cushion or pile butt if no cushion, in.	Medium driving, $p_1 = 1{,}000$ psi on head or cap, in.	Hard driving, $p_1 = 1{,}500$ psi on head or cap, in.	Very hard driving, $p_1 = 2{,}000$ psi on head or cap, in.
Head of timber pile....	0.05	0.10	0.15	0.20
3–4-in. packing inside cap on head of precast concrete pile....	0.05 + 0.07[b]	0.10 + 0.15[b]	0.15 + 0.22[b]	0.20 + 0.30[b]
½–1-in. mat pad only on head of precast concrete pile........	0.025	0.05	0.075	0.10
Steel-covered cap, containing wood packing, for steel piling or pipe...............	0.04	0.08	0.12	0.16
3⁄16-in. red electrical-fiber disk between two ⅜-in. steel plates, for use with severe driving on Monotube pile......	0.02	0.04	0.06	0.08
Head of steel piling or pipe...............	0	0	0	0

[a] Largely from A. Hiley, "Pile Driving Calculations with Notes on Driving Force and Ground Resistance," *The Structural Engineer*, vol. 8, July and August, 1930.[7] For a fuller discussion of the means of obtaining these values see this reference. For purpose of this article values represent average conditions and may be used.

[b] The first figure represents the compression of the cap and wood dolly or packing above the cap, whereas the second figure represents the compression of the wood packing between the cap and the pile head.

NOTE: Superior numbers (with or without letters) refer to the Bibliography, pp. 641*ff.*, in which the material is organized by subject.

TABLE II. TEMPORARY COMPRESSION VALUES OF C_2 FOR PILES

Type of pile	Easy driving, $p_2 = 500$ psi for wood or concrete piles, 7,500 psi for steel, net section, in.	Medium driving, $p_2 = 1{,}000$ psi for wood or concrete piles, 15,000 psi for steel, net section, in.	Hard driving, $p_2 = 1{,}500$ psi for wood or concrete piles, 22,500 psi for steel, net section, in.	Very hard driving, $p_2 = 2{,}000$ psi for wood or concrete piles, 30,000 psi for steel, net section, in.
Timber pile, based on value of $E = 1{,}500{,}000$. Proportion for other values of E given in Table VI[a]	$0.004 \times L^b$	$0.008 \times L^b$	$0.012 \times L^b$	$0.016 \times L^b$
Precast concrete pile ($E = 3{,}000{,}000$[a,c])	$0.002 \times L$	$0.004 \times L$	$0.006 \times L$	$0.008 \times L$
Steel sheet piling, Simplex tube, pipe pile, Monotube shell, Raymond steel mandrel[d] ($E = 30{,}000{,}000$)	$0.003 \times L$	$0.006 \times L$	$0.009 \times L$	$0.012 \times L$

[a] All other values in direct proportion to p_2 and inverse proportion to E.

[b] L should be considered as length to center of driving resistance, not necessarily full length of pile.

[c] May reach 6,000,000 for exceptionally good mix.

[d] When computing p_2 for a Raymond steel mandrel, it is suggested that the weight of the mandrel be divided by 3.4 × the effective length of pile in feet to obtain the average area.

TABLE III. TEMPORARY COMPRESSION OR QUAKE OF GROUND ALLOWANCE C_3[a]

All values of p_3 to be taken on projected area of pile tips or driving points for end-bearing piles and piles of constant cross section; on gross area of pile at ground surface in case of tapered friction piles; and on bounding area under H piles

	Easy driving, $p_3 = 500$ psi, in.	Medium driving, $p_3 = 1{,}000$ psi, in.	Hard driving, $p_3 = 1{,}500$ psi, in.	Very hard driving, $p_3 = 2{,}000$ psi, in.
For piles of constant cross section[b,c]	0 to 0.10	0.10	0.10	0.10

[a] Largely from A. Hiley, "Pile Driving Calculations with Notes on Driving Force and Ground Resistance," *The Structural Engineer*, vol. 8, July and August, 1930.[7] For a fuller discussion of the means of obtaining these values see this reference. For purpose of this article values represent average conditions and may be used.

[b] It is recognized that these values should probably be increased in the case of piles with battered faces, but insufficient test data are available at present time to cover this condition.

[c] If the strata immediately underlying the pile tips are very soft, it is possible that these values might be increased to as much as double those shown

GROUP II

OPERATING DATA ON HAMMERS, EXTRACTORS, AND RELATED EQUIPMENT

AMERICAN HAMMERS

TABLE IV.1. VULCAN DROP HAMMERS

Item	Type of hammer						
	K	M	P	S	T	V	W
Ram, weight (W_r), lb	500	700	1,000	.....	.....	2,000	
	600	800	1,200	1,500	1,800	2,500	3,000
Helmet for steel sheet pile, weight, lb	...	...	.....	.....	600	700	750
Pipe cap, weight, lb	280	300	520	720	760	900	970
Driving-head assemblies for use with Monotube piles,[a] weight, lb:							
Wood or fiber cushions	...	...	760	760	760	760	760
Steel and fiber cushions	...	...	814	814	814	814	814
Sheeting cap, weight, lb	Vary considerably, but about same weight as pile caps						

[a] These weights include an allowance of 30 lb for pilot ring weights, which vary from approximately 15 to 45 lb, depending on diameter of pile head.

Table IV.2. Eagle Drop Hammers[a]

Pile caps (follower blocks) lb,[b] for:	Weight of hammer, lb																			
	1,000	1,500	1,750	2,000	2,250	2,500	2,750	3,000	3,250	3,500	4,000	4,500	5,000	5,500	6,000	6,500	7,000	7,500	8,000	8,500
Wood piles:																				
9 in.	...	...	700	700	800	800	900	900												
10 in.[c]	...	...	...	...	870	870	950	950	950	1,000	1,000	1,000								
Wood sheeting	400	600	600	650																
H piles, 8 to 12 in.	...	...	...	900	900	950	950	1,000	1,000	1,050	1,050	1,050								
Steel sheet piles	...	...	...	900	900	900	950	950	950	1,000	1,000	1,000								
Steel pipe piles:																				
10 to 12 in.	...	...	...	900	900	900	950	950	950	1,000	1,000	1,000								
14 to 16 in.	...	...	...	...	...	1,000	1,000	1,050	1,050	1,100	1,100	1,100								
Precast concrete piles:																				
12 in. square	...	...	...	...	...	.....	.....	.....	.....	.....	1,000	1,000	1,050	1,050	1,100					
14 to 16 in. square[d]	...	...	...	...	...	.....	.....	.....	.....	.....	.....	.....	1,250	1,300	1,400	1,500	1,500	1,600	1,600	1,700

[a] Manufactured and sold by Eagle Iron Works, Des Moines, Ia. Hammers and follower block more or less tailor-made to fit purchaser's leads.
[b] Follower-block weights are approximate and will vary with different lead sizes. A wood cushion block required between hammer and follower block.
[c] Larger sizes furnished on request.
[d] Requires center filler for 16-in. size.

TABLE IV.3. McKIERNAN-TERRY SINGLE-ACTING STEAM HAMMERS

Item	Type of hammer					
	S3	S5	S8	S10	S14	S20
Normal stroke (h), in.	36	39	39	39	32	36
Maximum stroke, in.	39	42	42	42	36	38
Ram, weight (W_r), lb	3,000	5,000	8,000	10,000	14,000	20,000
Casing, weight (W_c), lb	5,000	6,000	8,500	10,000	14,400	15,500
Manufacturer's rated energy per blow (E_n), ft-lb	9,000	16,250	26,000	32,500	37,500	60,000
Strokes per min	65	60	55	55	60	60
Weight of anvil block, lb:						
Flat	800	1,375	1,600	2,200	3,200	3,100
Cup	800	1,400	1,650	2,250	3,300	
Pipe	940	1,550	1,900	2,700	3,600	
H-pile (approx)	900	1,500	1,750	2,500	3,400	
Sheet piling (approx)	900	1,500	1,750	2,500	3,400	
Concrete piles without extended bars	1,400	1,900	2,700	3,200	6,000	
Recommended steam pressure at boiler or air pressure at compressor, psi	100	100	100	100	125	165
Steam or air pressure at hammer, psi	80	80	80	80	100	150
Size of boiler, hp	25	40	55	65	90	150
Hose size, in.	1½	2	2½	2½	3	3
Size of exhaust hose for underwater driving, in.	3	3½	4	4	5	

TABLE IV.4. VULCAN SINGLE-ACTING STEAM HAMMERS

Item	Type of hammer[a]								
	2	1	06	08	010	014	020	0[b]	0R[b]
Normal stroke (h), in	29	36	36	39	39	36	36	39	39
Ram, weight (W_r), lb	3,000	5,000	6,500	8,000	10,000	14,000	20,000	7,500	9,300
Casing, weight (W_c), lb:									
With standard base	3,700	4,700	4,700	8,750	8,750	13,500	19,000		
With McDermid base	4,100	5,200	5,200						
Manufacturer's rated energy per blow (E_n), ft-lb	7,260	15,000	19,500	26,000	32,500	42,000	60,000	24,375	30,225
Strokes per min (normal stroke and no set)	70	60	60	50	50	60	60	80	80
Steam or air pressure at hammer, psi[c]	80	80	100	83	105	110	120	80	100
Size of boiler, nominal, hp	25	40	50	60	75	100	130	60	72
Boiler heating surface, sq ft[d]	300	480	600	720	900	1,200	1,580	720	864
Free air, cfm:									
Adiabatic	336	565	625	880	1,002	1,282	1,756	841	1,020
Isothermal	578	972	1,135	1,525	1,832	2,380	3,321	1,477	1,754
Hose size, in	1½	2	2	2½	2½	3	3	2½	2½

[a] Types Nos. 3 and 4 not now manufactured. Very few extant. Consult manufacturer for properties.

[b] Types Nos. 0 and 0R not manufactured after Aug. 1, 1959, but data are included for use with existing hammers.

[c] From 5 to 15 psi higher at boiler or air compressor.

[d] Based on 12 sq ft per hp. It is conservative for some types of boilers, particularly those with large fireboxes or horizontal tubes or for intermittent use.

TABLE IV.5. RAYMOND SINGLE-ACTING STEAM HAMMERS

Item	Type of hammer[a]						
	1	1S	0	00	000	0000	00000
Normal stroke (h), in.	36	36	39	39	39	39	39
Ram, weight (W_r), lb	5,000	6,500	7,500	10,000	12,500	15,000	17,500
Casing, weight (W_c), lb	6,000	6,000	8,500	8,500	8,500	8,800	8,950
Manufacturer's rated energy per blow, (E_n), ft-lb	15,000	19,500	24,375	32,500	40,600	48,750	56,875
Strokes per minute (when pile takes up)	60	58	50	50	50	46	44
Steam pressure at boiler, min, psi	110	125	125	125	135	140	150
Size of boiler, hp[b]	30	35	50	55	70	85	100
Free air, cfm	500	[c]	750	[c]	[c]	[c]	[c]
Hose size, in.[d]	1½	1½	2	2	2½	2½	2½–3
Raymond cap block:							
Diam., in.	11 7/16	11 7/16	13 11/16	13 11/16	13 11/16	13 11/16	13 11/16
Thickness, in.	6	6	6½	6½	6½	6½	6½

[a] Raymond Concrete Pile Company also has Vulcan No. 2 and No. 0R hammers to take Raymond cores.

[b] Boiler horsepower based on standard IBW rating of 10 sq ft of heating surface per horsepower. New packaged boiler units are rated at about 6 sq ft per hp and must be selected on basis of pounds of steam delivered per horsepower rather than on their rated horsepower.

[c] Pressure of 115 psi at the compressor furnished by ordinary portable compressor is not high enough.

[d] Steam piping should be one size larger than hose. About 50 ft of hose and 70 ft of steam piping may be considered normal; larger lengths require larger sizes or higher boiler pressure.

TABLE IV.6. SUPER-VULCAN DIFFERENTIAL-ACTING STEAM HAMMERS, OPEN TYPE

Item	Size of hammer							
	18C	30C	50C	65C	80C	140C	200C	400C
Normal stroke (h), in	10½	12½	15½	15½	16½	15½	15½	16½
Ram, weight (W_r), lb	1,800	3,000	5,000	6,500	8,000	14,000	20,000	40,000
Casing with standard base, weight (W_c), lb	2,339	4,036	6,782	8,386	9,885	13,984	19,050	43,000
Casing with McDermid base, weight (W_c), lb	2,364	4,090	6,845	8,449				
Manufacturer's rated energy per blow (E_n), ft-lb	3,600	7,260	15,100	19,200	24,450	36,000	50,200	113,488
Strokes per min	150	133	120	117	111	103	93	100
Standard plate, weight, lb	15	40	48	......	75			
McDermid plate, weight, lb	22	55	65	......	80			
Dished-cap plate, weight, lb	30	55	70	......	125			
Driving-head assembly for use with Monotube piles,[a] weight, lb:								
Wood or fiber cushions	......	......	......	......	1,205			
Steel and fiber cushions	......	......	......	......	1,249			
Steam or air pressure at hammer, psi[b]	120	120	120	150	120	140	142	150
Steam pressure at boiler or air pressure at compressor, psi[b]	125–130	125–130	125–130	......	125–130			
Size of boiler, nominal, hp	25	40	60	70	80	100	120	300
Boiler heating surface, sq ft[c]	300	480	720	807	960	1,200	1,440	3,600
Free air, cfm:								
Adiabatic	308	488	880	991	1,245	1,425	1,746	4,659
Isothermal	586	930	1,675	1,998	2,370	2,830	3,470	9,400
Hose size, in	1¼	1½	2	2	2½	3	4	5

[a] These weights include an allowance of 30 lb for pilot-ring weights, which vary from approximately 15 to 45 lb, depending on diameter of pile head.

[b] Steam pressure should be 5 to 10 lb higher at boilers. The standard portable-air-compressor pressure of 115 psi is insufficient. Portable air compressors in first-class condition can deliver these pressures by taking advantage of the overload capacity, but this is a practice the manufacturers do not like to recommend.

[c] Based on 12 sq ft per hp. It is conservative for some types of boilers, particularly those with larger fireboxes or horizontal tubes or for intermittent use.

TABLE IV.7. SUPER-VULCAN DIFFERENTIAL-ACTING STEAM HAMMERS, CLOSED TYPE[a]

Item	Size of hammer							
	600	1,100	1,800	3,000	5,000	8,000	14,000	20,000
Normal stroke (h), in.	7½	9	10½	12½	15½	16½	15½	15½
Ram, weight (W_r), lb	600	1,100	1,800	3,000	5,000	8,000	14,000	20,000
Casing, weight (W_c), lb	1,232	1,839	2,339	4,036	6,782	9,885	13,984	19,050
Manufacturer's rated energy per blow (E_n), ft-lb	1,125	2,180	3,600	7,260	15,100	24,450	36,000	50,200
Strokes per min	225	181	150	133	120	111	103	98
Flat anvil, weight, lb	.....	.....	200	392	588	877	1,460	1,450
Cup or bell anvil, weight, lb	55	101	135	214	358	592		
Driving-head assembly for use with Monotube piles:[c]								
Wood or fiber cushions, weight, lb	.....	.....	....	....	....	1,205		
Steel and fiber cushions, weight, lb	.....	.....	....	....	....	1,249		
Steam pressure at boiler or air pressure at compressor, psi	.....	.....	125–130	125–130	125–130	125–130		
Steam or air pressure at hammer, psi[c]	.....	.....	120	120	120	120	140	142
Size of boiler, nominal, hp	.....	.....	25	40	60	80	100	120
Boiler heating surface, sq ft[d]	.....	.....	300	480	720	960	1 200	1,440
Compressed air, cfm	.....	.....	308	488	880	1,245	1,425	1,746
Hose size, in.	.....	.....	1¼	1½	2	2½	3	3

[a] Manufacture of Super-Vulcan differential-acting closed-type hammer was discontinued in 1949, but data are included for use with existing hammers.

[b] These weights include an allowance of 30 lb for pilot-ring weights, which vary from approximately 15 to 45 lb, depending on diameter of pile head.

[c] At least 5 to 10 lb higher at boiler. For use submerged, add slightly less than ½ psi per ft of depth.

[d] Based on 12 sq ft per hp. It is conservative for some types of boilers, particularly those with large fireboxes or horizontal tubes or for intermittent use.

TABLE IV.8. VULCAN DIFFERENTIAL-ACTING MARINER HAMMERS[a]

Item	Size of hammer		
	3M	5M	8M
Normal stroke (h), in.	12½	15½	16½
Ram, weight (W_r), lb	3,000	5,000	8,000
Casing, weight (W_c), lb	4,590	6,300	9,900
Manufacturer's rated energy per blow (E_n), ft-lb	7,260	15,100	24,450
Strokes per min	133	120	111
Steam or air pressure, psi	120	120	120
Size of boiler, nominal, hp	40	60	80
Free air, cfm:			
Adiabatic	488	880	1,245
Isothermal	930	1,675	2,370
Hose size, in.	1½	2	2½

[a] Submersible.

TABLE IV.9. VULCAN DIFFERENTIAL-ACTING PILE AND STEEL SHEETING HAMMER

Item	Size of Hammer, DHG 900
Normal stroke (h), in	10
Ram, weight (W_r), lb	900
Casing and flat anvil, weight (W_c), lb	4,100
Manufacturer's rated energy per blow (E_n), ft-lb	4,000
Strokes per min	236
Steam or air pressure at hammer, psi	78
Size of boiler, nominal, hp	40
Free air, cfm:	
Adiabatic	580
Isothermal	1,171
Hose size, in	1½

TABLE IV.10. VULCAN DIFFERENTIAL-ACTING STEAM HAMMER, PORTABLE TYPE

Item	Size of Hammer, DGH 100
Normal stroke (h), in	6
Ram, weight (W_r), lb	100
Casing, weight (W_c), lb	686
Manufacturer's rated energy per blow (E_n), ft-lb	386
Strokes per min	303
Air pressure, psi	60
Free air, cfm:	
Adiabatic	74
Isothermal	118
Size of boiler, nominal, hp	5
Hose size, in	1

TABLE IV.11. RAYMOND DIFFERENTIAL-ACTING STEAM HAMMERS

Item	Size of hammer[a]		
	15M[b]	65C	150C
Normal stroke (h), in	18	16	18
Ram, weight (W_r), lb	5,000	6,500	15,000
Casing, weight (W_c), lb	5,305	8,175	17,500
Manufacturer's rated energy per blow (E_n), ft-lb[c]	15,060	19,500	48,750
Blows per min when pile takes up[c]	85	100–110	
Steam pressure at boiler, psi[d]	150	150	150
Steam pressure at hammer, min, psi[d]	120	120	
Size of boiler, hp	......	70	160
Air pressure at hammer, psi[e]	120	120	
Boiler heating surface, sq ft	600	700	1,600
Hose size, in	2	2	3
Raymond cap block:			
Diameter, in	11⅜	11⁷⁄₁₆	18
Thickness, in	Special	6	Special

[a] Raymond Concrete Pile Company also has Vulcan Nos. 50C and 80C hammers fitted to take Raymond cores.

[b] Special hammer fits inside core.

[c] These hammers will deliver the full rated striking energy if supplied with adequate steam pressure. However, the number of blows struck per minute will vary with the driving and will reach the maximum number listed only when the pile is practically at refusal.

[d] Steam pressure at the hammer should be great enough so that the cylinder and base will rise slightly with every blow. It should not be attempted to operate these hammers with boilers smaller or of lower pressure than indicated.

[e] These hammers will not run on the usual standard portable-air-compressor pressure of 115 psi. In order to obtain 120 psi at the hammer, the compressor should deliver air at 130 psi minimum, preferably at 135 or 140 psi. Portable air compressors in first-class condition can deliver these pressures by taking advantage of this full overload capacity, although this is a practice that the manufacturers do not like to recommend. If pressure is allowed to fall off, full driving energy will not be attained.

TABLE IV.12. MCKIERNAN-TERRY DOUBLE-ACTING STEAM HAMMERS
HAMMER SIZES 0 TO 5

Item	Size of hammer				
	0	1	2	3	5
Normal stroke (h), in	4¾	3¾	5¼	5¾	7
Ram, weight (W_r), lb	5¼	21	48	68	200
Casing, weight (W_c), lb	89¾	124	295	607	
Manufacturer's rated energy per blow (E_n), ft-lb	[a]	[a]	[a]	[a]	1,000
Strokes per min	....	...	...	...	300
Recommended steam pressure at boiler or air pressure at compressor, psi	125	125	125	125	125
Steam or air pressure at hammer, psi	100	100	100	100	100
Size of boiler, hp	5	10	10	15	20
Compressed air, cfm[b]	60	70	70	110	250
Hose size, in	¾	¾	¾	1	1¼

[a] Value not stated by manufacturer.

[b] Actual delivery.

TABLE IV.12. MCKIERNAN-TERRY DOUBLE-ACTING STEAM HAMMERS (*Continued*)
HAMMER SIZES 6, 6.5, 7, 9, 9-B-2, 10-B-2

Item	Size of hammer					
	6	6.5	7	9[a]	9-B-2[b]	10-B-2[b]
Normal stroke (h), in.	8¾	8⅜	9½	12	16	20
Ram, weight (W_r), lb	400	600	800	1,250	1,500	2,500
Casing, weight (W_c), lb	.....	3,790	3,872	5,850	4,815	6,662
Manufacturer's rated energy per blow (E_n), ft-lb	2,500	3,200	4,150	.....	8,200	15,000
Strokes per min	275	280	225	.....	140	115
Anvil, weight, lb:						
Flat	.....	160	328	400	445	838
Cup or bell	.....	.....	342	.....	420	850
Driving-head assemblies for use with Monotube piles, weight, lb:[c]						
Standard base, wood or fiber cushions	.....	.....	.....	.....	712	908
Same, adapter type	.....	.....	.....	.....	734	795
Standard base, steel and fiber cushions	.....	.....	.....	.....	767	996
Same, adapter type	.....	.....	.....	.....	787	848
Recommended steam pressure at boiler or air at air compressor, psi	125	125	125			
Steam or air pressure at hammer, psi	100	100	100			
Size of boiler, hp	25	35	35	.....	40	50
Compressed air, cfm[a]	400	450	450	.....	450	600
Hose size, in	1¼	1½	1½	.....	1½	2

[a] No. 9 not now manufactured; supplied on special order only.

[b] The B-2 series is not now manufactured but data are included for use with existing hammers.

[c] These weights include an allowance of 30 lb for pilot ring weights, which vary from approximately 15 to 45 lb, depending on the diameter of the pile head.

[d] Actual delivery.

TABLE IV.12. MCKIERNAN-TERRY DOUBLE-ACTING STEAM HAMMERS (*Continued*)
HAMMER SIZES 11-B-2, 9-B-3, 10-B-3, 11-B-3, C5

Item	Size of hammer				
	11-B-2[a]	9-B-3	10-B-3	11-B-3	C5
Normal stroke (h), in	20	17	19	19	18
Ram, weight (W_r), lb	3,625	1,600	3,000	5,000	5,000
Casing, weight (W_c), lb	8,570	5,070	7,225	8,182	6,880
Manufacturer's rated energy per blow (E_n), ft-lb	22,080	8,750	13,100	19,150	16,000
Strokes per min	120	145	105	95	100–110
Anvil, weight, lb:					
Flat	990	330	625	818	
Cup or bell	1,060	342	726	991	
Driving-head assemblies for use with Monotube piles,[b] weight, lb:					
Standard base, wood or fiber cushions	......	712	876		
Same, adapter type	......	373	757		
Standard base, steel and fiber cushions	......	767	965		
Same, adapter type	......	786	810		
Recommended steam pressure at boiler or air pressure at compressor, psi	125	125	125	125	125
Steam or air pressure at hammer, psi	100	100	100	100	100
Size of boiler, hp	60	45	50	60	45
Compressed air, cfm[c]	800	600	750	900	585
Hose size, in	2	2	2½	2½	2½

[a] The B-2 series is not now manufactured but data are included for use with existing hammers.
[b] These weights include an allowance of 30 lb for pilot ring weights, which vary from approximately 15 to 45 lb, depending on the diameter of the pile head.
[c] Actual delivery.

TABLE IV.13. VULCAN INTERNAL-COMBUSTION DOUBLE-ACTING HAMMER, OPEN TYPE

Item	*Size of Hammer, Mark V, Size* IC65
Normal stroke (h), in	15½
Ram, weight (W_r), lb	6,500
Casing, weight (W_c), lb	8,000
Manufacturer's rated energy per blow (E_n), ft-lb	19,565
Strokes per min	100
Over-all length, ft	4.25
Fuel consumption, gal per hr	5
Largest diameter of pile, in	26

TABLE IV.14. UNION IRON WORKS DOUBLE-ACTING STEAM HAMMERS
HAMMER SIZES 00, 0A, 0, 1, 1A

Item	Size of hammer				
	00	0A	0	1	1A
Normal stroke (h), in.	36	21	24	21	18
Ram, weight (W_r), lb	6,000	5,000	3,000	1,850	1,600
Casing, weight (W_c), lb	15,000[a]	12,000[a]	11,500[a]	8,650[a]	8,900[a]
Manufacturer's rated energy per blow (E_n), ft-lb	54,900	22,050	19,850	13,100	10,020
Strokes per min	85	90	110	130	120
Steam pressure at boiler, psi	100–125	100–125	100–125	100–125	100–125
Air pressure at compressor, psi	100	100	100	100	100
Steam or air pressure (mean effective) at hammer, psi	80	80	80	80	80
Size of boiler, hp[b]	125	60	40	40	35
Compressed air, cfm[c]	....	800	600	500	450
Hose size, in.	3	2½	1½	1½	1½

[a] Weights include average base.
[b] 100 to 125 psi.
[c] 100 psi actual delivery.

TABLE IV.14. UNION IRON WORKS DOUBLE-ACTING STEAM HAMMERS (*Continued*)
HAMMER SIZES 1½A, 2, 3, 3A, 4

Item	Size of hammer				
	1½A	2	3	3A	4
Normal stroke (h), in.	18	16	14	13½	12
Ram, weight (W_r), lb	1,500	1,025	700	820	370
Casing, weight (W_c), lb	7,700[a]	5,575[a]	4,000[a]	4,380[a]	2,430[a]
Manufacturer's rated energy per blow (E_n), ft-lb	8,680	5,755	3,660	4,390	2,100
Strokes per min	125	145	160	150	200
Steam pressure at boiler, psi	100–125	100–125	100–125	100–125	100–125
Air pressure at compressor, psi	100	100	100	100	100
Steam or air pressure (mean effective) at hammer, psi	80	80	80	80	80
Size of boiler, hp[b]	25	20	25	12	10
Compressed air, cfm[c]	400	300	350	150	100
Hose size, in.	1¼	1¼	1¼	1	1

[a] Weights include average base.
[b] 100 to 125 psi.
[c] 100 psi actual delivery.

TABLE IV.14. UNION IRON WORKS DOUBLE-ACTING STEAM HAMMERS (*Continued*)
HAMMER SIZES 5, 6, 7A, 8, 9

Item	Size of hammer				
	5	6	7A	8	9
Normal stroke (h), in	9	7	6	6½	4
Ram, weight (W_r), lb	210	100	80	40	25
Casing, weight (W_c), lb	1,415[a]	810[a]	460	160	75
Manufacturer's rated energy per blow (E_n), ft-lb	1,010	445	320	210	90
Strokes per min	250	340	400	Over 450	Over 550
Steam pressure at boiler, psi	100–125	100–125	100–125	100–125	100–125
Air pressure at compressor, psi	100	100	100	100	100
Steam or air pressure (mean effective) at hammer, psi	80	80	80	80	80
Size of boiler, hp[b]	8				
Compressed air, cfm[c]	75	70	60	45	
Hose size, in	¾	¾	¾	½	

[a] Weights include average base.
[b] 100 to 125 psi.
[c] 100 psi actual delivery.

TABLE IV.15. MCKIERNAN-TERRY DIESEL HAMMERS

Item	Size of hammer	
	DE20	DE30
Stroke (h), in	48–96	48–96
Ram, weight (W_r), lb	2,000	2,800
Casing weight (W_c), lb	3,750	5,325
Manufacturer's rated energy per blow (E_n), ft-lb:		
Min	12,000	16,800
Max	16,000	22,400
Strokes per min	48–52	48–52
Over-all length, ft	11.67	14
Diesel fuel, gal per hr	1.6	2

TABLE IV.16. LINK-BELT SPEEDER DIESEL HAMMERS

Item	Size of hammer		
	105	312	520
Maximum stroke (h), in	To 38	To 36	To 48
Idling strokes, no impact, in	14	16	22–24
Ram, weight (W_r), lb	1,460	3,800	5,000
Casing weight (W_c), lb	1,845	5,010	6,120
Manufacturer's rated energy per blow (E_n), ft-lb	Up to 7,500	Up to 18,000	Up to 30,000
Strokes per min	90–98	100–105	80–84
Anvil weight (standard equipment, for driving wood piles chamfered to dimension shown), lb	(9 in.) 395	(13 in.) 1,190	(13 in.) 1,180
Driving head weight, if used (additional to anvil weight), lb, for:			
Cylindrical pile	220	900–1,400	900–1,400
12- to 14-in. H piles, sheet piles, and 12- to 14-in.-o.d. pipe piles	....	775–795	775–795
H piles, sheet piles, and 10-in.o.d. pipe piles	420		
16-in. round concrete pile	....	1,000	1,000
Adapter (required with driving heads for all except wood piles), weight, lb	100	350	350
Over-all length, ft	10.17	10.75	13.50
Min width of leads usable with hammer, in	18.5	26.5	26.5
Diesel fuel, gal per hr	0.9	2.5	3.5

BRITISH AND GERMAN HAMMERS

TABLE IV.17. SEMIAUTOMATIC SINGLE-ACTING STEAM HAMMERS—BRITISH STEEL PILING COMPANY, LTD.

Item	Size of hammer[a]									
	4b	5b	6b	7b	9b	9c	10b	10c	12b	12c
Normal stroke (h), in.	54[b]	54[b]	54[b]	54[b]	54[b]	72[b]	54[b]	72[b]	54[b]	72[b]
Weight of falling mass (cylinder), cwt[c]	30	40	50	60	80	80	100	100	120	120
Total weight, cwt[c]	41	52	63	70	90	97	117	118	138	142
Boiler capacity required:										
Evaporation of water, lb per hr[d]	1,130	1,320	1,550	1,700	2,000	2,000	2,400	2,400	3,000	3,000
Size of boiler (Spencer-Hopwood)[d]	12	14	16	18	20	20	22	22	24	24
Free air, cfm	600	700	800	900	1,000	1,100	1,300	1,300	1,400	1,400

[a] Other sizes available in sizes ranging from 10 cwt to 10 tons.
[b] Longest stroke; length of stroke may be controlled by operator.
[c] 1 cwt = 112 lb.
[d] Recommended for normal working. When driving is hard and continuous, consumptions may increase up to about 30 per cent.

TABLE IV.18. INDUSTRIAL WORKS DOUBLE-ACTING STEAM HAMMER

Weight of falling mass, casing, lb	1,402
Total weight, lb	5,843
Manufacturer's rated energy per blow (E_n) at 120 strokes per min, ft-lb	10,379

TABLE IV.19. JOHNSON DIESEL PILE DRIVERS—C. H. JOHNSON (MACHINERY) LIMITED, STOCKPORT, ENGLAND

Size of hammer	Ram (W_r), weight, lb	Manufacturer's rated energy at given strokes			Over-all length, ft	Diam of beater plate, in.	Max diesel fuel, gal per 8 hr
		Stroke, in.	Strokes per min	Gross energy per blow (E_n), ft-lb			
D1	702	33	60	4,000	8.75	12	1.775
		63	40	7,200			
D3	1,764	34	55	7,842	10.17	15¾×25¼	1.6
		54	40	11,920		15¾×25¼	
	1,872[a]						
	1,980[a]						

[a] Using additional weights on falling cylinder.

TABLE IV.20. DELMAG DIESEL PILE HAMMERS—THE FOUNDATION EQUIPMENT CORPORATION, NEWCOMERSTOWN, OHIO

Item	Size of hammer		
	D-5	D-12	D-22
Total weight, lb	2,220	4,940	8,800
Ram, weight (W_r), lb	1,100	2,750	4,850
Manufacturer's rated energy (E_n), ft-lb per blow	9,100	22,500	39,700
Strokes per min	50–60	50–60	50–60
Over-all length, ft	11.25	12.25	12.83
Diesel fuel, gal per hr	0.67	1.75	3.33

PILE EXTRACTORS

TABLE IV.21. McKIERNAN-TERRY PILE EXTRACTORS

Item	Type number	
	E2	E4
Normal stroke (h), in	3	3
Ram, weight (W_r), lb	200	400
Total weight, lb	2,600	4,400
Manufacturer's rated energy per blow (E_n), ft-lb	700	1,000
Strokes per min	450	400
Max crane pull, tons	50	100
Steam, boiler hp	30	35
Air consumption, actual, cfm	400	450
Pressure delivered, psi	100–125	100–125

TABLE IV.22. VULCAN PILE EXTRACTORS

Item	Type number			
	200-A	400-A	800-A	1,200-A
Length of stroke, in	2	2	2	
Ram, weight, lb	200	400	800	1,200
Total weight, including connecting links, lb	1,500	2,850	5,460	9,200
Manufacturer's rated energy per blow, ft-lb	250	500	1,000	
Strokes per min. (100 psi at extractor)[a]	550	550	550	530
Over-all length, in.:				
Omitting connecting links	71	80	94	109½
Including connecting links	93¼	109½	127⅜	150⅝
Extreme width, over side bars, in	17	22½	29	34⅝
Extreme depth, front to back, in	13½	15½	20	24½
Width at lower end of connecting link, in	7	7½	8½	7¼
Height of opening in loop of lifting cap, in	7	7	8¼	10½
Width of opening in loop of lifting cap, in	3½	3½	4	5½
Max crane pull, tons	25	40	50	100
Size of connection for hose, in	1	1¼	1½	2
Steam or air pressure, psi	75–150	75–150	75–150	75–150
Boiler horsepower required, normal rating	18	25	40	65
Free air, cfm:				
Adiabatic	173	342	740	1,020
Isothermal	312	615	1,330	1,850

[a] Allowable air or steam pressure may vary from 75 to 125 psi.

TABLE IV.23. ZENITH EXTRACTORS FOR USE WITH WOOD PILES

Item	Type number		
	20	80	120
Total weight, lb	2,150	4,750	6,900
Ram, weight, lb	200	800	1,200
Strokes per min	500	350	350
Over-all length	7′8¼″	9′7″	10′6″
Width of grip	2′6″	3′0″	3′8″
Size of pile, in.:			
Max	10 × 10	13 × 13	16 × 16
Min	7 × 7	10 × 10	12 × 12
Max permissible crane pull, tons (long)	10	20	30
Manufacturer's rated energy per blow (at 90 psi pressure at extractor) (E_n), ft-lb	600	2,500	4,000
Volume of air required, cfm	250	550	850
Size of hose, in	1	1½	2
Evaporation required per hr, lb	500	1,100	1,700

TABLE IV.24. ZENITH EXTRACTORS FOR USE WITH STEEL PILES

Item	Type number		
	20	80	120
Total weight, lb	2,000	4,400	6,400
Ram, weight, lb	200	800	1,200
Strokes per min	500	350	350
Over-all length	6′7½″	8′5″	8′9″
Width of grip	1′5¾″	1′9″	1′11″
Size of pile, in.:			
Max thickness of web	¾	⅞	⅞
Min width of flat of web	5	7⅝	7⅝
Max permissible crane pull, tons (long)	10	20	30
Manufacturer's rated energy per blow (at 90 psi pressure at extractor) (E_n), ft-lb	600	2,500	4,000
Volume of air required, cfm	250	550	850
Size of hose, in	1	1½	2
Evaporation required per hr, lb	500	1,100	1,700

TABLE IV.25. BRITISH STEEL PILING CO. EXTRACTORS

Item	Type No.	
	HD-10	HD-15
Total weight, lb	6,600	10,000
Ram, weight, lb	1,650	3,850
Strokes per min, approx	160	120
Over-all length	15′10″	17′10″
Width of grip	2′0″	2′0″
Size of steel piling, in.:		
Max thickness	1	1
Min thickness	⅜	⅜
Min width of flat	6¼	6¼
Max permissible crane pull, tons (long)	20	20
Min crane capacity required, tons (long)	10	15
Manufacturer's rated energy per blow (at 100 psi pressure at extractor), E_n, ft-lb	Up to 8,000	Up to 12,000
Size of hose, in	2	2
Evaporation required, lb of steam per hr	1,500	2,000
Steam or air pressure		
Normal	90–110	90–110
Maximum	150	150

TABLE IV.26. MCKIERNAN-TERRY PILE EXTRACTORS—USING DOUBLE-ACTING STEAM HAMMERS

Item	Type number				
	2	5	6	7	9[a]
Normal stroke (h), in.	5¼	7	8¾	9½	12
Ram, weight (W_r), lb	48	200	400	800	1,250
Manufacturer's rated energy per blow (E_n), ft-lb	120[b]	800[b]	1,900[b]	2,900[b]	6,000[b]
For use with wood and concrete piles:					
Min size, in.	3 × 3	7 × 7	10 × 10	12 × 12	12 × 12
Max size, in.	3 × 8	10 × 10	13 × 13	16 × 16	16 × 16
Height	4′10″	9′6″	10′6″	12′3″	14′7″
Additional weight of attachments, lb	300	1,200	2,450	3,300	4,400
For use with steel piles:[c]					
Width of grip	10¼″	1′5¾″	1′8″	1′9⅝″	1′9⅝″
Height	4′10″	8′5″	9′4″	10′6″	12′10″
Additional weight of attachments, lb	280	1,050	2,150	2,800	3,900

[a] No. 9 is not now manufactured but is supplied to special order only.

[b] Based on 90 psi pressure at hammer.

[c] Using weights of attachment furnished by the British Steel Piling Co., Ltd. Similar values for attachments furnished by McKiernan-Terry Corp.

TABLE IV.27. UNION PILE EXTRACTORS—USING DOUBLE-ACTING STEAM HAMMERS

Item	Type number								
	0	1	1A	1½A	2	3	3A	4	5
Normal stroke (h), in.	24	21	18	18	16	14	13½	12	9
Ram, weight (W_r), lb.	3,000	1,600	1,600	1,500	1,025	700	820	370	210
Total weight, lb.	14,500	10,000	10,500	9,200	6,600	4,700	5,200	2,800	1,625
Manufacturer's rated energy per blow (E_n), ft-lb.	3,930	4,080	3,250	2,500	2,375	1,780	2,320	1,390	910
Strokes per min	80	95	95	100	115	120	120	120	190
Steam, boiler hp	50	40	40	35	25	20	25	12	10
Free air, cfm	750	600	500	450	400	300	350	150	100
Pressure delivered, psi	100	100	100	100	100	100	100	100	100

GROUP III

PILE DATA

CONCRETE AND STEEL PILES AND ACCESSORIES

Table V.1. Raymond Standard Piles

LENGTHS OF STANDARD PILES, FT[a]

Tip diam, in.	Nominal butt diam, in.[b]														
	9	10	11	12	13	14	15	16	17	18	19	20	21	22	23
8	2.5	5.0	7.5	10.0	12.5	15.0	17.5	20.0	22.5	25.0	27.5	30.0	32.5	35.0	37.5 (max)
10.8	...	...	...	3.0	5.5	8.0	10.5	13.0	15.5	18.0	20.5	23.0 (max)			

[a] Pile tapers uniformly at 0.4 in. per ft.

[b] Butt diameters of intermediate-length piles would increase at rate of 0.4 in. per ft.

RAYMOND STANDARD CORES

Tip diameter, in.	Weight, lb	Length, ft
8	9,675	37
	9,275	30
10	8,000	23

RAYMOND STANDARD SHELLS

Section No.	Outside diam,[a] top of section, in.	Approximate surface area,[b] sq ft	Approximate weight,[c,d] lb	Approximate cross-section area at top of section, sq in.
1	11.2	20	51	100
2[e]	14.1	25	64	156
3	17.0	30	78	227
4	19.9	35	92	310
5	22.8	42	108	408

[a] Based on 8-ft section lengths. Starting with 8-in. tip diameter. Four-ft sections are also made; check with manufacturer.

[b] Circumscribed area only; does not include projected end area due to taper.

[c] Based on 20-gage. Add 24 lb per pile for boot.

[d] Gages vary from Nos. 18 to 24.

[e] Stub standard (10.8-in tip) starts with No. 2 section.

Table V.2. Raymond Step-taper Piles

LENGTHS OF PILES USING 8- AND 12-FT STEPS[a–c]

Nominal tip diameter, in.[d,e]	Nominal butt diameter, in.[d,f]																			
	8		9		10		11		12		13		14		15		16		17[g]	
	8-ft steps	12-ft steps	8-ft steps	12-ft steps	8-ft steps	12-ft steps	8-ft steps	12-ft steps	8-ft steps	12-ft steps	8-ft steps	12-ft steps	8-ft steps	12-ft steps	8-ft steps	12-ft steps	8-ft steps	12-ft steps	8-ft steps	12-ft steps
8	8	12	16	24	24	36	32	48	40	60	48	72	56	84	64	96	72	108	80	120
9	...	...	8	12	16	24	24	36	32	48	40	60	48	72	56	84	64	96	72	108
10	...	...	...	...	8	12	16	24	24	36	32	48	40	60	48	72	56	84	64	96
11	...	...	...	...	...	...	8	12	16	24	24	36	32	48	40	60	48	72	56	84
12	...	...	...	...	...	...	...	...	8	12	16	24	24	36	32	48	40	60	48	72
13	...	...	...	...	...	...	...	...	...	...	8	12	16	24	24	36	32	48	40	60
14	...	...	...	...	...	...	...	...	...	...	...	...	8	12	16	24	24	36	32	48
Shell section No. (No. of top section)	000		00		0		1		2		3		4		5		6		7	

[a] Longer piles available using pipe extension below Nos. 1, 2, 3, or 4 shell section or increasing step lengths provided driving rig has adequate capacity.

[b] Other step lengths can be furnished for special conditions, *e.g.*, 4, 16, and 24 ft.

[c] Available step lengths can be combined to meet special requirements.

[d] Normal tip diameters used are 8- to 11-in. Tip diameters 12-in. and larger available for special conditions—check with manufacturer.

[e] Intermediate pile lengths would have the next-larger butt diameter.

[f] Actual outside diameter is ¼ to ½ in. greater.

[g] **No. 7 sections (17-in.-diameter) can be added to increase pile lengths.**

Table V.2. Raymond Step-taper Piles (*Continued*)

RAYMOND STEP-TAPER CORES

Section No.	Nominal o.d., in.	Average diam, in.	Approximate weights,[a] lb per lin ft	Approximate cross-sectional area of steel to take driving stresses, sq in.
000	8	7 7/16	75	22
00	9	8 5/16	90	26
0	10	9 3/16	105	31
1	11	10	120–200	35–60
2	12	11 1/8	120–220	35–65
3	13	12 1/8	125–200	37–60
4	14	13 1/8	135–165	40–50
5	15	14 1/8	150–165	45–50
6	16	15 1/8	160	47
7	17	16	200	60
Average.......	...	.....	130–160	38–47

[a] For each full core add 1,000 lb for core head, cap-block assembly, etc.

RAYMOND STEP-TAPER SHELLS

Section No.	O.D.,[a] in.	Nominal surface area,[b] sq ft		Approximate weight, lb per lin ft[d]	Nominal cross-sectional area, sq in.
		8-ft steps[c]	12-ft steps[c]		
000	8 5/8	18	27	8.5	58
00	9 1/2	20	30	9	71
0	10 3/8	22	33	10	85
1	11 1/4	24	36	11	100
2	12 1/4	26	39	12	118
3	13 1/4	28	42	13	138
4	14 1/4	30	45	14	160
5	15 1/4	32	48	15	182
6	16 1/4	34	51	16	208
7	17 1/4	36	54	17	234

[a] Diameters shown are at bottom of shell section. Top diameter of shell section is 1/8 in. larger.

[b] Circumscribed area only; does not include full contact surface of corrugated section nor projected end area due to taper.

[c] Sections are available also in lengths of 4, 16, and 24 ft—check with manufacturer.

[d] Gages range from Nos. 12 to 20. Manufacturer assumes responsibility for selection of proper gages.

TABLE V.3. UNION METAL MONOTUBE FLUTED PILES

	Type F, 1 in. taper in 7 ft 0 in.				Type J, 1 in. taper in 4 ft 0 in.				Type Y, 1 in. taper in 2 ft 6 in.				Type N, 5/8 in. taper in 20 ft 0 in.				
Nominal butt diam, in.	12	14	16	18	12	14	16	18	12	14	16	18		12	14	16	18
Length, ft	30	40	60	75	17	25	33	40	10	15	20	25		10 to 40	10 to 40	10 to 40	10 to 40
Concrete area, butt, sq in.	101	136	176	226	102	139	182	224	98	134	176	224		98	134	176	224
Estimated concrete vol, cu yd	0.55	0.95	1.68	2.59	0.32	0.58	0.95	1.37	0.18	0.34	0.56	0.86	Cu yd per foot	0.026	0.035	0.045	0.058
Surface area, sq ft	88	119	206.5	282.5	47.5	76.5	111.0	145.0	27.3	45.3	66.5	90.3	Sq ft per 20-ft length	66.6	76.7	88.6	100.3
11 gage, 0.1196 in.:																	
Properties at butt:																	
A, sq in.	4.54	5.26	5.98	6.77	4.56	5.31	6.06	6.72	4.46	5.22	5.97	6.72		4.46	5.22	5.97	6.72
I, in^4	76.2	118	174	252	77.1	122	181	247	72.5	116	173	247		72.5	116	173	247
S, in.^{3}	12.5	16.7	21.7	27.8	12.6	17.0	22.3	27.4	12.1	16.5	21.6	27.4		12.1	16.5	21.6	27.4
r, in.	4.09	4.74	5.39	6.10	4.11	4.80	5.47	6.06	4.03	4.71	5.38	6.06		4.03	4.71	5.38	6.06
W, lb	404	597	955	1289	235	372	529	690	142	227	323	432	Nom wt per ft	16	19	21	24
9 gage, 0.1495 in.:																	
Properties at butt:																	
A, sq in.	5.65	6.55	7.45	8.44	5.68	6.62	7.56	8.38	5.56	6.50	7.44	8.38		5.56	6.50	7.44	8.38
I, in^4	94.8	147	216	314	95.7	152	225	307	89.9	144	215	307		89.9	144	215	307
S, in.^{3}	15.5	20.9	27.0	34.6	15.6	21.3	27.8	34.1	15.0	20.5	26.9	34.1		15.0	20.5	26.9	34.1
r, in.	4.09	4.73	5.39	6.10	4.11	4.78	5.46	6.05	4.02	4.70	5.37	6.05		4.02	4.70	5.37	6.05
W, lb	500	749	1190	1624	289	460	656	857	173	278	399	535	Nom wt per ft	20	23	26	30
7 gage, 0.1793 in.:																	
Properties at butt:																	
A, sq in.	6.77	7.98	8.85	10.04	6.80	7.93	9.05	10.04	6.66	7.79	8.91	10.04		6.66	7.79	8.91	10.04
I, in^4	113	184	252	367	114	181	270	367	107	171	257	367		107	171	257	367
S, in.^{3}	18.6	25.7	31.7	40.8	18.6	25.3	33.2	40.8	17.8	24.5	32.1	40.8		17.8	24.5	32.1	40.8
r, in.	4.08	4.80	5.34	6.05	4.10	4.77	5.45	6.04	4.01	4.69	5.36	6.04		4.01	4.69	5.36	6.04
W, lb	589	895	1412	1957	338	541	775	1023	203	330	476	641	Nom wt per ft	24	28	32	36

Table V.3. Union Metal Monotube Fluted Piles (*Continued*)

	Type F, 1 in. taper in 7 ft 0 in.				Type J, 1 in. taper in 4 ft 0 in.				Type Y, 1 in. taper in 2 ft 6 in.					Type N, ⅝ in. taper in 20 ft 0 in.			
5 gage, 0.2093 in.:																	
Properties at butt:																	
A, sq in.	7.88	9.37	10.43	12.00	7.92	9.23	10.55	11.86	7.92	9.23	10.55	11.86		7.75	9.07	10.38	11.70
I, in⁴	130	219	301	460	132	209	312	444	132	209	312	444		124	199	298	426
S, in.³	21.3	30.3	37.5	49.8	21.6	29.4	38.4	48.7	21.6	29.4	38.4	48.7		20.7	28.3	37.2	47.3
r, in.	4.07	4.83	5.38	6.19	4.08	4.76	5.44	6.12	4.08	4.76	5.44	6.12		4.00	4.68	5.36	6.03
W, lb.	685	1055	1645	2307	342	629	902	1191	235	383	553	746	Nom wt per ft	28	33	37	42
3 gage, 0.2391 in.:																	
Properties at butt:																	
A, sq in.	8.98	10.78	11.90	13.77	9.02	10.52	12.02	13.52	9.02	10.52	12.02	13.52		8.83	10.34	11.84	13.34
I, in⁴	148	256	343	524	150	237	354	504	150	237	354	504		141	225	339	484
S, in.³	24.2	35.4	42.7	57.0	24.4	33.3	43.6	55.3	24.4	33.3	43.6	55.3		23.5	32.2	42.3	53.8
r, in.	4.06	4.87	5.37	6.22	4.07	4.75	5.43	6.11	4.07	4.75	5.43	6.11		3.99	4.67	5.35	6.02
W, lb.	778	1215	1872	2653	444	714	1025	1355	266	434	628	848	Nom wt per ft	32	37	42	48

Notes: Gages are Nos. 11, 9, and 7 U.S. Manufacturers' Standard. Other gages such as Nos. 5 and 3 can be provided.

Lengths of types F, J, and Y shown are extendible sections.

Sections for types F and N are available in 5-ft steps and for sections J and Y in 2- to 3-ft steps; for properties of these sections of intermediate lengths, interpolate or consult Union Metal Mfg. Co. catalogue.

Lengths over 40 ft are made in multiple sections, weld-assembled.

Physical properties are computed on the basis of a 16-flute section.

TABLE V.4. H PILES—UNITED STATES STEEL AND BETHLEHEM

	CBP146[a] BP14[b]	CBP146 BP14	CBP146 BP14	CBP146 BP14	CBP124 BP12	CBP124 BP12	CB124 B12c	CBP103 BP10	CBP103 BP10	CBP103 BP10	CB103 B10b	CBP83 BP8	CB83 B8b
Weight per lin ft, lb	117	102	89	73	74	53	65	57	44	42	49	36	33
Net area of steel, sq in	34.44	30.01	26.19	21.46	21.76	15.58	19.11	16.76	12.95	12.35	14.40	10.60	9.70
Gross area inside bounding rectangle, sq ft	1.47	1.44	1.41	1.38	1.03	0.97	1.01	0.71	0.68	0.68	0.69	0.45	0.45
Depth of section, in	14.23	14.03	13.86	13.64	12.12	11.78	12.12	10.01	9.76	9.72	10.00	8.03	8.06
Width of flange, in	14.88	14.78	14.70	14.59	12.22	12.05	12.00	10.22	10.10	10.08	10.00	8.16	8.01
Flange thickness, in	0.805	0.704	0.616	0.506	0.607	0.436	0.606	0.564	0.438	0.418	0.558	0.446	0.463
Web thickness, in	0.805	0.704	0.616	0.506	0.607	0.436	0.390	0.564	0.438	0.418	0.340	0.446	0.300
Radius of gyration:													
Axis 1–1	5.97	5.93	5.89	5.85	5.10	5.03	5.28	4.19	4.14	4.13	4.35	3.36	3.49
Axis 2–2	3.59	3.56	3.53	3.49	2.91	2.86	3.02	2.45	2.41	2.40	2.54	1.95	2.02
Surface area inside bounding rectangle per ft of height, sq ft	4.85	4.80	4.77	4.70	4.05	3.98	4.02	3.37	3.31	3.30	3.33	2.70	2.68

[a] United States Steel.
[b] Bethlehem.

TABLE V.5. PROPERTIES OF AMERICAN STEEL PIPE SECTIONS[a]

Nominal size, in.	Diam, in.[b]		Wall thickness, in.[b]		Weight per ft, lb	Gross area of metal, sq in.	Moment of inertia, I, in.^{4}	Section modulus, I/c, in.^{3}	Radius of gyration, r	Area of concrete		Effective area of metal,[c] sq in.
	Exterior	Interior	Decimal	Fraction						Sq in.	Sq ft	
6	6.00	**5.732**[d]	**0.109**[d]	7/64	6.86	2.02	8.74	2.91	2.08	26.25	0.1823	
		5.718	**0.141**	9/64	8.81	2.59	11.13	3.71	2.07	25.68	0.1783	1.42
		5.656	**0.172**	11/64	10.71	3.15	13.38	4.46	2.06	25.12	0.1744	1.98
6	6.625	**6.407**[d]	**0.199**[d]	7/64	7.16	2.23	11.81	3.57	2.31	32.24	0.2238	0.94
		6.343[d]	**0.141**[d]	9/64	9.54	2.87	15.07	4.55	2.30	31.60	0.2194	1.58
		6.249[d]	**0.1875**[d]	3/16	12.92	3.80	19.67	5.93	2.28	30.67	0.2129	2.51
		6.187[d]	**0.219**[d]	7/32	15.00	4.41	22.62	6.83	2.27	30.06	0.2080	3.12
		6.065	*0.28*	....	18.97	5.58	28.14	8.50	2.25	28.89	0.2006	4.29
		5.875	*0.375*	3/8	25.03	7.36	36.07	10.89	2.21	27.10	0.1882	6.07
		5.761	*0.432*	....	28.57	8.40	40.49	12.22	2.20	26.06	0.1810	7.11
8	8.00	**7.782**[d]	**0.109**[d]	7/64	9.20	2.70	21.03	5.28	2.80	47.58	0.3304	1.14
		7.718	**0.141**	9/64	11.83	3.48	26.88	6.72	2.79	46.78	0.3249	1.92
		7.656	**0.172**	11/64	14.39	4.23	32.42	7.97	2.78	46.03	0.3197	2.67
8	8.625	**8.407**[d]	**0.109**[d]	7/64	9.92	2.92	26.40	6.12	3.02	55.51	0.3855	1.24
		8.343[d]	**0.141**[d]	9/64	12.82	3.77	33.82	7.85	3.00	54.69	0.3797	2.09
		8.281[d]	**0.172**[d]	11/64	15.52	4.57	40.81	9.45	3.00	53.86	0.3740	2.89
		8.249	**0.1875**	3/16	16.93	4.98	44.36	10.28	2.99	53.44	0.3711	3.38
		8.187[d]	**0.219**[d]	7/32	19.65	5.78	51.12	11.85	2.98	52.64	0.3656	4.10
		8.125[d]	**0.250**[d]	1/4	22.37	6.58	57.72	13.40	2.97	51.85	0.3600	4.90
		8.071	*0.277*	....	24.69	7.26	63.35	14.69	2.95	51.16	0.3553	5.58
		7.981	*0.322*	....	28.55	8.40	72.49	16.81	2.94	50.02	0.3474	6.72
		7.625	*0.500*	1/2	43.38	12.76	105.7	24.51	2.88	45.66	0.3171	11.08

10	10.00	**9.782**[d]	**0.109**[d]	7/64	11.52	3.39	41.37	8.47	3.50	75.15	0.5209	1.52
		9.718	**0.141**	9/64	14.85	4.37	53.06	10.61	3.49	74.14	0.5148	2.45
		9.656	**0.172**	11/64	18.05	5.31	64.13	12.82	3.48	73.29	0.5085	3.36
10	10.75	**10.532**[d]	**0.109**[d]	7/64	12.39	3.64	51.58	9.59	3.77	87.12	0.6050	1.54
		10.468[d]	**0.141**[d]	9/64	15.93	4.70	65.95	12.27	3.75	86.06	0.5977	2.60
		10.406[d]	**0.172**[d]	11/64	19.43	5.72	79.92	14.87	3.74	85.05	0.5905	3.61
		10.374	**0.1875**	3/16	21.15	6.23	86.81	16.15	3.73	84.52	0.5870	4.14
		10.312[d]	**0.219**[d]	7/32	24.60	7.25	100.4	18.68	3.72	83.52	0.5800	5.15
		10.250	0.250	1/4	28.04	8.24	113.7	21.16	3.71	82.52	0.5730	6.15
		10.189	0.281	....	31.44	9.24	126.8	23.60	3.70	81.52	0.5665	7.14
		10.136	*0.307*	....	34.24	10.07	137.4	25.57	3.69	80.68	0.5603	7.97
		10.125	*0.3125*	5/16	34.83	10.25	139.6	25.98	3.69	80.52	0.5591	8.15
		10.020	*0.365*	....	40.48	11.91	160.7	29.90	3.67	78.85	0.5476	9.81
		10.000	*0.375*	3/8	41.55	12.22	164.7	30.64	3.67	78.54	0.5454	10.12
		9.875	*0.4375*	7/16	48.18	14.17	188.8	35.12	3.65	76.59	0.5318	12.07
		9.750	*0.500*	1/2	54.73	16.10	212.0	39.43	3.63	74.66	0.5184	14.02
12	12.00	**11.782**[d]	**0.109**[d]	7/64	13.84	4.07	71.98	12.00	4.21	109.03	0.7571	1.73
		11.718	**0.141**	9/64	17.86	5.25	92.38	15.39	4.20	107.87	0.7491	2.91
		11.656	**0.172**	11/64	21.73	6.39	111.8	18.64	4.19	106.70	0.7410	4.05
12	12.75	**12.532**[d]	**0.109**[d]	7/64	14.71	4.33	86.49	13.57	4.48	123.35	0.8566	1.84
		12.468[d]	**0.141**[d]	9/64	18.94	5.58	111.7	17.37	4.46	122.09	0.8479	3.09
		12.406[d]	**0.172**[d]	11/64	23.09	6.80	134.4	21.08	4.45	120.88	0.8394	4.30
		12.374	**0.1875**	3/16	25.16	7.42	146.1	22.91	4.44	120.26	0.8351	4.93
		12.312[d]	**0.219**[d]	7/32	29.28	8.62	169.1	26.53	4.43	119.06	0.8268	6.13
		12.250	**0.250**	1/4	33.38	9.82	191.8	30.09	4.42	117.86	0.8184	7.32
		12.188[d]	**0.281**[d]	9/32	37.45	11.01	214.3	33.61	4.41	116.67	0.8102	8.52
		12.126[d]	**0.3125**[d]	5/16	41.51	12.19	236.3	37.06	4.40	115.49	0.8018	9.70
		12.090	*0.330*	....	43.77	12.88	248.5	38.97	4.39	114.80	0.7972	10.38

Table V.5. Properties of American Steel Pipe Sections[a] (*Continued*)

Nominal size, in.	Diam, in.[b]		Wall thickness, in.[b]		Weight per ft, lb	Gross area of metal, sq in.	Moment of inertia, I, in.^{4}	Section modulus, I/c, in.^{3}	Radius of gyration, r	Area of concrete		Effective area of metal,[c] sq in.
	Exterior	Interior	Decimal	Fraction						Sq in.	Sq ft	
		12.000	*0.375*	3/8	49.56	14.58	279.3	43.82	4.38	113.10	0.7854	12.09
		11.875	*0.4375*	7/16	57.53	16.92	321.2	50.38	4.36	110.75	0.7691	14.43
		11.750	*0.500*	1/2	65.41	19.24	361.5	56.71	4.34	108.43	0.7530	16.75
14	14.00	**13.782**[d]	**0.109**[d]	7/64	16.17	4.75	114.7	16.39	4.92	149.18	1.0360	2.01
		13.718[d]	**0.141**[d]	9/64	20.82	6.14	147.0	21.00	4.90	147.80	1.0264	3.40
		13.656[d]	**0.172**[d]	11/64	25.38	7.47	178.5	25.50	4.90	146.47	1.0171	4.73
		13.625	**0.1875**	3/16	27.66	8.16	194.1	27.72	4.88	145.80	1.0125	5.42
		13.562[d]	**0.219**[d]	7/32	32.20	9.48	224.8	32.13	4.87	144.46	1.0032	6.74
		13.500	**0.250**	1/4	36.71	10.80	255.3	36.48	4.86	143.14	0.9940	8.06
		13.438[d]	**0.281**[d]	9/32	41.21	12.11	283.1	40.44	4.85	141.83	0.9849	9.37
		13.375[d]	0.3125[d]	5/16	45.68	13.42	314.9	44.98	4.84	140.48	0.9756	10.68
		13.250	*0.375*	3/8	54.56	16.05	372.8	53.25	4.82	137.89	0.9575	13.31
		13.125	*0.4375*	7/16	63.37	18.64	429.1	61.29	4.80	135.30	0.9395	15.90
		13.000	*0.500*	1/2	72.09	21.21	483.8	69.11	4.78	132.73	0.9217	18.47
16	16.00	**15.782**[d]	**0.109**[d]	7/64	18.50	5.44	171.8	21.47	5.64	195.62	1.3585	2.31
		15.718[d]	**0.141**[d]	9/64	23.88	7.03	220.9	27.61	5.62	194.04	1.3475	3.90
		15.656[d]	**0.172**[d]	11/64	29.06	8.55	267.7	33.46	5.60	192.51	1.3369	5.42
		15.625	**0.1875**	3/16	31.66	9.34	291.1	36.39	5.59	191.72	1.3316	6.21
		15.562[d]	**0.219**[d]	7/32	36.87	10.86	337.8	42.22	5.58	190.20	1.3209	7.73
		15.500	**0.250**	1/4	42.05	12.37	383.7	47.96	5.57	188.69	1.3104	9.24
		15.438[d]	**0.281**[d]	9/32	47.22	13.88	429.1	53.64	5.56	187.19	1.2999	10.75
		15.375[d]	0.3125[d]	5/16	52.36	15.38	473.9	59.24	5.55	185.69	1.2893	12.25
		15.250	0.375	3/8	62.58	18.41	562.1	70.26	5.53	182.65	1.2684	15.28

		15.125	*0.4375*	7/16	72.71	21.39	648.1	81.01	5.52	179.67	1.2477	18.26
		15.000	*0.500*	1/2	82.77	24.35	731.9	91.49	5.50	176.72	1.2271	21.22
18	18.00	**17.782**[d]	**0.109**[d]	7/64	20.83	6.13	245.1	27.24	6.35	248.34	1.7246	2.61
		17.718[d]	**0.141**[d]	9/64	26.89	7.91	315.5	35.06	6.33	246.56	1.7122	4.39
		17.656[d]	**0.172**[d]	11/64	32.74	9.63	382.8	42.53	6.32	244.84	1.7002	6.11
		17.625	**0.1875**	3/16	35.67	10.49	416.2	46.24	6.32	243.98	1.6943	6.97
		17.562[d]	**0.219**[d]	7/32	41.54	12.23	483.0	53.67	6.29	242.24	1.6822	8.71
		17.500	**0.250**	1/4	47.39	13.94	549.1	61.01	6.28	240.53	1.6703	10.42
		17.438[d]	**0.281**[d]	9/32	53.22	15.64	614.5	68.28	6.27	238.83	1.6585	12.12
		17.375[d]	0.3125[d]	5/16	59.03	17.34	679.2	75.47	6.25	237.13	1.6466	13.81
		17.250	0.375	3/8	70.59	20.76	806.6	89.62	6.23	233.71	1.6229	17.24
		17.125	*0.4375*	7/16	82.06	24.14	931.3	103.5	6.21	230.33	1.5995	20.62
		17.000	*0.500*	1/2	93.45	27.49	1,053.0	117.0	6.19	226.98	1.5762	23.97
		16.875	*0.5625*	9/16	104.75	30.81	1,172.0	130.3	6.17	223.65	1.5531	27.29
		16.750	*0.625*	5/8	115.97	34.12	1,289.0	143.2	6.15	220.35	1.5302	30.59
20	20.00	**19.782**[d]	**0.109**[d]	7/64	23.15	6.81	336.9	33.69	7.02	307.35	2.1343	2.84
		19.718[d]	**0.141**[d]	9/64	29.92	8.80	433.8	43.38	7.01	305.36	2.1206	4.83
		19.656[d]	**0.172**[d]	11/64	36.43	10.71	526.6	52.66	7.01	303.45	2.1073	6.74
		19.625	**0.1875**	3/16	39.68	11.67	572.7	57.27	7.00	302.48	2.1005	7.70
		19.562[d]	**0.219**[d]	7/32	46.27	13.61	665.7	66.57	6.99	300.55	2.0872	9.64
		19.500	**0.250**	1/4	52.73	15.51	756.5	75.65	6.98	298.65	2.0733	11.60
		19.438[d]	**0.281**[d]	9/32	59.23	17.41	847.1	84.71	6.97	296.75	2.0608	13.49
		19.375[d]	0.3125[d]	5/16	65.71	19.30	936.7	93.67	6.96	294.80	2.0474	15.38
		19.250	0.375	3/8	78.60	23.12	1,113.5	111.3	6.94	291.04	2.0210	19.20
		19.125	*0.4375*	7/16	91.40	26.89	1,287.0	128.7	6.92	287.27	1.9949	22.92
		19.000	*0.500*	1/2	104.13	30.63	1,457.0	145.7	6.90	283.53	1.9689	26.72
		18.875	*0.5625*	9/16	116.77	34.35	1,624.0	162.4	6.88	279.81	1.9431	30.43
		18.750	*0.625*	5/8	129.33	38.04	1,787.0	178.7	6.85	276.12	1.9174	34.13

TABLE V.5. PROPERTIES OF AMERICAN STEEL PIPE SECTIONS[a] (*Continued*)

Nominal size, in.	Diam, in.[b]		Wall thickness, in.[b]		Weight per ft, lb	Gross area of metal, sq in.	Moment of inertia, I, in.^{4}	Section modulus, I/c, in.^{3}	Radius of gyration, r	Area of concrete		Effective area of metal,[c] sq in.
	Exterior	Interior	Decimal	Fraction						Sq in.	Sq ft	
22	22.00	**21.718**[d]	**0.141**[d]	9/64	32.92	9.68	578.4	52.58	7.75	370.45	2.5726	5.36
		21.656[d]	**0.172**[d]	11/64	40.12	11.79	702.5	63.87	7.74	368.34	2.5579	7.47
		21.625[d]	**0.1875**[d]	3/16	43.69	12.85	764.2	69.47	7.73	367.28	2.5506	8.53
		21.562[d]	**0.219**[d]	7/32	50.95	14.99	888.7	80.79	7.72	365.15	2.5357	10.67
		21.500[d]	**0.250**[d]	1/4	58.07	17.08	1,010.3	91.84	7.71	363.05	2.5212	12.66
		21.438[d]	**0.281**[d]	9/32	65.18	19.17	1,130.2	102.7	7.69	360.96	2.5067	14.85
		21.375[d]	0.3125[d]	5/16	72.38	21.29	1,252.0	113.8	7.67	358.83	2.4919	16.97
		21.250[d]	0.375[d]	3/8	86.60	25.48	1,490.0	135.4	7.65	354.66	2.4628	21.16
		21.125	*0.4375*	7/16	100.75	29.64	1,723.0	156.6	7.62	350.50	2.4339	25.32
		21.000	*0.500*	1/2	114.81	33.77	1,952.0	177.5	7.60	346.36	2.4052	29.45
		20.875	*0.5625*	9/16	128.78	37.88	2,178.0	198.0	7.58	342.25	2.3767	33.56
		20.750	*0.625*	5/8	142.68	41.97	2,399.0	218.1	7.56	338.16	2.3483	37.65
24	24.00	**23.718**[d]	**0.141**[d]	9/64	35.83	10.57	752.1	62.67	8.45	441.82	3.0682	5.87
		23.656[d]	**0.172**[d]	11/64	43.76	12.88	913.3	76.11	8.44	439.51	3.0522	8.08
		23.625	**0.1875**	3/16	47.69	14.03	993.3	82.77	8.43	438.36	3.0441	9.33
		23.562[d]	**0.219**[d]	7/32	55.66	16.37	1,156.7	96.39	8.42	436.03	3.0280	11.67
		23.500	**0.250**	1/4	63.41	18.65	1,316.2	109.6	8.40	433.74	3.0121	13.95
		23.438[d]	**0.281**[d]	9/32	71.25	20.94	1,472.7	122.8	8.39	431.45	2.9962	16.24
		23.375[d]	0.3125[d]	5/16	79.06	23.22	1,631.7	135.9	8.38	429.17	2.9800	18.52
		23.250	0.375	3/8	94.62	27.83	1,942.0	161.9	8.35	424.56	2.9482	23.13
		23.125	*0.4375*	7/16	110.09	32.39	2,248.0	187.4	8.33	420.00	2.9166	27.69
		23.000	*0.500*	1/2	125.49	36.91	2,549.0	212.4	8.31	415.48	2.8852	32.21
		22.875	*0.5625*	9/16	140.80	41.42	2,846.0	237.1	8.29	410.97	2.8539	36.72
		22.750	*0.625*	5/8	156.03	45.90	3,137.0	261.4	8.27	406.49	2.8228	41.20

26	26.00	**25.500**[d]	**0.250**[d]	1/4	68.76	20.22	1,676.4	129.0	9.12	510.70	3.5466	15.12
		25.375[d]	**0.3125**[d]	5/16	85.75	25.22	2,085.3	160.4	9.10	505.71	3.5121	20.12
		25.250[d]	**0.375**[d]	3/8	102.65	30.19	2,478.4	190.6	9.08	500.74	3.4773	25.09
		25.000[d]	**0.500**[d]	1/2	136.19	40.06	3,257.0	250.5	9.04	490.86	3.4088	34.96
28	28.00	**27.500**[d]	**0.250**[d]	1/4	74.09	21.79	2,098.1	149.9	9.83	593.96	4.1247	16.30
		27.375[d]	**0.3125**[d]	5/16	92.41	27.18	2,605.0	186.1	9.81	588.57	4.0873	21.69
		27.250[d]	**0.375**[d]	3/8	110.66	32.55	3,105.1	221.8	9.79	583.21	4.0500	27.06
		27.000[d]	**0.500**[d]	1/2	146.88	43.20	4,084.8	291.8	9.75	572.56	3.9761	37.71
30	30.00	**29.500**[d]	**0.250**[d]	1/4	79.44	23.36	2,585.1	172.3	10.52	683.49	4.7465	17.49
		29.375[d]	**0.3125**[d]	5/16	99.08	29.10	3,211.8	214.1	10.50	677.60	4.7063	23.22
		29.250[d]	**0.375**[d]	3/8	118.65	34.90	3,829.9	255.3	10.47	672.00	4.6663	29.02
		29.000[d]	**0.500**[d]	1/2	157.53	46.34	5,043.0	336.1	10.43	660.52	4.5870	40.46
36	36.00	**35.500**[d]	**0.250**[d]	1/4	95.39	28.06	4,485.8	249.2	12.66	989.80	6.8736	21.00
		35.375[d]	**0.3125**[d]	5/16	119.11	34.98	5,578.0	309.9	12.62	982.90	6.8253	27.92
		35.250[d]	**0.375**[d]	3/8	142.68	41.97	6,658.7	369.9	12.60	975.91	6.7771	34.91
		35.000[d]	**0.500**[d]	1/2	189.57	55.76	8,786.0	488.1	12.55	962.12	6.6813	48.70

[a] Pipe furnished by The American Rolling Mill Co. conforms to *ASTM Standard Specifications for Electric-Fusion (Arc) Welded Steel Pipe* (sizes 8 in. to but not including 30 in.): A139, Grade B, except that hydrostatic testing will not be required, or *ASTM Standard Specifications for Welded and Seamless Steel Pipe Piles:* A252, for all sizes except 6, 8, 10, and 12 in. o.d. which conform to *ASTM Standard Specifications for Spiral-welded Steel or Iron Pipes:* A211 and *ASTM Standard Specifications for Low Tensile Strength Carbon-steel Plates of Structural Quality for Welding:* A78, Grade B.

Pipe furnished by the National Tube Co. conforms to *ASTM Standard Specifications for Welded and Seamless Steel Pipe Piles:* A252.

[b] Sizes in boldface type furnished, spiral welded pipe, by The American Rolling Mill Co. Sizes in italics furnished, seamless and welded pipe, by the National Tube Co. Other sizes furnished by both The American Rolling Mill Co. and the National Tube Co.

[c] After deducting the outer 1/16 in. of pipe wall.

[d] Indicates these sizes not carried in stock but available from mill of The American Rolling Mill Co. in minimum quantities of 10,000 lb. Other sizes indicated as furnished by The American Rolling Mill Co. are standard inventory sizes.

Table V.6. Properties of British Steel Pipe Sections
British Steel Piling Co., Ltd.[a]

O.d., in.	Dimensions	Thickness of casing										
		No. 11 gage, 0.116 in.	No. 10 gage, 0.128 in.	No. 9 gage, 0.144 in.	No. 8 gage, 0.160 in.	No. 7 gage, 0.176 in.	No. 6 gage, 0.192 in.	No. 5 gage, 0.212 in.	No. 4 gage, 0.232 in.	¼″ thick, 0.25 in.	5/16″ thick, 0.3125 in.	⅜″ thick, 0.375 in.
10	Weight, lb/lin ft.........	12.247	13.497	15.160	16.817	18.468	20.115					
	Area of steel, sq in.......	3.60	3.97	4.46	4.95	5.43	5.92					
	Area of concrete, sq in....	75	75	74	74	73	73					
12	Weight, lb/lin ft.........	14.725	16.232	18.236	20.235	22.228	24.216	26.693				
	Area of steel, sq in.......	4.33	4.77	5.36	5.95	6.54	7.12	7.85				
	Area of concrete, sq in	109	108	108	107	107	106	105				
14	Weight, lb/lin ft.........	17.203	18.966	21.312	23.653	25.988	28.318	31.222	34.118	36.717		
	Area of steel, sq in.......	5.06	5.58	6.27	6.96	7.64	8.33	9.18	10.04	10.80		
	Area of concrete, sq in....	149	148	148	147	146	146	145	144	143		
16	Weight, lb/lin ft	19.681	21.700	24.388	27.071	29.748	32.419	35.751	39.074	42.058	52.364	
	Area of steel, sq in.......	5.79	6.38	7.17	7.96	8.75	9.54	10.52	11.49	12.37	15.40	
	Area of concrete, sq in....	195	195	194	193	192	192	191	190	189	186	
18	Weight, lb/lin ft	22.159	24.435	27.465	30.489	33.508	36.521	40.280	44.031	47.399	59.040	70.597
	Area of steel, sq in.	6.52	7.19	8.08	8.97	9.86	10.74	11.85	12.95	13.94	17.37	20.76
	Area of concrete, sq in.	248	247	246	246	245	244	243	242	241	237	234
20	Weight, lb/lin ft	24.637	27.169	30.541	33.907	37.268	40.623	44.809	48.987	52.739	65.716	78.608
	Area of steel, sq in.	7.25	7.99	8.98	9.97	10.96	11.95	13.18	14.41	15.51	19.33	23.12
	Area of concrete, sq in.	307	306	305	304	303	302	301	300	299	295	291
22	Weight, lb/lin ft	27.115	29.904	33.617	37.325	41.027	44.724	49.338	53.943	58.080	72.391	86.619
	Area of steel, sq in.	7.98	8.80	9.89	10.98	12.07	13.15	14.51	15.87	17.08	21.29	25.48
	Area of concrete, sq in.	372	371	370	369	368	367	366	364	363	359	355
24	Weight, lb/lin ft	29.593	32.638	36.693	40.743	44.787	48.826	53.867	58.899	63.421	79.067	94.631
	Area of steel, sq in.	8.70	9.60	10.79	11.98	13.17	14.36	15.84	17.32	18.65	23.25	27.83
	Area of concrete, sq in.	444	443	442	440	439	438	437	435	434	429	425
26	Weight, lb/lin ft	32.071	35.373	39.770	44.161	48.547	52.928	58.396	63.855	68.761	85.743	102.642
	Area of steel, sq in.	9.43	10.40	11.70	12.99	14.28	15.57	17.18	18.78	20.22	25.22	30.19
	Area of concrete, sq in.	521	521	519	518	517	515	514	512	511	506	501
28	Weight, lb/lin ft	34.549	38.107	42.846	47.579	52.307	57.030	62.925	68.812	74.102	92.419	110.653
	Area of steel, sq in.	10.16	11.21	12.60	13.99	15.38	16.77	18.51	20.24	21.80	27.18	32.55
	Area of concrete, sq in.	606	605	603	602	600	599	597	596	594	589	583

[a] Spiral-welded steel pipe manufactured and driven by the British Steel Piling Co., Ltd. Other diameters and thicknesses available on order.

TABLE V.7. PROPERTIES OF TYPICAL SQUARE PRETENSIONED CONCRETE BEARING PILES[a]

Side dim, in.	Hole diam, in.	Min No. 7-wire strands, strand diam, in.[b]	Initial tension in each strand, lb	A_c, sq in.[c]	Section modulus, in.^{3}	Radius of gyration, in.	Max length, ft	
							Single-point pickup	Double-point pickup
12	Solid	12, 5/16	10,500	144	288	3.47	50	75
14	Solid	12, 3/8	14,000	196	457	4.05	60	85
18	Solid	16, 7/16	18,900	324	972	5.20	70	95
20	Solid	20, 7/16	18,900	400	1,333	5.78	75	107
24	12	24, 7/16	18,900	576	2,220	7.60	90	125

[a] Florida State Road Department, Bridge Division. Spiral ties No. 5 gage, pitch beginning at each end: 5 turns at 1 in., 16 at 3 in., balance at 9 in. through central portion. Set 2¼ in. clear from face.

[b] More strands may be required for severe driving conditions or to handle long lengths.

[c] 1-in. chamfers on corners disregarded.

TABLE V.8. PROPERTIES OF PRETENSIONED CONCRETE BEARING PILES[a]

Diam, in.	Shape	Solid or hollow core	No. of strands, and strand diam, in.	Effective prestress in concrete, psi[b]	A_c sq in.[c]	I, in.^{4}	I/c, in.^{3}	Perimeter, in.	Weight per lin ft, lb	Allowable design load[d]		Allowable moment, kip-in.[e]	Allowable moment for earthquakes, kip-in.[f]	Allowable unsupported length, ft[g]	Normal max length, ft
										Kips	Tons				
10	Octagonal	Solid	6, ⅜	839	83	547	109	33	86	91	46	124	157	32	75
12	Octagonal	Solid	8, ⅜	780	119	1,135	189	40	124	131	65	204	261	38	100
14	Octagonal	Solid	10, ⅜	716	162	2,103	300	46	169	178	89	305	395	45	130
16	Octagonal	Solid	13, ⅜	711	212	3,587	448	53	221	233	117	453	587	51	140
18	Octagonal	Solid	16, ⅜	701	268	5,746	638	60	279	295	147	637	828	58	150
20	Octagonal	Solid	20, ⅜	701	331	8,758	876	66	345	364	182	877	1,140	64	160
20	Octagonal	11″ HC	16, ⅜	786	236	8,039	804	66	246	260	130	873	1,114	73	160
20	Square	Solid	19, 7⁄16	749	398	13,146	1,315	78	415	438	219	1,379	1,774	72	160
20	Square	11″ HC	20, ⅜	766	303	12,427	1,243	78	316	333	167	1,325	1,698	80	160
26	Square	16″ HC	23, 7⁄16	760	475	32,272	2,482	97	495	523	261	2,631	3,376	103	160
30	Round	20″ HC	19, 7⁄16	759	393	31,907	2,127	94	409	431	216	2,252	2,891	113	180
36	Round	26″ HC	24, 7⁄16	774	487	60,016	3,334	113	507	536	268	3,581	4,581	138	200
48	Round	38″ HC	32, 7⁄16	744	675	158,222	6,593	151	703	743	371	6,883	8,861	191	225
54	Round	44″ HC	36, 7⁄16	734	770	233,409	8,645	170	802	847	424	8,939	11,532	217	250

[a] From Ben C. Gerwick, Inc.

[b] Effective prestress is based on a final effective force of 11,600 lb for ⅜-in.-diameter strand and 15,700 lb for 7⁄16-in.-diameter strand, or 145,000 psi.

[c] All holes are circular; 1-in. chamfer on 20 in. square pile corners and 3-in. chamfer on 26 in. square pile corners.

[d] Allowable design load is based on 1,100 psi on the concrete section. Where driving and soil conditions are favorable, this may be raised accordingly.

[e] Allowable moment is based on a tension of 300 psi with an effective prestress as given in the table. Where bending resistance is critical, the allowable moment may be increased by using more strands to raise the effective prestress to about 1,200 psi maximum.

[f] Allowable moment for earthquake or similar loads is based on a tension of 600 psi with an effective prestress as given in the table.

[g] Allowable unsupported length is computed for $E_c = 5{,}000{,}000$ psi, with a factor of safety of 2 on the allowable direct load, assuming pin-ended at both ends. If the external direct load is smaller, the length can be increased. Note that this length is for transient loads; for sustained loads the value must be revised for a modulus E_c of 2,000,000 psi. If eccentricity is expected, allowable length should be reduced.

TABLE V.9A. PROPERTIES OF PRETENSIONED CONCRETE SHEET PILES[a]

Section No.	Shape	Size, in.	Solid or cored	No. of strands, and strand diam, in.	A_c, in.^{2}	Weight, psf of wall, lb	I/C		I about minor axis, in.^{4}	Effective prestress, psi[b]	Allowable moment	
							Per pile, in.^{3}	Per foot of wall, in.^{3}			300 psi tension, kip-in. per ft of wall	600 psi tension, kip-in. per ft of wall
BG-6	Rect.	6 × 24	Solid	12, ⅜	144	75	144	72	432	967	91	113
BG-9	Rect.	9 × 36	Solid	28, ⅜	324	112	487	162	2,190	1,002	211	259
BG-12	Rect.	12 × 36	Solid	28, 7/16	432	150	864	288	5,180	1,014	379	465
BG-18	Rect.	18 × 36	2-10″ HC	32, 7/16	491	170	1,850	617	16,600	1,020	814	999

TABLE V.9B. PROPERTIES OF PRETENSIONED CONCRETE FENDER PILES[a]

Section No.	Shape	Size, in.	Solid or cored	No. of strands, and strand diam., in.	A_c, in.^{2}	Weight, lb per lin ft	I, in.^{4}	I/C, in.^{3}	Effective prestress, psi	Allowable moment	
										300 psi tension, kip-in.	600 psi tension, kip-in.
FP 12	Square	12 × 12	Solid	12, ⅜	144	150	1,728	288	967	365	451
FP 14	Square	14 × 14	Solid	16, ⅜	196	204	3,200	457	947	570	706

[a] From Ben C. Gerwick, Inc.

[b] The effective prestress is based on a uniform distribution of strands resulting in a uniform prestress. For special applications of sheet piles, eccentric prestress may be desirable and economical.

TABLE V.10. RAYMOND PRESTRESSED CONCRETE CYLINDER PILES

PROPERTIES FOR DESIGN

O.D., in.	Size, I.D., in.	Well thick-ness, in.	Area, in.^{2}	I, in.^{4}	S, in.^{3}	r, in.	Circum-ference, in.	Volume per foot, ft^3	Weight per foot, lb	Concrete design stress per cable, psi	No. of prestressing cables
	16	4	251	13,060	1,089	7.2	75	1.74	262	166.0	
24	15	4½	276	13,810	1,150	7.1	75	1.92	288	150.5	
	14	5	298	14,380	1,198	6.9	75	2.07	310	139.5	
	22	4	327	28,200	1,880	9.3	94	2.27	341	127.2	
30	21	4½	361	30,170	2,010	9.1	94	2.50	376	115.0	
	20	5	393	31,800	2,120	9.0	94	2.73	410	106.0	
	28	4	402	52,200	2,900	11.4	113	2.79	419	103.5	
36[a]	27	4½	445	56,400	3,130	11.3	113	3.09	463	93.6	8 to 12 or 16
	26	5	487	60,000	3,330	11.1	113	3.38	507	85.5	
	34	4	477	87,000	4,140	13.5	132	3.32	497	87.3	
42	33	4½	530	94,800	4,520	13.4	132	3.68	551	78.6	
	32	5	581	102,000	4,850	13.2	132	4.04	605	71.7	
	30	6	679	113,300	5,390	12.9	132	4.72	708	61.3	
	40	4	553	134,900	5,620	15.6	151	3.84	576	75.3	
48	39	4½	615	146,700	6,100	15.4	151	4.27	641	67.7	
	38	5	675	158,100	6,600	15.3	151	4.69	703	61.7	
	36	6	792	178,000	7,400	15.0	151	5.50	826	52.6	
	46	4	628	197,400	7,320	17.7	170	4.36	655	66.4	
54[a]	45	4½	700	216,000	8,000	17.6	170	4.86	728	59.5	12 to 16 or 24
	44	5	770	233,000	8,630	17.4	170	5.35	802	54.2	
	42	6	904	264,000	9,770	17.1	170	6.28	941	46.1	

[a] Output originally standardized to the two indicated sizes; other sizes and wall thicknesses can be furnished, provided the project is of sufficient size to warrant the purchase of the special equipment required.

TABLE V.11. TYPICAL WELDED SPLICES FOR H PILES[a]

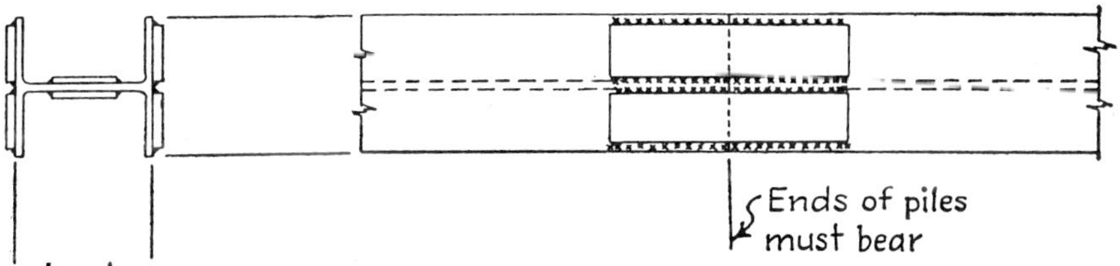

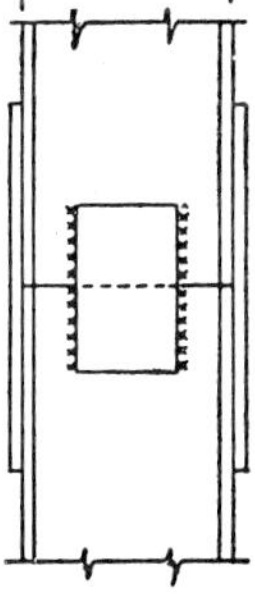

The upper pile section, with splice plates attached, should be set in place on top of the driven section and tapped several times with the hammer to improve bearing contact, after which the welding should be completed. The 33⅓ per cent weld develops ⅓ of the strength of the pile section and the 100 per cent weld develops the full strength of the pile section. All welding to be ⅜-in. continuous fillets.

H pile section	Flange plates			Web plates			Total weight, lb[c]
	Size, in.	Length[b]	Total lin in. weld	Size, in.	Length	Total lin in. weld	
33⅓ per cent weld							
B8b × 33	3 × ⅜	0′6″	48	4 × ⅜	0′6″	24	13
HP8 × 36	3 × ⅜	0′6″	48	4 × ⅜	0′6″	24	13
HP10 × 42	4 × ⅜	0 6 ′	48	4 × ⅜	0′6″	24	15
B10b × 49	4 × ⅜	0′6″	48	4 × ⅜	0′6″	24	15
HP10 × 57	4 × ⅜	0′7″	56	4 × ⅜	0′6″	24	17
HP12 × 53	5 × ⅜	0′7″	56	4 × ⅜	0′6″	24	20
B12c × 65	5 × ⅜	0′9″	72	4 × ⅜	0′6″	24	24
HP12 × 74	5 × ⅜	0′9″	72	4 × ⅜	0′6″	24	24
HP14 × 73	6 × ⅜	0′9″	72	6 × ⅜	0′6″	24	31
HP14 × 89	6 × ⅜	0′11″	88	6 × ⅜	0′7″	28	37
HP14 × 102	6½ × ⅜	1′0″	96	6 × ⅜	0′8″	32	43
HP14 × 117	6½ × ⅜	1′2″	112	6 × ⅜	0′9″	63	50
100 per cent weld							
B8b × 33	3 × ⅝	1′0″	96	4 × ⅜	0′6″	24	31
HP8 × 36	3 × ⅝	1′0″	96	4 × ½	0′8″	32	35
HP10 × 42	4 × ⅝	1′3″	120	5 × ½	0′9″	36	55
B10b × 49	4 × ¾	1′6″	144	5 × ⅜	0′8″	32	70
HP10 × 57	4 × ¾	1′7″	152	6 × ½	1′1″	52	88
HP12 × 53	5 × ⅝	1′6″	144	5 × ⅝	1′0″	48	85
B12c × 65	5 × ¾	1′11″	184	5½ × ½	0′11″	44	115
HP12 × 74	5 × ⅞	2′0″	192	6 × ⅝	1′4″	64	153
HP14 × 73	6 × ¾	2′2″	208	6 × ½	1′1″	52	155
HP14 × 89	6 × ⅞	2′6″	240	7 × ⅝	1′6″	72	223
HP14 × 102	6½ × ⅞	2′10″	272	7 × ¾	1′8″	80	279
HP14 × 117	6½ × 1	3′2″	304	8 × ¾	1′11″	92	358

[a] The above splices are Bethlehem Steel Co. designs. [b] Lengths of welding given in tables include both sides of the joint. [c] Weights given in tables do not include welding wire. *Champion steel H-beam bearing pile splicers* consist of channel-shaped bent plates designed and prefabricated to develop fully the bending strength of the pile with a minimum of field welding. A table of dimensions may be obtained from the manufacturer, Associated Pipe and Fitting Co., Clifton, N.J.

Table V.12. Followers for H Piles

Length of follower section must be given. Bolts to have square heads and hexagonal nuts

BETHLEHEM FOLLOWERS

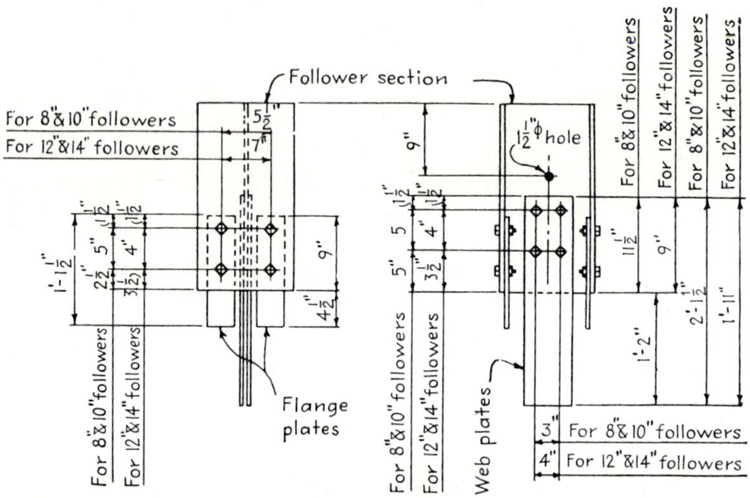

	H pile			
	8 in., BP8	10 in., BP10	12 in., BP12	14 in., BP14
Follower section	B8b × 58 lb	B10b × 89 lb	B12c × 106 lb	B14d × 136 lb[a] B14e × 176 lb
Flange plates:				
No.	4		4	
Size	3" × ½" × 1'½"		4" × ½" × 1'1½"	
Web plates:				
No.	2		2	
Size	6" × ½" × 2'1½"		8" × ½" × 1'11"	
Bolts:				
No.	12		12	
Size	⅞" diam × 2¾"		⅞" diam × 3"	
Total weight[b]	79 lb		96 lb	

[a] To be used with 14-in. H pile sections having a web thickness not greater than ⅝ in.

[b] Not including follower section.

TABLE V.12. FOLLOWERS FOR H PILES (*Continued*)

UNITED STATES STEEL CORPORATION FOLLOWERS

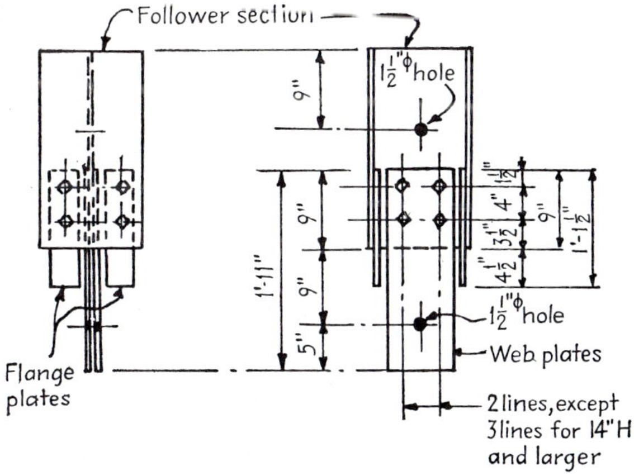

	H pile				
	8 in., CBP83	10 in., CBP103	12 in., CBP124	14 in., CBP145	14 in., CBP146
Follower section........	CB83 × 58 lb	CB103 × 89 lb	CB124 × 106 lb	CB145 × 136 lb	CB146 × 176 lb
Flange plates:					
No.........	4		4	4	
Size........	4″ × ¾″ × 1′1½″		4″ × ¾″ × 1′1½″	4″ × ¾″ × 1′1½″	
Web plates:					
No.........	2		2	2	
Size........	6″ × ¾″ × 1′11″		7″ × ¾″ × 1′11″	9″ × ¾″ × 1′11″	
Bolts:					
No.........	12		12	14	
Size........	⅞″ diam × 3″		⅞″ diam × 3¼″	⅞″ diam × 3¼″	
Total weight[a]	116 lb		127 lb	149 lb	

[a] Not including follower section.

STEEL-BOX-PILING DATA

TABLE V.13. PROPERTIES OF LARSSEN BOX PILES

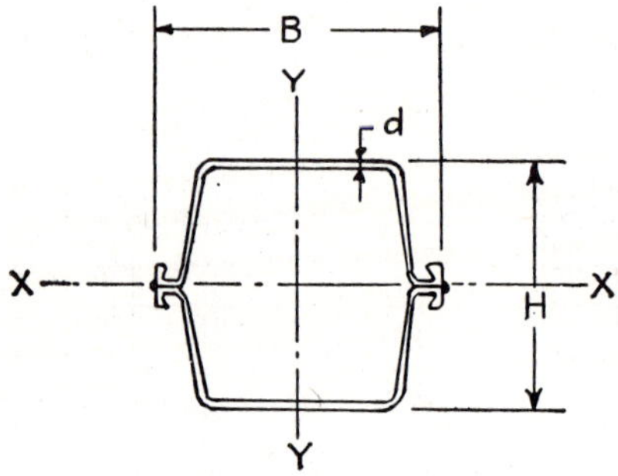

Section No.[a]	Larssen section	*B*, in.	*H*, in.	*d*, in.	Weight, lb per lin ft	Section modulus, in.³	
						X-X	*Y-Y*
BP1	1GB	17	6½	0.31	49	27.8	40.5
BP1U	1U	17	6½	0.375	58	32.8	50.1
BP2	2	17¼	9 7/16	0.41	66	52.0	71.8
BP3	3	17¼	11⅜	0.55	84	81.5	82.9
BP3B	3B	17¼	13½	0.53	83	91.7	84.2
BP4	4B	18	15½	0.63	114	141.2	126.8
BP5	5	18	15½	0.87	135	180.0	137.4
BP6	6	18¼	19¾	0.87	164	257.3	177.7

[a] British Steel Piling Company, Ltd., numbers.

Table V.14. Properties of Algoma Box Piles

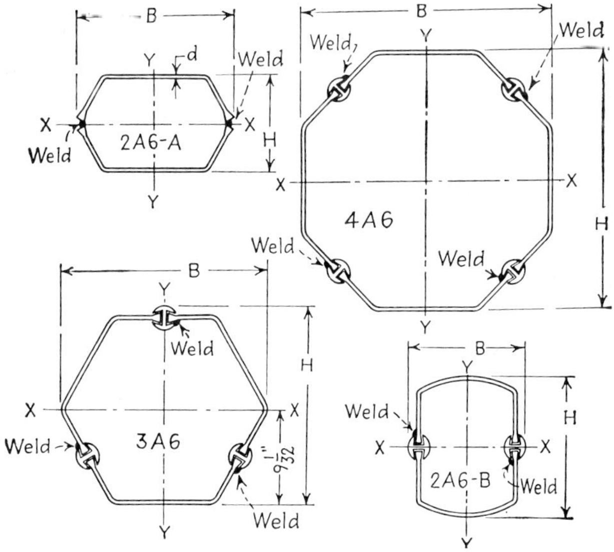

Section No.	B, in.	H, in.	d, in.	Weight, lb per lin ft	Section modulus, in.^{3}	
					X-X	Y-Y
2A6-A	$15\frac{5}{8}$	$9\frac{5}{16}$	$\frac{3}{8}$	53.3	44.81	50.57
2A6-B[a]	$11\frac{29}{32}$	$19\frac{3}{64}$	$\frac{3}{8}$	65.5	49.1	57.8
3A6	$20\frac{3}{4}$	$18\frac{31}{32}$	$\frac{3}{8}$	98.25	129.0	121.0
4A6	$25\frac{9}{32}$	$25\frac{9}{32}$	$\frac{3}{8}$	131.0	245.0	245.0

[a] User must arrange for fabricating this from standard sections.

TABLE V.15. PROPERTIES OF RENDHEX PILES

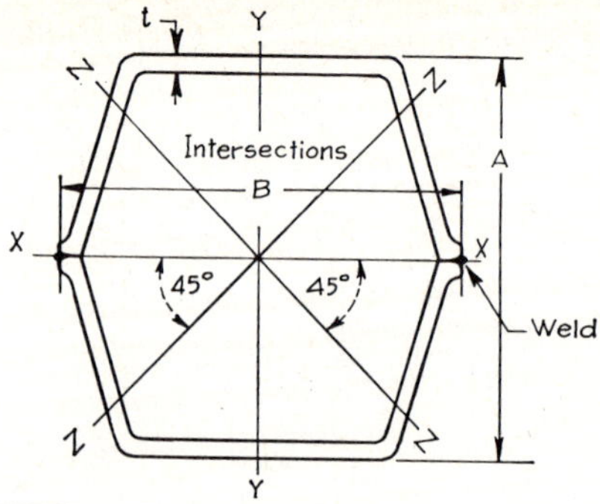

Section No.	A, in.	B, in.	t, in.	Weight, lb per lin ft	Steel area, in.²	Section modulus, in.³			Radius of gyration, in.			Moment of inertia, in.⁴		
						X-X	Y-Y	Z-Z	X-X	Y-Y	Z-Z	X-X	Y-Y	Z-Z
3	13	12⅞	½	72.50	21.26	72.20	58.70	57.30	4.69	4.23	4.47	469	381	425
4	16	16$\frac{5}{16}$	⅝	117.75	34.3	151.75	125.82	116.15	5.95	5.50	5.73	1,214	1,038	1,126
6	20	20	0.7	162.80	47.47	265.80	218.60	207.20	7.49	6.81	7.16	2,658	2,203	2,431

TABLE V.16. PROPERTIES OF FRODINGHAM OCTAGONAL BOX PILES

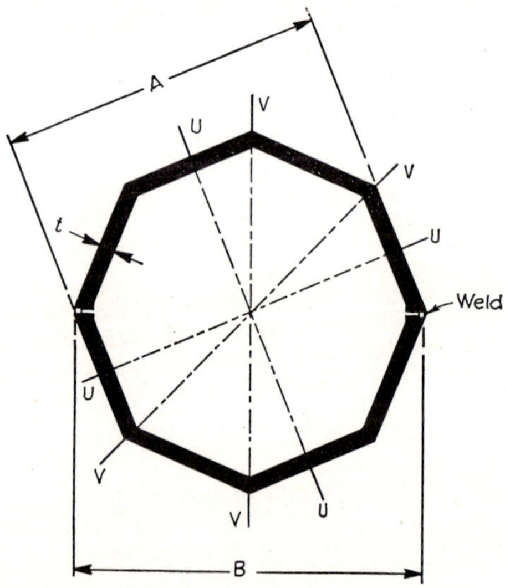

Section No.	A, in.	B, in.	t, in.	Weight, lb per lin ft	Section modulus, in.³		Radius of gyration, in.	Moment of inertia, in.⁴
					U-U	V-V		
4	16½	17¾	⅝	112.6	133.2	123.6	5.75	1,096

TABLE VI. APPROXIMATE VALUES OF WEIGHTS AND STRENGTHS OF WOOD[a]

Kind of wood	Green (untreated)[b]					Air seasoned (untreated)[b]				
	Weight, lb per cu ft[c]	Modulus of elasticity E psi [d]	Max crushing strength, psi[d,e]	Max bending strength, psi	Per cent moisture[f]	Weight, lb per cu ft[c]	Modulus of elasticity E psi [d]	Max crushing strength, psi[d]	Max bending strength, psi	Per cent moisture[f]
North American Woods										
Beech	54	1,378,000	3,550	8,600	54	45	1,720,000	7,300	14,900	12
Birch	57	1,560,000	3,510	8,700	62	44	2,070,000	8,310	16,700	12
Cedar:										
Alaska	36	1,140,000	3,050	6,400	38	31	1,420,000	6,310	11,100	12
Eastern red	37	650,000	3,570	7,000	35	33	880,000	6,020	8,800	12
Northern white	28	640,000	1,990	4,200	55	22	800,000	3,960	6,500	12
Port Orford	56	1,420,000	3,130	6,200	43	29	1,730,000	6,470	11,300	12
Southern white	26	750,000	2,390	4,700	35	23	930,000	4,700	6,800	12
Western red	27	920,000	2,750	5,100	37	33	1,120,000	5,020	7,700	12
Chestnut	55	1,230,000	5,320	5,600	122	30	930,000	2,470	8,600	12
Cottonwood:										
Eastern	49	1,010,000	2,280	5,300	111	28	1,370,000	4,910	8,500	12
Northern black	46	1,070,000	2,160	4,800	132	24	1,260,000	4,420	8,300	12
Cypress, southern	51	1,180,000	3,580	6,600	91	32	1,440,000	6,360	10,600	12
Douglas fir:										
Coast region	38	1,550,000	3,890	7,600	36	34	1,920,000	7,420	11,700	12
Inland empire	36	1,340,000	3,240	6,800	42	31	1,610,000	6,700	11,300	12
Rocky Mountain	35	1,180,000	3,000	6,400	38	30	1,400,000	6,060	9,600	12
Elm										
American	54	1,110,000	2,910	7,200	89	35	1,340,000	5,520	11,800	12
Rock	53	1,190,000	3,780	9,500	48	44	1,540,000	7,050	14,800	12
Slippery	56	1,230,000	3,320	8,000	85	37	1,490,000	6,360	13,000	12

TABLE VI. APPROXIMATE VALUES OF WEIGHTS AND STRENGTHS OF WOOD[a] (*Continued*)

Kind of wood	Green (untreated)[b]					Air seasoned (untreated)[b]				
	Weight, lb per cu ft[c]	Modulus of elasticity E psi [d]	Max crushing strength, psi[d,e]	Max bending strength, psi	Per cent moisture[f]	Weight, lb per cu ft[c]	Modulus of elasticity E psi [d]	Max crushing strength, psi[d]	Max bending strength, psi	Per cent moisture[f]
Fir:										
Balsam	45	960,000	2,400	4,900	117	25	1,230,000	4,530	7,600	12
White	46	1,120,000	2,810	5,800	108	27	1,470,000	5,380	9,300	12
Gum:										
Black	45	1,030,000	3,040	7,000	55	35	1,200,000	5,520	9,600	12
Red	50	1,150,000	2,840	6,800	81	34	1,490,000	5,800	11,900	12
Tupelo	56	1,050,000	3,370	7,300	97	35	1,260,000	5,920	9,600	12
Hemlock:										
Eastern	50	1,070,000	3,080	6,400	111	28	1,200,000	5,410	8,900	12
Western	41	1,220,000	2,990	6,100	74	29	1,490,000	6,210	10,100	12
Hickory:										
Pecan	62	1,380,000	4,320	9,900	68	45	1,780,000	8,280	16,300	12
True	63	1,570,000	4,570	11,300	57	51	2,180,000	8,970	19,700	12
Honey locust	61	1,290,000	4,420	10,200	63	..	1,630,000	7,500	14,700	12
Ironwood	59	1,580,000	3,790	10,400	47	50	2,080,000	8,690	19,400	11
Larch, western	48	1,350,000	3,800	7,500	58	36	1,710,000	7,490	11,900	12
Locust, black	58	1,850,000	6,800	13,800	40	48	2,050,000	10,180	19,400	12
Mangrove (*Rhizophora mangle*)	77	....	5,400	......	...	67	....	11,300		
Maple:										
Big leaf	47	1,100,000	3,240	7,400	72	34	1,450,000	5,950	10,700	12
Black	54	1,330,000	3,270	7,900	65	40	1,620,000	6,680	13,300	12
Red	50	1,390,000	3,280	7,700	63	38	1,640,000	6,540	13,400	12
Silver	45	940,000	2,490	5,800	66	33	1,140,000	5,220	8,900	12
Sugar	56	1,550,000	4,020	9,400	58	44	1,830,000	7,830	15,800	12

Oak:										
Red	64	1,360,000	3,520	8,500	80	44	1,810,000	6,920	14,400	12
White	63	1,200,000	3,520	8,100	70	47	1,620,000	7,040	13,900	12
Osage orange	..	....	......	......	...	48	1,300,000	......	16,000	
Palmetto, cabbage	54	....	1,250	......	...	27	....	2,600		
Pine:										
Jack	38	1,260,000	3,000	6,300	42	31	1,430,000	5,260	10,100	13
Loblolly	53	1,410,000	3,490	7,300	81	36	1,800,000	7,080	12,800	12
Lodgepole	39	1,080,000	2,610	5,500	65	29	1,340,000	5,370	9,400	12
Longleaf	55	1,600,000	4,300	8,700	63	41	1,990,000	8,440	14,700	12
Northern white	36	1,020,000	2,490	5,000	68	25	1,280,000	4,840	8,800	12
Norway	42	1,380,000	3,080	6,400	54	34	1,800,000	7,340	12,500	12
Ponderosa	45	970,000	2,400	5,000	91	28	1,260,000	5,270	9,200	12
Shortleaf	52	1,390,000	3,430	7,300	81	36	1,760,000	7,070	12,800	12
Sugar	52	940,000	2,530	5,100	137	25	1,200,000	4,770	8,000	12
Western white	35	1,170,000	2,650	5,200	54	27	1,510,000	5,620	9,500	12
Western yellow	47	1,260,000	2,840	5,700	70	32	1,480,000	6,050	10,200	13
White	42	1,260,000	2,450	4,700	97	24	1,590,000	4,980	8,500	11
Poplar, yellow	38	1,090,000	2,420	5,400	64	28	1,500,000	5,290	9,200	12
Redwood	50	1,180,000	4,200	7,500	112	28	1,340,000	6,150	10,000	12
Spruce:										
Black	37	1,520,000	2,860	5,800	48	30	1,790,000	5,850	10,300	13
Eastern	34	1,110,000	2,600	3,300	46	28	1,440,000	5,590	6,500	12
Englemann	39	830,000	1,980	4,200	100	23	1,160,000	4,580	8,500	12
Red	35	1,470,000	2,800	5,100	49	28	1,730,000	5,300	10,100	13
Sitka	33	1,230,000	2,670	5,700	42	28	1,570,000	5,610	10,200	12
White	36	1,260,000	2,450	5,100	69	26	1,590,000	4,980	8,600	13
Sycamore	52	1,060,000	2,920	6,500	83	34	1,420,000	5,380	10,000	12
Tamarack	47	1,240,000	3,480	7,200	52	37	1,640,000	7,560	11,600	12
Walnut, black	58	1,420,000	4,300	9,500	81	38	1,680,000	7,580	14,600	12

TABLE VI. APPROXIMATE VALUES OF WEIGHTS AND STRENGTHS OF WOOD[a] (*Continued*)

Kind of wood	Green (untreated)[b]					Air seasoned (untreated)[b]				
	Weight, lb per cu ft[c]	Modulus of elasticity E psi [d]	Max crushing strength, psi[d,e]	Max bending strength, psi	Per cent moisture[f]	Weight, lb per cu ft[c]	Modulus of elasticity E psi [d]	Max crushing strength, psi[d]	Max bending strength, psi	Per cent moisture[f]
European Woods										
Ash, English	48	1,500,000	3,750	8,900	47	42	1,860,000	6,990	15,100	12
Beech, English	..	1,520,000	3,860	8,900	88	43	1,950,000	7,870	16,200	12
Elm, English (*Ulmus procera*)	..	810,000	2,360	5,500	140	32	1,090,000	4,740	9,300	12
Hickory	..	1,570,000	4,480	11,100	59	51	2,220,000	8,940	19,200	12
Larch, European	44	1,130,000	3,480	7,350	50	34	1,480,000	6,920	12,700	12
Oak, British (*Quercus robur, Quercus petraea*)	..	1,290,000	3,850	8,100	89	43	1,560,000	7,210	13,300	12
Redwood, Baltic	..	1,240,000	3,140	6,500	30	30	1,420,000	6,360	11,200	12
South American and Caribbean Woods										
Angelique	67	1,840,000	5,590	11,400	79					
Greenheart	78	3,700,000	10,690	20,900	38	66	4,140,000	13,040	25,500	14.4
Mangrove (*Rhizophora mangle*)	77	....	5,400	......	...	67	....	11,300		
Mora	..	....	......	......	...	60–70	3,370,000	10,870	21,000	12
Palmetto, cabbage	54	....	1,250	......	...	27	....	2,600		
Purpleheart	..	2,170,000	7,890	14,400	62	55	2,580,000	10,960	20,200	12
Wallaba	76	2,860,000	8,270	16,000	57	50–60	2,870,000	10,230	22,000	13.1

[a] Tabular values obtained largely from *Wood Handbook*, Forest Products Laboratory, Forest Service, 1935; *Reports of Sea-action Committee of the Institution of Civil Engineers; Report of the Committee on Marine Piling Investigation*, National Research Council; and "The Mechanical Properties of Canadian Woods," *Bull.* 82, Canadian Forest Service.

[b] For treated air-seascned piling, add weight of preservative. For green piling, preliminary steaming or vacuum treatment may reduce weight more than weight of preservative added.

[c] Weights include weight of contained water and are values desired in pile-driving formulas; values may differ from those found in many reference tables, which state only weight of wood, based on specific gravity.

[d] Values of E and crushing strength are taken parallel to grain.

[e] *Reduce maximum crushing strength by approximately one-third for green wood seasoned by steaming or boiling under vacuum.* If temperature, pressure, and time of seasoning are not controlled within proper limits, much greater weakening may occur. Value of E should also be reduced somewhat. Also *reduce maximum crushing strength by approximately one-third for green wood submerged in creosote heated above the boiling point of water, under atmospheric pressure only.*

[f] For air-seasoned woods, without artificial heating and sheltered from precipitation, the moisture content reaches approximately 12 per cent in the north central United States. In any lot, the variation is usually not over 10 per cent. In green material, the variation may occasionally be as great as 20 per cent. Particularly in species which have a high moisture content in the sapwood, large variations in weight may occur when the wood is green. Since young softwood trees contain a larger proportion of sapwood than old trees, their wood is heavier on the average when green. The greatest changes in weight are those which occur in the early stages of the drying of green wood Changes in the moisture content of air-seasoned wood are attended by only relatively small changes in weight per cubic foot owing to the countereffect of change in volume. For estimating changes in weight for moisture contents in the vicinity of 12 per cent, assume ½ per cent change in weight as occurring for every 1 per cent change in moisture content.

A formula,[2b] known as the "exponential formula," has been devised by the U.S. Forest Products Laboratory for adjusting strengths to water contents:

$$\log S_3 = \log S_1 + \frac{M_1 - M_3}{M_1 - M_2} \log \frac{S_2}{S_1} \tag{T1}$$

where S_1 and M_1 are one pair of corresponding strength and moisture-content values as found from test, S_2 and M_2 are another pair, and S_3 is the strength value adjusted to the moisture content M_3. If one strength value is for green wood, the following experimentally determined values (M_p) must be used for the corresponding moisture content.

Species	*M_p, Per Cent*	*Species*	*M_p, Per Cent*
Ash, white	24	Pine:	
Birch, yellow	27	Loblolly	21
Chestnut	24	Longleaf	21
Douglas fir	24	Norway	24
Hemlock, western	28	Redwood	21
Larch, western	28	Spruce:	
		Red	27
		Sitka	27
		Tamarack	24

For all other species, assume M_p equals 24 per cent. This formula is not applicable where there is a large variation in moisture content from one part of the cross section to another.

GROUP IV

ENGINEERING DATA FOR JETTING

TABLE VII.1. HOSE DELIVERY TABLE[a]
Necessary Pounds Pressure on Intake End of 100 Ft of Rubber Hose to Deliver Gallons per Minute Desired

U.S. gal per min	Inside hose diameter						
	1¼″	1½″	2″	2½″	3″	4″	5″
100	88	37	12.5	3.5	1.7	0.4	0.1
150	183	78	27	7.5	3.5	0.7	0.3
200	...	133	46	13	5.9	1.4	0.5
250	...	...	70	19	9.1	2.1	0.7
300	...	...	95	27	12	2.9	1.0
350	...	...	126	36	17	4	1.3
400	...	...	.....	46	21	5.1	1.7
450	...	...	.... .	57	26	6.3	2.1
500	...	...	.....	70	32	7.4	2.6
1,000	...	...	.....	....	116	27.0	9.6
2,000	...	...	....	....	.....	100.0	32.0

[a] All values in this table furnished by the Griffin Equipment Corp.

TABLE VII.2. THEORETICAL DISCHARGE OF NOZZLES, IN GALLONS PER MINUTE[a]

Pounds pressure, psi	1″ nozzle	1⅛″ nozzle	1¼″ nozzle	1½″ nozzle	2″ nozzle	2½″ nozzle
50	211	267	330	475	845	1,320
75	259	327	404	582	1,036	1,618
100	299	378	467	672	1,196	1,870
125	334	423	522	751	1,338	2,090
150	366	463	572	824	1,466	2,290
175	395	500	618	890	1,582	2,473
200	423	535	660	950	1,691	2,645
250	473	598	739	1,063	1,891	2,955
300	517	655	808	1,163	2,070	3,235

[a] All values in this table furnished by the Griffin Equipment Corp.

TABLE VII.3. PRESSURES IN POUNDS PER SQUARE INCH WITH EQUIVALENT FEET HEAD[a]

Pressure, psi	0	1	2	3	4	5	6	7	8	9
	Equivalent Feet Head[b]									
10	230.4	232.7	235.0	237.3	239.6	241.9	244.2	246.5	248.8	251.1
12	276.4	278.7	281.0	283.3	285.6	287.9	290.2	292.5	294.8	297.1
14	322.5	324.8	327.1	329.4	331.7	334.0	336.3	338.6	340.9	343.2
16	368.6	370.9	373.2	375.5	377.8	380.1	382.4	384.7	387.0	389.3
18	414.7	417.0	419.3	421.6	423.9	426.2	428.5	430.8	433.1	435.4
20	460.8	463.1	465.4	467.7	470.0	472.3	474.6	476.9	479.2	481.5
22	506.8	509.1	511.4	513.7	516.0	518.3	520.6	522.9	525.2	527.5
24	552.9	555.2	557.5	559.8	562.1	564.4	566.7	569.0	571.3	573.6
26	599.0	601.3	603.6	605.9	608.2	610.5	612.8	615.1	617.4	619.7
28	645.1	647.4	649.7	652.0	654.3	655.6	658.9	661.2	663.5	665.8
30	693.0	695.3	697.6	699.9	702.2	704.5	706.8	709.1	711.4	713.7
32	739.2	741.5	743.8	746.1	748.4	750.7	753.0	755.3	757.6	759.9
34	785.4	787.7	790.0	792.3	794.6	796.9	799.2	801.5	803.8	806.1
35	808.5	810.8	813.1	815.4	817.7	820.0	822.3	824.6	826.9	829.2
40	924.0	926.6	928.6	930.9	933.2	935.5	937.8	940.1	942.4	944.7
45	1,039.5	1,041.8	1,044.1	1,046.4	1,048.7	1,051.0	1,053.3	1,055.6	1,057.9	1,060.2
50	1,155.0	1,157.3	1,159.6	1,161.9	1,164.2	1,166.5	1,168.8	1,171.1	1,173.4	1,175.7

[a] All values in this table furnished by the Griffin Equipment Corp.

[b] Example: In order to find the equivalent feet head for 126 lb, follow down the first column to the figure 12, then across on the same horizontal line until under the figure 6, which gives 290.2 ft as the equivalent to 126 lb pressure.

TABLE VII.4. LOSS OF PRESSURE BY FRICTION IN JET PIPE AND HOSE

Size of pipe, in.	Gal per min								
	100	150	200	250	300	350	400	450	500
	Friction loss, lb per ft of length								
2	0.14	0.30	0.55	0.85	1.20				
2½	0.05	0.10	0.18	0.28	0.40	0.54	0.72	0.90	1.12
3	0.02	0.04	0.07	0.12	0.16	0.22	0.30	0.40	0.45
3½	....	0.02	0.03	0.05	0.08	0.10	0.13	0.16	0.20
4	....	0.01	0.02	0.03	0.04	0.05	0.07	0.08	0.11
5	....	....	....	....	0.01	0.02	0.02	0.03	0.04
6	....	....	....	....	....	....	....	0.01	0.02

APPENDIX

APPENDIX I

FORMULAS

Derivation of Dynamic Formula

Following is the derivation of formulas (2.1*a*) and (2.1*b*), where

g = 32.2 ft per sec^2;
v = velocity of ram due to free fall, at moment of impact;
v_c = velocity of ram and pile, at end of compression period;
v_r = velocity of ram at end of period of restitution;
v_p = velocity of pile at end of period of restitution;
M_t = amount of impulse causing compression; and
eM_t = amount of impulse causing restitution.

The momentum of the ram at the moment of impact is W_rv/g. At the end of the period of compression the momentum of the ram is $(W_rv/g) - M_t$, and the velocity $v_c = [(W_rv/g) - M_t]/(W_r/g)$. Assuming that the pile is able to move a short distance, and that the hammer blows and rebounds of the pile result in a looseness between the pile and the earth, then the momentum of the pile at the end of the period of compression may be taken as M_t, and the velocity of the pile becomes $v_c = M_t/(W_p/g)$.

Since the velocities of the ram and pile are equal at the end of the compression period, at that time $M_t = vW_rW_p/g(W_r + W_p)$.

At the end of the period of restitution, the momentum of the ram is $(W_rv/g) - M_t - eM_t = W_rv_r/g$; therefore,

$$v_r = v - \frac{M_t(1 + e)}{W_r/g} = v - \frac{W_p}{W_r + W_p}v(1 + e) = \frac{W_r - eW_p}{W_r + W_p}v$$

At the end of the period of restitution, the momentum of the pile is $M_t + eM_t = W_pv_p/g$; therefore,

$$v_p = \frac{M_t(1 + e)}{W_p/g} = \frac{W_r}{W_r + W_p}(1 + e)v = \frac{W_r + eW_r}{W_r + W_p}v$$

Using the above values for v_r and v_p, it is possible to determine the sum of the gross energies in the ram and pile, at the end of the period of restitution, available for expenditure in overcoming ground resistance to driving, and in causing temporary elastic compressions in the driving cap, pile, and soil.

The energy available in the ram and pile at the close of the period of restitution will be equal to

$$\frac{W_r}{2g}(v_r)^2 + \frac{W_p}{2g}(v_p)^2 = \frac{W_r v^2}{2g}\frac{(W_r - eW_p)^2}{(W_r + W_p)^2} + \frac{W_p v^2}{2g}\frac{(W_r + eW_r)^2}{(W_r + W_p)^2}$$
$$= \frac{W_r v^2}{2g}\frac{(W_r + e^2 W_p)}{(W_r + W_p)}$$
$$= W_r h \frac{W_r + e^2 W_p}{W_r + W_p} \tag{A1}$$

If the above expression for energy available at the close of the period of restitution is written as

$$\frac{W_r v^2}{2g}\left[1 - \frac{W_p(1 - e^2)}{W_r + W_p}\right]$$

then the loss in energy due to impact is

$$\frac{W_r v^2}{2g}\frac{W_p(1 - e^2)}{W_r + W_p} = W_r h \frac{W_p(1 - e^2)}{W_r + W_p} \tag{A2}$$

and the efficiency of the hammer blow is

$$\frac{W_r + e^2 W_p}{W_r + W_p} = \frac{1}{1 + W_p/W_r} + \frac{e^2}{1 + W_p/W_r} \tag{A3}$$

(Because of the smallness of the term e^2 compared with 1, the efficiency has been taken as $\frac{1}{1 + W_p/W_r}$ to serve as the basis of the expression W_p/W_r in the denominators of the Eytelwein and Navy-McKay formulas, in each of which, however, it has been modified empirically.)

If no impact or elastic losses occurred, and the mechanical efficiency of the hammer were 100 per cent, the following expression could be written

$$R_u s = W_r h \tag{A4}$$

Denoting the mechanical efficiency by the term e_f, the above expression becomes

$$R_u s = e_f W_r h \tag{A5}$$

Replacing the terms $W_r h$ by the expression derived for energy available at the close of the period of restitution,

$$R_u s = e_f W_r h \frac{W_r + e^2 W_p}{W_r + W_p} \tag{A6}$$

which may be transposed to read

$$R_u = \frac{e_f W_r h}{s}\frac{W_r + e^2 W_p}{W_r + W_p} \tag{A7}$$

However, while the tip of the pile moves downward a distance s, the top of the driving cap moves downward an additional distance $C_1 + C_2 + C_3$ owing to temporary elastic compressions in the cap, pile, and soil. Within elastic limits, the deformation of each of these materials may be assumed to vary with load. For the cap and pile, the amounts of temporary compression may be computed from the expression $C = R_u l/AE$. The work obtained from the kinetic energy of the blow may be written, instead of $R_u s$, as $R_u(s + C/2)$, and the above expression becomes

$$R_u = \frac{e_f W_r h}{s + C/2} \frac{W_r + e^2 W_p}{W_r + W_p} \tag{A8}$$

and if the temporary compression C is subdivided into its component elements the following expression results:

$$R_u = \frac{e_f W_r h}{s + \frac{1}{2}(C_1 + C_2 + C_3)} \frac{W_r + e^2 W_p}{W_r + W_p} \tag{2.1a}$$

Formulas (2.1*a*) and (2.1*b*), while not mathematically quite true for the special case of driving to refusal on impenetrable material or rock, where the point of the pile is not free to move on, give results which differ only inconsiderably from those which would be obtained from a more correct, but more complicated, formula.

Relationship between Various Dynamic Formulas

The relation of the various most frequently used formulas to each other and a brief indication of their common basis may be of interest.

If there were no losses, the driving resistance could be expressed by formula (A4). Since there are losses due to efficiency, impact, and elastic compressions of the cap, pile, and soil, these items are deducted as shown in the following expression:

$$R_u = \frac{e_f W_r h}{s} - \frac{e_f W_r h \dfrac{W_p(1 - e^2)}{W_r + W_p}}{s} - \frac{R_u C_1}{2s} - \frac{R_u{}^2 l}{2AEs} - \frac{R_u C_3}{2s} \tag{2.5a}$$

and, calling $\dfrac{R_u l}{AE} = C_2$, by combining terms, we obtain

$$R_u = \frac{e_f W_r h}{s + \frac{1}{2}(C_1 + C_2 + C_3)} \frac{W_r + e^2 W_p}{W_r + W_p} \qquad \textit{Hiley formula} \tag{2.1a}$$

If the rebound coefficients in formula (2.1*a*) are modified empirically, and a factor of safety of 3 is assumed, the following Canadian National Building Code formula is obtained:

$$R = \frac{4nW_r H}{s + C/2} \qquad \textit{Canadian National Building Code formula} \tag{A9}$$

where $n = \dfrac{W_r + e^2W_p}{W_r + W_p}$ for friction piles

$$n = \frac{W_r + 0.5e^2W_p}{W_r + W_e} \text{ for refusal}$$

$$C = \frac{3R}{A}\left(\frac{l}{E} + 0.0001\right)$$

The resulting value is used for drop hammers with triggers, and should be multiplied by 0.9 for single-acting hammers, and by 0.8 for drop hammers with winch drag.

If it is assumed that there are no elastic losses in the cap or soil quake, formula (2.5a) reduces to

$$R_u = \frac{e_fW_rh}{s} - \frac{e_fW_rh\,\dfrac{W_p(1 - e^2)}{W_r + W_p}}{s} - \frac{R_u{}^2l}{2AEs} \qquad \text{(A10)}$$

Now assume the hammer to be mechanically 100 per cent efficient, thus omitting the term e_f, and solve for R_u, in which case,

$$R_u = \frac{AE}{l}\left[-s + \sqrt{s^2 + W_rh\left(\frac{W_r + e^2W_p}{W_r + W_p}\right)\frac{2l}{AE}}\right] \quad \textit{Universal or Stern formula} \quad \text{(A11)*}$$

If the impact is assumed to be perfectly inelastic instead of semielastic, then $e = 0$, and formula (A11) becomes

$$R_u = \frac{AE}{l}\left[-s + \sqrt{s^2 + \left(\frac{W_r{}^2h}{W_r + W_p}\right)\frac{2l}{AE}}\right] \quad \textit{Redtenbacher formula} \quad \text{(A12)}$$

If the temporary elastic shortening of the ground is included, as well as of the pile, and these are measured from a load-test diagram near to or beyond the failure point, the above formula becomes

$$R_u = \frac{s}{\tan\phi_e}\left[-1 + \sqrt{1 + \left(\frac{W_r{}^2h}{W_rW_p}\right)\frac{2\tan\phi_e}{s^2}}\right] \quad \textit{Schenk formula} \quad \text{(A13)}$$

where $\tan\phi_e$ = tangent of angle between horizontal and rebound line (in Fig. 15.8, the lowest inclined line).

If the impact loss is entirely neglected, formula (A12) becomes

$$R_u = -\frac{sAE}{l} + \sqrt{\frac{2W_rhAE}{l} + \left(\frac{sAE}{l}\right)^2} \quad \textit{Weisbach formula} \quad \text{(A14)}$$

In formulas (A11), (A12), and (A14) the value of l was intended by the authors to be the full length of the pile.

* This formula was published for the first time, as fas as has been ascertained, in a book by Krapff, entitled *Formeln und Versuche über die Tragfähigkeit eingerammter Pfähle*, Fortschr. der Ingineur-Wissensch., Zweite Gruppe, 12. Heft, Leipzig, 1906.

If the hammer is assumed to be mechanically 100 per cent efficient, thus omitting the term e_f, and if, instead of considering the elastic losses in the cap or soil quake, twice the average elastic loss is used, taking into account the full length of the pile, and if fixed values are assumed for e, then formula (2.1a) becomes

$$R_{ut} = \frac{12W_{rt}H \dfrac{W_{rt} + KW_{pt}}{W_{rt} + W_{pt}}}{s + \dfrac{24{,}000R_{ut}L}{AE}} \qquad \textit{International Conference Uniform Building Code formula} \qquad \text{(A15)}$$

where R_{ut} = ultimate driving resistance, in tons (to which it is specified that a factor of safety of 4 should be applied to obtain the working load, in tons);
W_{rt} = weight of falling mass, in tons;
W_{pt} = weight of pile, in tons; and
K = 0.25 for steel piles and 0.10 for other piles.

On the other hand, assume the impact to be perfectly elastic, and also assume that the pile is fully embedded in the ground and is a friction pile without end bearing, so that the distance from the butt to the center of resistance is $l/2$, and formula (A11) becomes

$$R_u = \frac{2AEs}{l}\left(\sqrt{1 + \frac{W_r hl}{s^2EA}} - 1\right) \qquad \textit{Rankine formula} \qquad \text{(A16)}$$

By taking formula (2.1a) and assuming that the mechanical efficiency is 100 per cent (e_f = 1.0), that the impact is perfectly inelastic (e = 0), and that there are no elastic losses in the cap, pile, or soil, we obtain

$$R_u = \frac{W_r h}{s}\,\frac{W_r}{W_r + W_p} \qquad \textit{Dutch formula} \qquad \text{(A17)}$$

With the Dutch formula it is customary to use a factor of safety of 10 when driving with a drop hammer, and of 6 with a steam hammer. This formula and the Hiley formula are the best known in Great Britain.

The Ritter formula is the same as the Dutch formula, with the inclusion of terms to add the weights of the ram and pile:

$$R_u = \frac{W_r h}{s}\,\frac{W_r}{W_r + W_p} + W_r + W_p \qquad \textit{Ritter formula} \qquad \text{(A18)}$$

By writing the Dutch formula in the following form, taking H in feet, and assuming a factor of safety of 6, we obtain a formula for drop hammers:

$$R = \frac{2W_rH}{s\left(1 + \dfrac{W_p}{W_r}\right)} \qquad \textit{Eytelwein formula} \qquad \text{(A19}a\text{)}$$

The Eytelwein formula is modified as follows for single-acting and double-acting steam hammers:

Single-acting:

$$R = \frac{2W_rH}{s + 0.1\,\dfrac{W_p}{W_r}} \tag{A19b}$$

Double-acting:

$$R = \frac{2(W_rH + Ap)}{s + 0.1\,\dfrac{W_p}{W_r}} \tag{A19c}$$

where A = effective area of piston, in square inches; and
p = mean effective pressure of steam or air, in pounds per square inch.

If H is taken in feet, a factor of safety of 6 assumed, and the ratio W_p/W_r in formula (A19*b*) modified by a factor of 0.3*s* instead of 0.1, we have

$$R = \frac{2W_rH}{s\left(1 + 0.3\,\dfrac{W_p}{W_r}\right)} \qquad \textit{Navy-McKay formula} \quad \text{(A20)}$$

(*not now used by Navy*). If in formula (2.1*a*) the impact loss is entirely neglected, the mechanical efficiency taken as 100 per cent, the elastic losses in the cap, pile, and soil represented by a constant term of 1.0, H taken in feet and then multiplied by 12, and a factor of safety of 6 is assumed, the following expression is obtained for use with drop hammers, giving working loads instead of ultimate loads:

$$R = \frac{2W_rH}{s + 1.0} \qquad \textit{Engineering News formula} \quad \text{(A21a)}$$

For use with single-acting steam hammers, formula (A21*a*) was modified by its author by changing the term 1.0 to 0.1, and in this form as given below has been widely used for single-acting, double-acting, and differential-acting steam hammers.

Single-acting:

$$R = \frac{2W_rH}{s + 0.1} \tag{A21b}$$

Double- and differential-acting:

$$R = \frac{2E_n}{s + 0.1} \tag{A21c}$$

Formulas (A21*b*) and (A21*c*) may be expressed as follows, n being the number of blows per foot of penetration:

Single-acting:

$$R = \frac{20n}{120 + n} \times W_r H \qquad \textit{Vulcan Iron Works formula} \quad \text{(A22}a\text{)}$$

Double- and differential-acting:

$$R = \frac{20n}{120 + n} \times E_n \qquad \text{(A22}b\text{)}$$

The United States Steel Co. modifies the *Engineering News* formula by varying the constant in the numerator, as follows:

Drop hammers:

$$R = \frac{FW_r H}{s + 1.0} \qquad \textit{United States Steel formula} \quad \text{(A23}a\text{)}$$

Single-acting steam hammers:

$$R = \frac{FW_r H}{s + 0.1} \qquad \text{(A23}b\text{)}$$

Double- and differential-acting:

$$R = \frac{FH(W_r + Ap)}{s + 0.1} \qquad \text{(A23}c\text{)}$$

where F = 2 for piles driven to refusal or practical refusal in all materials;
= 6 for piles driven easily in sands and for gravels;
= 4 for piles driven easily in hard or sandy clays;
= 3 for piles driven easily in mixed medium clays and sand or sand and silt;
= 2 for piles driven easily in alluvial deposits, soft clays, and silts;
A = effective area of piston, in square inches; and
p = mean effective pressure of steam or air, in pounds per square inch.

Another modification of formula (A21*b*) is

$$R = \frac{2W_r H}{s + 0.3} \qquad \textit{Bureau of Yards and Docks formula} \quad \text{(A24)}$$

(*not now used by Navy, except for precast concrete piles weighing over* 10,000 *lb driven with hammer energies between* 19,000 *and* 36,000 *ft-lb.*)

The Benabencq formula is

$$R = \frac{W_r h}{2s} + W_r + W_p \qquad \textit{Benabencq formula} \quad \text{(A25)}$$

The Sanders formula, proposed in 1851, was obtained by applying a purported factor of safety of 8 to formula (A4),

$$R = \frac{W_r h}{8s} \qquad \textit{Sanders formula} \quad \text{(A26)}$$

and Merriman used the same terms with a purported factor of safety of 6,

$$R = \frac{W_r h}{6s} \qquad \textit{Merriman formula} \quad \text{(A27)}$$

The Goodrich formula is a simplification of a comprehensive formula which contains 25 terms covering conditions of the pile, hammer, cap, and ground, and was intended for use only with wood piles and drop hammers with a fall of about 15 ft and set of 1.0 in. Under these conditions it was believed by its author to have an accuracy within 10 per cent of that of the comprehensive formula.

$$R_u = \frac{10 W_r H}{3s} \qquad \textit{Goodrich formula} \quad \text{(A28)}$$

The Kafka formula[80,81] is the earliest form in which the elastic rebounds of the pile and soil, measured from a graph taken on the pile, have been found. This formula is

$$R_u = X\left[-1 + \sqrt{1 + \frac{Y}{X(2s + \lambda')}}\right] + W_r + W_p \qquad \textit{Kafka formula} \quad \text{(A29)}$$

where $X = (2s + \lambda')\dfrac{AE}{l}$

$$Y = 6W_r h \frac{W_r + e^2 W_p}{W_r + W_p}$$

$$\lambda' = s + C_2 + C_3$$

Kreuter's formula can be used only with drop hammers, and is

$$R_u = \frac{h_1 - h_2}{s_1 - s_2} \times W_r \qquad \textit{Kreuter formula} \quad \text{(A30)}$$

where h_1 and h_2 = different heights of fall of hammer; and

s_1 and s_2 = average penetration of pile under one blow of the respective sets of blows, corresponding to h_1 and h_2.

In order to obtain reasonable results, h_1 and h_2 must not be too widely different, and must be close to the maximum value of h which can be used without causing any set. Great values of h_1 and h_2, under which the pile advances considerably, cannot be used, even if the difference between them is small. This is because loss in energy must always increase with increased energy of blows, but the formula is based on losses in energy being nearly the same for both h_1 and h_2, so that the losses are canceled out in the derivation of the formula. This formula gives the ultimate driving resistance. The tip should be in the same stratum and the groups of blows with different falls consecutive in order to avoid other variables. If under the first set of blows the pile advances considerably, let the blows of the second set be weaker; if the advance of the pile is small, let the second set of blows be heavier. This formula dispenses with the need for measuring or assuming values for C_1, C_2, and C_3 or including a term for impact loss, as is necessary in formulas (2.1*a*) and (2.1*b*), and has been found to be reliable when the above limitation conditions are met. It is founded for each case upon observations for the special work in question. A suitable factor of safety must be selected but it can be of a fairly low value, such as is recommended for use with formulas (2.1*a*) and (2.1*b*). The chart in Fig. 2.4 is based on the same theory as this formula.

By a consideration of the relative magnitudes and effects of the terms omitted or approximated, using different types and lengths of piles, and types of hammers and caps, an idea may be gained of the relative accuracies of the different formulas in different ranges of penetrations. For instance, inspection of formula (A17) indicates that when s approaches zero, R_u approaches infinity, thus showing that this formula is of no value in the range of small sets. Another evident fact in formula (A21*a*) is that no means is provided of taking into account the weight of the pile, although it is evident that the penetrations obtained by driving a wood pile would be far different from those when driving a heavy steel or concrete pile weighing many times as much. Similar inspections will reveal many other interesting and pertinent points regarding the values of the various formulas.

APPENDIX II

NUMERICAL EXAMPLES USING ASSUMED DATA

Application of Dynamic Formulas to Determine Set for Given Working Load and Analyze Force Losses

Assumed Conditions:

Soil conditions:
- 5 ft topsoil and recent fill
- 20 ft soft clay
- 2 ft sand
- 20 ft gravel
- 5 ft hardpan over rock

Hammer—McKiernan-Terry No. 9-B-2
Blows per min—140 (Table IV.12)
Energy (E_n) = 8,200 ft-lb (Table IV.12)
Pile—steel pipe 10 in. OD × ¼ in. thick × 30 ft long, weight 790 lb (to be filled with concrete after driving)
Cap—650 lb (Table IV.12, unless found otherwise)
$W_p = 790 + 650 = 1{,}440$ lb
$W_r = 1{,}500$ lb (Table IV.12)
$e_f = 0.85$ (page 30)
$e = 0.4$ (page 32)
$R = 30$ tons (working load)
Factor of safety assumed—2.5
$R_u = 30 \times 2{,}000 \times 2.5 = 150{,}000$ lb

$$C_1 = 0.16 \text{ in. (Table I) } (p_1 = 150{,}000 \div 78 \text{ sq in.} = 2{,}000 \text{ psi}) \qquad (2.2)$$

Net area of pile = 9.87 in. × 3.1416 × 0.25 in. = 7.73 sq in.

$$C_2 = \frac{19{,}400}{15{,}000} \times 0.006 \times 30 = 0.23 \text{ in. (Table II)}$$

$$(p_2 = 150{,}000 \div 7.73 = 19{,}400 \text{ psi}) \qquad (2.3)$$

$$C_3 = 0.05 \text{ in. (Table III, value increased to 0.10 in this edition)}$$
$$(p_3 = 150{,}000 \div 78 \text{ sq in.} = 2{,}000 \text{ psi}) \qquad (2.4)$$

$$150{,}000 = \frac{12 \times 0.85 \times 8{,}200}{s + (\tfrac{1}{2})(0.16 + 0.233 + 0.05)} \times \frac{1{,}500 + 0.4^2 \times 1{,}440}{1{,}500 + 1{,}440} \tag{2.1b}$$

$$150{,}000 = \frac{83{,}500}{s + 0.2215} \times 0.589$$

$$s + 0.2215 = 0.328$$

$s = 0.328 - 0.2215 = 0.1065$ in. penetration ($9\frac{1}{3}$ blows per inch)

Then plot the curves in Figs. 2.1 and 2.2, from which the safe working load for any other set can be read in the field, by repeating the solution of formula (2.1*b*) for several assumed values of R_u. The C_1, C_2, and C_3 values may be varied directly with R_u. In the case of double-acting or drop hammers, where it is possible for the operator to vary the amount of applied energy, curves should be plotted for several likely speeds of double-acting hammers, or heights of drop for drop hammers, in order that no field delays will occur for lack of this information.

Analyzing force losses:

Total kinetic energy applied by hammer = 83,600 in.-lb (2.5*b*)

Impact loss:

$$12 \times 0.85 \times 8{,}200 \times \frac{1{,}440(1 - 0.4^2)}{1{,}500 + 1{,}440} = 34{,}400 \text{ in.-lb } (41\%)$$

Cap elastic loss:

$$\frac{150{,}000 \times 0.16}{2} = 12{,}000 \text{ in.-lb } (14\%)$$

Pile elastic loss:

$$\frac{(150{,}000)^2 \times 360}{2 \times 7.73 \times 30{,}000{,}000} = 17{,}480 \text{ in.-lb } (21\%)$$

Soil elastic loss:

$$\frac{150{,}000 \times 0.05}{2} = 3{,}750 \text{ in.-lb } (5\%)$$

Net effective energy available for driving $150{,}000 \times 0.1065$ = 15,970 in.-lb (19%)

(check) 83,600 in.-lb (100%)

Application of Dynamic Formula to Determine Safe Maximum Sets and Proper Size of Hammer

These computations illustrate the actual mathematical steps involved in preparing set-resistance curves (which would be set-bearing-value curves when driving in cohesionless soils) (Fig. A.1), for long tapered creosoted wood piles carrying the load in friction in fairly firm clay in the bottom 15 to 20 ft. The problem is to drive the piles, which necessarily have fairly small cross-sectional area near the bottom, into the friction-load-carrying stratum a sufficient depth not to exceed a safe working value of skin friction, without breaking the piles by overdriving, and to select the proper size of hammer for this purpose and specify the minimum allowable sets. This problem illustrates the use of the graphs in determining the stress in the piles during driving. If the supporting

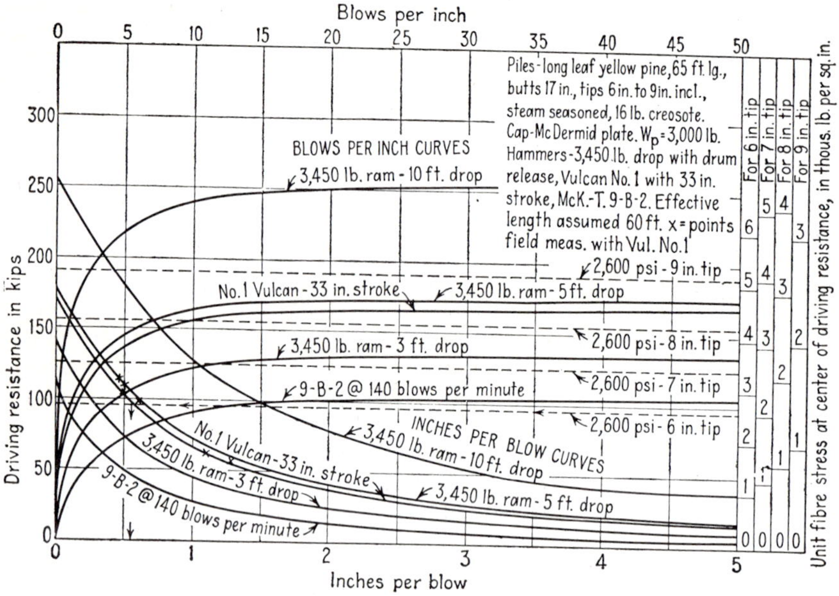

FIG. A.1. Set–resistance curves.

stratum had been of cohesionless material, the driving resistances divided by a reasonable factor of safety would also have given the safe working loads which would have been tabulated on the left-hand side of the graph, as in Figs. 2.1 and 2.2. However, when driving in clays, the temporary resistance to driving has no known relation to the permanent load-carrying capacity.

Assumed Conditions:

Hammers available—McKiernan-Terry 9-B-2, Vulcan No. 1, and 3,450-lb drop hammer

Hammer data—for 9-B-2, blows per min. = 140, energy (E_n) = 8,200 ft-lb (Table IV.12), e_f = 0.85 (page 30)

—for Vulcan No. 1, measured stroke = 33 in., e_f = 0.75 (page 30)

—for drop hammer assume any drop from 3 to 10 ft, drum release, e_f = 0.75 (page 29)

Cap—McDermid plate

Piles—Longleaf yellow pine, steam seasoned, 16-lb creosote treatment, 7-in. min tips specified, 16¾ in. ave 3 ft from butts, 65 ft long, weight per cu ft = 41 + 16 = 57 lb

W_p = 3,000 lb including plate

Soil conditions:

5 ft fill

30 ft river mud and silt

50 ft fairly firm clay

Cutoff is 10 ft aboveground

Assumed distance to center of driving resistance—60 ft

e = 0.25 (page 32)

Working design load—15 tons

It is most convenient to compute C_2, and take C_1 and C_3 from Tables I and III, for an arbitrary value of R_u = 100,000 lb, and obtain the value of $\frac{1}{2}(C_1 + C_2 + C_3)$ for this figure, then obtain the value of $\frac{1}{2}(C_1 + C_2 + C_3)$ for other values of R_u by direct proportion to 100,000 lb.*

C_1 = 0.10 in. (Table I)

(p_1 = 100,000 ÷ 103 sq in. area of trimmed butt = 950 psi) (2.2)

Average of cross-sectional areas of pile at butt and at center of driving resistance:

$$\left[\frac{(5/65 \times 10.0 \text{ in.} + 7 \text{ in.})^2 + 17.0 \text{ in.}^2}{2}\right] \times \frac{\pi}{4} = 137 \text{ sq in.}$$

$$C_2 = \frac{R_u l}{AE} = \frac{100{,}000 \times (60 \times 12)}{137 \times 1{,}600{,}000} = 0.33 \text{ in.} \tag{2.6}$$

C_3 = 0.10 in. (Table III) (p_3 = 100,000 ÷ [(17 in. − 10/65 × 10 in.)^2 × π/4] sq in. at ground line = 530 psi (2.4b)

For each hammer or drop, solve formula (2.1*a*) or (2.1*b*) for values of s to use in plotting the curves. Starting with the 9-B-2 hammer,

$$100{,}000 = \frac{12 \times 0.85 \times 8{,}200}{s + \frac{1}{2}(0.10 + 0.33 + 0.10)} \times \frac{1{,}500 + 0.25^2 \times 3{,}000}{1{,}500 + 3{,}000} \tag{2.1b}$$

$$= \frac{83{,}600}{s + 0.265} \times 0.375 = \frac{31{,}300}{s + 0.265}$$

and

$$s + 0.265 = \frac{31{,}300}{100{,}000} = 0.313; \qquad s = 0.05 \text{ in.}$$

* In future work, recommend assuming C_3 constant at 0.1.

then

$$125{,}000 = \frac{31{,}300}{s + 0.265 \times {}^{125}\!/_{100}}; \qquad s = \text{refusal}$$

$$75{,}000 = \frac{31{,}300}{s + 0.265 \times {}^{75}\!/_{100}}; \qquad s = 0.22 \text{ in.}$$

$$50{,}000 = \frac{31{,}300}{s + 0.265 \times {}^{50}\!/_{100}}; \qquad s = 0.49 \text{ in.}$$

$$25{,}000 = \frac{31{,}300}{s + 0.265 \times {}^{25}\!/_{100}}; \qquad s = 1.19 \text{ in.}$$

From these values, and as many more as desired to clearly locate the curve, including the ends, the curve for inches per blow for the 9-B-2 hammer is plotted. With this curve it is easy to pick off sufficient points to plot the blows-per-inch curve for this hammer. It has been found confusing to plot both sets of curves on one graph sheet, and it is recommended that two separate sheets be used.

Repeating the above operations for the Vulcan No. 1 hammer,

$$100{,}000 = \frac{0.75 \times 5{,}000 \times 33}{s + 0.265} \times \frac{5{,}000 + 0.25^2 \times 3{,}000}{5{,}000 + 3{,}000} \tag{2.1a}$$

$$= \frac{124{,}000}{s + 0.265} \times 0.649 = \frac{80{,}300}{s + 0.265}$$

and

$$s + 0.265 = \frac{80{,}300}{100{,}000} = 0.803; \qquad s = 0.54 \text{ in.}$$

then

$$150{,}000 = \frac{80{,}300}{s + 0.265 \times {}^{150}\!/_{100}}; \qquad s = 0.13 \text{ in.}$$

$$125{,}000 = \frac{80{,}300}{s + 0.265 \times {}^{125}\!/_{100}}; \qquad s = 0.31 \text{ in.}$$

$$75{,}000 = \frac{80{,}300}{s + 0.265 \times {}^{75}\!/_{100}}; \qquad s = 0.87 \text{ in.}$$

$$50{,}000 = \frac{80{,}300}{s + 0.265 \times {}^{50}\!/_{100}}; \qquad s = 1.47 \text{ in.}$$

$$25{,}000 = \frac{80{,}300}{s + 0.265 \times {}^{25}\!/_{100}}; \qquad s = 3.14 \text{ in.}$$

From these values the inches-per-blow and then the blows-per-inch curves for this hammer are plotted.

Repeating these operations for a 5-ft drop of the drop hammer,

$$100{,}000 = \frac{0.75 \times 3{,}450 \times 5 \times 12}{s + 0.265} \times \frac{3{,}450 + 0.25^2 \times 3{,}000}{3{,}450 + 3{,}000} \tag{2.1a}$$

$$= \frac{155{,}000}{s + 0.265} \times 0.564 = \frac{87{,}300}{s + 0.265}$$

and

$$s + 0.265 = \frac{87{,}300}{100{,}000} = 0.873; \qquad s = 0.61 \text{ in.}$$

Then

$$150{,}000 = \frac{87{,}300}{s + 0.265 \times {}^{150}\!/_{100}}; \qquad s = 0.18 \text{ in., etc.}$$

The above step is repeated for 3-, 4-, 6-, 8-, and 10-ft drops of the drop hammer, and as few of the inches-per-blow and blows-per-inch curves as desired are plotted. In order to avoid confusing the illustration not all are plotted. Usually if a few drops are plotted, the rest can be interpolated.

The scale of values of driving resistance, R_u, on the left side of the graph will, when divided by the cross-sectional area of the pile at the center of resistance, give the fiber stress in the pile. Scales have been shown for 6-, 7-, 8-, and 9-in. tip piles (having areas at center of driving resistance of 37, 48, 60, and 73 sq in., respectively). The ultimate fiber stress, in crushing, of the green wood is 4,300 psi (Table VI), and this value is assumed to be reduced by one-third to 2,900 psi, for the steamed and treated wood, although the reduction may be slightly greater or much less. No set should be used which will result in a greater figure than this, and to be safe, some lower figure such as 2,600 psi will be assumed as the maximum allowable. The smallest sets which should be used in the field can now be read from the graph and listed for the use of the inspector in signaling the stop of driving to avoid damaging the pile. Horizontal lines have been drawn at the 2,600-psi value for each tip size, and the limiting set is found at the intersection of such a line with the curve of the hammer used. These values from the graphs for the cases listed are found in the accompanying table.

Maximum Number of Blows per Inch Allowable

Hammer	For pile with tip diameter of—			
	6 in.	7 in.	8 in.	9 in.
10-ft drop	⅔	1	1⅓	2½
5-ft drop	1½	3	7	Unlimited
3-ft drop	3	10	Unlimited	Unlimited
Vulcan No. 1	2	3	10	Unlimited
9-B-2	15	Unlimited	Unlimited	Unlimited

If the amount of force required to obtain a satisfactory resistance when driving in cohesionless materials

$$(R_u = 15 \text{ tons} \times 2{,}000 \times \text{F.S. } 2.5 = 75{,}000 \text{ lb})$$

is assumed to be the *least* that could be required to drive in cohesive materials, it is seen that the curve for blows per inch shows the 9-B-2 hammer to be too small, and that although it would be practically impossible to damage a pile with it, to obtain the desired depth of penetration (if this is possible at all), might require an excessive number of blows which would consume too much time. Greater resistance to driving may be exerted by the fairly firm clay than would be exerted by sand. The No. 1 hammer can exert a force of about 125,000 to 150,000 lb without requiring too small sets and too much time, whereas the limit of force for the 9-B-2 hammer is 110,000 lb. The smallness of the impact factor for the 9-B-2 hammer in the solution of formula (2.1*b*) also shows a very large impact loss, indicating that the pile is undesirably heavy for this hammer. Comparison of this term with those obtained when using the other hammers will make this unnecessary loss in efficiency in driving apparent.

From the curves and the above tabulation, it appears that it would be very easy to damage the piles with a 10-ft drop of the 3,450-lb ram, and that it would be wiser to limit the drop to 5 ft, as it may be difficult to obtain such large sets as are required to prevent overstress. Sudden heavy impacts on long slender piles have a tendency to snap them before the required set can occur. It also appears from the curves that, although it will be fairly easy to overstress the piles having 6- and 7-in. tips, it will be less easy to hurt the 8-in. tip piles, and the 9-in. tip piles will stand almost any number of blows from the Vulcan No. 1 hammer, or from the drop hammer for drops up to 6 ft.

The choice between the No. 1 single-acting steam hammer and the drop hammer lies with the No. 1 hammer on account of the greater number of piles per shift it can drive—its speed being about 55 blows per minute against about 8 for the drop hammer on which the operator must work the drum controls.

By taking field graphs of the rebounds, the true sum of $C_2 + C_3$ is obtained. These should be taken for several different values of s, both small and large, and the distances scaled plus the previously assumed C_1 values, used to plot new points on the curves. These points usually lie so close to the curves already computed that slight or no change is necessary, if the effective length has been assumed wisely, as it readily can be if boring logs have been made, as they should have been. Field-measured points for the above examples, when using the No. 1 hammer, are shown in Fig A.1.

Pile inspectors' reports revealed that some of the piles with small tips were actually driven slightly harder than permitted by the above table, in the endeavor to obtain the 15- to 20-ft penetration desired in the firm clay, resulting in a somewhat higher fiber stress than desirable, and consequently some breakage. The results for piles driven with a No. 1 Vulcan hammer were as follows:

	Tip diameter, in.							
	7	7½	8	8½	9	9½	10	11
Specified min set, in...	0.33	0.20	0.10	Unlimited				
Average set reported, in.	0.42	0.30	0.25	0.27	0.25	0.15	0.19	0.19
Average direct fiber stress at center of driving resistance, psi, corresponding to reported sets, from curves[a]	2,550 (1,650–3,350)	2,600 (1,500–2,900)	2,400 (1,900–2,650)	2,200 (1,450–2,400)	2,000 (1,250–2,200)	1,950	1,700	1,600
Number of piles driven	101	64	130	38	48	9	18	5
Number of piles broken	9	6	8	2	2	0	1	0

[a] Minimum–maximum range is given in parentheses.

For piles driven with a 3,450-lb drop hammer having a 5-ft fall, the percentages of breakage were the same or smaller, with no breakages occurring when the stress was 2,450 psi or less.[54]

It is seen that when the stresses were occasionally allowed to exceed the average values, which were slightly under the 2,600 psi used as the basis of the table of limiting sets, the percentage of breakage tended to rise. This trend appears, although not all overdriven piles broke, and the direct fiber stress in the piles which broke did not always exceed the limit proposed. These individual deviations and occasional breakages in the larger piles are attributable to defects in the piles, such as curvature, knots, differences from average butts or taper, seasoning damage, or eccentric driving, as well as the variations in natural strength of the different piles and effects of seasoning and treatment.

APPENDIX III

NUMERICAL EXAMPLES USING FIELD DATA

Determination of Center of Driving Resistance

If no test piles with continuous driving records have been driven, the center of driving resistance can only be estimated from inspection of the boring log and samples.

After a test record has been made, the center of driving resistance may be taken as the approximate center of gravity of frictional driving resistance and final end-bearing resistance, measured in kips and not in number of blows per inch (Fig. A.2). The graph shown is for the previous example, given in Appendix II.

Various impressions about the center of resistance of the soil can be obtained from visual inspection of the number of blows required to drive a unit depth with different sized hammers. A light hammer will show a large increase in number of blows as its capacity is neared, whereas a heavier hammer will find no difficulty in penetrating the same strata. Observation of the manner in which the blows-per-inch curves in Fig. A.1 flatten out, in relation to a given driving resistance, at different points for different hammers, will make this clear.

Translation of sets into driving resistances is done by reading values of resistances from set-bearing-value curves such as those in Fig. A.1. The amount of end bearing included in the resistance at each foot of driving is not cumulative and should be deducted as well as possible by considering what portion of the increases from foot to foot are due to end bearing. Final end bearing is active, however, with friction on the sides of the pile. Approximately, it may be satisfactory to consider the increase in resistance to driving between the next-to-last foot and the last driving as end bearing, if more definite data are lacking.

In Fig. A.2, the continuous-driving-resistance records of two identical piles driven with a light and a medium-sized hammer are plotted. Both give the same driving resistance when numbers of blows per foot are translated into pounds. The end-bearing resistance portion is estimated and shown graphically. The shaded area between this end-bearing resistance line and the total resistance graph represents frictional resistance. The center of gravity of the shaded area and the end resistance at final penetration may be satisfactorily chosen by visual inspection as a rule, or it may be computed. This can be done by figuring the center of gravity of the shaded area, which will be that of frictional resistance, by taking moments about the tip, and then figuring the center of gravity of the frictional resistance and end resistance together by again taking moments about the tip and dividing by the total R_u to find the

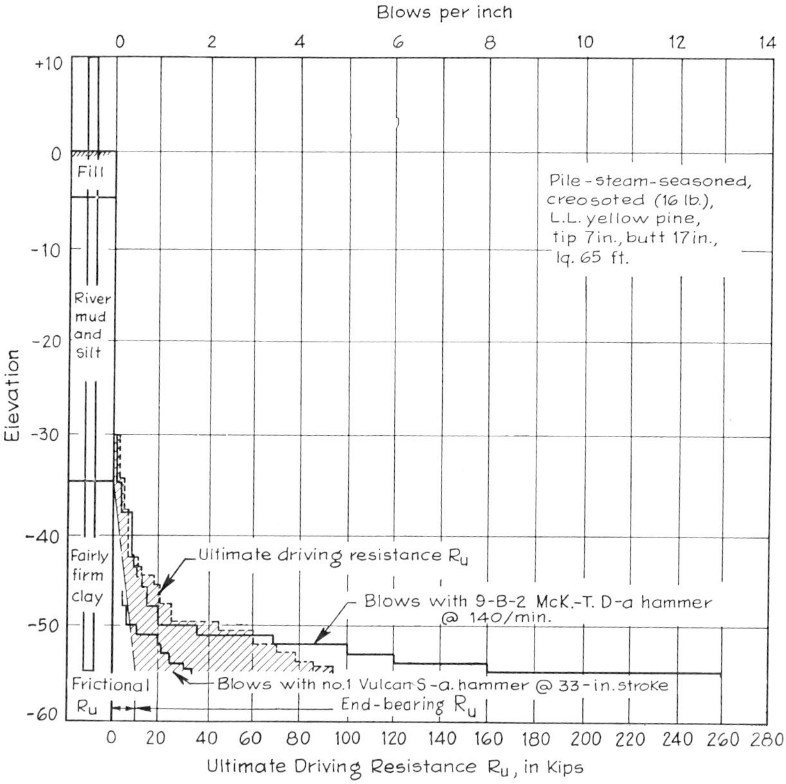

FIG. A.2. Continuous-driving-record sets compared with ultimate driving resistances.

center of driving resistance of both friction and end bearing. The distance from this center of gravity of driving resistance to the pile butt is the effective length, L.

Determination of Stresses in Piles from Field Measurements

Observed Data:

Hammer—Vulcan No. 1 single-acting steam hammer
W_r = 5,000 lb (Table IV.4)
h = 36 in. (Table IV.4, verified by field measurement)
e_f = 0.85 (pages 29*n*, 30)

Pile—8-in. tip, 12-in. butt, 40 ft long, longleaf yellow pine, steam-seasoned at not over 20 psi pressure, 16-lb creosote treatment. Weight of green wood 55 lb per cu ft; assumed that loss of weight in steaming equals weight of creosote, which is 16 lb per cu ft, and that net weight is 55 lb per cu ft

$$W_p = \left(\frac{8+12}{2}\right)^2 \times \frac{\pi}{4} \times \frac{40}{144} \times (55 - 16 + 16)$$

$$= 1{,}200 \text{ lb}$$

Driving cap—none
Follower—none
Soil strata—30 ft soft clay, 10 ft sand
l = 37.5 ft × 12 = 450 in. (assuming center of driving resistance 37.5 ft from butt)
E = 1,600,000
Compute the stress in the pile when the set measured from field graph (a) in Fig. A.3 is 0.5 in., and the rebound is 0.75 in.

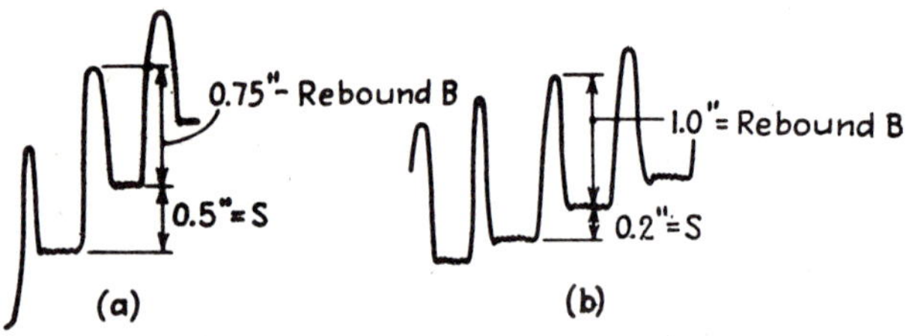

FIG. A.3. Field-measured rebound graphs.

Assuming that no resistance to driving is being exerted by the soft clay (on the safe side), the full amount of R_u occurs in the pile at the top of the sand stratum, at which point the area of the pile is

$$A_p = \left(8 + \frac{10}{40} \times 4\right)^2 \times \frac{\pi}{4} = 63.5 \text{ sq in.}$$

$$R_u = \frac{0.85 \times 5{,}000 \times 36}{0.5 + \frac{1}{2}(0 + 0.75)} \times \frac{5{,}000}{5{,}000 + 1{,}200}$$

$$= \frac{154{,}000}{0.875} \times 0.81 = 144{,}000 \text{ lb} \qquad (3.1a)$$

$$p = \frac{144{,}000 - 0}{63.5} = 2{,}270 \text{ psi} \qquad (3.2a)$$

Assuming the center of driving resistance is 2.5 ft from the tip and investigating the stress at that point,

$$A_p = \left(8 + \frac{2.5}{40} \times 4\right)^2 \times \frac{\pi}{4} = 53.2 \text{ sq in. at center of driving resistance}$$

$$R_u = \frac{0.85 \times 5{,}000 \times 36}{0.5 + \frac{1}{2}(0 + 0.75)} \times \frac{5{,}000}{5{,}000 + 1{,}200}$$

$$= \frac{154{,}000}{0.875} \times 0.81 = 144{,}000 \text{ lb} \qquad (3.1a)$$

$$p = \frac{144{,}000 - 72{,}000}{53.2} = 1{,}350 \text{ psi} \qquad (3.2b)$$

The ultimate fiber stress in longleaf yellow pine is as follows: green, 4,300 psi; air-seasoned, 8,440 psi; steam-seasoned, approximately two-thirds of the green value, or 2,900 psi, if steam pressure does not exceed 20 psi and is not applied longer than necessary in order to prevent much more damage to the wood. The pile stress has a more than adequate factor of safety at this amount of set with this hammer.

The above value of $R_u = 144{,}000$ is as great a value as would be required when divided by a factor of safety on a wood pile of this size. However, if the desired tip grade can not be obtained without jetting or harder driving than represented by a value of $s = 0.5$ in., the fiber stresses due to harder driving should be investigated.

Assume that the final set is only 0.2 in. at the time the desired tip grade is reached, and that the measured graph is (*b*) in Fig. A.3. Again assuming that the clay provides no resistance to driving, the fiber stress in the pile at the top of the sand stratum becomes

$$R_u = \frac{154{,}000}{0.2 + \frac{1}{2}(0 + 1.0)} \times 0.81 = 179{,}600 \text{ lb} \qquad (3.1a)$$

$$p = \frac{179{,}600 - 0}{63.5} = 2{,}830 \text{ psi} \qquad (3.2b)$$

This fiber stress is approximately the same as the yield point of the steam-seasoned material. If air-seasoned piles had been used, the stress would have had a very good factor of safety during driving. If green untreated piles had been used, the factor would have been 1.5, which is low but doubtless satisfactory. The same procedure may be followed for any other observed sets and rebounds.

APPENDIX IV

SUMMARY OF COMPARATIVE RESULTS OF TESTS

To illustrate the possible effects of formulas (2.1*a*) and (2.1*b*) on the matter of reaching a given depth or stratum and on the question of economies when determining pile lengths, the following driving test results are presented:

A. Different Types of Piles with Same Hammer to Same Depth

In this case, three different types of piles but with fairly comparable friction areas and volumes of soil displacement below ground surface were driven with the same hammer. Eleven piles were driven near each other in the same strata with a Raymond No. 1 (similar to Vulcan No. 1) single-acting steam hammer having an observed 34-in. stroke. Three piles were standard Raymond piles 29 ft 4 in. long; three were No. 3 gage Monotube shells 25 ft long with 8-in. tips and 14¼-in. butts; two were No. 7 gage Monotube shells 25 ft long with 9-in. tips and 15⅝-in. butts; and three were Monotube shells 25 ft long with 8-in. tips and 14¼-in. butts filled with Incor cement 6 days before driving. Soil conditions consisted of 4 ft of cinder fill on 2 to 4 ft of marsh over fully inundated noncohesive lake sand. The ground-water level was about 5 ft below the surface. A pit about 2 ft deep was dug for each pile to remove a frozen top layer of the fill.

A driving cap about 11½ in. in diameter, consisting of a 2-in. steel plate over a 6-in. hardwood block over a 1-in. steel plate, weighing about 150 lb, was used on each pile. In addition, a 4-in. hardwood block was placed on the concrete of the prefilled piles under the lower plate. The Monotube piles were all driven with a follower in addition, consisting of a piece of Raymond mandrel weighing 540 lb.

Piles, mandrels, caps, and followers weighed as follows, in pounds:

	Mandrel	Pile	Cap	Follower	Total weight (W_p)
For Raymond piles	9,425	250	150	...	9,825
For 8-in. tip Monotube shells	.....	780	150	540	1,470
For 9-in. tip Monotube shells	.....	600	150	540	1,290
For 8-in. tip prefilled Monotube shells	.....	2,560	165	540	3,265

For all piles driven to tip elevations 15 ft below ground surface, the following safe bearing values were computed from the observed sets at this depth (see Table *A*, Example *a*, page 584).

The term P in the Eytelwein and Navy-McKay formulas was taken as the weight of the pile only, in accordance with customary practice when using those formulas.

In computing values by formulas (2.1*a*) and (2.5*a*), values of L were taken as the distances from the heads of the piles to a point at half the distance of embedment in the sand stratum, the value of e_f taken as 0.8, and the value of C_3 as 0.05. From the above table, it appears that formula (2.1*a*) results in a maximum difference of only 13 per cent in bearing values for the different types of piles. If some allowance were made for end bearing during driving, and the value of L therefore taken somewhat longer, the load-carrying capacities of the compressible piles, such as the Monotube empty shells, would decrease by a ton or two, and even closer agreement would be noted between the various types of piles. It should also be noted that the surface area and end area of the 9-in. tip Monotube shells are approximately 10 per cent greater than for the 8-in. tip Monotube shells, resulting in an indicated increase in safe bearing value of about 10 per cent over the 8-in. tip Monotube shells. Actually the 9-in. tip Monotube shells need not have been driven 15 ft into the sand, as for the 8-in. tip Monotube shells, on this account. These agreements are close enough for all practical purposes.

To determine the depths to which the Monotube shells would have to be driven to give the same indicated bearing value as the *Engineering News* value of 63 tons for the Raymond piles, it was necessary to observe the tip elevations on the driving records at which the set of 0.125 in. was obtained for these piles. It was necessary to drive an 8-in. Monotube shell 22 ft 6 in. below ground level to obtain this set, thus having the effect of increasing the length of all the piles on the job 50 per cent. This added length is of no practical value, since the 15-ft length was sufficient to transmit the load safely to the bearing stratum of sand. It will be noted, however, that by the use of formula (2.1*a*), practically identical depths of penetration would be required for any of the types of piles to obtain the same safe bearing value.

B. Different Types of Piles with Same Hammer to Same Capacity by Engineering News Formula

A further comparison is given by the following case in which three Monotube shells, 40 ft long, having 8-in. tips and a taper of 1 in. in 4 ft, and five Raymond shells 31 ft 6 in. long, having 8-in. tips and a taper of 1 in. in 2 ft 6 in., were driven to the same tip grade in the same strata by a No. 1 Vulcan single-acting steam hammer. A Monotube pile weighed 600 lb, the bathtub follower, 2 ft long, weighed 800 lb, and the driving cap (a hardwood block 11½ in. in diameter by 6 in. thick, with 2-in. steel plate on top and 1-in. steel plate below fitted into a shield) weighed 225 lb. A Raymond mandrel weighed 12,300 lb and the 225-lb cap only was used with it. The strata consisted of 10 ft of cinder, gravel, and ash fill, 5 ft of fine sand, 11 ft of peat and sand, 8 ft of loose silty sand, 4 ft of coarse sand and gravel in which the final tip readings were taken, and below that 6 ft of medium gray clay, 12 ft of firm fine sand and a little clay, and a deep bed of firm fine sand. The comparative results are tabulated, showing the safe bearing values computed by formulas (2.1*a*) and (2.5*a*) and by the *Engineering News* formula (see Table *A*, Example *b*, page 584).

The value of L in computing C_2 for the Monotube piles was taken as 30 ft, being based on a center of resistance about two-thirds of the distance from the ground surface to the tip, and for the 37-ft 4-in.-long Raymond mandrels as 20 ft, being based on a center of resistance about half the embedment of the casing, on account of the greater taper. Owing to the small elastic loss in the Raymond mandrels, it makes practically no difference as to the value assumed for L for them. The value of C_3 was taken as 0.10 for driving the Monotube piles, and 0.15 for the Raymond piles.

In order to secure for the Monotube shells an equal 31-ton capacity by the *Engineering News* formula, it was necessary to drive them an additional 15 ft. On the other hand, the Raymond-pile tips stopped just short of the 6-ft bed of clay, and the Monotube-pile tips carried through this clay into the firm sand below. The stratum in which the tips should rest should be determined from the borings, however, and not be governed by finding adequate driving resistance just above the clay, if detrimental settlement of the clay bed is expected. In the previous example *A*, it appeared that use of the *Engineering News* formula required the Monotube piles to be driven to a greater depth than actually necessary, whereas in this example it would appear that use of the *Engineering News* formula, without consideration of the relation of the tip elevation to the strata, would result in stopping the Raymond piles at too high an elevation.

C. Same Types of Piles Driven with Different Types of Hammers (Double-acting and Differential-acting)

The preceding paragraphs have illustrated the computation of safe loads in the cases of different types of piles driven with the same hammer. The following example shows the computation of safe loads for a case in which the same type of pile was driven with two different kinds of hammers. The piles were 8 in. H 36 lb by 46 ft long, driven to a length of 44 ft 6 in. into the ground. The strata consisted of 4 ft of soft clay with some sand, 19 ft of soft mud with some sand, 8 ft of soft mud and fine sand, slightly firmer than above, 5 ft of fine silty sand and mud, 10 ft of silty coarse sand and gravel, 7 ft of coarse sand and gravel in which the tips rested, and coarse sand and clay below. Seven piles were driven with a McKiernan-Terry 9-B-3 double-acting steam hammer and nine piles with a Vulcan 50-C differential-acting steam hammer. The speed of the 9-B-3 hammer was observed as only 120 blows per minute. The value of L for use in determining C_2 was taken as 27 ft. The value of C_1 was taken as zero, since no cap was used and the piles were steel. The value of C_3 was taken as 0.05 in both cases (see Table *A*, Example *c*, page 584).

D. Same Types of Piles Driven with Different Types of Hammers (Double-acting and Drop)

The following case also illustrates the computation of safe loads for a case in which piles of the same type were driven in the same building in the same strata, and to the same depth, with different kinds of hammers. This case, in which one hammer was a drop hammer and the other a double-acting steam hammer, permits comparison of the results obtained by both forms of the *Engineering News* formula with those obtained by means of formulas (2.1*a*) and (2.1*b*). It is generally thought that the *Engineering News* formulas give good results when used with wood piles, but this illustration shows that such is not always the case.

Twenty-seven green untreated white-oak piles with an average length of 24 ft, and with average 8½-in.-diameter tips and 11-in.-average-diameter butts were driven under a building, sixteen by a 9-B-2 McKiernan-Terry double-acting steam hammer operating at a speed of 140 strokes per minute, and eleven by a 1,750-lb drop hammer falling 10 ft and having a 250-lb driving head and 60 ft of cable on a winch. The comparative results, showing safe bearing values computed by formulas (2.1*a*) and (2.1*b*) and by the *Engineering News* formula, are tabulated as shown in Table *A*, Example *d*, page 584.

The efficiency of the drop hammer was taken as 75 per cent and of the double-acting steam hammer as 85 per cent.

The soil consisted of approximately 3 ft of cinder fill, 8 ft of soft silt with a little fine sand, then hard coarse sand and gravel in which the tips came to rest after a penetration of about 13 ft.

E. Correspondence of Computed and Observed Temporary Compressions and Stress in Pile

The following example illustrates the correspondence of the computed and measured value of C_2, and of the theoretical and actual yield points

TABLE A. COMPARISON OF COMPUTED SAFE BEARING VALUES

	Set (av value), in.	Formula: (2.1a) and (2.5a): Total applied energy, in.-lb	Per cent energy losses due to: Impact	Cap	Elasticity	Quake	Net ultimate load, tons	Net working load, tons (F.S. = 2.5)	*Eng. News*, $\frac{2W_rH}{s+0.1}$,* $\frac{2E_n}{s+0.1}$,† $\frac{2W_rH}{s+1.0}$,‡ tons	Eytelwein, $\frac{2W_rH}{s+0.1\frac{W_p}{W}}$,* $\frac{2E_n}{s+0.1\frac{W_p}{W_r}}$,† $\frac{2W_rH}{s+\frac{W_p}{W_r}}$,‡ tons	Navy-McKay, $\frac{2W_rH}{s\left(1+0.3\frac{W_p}{W_r}\right)}$, tons	Bureau of Yards and Docks, $\frac{2W_rH}{s+0.3}$, tons
							Example a					
Type of pile:												
Raymond piles	0.125	136,000	61	14	2	4	102	41	63	45	72	33
8-in. tip Monotube shells	0.30	136,000	21	14	15	4	106	42	35	43	44	24
9-in. tip Monotube shells	0.23	136,000	19	15	24	4	112	45	43	58	58	27
8-in. tip Monotube prefilled shells (assumed $n = 10$ for concrete)	0.26	136,000	35	18	5	4	98	39	39	45	46	25
							Example b					
Monotube shells	0.85	135,000	21	3	13	4	47	19	16	17	17	13
Raymond piles	0.375	135,000	64	1	1	6	51	20	31	24	23	22
							Example c					
Type of hammer:												
9-B-3 (at 120 blows per min)	0.116	55,100§	39§	0§	22§	7§	77§	31	25	25	36	13
50-C	0.563	135,000	18	0	12	3	80	32	23	22	20	17
							Example d					
Method of driving:												
McKiernan-Terry 9-B-2	0.062	83,800§	35§	0§	43§	10§	82.5§	33	51	62	109	23
1,750 lb hammer, 10-ft drop	0.275	157,500	35	0	29	5	85	34	14	18	52	30

* For single-acting steam hammers.
‡ For drop hammers.
† For double- and differential-acting steam hammers.
§ Also for formulas (2.1b) and (2.5b).

of the piles. Piles consisting of 10-in. 42-lb H's 50 ft long were driven through soft material consisting of 10 ft of slag and cinder fill, 19 ft of river muck, 20 ft of sand, clay, and gravel, and 1 or 2 ft of weathered shale or hard clay to hard shale rock, with a No. 0 Vulcan single-acting steam hammer. The weight of the driving cap, containing a wood block, was 700 lb. The ultimate driving resistance equaled the theoretical yield point of the pile, and was 406,000 lb. The computed elastic compression C_2 at this load was 0.67 in. This condition should have occurred when driving resistance reached approximately 30 blows per inch. Because of the solidity of the rock resistance compared with any frictional resistance, the value of L was taken as the full length of the pile. These results were checked by field observation, a value of ⅝ in. being measured for C_2, and failure of the pile metal occurring when driving required slightly over 30 blows per inch. As ultimate bearing values are reached, the formula indicates that the number of blows per inch increases rapidly, with almost no increase in the value of R_u (see Fig. 2.2 for typical curves).

F. Correspondence of Test Loads and Computed Pile Carrying Capacities

The comparison revealed in Fig. A.4 illustrates the correspondence between test loads and pile carrying capacities in cohesionless or principally cohesionless soils, computed by means of formulas (2.1*a*) and (2.1*b*). Carrying capacities computed from the *Engineering News* formula are also shown. Other load tests in the above series, made on cohesive or predominantly cohesive soils, showed lower values than those computed by the formulas, the driving resistances computed in these cases having no use except as a means of checking unit stresses in the piles during driving.

The descriptions of the piles, driving equipment, and soil conditions in these tests are tabulated in Table *B*, pages 588–599.

The good correspondence between test loads and pile carrying capacities computed by means of formulas (2.1*a*) and (2.1*b*) should be noted, and the fact observed that the trend of the computed values follows the trend of the load tests, with wide discrepancies avoided. For comparison, the load-carrying capacities computed by means of the *Engineering News* formula are also plotted in Fig. A.4, where the characteristic occasional divergence from more accurately computed values, and from the load tests, is apparent. It will be noted that many of the *Engineering News* values are in good agreement with the other values, and the fact that good results are often obtained by its use frequently leads to a belief that it is still a satisfactory and sufficiently reliable tool for use in all cases.

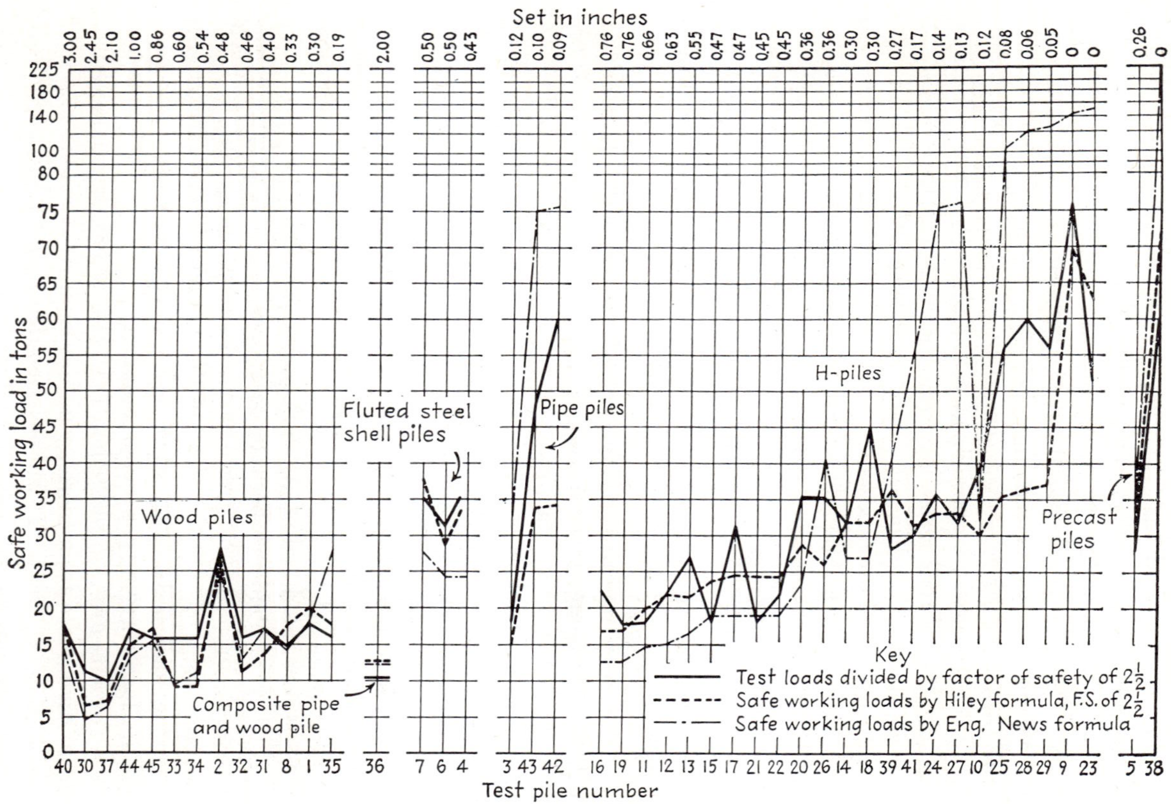

FIG. A.4. Graphic relation between pile formula and load-test results.

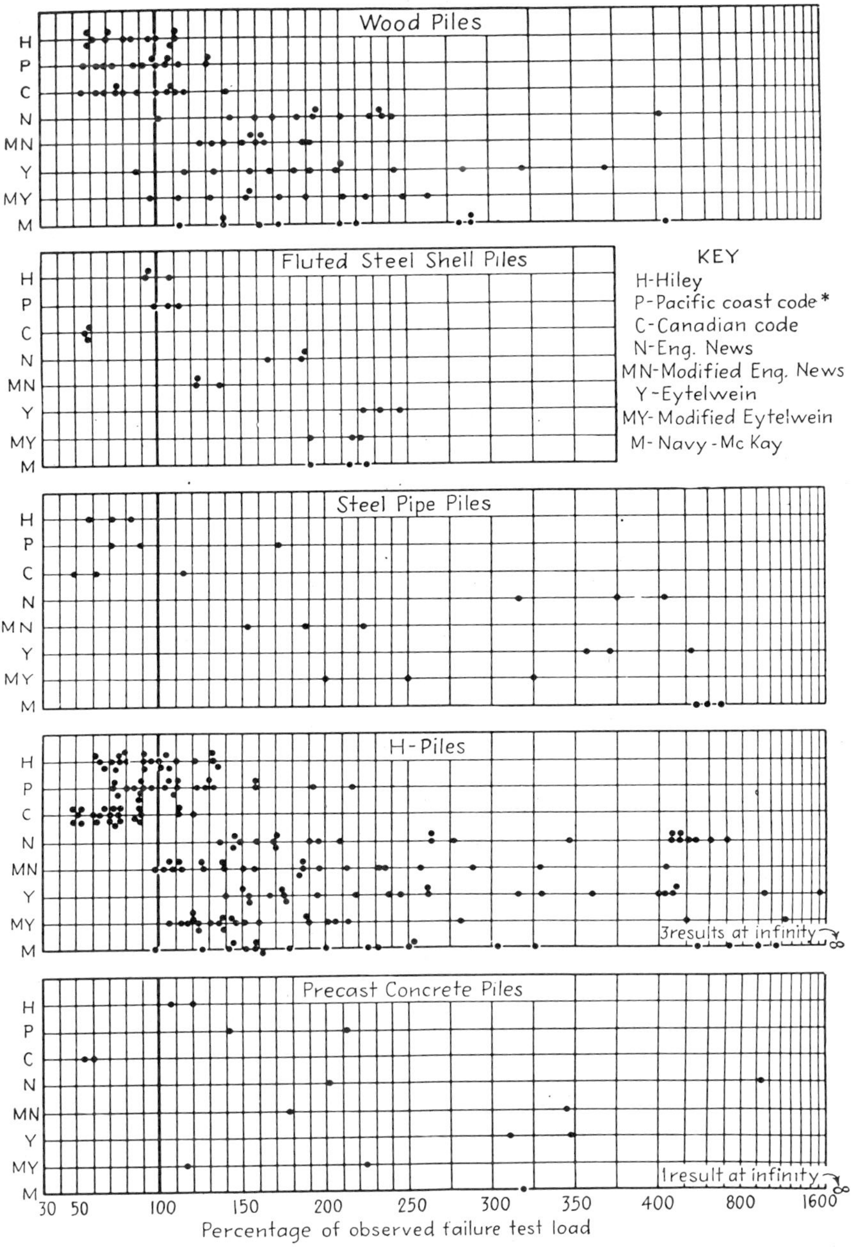

* Now International Conference of Building Officials Uniform Building Code.

FIG. A.5. Scattering of the ratios between resistances computed from various pile formulas and observed failure test loads.

TABLE *B*. DATA ON

Test no.	Location	Pile: Type	Pile: Length, ft	Pile: Embedment, ft	Pile: Butt, in.	Pile: Tip, in.	Set (s), in.	Hammer	Pile cap, lb	W_p (including cap), lb
	Wood Piles									
40	San Francisco Harbor, Calif.	Douglas fir, creosoted	92	73	16	8	3.00	Drop 2,900 lb (20-ft drop)	....	3,700
30	Lavaca Bay Bridge Port Lavaca, Tex.	Wood—upper 32 ft, 24 lb creosoted, lower 52 ft untreated	84	63	12 (upper sect), 9–13 (lower sect)	9 (upper sect), 8–9 (lower sect)	2.45	Drop 3,290 lb, cable drag on drum, 5 ft	Wood dolly in steel cap, rope coil on pile 1,080	3,130
37	Marathon, Ont., Canada	Douglas fir, green	68.8	64.8	15	10.25	2.10	Vulcan No. 1	....	2,750
44	Norfolk, Va., waterfront	Longleaf yellow pine, green, untreated	60	57.7	13½	8	1.00	Vulcan No. 1	65	3,025
45	Norfolk, Va., waterfront	Longleaf yellow pine, green, untreated	70	67	17	7	0.86	Vulcan No. 1	65	3,115
33	Marathon, Ont., Canada	Douglas fir, green	63	59.5	14	7.5	0.6	9-B-2 at 128 per min	....	2,200
34	Marathon, Ont., Canada	Douglas fir, green	61	59.5	14	8	0.54	9-B-2 at 132 per min	....	2,200
2	Pee Pee Creek Bridge, Pike County, Ohio	Wood, green, untreated	45	42	14	6	0.48	Vulcan No. 1 (35-in. stroke)	100	1,400

PILES, DRIVING AND TESTS

Failure test load, tons	Ultimate resistances from formulas, tons								Soil conditions	Source*
	Hiley	Pacific Coast[a]	Canadian National (working load × 3)	*Eng. News* (working load × 6)	Eytelwein (working load × 6)	Modified *Eng. News* (working load × 6)	Modified Eytelwein (working load × 6)	Navy-McKay (working load × 6)		
45	45	47	45	88	52	...	...	.....	Harbor mud	Squire
28	17	24.5	21.3	28.5	24.6	...	...	.....	Silt, some shell strata	King
25	18	23	22	40	42	38	39	35	5.5 ft sand, 37 ft highly plastic clay, 44 ft silty sand under hydrostatic pressure	
44 (compr, day of driving)	37.5	47	49	81	85	69	76	76	9 ft soft gray sandy silt, 12 ft soft gray silt, 4 ft med. gray sand and silt, 2 ft med. gray sand and little clay, 8 ft soft gray and blue clay, 5 ft med. gray and blue sandy clay, 5 ft soft gray and blue clay and sand traces, 4 ft soft gray silt and some peat and sand traces, 4 ft med. gray silty sand, 75 ft fine green sand and shells and some clay	
40 (compr, 30 days duration)	43.5	52.5	57	94	98	77	85	88		
40	23	22.5	22	58	54	45	39	46	5 ft sand, 39.5 ft highly plastic clay, 3 ft slightly plastic silt, 12 ft very fine sand under hydrostatic pressure	
40	23	26	25.5	68	63	53	45	56	5 ft sand, 39.5 ft highly plastic clay, 3 ft slightly plastic silt, 12 ft very fine sand under hydrostatic pressure	
71	68	71	75	150	148	134	188	204	Sand and gravel	Rabo

TABLE *B*. DATA ON PILES,

Test no.	Location	Pile: Type	Pile: Length, ft	Pile: Embedment ft	Pile: Butt, in.	Pile: Tip, in.	Set (s), in.	Hammer	Pile cap, lb	W_p (including cap), lb
32	Marathon, Ont., Canada	Douglas fir, green	63	59.5	13.5	8	0.46	9-B-2 at 132 per min	....	2,200
31	Marathon, Ont., Canada	Douglas fir, green	75	71	13.5	8	0.4	9-B-2 at 140 per min	....	1,820
8	Kokasing River Bridge, Morrow County, Ohio	Oak, green, untreated	30	29	12	7	0.333	Vulcan No. 2 (25-in. stroke)	40	780
1	Crooked Creek Bridge, Pike County, Ohio	Wood	39	25	13	8	0.3	Vulcan No. 2 (29-in. stroke)	100	1,300
35	Marathon, Ont., Canada	Douglas fir, green	63	59.5	14	7.5	0.19	9-B-2 at 140 per min	....	2,100
	Composite Pipe and Wood Pile									
36	Marathon, Ont., Canada	Douglas fir, creos, driven through casing extending to lower sand stratum, using pipe mandrel 10¾ in. × ⅜ in. × 50 ft	40	25	12.4	10	2.00	Vulcan No. 0 (38-in. stroke)	....	1,800
	Monotube Piles									
7	Scioto River Bridge, Pike County, Ohio	Monotube Type 11-Y	25	25	18	8	0.429	Vulcan No. 1 (35-in. stroke)	670	1,100
6	Scioto River Bridge, Pike County, Ohio	Monotube Type 11-JN18	45	26	18	8	0.5	Vulcan No. 1 (35-in. stroke)	670	1,500

DRIVING AND TESTS *(Continued)*

Failure test load, tons	Ultimate resistances from formulas, tons								Soil conditions	Source*
	Hiley	Pacific Coast[a]	Canadian National (working load × 3)	*Eng. News* (working load × 6)	Eytelwein (working load × 6)	Modified *Eng. News* (working load × 6)	Modified Eytelwein (working load × 6)	Navy-McKay (working load × 6)		
40	28	27	27	78	73	57	47	65	5 ft sand, 39.5 ft highly plastic clay, 3 ft slightly plastic silt, 12 ft very fine sand under hydrostatic pressure	
43	34.5	32	34.5	98	95	70	65	90	10 ft sand, 5 ft silt, 2 ft sand, 38.5 ft highly plastic clay, 8 ft slightly plastic silty sand, 8 ft very fine sand under hydrostatic pressure	
37	45	48	49	87	105	60	92	104	12 ft silt, fine sand and gravel; then sand and gravel	Rabe
45	50	51	53	109	144	72	102	129	Fine sand and gravel	Rabe
40	44.5	39	43.5	170	147	100	76	176	5 ft sand, 39.5 ft highly plastic clay, 3 ft slightly plastic silt, 12 ft very fine sand under hydrostatic pressure	
26	32	45.5	32	75	78	68	75	68	14 ft sand, 32 ft highly plastic clay, 60 ft very fine sand under hydrostatic pressure	
88	94	100	50	166	206	108	192	198	25 ft silt, sand and gravel; then sand and gravel	Rabe
78	72	76	46	146	174	108	174	168	Sand and gravel	Rabe

TABLE *B*. DATA ON PILES,

Test no.	Location	Pile: Type	Length, ft	Embedment, ft	Butt, in.	Tip, in.	Set (*s*), in.	Hammer	Pile cap, lb	W_p (including cap), lb
4	Pee Pee Creek Bridge, Pike County, Ohio	Monotube Type 11-Y	25	19	18	8	0.502	Vulcan No. 1 (35-in. stroke)	670	1,100
	Pipe Piles									
3	Ohio Brush Creek Bridge, Adams County, Ohio	Pipe, ⅜ in.	50	47	12.8	12.8	0.125	Vulcan No. 2 (29-in. stroke)	600	3,100
43	Cuyahoga River, turning basin, Cleveland, Ohio	Pipe, 12¾ in. OD × ¼ in.	80	78	12¾	12¾	0.10	Vulcan No. 1 (36-in. stroke)	1,080	3,250
42	Cuyahoga River, turning basin, Cleveland, Ohio	Pipe, 12¾ in. OD × ¼ in.	80	70	12¾	12¾	0.09	Vulcan No. 1 (36-in. stroke)	690	3,250
	Steel H Piles									
16	Wharf, Bremerhaven, Germany	IP 24 (10 in. 60 lb)	39.3	35.4	10	10	0.76	10,850 ft-lb	....	......
19	Wharf, Bremerhaven, Germany	IP 24 (10 in. 60 lb)	39.3	35.4	10	10	0.76	10,850 ft-lb	....	......
11	Wharf, Bremerhaven, Germany	IP 24 (10 in. 60 lb)	39.3	35.4	10	10	0.66	10,850 ft-lb	....	......
12	Wharf, Bremerhaven, Germany	IP 24 (10 in. 60 lb)	39.3	35.4	10	10	0.63	10,850 ft-lb	....	......

DRIVING AND TESTS (*Continued*)

Failure test load, tons	Ultimate resistances from formulas, tons								Soil conditions	Source*
	Hiley	Pacific Coast[a]	Canadian National (working load × 3)	*Eng. News* (working load × 6)	Eytelwein (working load × 6)	Modified *Eng. News* (working load × 6)	Modified Eytelwein (working load × 6)	Navy-McKay (working load × 6)		
88	83	93	48	146	177	108	168	170	12 ft silt; then sand and small gravel	Rabe
46	38	79	53	194	164	102	115	276	14 ft silt, clay and fine sand; 10 ft clay; then gravel, fine sand and broken stone	Rabe
120 (compr, 7 days), 140 (pull, 17 days)	85	107	74	450	626	225	390	798	7.5 ft sand and loam, 6.5 ft fine brown sand, 20 ft fine gray sand, 1 ft gravel, 9 ft silty damp sand, 50 ft silty sand and stiff dry clay	Sowers
150 (compr, 10 days), 80 (pull, 16 days)	86	108	74	474	556	230	300	822	7.5 ft sand and loam, 6.5 ft fine brown sand, 20 ft fine gray sand, 1 ft gravel, 9 ft silty damp sand, 50 ft silty sand and stiff dry clay	Sowers
56	42.5	48	38	76	78	61	64	70	6 ft sand, 4 ft clayey sand, 2 ft sand, 1 ft clayey sand, 1 ft earth, 10 ft sand, 1 ft clay, 3 ft clay and sand, 7.5 ft fine sand	Agatz and Cummings
44.5	42.5	48	38	76	78	61	65	70	5.5 ft sand, 4.5 ft dirty sand, 12.5 ft very fine sand, 1.5 ft clay, 1 ft sand and clay, 3 ft dirty sand, 6.5 ft sharp sand	Agatz and Cummings
45	50	54	40	86	88	68	72	80	8 ft sand, 4 ft clayey sand, 16 ft sand, 0.5 ft. hard clay, 2 ft clay with sand, 5 ft fine sand	Agatz and Cummings
55	55	57	42	90	92	70	75	84	6 ft sand, 4 ft clayey sand, 14 ft very fine sand, 2 ft clay, 0.5 ft peat, 1 ft clay, 5.5 ft clayey sand, 2.5 ft sharp sand	Agatz and Cummings

TABLE *B*. DATA ON PILES,

Test no.	Location	Pile: Type	Pile: Length, ft	Pile: Embedment ft	Pile: Butt, in.	Pile: Tip, in.	Set (*s*), in.	Hammer	Pile cap, lb	W_p (including cap), lb
13	Wharf, Bremerhaven, Germany	IP 24 (10 in. 60 lb)	39.3	35.4	10	10	0.55	10,850 ft-lb	....	......
15	Wharf, Bremerhaven, Germany	IP 24 (10 in. 60 lb)	39.3	35.4	10	10	0.47	10,850 ft-lb	....	......
17	Wharf, Bremerhaven, Germany	IP 24 (10 in. 60 lb)	39.3	35.4	10	10	0.47	10,850 ft-lb	....	......
21	Wharf, Bremerhaven, Germany	IP 24 (10 in. 60 lb)	29.3	25.4	10	10	0.45	10,850 ft-lb	....	......
22	Wharf, Bremerhaven, Germany	IP 24 (10 in. 60 lb)	29.3	25.4	10	10	0.45	10,850 ft-lb	....	......
20	Wharf, Bremerhaven, Germany	IP 24 (10 in. 60 lb)	39.3	35.4	10	10	0.36	10,850 ft lb	....	......
26	Bay Bridge falsework, San Francisco, Calif.	12 in. H 65 lb	99.2	43.2	12	12	0.364	No. 11-B-2 at 110 blows per min	....	7,200
14	Wharf, Bremerhaven, Germany	IP 24 (10 in. 60 lb) with 4 sets projecting fins tilted from horizontal	27	23	10	10	0.30	10,850 ft-lb	....	......

Driving and Tests (*Continued*)

Failure test load, tons,	Ultimate resistances from formulas, tons								Soil conditions	Source*
	Hiley	Pacific Coast[a]	Canadian National (working load × 3)	*Eng. News* (working load × 6)	Eytelwein (working load × 6	Modified *Eng. News* (working load × 6)	Modified Eytelwein (working load × 6)	Navy-McKay (working load load × 6)		
67.5	54	61	52	100	104	77	82	96	6 ft sand, 5 ft fine dirty sand, 12 ft very fine sand, 1.5 ft clay, 0.5 ft peat, 0.5 ft clay, 10 ft sand	Agatz and Cummings
45	60	71	50	114	118	84	92	113	7 ft sand, 3 ft clayey sand, 2.5 ft sand, 0.5 ft clayey sand, 2 ft earth, 10 ft fine sand, 1.5 ft clay, 6 ft clay and sand, 3 ft fine sand	Agatz and Cummings
79	62	71	50	114	118	84	93	113	7 ft sand, 3 ft clayey sand, 15 ft very fine sand, 2 ft clay, 0.5 ft peat, 1 ft clay, 7 ft clayey sand	Agatz and Cummings
45	61	71	50	114	118	84	93	113	5 ft fine sand, 4 ft dirty fine sand, 14 ft very fine sand, 1 ft clay, 1 ft peat, 0.5 ft sandy clay	Agatz and Cummings
67	61	71	50	114	118	84	93	113	4 ft fine sand, 4 ft dirty sand, 10 ft very fine sand, 4 ft sharp sand and silt, 1 ft clay, 1 ft sandy clay, 1.5 ft dirty sand	Agatz and Cummings
89	72.5	84	48	141	154	99	109	144	5 ft sand, 5 ft dirty sand, 13 ft very fine sand, 1 ft clay, 1 ft dirty sand, 3 ft sandy clay, 6.5 ft sharp sand	Agatz and Cummings
88	65	78	57	244	210	171	126	204	40 ft water, 13 ft surface mud, 20 ft clayey sand, 18 ft very sandy silty clay, 37 ft silty clay	Agatz and Cummings
78.8	80	100	57	162	171	109	120	177	4 ft fine sand, 7 ft fine dirty sand, 12 ft very fine sand	Agatz and Cummings

TABLE *B*. DATA ON PILES,

Test no.	Location	Pile: Type	Pile: Length, ft	Pile: Embedment, ft	Pile: Butt, in.	Pile: Tip, in.	Set (s), in.	Hammer	Pile cap, lb	W_p (including cap), lb
18	Wharf, Bremerhaven, Germany	IP 24 (10 in. 60 lb)	39.3	35.4	10	10	0.30	10,850 ft-lb	....	......
39	Cowen Park Bridge, Seattle, Wash.	10 in. HP 42 lb	46	32	10	10	0.27	Vulcan No. 1	750	2,680
41	Cuyahoga River, turning basin, Cleveland, Ohio	10 in. H 54 lb	80	78	10	10	0.17	Vulcan No. 1 (36-in. stroke)	660	4,900
24	Bay Bridge falsework, San Francisco, Calif.	12 in. H 65 lb	98.8	40.5	12	12	0.141	No. 11-B-2 at 110 blows per min	....	7,200
27	Bay Bridge falsework, San Francisco, Calif.	12 in. H 65 lb	99.5	44.7	12	12	0.131	No. 11-B-2 at 110 blows per min	....	7,200
10	Alameda Creek Bridge, Niles, Calif.	10 in. HP 42 lb	38	35	10	10	0.125	Vulcan No. 2	....	2,300
25	Bay Bridge falsework, San Francisco, Calif.	12 in. H 65 lb	99	44.4	12	12	0.083	No. 11-B-2 at 110 blows per min	....	7,200
28	Bay Bridge falsework, San Francisco, Calif.	12 in. H 65 lb	99.6	40.4	12	12	0.056	No. 11-B-2 at 110 blows per min	....	7,200
29	Bay Bridge falsework, San Francisco, Calif.	12 in. H 65 lb	100.1	41.4	12	12	0.05	No. 11-B-2 at 110 blows per min	....	7,200
9	Strip mill foundations, Lackawanna, N.Y.	10 in. HP 57 lb	30.75	24.7	10	10	0	Vulcan No. 1	....	2,500

Driving and Tests (*Continued*)

Failure test load, tons	Ultimate resistances from formulas, tons								Soil conditions	Source*
	Hiley	Pacific Coast[a]	Canadian National (working load × 3)	*Eng. News* (working load × 6)	Eytelwein (working load × 6)	Modified *Eng. News* (working load × 6)	Modified Eytelwein (working load × 6)	Navy-McKay (working load × 6)		
Over 112	80	100	57	162	171	109	120	177	6 ft sand, 4.5 ft clayey sand, 12.5 ft sand, 2 ft clay, 0.5 ft peat, 0.5 ft clay, 3 ft clayey sand, 5.5 ft sharp sand	Agatz and Cummings
Over 70	92	152	84	243	291	300	780	∞	Fine sand	Bethlehem
75 (compr, 9 days), 85 (pull, 19 days)	79	100	66	336	353	192	210	426	7.5 ft sand and loam, 6.5 ft fine brown sand, 20 ft fine gray sand, 1 ft gravel, 9 ft silty damp sand, 50 ft silty sand and stiff dry clay	Sowers
90	83	100	67	472	360	258	168	534	40 ft water, 13 ft surface mud, 20 ft clayey sand, 18 ft very sandy silty clay, 37 ft silty clay	Bethlehem and Cummings
80	84	104	70	492	372	264	172	564		Bethleham and Cummings
99	75	72	60	194	244	102	123	300	Gravel and small boulders	Bethlehem and Cummings
140	89	108	69	636	444	258	170	528	40 ft water, 13 ft surface mud, 20 ft clayey sand, 18 ft very sandy silty clay, 37 ft silty clay	Bethlehem and Cummings
150	92	110	72	729	496	318	194	1,332		Bethlehem and Cummings
140	93	114	75	755	504	324	195	1,494	40 ft water, 13 ft surface mud, 20 ft clayey sand, 18 ft very sandy silty clay, 37 ft silty clay	Bethlehem and Cummings
190	175	210	144	875	1,800	300	360	∞	6 ft clay, 2 ft sand and gravel, 13 ft soft clay, 4.7 ft sand and gravel to rock	Bethlehem and Cummings

TABLE *B*. DATA ON PILES,

Test no.	Location	Pile					Set (s), in.	Hammer	Pile cap, lb	W_p (including cap), lb
		Type	Length, ft	Embedment, ft	Butt, in.	Tip, in.				
23	Strip mill foundations, Lackawanna, N.Y.	10 in. HP 57 lb	40	30	10	10	0	Vulcan No. 1	....	3,000
	Precast Piles									
5	Scioto River Bridge, Pickaway County, Ohio	Precast concrete, 12 × 17 in.	14	11	12 × 17	12 × 17	0.261	Vulcan No. 1 (28-in. stroke)	1,480	4,460
38	Naval Operating Base, San Diego, Calif.	20 × 20-in. Concrete	53	....	20	20	0	Vulcan No. 0 (34-in. stroke)	....	22,550

On account of the occasional wide divergences from more accurate results, this is not the case.

The values in Fig. A.4 have been grouped according to the type of pile, and within each group have been placed in descending order of set values. The increasing divergence of the *Engineering News* values as sets decrease should be noted. There are many other variables involved, but this trend is apparent. The divergences appear to start at larger sets with the heavier piles.

The reasonable coincidence of the graphs in Fig. A.4 for wood piles, for the easier drivings, indicates that the actual factor of safety provided by the *Engineering News* formula is nearer 2½ than 6, although a larger value might be necessary with hard driving. On the other hand, harder driving than 0.3-in. set with a heavy hammer might damage a wood pile. As piles become heavier, it appears that use of the *Engineering News* formula becomes more dangerous with small sets.

The scattering of ratios between the ultimate driving resistances, computed by means of the various dynamic formulas in most common use, and the failure points of test loads is shown in Fig. A.5. Since the *Engineering News* and Eytelwein formulas are intended to give working loads with a factor of safety of 6, the results from these formulas have been multiplied by 6 to obtain ultimate resistances. Since the Canadian National Building Code formula contains a factor of safety of 3, the

Driving and Tests (*Continued*)

Failure test load, tons	Ultimate resistances from formulas, tons: Hiley	Pacific Coast[a]	Canadian National (working load × 3)	*Eng. News* (working load × 6)	Eytelwein (working load × 6)	Modified *Eng. News* (working load × 6)	Modified Eytelwein (working load × 6)	Navy-McKay (working load × 6)	Soil conditions	Source*
128	157	248	90	900	1,920	300	654	∞	7.5 ft clay, 3 ft sand, 10.5 ft soft clay, 6 ft sand and gravel to rock	Bethlehem and Cummings
70	75	153	42	192	218	124	158	225	Sand and gravel, relatively loose	Rabe
Over 150	180	214	82	1,380	523	520	174	∞	Water, mud, and sandy clay to firm sand	*Bull. No. 36*

[a] Now International Conference of Building Officials.
* References.
Squire, *Proc. Inst. Civil Eng. (London)*, vol. 226.
King, *Eng. News-Record*, vol. 134, No. 26.
W. H. Rabe, Bureau of Bridges, State of Ohio.
G. B. Sowers, Drilled-In Caisson Corp.
Agatz, *Die Bautechnik*, Feb. 2 and 9, 1934.
E. M. Cummings, Bethlehem Steel Co.
Bethlehem H-Piling, Bethlehem Steel Co.
Bull. 36, Public Works of the Navy, October, 1927.

results have been multiplied by 3. The International Conference (Pacific Coast) Uniform Building Code formula gives ultimate resistances directly.

The best and safest range of values in Fig. A.5 appears to be obtained from formulas (2.1*a*) and (2.1*b*), and no dangerously high percentages appear. The International Conference (Pacific Coast) formula shows a somewhat wider range. The Canadian-formula range falls into somewhat lower figures. The *Engineering News* and Eytelwein ranges are high and widely scattered.

It may be observed in Fig. A.5 that the results of formulas (2.1*a*) and (2.1*b*), the International Conference (Pacific Coast), and the Canadian formulas are grouped in some proximity to the 100 per cent line, so that with the factors of safety assumed, none would actually be unsafe and none very wasteful. On the other hand, the scattering of results from the *Engineering News* and Eytelwein formulas is too wide to be comprehended economically within any one factor of safety, and even a factor of 6 would not be adequate for a number of the results.

APPENDIX V

PILE INSPECTOR'S REPORTS AND DUTIES

Qualifications and Instruction of Inspector

A pile-driving inspector must be able to read plans, to keep neat and accurate records, and to understand the principles of the specification. If he is to be in charge, previous experience is essential.

Inexperienced inspectors may permit failure in meeting the objects of the specifications through lack of understanding, "stage fright," or poorly exercised authority.

It is not reasonable to expect an inspector, who, almost certainly, is far less experienced in pile-foundation design than the engineer, to be able to represent the engineer adequately and to exercise judgment and control over many varying or unforeseen conditions unless he is informed of the background of the design and of the tolerances permissible.

It is recommended that the engineer gather together the superintendent, field engineers, pile inspectors, and pile-driving contractor's foremen, prior to the start of driving, and outline to them the history of the pile design. Any soil mechanics study should be discussed. The general geological history of the site should be described. The factors leading to the selection of the particular type and length of pile and size of hammer should be stated. Minimum and maximum driving requirements should be given, and the reasons for their adoption stated. Matters which would tend to defeat the purpose of the specification should be mentioned, such as lack of care in handling piles, overdriving, hitting obstructions, driving out-of-plumb, retardation of stroke, variations in cushioning material, and sequence of driving. Sketching the soil stratification, with piles, on a large sheet of paper on the wall is most helpful. Explanation of the characteristics of the graphs in Figs. 2.1 and 2.2 is advisable.

It is recommended that the results of borings be plotted to scale to show profiles of the underground stratification, and that copies be in the hands of all inspectors. An exaggerated vertical scale may be necessary.

Printed manuals or typewritten instructions embodying the above information for the specific project are suggested as helpful.

Pile Reports. The forms in Figs. A.6, A.7, and A.8 are useful in recording essential continuous driving information. They should be used for the first piles, for occasional later typical piles in quite constant site

CONTINUOUS PILE-DRIVING RECORD (FOR USE WITH SINGLE-ACTING AND DROP HAMMERS ONLY)

▲ 5040.3A

STONE & WEBSTER ENGINEERING CORPORATION PILE NO.________

CLIENT	STRUCTURE	CONTRACTOR	HAMMER NO.	J. O. NO.	DATE

PILE

TYPE*

TIP DIAMETER IN. — BUTT DIAMETER IN.

LENGTH DRIVEN FT. IN. — WEIGHT† LB.

HAMMER

MAKE AND MODEL

STROKE: RATED IN. MEAS.‡ IN.

WEIGHT OF RAM LB.

MANDREL (IF USED)

DESCRIPTION

LENGTH FT. IN. — WEIGHT LB.

DRIVING CAP, ANVIL, HEMLET, ETC.

WEIGHT LB. — DESCRIPTION (MAKE SKETCH ON BACK)

FOLLOWER (IF USED)

DESCRIPTION

LENGTH FT. IN. — WEIGHT LB.

PENETRATION

ELEVATION OF CUTOFF — ELEVATION OF GROUND

ELEVATION OF TIP — LG., CUTOFF TO TIP FT. IN.

REMARKS¶

TIME

STARTED DRIVING: A.M. P.M. — FINISHED DRIVING: A.M. P.M. — DRIVING TIME: MIN.

Ft.	No. of Blows	Ft.	No. of Blows	Ft.	No. of Blows	Ft.	No. of Blows	Ft.	No. of Blows	Ft.	No. of Blows	Ft.	No. of Blows	Ft.	No. of Blows
0		10		20		30		40		50		60		70	
1		11		21		31		41		51		61		71	
2		12		22		32		42		52		62		72	
3		13		23		33		43		53		63		73	
4		14		24		34		44		54		64		74	
5		15		25		35		45		55		65		75	
6		16		26		36		46		56		66		76	
7		17		27		37		47		57		67		77	
8		18		28		38		48		58		68		78	
9		19		29		39		49		59		69		79	
10		20		30		40		50		60		70		80	

Record number of blows required for each 6 in. of penetration. Note points at which stoppages occur, with times of stopping and starting.

*If wood, state kind, seasoning and treatment. If concrete, state mix and age.
†For wood piles determine actual weight per cubic foot of the wood by weighing a butt section (can determine volume by measuring quantity of water displaced by section).
‡Note any falling off in rated stroke during driving.
¶Jetting, cause and duration of delays in driving, boulders, bark, condition of cushions, plumbness, banding, damage, driving shoe, etc.

____
Pile Inspector

FIG. A.6. Continuous pile-driving record (for use with single-acting steam and drop hammers only).

CONTINUOUS PILE-DRIVING RECORD (FOR USE WITH DOUBLE-ACTING AND DIFFERENTIAL-ACTING HAMMERS ONLY)

▲ 8040.8A

STONE & WEBSTER ENGINEERING CORPORATION PILE NO.________

CLIENT	STRUCTURE	CONTRACTOR	HAMMER NO.	J. O. NO.	DATE

PILE

TYPE*

TIP DIAMETER IN. | BUTT DIAMETER IN.

LENGTH DRIVEN FT. IN. | WEIGHT† LB.

HAMMER—MAKE AND MODEL

DRIVING CAP, ANVIL, HELMET, ETC.: WEIGHT LB. | DESCRIPTION (MAKE SKETCH ON BACK)

MANDREL (IF USED)

DESCRIPTION

LENGTH FT. IN. | WEIGHT LB.

FOLLOWER (IF USED)

DESCRIPTION

LENGTH FT. IN. | WEIGHT LB.

PENETRATION

ELEVATION OF CUTOFF | ELEVATION OF GROUND

ELEVATION OF TIP | LG., CUTOFF TO TIP FT. IN.

REMARKS‡

TIME

STARTED DRIVING: A.M. P.M. | FINISHED DRIVING: A.M. P.M. | DRIVING TIME: MIN.

Ft.	No. of Blows	Obs. Speed, Blows Per Min.	Ft.	No. of Blows	Obs. Speed, Blows Per Min.	Ft.	No. of Blows	Obs. Speed, Blows Per Min.	Ft.	No. of Blows	Obs. Speed, Blows Per Min.	Ft.	No. of Blows	Obs. Speed, Blows Per Min.	Ft.	No. of Blows	Obs. Speed, Blows Per Min.
0			10			20			30			40			50		
1			11			21			31			41			51		
2			12			22			32			42			52		
3			13			23			33			43			53		
4			14			24			34			44			54		
5			15			25			35			45			55		
6			16			26			36			46			56		
7			17			27			37			47			57		
8			18			28			38			48			58		
9			19			29			39			49			59		
10			20			30			40			50			60		

Record number of blows required for each 6 in. of penetration. Note points at which stoppages occur, with times of stopping and starting.

*If wood, state kind, seasoning and treatment. If concrete, state mix and age.
†For wood piles determine actual weight per cubic foot of the wood by weighing a butt section.
‡Jetting, cause and duration of delays in driving, boulders, bark, condition of cushions, plumbness, banding, damage, driving shoe, etc.

____
Pile Inspector

FIG. A.7. Continuous pile-driving record (for use with double-acting and differential-acting steam hammers only).

CONTINOUS PILE-DRIVING RECORD (FOR USE WITH DIESEL HAMMERS ONLY)

▲ 5040.9

STONE & WEBSTER ENGINEERING CORPORATION PILE NO.______

CLIENT	STRUCTURE	CONTRACTOR	HAMMER NO.	J. O. NO.	DATE

PILE
- TYPE*
- TIP DIAMETER ____ IN.
- BUTT DIAMETER ____ IN.
- LENGTH DRIVEN ____ FT. ____ IN.
- WEIGHT† ____ LB.

HAMMER: MAKE AND MODEL; RATED ENERGY, FT. LB.
- WEIGHT OF IMPACT BLOCK ____ LB.
- WEIGHT OF HELMET OR ANVIL ____ LB.
- DESCRIPTION (MAKE SKETCH ON BACK)

MANDREL (IF USED)
- DESCRIPTION
- LENGTH ____ FT. ____ IN.
- WEIGHT ____ LB.

FOLLOWER (IF USED)
- DESCRIPTION
- LENGTH ____ FT. ____ IN.
- WEIGHT ____ LB.

PENETRATION
- ELEVATION OF CUTOFF
- ELEVATION OF GROUND
- ELEVATION OF TIP
- LG., CUTOFF TO TIP ____ FT. ____ IN.

REMARKS¶

TIME
- STARTED DRIVING: ____ A.M. ____ P.M.
- FINISHED DRIVING: ____ A.M. ____ P.M.
- DRIVING TIME: ____ MIN.

Ft.	No. of Blows	Ft.	No. of Blows	Ft.	No. of Blows	Ft.	No. of Blows	Ft.	No. of Blows	Ft.	No. of Blows	Ft.	No. of Blows	Ft.	No. of Blows
0		10		20		30		40		50		60		70	
1		11		21		31		41		51		61		71	
2		12		22		32		42		52		62		72	
3		13		23		33		43		53		63		73	
4		14		24		34		44		54		64		74	
5		15		25		35		45		55		65		75	
6		16		26		36		46		56		66		76	
7		17		27		37		47		57		67		77	
8		18		28		38		48		58		68		78	
9		19		29		39		49		59		69		79	
10		20		30		40		50		60		70		80	

Record number of blows required for each 6 in. of penetration. Note points at which stoppages occur, with times of stopping and starting.

*If wood, state kind, seasoning and treatment. If concrete, state mix and age.
†For wood piles determine actual weight per cubic foot of the wood by weighing a butt section (can determine volume by measuring quantity of water displaced by section).
¶Jetting, cause and duration of delays in driving, boulders, bark, condition of cushions, plumbness, banding, damage, driving shoe, etc.

____
Pile Inspector

FIG. A.8. Continuous pile-driving record (for use with diesel hammers only).

conditions, and for each type of driving, at least. It is desirable to keep these continuous records on all piles, since this permits the best control over the job and may provide data to detect or explain oddities or damage that otherwise could not be diagnosed. Pile behavior indicates resistance encountered and should be intelligently watched, and not ignored or merely recorded. By checking driving against the boring logs, normal action of the pile can be observed and variations quickly detected, such

FIG. A.9. Markings on pile frame for use in reading foot-by-foot driving records. Hammer is No. 1 Vulcan single-acting steam hammer driving 10-in.-o.d. core for pedestal pile.

as breaks, breakthroughs, or obstructions. If causes are not certain, a pile may be pulled. Since the inspector should be present all the time, keeping these records will not involve extra cost, and they will also familiarize him with conditions more fully.

If individual pile report sheets are not used for all piles, a summary of the final driving results should be reported. A typical form for this purpose is shown in Fig. A.10. Under "Remarks" should be stated such items as redriving, heaving, weaving, obstructions, jetting, spudding, stops, etc. Piles should be entered on the summary report in the order driven, since study of sequence is sometimes valuable.

Cushioning Material. The inspector should be cautioned against the use of inadequate and unsuitable cap blocks such as "wood chips" or

PILE INSPECTOR'S SUMMARY REPORT
▲ 5040.7A

SHEET______OF______

STONE & WEBSTER ENGINEERING CORPORATION

CLIENT	STRUCTURE	J. O. NO.	DATE
PILE DRIVER NO.	MAKE & MODEL OF HAMMER	CONTRACTOR	
TYPE & MATERIAL OF PILE	WT. OF RAM (SINGLE-ACTING & DROP HAMMERS) LB.	TYPE & WT. OF DRIVING CAP LB.	SAFE LOAD BY FORMULA OR GRAPH TONS

PILE MARK		SIZE OF PILE			ELEVATION			LENGTH, CUTOFF	AT FINAL PENETRATION				FOR DROP AND SINGLE-ACTING HAMMERS ONLY	FOR DOUBLE-ACTING AND DIFFERENTIAL-ACTING HAMMERS ONLY, AT FINAL PENETRATION			REMARKS
FOOTING OR BENT	PILE NO.	TIP, IN.	BUTT, IN.	LENGTH FT.	TIP	GROUND	CUT OFF	TO TIP, FT.	NO. OF BLOWS	DIST. DRIV., IN.	AVE. FIN. SET, IN. PER BLOW	SAFE LOAD, BY GRAPH OR FORM., T.	ACTUAL STROKE IN.	BLOWS PER MIN.*	OBSERVED NO. BLOWS FOR ANY TIME: NO. BLOWS	OBSERVED NO. BLOWS FOR ANY TIME: TIME, SEC.	

*BLOWS PER MINUTE MAY BE COUNTED DIRECTLY FOR ONE MINUTE, OR IF MORE CONVENIENT MAY BE READILY OBTAINED BY FILLING IN THE TWO FOLLOWING COLUMNS AND CONVERTING LATER BY USE OF THE CHART ON THE BACK OF THIS SHEET. IT IS ESSENTIAL THAT THE "BLOWS PER MINUTE" COLUMN BE FILLED OUT FOR DOUBLE-ACTING AND DIFFERENTIAL-ACTING HAMMERS BEFORE SUBMITTING THIS REPORT.

LOAD COMPUTATIONS BY ______________________ SIGNED ______________________
INSPECTOR

FIG. A.10. Pile inspector's summary report.

random blocks of wood. The practice of adding fresh cap-block material, such as tossing wood chips under the ram during the final driving of the pile, should be prohibited. The inspector should note on the driving log when fresh cap-block material is placed under the hammer and should discount the measured resistance to driving immediately thereafter. The effects of soft, renewed blocking can make a very large difference in the set and pile length, which can become particularly dangerous if the hammer is finding hard driving because it is driving close to its capacity.

Rebound-set Graphs. The inspector should be instructed how to make rebound-set graphs on the piles during driving. These should generally be taken at several points and at final set on the first few piles of each type or length. The distances from tips or heads should be stated.

Piles Driven Out of Position

It sometimes becomes necessary to move the locations of piles somewhat, or piles become driven out of position. In such cases, sketches should be prepared and sent to the engineer immediately. A procedure should be developed regarding acceptance or redriving of piles out of line, and regarding locations of alternate piles when required, so that it will not be necessary to bring the rig back. The procedure will depend somewhat upon the availability of the engineer.

Pile Numbering Plan

If a pile numbering plan is not provided for field use, one should be prepared before driving. Piles should be entered on the pile inspector's report in the order driven, so that the sequence of driving may thus be preserved for study if necessary. The sequence should be stated.

Plumbness

Both sunlight and headlights may be reflected down a hollow pile by a mirror, or lamps lowered. If curvature is so great that lights cannot be seen, out-of-plumbness must be estimated, or an inclinometer[147] or an electronic plumb bob used that gives such measurements.*

A manometric inclinometer for measuring the out-of-plumbness of pipe piles when a lowered light can no longer be observed around the bend has been devised.† This consists of a frame on wheels which fits the inside of the pipe. A three-tube manometer is connected to a single stopcock located at the lower end. A nylon fish line fastened to the stopcock can be jerked from the surface to close the cock at a predetermined depth. By reading the manometer and observing the difference in liquid levels, the amount and direction of the slope can be observed at desired elevations.

* Manufactured by The Hinchman Corporation, Detroit, Mich.

† H. P. Eddy, Jr., *J. Boston Soc. Civil Engrs.*, vol. 46, No. 5, July, 1959, p. 261.

APPENDIX VI

STANDARD SPECIFICATIONS

Standard Specifications for Round Timber Piles of the American Society for Testing Materials (D25–37) and of the American Standards Association (ASA 06-1939)

Scope

1. These specifications cover round timber piles to be used untreated, or treated by standard preservatives.

Note: Where sawed timber is used as piling, such as heavy sheet piles, appropriate specifications should be selected from the "*Standard Specifications for Structural Wood Joist and Plank, Beams and Stringers, and Posts and Timbers* (*ASTM* D245)" of the American Society for Testing Materials.

Kinds of Wood

2. (*a*) The Purchaser shall specify the kind or kinds of wood he desires, and shall designate the kinds he desires for preservative treatment.

Note: Commonly used species are cedars, chestnut, cypress, Douglas fir, larch, oaks, pines, spruces, and tamarack.

(*b*) Piles of different kinds of wood shall be delivered in separate lots.

Use Classification

3. Timber piles are classified in these specifications under three general divisions according to the use intended, as follows:

(*a*) Class *A*. Piles suitable for use in heavy railway bridges and trestles. The minimum diameter of butt assumes the use of load-bearing timber caps 14 in. in width.

(*b*) Class *B*. Piles suitable for use in docks, wharves, highway work, and general construction. The minimum diameter of butt assumes the use of load-bearing timber caps 12 in. in width. When timber caps are not used, as in the case of piles under masonry foundations, the sizes given for Class *C* piles are recommended.

(*c*) Class *C*. Piles suitable for use in foundations which will always be completely submerged, for coffer-dams, false-work, and sundry temporary work.

CLASS *A* AND CLASS *B* PILES

General Quality

4. Except as hereinafter provided, Class *A* and Class *B* piles shall be free from any defects which may impair their strength or durability as piling, such as decay, red heart, splits in piles to be treated, or splits longer than the measured butt diameter of piles not to be treated, twist of grain exceeding one-half of the circumference in any 20 ft of length, unsound knots, numerous knots or holes, or shake more than one-third of the diameter of the pile. Piles which have been scored for turpentine shall be

accepted, provided such scar does not exceed 36 in., and provided the scoring is of recent date showing the scar to be entirely sound and free from insect damage.

Knots

5. Sound knots will be permitted in Class *A* and Class *B* piles, provided they are not in clusters. The diameter of a sound knot shall not be greater than one-third of the minimum diameter of the pile at the section where it occurs, and shall not exceed 4 in. for piles 50 ft and under in length. For piles over 50 ft in length, knots between the section at mid-length and the butt shall conform to the limitation prescribed for piling under 50 ft. Between mid-length and the tip, single knots up to 5 in. in diameter will be permitted, provided they do not exceed one-half the minimum diameter of the pile at the section where they occur. The diameter of a knot shall be measured at right angles to the length of the pile.

General Requirements

6. (*a*) *Sound Timber.* Class *A* and Class *B* piles shall be cut from sound, live trees, except that fire-killed, blight-killed, or wind-felled timber may be used if not attacked by decay or insects. Piles shall be cut above the ground swell.

(*b*) *Tip.* The tip shall be sound.

(*c*) *Butt End.* The butt end shall be sound except in cedar piles, which may have a pipe or stump rot hole not more than 1½ in. in diameter.

(*d*) *Taper.* Piles shall have a gradual taper from the point of butt measurement to the tip.

(*e*) *Surface Finish.* All knots and limbs shall be trimmed or smoothly cut flush with the surface of the pile. The butt and tip shall be sawed square with the axis of the pile, or the tip may be tapered to a point not less than 4 in. in diameter if directed by the engineer in charge.

Sapwood

7. (*a*) *Piles for Use Untreated.* Piles for use without preservative treatment shall have as little sapwood as possible, and when used in exposed work, the diameter of the heartwood shall not be less than eight-tenths of the actual diameter of the pile at the butt.

(*b*) *Piles for Treatment.* Piles for use with preservative treatment shall have no sapwood restrictions, but preferably shall contain as much sapwood as possible. In southern pine the sapwood thickness shall not be less than 1½ in. and in Douglas fir and larch not less than 1 in. on the butt end.

Peeled Piles

8. (*a*) Piles shall be peeled of bark, including the inner skin, soon after cutting so that the piles are smooth and clean. Care shall be taken to remove as little sapwood as possible while peeling the bark. The sapwood shall not be injured by unnecessary axe cuts. These piles shall be designated as piles for treatment.

(*b*) No pile shall be considered as thoroughly peeled unless all of the rough bark and at least 80 per cent of the inner bark which remains on the pile shall have been removed. In no case shall any piece of inner bark be over ¾ in. in width or over 8 in. in length, and there shall be 1 in. of clean wood surface between any two strips of inner bark.

Diameter

9. (*a*) It is recommended that the diameters of piles measured under the bark shall conform to the requirements shown in the table (see p. 610) subject to a permissible

variation of minus ½ in. in any diameter in not more than 25 per cent of the piles of that diameter.

(*b*) The diameter of a pile in cases where the tree is not exactly round shall be determined either by measuring the circumference and dividing the number of inches by 3.14, or by taking the average of the maximum and minimum diameters of the location specified.

Length

10. All piles shall be furnished cut to any of the following lengths, as specified: 16 to 40 ft in multiples of 2 ft, and over 40 ft in multiples of 5 ft. A variation of 6 in. in length shall be allowable, but the average length in any shipment shall be equal to, or greater than, the specified lengths. The length of each pile shall be legibly marked on the butt end with white or black paint.

Straightness

11. Piles shall be free from short or reversed bends, and free from crooks greater than one-half of the diameter of the pile at the middle of the bend. In short bends, the distance from the center of the pile to a line stretched from the center of the pile above the bend to the center of the pile below the bend shall not exceed 4 per cent of the length of the bend, or 2½ in. A line drawn from the center of the butt end to the center of the tip shall lie within the body of the pile.

CLASS *C* PILES

General Quality

12. Class *C* piles shall be of sound, live timber that will stand driving, and need not be peeled if they are to be used without preservative treatment. They shall be free from decay and other imperfections such as bad knots and shakes which will materially affect their strength. Piles which have been scored for turpentine shall be accepted, provided such scar does not exceed 36 in., and provided the scoring is of recent date showing the scar to be entirely sound and free from insect damage.

General Requirements

13. (*a*) *Tip.* The tip shall be sound.

(*b*) *Taper.* Piles shall have a gradual taper from the point of butt measurement to the tip.

(*c*) *Surface Finish.* All knots and limbs shall be trimmed or smoothly cut, flush with the surface of the pile. The butt and tip shall be sawed square with the axis of the pile, or the tip may be tapered to a point not less than 4 in. in diameter if directed by the engineer in charge.

Sapwood

14. Sapwood requirements shall be identical with those for Class *A* and Class *B* piles as specified in section 7.

Peeled Piles

15. The requirements for peeled piles shall be identical with those for Class *A* and Class *B* piles as specified in section 8.

Diameter

16. The methods of measurement of diameter and the permissible variations in diameter shall be identical with those for Class *A* and Class *B* piles as specified in section 9. The recommended sizes are given in the table (see p. 610).

LIMITING DIMENSIONS OF PILES, INCHES

Place measured	Southern pine and Douglas fir[a]					Oak, cypress, and chestnut[b]			Cedar		
	Less than 40 ft long	40–50 ft long	51–70 ft long	71–90 ft long	Over 90 ft long	Less than 30 ft long	30–40 ft long	Over 40 ft long	Less than 30 ft long	30–40 ft long	Over 40 ft long
ASCE, ASTM, ASA, and CESA Class *A* piles AREA First-class piles for railway bridges											
3 ft from butt:											
Minimum	14	14	14	14	14	14	14	14	14	14	14
Maximum	18	18	18	20	20	18	18	18	22	22	22
Tip (minimum)	10	9	8	7	6	10	9	8	10	9	8
ASCE, ASTM, ASA, and CESA Class *B* piles AREA First-class piles for highway bridges											
3 ft from butt:											
Minimum	12	12	13	13	13	12	13	13	12	13	13
Maximum	20	20	20	20	20	18	20	20	22	22	22
Tip (minimum)	8	7	7	6	5	8	8	7	8	8	7
ASCE, ASTM, ASA, and CESA Class *C* piles (Values in boldface are for ASTM, ASA, and CESA)[c] AREA Second-class piles											
3 ft from butt:											
Minimum	12 **12**	12 **12**	12 **12**	12 **12**	12 **12**	12 **12**	12 **12**	12 **12**	12 **12**	12 **12**	12 **12**
Maximum	18 **20**	18 **20**	18 **20**	18 **20**	18 **20**	18 **20**	18 **20**	18 **20**	18 **22**	18 **22**	18 **22**
Tip (minimum)	8 **8**	6 **6**	6 **6**	6 **6**	6 **5**	8 **8**	8 **8**	6 **6**	8 **8**	8 **8**	6 **7**

[a] For ASTM and ASA, where larch, lodgepole or Norway pine, spruce or tamarack piles are specified, their dimensions shall correspond to the requirements shown for Douglas fir and southern pine. For CESA, where spruce, pine, larch, tamarack, or other softwoods are specified, their dimensions shall correspond to the requirements shown for Douglas fir. No values for southern pine are given for CESA.

[b] For ASTM and ASA include black-oak, pin-oak, post- or burr-oak, red-oak, white-oak, and willow-oak piles. For CESA, include birch, beech, and maple, and omit cypress and chestnut.

[c] For ASTM and ASA Class *C* piles a minimum diameter (at cutoff) of 10 in. may be specified for lengths of 20 ft and under.

Length

17. The requirements for length shall be identical with those for Class *A* and Class *B* piles as specified in section 10.

Straightness

18. The requirements relating to straightness shall be identical with those for Class *A* and Class *B* piles as specified in section 11.

Specifications for Driving Wood Piles of the American Railway Engineering Association, 1940

1. *Scope*

These specifications cover the driving of wood piles in trestles, foundations, and for protection work. For the driving of concrete piles and steel piles, see the masonry chapter.*

2. *Tests*

A thorough exploration shall be made at the proposed site by borings or preliminary test piles as indicated in the Design of Pile Trestles in this chapter and in the Specifications for Pile Foundations in the masonry chapter.*

3. *Materials*

The kinds of wood, physical requirements, dimensions, and manufacture are specified in this chapter.

4. *Handling of Material*

Treated piles shall be handled with rope slings, taking care to avoid dropping, bruising or breaking of outer fibers, or penetrating the surface with tools. Sharp pointed tools shall not be used in handling treated piles or turning them in the leads.

The surface of treated piles below cutoff elevation shall not be disturbed by boring holes or driving nails or spikes into them to support temporary material or staging. Staging may be supported in rope slings carried over the tops of piles or attached to pile clamps of an approved design.

5. *Selection and Preparation of Piles*

(*a*) *Size.* The piles in each bent of a pile trestle shall be selected for uniformity of size to facilitate placing of the brace timbers.

(*b*) *Pointing.* Piles may be pointed where soil conditions make it desirable.

(*c*) *Pile Shoes.* Where the driving of a test pile or former experience at the site indicates that difficult driving will be encountered, metal shoes of an approved design may be attached to the tips of the piles.

(*d*) *Collars.* Where the heads of the piles tend to check or split under the hammer, the heads shall be wrapped with wire, or metal bands applied to obviate this condition.

(*e*) *Driving Cap.* The heads of all piles shall be protected, while being driven, with a cushion cap of approved design. Care shall be exercised to ensure full bearing of the driving cap on the pile for proper distribution of the hammer blow.

6. *Type of Hammers*

Pile driving shall not be started on any project until approval is secured from the Engineer as to the type and weight of the hammer to be used.

* These portions of the AREA specifications not quoted in this volume. See AREA specifications.

Piles shall be driven with the heaviest hammer that, in the judgment of the Engineer, can be used to secure maximum penetration without appreciable damage to the pile.

Where a drop hammer is used, the striking ram shall weigh not less than 3,000 lb. The fall shall be so regulated as to avoid injury to the pile.

A steam hammer shall be used where the shock to surrounding material may cause damage to an adjacent structure.

7. *Driving*

(*a*) *Leads.* Pile driver leads shall be constructed in such a manner as to afford freedom of movement of the hammer, and they shall be held in position by guys or stiff braces to ensure support for the pile during driving.

(*b*) *Followers.* The use of followers shall be avoided if practicable and shall be used only with the written permission of the Engineer.

(*c*) *Line.* Piles shall be driven as accurately as possible in the correct location, true to line both laterally and longitudinally, and to the vertical or batter lines as indicated on the plans. On sloping ground or under difficult conditions of driving, the pile shall be started in a hole or guiding templet or other necessary means provided to ensure driving in the proper location. In case a pile works out of line in driving, it shall be properly aligned before it is cut off or braced, and the distance that it may be pulled shall be determined by the Engineer.

(*d*) *Jetting.* When water jets are used, the number of jets and the volume and pressure of water shall be sufficient to freely erode the material adjacent to the pile. The plant shall have sufficient capacity to deliver at least 100 psi pressure at two ¾-in. nozzles. Before the desired penetration is reached, the jets shall be removed and the pile finally set under normal driving by at least 50 blows from a gravity or single-acting steam hammer or 200 blows from a double-acting steam hammer.

(*e*) *Drilling.* When it has been satisfactorily demonstrated to the Engineer that piling can not be driven in the regular manner or by jetting, holes may be drilled to facilitate the driving.

Where drilling is permitted, the holes drilled shall have a diameter not more than 1 in. larger than the tip diameter of the pile and the drilling will continue only through the strata of hard material obstructing the driving. Where the hard material extends below the desired penetration, the drilling shall be stopped above that penetration level and the pile finally set under normal driving by at least 50 blows from a gravity or single-acting steam hammer or 200 blows from a double-acting steam hammer.

(*f*) *Drilling and Shooting.* Where it is impossible to drive, jet, or drill and drive the piles, the Engineer will determine whether shooting the holes with explosives or redesign of the structure is necessary. Shooting will not be permitted except by written permission of the Engineer.

(*g*) *Penetration.* It is expected that piles shall be driven, jetted, or drilled and driven to the full penetration shown on the plans or as otherwise required. This shall not be construed to mean that driving may stop when such penetration as shown on the plans has been secured, but on the contrary, driving shall continue in every case until the total penetration obtained is satisfactory to the Engineer, regardless of the fact that sufficient bearing capacity as determined by formula may be obtained at a lesser depth.

(*h*) *Bearing Capacity.* Where possible, test piles shall be driven and loading tests made before construction is started. However, if these test data are not available, the *Engineering News* formulas may be used to determine the approximate bearing capacity of piles.

These formulas are applicable only when the hammer has a free fall, the head of the pile is not broomed or crushed, the penetration is reasonably uniform, and there is no appreciable bounce of the hammer. The character of the soil penetrated; conditions of driving; spacing, size, and length of piles; and experience under similar conditions shall be given due consideration in determining the value of piles by formula.

For jetted piles, the same formulas will apply and the test shall be made when driving is resumed after removal of the jets. For piles driven in drilled holes, the tests shall be made after the tip of the pile has passed the bottom of the hole.

(*i*) *Delay.* When driving is interrupted before final penetration is reached, record for bearing capacity shall not be taken until at least 12 in. penetration or refusal has been obtained after driving has been resumed.

(*j*) *Overdriving.* When the point of refusal is reached, care shall be taken to avoid damaging the pile by overdriving. This condition is indicated when the hammer begins to bounce, or when the energy of the blow is dissipated in the bending or kicking of the pile.

(*k*) *Replacing.* Any pile driven too far out of line, driven below cutoff elevation, or so injured in driving or straightening as to impair its structural value as a pile under the conditions of use, shall be pulled and replaced by a new pile.

8. *Framing*

(*a*) *Cutoff.* The tops of piles shall be pulled into line if necessary, fixed in position, cut off to a true plane as shown on the plans, and at the elevation established by the Engineer. Piles shall show a solid head at the plane of the cutoff.

(*b*) *Treatment.* After the cutoff has been made, the tops of treated piles shall be saturated with hot preservative, followed by two coats of hot sealing compound. The sealing compound shall be a mixture of creosote coal tar pitch, mixed to about the consistency of vaseline, and brushed thoroughly into the wood. For detailed description of kinds of preservatives, methods of application, and handling of treated piles and timber, see Section V of Chapter 117, Wood Preservation.*

(*c*) *Pile Covering.* Where the plans call for a heavy roofing material or metal pile covering, it shall be placed on the tops of the piles immediately after treatment, the edges bent down over the sides of the pile, neatly trimmed, and fastened with roofing nails.

(*d*) *Placing Caps.* Caps shall be placed while the piles are held in correct position. Where drift bolts are used for making the connection, the caps and tops of piles shall be bored the same diameter as the drift bolt and to a depth of 3 in. less than its length. Where the connection is made with straps and bolts, see paragraph (*f*) for boring and treatment of holes.

(*e*) *Bracing.* Piling shall not be trimmed or cut to facilitate the framing of sway or longitudinal bracing. Where necessary, filler blocks shall be used between the pile and brace to establish the bracing in a true plane.

(*f*) *Holes for Bolts.* Holes shall be bored the same diameter as the bolt.

When holes are bored in treated piles, caps, or bracing in the field, the entire hole shall be pressure treated or swabbed with hot preservative and sealing compound just before the bolt is placed. Bolts shall be cleaned of rust and scale, and dipped in hot sealing compound before placing. All unused holes shall be plugged at each end with tight-fitting treated wooden plugs.

(*g*) *General Field Treatment.* Where it is necessary to disturb the surface of treated piles or timber, or where the surface has been damaged in handling, such surfaces shall be treated with a liberal quantity of hot preservative followed by two applications of hot sealing compound.

* These portions of the AREA specifications not quoted in this volume. See AREA specifications.

9. *Foundation Piles*

For the design of pile foundations, exploration at the site, and test pile loading, see the masonry chapter.*

The general specifications above shall apply to the driving of wood foundation piles.

Pile driving shall not be started until foundation excavation has been carried to plan depth.

After all of the piles are driven, tests shall be made to determine if any of the piles have raised due to driving of adjacent piles. Any piles that have raised shall be driven down again.

After driving is completed, the piles shall be cut off as shown on the plans and at the elevation established by the Engineer. All loose and displaced materials down to the level of original excavation shall be removed from the foundation pit, leaving a clean solid surface on the piles, and bottom and walls of the pit.

10. *Protection Work*

The general specifications above shall apply to the driving of wood piles for protection work. It is essential that protection work be constructed as securely as possible, accurately located as shown on the plans, and the piles driven to a fixed penetration or to refusal as may be determined by the Engineer.

Standard for the Purchase and Preservation of Forest Products—Specification M1 of the American Wood-Preservers' Association, 1954

The following recommendations are intended as guides to purchasers of treated forest products in order that a better product will result from proper application of the standards of the AWPA.

1. *General Requirements*

1.1. Forest products to be used in permanent locations where exposed to decay or attack by other wood-destroying organisms should be given preservative treatment.

1.2. Only wood free from defects which will render it unfit for its use should be treated. Preservative treatment will not restore any loss of strength resulting from defects of any kind.

1.21. The heartwood of certain species resists penetration, but its life is extended by treatment, even though the penetration is shallow. Such species should be specified with as much sapwood present in the piece as possible to ensure maximum treatment.

1.22. The supplier of Douglas fir material to be treated should not mix species of any parcel without suitable identification and should advise the treating plant in advance of unloading or storing, with all necessary information concerning the origin of the timber.

1.3. *Conditioning.* Material to be treated should be conditioned by a method indicated in the specification selected to govern the treatment.

1.31. *Air Seasoning.* When circumstances permit, material should be air-seasoned. When there is not sufficient time for proper air seasoning, or in cases of large material which in some localities during all seasons of the year will not air season successfully without deterioration, artificial conditioning should be used.

1.311. The seasoning yard should be in the open where the air current will circulate freely; should not be in a low, humid situation if it can be avoided; should have good drainage; and should be kept free from vegetation and debris, especially from wood already infected with decay.

* These portions of the AREA specifications not quoted in this volume. See AREA specifications.

1.312. All stacks of seasoning material should be supported on treated or other nondecaying sills, and in all cases there should be at least 12 in. of unenclosed air space underneath the lowermost layer of material with more in warm humid climates or localities. To allow proper air circulation, the alleys between the stacks should extend in continuous lines across the seasoning yard and should be not less than 3 ft wide in the working spaces and 1 ft wide in other directions, to obtain full advantage of the prevailing winds.

1.313. Material will season faster in fairly open stacks. Crossties should be stacked in layers of eight to ten with one tie as a stringer at every other end. Timbers 5 in. or more thick should be stacked with at least 2 in. of air space between layers. Lumber less than 5 in. thick should be stacked with at least 1 in. of air space between layers. Round material will season faster and with less deterioration in layers separated by at least 1 in. of air space. Crossties and timbers should be stacked at least 2 in. apart within layers. Stickers should be of treated wood.

1.314. Material being air-seasoned should be held in stacks until the amount of moisture in the wood will not prevent the adequate penetration and retention of the preservative. The length of time required varies with the species, dimensions, locality, climatic conditions at the storage yard, and the moisture content of the material when it is stacked for seasoning. The following ranges of time are approximate and depend on the variables listed above:*

Material	Kind of wood	Time seasoning, months
Piles..........	Douglas fir	6–12
	Southern pine	3–9
	Oak	9–15

1.315. While the foregoing limits have been found satisfactory in general, all material should be treated before it commences to deteriorate. Deterioration because of decay is more apt to occur during humid weather and evidences of infection will first be noticed in the form of discolored areas at points of contact and later in the form of fruiting bodies of fungi. Destruction by insects may be noticed in the form of sawdust from their borings.

1.316. Antisplitting devices should be applied as soon as possible after they are stacked in the seasoning yard, to the ends of ties and timber of hardwood species which have a tendency to split during conditioning.

1.32. *Steam Conditioning.* The conditioning of material in steam is particularly adapted for woods which do not check materially, split, or warp when subjected to high temperatures, such as southern yellow pine. Steaming periods should be no longer than necessary to assure the required penetration, and in no case should they exceed that allowed in the specifications for the individual species. Steaming is applied to Douglas fir and to hardwood species only when the material is to be given a salt treatment.

1.33. *Heating in Preservative at Atmospheric Pressure.* This method is used when the purpose is to precondition the wood to facilitate penetration of preservative under pressure, on the assumption that the removal of appreciable quantities of water from the wood is not necessary to obtain the desired penetration of preservative. It is used chiefly in the conditioning of all types of material of Douglas fir and of hardwood species.

* Values for crossties, poles, posts, lumber and timber omitted. See AWPA specifications.

1.34. *Heating in Oil under Vacuum.* For conditioning unseasoned material of Douglas fir and most hardwood species which split, check, or warp when subjected to high temperatures. To condition material properly in this manner plants should be equipped with adequate condensers and measuring tanks. This procedure is not applicable when material is to be given a salt treatment.

1.4. *Peeling.* Bark on material to be treated retards the penetration of preservative. For proper treatment no strip of inner bark wider than ½ in. should remain.

1.5. *Machining.* Wherever possible, the material to be treated should be completely manufactured in its final form to obviate any necessity for cutting into treated wood. The preboring of holes to accommodate fastening devices ensures the introduction of preservative to the surrounding wood.

1.6. *Incising.* Woods which are difficult to penetrate and exposed heartwood resistant to treatment should be incised.

2. *Treatment*

2.1. For best results the treatment of material should be in accordance with the standards of the American Wood-Preservers' Association.

2.2. *Processes*

2.21. *Empty-cell Processes.* When the objective of the treatment is to obtain as deep and uniform penetration as possible with the retention of preservation stipulated, either the Lowry or Rueping process should be specified.

2.22. *Full-cell Processes*

2.221. *Bethell Process.* For material of any species, type, or condition to be treated with oil where a maximum retention of preservative is desired, and used chiefly for the treatment of marine piling and timbers.

2.222. *Burnett Process.* For material of any species, type, or condition to be treated with zinc chloride where it is possible to properly air-season or kiln-dry the wood after treatment.

2.23. A description of each process is in Standards C1 and M5.

3. *Results of Treatment*

3.1. *Retentions.* The retentions given in the treating standards are those for the use requirements designated. When the material is to be used where conditions are especially favorable to leaching or rapid attack by decay or other organisms, retentions higher than the minima should be stipulated.

3.2. *Penetrations.* The penetrations listed in the treating standards are those consistent with the retentions given.

4. *Preservatives*

4.1. *Creosote.* Creosote is the standard wood preservative for all uses where a colorless, odorless, and paintable product is not required. It has a high degree of permanence and is highly toxic to fungi, insects, and marine borers.

4.11. Creosote should be purchased in accordance with Standard P1.

4.12. Preservative oils for nonpressure treatments should be purchased in accordance with Standards P1 and P7.

4.2. *Creosote Solutions.* Creosote is often mixed with coal tar or petroleum to decrease the cost, increase the water-repelling properties, and retard the evaporation of the preservative in hot climates. Treatment with such mixtures or solutions is adapted for ties, piles, posts, and structural lumber, except that creosote petroleum mixtures should not be used in marine structures.

4.21. Creosote coal tar solutions should be purchased in accordance with Standard P2.

4.22. Petroleum for blending with creosote should be purchased in accordance with Standard P4.

4.23. Creosote-petroleum solutions should be purchased in accordance with Standard P3.

4.3. *Water-borne Preservatives.* Chromated zinc chloride and Tanalith are the standard wood preservatives for use where a colorless, odorless, and paintable product is required. They are forced into the wood in water solutions. They are highly toxic to fungi and insects, but will leach out of the wood faster than creosote in localities where the wood is in contact with wet soil or subject to periodic wetting. Material treated with zinc chloride should be kiln- or air-dried after treatment and before installation.

4.31. Water-borne preservatives should be purchased in accordance with Standard P5.

4.4. *Oil-borne Preservatives.* Pentachlorophenol and copper napthenate are standard oil-borne wood preservatives. They are forced into the wood in oil solutions. They are highly toxic to fungi and insects and are insoluble in water, so are resistant to leaching. They are not recommended for use in coastal waters.

4.41. Oil-borne preservatives should be purchased in accordance with Standard P5.

4.42. Petroleum for use with oil-borne preservatives should be purchased in accordance with Standard P9.

5. *Inspection*

The inspection of preservative treatment should be in accordance with Standard M2.

6. *Care after Treatment*

6.1. To assure best results it is necessary to protect treated material from mechanical injury both in handling and under service conditions. Cutting of treated material should be avoided whenever possible.

6.2. The field treatment of material after treatment should be in accordance with Standard M4.

Standard for Preservative Treatment by Pressure Processes—All Timber Products—Specification C1 of the American Wood-Preservers' Association, 1960

1. *General Requirements*

1.1. The following requirements, except as modified or supplemented by the other Commodity Standards, for the various species and types of material, apply to each of the treating processes and to all species and types of material. If these requirements are to be otherwise modified to meet special conditions, complete detailed instructions shall be given.

Maximum time duration (total elapsed time of a treating phase), maximum temperature, and maximum pressure limits shall not be exceeded. A phase shall begin when a change in conditions within the cylinder is initiated and shall end when either new conditions are imposed or the cylinder is emptied of preservative.

The minimum time duration stipulated for each phase shall be the period of time after the minimum condition has been attained and until the end of that phase.

1.2. *Plant Equipment.* Treating plants shall be equipped with the thermometers and gages necessary to indicate and record accurately the conditions at all stages of treatment, and all equipment shall be maintained in acceptable, proper working condition. The apparatus and chemicals necessary for making the analyses and tests required by the purchaser shall also be provided by plant operators, and kept in condition for use at all times.

1.3. *Conditioning.* Material shall be conditioned by air-seasoning, by kiln drying, by steaming, by heating in the preservative either at atmospheric pressure or under vacuum, or by a combination of them as agreed upon, in such a manner as will not cause damage for the use intended. Ice-coated or frozen material may be steamed prior to conditioning or treatment for a total period not to exceed 2 hr; the temperature shall not exceed 240°F.

1.31. When air-seasoning is used, it shall be done, as far as practicable, according to Standard M1.

1.32. When steam-conditioning is used, material shall be steamed in the cylinder at the temperature specified for the individual type of material or species, but in any case the maximum temperature specified shall not be reached in less than one hour. The cylinder shall be provided with vents to relieve it of air and insure proper distribution of steam. A vacuum may be drawn to remove the air prior to the introduction of steam. After steaming is completed, a vacuum as specified for the individual type of material or species, shall be created in the cylinder. The cylinder shall be drained continuously or frequently enough to prevent condensate from accumulating in sufficient quantity to reach the wood. Before the preservative is introduced, the cylinder shall be drained of condensate.

1.33. When conditioning by heating in the preservative is used the preservative shall cover the material in the cylinder. The temperature of the preservative during the conditioning period shall not exceed the maximum specified for the individual type of material or species.

If a vacuum is drawn during the conditioning period it shall be of sufficient intensity to evaporate water from the material at the temperature of the preservative. The intensity of the vacuum, or the temperature of the preservative, or both, shall be adjusted so as to regulate the evaporation of the water satisfactorily. The conditioning shall continue until the material is sufficiently heated and enough water removed to permit proper penetration. The preservative shall be removed from the cylinder and air admitted before an empty-cell process is applied.

1.4. *Sorting and Spacing.* Whenever it is practicable the material in any charge shall consist of pieces of the same species similar in form and size, moisture content and receptivity to treatment, and so separated as to insure contact of treating medium with all surfaces.

1.5. *Machining.* So far as practicable, all adzing, boring, chamfering, framing, gaining, surfacing, trimming, etc., shall be done prior to treatment. Gaining and boring bolt holes and step holes shall be permitted after treatment on poles with 100 per cent sapwood penetration, provided the surfaces of such gains and holes are protected as described in Standard M4.

1.6. *Incising.* Woods which are difficult to penetrate should be incised. When required or recommended in subsequent standards, material shall be incised prior to treatment by a method that will provide at least the minimum penetration specified without damage and with the least loss in strength with the exception that incising shall be waived when it will make the material unfit for the use intended.

2. *Treatment*

2.1. *Manner of Treatment.* The material shall be impregnated with preservative by a combination of such processes and under such conditions as will produce a satisfactory product for the use intended.

2.11. *Oil Treatment.* Following the conditioning period, the material shall be treated by an empty-cell process whenever practicable, unless otherwise specified, to obtain as deep and uniform penetration as possible with the retention of preservative stipulated. Material shall be treated by the full-cell process only when the maximum net retention is desired and where pressure is held to refusal, or when the stipulated retention is greater than can be obtained by the use of an empty-cell process.

2.12. *Salt Treatment.* Following the conditioning period, the material shall be treated by an empty-cell process or by the full-cell process. The treating solution shall be of uniform concentration and no stronger than necessary to obtain the required retention of preservative with the largest volumetric absorption practicable with the process used.

2.2. *Standard Processes*

2.21. *Initial Air Pressure or Vacuum.* Initial conditions shall be applied prior to filling and shall be maintained while the cylinder is being filled with preservative.

2.211. *Empty-cell.* Material shall be subjected to atmospheric air pressure (Lowry) or to higher air pressures (Rueping) of the necessary intensity and duration.

2.212. *Full-cell.* Material shall be subjected to a vacuum of not less than 22 in. at sea level for not less than 30 min either before the cylinder is filled or during the period of heating in preservative.

2.213. When refusal treatment is specified, material shall be treated by the full-cell process; subsequent pressure and temperature conditions shall be maintained as recommended in Standard M1, paragraph 3.1.

2.22. *Pressure Period.* Pressure shall be increased to at least the minimum but not higher than the maximum specified and shall be maintained until the desired volumetric injection has been obtained. Pressure shall be reduced to atmospheric either before or while the cylinder is emptied of preservative. A vacuum of not less than 22 in. at sea level shall be created and maintained until the wood can be removed free of dripping preservative, except that a vacuum need not be used after a full-cell or refusal treatment when the maximum possible retention is desired.

2.221. *Temperature of Preservative.* The temperature of the preservative during the entire pressure period shall not exceed the maximum temperatures but shall average at least the minimum average temperatures specified below:

	Average, °F	Maximum, °F
Creosote or creosote solutions:		
Western red cedar	180	190
All other species	180	210
Oil-borne preservatives:		
Southern, ponderosa, jack, red, and lodgepole pines	140	210
All other species	180	210
*Water-borne preservatives:**		
Acid copper chromate (ACC)	...	120
Ammoniacal copper arsenite (ACA)	...	150
Chromated copper arsenate (CCA)	...	120
Chromated zinc arsenate (CZA)	...	120
Chromated zinc chloride (CZC)	...	140
Copperized chromated zinc chloride (CuCZC)	...	140
Fluor chrome arsenate phenol (FCAP)	...	140

* A list of trade names for water-borne preservatives is shown in Standard M9.

2.23. *Expansion Bath.* When permitted by Standards C2, C3, C4, C5, C6 or C11, an expansion bath may be applied after pressure of an oil treatment is completed and before removal of preservative from the cylinder, by quickly reheating the oil surrounding the material to the maximum temperature permitted by the individual species specification, either at atmospheric pressure or under vacuum, the steam to be turned off the heating coils immediately the maximum temperature is reached. The cylinder shall then be emptied speedily of preservative, and a vacuum of not less than 22 in. at sea level created promptly and maintained until the wood can be removed from the cylinder free of dripping preservative.

2.24. *Final Steaming.* At the completion of an oil treatment, material may be cleaned by final steaming as specified for the individual type of material or species.

3. *Results of Treatment*

3.1. *Retention of Preservative.* Unless otherwise specified, the net retention in any charge shall be not less than 90 per cent of the quantity of preservative that may be specified; but the average retention by the material treated under any contract or order of 5 charges or more and the average retention of any 5 consecutive charges shall be at least 100 per cent of the quantity specified, except when the character of the wood in any charge makes these requirements impracticable, despite treatment to refusal, in which latter case allowance shall be made for the difference between the quantities of preservative specified and retained. When the contract or order comprises less than 5 charges the average retention shall be not less than 95 per cent of the quantity specified. The amount of preservative solution retained shall be determined from readings of working-tank gauges or scales, or from weights before and after treatment of loaded trams on suitable track scales, with the necessary corrections for changes in moisture content. The retention of preservative shall be calculated after correcting the volume of preservative solution retained to 100°F. Corrections of volume or specific gravity shall be made using the factors contained in Section F.

As an alternate method, preservative retention in a specified zone of the treated product may be determined by extraction or analysis as described in Section A, Analysis Method. Retentions determined by extraction or analysis and specified in subsequent standards shall be minimum retentions and no tolerances shall apply.

3.11. *Creosote and Creosote Solutions.* The retention of creosote or creosote solutions shall be calculated as pounds of the preservative as defined in Standards P1, P2 and P3.

3.12. *Oil-borne Preservatives.* The retention of oil-borne preservative shall be calculated as pounds of the preservative as defined in Standard P8. Concentration of preservative in solution shall be determined by analysis in accordance with Standard A5. Unless otherwise specified the concentration of pentachlorophenol in the treating solution shall be between 4.5 and 5.5 per cent, by weight; the petroleum used shall meet the requirements of Standard P9 for heavy solvent.

3.13. *Water-borne Preservatives.* The retention of water-borne preservative shall be calculated as pounds of the preservative as defined in Standard P5. Concentration of preservative in solution shall be determined by analysis in accordance with Standard A2.

3.2. *Penetration.* The penetration shall be specified by the purchaser in accordance with use requirements, but shall not be less than that specified for the individual type of material or species.

3.21. *Determination of Penetration.* Penetration shall be determined by boring a representative number of pieces that are well distributed throughout each charge as specified for the individual type of material and species.

3.22. *Plugging Penetrance Test Holes.* All holes made for determining penetration of preservative shall be filled with tight-fitting, treated cylindrical plugs.

3.3. *Condition of Material.* When minimum retentions are specified for creosote, creosote solutions, or oil-borne preservatives, material shall be supplied reasonably free of exudate and surface deposits. Such surface conditions cannot be required for heavier retentions of these preservatives. Material treated with water-borne preservatives and fire retardants shall be supplied free of excessive dust. All material shall be processed in such a manner as to prevent damage and minimize degrade.

4. *Preservatives*

The preservative used shall be whichever of the following standards of the American Wood-Preservers' Association is stipulated:

4.1. *Creosote and Creosote Solutions*

4.11. Creosote—Standard P1.

4.12. Creosote-coal tar solutions—Standard P2.

4.13. Creosote-petroleum solution—Standard P3.

4.2. *Oil-borne Preservatives*

4.21. Pentachlorophenol—Standards P8 and P9.

4.3. *Water-borne Preservatives**

4.31. Acid copper chromate (ACC)—Standard P5.

4.32. Ammoniacal copper arsenite (ACA)—Standard P5.

4.33. Chromated copper arsenate (CCA)—Standard P5.

4.34. Chromated zinc arsenate (CZA)—Standard P5.

4.35. Chromated zinc chloride (CZC)—Standard P5.

4.36. Copperized chromated zinc chloride (CuCZC)—Standard P5.

4.37. Fluor chrome arsenate phenol (FCAP)—Standard P5.

5. *Inspection*

Inspection of material for conformity to the requirements of this specification shall be in accordance with American Wood-Preservers' Association Standard M2, Standard Instructions for the Inspection of Preservative Treatment of Wood.

6. *Re-treatment*

6.1. Material not conforming to the stipulated minimum requirements may be re-treated and may be re-offered for acceptance, under the following conditions:

6.11. Material shall not be re-treated more than twice.

6.12. The limits for conditioning and treatment stipulated in this Standard, C1, and in the other Commodity Standards, shall not be exceeded during re-treatment.

6.13. When material is re-treated in a charge with untreated material, the volume of the re-treatable material shall not exceed 10 per cent of the total volume of the charge, and in the computation of the required minimum net retention of preservative, all material in the charge shall be considered as untreated.

6.14. When a charge as a whole is re-treated, the total retention as a result of all treatments shall be sufficient to satisfy the specified requirements for both net retention and penetration.

6.15. When a charge made up of pieces rejected for insufficient penetration only is re-treated, the amount of preservative injected during re-treatment shall be sufficient to produce the required penetration.

* A list of trade names for water-borne preservatives is shown in Standard M9.

The table below is part of Standard C3, which starts on the following page.

	Southern pine, ponderosa pine		Pacific Coast Douglas fir		Oak	
1.3. Conditioning	Air-seasoning or steaming or heating in the preservative or a combination		Air-seasoning or heating in the preservative or a combination		Air-seasoning or heating in the preservative or a combination	
1.32. Steaming			Not permitted		Not permitted	
Temp., °F, max	259					
Duration, hr, min	6					
Max	18					
Vacuum						
Inches at sea level, min	22					
Duration, hr, min	Optional					
1.33. Heating in preservative						
Temp., °F, max	220		Seasoned: 210° and 6 hr. Green or partially seasoned: 220° and no time limit		220	
Duration, hr, max	Optional				Optional	
2. *Treatment*						
2.22. Pressure, lb., min	125		75		150	
Max	200		150		200	
2.23. Expansion bath						
Temp., °F, max	220		220		Not permitted	
2.24. Final steaming	Not permitted		Not permitted for piles for coastal waters		Not permitted	
Temp., °F, max			240			
Duration, hr, max			3			
3. *Results of Treatment*						
3.1. Retention, lb per cu ft, min						
Creosote and creosote solutions	*General Use*	*Coastal Waters*	*General Use*	*Coastal Waters*	*General Use*	*Coastal Waters*
Creosote						
by gauge (or scale weight)	12	20	8, 10[a]	12, 14, 16[a] full-cell	10	Refusal—full-cell
by extraction		18				
Creosote–coal tar	12	20	8, 10[a]	Not recommended	10	Refusal—full-cell
Creosote—petroleum	Not recommended		8, 10[a]	Not recommended	Not recommended (*Note:* Above retentions apply to red oak only. White oak to be treated to refusal for all uses.)	
Oil-borne preservatives						
Pentachlorophenol	Not approved		0.40, 0.50[a]	Not recommended	Not approved	
3.2. Penetration in inches or per cent of sapwood, min	3.5 or 90 for 12 to 16 lb retentions 4.0 or 100 for 17 to 24 lb retentions		¾, ⅞, and 1 for 12-, 14-, and 16-lb retentions, respectively, when treated by the full-cell process. For 8-lb creosote and 0.40-lb pentachlorophenol retentions, ¾. For 10-lb creosote and 0.50-lb pentachlorophenol retentions, ¾ and 85 up to a maximum of 1⅝		White oak—100% sapwood Red oak—65% of annual rings: charges of recalcitrant wood with less penetration may be accepted if wood is conditioned properly before treatment and if treatment is continued to refusal.	
3.21. Determination of penetration	A borer core shall be taken midway between the butt and top of each pile in each charge. Only those piles meeting the penetration requirements shall be accepted		A borer core shall be taken midway between the butt and top of each pile in each charge. Only those piles meeting the penetration requirements shall be accepted.[b]		A borer core shall be taken approximately 4 ft from the butt end of each pile in each charge, and only those piles meeting the penetration requirements shall be accepted	
4. *Preservatives*	All standard preservatives listed above		All standard preservatives listed above		All standard preservatives listed above	

[a] The higher retentions and corresponding penetrations are recommended for severe service conditions.

[b] Effective penetration as measured for piles to be used in coastal waters must be continuously black and concentrated with both summerwood and springwood penetrated.

Standard for the Preservative Treatment of Piles by Pressure Processes—Specification C3 of the American Wood-Preservers' Association, 1960*

Note: This Standard, C3, is to be used in conjunction with the Association Standard C1, "Standard for Preservative Treatment by Pressure Processes—All Timber Products," and C1 is hereby made a part of this Standard. On succeeding pages are given Specific Requirements for the treatment of piles of the following species:

1. *Specific Requirements*

1.1. Piles shall be treated in accordance with the requirements of American Wood-Preservers' Association Standard C1, "Standard for Preservative Treatment by Pressure Processes—All Timber Products" except as modified or supplemented by the table shown on page 622.

Standard for Creosoted-wood Foundation Piles—Specification C12 of the American Wood-Preservers' Association, 1951

General. A foundation pile is one which is entirely embedded in the ground and capped with masonry and should be pressure creosoted. Pacific Coast Douglas fir, southern yellow pine and red (Norway) pine are acceptable for this purpose.

1. *Specific Requirements*

1.1. Foundation piles shall be treated in accordance with the requirements of American Wood-Preservers' Association Standard C1 "Standard for Preservative Treatment by Pressure Processes—All Timber Products" except as modified or supplemented in paragraphs below.

2. *Treatment.* Foundation piles should be treated in accordance with Standard C3.

3. *Retention.* Not less than 12 lb per cu ft of wood.

4. *Preservative.* Creosote.

6. *Care after Treatment.* Pile heads, after cutoffs are made to final elevation, shall be brushed liberally with two coats of hot creosote, followed by the application of a coat of coal tar pitch. There shall be sufficient interval between applications to permit absorption of each coat before the succeeding one is applied.

Standard for Pressure Treated Piles and Timbers in Marine Construction—Specification C18 of the American Wood-Preservers' Association, 1959

1. *Specific Requirements*

1.1. Piles and timbers in marine construction shall be treated in accordance with the requirements of American Wood-Preservers' Association Standard C1 "Standard for Preservative Treatment by Pressure Processes—All Timber Products," except as modified or supplemented by the table below.

1.5. Pile cutoffs, bolt holes and field cuts shall be protected in accordance with AWPA Standard M4.

1.51. The lower substructure bracing timbers shall be attached to the piles at a minimum height of 3.5 ft above mean low water for marine structures at sites where the tide range is 6 ft or less, and at middle elevation for tidal ranges exceeding 6 ft.

* Sections covering jack pine, lodgepole pine, red pine, and western larch not quoted.

3. *Results of Treatment*

3.1. Retention-lb per cu ft—min.

	Coastal Waters		Fresh Water or ground contact		AWPA Manual Standards
	Creosote	Creosote–coal-tar solution	Creosote	Creosote–coal-tar solution	
Round timber piles:					
Southern yellow pine......	20	20	12	12	C3
Red pine.................	...	...	12	12	C3
Douglas fir...............	14	14	12	12	C3
Red oak.................	...	...	12	12	C3
Timber (substructure exposed to tides, running water or wave action)					
Southern yellow pine......	16	16	12	12	C2
Douglas fir...............	12	12	10	10	C2
Gum.....................	...	...	10	10	C2
Red oak.................	...	...	10	10	C2

Note: Sections on treatments for substructure members out of water but subject to salt-water splash, and superstructure members, not quoted here.

Standard Instructions for the Care of Pressure-treated Wood after Treatment—Specification M4 of the American Wood-Preservers' Association, 1954

1. *General Requirements*

1.1. Insofar as practicable, all adzing, boring, chamfering, framing, gaining, incising, surfacing, trimming, etc., shall be done prior to treatment.

1.2. To prevent damage to the treated portions, cant hooks, peavies, pickaroons, and end hooks shall not be used on the side surfaces of treated material. All handling of treated piles, poles, ties, lumber, or timbers with pointed tools shall be confined to the ends only.

1.3. When pressure-treated materials have been accidentally damaged, or when it has been absolutely necessary to cut or bore into them after treatment in such a way as to expose, or nearly expose, the untreated wood, such injuries, cuts or holes shall be carefully field treated with hot preservative solution by brushing, spraying, or dipping, in the manner described below, so as to minimize, as far as possible, the danger of decay, insect, or borer attack.

1.31. Holes bored in pressure-treated material shall be poured full of hot preservative per Paragraphs 1.33 and 1.35. Horizontal holes, such as those for sway brace bolts, may be filled by pouring creosote into them through a bent funnel. The use of equipment to apply creosote under pressure in holes bored in the field is recommended. Holes shall not be bored or spikes driven into piles to support scaffolding, and such holes and spikes in lumber or timbers for temporary use shall be held to a minimum. Holes bored in treated material and not used for bolts shall not be left open, but shall be poured full of hot preservative and plugged with tight-fitting treated plugs.

1.32. All pressure-treated material that has been damaged or cut into after treat ment shall have the exposed surfaces covered with at least two coats of hot preservative solution immediately after cutting. Where maximum protection is desired, two coats of preservative shall be applied, followed by a heavy application of coal tar pitch, flashing cement, or other sealers.

1.33. Preservatives for field treatments should be the same as initially used and conform to the same specification except that creosote meeting the requirements of P1 or P7 may be used on material initially treated with creosote solutions. The strength of oil-borne preservatives in solutions for field treatments should be the same as or stronger than that initially used. The strength of salt solutions should be a concentrate not less than three times and not greater than five times the strength of the original treating solution.

1.34. Preservative solution for field applications should be obtained from the treating plant at the time the treated wood is purchased. Coal tar pitch or other sealers must be purchased elsewhere.

1.35. Creosote or creosote solutions shall be heated to temperatures from 150 to 200°F, before application. Coal tar pitch should be heated as necessary for ease in applying. Proprietary sealers should be applied in accordance with the manufacturers' recommendations. Salt or oil-borne preservatives shall not be heated in excess of the temperatures given in Paragraph 2.4 of Standard C1.

2. *Specific Requirements*

2.1. *Piles.* Immediately after making the final cutoff, the cut area shall be given two applications of hot creosote followed by a heavy application of coal tar pitch, flashing cement, or other sealer. Piles should be cut square except in the case of piles to be capped with masonry, cutoffs should be further protected by the application of two thicknesses of tar-saturated fabric over the cutoff, and overlapping the sides of the pile at least two inches. The overlap shall be folded down along the sides and glued in place with coal tar pitch, flashing cement, or other sealer. The fabric shall then be coated with one coat of coal tar pitch, flashing cement, or other sealer. Pressure treated piles must not be dapped in the field for sway bracing; piles of uniform size should be selected for each bent and, where necessary, pressure treated filler blocks used to fill in between piles or caps and sway bracing. A satisfactory plastic compound for the protection of piles at cutoff may be made with 10 to 20 per cent of creosote and 90 to 80 per cent pitch.

2.11. Pile cutoff protection: A sheet metal ring of 12 gauge metal 4 inches in height shall be lightly driven into the pile at final cutoff to an oil tight seat. The diameters of the rings shall be slightly less than that of the pile, so that the untreated center of the pile is enclosed by the ring. The ring shall then be filled to a depth of at least 2 in. with hot creosote and left in place until the creosote has been absorbed by the pile through end penetration. After use the ring shall be removed for reuse.

2.12. An alternate method consists of encircling the pile at final cutoff with roofing felt, or thin metal tightly banded to the pile forming a cup extending 3 in. above the point of cutoff. This cup is filled with hot creosote to a depth of 2 in. and left in place until the oil has been absorbed through end penetration.

NOTE: Paragraph on Ties omitted.

SPECIFICATIONS FOR PILES JACKETED BY PNEUMATICALLY APPLIED CONCRETE OF BEN C. GERWICK, INC.

1. *Materials*

a. Wood piles shall be in accordance with or equal to Navy Specification No. 39P14a, Type B (Douglas fir), Class I (untreated) with particular attention to requirements for straightness and bends. Piles shall be rough peeled.

b. Cement shall be portland cement which conforms to the specifications of the ASTM or to Navy Specification No. SS-C-191b as amended.

c. Sand employed shall be clean, sharp, and reasonably free from clay, loam, and silt. Grading shall be as follows:

U.S. Standard Sieve No.	*Per Cent Passing (by Weight)*
4	96–100
8	75– 90
16	55– 75
30	35– 50
50	10– 17
100	0– 5

d. Reinforcement shall be galvanized electrically-welded wire fabric of 12-gage (0.1055 in. diam.), 2- by 2-in. mesh, in accordance with ASTM A185-37.

2. *Preparation of Piles*

a. Prior to placing reinforcing mesh, piles shall be notched or keyed, and all bark and objectionable knots removed. Notches or keys, 1 in. deep and 4 in. square, shall be cut in rows on three points of the circumference of the pile. Notches 1 in. deep and 4 in. in diameter may be used in lieu of squares at the option of the Contractor. Notches of either style in each row shall be 6 ft apart except for the top 15 ft where they shall be 3 ft apart. The rows shall be arranged in such a manner that the notches will be staggered on the face of the pile.

b. Immediately prior to encasement, piles shall be thoroughly wetted and all dirt or other materials that would impair the bond of the concrete jacket shall be removed.

3. *Reinforcement Installation*

a. Wire mesh shall be held away from the piles by metal spacers at fifth points on the circumference of the pile, starting from a point not more than 2 in. above the lower end of the reinforcing, and extending at 18-in. intervals to a point not more than 2 in. below the upper end of the reinforcing. Metal squares 1 by 1 in., 16 gage (0.0625 in.) in thickness, approximately and secured with 2.25-in. wire staples or combined spacer and fasteners of the same material and approved design (or both types), may be used at the Contractor's option.

b. Mesh shall be placed not less than ¾ in. nor more than ⅞ in. from the surface of the pile.

c. At points of splicing mesh shall be lapped not less than 2 squares or 4 in., whichever may be greater.

d. The mesh shall conform as nearly as possible to the natural lines of the pile.

e. Extra reinforcement of 8-gage (0.162 in. diam) steel wire shall be placed at the upper and lower ends of the mesh. The wire shall be installed by spiral wrapping over the mesh starting at the extreme ends of the mesh and wrapping toward the opposite end, a distance of not less than 8 in., using not less than 6 complete wraps of wire.

4. *Concrete and Its Application*

a. Concrete shall be placed pneumatically with machines of approved design. Only experienced nozzlemen shall be employed. No nozzleman shall be deemed experienced unless he has performed a minimum of 3 months' work of a similar nature and has proved his qualifications to the satisfaction of the Engineer.

b. Concrete mixture measured by volume shall be one part cement to four parts fine aggregate, mixed dry. With a normal rebound of 40 to 50 per cent this will give a 1:3 mix on the pile. "Dry" as applied to the sand means that it shall have a normal moisture content of 3 to 6 per cent by weight. Sand with an upper limit of 10 per cent moisture by weight may be used at the option of the Contractor if the use of a suitable and approved type of heater can be demonstrated to produce a satisfactory mixture leaving the nozzle. A satisfactory mixture shall be one similar in all respects to that produced with less than 6 per cent moisture. No lumps, balling of the mixture, or burning of the cement will be allowed.

c. Rebound may be reused only when screened, counted as sand, and remixed with cement in the aforementioned proportions.

d. Capacity of equipment shall be such that a minimum of 1½ yd of mix per hour is placed per pile, exclusive of rebound.

e. The concrete shall be applied in layers of thickness that will not sag or separate. If succeeding layers are necessary, the concrete shall have attained its initial set and shall be cleaned thoroughly with water and air blast.

f. Any spalling or chipping of the concrete jacket which occurs during or after encasement shall be repaired by and at the expense of the Contractor. The area around the defect shall be tapered or chamfered for a distance of 3 in. in all directions, thoroughly cleaned of loose material, and the entire area reshot.

g. The top 3 to 5 ft at the butt end of the wood pile shall be left uncoated until after driving.

5. *Support and Handling of Piles*

a. Piles shall be supported horizontally while concrete is being applied in a manner to produce a uniform, straight, finished product.

b. While the concrete is being applied to piles and during curing, where any portion of the pile to be jacketed is located between supports, the maximum span between supports shall be 40 ft.

c. Where pile is supported within the length of the gunite jacket, the support shall be such as to prevent dislocation of the wire mesh and spalling of the gunite. Encasement at the point of support shall be equal in all respects to that in the remainder of the jacket.

d. If pile is rolled during application of gunite, care shall be exercised so that at no time will the pile being treated receive a jerk, twist, or sudden movement that might disturb the encasement already in place.

e. Piles shall not be picked up or rolled during the period of time starting 1 hr after encasement is complete and extending 7 days. However, piles may be skidded during this period.

6. *Curing*

a. Concrete jackets shall be cured by a continuous moisture blanket provided by sprinklers or burlap wrapping regularly dampened. An approved membrane type of cure (Hunt process "clear" or equal) may be substituted at the option of the Contractor.

b. Curing by either method shall commence not later than 90 min after the jacket is complete. If water curing is used, it shall be continuous for 7 days.

Specifications for Manufacture and Driving of Precast Concrete Piles, of the Portland Cement Association

Materials

1. *Portland Cement*

Portland cement shall comply with the "Standard Specifications for Portland Cement" (ASTM C150) or the "Tentative Specifications for Air-entraining Portland Cement" (ASTM C175) and shall be Type ______.

NOTE: These specifications cover the types of Portland cement listed below and provide that when no type is specified, the requirements of Type I or Type IA are to govern.

The letter A after type number designates air-entraining portland cement. Attention is called to the fact that cements conforming to the requirement for Type IV and Type V are not usually carried in stock. In advance of specifying their use, purchasers or their representatives should determine whether these types of cement are, or can be made, available.

Type I or IA—for use in general concrete construction when the special properties of other types are not required.

Type II or IIA—for use in general concrete construction exposed to moderate sulfate action or where moderate heat of hydration is required.

Example: concrete exposed to sulfates where the concentration is not unusually severe, or heavy sections of concrete placed in warm weather.

Type III or IIIA—for use when high-early-strength is required.

Type IV—for use where a low heat of hydration is required. *Example:* large masses of concrete such as gravity dams.

Type V—for use when high sulfate resistance is required. *Example:* concrete exposed to soils or waters of high alkali content encountered in a few locations, principally in some western states.

2. *Concrete Aggregates*

(*a*) Concrete aggregates shall conform to the "Standard Specifications for Concrete Aggregates" (ASTM C33). Where aggregates conforming to these specifications are not obtainable, aggregates that have been shown by test or actual service to produce concrete of the required strength, durability, and watertightness may be used where authorized by the Engineer.

(*b*) Maximum size of the aggregate shall be not larger than 1½ in. nor more than three-fourths of the minimum clear spacing between reinforcing bars.

3. *Water*

Water used in mixing concrete shall be clean, and free from deleterious amounts of acids, alkalis, or organic materials.

4. *Metal Reinforcement*

(*a*) Metal reinforcement shall conform to the requirements of the "Standard Specification for Minimum Requirements for the Deformations of Deformed Steel Bars for Concrete Reinforcement" (ASTM A305) and of the "Standard Specifications for Billet-steel Bars for Concrete Reinforcement" (ASTM A15), or of "Standard Specifications for Rail-Steel Bars for Concrete Reinforcement" (ASTM A16).

(*b*) Cold-drawn wire for concrete reinforcement shall conform to the requirements of the "Standard Specifications for Cold-Drawn Steel Wire for Concrete Reinforcement" (ASTM A82).

5. *Storage of Materials*

Cement and aggregates shall be stored at the work in such a manner as to prevent deterioration or intrusion of foreign matter. Any material which has deteriorated or which has been damaged shall not be used.

CONCRETE QUALITY

6. *Concrete Quality*

The concrete shall contain not more than 6 gal of water per sack of cement, including the surface moisture carried by the aggregates. Piles subjected to sea water or other severe exposure shall contain not more than 5½ gal of water per sack of cement. Moisture in the aggregates shall be determined by methods which will give results within 1 lb per 100 lb aggregate, and proper deduction shall be made in the amount of water added to each batch.

7. *Tests on Concrete*

(*a*) During the progress of the work, compression tests shall be made in accordance with the "Standard Method of Making and Storing Compression Test Specimens of Concrete in the Field" (ASTM C31) Each test shall consist of one laboratory control cylinder and one field control cylinder.

(*b*) At least one test shall be made for every 25 piles cast and not less than one test shall be made for any one day's operation.

(*c*) The standard age of test shall be 28 days, but 7-day tests may be used, provided that the relation between the 7- and 28-day strengths of the concrete is established by test for the materials and proportions used.

(*d*) In all cases where the average strength of the laboratory control cylinders shown by these tests falls below an ultimate compressive strength of 3,500 psi, the Engineer shall have the right to order a change in the mix or in the water content for the remaining piles. In cases where the average strength of the cylinders cured on the job falls below the required strength, the Engineer shall have the right to require conditions of temperature and moisture at the job necessary to secure the required strength.

(*e*) In the event that the Engineer changes the water content specified, adjustment, covering amount of cement and aggregates affected, will be made as an extra or a credit under the provisions of the contract.

8. *Concrete Proportions and Consistency*

(*a*) The proportions of aggregate to cement for any concrete shall be such as to produce a mixture which will work readily into the corners and angles of the forms and around reinforcement with the method of placing employed on the work, but without permitting the materials to segregate or excess free water to collect on the surface. The combined aggregates shall be of such composition of sizes that when separated on the No. 4 standard sieve, the weight passing the sieve (fine aggregate) shall not be less than 30 per cent nor greater than 50 per cent of the total unless otherwise required by the Engineer.

(*b*) The methods of measuring concrete materials shall be such that the proportions can be accurately controlled and easily checked at any time during the work. Measurement of materials for ready-mixed concrete shall conform to the "Standard Specifications for Ready Mixed Concrete" (ASTM C94).*

(*c*) The slump of the concrete shall be not greater than 4 in. when concrete is placed by hand, or 2 in. when placed by vibration.

MIXING AND PLACING CONCRETE

9. *Preparation of Equipment and Place of Deposit*

(*a*) Before placing concrete, all equipment for mixing and transporting the concrete shall be cleaned, all debris and ice shall be removed from the forms, which shall be thoroughly wetted (except in freezing weather) or oiled, and the reinforcement shall be thoroughly cleaned of ice or other coatings.

10. *Mixing of Concrete*

(*a*) The concrete shall be mixed until there is a uniform distribution of the materials and shall be discharged completely before the mixer is recharged.

* Wherever practicable the Engineer should require measurement by weight rather than by volume.

(*b*) For job-mixed concrete, the mixer shall be rotated at a speed recommended by the manufacturers and mixing shall be continued for at least 1 min after all materials are in the mixer.

(*c*) Ready mixed concrete shall be mixed and delivered in accordance with the requirements set forth in the "Standard Specifications for Ready Mixed Concrete" (ASTM C94).

11. *Conveying*

(*a*) Concrete shall be conveyed from the mixer to the place of final deposit by methods which will prevent the separation or loss of the materials.

12. *Depositing*

(*a*) When concreting is once started, it shall be carried on as a continuous operation until the pile is completed, beginning at the head and working toward the point of the pile. The top surface shall be screened and brushed to a uniform even texture similar to that produced by the forms. No concrete that has partially hardened or been contaminated by foreign material shall be deposited in the forms, nor shall retempered concrete be used.

(*b*) All concrete shall be thoroughly compacted by vibrating and/or spading and rodding during the operation of placing and shall be thoroughly worked around reinforcement and into the corners of the forms.

(*c*) The frequency of vibrators shall be not less than 3,600 per min. The intensity of vibration shall be sufficient to cause the concrete to flow and settle into place, and to make the effect on the concrete visible over a radius of at least 2 ft. Vibrators shall be applied at points not over 2 ft apart and there shall be an average of not less than 20 sec of vibration per foot of pile. In general, vibration shall be of sufficient duration to accomplish thorough compaction and complete embedment of reinforcement. To secure even and dense surfaces free from honeycomb, vibration shall be supplemented by spading or rodding by hand while concrete is plastic under the vibrating action.

13. *Protecting and Curing*

Side forms may be removed 24 hr after concrete is placed, provided the concrete has hardened sufficiently.

Provision shall be made for maintaining the surfaces of concrete made with normal portland cement moist for at least 7 days and for that made with high-early strength portland cement at least the first 3 days after the placement of the concrete, or until the concrete has attained a compressive strength of 2,500 psi as shown by test cylinders under like curing conditions.

14. *Cold-weather Requirements*

(*a*) Adequate equipment shall be provided for heating the concrete materials and protecting the concrete during freezing or near-freezing weather. No frozen materials or materials containing ice shall be used.

(*b*) All concrete materials and all reinforcement and forms with which the concrete is to come in contact shall be free from frost. Whenever the temperature of the surrounding air is below 40°F, all concrete placed in the forms shall have a temperature of between 70°F and 80°F, and adequate means shall be provided for maintaining a temperature of 70°F for not less than 3 days after placing except when high-early strength portland cement or concrete is used the temperature shall be maintained at not less than 70°F for 2 days, or for as much more time as is necessary to ensure proper curing. The addition of salt or other chemicals to the mix for the prevention of freezing shall not be permitted.

FORMS AND REINFORCEMENT

15. *Design of Forms*

Forms may be of wood or metal and shall conform to the shape, lines and dimensions of the pile as called for on the drawings, and shall be substantial and sufficiently tight to prevent leakage of mortar. They shall be properly braced or tied together so as to maintain position and shape.

16. *Cleaning and Bending Reinforcement*

Metal reinforcement, at the time concrete is placed, shall be free from rust scale or other coatings that will destroy or reduce the bond. Bends for ties shall be made around a pin having a diameter not less than two times the minimum thickness of the bar. All bars shall be bent cold.

17. *Placing Reinforcement*

Metal reinforcement shall be accurately placed in accordance with the plans and shall be adequately secured in position by concrete or metal chairs and spacers. A minimum of 1½ in. of concrete shall be provided over all reinforcing, except that for piles subjected to sea water or other severe exposure 3 in. of concrete shall be provided.

18. *Splices and Offsets in Reinforcement*

Splices of reinforcement at points of maximum stress shall generally be avoided. Splices shall provide sufficient lap to transfer the stress between bars by bond and shear.

HANDLING AND DRIVING

19. *Marking*

Each pile shall be stamped or marked with the date of its manufacture. Lifting points indicated on the drawings shall be plainly marked.

20. *Handling*

Piles shall be handled carefully to avoid dropping or severe jarring while in a horizontal position. Piles shall be handled only when the concrete has obtained a strength of 3,500 psi as determined by field control test cylinders as required in section 7, unless special provision is made for handling to reduce the stresses in proportion to strength of the concrete.

21. *Driving Cap*

Piles shall be protected with an approved cushion and cap while being driven.

22. *Hammer*

A steam hammer shall be used unless other equipment is permitted by the Engineer. The weight of the striking parts of the hammer shall not be less than one-third the weight of the pile.

23. *Leads*

Piles shall be secured against lateral movement during driving by leads or other suitable means.

24. *Jetting*

Piles may be driven with a hammer and water jets or by water jets alone. Jet pipes may be either separate or cast in the pile. An ample supply of water at adequate pressure shall be provided. Piles shall be driven by a hammer alone for the last 5 ft of penetration.

25. *Penetration*

Piles shall be driven to depths or to penetrations per blow as directed by the Engineer. Accurate records of the penetration per blow for the last foot shall be kept for the guidance of the Engineer in determining allowable loads on the pile. Where driving is interrupted before final penetration is reached, the record for penetration shall not be taken until after at least 12 in. penetration has been obtained on resumption of driving.

26. *Replacing*

Any pile so injured in driving or handling that its structural integrity as a pile under the conditions of use is impaired, shall be replaced by a new pile, or the injured part replaced by splicing or otherwise repaired as directed by the Engineer.

27. *Alignment*

Unless otherwise called for on the plans, piles shall be driven as nearly as possible in plumb position. Any pile so out of line or plumb as to impair its usefulness shall be pulled and redriven or an additional pile driven as directed by the Engineer.

28. *Splicing*

Heads of piles shall be cut off or the concrete stripped from the reinforcement as directed by the Engineer. If proper resistance to driving is not attained at contemplated level of cutoff, the driving shall be continued and an additional length of pile required shall be supplied by splicing in such a way as to develop the full strength of the section of the pile.

29. *Test Piles*

Test piles shall be of the same size and materials as the permanent piles and shall be driven with the same equipment and in the same manner as specified for such pile. Test piles shall be driven in advance of final driving of permanent piles so that lengths for casting may be determined. During driving, an accurate record of the penetration shall be kept. Load tests shall be made with equipment approved by the Engineer. Compensation for driving and loading test piles shall be made at a unit price agreed upon in the contract.

Suggested Specifications for Pretensioned Prestressed Concrete Piles*

1. *General*

(*a*) Piles, which may be either solid or hollow, shall be cast as monolithic units of high-strength concrete stressed with high-tensile cold-drawn steel strands.

(*b*) Cored holes shall be concentric with the pile section. Forms shall be secured against both flotation and sinking, with deflections not to exceed ¼ in.

(*c*) Piles may be cured in controlled steam chambers at temperatures not exceeding those set by the Portland Cement Association for prestressed concrete, or cured by covering with wet burlap kept saturated.

* Based in part on suggested specifications of Ben C. Gerwick, Inc., which, in turn, were based partly on *Criteria for Prestressed Concrete Bridges* by the Bureau of Public Roads.

(*d*) The head of the pile shall be a plane normal to the pile axis.

(*e*) Solid heads and tips may be provided for hollow piles, provided the ends of the hollow core are rounded.

2. *Concrete*

(*a*) Type II cement shall normally be used for pretensioned piling exposed to sea water. Calcium chloride shall not be used, except for a trace in admixtures used to accelerate strength gain and reduce shrinkage. Type I cement may be used in piles where low alkali content is not required.

(*b*) Grading and composition of fine and coarse aggregates shall conform to the latest ASTM specifications and be proportioned to produce a smooth, dense, workable mixture of high strength and low shrinkage. Maximum size of aggregate shall be as large as may properly be placed.

(*c*) Concrete shall be compacted by high-frequency internal or external vibrators, keeping contact with strands to a minimum.

(*d*) The full amount of tensioning shall be maintained until concrete cylinders, cast and cured under the same conditions as the piles, indicate a strength of at least 3,500 psi.

(*e*) Concrete shall have a minimum strength of 5,000 psi at 7 days if steam-cured or at 28 days if cured by other means and shall contain between 6 and 8 bags of cement per cu yd.

3. *Strand*

(*a*) Strand shall be of the ungalvanized 7-wire high-tensile cold-drawn type, stress-relieved as a unit after the wires have been formed into a strand. Strand properties shall conform to tables as published by an acceptable recognized manufacturer of strand.

(*b*) Strands shall be free of dirt, oil, grease, and loose rust and be accurately positioned. Minor surface corrosion from brief exposure to the weather is permissible.

(*c*) Strands shall be held in position and stressed uniformly by hydraulic jack having an accurate calibrated gage to permit strand stress to be computed at any time. Elongation shall be measured at completion of stressing, prior to pouring concrete, and shall conform to elongation tables furnished by the strand manufacturer.

(*d*) Transfer of prestress shall be accomplished by means of gradual release of all strands by the hydraulic jack. Strands shall not be cut until prestress has been transferred.

4. *Reinforcing Steel*

(*a*) All other reinforcing steel shall meet the standards of the ASTM specifications, and shall be accurately positioned.

(*b*) Cold-drawn wire spirals or square ties shall be used throughout the pile, with close spacing near head and tip. Adequate lateral ties are essential.

5. *Design Factors*

(*a*) Initial stress in the strand, prior to pouring concrete, shall not exceed 70 per cent of minimum ultimate strength.

(*b*) Loss in initial prestress due to creep and shrinkage of concrete and creep in steel shall, under normal conditions, be assumed not less than 16 per cent of the initial tensioning force.

(*c*) Average working stress in the strand shall not exceed 60 per cent of the ultimate strength. Working stress and working force will be considered as those remaining in the strand after creep and shrinkage have taken place.

(*d*) Minimum spacing shall be three times strand diameter measured on centers, nor the clear distance less than 1½ times the maximum size of coarse aggregate. Minimum cover shall be 2 in. for piles in water and 1½ in. for piles in soil.

6. *Handling and Driving*

(*a*) Piles may be moved when the compressive strength has reached 3,500 psi, but are not to be driven until 4,500 psi.

(*b*) Piles shall be lifted or supported only at the designated points and handled and driven in such a manner as to avoid excessive bending stresses, cracking, or spalling.

(*c*) Pile heads shall be protected from direct impact of the hammer by a cushion head block so arranged that strands or bars projecting from the pile head will not be displaced or deformed during driving.

(*d*) Jetting will be permitted or required where necessary to reach the desired depth.

(*e*) Piles shall be cut off where necessary. Piles driven below cutoff grade may be extended as shown on the plans or in a manner approved by the Engineer. In cutting off, a circumferential cut shall first be made with a diamond saw to prevent spalling.

(*f*) Minimum energy per blow of the hammer to be used will be established by the Engineer.

Suggested Specifications for Rotary-drilled Caissons of the Case Foundation Co.

1. *Scope*

The work covered by this section of the specifications consists in furnishing all plant, labor, equipment, appliances and materials, and in performing all operations in connection with the installation of drilled caissons, complete, in strict accordance with this section of the specifications and the applicable drawings, and subject to the terms and conditions of the contract.

2. *Installation Equipment*

Caissons shall be formed by means of power-driven rotary bucket type foundation drilling rigs.

3. *Caisson Installations*

(*a*) All caissons shall be installed from the ground surface as existing after top soil has been removed and general excavation work completed. The maximum variation of the center of any caisson from the required location shall be 1 in. for 18-in.-diameter shafts and up to 3 in. for 48-in.-diameter shafts at the ground surface, and no caisson shall be out of plumb more than 1 per cent. If these tolerances are exceeded, proper additional construction as required by the engineer shall be provided without additional cost to the owner.

(*b*) A shaft of the diameter specified shall be drilled from the ground surface to the required depth. Material excavated by drilling shall be removed from within the area of operation and disposed of as directed by the engineer. A protective casing shall be installed in the hole, if necessary, for the protection of personnel and to prevent cave-ins and displacement of earth, and for the retention of groundwater. Before any concrete is poured, the bottom of the hole shall be cleaned of mud and any extraneous matter and dewatered by pumping for inspection purposes.

(*c*) Concrete having ultimate strength of 3,000 psi at 28 days is to be used.

(*d*) All concrete materials conform to the applicable requirements of Section ______ "Concrete."

(*e*) The caissons shall be filled with concrete as specified below and care shall be taken to maintain a sufficient head of concrete to prevent reduction in diameter of the caisson by earth pressure on the fresh concrete. The concrete in any caisson shall be stopped at the cut-off elevation indicated on the drawings within a tolerance of 1 in. If the cutoff elevation is above the ground elevation from which the caisson is installed, the caisson shaft shall be extended to the cut-off elevation by suitable means approved by the engineer.

(*f*) The excavation of any caisson or group of caissons having been completed, the concrete shall be placed in a manner that will prevent separation of its constituent materials. Where any inflow of water from the bottom or sides of the excavation is encountered, the concrete shall be placed through still water by means of a tremie or bottom-dump bucket to a height sufficient to permit the balance of the concrete to be placed as specified below. All other concreting shall be done by an approved method which will provide a continuous flow, without segregation, from bottom to top of caisson. When any caisson is concreted in such manner as to allow part of the concrete to set for more than one hour before the balance is placed, the resulting stoppage surface shall be left approximately level. Such surface shall then be thoroughly cleaned of all laitance, roughened (if required) and slushed with a 1 to 1 cement grout before more concrete is placed.

(*g*) If steel cylinders are used for lining, they may be withdrawn as the concrete is deposited, in which case a sufficient head of concrete shall be maintained to insure that no extraneous material enters the concreted caisson.

(*h*) A complete report of each caisson installed shall be made for the engineer on forms furnished by him. The report shall contain all dimensions, location of caisson, elevation of bottom and top as actually poured, and any other pertinent data.

(*i*) The contractor shall state in his proposal as a unit price the cost per lineal foot of casing for providing casings in place and removing same as concrete is poured. The base bid shall not include any casings.

(*j*) If rock or boulders are encountered which cannot be removed by the standard caisson drilling procedure or if the character of the obstruction necessitates the use of hand labor and/or air tools, it is agreed that the contractor will proceed on a time and material basis until such time as he can return to the regular drilling methods.

Suggested Addendum

Belled Caissons. Under this option concrete belled caissons shall be carried to hardpan as defined by the Local Building Code. The minimum bearing capacity of the hardpan on which the caissons are to rest shall be determined by tests as provided under Section ______ of the Local Building Code. If this option is taken, the contractor shall provide a complete design of alternate foundations, which shall be approved by the owner before work is started. The cost of the redesign of alternate foundations and caissons, including the load test specified herein for each building, shall be borne by the contractor without any additions to the contract price.

Suggested Specifications for Steel Bearing Piles of the United States Steel Corporation

Description

All piles shall be United States Steel Co. rolled steel CBP sections or CB sections of the section number, size, and weight per lineal foot as indicated on the plans. Piles shall conform at time of driving to camber and sweep as permitted by allowable mill tolerances.

Material

The material in rolled steel piles and splices shall be standard structural grade open-hearth steel, conforming to ASTM Standard Specifications for Bridges A7.

Where a considerable portion of the pile structure is exposed to the atmosphere, such as in the case where a pile also forms part of a trestle bent, it is well to add the following clause to the steel material specification: "All steel shall have a copper content of 0.2 per cent minimum."

Pile Lengths

Pile lengths for estimating purposes, as shown on the plans, are based upon probable lengths remaining in place in the completed structure. The Engineer will determine the final lengths of piles required to develop the bearing values specified for the minimum penetrations acceptable, or to develop both the bearing value and minimum penetration, by means of tests specified herein.

Test Piles

Test piles shall be driven under the observation of the Engineer.

Option (*a*) The contractor shall drive (here state approximate lengths indicated by test boring, soil data, local conditions, etc., for example: 3—35 ft lengths and 2—45 ft lengths of steel test piles) at the points indicated on the plans, or where and as directed by the Engineer. They shall be driven at such points that they may be left in place, cut off, and become a part of the permanent structure. From their performance under driving, the Engineer will determine the lengths of piles required.

Or (*b*) The Engineer shall select length and number of test piles on which loading tests shall be conducted in accordance with (here describe method, test loadings, and permissible net settlements). From the results of the pile tests, the Engineer will determine the lengths of piles required.

Or (*c*) The piles shall be driven to obtain a bearing power of ______ tons based on the use of the following formula (here state formula) and if indicated on the plans, to a minimum depth of penetration of (depth shown).

Splicing

Should it become necessary or desirable to splice the piles, the splices shall be made in accordance with details shown on the plans.

Welding

In the case of welded connections, splices, etc., all work shall be done with approved methods, materials, and experienced personnel whose ability and qualifications to do acceptable welding shall be fully demonstrated to the satisfaction of the Engineer.

Bolting

Details of bolted splices contemplate the use of milled ends of the sections of piles. Where the ends are cut by other means, the bolted splices must be proportioned to develop the full capacity of the pile on its entire cross-sectional area, in order to withstand the forces encountered in driving.

If permitted by the Engineer in writing, the rough ends of pile sections may be brought into contact by means of adequate deposits of weld metal, applied between butt ends of adjacent sections after sections have been bolted in alignment.

Pile Caps

All piles shall be cut off at elevations shown on plans and capped as indicated on the structural details. The ends of the piles shall be cut off level and surface made as smooth as practical before cap is welded in place.

Pile Bent Bracing Members

Structural-steel sway and cross bracing, and other channel and angle-tie bracing, shall be placed on steel pile bents where indicated on plans. All of this bracing material shall be welded in place as shown on plans. Where piles are not driven in the exact position and to the alignment specified, it will be necessary to use fills and shims between the bracing and the flanges of the piles. All fills and shims required to square and line up faces of flanges for cross bracing shall be furnished and placed by the contractor without cost to the owner. Weight of fills and shims used on piles will not be included in the weight of any item of structural steel paid for by the Owner.

Driving

Steel piles, including test piles, shall be driven with steam or air hammers, developing an energy per blow of not less than 7,250 ft-lb, nor more than 15,000 ft-lb. (The preceding stipulations as to hammer energy may be lowered for the lightest sections and increased for very heavy sections.) The hammer shall be operated at all times at steam or air pressures, and the speed recommended by the manufacturer.

An accurate record shall be kept of the date, time, total depth of penetration, rate of penetration and number of blows for every foot penetration under last five blows of hammer, steam or air pressure, and kind and size of hammer used in the driving. Any unusual phenomena shall also be recorded.

A cast- or structural-steel driving head shall be used for driving steel piles, if required, to keep the pile heads from upsetting excessively under extremely hard driving conditions.

Piles shall be driven as nearly as possible in the exact position specified on design plans; however, a maximum deviation of 1½ in. from exact position will be permissible in combination pile and trestle bents, and a maximum deviation of 3 in. from design plan position will be allowed for all piles in footings of piers or abutments. These deviations from required location will be permitted only if the Contractor at his own expense widens the footing so that the minimum distance from face of pile to face of cap is not less than 3 in. Care shall be taken during driving to prevent and correct any tendency of the steel piles to twist or rotate.

The rows of piles around the perimeter of the bases or footings shall be driven before those in the middle. Piles shall be driven and sunk vertically or to the batters shown by the plans. For batter piles, the pile driver leads shall be inclined so as to be in line with the desired position of the piles.

Piles shall be driven for their full effective length. Excavation below the bottom of concrete footings will not be permitted.

Pile drivers shall have firmly supported leads extending down to the lowest point the hammer must reach; short leads suspended from a line and braced only by lines will not be acceptable unless the piles are rigidly braced and held in alignment by suitable guide frames. Underwater hammers may be used only where held in rigid leads extending to full depth.

Jetting shall not be permitted unless special permission has been given in writing by the Engineer.

If the material be such that cavities remain about the piles after driving, the cavities shall be filled with sand or other approved material, deposited with water.

Protective Encasement

Concrete

(*a*) After driving piles, an encasement of (here specify dense concrete mix) vibrated concrete ______ in. thick measured from the exterior projection of steel surfaces shall be placed, extending ______ ft above and ______ ft below grade or mean high or low water level.

Or (*b*) If permitted by the Engineer, in writing, the Contractor may apply a gunite concrete to the piles before driving. This gunite must be reinforced with an approved mesh wrapped around the pile and spaced at least 1 in. away from the steel surfaces. The protection shall be applied so that after driving is completed it will extend over the full lengths of zone or zones for which a protective encasement is specified.

Combination Types

These should be specified by a comprehensive description of the methods desired and by reference to complete details on the drawings.

Basis of Payment

Use suitable paragraph from following group:

(*a*) Payment for steel bearing piles will be considered as completely covered by the contract price per pound for the steel piling in place, which will include all material, tools, equipment, labor, and work incidental thereto. Estimates will be made on the basis of the theoretical weight of metal in beam sections which remain as a permanent part of the structure after cutoff is removed, and will not include weight of cutoff, splices, caps, or other miscellaneous material. Payment for cutoffs, or other waste material, and for splices including all material necessary for making splices in accordance with design details shall be considered as completely paid for under estimates made in accordance with the preceding paragraphs.

(*b*) Payment for steel bearing piles will be made at a price per lineal foot for the length of pile extending below the bottom of the cutoff line indicated on the plans. The length of pile ordered shall provide suitably for variation of depth, and for cutoffs and the part of pile cutoff will be paid for at the cost per lineal foot of the pile delivered at the site before driving as determined by the Engineer. The prices per lineal foot applied to piles driven on a batter shall be the same as piles driven vertically. The prices so paid for piles shall cover and include all material, plants, equipment, tools, and work incidental to furnishing and preparing the permanent structure.

Splices will be paid for at the price of (here state price per splice), which price shall include full compensation for furnishing all materials, labor, tools, and equipment required to construct one splice in accordance with the details shown on the plans. No splices will be permitted in piles less than 25 ft in length, and not more than one splice will be permitted in any pile under 40 ft long.

(*c*) Payment for steel bearing piles will be made at a price per lineal foot for the length of pile ordered into the driving leads by the Engineer. Balance of this paragraph made up of suitable clauses from (*a*) or (*b*).

(*d*) Measurement of pay quantities for furnishing steel piles will be based on the lengths of piles ordered by the Engineer, and no deductions will be made for cutoff lengths. Balance of this paragraph made up of suitable clauses from (*a*) or (*b*).

(*e*) Payment for (here state number) pile tests as specified in section or paragraph ______ shall be considered as covered by the contract price for piling in place.

(*Or*) Payment for pile tests as specified in section or paragraph ______ shall be made at the rate of (here state lump sum) for test for each complete test ordered by the Engineer.

Suggested Specification Clauses Appertaining to Various Types of Piles

General Clauses for All Types of Piles

The piles (or pipes, pile-forming apparatus, timber tips, lower pipe sections, or wood piles, as the case may be, depending on what is the driven item) shall be driven to an indicated working load of not less than ______ tons, by a formula or graph provided or approved by the Engineers, after the type of pile and the driving equipment have been selected. Pile tips shall be driven down to at least Elevation ______, or shall extend ______ ft into ______. (In the case where piles are not desired to break through a distributing bed of firm material overlying soft material, state: Pile tips shall not extend closer than ____ ft to the underside of the ______.) (In the case of composite piles having wood lower sections, state: The top of the wood section shall not be higher than Elevation _____.)

A wood cap or cushion block shall be a one-piece hardwood block with grain parallel to the pile axis and enclosed in a close-fitting steel housing. A steel plate of 2-in. minimum thickness shall be used on top of the wood block to distribute the hammer blows. If capblock material other than wood is used, it shall be at least equal to wood in transmitting hammer energy to the pile. The use of wood chips, small wood blocks, or other materials that allow excessive loss of energy is prohibited. Capblock material shall not be replenished just before or during measurements of final driving resistance.

After driving casings for cast-in-place concrete piles, they shall be inspected internally and the length checked, and, if satisfactory, inspected again just before concrete is placed. Prior to pouring concrete, 1:2 cement grout shall be deposited in the pile point to a depth of approximately 1 ft. The piles shall then be filled with concrete having an ultimate strength at 28 days of not less than 2,500 psi. At least 6 bags* of portland cement shall be used per cubic yard of concrete. The concrete shall have sufficient slump to drop into place properly, but shall not contain more than 6 gal† of water per bag of cement, including the water contained in the aggregates. Concrete shall be deposited through a steel funnel having a discharge diameter of approximately 8 in. (*This paragraph applies only to piles containing concrete poured in the field.*)

During the first day of driving, a pile shall be formed using quick-setting cement. This pile shall be tested to a load 50 per cent in excess of the working pile load called for in the specifications. The test shall be made within 3 days after driving the first pile. (*This paragraph to be used at the discretion of the Engineer.*)

Driving Tolerance Clauses (for Use Where Applicable to Type)

The maximum bow shall not exceed one-half of the outside diameter of the pile, nor be greater than $\frac{1}{200}$ times the length. The deviation from the vertical shall not exceed 2 per cent on any section of the length for vertical piles. For batter piles, the deviation from the axis shall not exceed 4 per cent of the length in any section. (These requirements have been often specified, but sometimes all cannot be applied to wood or H piles and therefore might be reconsidered for cast-in-place piles, after studying the character of the soil strata, mathematical analysis, and load tests on questionable piles.)

Reduction of the gross cross-sectional area of the pile shall not exceed 10 per cent.

The tops of all piles shall be within 6 in. of the locations shown on the design drawings.

* U.S. bags of 94 lb.

† U.S. gal.

An elevation shall be taken on each pile as soon as it has been driven, and elevations shall again be taken on all piles after completion of driving. Any pile found to have heaved shall be redriven to the original elevation, or to the satisfaction of the Engineer.

Rejected piles shall be pulled, left in place as is, or filled if open shells, as directed by the Engineers, and additional piles driven to replace them where directed by the Engineers, without additional expense to the Purchaser.

Precast Concrete Piles

Piles shall be designed for handling stresses and selected points of support when stored. Lifting loops shall be embedded (if specified).

Piles shall have a ______ cross section (Engineer to specify whether constant, tapered, or with point). Lengths shall be (specify).

Where piles bring up not over 6 ft below cutoff and do not number over 25 per cent of the total number in any footing, they may be spliced by clamping sectional or corrugated forms larger than the pile, cleaning exposed bars of proper length for bond, and placing vibrated concrete.

Cased Pedestaled Concrete Piles

The piles shall be of corrugated cased, pedestaled, cast-in-place type. The pedestal section shall be not less than 4 ft in length. The cased shaft shall not be less than 12 in. in diameter, the shell consisting of a corrugated-steel shell of 18-gage metal.

The piles shall be formed by driving into the ground an apparatus consisting of a heavy steel core and casing, removing the core, placing 4 ft of concrete inside the casing, replacing the core in contact with the concrete, drawing the casing up over the core, redriving the apparatus through the deposited concrete to form a pedestal, placing a corrugated shell in the casing, withdrawing the drive casing, and filling the shell with concrete. The shell may be filled before the casing is withdrawn. If filling of the entire shell is to be deferred some time, at least 8 cu ft of concrete shall be deposited immediately.

Compressed Concrete Piles

Piles shall be of the cast-in-place compressed concrete type. They shall be cylindrical in form and shall have a minimum diameter of 12 in.

In the formation of the pile, positive means shall be employed to assure that the concrete shaft will be continuous and of a minimum diameter greater than the diameter of the forming apparatus used.

Cased Concrete Piles with Compressed Base Section

Piles shall be of the cast-in-place concrete type, having a compressed concrete-base section and a cased concrete shaft. They shall be cylindrical in form and shall have a minimum diameter of 12 in. in the cased section and 16 in. in the compressed section.

In the formation of the compressed section, positive means shall be employed to assure that the concrete shaft will be continuous and of a minimum diameter greater than the diameter of the forming apparatus used.

The cased shaft shall be formed by placing within the driven casing a continuous corrugated metal shell 12 in. in diameter, made of 22-gage steel, and filling this shell with concrete of the same proportion specified for the base section.

Compressed Concrete Pedestal Piles

Piles shall be of the cast-in-place compressed concrete-pedestal type. They shall be cylindrical in form and shall have a minimum diameter of not less than ___ in.

In the formation of the pile, positive means shall be employed to assure that the concrete shaft will be continuous and of a minimum diameter greater than the diameter of the forming apparatus used.

A pedestal base shall be formed on each pile.

Compressed Concrete Piles with Mushroom Base

Piles shall be of the cast-in-place compressed concrete type. They shall be cylindrical in form and shall have a minimum diameter of 12 in.

In the formation of the pile, positive means shall be employed to assure that the concrete shaft will be continuous and of a minimum diameter greater than the diameter of the forming apparatus used.

A mushroom base shall be formed on the piles.

Composite Piles, Cased Concrete and Wood

Piles shall be of the composite wood and concrete type. They shall have:

1. *A Lower Wood Section.* This section shall be a wood pile which shall be not less than 6 in. at the tip and 12 in. at the butt before forming of the tenon. The wood section may be cypress, yellow pine, fir, or mixed hardwoods. Piles shall be sufficiently straight so that a straight line drawn from the center of the butt to the center of the tip shall lie wholly within the pile. Piles shall be free from dry rot and from short bends.

2. *A Splice.* The splice shall be made by forming a tenon at the head of the wood pile, 18 in. long, 9 in. at the shoulder, and 8 in. at the top. Tenons shall be measured and cut accurately, using a saw, in order to obtain equal and even bearing on the shoulders and top of the wood section. This tenon shall be wrapped with No. 9 wire at ¾-in. spacing, well stapled in place. Surrounding the tenon and extending 18 in. into the upper concrete section, there shall be placed a reinforcing cage consisting of six vertical ⅜-in. bars and a 3⁄16-in. spiral on 4-in. pitch, or their equivalents. This reinforcement shall be wired in place in the shell, which will later act as a form for the concrete of the upper section. The section of the splice from the shoulder of the tenon to the head of the wood pile shall be filled with a grout consisting of one part of portland cement to two parts of sand by volume, immediately after driving. Provision must be made to prevent the concrete section from lifting off the wood section, due to upheaval of the ground or any other cause, before the concrete takes its initial set.

3. *An Upper Concrete Section.* The concrete section shall consist of a 22-gage continuous-sheet steel pipe having ½-in. deep corrugations, not less than 13 in. in diameter, or a laminated spirally-wound fiber shell, which shall be filled above the head of the wood pile with concrete. If fiber shells are used, they shall be filled with concrete immediately.

The pile shall be formed in such a way as to exclude water and mud from the splice and shell of the upper section at all times without recourse to pumping or siphoning.

After the pile is driven and before the concrete is deposited, the head of the wood section shall be examined with a light and if there is mud or water surrounding the tenon, the pile shall be rejected and a new pile driven.

To assure that the two sections of the pile shall be in proper alignment at the splice, at least 3 ft of the butt of the wood pile shall be within the guide casing at the time of the forming of the splice.

If the diameter of the cased section is less than that of the butt of the lower section, soil shall be jetted or sand sluiced, as directed by the Engineers, around the upper section, prior to driving adjacent piles.

Composite Piles, Projectile Type

Piles shall be of the projectile composite type. They shall have:

1. *A Lower Steel-pipe Section.* This section shall consist of one or more sections of 10 in. inside diameter, ¼-in. wall, steel pipe, the lower end of which shall be closed by means of a cast-iron point or steel plate. Where it is necessary to use more than one section of steel pipe, the joint between sections shall be made with a 10-in. cast-steel drive sleeve.

2. *A Splice.* The splice between the upper cased-concrete section and the lower steel-pipe section shall be made by carrying the shell of the upper section not less than 12 in. below the upper end of the pipe section and concreting the two sections as a unit. (Where reinforcing is used in the upper section, the length of the splice will be governed by the necessity of carrying the vertical bars of the reinforcing cage a minimum of 40 diam beyond the upper end of the steel pipe.)

3. *An Upper Cased-concrete Section.* The concrete section shall consist of a 12-in. diam 22-gage continuous-sheet steel pipe having ½-in.-deep corrugations. This shell and the projectile pipe section shall be filled with concrete.

The forming apparatus shall be first driven to a point such as to make the projectile section as short as is economically possible. The projectile section shall then be driven independently to a refusal of at least four blows of a No. 1 Vulcan hammer, or equivalent, to 1 in. of penetration.

The pile shall be formed in such a way as to exclude water and mud from the splice and shall be kept dry at all times without recourse to bailing and pumping.

To assure that the two sections of the pile shall be in proper alignment at the splice, at least 3 ft of the steel pipe section shall be within the guide casing at the time of the forming of the splice.

Button-bottom Cased Concrete Piles

The piles shall be of the cased, cast-in-place concrete type. The point of the pile shall be larger than the shaft of the pile and the driving casing. The outside diameter of the permanent shell shall be not less than 12 in.

After the pile has been driven to final bearing, the shell shall be filled with concrete.

The tops of the piles shall be cut off to the level shown on the plans.

Solid-point Steel-pipe Piles

Piles shall be of the solid-point steel-pipe type.

Pipe shall be of the diameters and wall thickness shown on the plans and shall be: welded or seamless steel pipe conforming to the latest revision of the ASTM Standard Specifications for Welded and Seamless Steel Pipe Piles (A252) for grade ____: spiral-seam welded pipe conforming to the latest revision of the ASTM Standard Specifications for Electric-Fusion (Arc)-Welded Steel Pipe (A139 for sizes 8-in. to but not including 30-in. for spiral-seam grade *B* pipe, or A134 for sizes 30-in. and over) except that hydrostatic testing is not required, the ASTM Standard Specifications for Spiral-Welded Steel or Iron Pipe A211 for pipe and the ASTM Standard Specification for Low and Intermediate Tensile Strength Carbon-steel Plates of Structural Quality A283 (grade *B*) for material, for the 6-, 8-, 10-, and 12-in. outside diameter pipe sizes only; or seamless steel pipe conforming to the applicable parts of the latest revisions of the API Specifications 5L for Line Pipe for grade ____ pipe.

The pipe shall (not) be thoroughly cleaned and coated before shipment with ____ coat of __________.

Field splices shall be of the self-tightening type, unwelded, with the outer surfaces coated with bituminous cement for watertightness if required by the Engineers, or

Split Chill Rings No. P-421 as manufactured by Wedge Protectors, Inc., Cleveland, Ohio, field welded. Shielded electric arc welding shall be used.

Pipe ends shall be square cut for driving and bearing surfaces, and beveled for welding ends.

The points shall be gray cast iron, cast steel, or steel plates of sufficient strength to prevent cracking under driving.

The tops of the pipes shall be cut off to the level shown on the plans.

Open-end Steel Pipe Piles

The piles shall be of the open-end steel pipe type.

The specifications for pipe are the same as given previously for solid-point steel pipe piles.

The pipe shall (not) be thoroughly cleaned and coated before shipment with ____ coat of __________.

Field splices shall be the same as specified for solid-point steel pipe piles. (In case of exceptionally long or large diameter piles, specify lengths of sleeves, or other type of connection desired.)

After the pipe has been driven to rock, all soil and other material shall be cleaned out of it. The empty pipe shall then be again driven to a refusal indicating a bearing value of ____ tons.

The pipe shall be examined by means of a light or a sounding rod to make sure that it has not crumpled at the bottom and that a full bearing has been obtained on rock.

The pipe shall then be filled with concrete. No concrete shall be dumped through water. If it is not found possible to exclude all water from the pipe, then the concrete shall be deposited by means of a bottom dump tremie.

The pipe shall be cut off to the level indicated on the plans. This cut must be clean and true so as to obtain a full contact between the pipe and the bearing plate.

The bearing plates will consist of steel plates of the dimensions shown on the plans. They shall have two 1½ in. in diameter grouting holes. They shall be grouted in place at the time the pile caps are poured.

Union Metal Monotube Piles

Piles shall be steel-encased concrete piles of the type wherein steel casings are driven, left in place, and filled with concrete. Casings shall be of such gage as to withstand driving and collapsing forces. Joints shall be welded or lock-seamed. Points shall be steel. Piles shall be (specify straight, tapered or combination, also diameters and lengths). Minimum tip diameter shall be 8 in.

Cased Mandrel-driven-shell Concrete Piles

Cased mandrel-driven-shell concrete piles shall have steel shells driven by use of an internal mandrel full length to the required bearing in direct contact with the soil. Shells shall be left in place and filled with concrete. Shells shall have sufficient strength to withstand driving to the required resistance without injury and to resist harmful distortion or buckling due to soil pressures. Shells shall be sufficiently water-tight to exclude water during placing of concrete. Piles shall be of constant diameter, uniformly tapered, or step-tapered. Minimum tip diameter shall be 8 in.

Drilled-In Caissons

Caissons shall be not over 1⅓ per cent out of plumb. The centers shall not vary from a line connecting top and bottom centers by over 0.5 per cent.

Splices shall be external sleeves at least 12 in. long. Ends of pipe sections shall have full bearing and be welded continuously to the sleeves. Splices in the structural steel core shall be formed by welded cover plates, and be strong enough to withstand driving. Butt ends shall be milled.

Pipe shall conform to ASTM A134.

Drilled Caisson Piles

Caissons shall be drilled by a power machine to neat diameters not over 1 in. from those specified, and not over 1 in. in 10 ft from the vertical.

A collar shall be used at the top to keep dirt from falling in.

Open Specification

Piles shall be of one of the following types: (list acceptable types, giving appropriate specification clauses for each).

In the event that the bidders wish to submit proposals on other types of piles or caissons than those specified herein, they shall communicate with the Engineers before preparing their proposals to determine if the proposed methods are satisfactory for this project.

INFORMATION WHICH SHOULD BE CONTAINED IN SPECIFICATIONS OR INVITATIONS FOR BIDS

In order to permit the bidders to submit intelligent proposals, avoid delays in obtaining bids, and obtain uniformity in the basis of bids, the following information should be stated:

Site. (1) Location of project, and of structure on site. (2) Railroad facilities, distance to structure, availability or restrictions on use. (3) Surface water and disposal, pumping in pits and excavations, by whom, and on what basis. (4) Overhead obstructions such as wires, bridges, pipes, or guys. (5) Underground obstructions such as pipes to be cleared, removed, or kept in service. (6) Distance to adjacent structure if close enough to affect pile-driving operations, design of piles, or stability of adjacent structure. If affected, give design or details of existing foundation. (7) If previous pile foundation was built on site, give details and method of handling. (8) A statement should generally be included to the effect that bidders shall inspect the site before submitting their proposals, and that ignorance of conditions will not be accepted as the basis of a claim for additional compensation.

Soil. (1) Information regarding borings or test pits, if available. (2) Groundwater level, with seasonal or long-term variations if known, and expected future low for basis of design.

Lines and Grades. (1) State who will furnish lines and grades, or the exact division of such work.

Facilities. (1) Is power available for power and lighting? If so, state voltage, alternating or direct current, cycles, location of outlet, limitations on use if any, and rate if not free. (2) Who furnishes night lighting if required, and lights for inspection of open piles? (3) Location of water outlet, limitations on use if any, and rate if not free. (4) Compressed air and steam if available, limitations on use if any, and rate if not free. (5) Watchman service, whether supplied or not. (6) Toilet and telephone facilities if available.

Permits. (1) State who obtains and pays for various permits required. (2) Boiler operator's license if required.

Insurance. (1) Types and amounts required. State if necessary to be carried in particular companies.

Structure. (1) Send drawings of pile location plan and of footings if available. Otherwise state dimensions of structure, column spacing and loads, maximum load per pile, and depth of excavation.

Completion. (1) State starting date. (2) State completion date required for all, or for various portions of the work. (3) State sequence of operations, if necessary.

Price. (1) State whether unit prices only are desired, giving basis of payment and measurement, or whether a lump-sum price with units for additions to or deductions from the scope of the work is wanted. Lengths of piles upon which bids are to be based should be stated, where practicable, with a provision for an adjustment in price for variation from such lengths. Whether or not this policy will generally result in a money saving is conjecturable, but it should impress contractors with its fairness, and reduce the probability of more cautious firms being underbid by inexperienced, reckless, or gambling concerns, with possibilities of delay, poor work, and litigation. (2) State whether prices are to be based on single shift, straight-time basis, or whether alternate prices are desired for overtime or additional rigs or additional shifts.

Special Conditions. State any other special conditions which might affect the price, starting or completion date, or price of the work.

Prequalification. A prequalification clause requiring the bidder to have had sufficient experience in installing the specified type of piling is suggested.

Tentative Method of Test for Load-Settlement Relationship for Individual Piles under Vertical Axial Load of the American Society for Testing Materials
(Designation: D 1143-57 T)

Scope

1. This method covers a procedure for testing individual vertical foundation piles to determine the relationship between the vertical load (applied at the top of the pile on the center or vertical axis of the pile) and the settlement of the pile.

Note 1. This method describes only a procedure for testing a single pile. It does not cover the application of the test results to the carrying capacity of a group of piles or to foundation design in general.

Apparatus

2. The apparatus required for the test shall consist of apparatus for applying known vertical loads to the top of the pile (Section 3) and apparatus for measuring the settlement of the pile (Section 4).

Loading Device

3. The apparatus for applying the vertical loads shall consist of one of the following devices:

(*a*) *Load Supported Directly by Pile.* A box shall be supported on top of the pile and loaded with earth, sand, cement, pig iron, or other suitable material. The construction of the box and the application of the loads shall be such that no lateral forces will be applied to the top of the pile and no impact will occur as the loads are placed. A suitable type of construction is shown in Fig. 1. Loads shall be so distributed that all wedges will remain loose as settlement occurs. The weight of the box shall be included in the calculated load on the pile and the supporting beams and the box shall be in place on the pile when the "no-load" reading is made. In cases where the test pile is in an excavation below the natural ground surface, an extension column of structural steel or steel pipe may be used to extend from the pile head up to

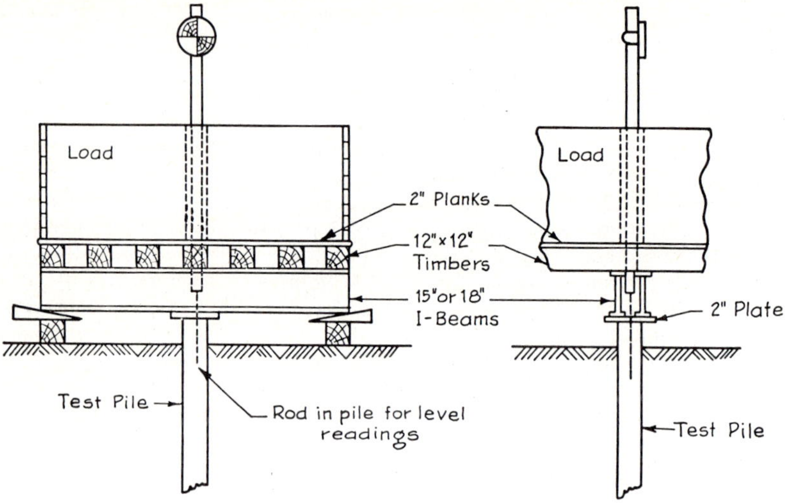

FIG. 1. Pile test with load from weighted box or platform supported directly by the pile.

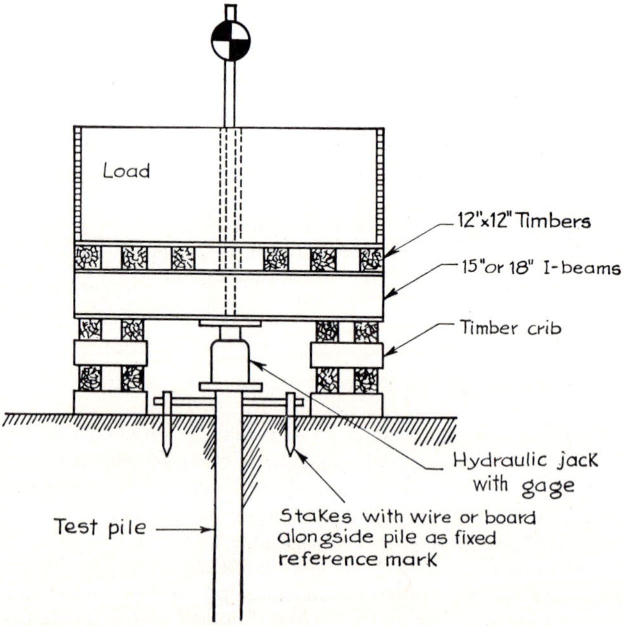

FIG. 2. Test pile with load from weighted box or platform applied to pile by means of a hydraulic jack.

the test box. Special precautions must be taken to avoid tilting of the box when heavy test loads are to be applied.

(*b*) *Load from Weighted Box or Platform Applied to Pile by Hydraulic Jack.* A test box or test platform resting on cribbing shall be constructed over the pile and loaded with earth, sand, cement, pig iron or other suitable material with a total weight greater than the anticipated maximum test load. A recently calibrated hydraulic jack with a pressure gage shall be interposed between the pile head and the load box, and load applied to the pile by operating the jack. A typical scheme for a test of this type is shown in Fig. 2.

(*c*) *Load Applied to Pile by Hydraulic Jack Acting against Anchored Reaction Member.* Two or more piles to be used as anchor piles shall be driven as far from the test

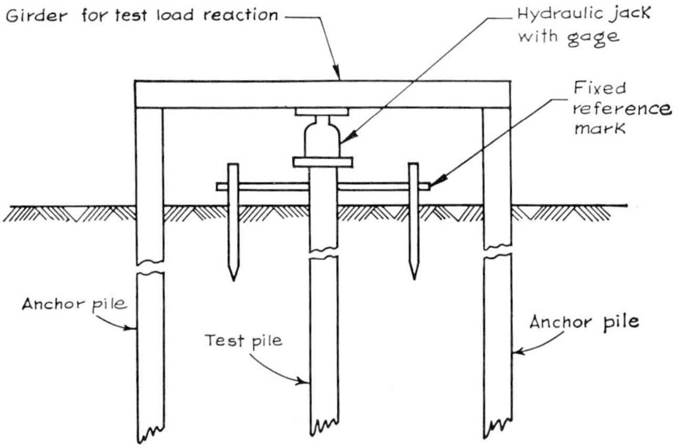

FIG. 3. Pile test with load applied to pile by means of a hydraulic jack acting against a reaction member held down by anchor piles.

pile as practicable. A girder of sufficient strength to act as a reaction beam shall be attached to the upper ends of the anchor piles. A recently calibrated hydraulic jack with pressure gage shall be interposed between the head of the test pile and the underside of the reaction beam and the test load applied to the pile by operating the jack. This general type of test is illustrated in Fig. 3.

Apparatus for Measuring Settlement

4. The apparatus for measuring settlement shall consist of one of the following devices:

(*a*) *Surveyor's Level and Target Rod.* A surveyor's level and target rod reading to 0.001 ft may be used as illustrated in Fig. 1. Two bench marks shall be established on permanent objects near the test pile location and settlements shall be determined by readings made on these bench marks and on the bolt or rod set in the pile head.

(*b*) *Wire and Scale.* A wire shall be stretched between two stakes each driven into the ground at a distance not less than 8 ft from the center line of the test pile. The wire shall pass across the face of a scale attached to the pile so that settlement readings can be made directly from the scale.

(*c*) *Dial Gage.* A beam shall be attached to two stakes each driven into the ground at a distance not less than 8 ft from the nearest point of the test pile. A dial gage with its stem resting on top of the pile or on lugs or similar reference points on the pile, shall be attached to this fixed beam to record the movement of the pile head. Alternatively, the gage may be attached to the pile with its stem bearing against the underside of the beam or against lugs attached thereto. While a dial gage may provide the greatest accuracy in consecutive settlement readings, a check settlement observation shall be made at selected intervals during the test by level rod or scale referred to a fixed elevation which is independent of the reference beam to which the dial gage is attached.

Procedure

5. (*a*) The head of the pile shall be cut off level or shall be capped in such a manner as to produce a horizontal plane bearing surface. A steel plate shall be set on top of the pile. If the method of loading is that shown in Fig. 1, the plate shall be provided with a hole in its center through which the head of the bolt or reinforcing rod may project to serve as a reference point for the level rod.

(*b*) The total test load shall be twice the anticipated working load on the pile and shall be applied in increments amounting to 25, 50, 75, 100, 125, 150, 175, and 200 per cent of the anticipated working load. Settlement readings made to an accuracy of 0.001 ft shall be taken before and after the application of each new load increment. Additional load shall not be applied until the rate of settlement under the previous increment is less than 0.001 ft in 1 hr or until 2 hr have elapsed, whichever occurs first. When loading has been completed, the full test load shall remain on the pile for 24 hr, or for a longer period if the necessity therefor is indicated by the rate of settlement of the pile, and settlement readings shall be taken during and at the end of that period. As an alternative method of loading, the specified load increments may be added in constant time intervals of not less than 30 min and preferably of 1 hr. Settlement readings shall be made immediately before and after the addition of each load increment and at not less than three specified times between load increments.

NOTE 2. Instead of applying only a predetermined amount of load to the test pile, it is recommended that, whenever practicable, loading of the test pile be continued until rapid progressive settlement occurs.

(*c*) During the unloading of the pile, the rebound shall be measured when the load remaining on the pile amounts to 75, 50, 25, 10, and 0 per cent of the full test load, with decrements of load released at not less than half-hour intervals, and with measurements of the rebound being made immediately before and after each decrement. The final rebound shall be recorded 24 hr after the entire test load has been removed.

Report

6. The report of the load test shall include the following information:

(1) A description of soil conditions at the location of the test pile,

(2) A description of the pile and its driving record, including the number of hammer blows per foot throughout the pile length and the final driving resistance in blows per inch for the last 3 in. of driving,

(3) A description of the hammer and its actual rate of operation during the driving of the test pile,

(4) A tabulation of the loads and settlement readings during the loading and unloading of the pile,

(5) A graphic representation of the test results in the form of a time-load-settlement curve, and

(6) Remarks concerning any unusual occurrences during the driving or loading of the pile.

BIBLIOGRAPHY

PILE-DRIVING THEORY AND PRACTICE

1. *Piles and Pile-driving*, edited by A. M. Wellington, Engineering News Publishing Co., New York, 1893.
2. E. P. Goodrich, "The Supporting Power of Piles," *Trans. ASCE*, vol. 48, 1902.
3. F. Redtenbacher, *Prinzipien der Mechanik und des Maschinenbaues*, Karlsruhe, 1852.
4. J. Benabencq, *Ann. ponts et chaussées*, vols. 5 and 6, 1911, and vol. 3, 1913.
5. J. Boussinesq, *Application des potentiels à l'étude de l'équilibre et du mouvement de solides élastiques*, Paris, 1885.
6. C. Terzaghi, "The Science of Foundations—Its Present and Future," *Trans. ASCE, Paper* 1704, 1929, pp. 270–405.
7. A Hiley, "Pile Driving Calculations with Notes on Driving Forces and Ground Resistance," *The Structural Engineer*, vol. 8, July and August, 1930.
8. J. S. Crandall, "Piles and Pile Foundations," *J. Boston Soc. Civil Engr.*, vol. 18, No. 5, May, 1931, pp. 176–189.
9. R. D. Chellis, "A Consideration of Pile Driving with Application of Pile Loading Formula," *J. Boston Soc. Civil Engrs.*, vol. 38, No. 1, January, 1941, pp. 1–30.
10. Progress Report of the Committee on the Bearing Value of Pile Foundations, "Pile-driving Formulas," *Proc. ASCE*, vol. 67, No. 5, May, 1941, pp. 853–866.
11. ——— Discussion by R. D. Chellis, vol. 67, No. 8, October, 1941, pp. 1517–1537.
12. ——— Discussion by H. T. Evans, vol. 67, No. 9, November, 1941, pp. 1784–1789.
13. G. C. Boonstra, "Eenige Beschouwingen over den Puntweerstand van Palen," *Ingenieur (Utrecht)*, vol. 55, No. 8, Feb. 23, 1940, pp. B33–B45.
14. C. Franx, "Proefbelasting van Heipalen te Rotterdam," *Polytech. Weekblad*, Nos. 23, 24, and 25, 1934.
15. F. M. Masters, "Timber Friction Pile Foundations," *Trans. ASCE, Paper* 2174, vol. 108, 1943, pp. 115–173 (including discussion).
16. I. F. Morrison, "The Fundamentals of Pile Foundations," *Eng. J.*, October, 1939, pp. 1–4.
17. G. Gilboy, "The Scientific Method in Earthwork," *Civil Eng.*, vol. 7, No. 12, December, 1937, pp. 827–830.
18. R. M. Miller, "Foundation Methods Involving Piling," *Civil Eng.*, vol. 2, No. 5, May, 1932, pp. 304–308.
19. M. K. Khalifa, "Model Studies of Pile Foundations," *Civil Eng. Research Laboratories, Bull.* 7, Columbia University, 1940.
20. T. C. Bruns, "Don't Hit Timber Piles Too Hard," *Civil Eng.*, vol. 2, No. 12, December, 1941, p. 726.
21. W. H. Glanville, G. Grime, E. N. Fox, and W. W. Davies, "An Investigation of the Stresses in Reinforced Concrete Piles During Driving," *Dept. Sci. Ind. Research (Brit.), Bldg. Research Tech. Paper* 20, 1938.
22. H. W. Bolin, "The Pile Efficiency Formula of the Uniform Building Code," *Building Standards Monthly*, vol. 10, No. 1, January, 1941.

23. V. Hansen and F. N. Kneas, "Static Load Tests for Bearing Piles," *Civil Eng.*, vol. 12, No. 10, October, 1942, pp. 545–547.

24. W. F. Swiger, "Foundation Tests for Los Angeles Steam Plant," *Civil Eng.*, vol. 11, No. 12, December, 1941, pp. 711–714.

25. R. M. Miller, "Soil Reactions in Relation to Foundations on Piles," *Trans. ASCE, Paper* 2004, vol. 103, 1938, pp. 1193–1236.

26. J. H. Gregory, R. A. Allton, and J. H. Blodgett, "Holding-down Power of Concrete Piles," *Civil Eng.*, vol. 3, No. 2, February, 1933, pp. 66–68.

27. "Driving and Loading of Concrete Test Piles at the Naval Supply Depot, Naval Operating Base, San Diego, Calif.," *Public Works of the Navy, Bull.* 36, October, 1927.

28. G. H. Gardner, "Practical Hints on Piling," *Soc. Engrs. (London), Trans.*, vol. 33, No. 2, July to December, 1942, pp. 68–84.

29. G. F. Stevenson, "Final Report—Morganza Floodway Bridge at Lottie, Louisiana, with Test Pile Data on Other Crossings of the Morganza Floodway," Louisiana State Highway Department, 1943.

30. O. S. Bray, "Power Plant Requires Variety of Foundations," *Eng. News-Record*, vol. 131, No. 27, Dec. 30, 1943, pp. 966–970.

31. E. J. Christenson, "Dam Foundation Compacted by Pile Driving," *Civil Eng.*, vol. 9, No. 2, February, 1939, pp. 92–93.

32. W. Z. Lidicker, "Pile Bearing Increased by Close Driving," *Eng. News-Record*, vol. 125, No. 5, Aug. 1, 1940, pp. 172–174.

33. A. Engelund, "A Special Foundation Method for Bridge Piers, Adopted in Danish Fjords," *Proc. Intern. Conf. Soil Mech. and Foundation Eng.*, Cambridge, Mass., vol. 1, 1936, pp. 291–294.

34. Whangpoo Conservancy Board, S.H.T., Series 1, No. 7, various reports to the Engineer-in-Chief on Special Investigations, Shanghai, China, 1921; report of the Engineer-in-Chief on Pile Tests.

35. L. B. Feagin, "Lateral Pile-loading Tests," *Trans. ASCE, Paper* 1959, vol. 102, 1937, pp. 236–288 (including discussion).

36. J. H. Davies, "Steel Piles to Resist Lateral Loads," *Eng. News-Record*, vol. 122, No. 3, Jan. 19, 1939, p. 103.

37. J. W. Pearl, "Foundation Pile Spacing for Eccentric Loads," *Eng. and Contr.*, vol. 53, No. 4, Jan. 28, 1920, p. 91.

38. "Construction of Repair and Fitting-out Pier, Navy Yard, Puget Sound, Wash.," *Public Works of the Navy, Bull.* 36, October, 1927.

39. G. B. Sowers, "Load Required to Compress Concrete and Rock," *Civil Eng.*, vol. 13, No. 8, August, 1943, pp. 389–390.

40. M. P. Hite, "Special Jet Clears Holes for Wood Piling," *Eng. News-Record*, vol. 102, No. 19, May 9, 1929, p. 764.

41. "Home-made Muffler Stops Exhaust Noise of Large Steam Hammer," *Eng. News-Record*, vol. 98, No. 17, Apr. 28, 1927, pp. 700–701.

42. J. F. Seiler and W. D. Keeney, "The Efficiency of Piles in Groups," *Wood Preserving News*, vol. 22, No. 11, November, 1944, pp. 109–118.

43. "Wrapping Pile Butts for Hard Driving," *Wood Preserving News*, vol. 23, No. 8, August, 1945, p. 79.

44. W. A. King, "Pile Tests Justify Increased Loading," *Eng. News-Record*, vol. 134, No. 26, June 28, 1945, pp. 73–75.

45. "Horizontal Pile Hammer," *Construction Methods*, vol. 27, No. 10, October, 1945, pp. 82–85, 124, 126, 132, 136.

46. E. A. Prentis and L. White, *Underpinning,* 2d ed., Columbia University Press, 1950.
47. A. Di Giacinto, "Factory Underpinned with Three Types of Piles to Prevent Settlement of Walls and Columns," *Construction Methods,* vol. 27, No. 10, October, 1945, pp. 92–94.
48. F. D. Hartford, "Notes on Driving Timber Piles," *Eng. News-Record,* vol. 135, No. 22, Nov. 29, 1945, pp. 735–737.
49. "Two-pipe Pile Jetting Rig Speeds Driving on Navy Pier Job," *Eng. News-Record,* vol. 128, No. 5, Jan. 29, 1942, p. 200.
50. "Concrete Piles Jetted to Place by Three-lead Driver," *Eng. News-Record,* vol. 135, No. 12, Sept. 20, 1945, pp. 371–372.
51. G. B. R. Pimm, "Recent Developments in Deep Ground Testing," *The Structural Engineer,* July, 1938.
52. A. E. Cummings, "Stability of Foundation Piles against Buckling under an Axial Load," *Proc., Highway Research Board,* 1938.
53. G. P. Manning, "Column Strength of Long Piles of Steel, Timbers, Cast-iron, and Reinforced Concrete," *Civil Eng. and Public Works Rev.,* vol. 61, No. 478, April, 1946, pp. 145–148, 150.
54. R. D. Chellis, "Piles and Pile Foundations," *Eng. News-Record,* vol. 136, 1946: Part I, "Avoiding Detrimental Settlement," No. 20, May 16, pp. 774–778; Part II, "Field Determination of Damaging Stresses during Driving," No. 22, May 30, pp. 863–865; Part III, "Making and Interpreting Pile Load Tests," No. 24, June 13, pp. 914–919.
55. "Loading with Water to Design Capacity Forces Steel Cylinders below Driven Depth," *Eng. News-Record,* vol. 136, No. 26, June 27, 1946, pp. 974–977.
56. "Pipe Piles Jetted for Tower Supports," *Construction Methods,* vol. 28, No. 6, June, 1946, pp. 99, 159–160.
57. E. H. Harlow, "Piers and Wharves on Steel Cylinders," *Eng. News-Record,* vol. 137, No. 2, July 11, 1946, pp. 33–35.
58. "Special Rig Pulls Long Concrete Piles," *Construction Methods,* vol. 28, No. 7, July, 1946, pp. 92–93.
59. "The Actual Bearing Value of Piles," *Civil Eng. and Public Works Rev.,* vol. 30, July, 1935, pp. 210–211.
60. "Quick-tilting Leads on Floating Pile Driver," *Construction Methods,* vol. 28, No. 8, August, 1946, p. 85.
61. E. E. Howard, "Air Jetting Aids Pile Sinking," *Civil Eng.,* vol. 16, No. 9, September, 1946, pp. 402–403.
62. E. Rose, "Thrust Exerted by Expanding Ice Sheet," *Trans. ASCE, Paper* 2314, vol. 112, 1947, pp. 871–900 (including discussion).
63. C. L. McGuinness, "Recharge and Depletion of Ground-water Supplies," *Trans. ASCE, Paper* 2318, vol. 112, 1947, pp. 972–994 (includes bibliography).
64. W. H. Rabe, "Dynamic Pile-bearing Formula with Static Supplement," *Eng. News-Record,* vol. 137, No. 26, Dec. 26, 1946, pp. 868–871.
65. "Barge-mounted Multiple-lead Driver Handles 13 Piles without Moving," *Construction Methods,* vol. 28, No. 12, December, 1946, pp. 88–90.
66. P. E. Gisiger, "Safeguarding Hydro Plants Against Ice Menace," *Civil Eng.,* vol. 17, No. 1, January, 1947, pp. 24–27.
67. S. Sehested, "Fundamental Application of Soil Mechanics to Piled Foundations," *J. Eng. Assoc. Malaya,* April, 1941, pp. 4–24.
68. E. M. Young, Jr., "How Good are Diesel-powered Hammers?" *Eng. News-Record,* vol. 160, No. 20, May 15, 1958, pp. 54–56, 58, 62.

69. A. J. Hunter, "A Critical Examination of Pile-driving Formulae," *Proc., New Zealand Soc. Civil Eng.*, vol. XIII, pp. 44–83.
70. J. H. Nicholson, "Pile Formulas," *Selected Engineering Papers,* No. 62, The Institution of Civil Engineers, London, 1928.
71. "Roslyn Viaduct Piledriving—Prejetting for Long Piles Keeps Driving on Fast Schedule," *Construction Methods,* vol. 29, No. 9, September, 1947, pp. 97–99.
72. E. E. Howard, "Tubular Steel Pile Piers Withstand Rigid Test," *Civil Eng.*, vol. 17, No. 10, October, 1947, pp. 14–17, 76.
73. G. R. Wadleigh, "Machinery Foundations Are Tricky," *Southern Power and Ind.*, vol. 65, No. 12, December, 1947, pp. 51, 52, 118, 119.
74. F. A. Ciampa, "Pier on 200-ft. Piles in N.Y. Harbor," *Eng. News-Record,* vol. 140, No. 4, Jan. 22, 1948, pp. 104–108.
75. O. A. Zimmerman, "From Any Angle—Special Rig Drives 2-way Battered Piles," *Construction Methods,* vol. 30, No. 3, March, 1948, pp. 102–105.
76. "Plenty of Snow—But Footings Go In at Boston Airport," *Construction Methods,* vol. 30, No. 3, March, 1948, pp. 86–88.
77. C. H. Wood, "Battered Piles in Bridge Abutments," *Eng. News-Record,* vol. 138, No. 26, June 26, 1947, pp. 1018–1019.
78. R. F. Kafka, "Die Berechnung der Tragfähigkeit gerammter Pfähle," *Armierter Beton,* Nos. 11 and 12, 1910.
79. O. Stern, "Das Problem der Pfahlbelastung," W. Ernst und Sohn, Berlin, 1908.
80. R. F. Kafka, "Praktische Anwendungen der Methode zur Bestimmung der zulaessigen Pfahlbelastung," *Beton u. Eisen,* Nos. 7, 8, and 9, 1909.
81. R. F. Kafka, "Die Theorie der Pfahlgruendungen," Springer-Verlag, Berlin, 1912.
82. "Steel Tubes for 100-ton Loads," *Eng. News-Record,* vol. 140, No. 14, Apr. 1, 1948, pp. 511–513.
83. "Chicago Subway's Deepest Open Cut," *Excavating Engineer,* vol. 42, No. 4, April, 1948, pp. 18, 19, 34.
84. R. H. Mann, "Man-made Island Enclosed by Creosoted Piles," *Wood Preserving News,* vol. 26, No. 4, April, 1948, pp. 46–48, 52.
85. *The Gunite-jacketed Pile,* B. C. Gerwick, Inc., Engineers, Contractors, San Francisco, Calif., 1948.
86. G. F. Nicholson, "War Time Waterfront Construction Applicable to Commercial Ports," *World Ports,* vol. 8, No. 11, August, 1946, pp. 31–32.
87. C. F. Dodge and W. F. Swiger, "Vibration Testing of Friction Piles," *Eng. News-Record,* vol. 140, No. 20, May 13, 1948, pp. 716–720.
88. W. L. Pugh, "Test Piles for Saguenay Terminals Wharf," *Eng. J.*, vol. 31, No. 6, June, 1948, pp. 325–334.
89. W. M. Angas, "New Foundation for 350-ton Crane," *Eng. News-Record,* vol. 141, No. 4, July 22, 1948, pp. 63–67.
90. L. Velie, "What Are We Going to Do for Water?" *Collier's,* May 15, 1948, and *Reader's Digest,* August, 1948.
91. "Pile Driving," *Safe Practices Pamphlet Industrial Safety Series No. Con.* 4, National Safety Council, Chicago, Ill.
92. H. B. Thornley, "Pile Tests for Design and Construction Economy," *Civil Eng.*, vol. 18, No. 9, September, 1948, pp. 32–35.
93. "Diesel Pile-driving Hammers," *Concrete and Constructional Eng.*, vol. 43, No. 9, September, 1948, p. 277.
94. "Well-points on Bridge Job Progress from Pier to Pier," *Eng. News-Record,* vol. 141, No. 6, Aug. 5, 1948, p. 85.

95. "Bridge Piles Driven by Precision Methods," *Eng. News-Record*, vol. 141, No. 6, Aug. 5, 1948, pp. 80–83.
96. R. D. Chellis, "The Relationship between Pile Formulas and Load Tests," *Trans. ASCE, Paper* 2369, vol. 114, 1949, pp. 290–320.
97. S. W. Oberg, "An Island of Steel for Offshore Well Drilling," *Eng. News-Record*, vol. 141, No. 24, Dec. 9, 1948, pp. 80–83.
98. "Pile Lead Lengthener," *Eng. News-Record*, vol. 141, No. 4, July 22, 1948, p. 79.
99. R. D. Chellis, "A Study of Pile Friction Values," *Proc. Second Intern. Conf. Soil Mech. and Foundation Eng.*, Rotterdam, vol. 5, 1948, pp. 142–146.
100. W. F. Swiger, "Effect of Vibration on Piles in Loose Sand," *Proc. Second Intern. Conf. Soil Mech. and Foundation Eng.*, Rotterdam, vol. 2, 1948, pp. 191–194.
101. "Multiple Jet Gnaws through Tough Clay," *Construction Methods*, vol. 31, No. 3, March, 1949, pp. 62–63.
102. J. H. Thornley, "New Pile Code Sets Load-test Criteria," *Eng. News-Record*, vol. 142, No. 21, May 26, 1949, pp. 56–59.
103. T. T. Knappen, "Oil-loading Docks in Venezuela Built of Welded Steel," *Civil Eng.*, vol. 19, No. 3, March, 1949, pp. 33–37.
104. "Steam Generator Cuts Pile Driving Costs," *Eng. News-Record*, vol. 143, No. 19, Nov. 10, 1949, pp. 44–45.
105. P. W. Roberts and F. A. F. Cooke, "Arctic Tower Foundations Frozen into Permafrost," *Eng. News-Record*, vol. 144, No. 6, Feb. 9, 1950, pp. 38–39.
106. A. B. Hunicke, "Penetrometer, New Electronic Device, Records Pile Penetration under Hammer," *Civil Eng.*, vol. 20, No. 3, April, 1950, pp. 32–33.
107. "Rail Rig Drives Miles of Piles for Garrison Dam Cutoff," *Construction Methods and Equipment*, vol. 32, No. 6, June, 1950, pp. 54–55.
108. W. Loos, "Die Beziehung zwischen Pfahlformeln und Belastungsversuchen," *Der Bauingenieur*, vol. 25, No. 2, February, 1950, pp. 63–66.
109. C. H. Harned, "New Tools, Techniques and Talents Used in 1951 Model Bridge Foundation Studies," *California Highways and Public Works*, vol. 29, Nos. 11 and 12, November and December, 1950, pp. 36–42.
110. "Sheeting Jig Insures Fast Sewer Job," *Construction Methods and Equipment*, vol. 33, No. 4, April, 1951, pp. 54–57.
111. W. H. James, "The Structures of Osborne 'B' Power Station," *J. Inst. Engrs., Australia*, vol. 22, No. 12, December, 1950, pp. 257–268.
112. "Records Action of Pile Hammers," *Construction Methods and Equipment*, vol. 31, No. 12, December, 1949, p. 109.
113. B. Lofquist, "Lyftkraft och Bärförmåga hos ett Istacke," *Tek. Tid.*, No. 25, Stockholm, 1944 ("Lifting Force and Bearing Capacity of an Ice Sheet," *Technical Translation* TT-164, National Research Council of Canada, Ottawa, 1951).
114. Joint Committee, San Francisco Section, ASCE and Structural Engineers Association of Northern California, "Lateral Forces of Earthquake and Wind," *Trans. ASCE, Paper* 2514, vol. 117, 1952, pp. 716–780. (Includes extensive bibliography on lateral forces.)
115. "British Oil Dock Incorporates Unusual Fenders, Foundations," *Eng. News-Record*, vol. 146, No. 21, May 24, 1951, p. 42.
116. D. H. Little, "Some Dolphin Designs," *J. Inst. Civil Engrs.* (London), No. 1, 1946–1947, November, 1946, pp. 48–66, also Supplement to No. 8, 1946–1947, October, 1947, for discussion.
117. R. J. Wilkins, "The Bending of a Vertical Pile under Lateral Forces," *Civil Eng. and Public Works Rev.*, vol. 46, No. 539, May, 1951, pp. 355–357.
118. "Piledriving Leads go Modern—With Pipe," *Construction Methods and Equipment*, vol. 33, No. 6, June, 1951, pp. 51–53.

119. H. G. Schlitt, *Steel Pile Tests—Q Street Viaduct—Omaha, Nebr.*, Dept. of Roads and Irrigation, Bridge Design Section, Lincoln, Nebr.

120. Hamilton Gray, "Field Vane Shear Tests of Sensitive Cohesive Soils," *Trans. ASCE, Paper* 2887, vol. 122, 1957, pp. 844–863.

121. E. A. L. Smith, "Pile-driving Analysis by the Wave Equation," *J. Soil Mech. and Foundations Div.*, No. SM 4, *Proc. ASCE, Paper* 2574, August, 1960, pp. 35–61 (also see later discussions).

122. "Easier Pile Driving," *Eng. News-Record*, vol. 160, No. 16, Apr. 18, 1958, p. 72.

123. "Horizontal Piles Brace Cofferdam Wall," *Eng. News-Record*, vol. 160, No. 17, Apr. 24, 1958, pp. 55, 58.

124. C. I. Mansur and R. I. Kaufman, "Pile Tests, Low-sill Structure, Old River La.," *J. Soil Mech. and Foundations Div.*, No. SM 4, *Proc. ASCE, Paper* 1079, October, 1956, pp. 1–33.

125. J. D. Parsons and S. D. Wilson, "Safe Loads on Dog-leg Piles," *Trans. ASCE, Paper* 2816, vol. 121, 1956, pp. 695–721.

126. "Tappan Zee Bridge: A Foundation Triumph," *Eng. News-Record*, vol. 154, No. 15, Apr. 14, 1955, pp. 44–46.

127. S. R. Medvedev, "The Use of Vibrators for Driving Steel Sheet Piling," *Civil Eng. and Public Works Rev.*, vol. 48, No. 559, January, 1953, pp. 61, 62.

128. R. O. Reid and C. L. Bretschneider, "Surface Waves and Offshore Structures: The Design Wave in Deep or Shallow Water, Storm Tide, and Forces on Vertical Piles and Large Submerged Objects," Texas A. & M. Research Foundation, Dept. of Oceanography, College Station, Tex., October, 1953. (A technical report to the Bureau of Yards and Docks, Beach Erosion Board, Office of Naval Research. Contains bibliography.)

129. P. T. Gavaris, "Computer Checks As-driven Pile Groups," *Eng. News-Record*, vol. 160, No. 21, May 22, 1958, p. 54.

130. Symposium on Vane Shear Testing of Soils, *ASTM Spec. Tech. Publ.* 193, 1956.

131. M. A. Mason, "Surface Water Wave Theories," *Trans. ASCE, Paper* 2556, vol. 118, 1953, pp. 546–574. (Includes extensive bibliography.)

132. W. W. Moore and R. D. Darragh, "Some Considerations in the Design of Foundations for Earthquakes," *Proc. World Conf. on Earthquake Eng., Paper* 28, Earthquake Engineering Research Institute, Berkeley, Calif., June, 1956.

133. S. Okamoto, "Bearing Capacity of Sandy Soil and Lateral Earth Pressure during Earthquakes," *Proc. World Conf. on Earthquake Eng., Paper* 27, Earthquake Engineering Research Institute, Berkeley, Calif., June, 1956.

134. H. J. Brunnier, "The Development of Asiesmic Construction in the United States," *Proc. World Conf. on Earthquake Eng., Paper* 26, Earthquake Engineering Research Institute, Berkeley, Calif., June, 1956.

135. D. H. Little, "Some Designs for Flexible Fenders," *Proc. Inst. Civil Engrs.*, Part II, *Maritime Paper* 21, February, 1953, pp. 42–105.

136. R. R. Minikin, *Winds, Waves and Maritime Structures*, Charles Griffin & Co., Ltd., London, 1950.

137. A. M. Robertson, "Fendering, Lead-in Jetties and Dolphins," *Dock and Harbour Authority*, vol. 34, No. 391, May, 1953, pp. 15–20, 25.

138. A. Eggink, Paper, *Proc. Eighteenth Intern. Navigation Congr., Sect.* 2, *Question* 2, Rome, 1953, pp. 167–187.

139. A. L. L. Baker, Paper, *Proc. Eighteenth Intern. Navigation Congr., Sect.* 2, *Question* 2, Rome, 1953, pp. 111–142.

140. H. Ridenhalgh, "A Berthing Beam for Large Vessels," *Dock and Harbour Authority*, vol. 36, No. 415, May, 1955, pp. 9–14.

141. D. A. Hopkins, "The Design of Piers, Jetties and Dolphins," *Trans. ASCE, Paper* 2846, vol. 122, 1957, pp. 22–38.

142. "British Oil Dock Incorporates Unusual Fenders, Foundations," *Eng. News-Record,* vol. 146, No. 21, May 24, 1951, p. 42.

143. A. deF. Quinn, *Design and Construction of Ports and Marine Structures,* McGraw-Hill Book Company, Inc., New York, 1961.

144. L. A. Volse, "Docking Fenders: Key to Pier Protection," *Eng. News-Record,* vol. 160, No. 19, May 8, 1958, pp. 40–42, 44.

145. "Ice Pressure against Dams: A Symposium," *Trans. ASCE, Paper* 2656, vol. 119, 1954, pp. 1–42.

146. "Mackinac Bridge—Incredible but True," *Eng. News-Record,* vol. 154, No. 4, Jan. 27, 1955, pp. 35–44.

147. L. Bjerrum, "Norwegian Experiences with Steel Pile Foundations to Rock," *J. Boston Soc. Civil Engrs.,* vol. 41, No. 3, July, 1957, pp. 155–198. (Includes bibliography.)

148. "Accuracy of Driving Hollow Piles Checked by Electronic Device," *Contractors and Engrs. Monthly,* July, 1951.

149. "Fins Guide Piles to Firm Bearing," *Eng. News-Record,* vol. 154, No. 14, Apr. 5, 1956, pp. 85, 86.

150. M. E. Essaglou, "Piling Operations in Alaska," *Military Engr.,* vol. 49, No. 330, July–August, 1957, pp. 282–287.

151. L. A. Nees, "Pile Foundations for Large Towers on Permafrost," *Trans. ASCE, Paper* 2528, vol. 117, 1952, pp. 935–947.

152. K. Terzaghi, "Permafrost," *J. Boston Soc. Civil Engrs.,* vol. 39, No. 1, January, 1952, pp. 1–50. (Includes bibliography.)

153. L. J. Ritter, Jr., "How to Save Money on Pile Construction," *Modern Highways,* Oct. 15, 1957, pp. 7–12, and Dec. 15, 1957, pp. 9–15.

154. Yang, Nai-Chen, "Redriving Characteristics of Piles," *J. Soil Mech. and Foundations Div.,* No. SM 3, *Proc. ASCE, Paper* 1026, vol. 82, July, 1956, pp. 1–19.

155. A. B. Seed and L. C. Reese, "The Action of Soft Clay along Friction Piles," *Proc. ASCE, Paper* 842, vol. 81, December, 1955, and discussions in *J. Soil Mech. and Foundations Div.,* No. SM 3, *Proc. ASCE, Paper* 1028, vol. 82, July, 1956, pp. 5–13.

156. M. F. Ghanem, "Bearing Capacities of Friction Piles in Deep Soft Clays," Thesis, University of Illinois, Urbana, Ill., 1953.

157. D. D. Barkan, "Foundation Engineering and Drilling by the Vibration Method," *Proc. Fourth Intern. Conf. Soil Mech. and Foundation Eng., London,* vol. 2, 1957, pp. 3–7. (Includes bibliography.)

158. Yi-Sheng, T. E. Mao, "The Yangtze River Bridge at Hankow, China," *Civil Eng.,* vol. 28, No. 12, December, 1958, pp. 54–57.

159. H. R. Reynolds, "Notes on the Effects of Vibration on Soils," *Civil Eng. and Public Works Rev.,* Part 1, vol. 49, No. 578, August, 1954, pp. 835–837, and Part 2, vol. 49, No. 580, October, 1954, pp. 1077–1079. (Includes bibliographies.)

160. A. M. Klein and J. H. A. Crockett, "Design and Construction of a Fully Vibration-controlled Forging Hammer Foundation," *J. ACI,* vol. 24, No. 5, January, 1953, pp. 421–444. (Includes bibliography.)

161. L. A. Palmer and J. B. Thompson, "The Earth Pressure and Deflection along the Embedded Lengths of Piles Subjected to Lateral Thrusts," *Proc. Second Intern. Conf. Soil Mech. and Foundation Eng.,* Rotterdam, vol. 5, 1948, pp. 156–161.

162. S. M. Gleser, "Lateral Load Tests on Vertical Fixed-head and Free-head Piles," *ASTM Spec. Tech. Publ.* 154, 1953, pp. 75–93.

163. A. A. Wagner, "Lateral Load Tests on Piles for Design Information," *ASTM Spec. Tech. Publ.* 154, 1953, pp. 59–74.

164. H. G. Mason and J. A. Bishop, "Piles Subject to Lateral Earth Thrust. Part I, Measurement of Earth Pressure and Deflection along the Embedded Portion of a 40-ft Steel Pile," *ASTM Spec. Tech. Publ.* 154-A, 1954, pp. 1–21.

165. L. A. Palmer and P. P. Brown, "Piles Subject to Lateral Thrust. Part II, Analysis of Pressure Deflection, Moments and Shears by the Method of Difference Equations," *ASTM Spec. Tech. Publ.* 154-A, 1954, pp. 23–32.

166. B. McClelland and J. A. Focht, Jr., "Soil Modulus for Laterally Loaded Piles," *J. Soil Mech. and Foundations Div.*, No. SM 4, *Proc. ASCE, Paper* 1081, vol. 82, October, 1956, pp. 1–22.

167. J. F. McNulty, "Thrust Loading on Piles," *J. Soil Mech. and Foundations Div.*, No. SM 2, *Proc. ASCE, Paper* 940, vol. 82, April, 1956, pp. 1–25; also discussion in SM 4, vol. 82, October, 1956, pp. 39–46.

168. J. R. Morison, M. P. O'Brien, J. W. Johnson, and S. A. Schaff, "The Force Exerted by Surface Waves on Piles," *Petroleum Trans., AIME*, vol. 189, 1950, pp. 149–154.

169. E. E. Howard, "Machine Foundation Supported by Steel Rods Driven through Clay," *Civil Eng.*, vol. 23, No. 4, April, 1953, pp. 60, 61.

170. L. Van Rijsselberghe and L. Descans, "Ducs d'Albe en palplanches métalliques—Application au Port de Zeebrugge," *Ann. travaux publics de Belgique*, 104e Année, 3e Fascicule, June, 1951, pp. 423–484. (Dolphins and steel sheet piling.)

171. J. R. Morison, "Design of Piling," *Coastal Engineering, Proc. First Conf. Coastal Eng.*, The Engineering Foundation, Council on Wave Research, Long Beach, Calif., 1951, pp. 254–258.

172. M. Banal, "The Effect of a Tidal Bore on Ships' Moorings," *Bull. Permanent Intern. Assoc. Navigation Congresses*, Brussels, vol. 1, No. 47, 1958, pp. 159–176.

173. G. W. Rathjens, "Arctic Engineering Requires Knowledge of Permafrost Behavior," *Civil Eng.*, vol. 21, No. 11, November, 1951, pp. 645–647.

174. G. P. Tschebotarioff, "How Russians Drive Piles by Vibration," *Eng. News-Record*, vol. 163, No. 3, July 16, 1959, pp. 53, 54.

175. P. W. Roberts and Virgil Blancato, "New Retractable Marine Fender System," *J. Waterways and Harbors Div.*, No. WW 1, *Proc. ASCE, Paper* 1513, vol. 84, January, 1958.

176. T. J. Risselada, "Flexible Steel Dolphins and Kindred Structures," *Dock and Harbour Authority*, vol. 39, No. 453, July, 1958, pp. 93, 94.

177. ———, "Dolphins at the Port of Antwerp," *Dock and Harbour Authority*, vol. 35, No. 405, July, 1954, pp. 88, 89.

178. R. J. Howe, "Design of Offshore Drilling Structures," *Trans. ASME*, vol. 77, No. 6, August, 1955, pp. 827–851.

179. *Shore Protection Planning and Design, Tech. Rept.* 4, Department of the Army, Corps of Engineers, Office of the Chief of Engineers, Beach Erosion Board, June, 1954. (Contains extensive bibliography and glossary of terms. Also Corrections, Revisions, and Addenda of February, 1957, and August, 1957.)

180. G. G. Meyerhoff, "Penetration Tests and Bearing Capacity of Cohesionless Soils," *J. Soil Mech. and Foundations Div.*, No. SM 1, *Proc. ASCE, Paper* 866, vol. 82, January, 1956, pp. 1–19. (Contains extensive bibliography.)

181. R. L. Wiegel, *Gravity Waves—Table of Functions*, The Engineering Foundation, Council on Wave Research, Berkeley, Calif., February, 1954.

182. J. R. Morison, J. W. Johnson, and M. P. O'Brien, "Experimental Studies of Forces on Piles," *Coastal Engineering, Proc. Fourth Conf. Coastal Eng.*, The Engineering Foundation, Council on Wave Research, Berkeley, Calif., 1956, pp. 340–370.
183. C. L. Bretschneider, "Selection of Design Wave for Offshore Structures," *Trans. ASCE, Paper* 3026, vol. 125, 1960, pp. 381–416.
184. R. L. Wiegel and R. E. Skjei, "Breaking Wave Force Prediction," *J. Waterways and Harbors Div.*, No. WW 1, *Proc. ASCE, Paper* 1573, vol. 84, March, 1958, pp. 1573-1, 1573-14.
185. ———, K. E. Beebe, and James Moon, "Ocean Wave Forces on Circular Cylindrical Piles," *J. Hydraulics Div.*, No. HY 2, *Proc. ASCE, Paper* 1199, vol. 84, April, 1957, pp. 1199-1, 1199-36. (Includes extensive bibliography.)
186. D. F. Denny, "Further Experiments on Wave Pressures," *J. Inst. Civil Engrs.* (*London*), vol. 35, No. 4, 1951, pp. 330–345.
187. R. Y. Hudson, "Wave Forces on Breakwaters," *Trans. ASCE, Paper* 2556, vol. 118, 1953, pp. 653–674.
188. E. N. Praeger, "Energy Absorption by the Pier or Wharf Structure," *Princeton Univ. Conf. on Berthing and Cargo Handling in Exposed Locations*, Oct. 20, 21, 1958, pp. 111–114.
189. R. L. Wiegel and K. E. Beebe, "The Design Wave in Shallow Water," *J. Waterways and Harbors Div.*, No. WW 1, *Proc. ASCE, Paper* 910, vol. 82, 1956, pp. 1–21. (Wave pressures on pile-supported structures.)
190. R. H. Fleming and C. C. Bates, "The Engineering Application of Sea and Swell Data," *Coastal Engineering, Proc. First Conf. Coastal Eng.*, The Engineering Foundation, Council on Wave Research, Berkeley, Calif., 1951, pp. 88–97.
191. Thorndike Savile, Jr., "Wave Run-up on Shore Structures," *Trans. ASCE, Paper* 2915, vol. 123, 1958, pp. 139–150.
192. B. W. Wilson, "The Energy Problem in the Mooring of Ships Exposed to Waves," *Princeton Univ. Conf. on Berthing and Cargo Handling in Exposed Locations*, Oct. 20–21, 1958, pp. 1–67. (Contains extensive bibliography.)
193. Zusse Levinton, "Elastic Fender System for Wharves," *Princeton Univ. Conf. on Berthing and Cargo Handling in Exposed Locations*, Oct. 20–21, 1958, pp. 87–95.
194. F. H. Stracke, "Offshore Mooring Facilities for Tankers up to 100,000 Dead Weight Ton Capacity," *Princeton Univ. Conf. on Berthing and Cargo Handling in Exposed Locations*, Oct. 20–21, 1958, pp. 157–173.
195. C. L. Bretschneider, "Revisions in Wave Forecasting: Deep and Shallow Water," *Coastal Engineering, Proc. Sixth Conf. Coastal Eng.*, The Engineering Foundation, Council on Wave Research, Berkeley, Calif., 1958, pp. 30–67.
196. R. O. Reid, "Correlation of Water Level Variation with Wave Forces on a Vertical Pile for Nonperiodic Waves," *Coastal Engineering, Proc. Sixth Conf. Coastal Eng.*, The Engineering Foundation, Council on Wave Research, Berkeley, Calif., 1958, pp. 749–786.
197. E. E. Gesler, R. O. Eaton, and J. V. Hall, Paper, *Proc. Eighteenth Intern. Navigation Congr., Sect.* 2, *Question* 1, Rome, 1953, pp. 67–90.
198. G. Plantema, "Construction and Method of Operation of a New Deep-sounding Device," *Proc. Second Intern. Conf. Soil Mech. and Foundation Eng.*, Rotterdam, 1948, pp. 277–287.
199. C. Van der Veen and L. Boersma, "The Bearing Capacity of a Pile Pre-determined by a Cone Penetration Test," *Proc. Fourth Intern. Conf. Soil Mech. and Foundation Eng.*, London, vol. 2, 1957, pp. 72–75.

200. E. V. Lewis and S. F. Borg, "Energy Absorption by the Ship," *Princeton Univ. Conf. on Berthing and Cargo Handling in Exposed Locations*, Oct. 20, 1958, pp. 69–86.

201. T. K. Huizinga, "Application of Results of Deep Penetration Tests to Foundation Piles," *Proc. Building Research Congr.*, London, vol. 1, 1951, pp. 173–179.

202. J. Darbyshire, "An Investigation of Storm Waves in the Atlantic Ocean," *Proc. Royal Soc. (London)*, vol. 230A, No. 1183, July 7, 1955, pp. 560–569.

203. "New Oil Jetties for Thames Haven," *Dock and Harbour Authority*, vol. 40, No. 466, August, 1959, pp. 97–103.

204. A. R. Striegl, "Ice on the Great Lakes and Its Engineering Problems," paper presented at Joint Meeting of American Meteorological Society and American Geophysical Union, Chicago, Sept. 11, 1952.

205. J. A. Pihlainen, "Pile Construction in Permafrost," *J. Soil Mech. and Foundations Div.*, No. SM 6, *Proc. ASCE*, vol. 85, December, 1959, pp. 75–95.

206. G. G. Meyerhoff, "Compaction of Sands and Bearing Capacity of Piles," *J. Soil Mech. and Foundations Div.*, No. SM 6, *Proc. ASCE*, vol. 85, December, 1959, pp. 1–29.

207. J. J. Hamilton, "Annotated Bibliography on Vane Testing in Soils," National Research Council, Division of Building Research, Ottawa, Canada, September, 1959.

208. K. Terzaghi, "Evaluation of Coefficients of Subgrade Reaction," *Géotechnique*, vol. 5, 1955, pp. 297–326; also *Harvard Soil Mechs. Ser.*, No. 51.

209. L. J. Phares, "Wet Rotary Pre-excavation for Piles," *Civil Eng.*, vol. 30, No. 2, February, 1960, pp. 64–66.

210. *Engineering Manual for Military Construction*, U.S. Army, Corps of Engineers, October, 1954.

211. J. H. Thornley, "Ship Anchorage on Rock," *Civil Eng.*, vol. 29, No. 3, March, 1959, pp. 50–52.

212. Hudson Matlock and L. C. Reese, "Generalized Solutions for Laterally Loaded Piles," *J. Soil Mechanics and Foundations Div.*, No. SM 5, *Proc. ASCE, Paper* 2626, vol. 86, October, 1960, pp. 63–91.

213. J. M. Weiss and V. Blancato, "A Breasting Dolphin for Berthing Supertankers," *J. Waterways and Harbors Div.*, No. WW 3, *Proc. ASCE*, September, 1959, pp. 183–195.

214. W. C. Gribble, Jr., "Foundations in Permafrost," *Military Engineer*, vol. 52, No. 350, November-December, 1960, pp. 445–447.

215. "Two Ways to Plant a Power Line," *Eng. News-Record*, vol. 165, No. 3, July 21, 1960, p. 32.

216. T. J. Flynn, "How to Build Modern Roads over Marshes," *Wood Preserving News*, vol. 38, No. 8, August, 1960, pp. 9, 10.

217. L. E. Hunter, *Underpinning and Strengthening of Structures*, Contractors Record and Municipal Engineering, London, 1942.

GENERAL PUBLICATIONS*

1*a*. E. Noë and L. Troch, *Pieux et sonnettes*, Gauthier-Villars & Cie, 1920.

1*b*. "Engineering and Contracting Procedure for Foundations," *ASCE Manuals of Engineering Practice No*. 8, 1934.

1*c*. A. C. Dean, *Piles and Pile Driving*, Crosby, Lockwood & Son, 1935.

1*d*. F. E. Wentworth-Sheilds and W. S. Gray, *Reinforced Concrete Piling*, Concrete Publications Ltd., 1938.

1*e*. *Specification for Concrete Pile-driving*, Institution of Structural Engineers, October, 1936.

1*f*. "Semipermanent Highway and Railway Trestle Bridges," *War Department Technical Manual TM*5-286, April, 1945.

1*g*. D. H. Lee, *Sheet Piling, Cofferdams, and Caissons*, Concrete Publications Ltd., 1945.

1*h*. "Pile Foundations and Pile Structures," *ASCE Manuals of Engineering Practice No*. 27, 1946.

1*i*. R. R. Minikin, "Bearing Piles," *Civil Eng. and Public Works Rev*., vol. 40, 1945: Part I, March, pp. 64–72; Part II, April, pp. 84–88; Part III, May, pp. 112–118; Part IV, June, pp. 132–134; Part V, July, pp. 158–165.

1*j*. F. Costa, *Estacas para Fundaçoes*, Instituto Superior Tecnico, Lisbon.

1*k*. R. R. Minikin, *Piling for Foundations*, Crosby, Lockwood & Son, 1948.

1*l*. *Codes of Practice for Civil Engineering, Public Works Building and Constructional Work*, Codes of Practices Committee: Code I, Site Investigations (General Investigations and Soil Surveys); Code II, Earthworks (Excavation, Embankments, Dredging, Techniques for Improving Ground Conditions); Code III, Foundations (Principles, Spread and Piled Foundations, Cofferdams, Caissons and Underwater Work); Code VII, Earth—Retaining Structures (Retaining Walls, Revetments, Sheet Piling). H. M. Stationery Office.

1*m*. W. Schenck, *Der Rammpfahl-Neue Erkenntnisse aus Theorie und Praxis*, Wilhelm Ernst & Sohn, 1951.

1*n*. K. Terzaghi and R. B. Peck, *Soil Mechanics in Engineering Practice*, John Wiley & Sons, Inc., New York, 1948.

1*o*. J. H. Thornley, *Foundation Design and Practice—An Economic View*, Columbia University Press, New York, 1959.

*Also see items 25, 46, 54, 2*ca*, 5*a*, 3*aw*, 11*c*, 11*m*, 6*f*, and 11*s*.

WOOD PILES, INCLUDING DETERIORATION AND PRESERVATION*

2*a*. "Timber Piles and Construction Timbers," *ASCE Manuals of Engineering Practice No.* 17, 1939.

2*b*. *Wood Handbook,* Forest Products Laboratory, Forest Service, U.S. Department of Agriculture, 1955.

2*c*. J. D. McLean, "Manual on Preservative Treatment of Wood by Pressure," *U.S. Dept. Agr., Misc. Pub.* 224.

2*d*. D. F. Holtman, *Wood Construction* (A Project of the National Committee on Wood Utilization), McGraw-Hill Book Company, Inc., 1929.

2*e*. G. M. Hunt and G. A. Garratt, *Wood Preservation,* McGraw-Hill Book Company, Inc., 1938.

2*f*. *Manual of Recommended Practice,* American Wood-Preservers' Association.

2*g*. T. R. C. Wilson, "The Effect of Creosote on the Strength of Fir Timbers," *Timberman* 31, 1930, pp. 50–56.

2*h*. W. K. Hatt, "Report on the Effect of Steaming Wood," *Proc. AREA,* 1927, pp. 1164–1186.

2*i*. L. W. Rees and S. J. Buckman, "Effect of Steaming on the Physical Properties of Southern Yellow Pine," *Proc. AWPA,* vol. 36, 1940, pp. 331–360.

2*j*. *Are Creosoted Timber Foundation Piles Economical and Permanent?* Service Bureau, American Wood-Preservers' Association, December, 1938.

2*k*. Various publications relating to marine borers by the William F. Clapp Laboratories, Duxbury, Mass.

2*l*. W. F. Clapp, "The Activities of Marine Borers," *Civil Eng.,* vol. 7, No. 2, February, 1937, pp. 105–108.

2*m*. W. F. Clapp, "Recent Increases in Marine-borer Activity," *Civil Eng.,* vol. 7, No. 12, December, 1937, pp. 836–838.

2*n*. W. G. Atwood and A. A. Johnson, "Marine Structures, Their Deterioration and Preservation," Report of Committee on Marine Piling Investigation, National Research Council, Washington, D. C., 1924.

2*o*.† C. L. Hill and C. A. Kofoid, *Marine Borers and Their Relation to Marine Construction on the Pacific Coast,* San Francisco Bay Marine Piling Committee, 1927.

2*p*.† T. Iredale, R. A. Johnson, and F. A. McNeill, *Destruction of Timber by Marine Organisms in the Port of Sydney,* Sidney Harbor Trust, 1932.

2*q*.† R. H. McGonigle, *Marine Borers of the Atlantic Coast of Canada, Report* 15, National Research Council of Canada, 1925.

2*r*.† R. H. McGonigle, *A Further Consideration of the Relations between the Distribution of Teredo Navalis (Linne) and the Temperature and Salinity of its Environment, Report* 20, National Research Council of Canada, 1926.

* Also see items 43, 48, 63, 84, 85, 111, 3*k*, 3*q*, 3*av*, and 5*aq*.

† For complete references to the literature on the subject to date, consult bibliographies in these publications.

2*s*. H. Titus, "Will There Be Enough Water?" *Saturday Evening Post*, vol. 216, No. 48, May 27, 1944, pp. 28, 105.

2*t*. H. W. Young, "Timber Copperplated to Frustrate Worms," *Compressed Air Mag.*, vol. 48, No. 6, June, 1943, pp. 7060–7062.

2*u*. C. A. P. Turner, "Wood Borers Remove Piles from Building Foundations," *Eng. News-Record*, vol. 111, Aug. 10, 1933, pp. 159–160.

2*v*. B. F. Perry, "Piles Cut by Blasting," *Eng. News-Record*, vol. 120, No. 2, Jan. 13, 1938, p. 78.

2*w*. W. E. Wheat, "Jacketed Creosoted Piles Repel Marine Borers," *Eng. News-Record*, vol. 108, Feb. 11, 1932, pp. 208–209.

2*x*. "Concrete Armored Piles in Wharf at Port Orford, Ore.," *Eng. News-Record*, vol. 87, No. 26, Dec. 29, 1921, p. 1061.

2*y*. T. H. Barnes, "The Reinforced Concrete Wharf of the United Fruit Company at Bocas Del Toro, Panama," *Trans. ASCE, Paper* 1137, vol. 66, Mar. 5, 1910, pp. 287–305.

2*z*. "Wood Piles with Mortar Coating Save Time and Critical Materials," *Eng. News-Record*, vol. 129, No. 19, Nov. 5, 1942, pp. 624–626.

2*aa*. "Mortar-coated Piles Resist Hard Treatment," *Eng. News-Record*, vol. 130, No. 12, Mar. 25, 1943, pp. 425–427.

2*ab*. *"Gunite" for Piling, Bull.* 2400, Cement-Gun Co., Inc., 1949.

2*ac*. "Concrete Jackets Make Up for Unequal Pile Lengths," *Eng. News-Record*, vol. 129, No. 17, Oct. 22, 1942, pp. 544–546.

2*ad*. "Pulled Piles Reveal Effectiveness of Asphalt-treated Wrapping," *Eng. News-Record*, vol. 130, No. 14, Apr. 8, 1943, pp. 480–481.

2*ae*. G. W. Davis, "Foundation Pile-head Bond and Anchorage Tests," *Public Roads*, vol. 9, No. 9, November, 1928, pp. 169–183.

2*af*. *Report on the San Francisco Bay Marine Piling Survey*, San Francisco Bay Marine Piling Committee of the American Wood-Preservers' Association, Jan. 24, 1921.

2*ag*. W. A. Stacey, "Proper Treatment of Field Cuts Will Prevent Needless Decay," *Wood Preserving News*, vol. 22, No. 10, October, 1944, pp. 102–108.

2*ah*. G. Smitt and A.-G. P. Palén, "Arsenic-cement," *Bull. Permanent Intern. Assoc. Navigation Congresses*, 11th year, No. 21, January, 1936, pp. 93–103.

2*ai*. "Comparative Value of Timber Cut from Live and Dead Trees," *U.S. Dept. Agr., Forest Products Lab., Technical Note* 101.

2*aj*. "Effect of Time of Cutting Timber on Its Durability," *U.S. Dept. Agr., Forest Products Lab., Technical Note F*-15.

2*ak*. "Comparative Durability of Green and Seasoned Timber," *U.S. Dept. Agr., Forest Products Lab., Technical Note F*-33.

2*al*. "Differences between Heartwood and Sapwood," *U.S. Dept. Agr., Forest Products Lab., Technical Note* 189.

2*am*. "Comparative Decay Resistance of Heartwood of Different Native Species when Used under Conditions That Favor Decay," *U.S. Dept. Agr., Forest Products Lab., Technical Note* 229.

2*an*. T. R. C. Wilson, T. A. Carlson, and R. F. Luxford, "Effect of Partial Seasoning on the Strength of Wood," *Proc. AWPA*, 1930, pp. 349–379.

2*ao*. "Timehri," *J. Roy. Agr. Comm. Soc. British Guiana*, June, 1917.

2*ap*. *Canadian Woods—Their Properties and Uses*, Forest Products Laboratories of Canada, 1940.

2*aq*. *Factors that Influence the Decay of Untreated Wood in Service and Comparative*

Decay Resistance of Different Species, U.S. Department of Agriculture, Forest Products Laboratory, 1941.

2*ar*. *Life of Creosoted Wooden Piling when Used for Building Foundations to Support Masonry Footings*, Forest Products Laboratories of Canada, 1934.

2*as*. *The Effect of Partial Seasoning on the Strength of Wood*, U.S. Department of Agriculture, Forest Products Laboratory, 1930.

2*at*. J. D. McLean, *Results Obtained in Marine Piling Experiments*, report presented at 27th annual meeting of the American Wood-Preservers' Association, 1931.

2*au*. J. D. McLean, *Information Obtained from Marine Piling Experiments on the Gulf Coast*, report presented at 37th annual meeting of the American Wood-Preservers' Association, 1941.

2*av*. "Field Treatment of Pile Cut-offs," *Wood Preserving News*, vol. 23, No. 4, April, 1945, pp. 39, 41.

2*aw*. "Thousands of Creosoted Piles in Gulf Coast Refinery," *Wood Preserving News*, vol. 23, No. 5, May, 1945, pp. 47–50.

2*ax*. *Presscrete Pile Encasing*, The Presscrete Co., Inc.

2*ay*. "Shotcrete Protects Wood Piles from Borers," *Concrete for Railways*, vol. **9**, No. 2, Portland Cement Association, pp. 11–14.

2*az*. C. J. J. Watson, F. A. McNeill, R. A. Johnson, and T. Iredale, "Destruction of Timber by Marine Organisms in the Port of Brisbane," *Bull.* 12, Subdepartment of Forestry, Dept. of Public Lands, July, 1936, pp. 107.

2*ba*. R. H. Mann, "Marine Structures on the North Atlantic Seaboard," *Proc. AWPA*, 1939.

2*bb*. *A Handbook of Empire Timbers*, Dept. of Scientific and Industrial Research, Forest Products Research Laboratory, H. M. Stationery Office, 1939; also *War Emergency Supplement*, 1943.

2*bc*. "Protection from Injurious Insects," *U.S. Dept. Agr., Farmers' Bull.* 1582, October, 1943.

2*bd*. "Preventing Damage to Buildings by Subterranean Termites and Their Control," *U.S. Dept. Agr., Farmers' Bull.* 1911, March, 1946.

2*be*. "Preventing Damage by Termites or White Ants," *U.S. Dept. Agr., Farmers' Bull.* 1472, August, 1939.

2*bf*. "Effectiveness of Wood Preservatives in Preventing Attack by Termites," *U.S. Dept. Agr., Circular* 683, September, 1943.

2*bg*. *A Report*, Marine Borer Research Committee, New York Harbor, December, 1946.

2*bh*. W. F. Clapp, "Tests Assure Adequate Warning of Marine Borer Attack in New York Harbor," *Civil Eng.*, vol. 17, No. 6, June, 1947, pp. 38–41, 84.

2*bi*. W. H. Johnson and C. J. Beer, "Encasing Bridge Piles in Concrete," *Concrete*, vol. 55, No. 7, July, 1947, pp. 3–5.

2*bj*. W. H. Johnson and C. J. Beer, "Unique Method for Pile Encasement," *California Highways and Public Works*, January–February, 1947, pp. 37–39.

2*bk*. R. D. Chellis, "Finding and Fighting Marine Borers," *Eng. News-Record*, vol. 140, 1948: Part I, "Destructive Borers and How They Work," No. 10, Mar. 4, 1948, pp. 344–347; Part II, "Geographic Range and Natural Controls," No. 12, Mar. 18, 1948, pp. 422–424; Part III, "Metal Armor and Concrete Jacketing," No. 14, Apr. 1, 1948, pp. 493–496; Part IV, "Creosote Impregnation and Poisoning," No. 16, Apr. 15, 1948, pp. 555–558.

2*bl*. R. G. Dyktor, "Concrete-spliced Wood Piles Answer Need for 55-ft. Lengths in Venezuela Dock," *Civil Eng.*, vol. 19, No. 1, January, 1949, pp. 36–37.

2*bm*. "Hydraulic Rig Removes Wood Piles," *Construction Methods*, vol. 31, No. 1, January, 1949, p. 82.

2*bn*. R. H. Mann, "Good Design and Chemical Treatment Give Long Life to Wood Waterfront Structures," *Civil Eng.*, vol. 19, No. 4, April, 1949, pp. 22–26, 80.

2*bo*. "Hydraulic Pile Puller," *Eng. News-Record*, vol. 142, No. 7, Feb. 17, 1949, p. 79.

2*bp*. W. R. Bond, "34-year-old Creosoted Piles Salvaged for Re-use," *Wood Preserving News*, March, 1949, pp. 25–27, 36.

2*bq*. J. Stenbuck, "Termites of the Sea," *Science Illustrated*, vol. 2, No. 7, July, 1947, pp. 42–45, 64.

2*br*. J. F. Jelley, "More Basic Research Vital in Fight against Costly Marine Borers," *Civil Eng.*, vol. 21, No. 4, April, 1951, pp. 28–29.

2*bs*. "Big-end-down Piles in Jersey Meadows Save Half-million Cost for Warehouse," *Eng. News-Record*, vol. 146, No. 24, June 21, 1951, p. 30.

2*bt*. *Harbor Reports on Marine Borer Activities*, NavDocks P-43, Dept. of the Navy, Bureau of Yards and Docks, Research Division, June, 1950. (Compiled by The William F. Clapp Laboratories, Inc.)

2*bu*. *Report on Marine Borers and Fouling Organisms in* 56 *Important Harbors and Tabular Summaries of Marine Borer Data from* 160 *Widespread Locations*, NavDocks TR Re-1, Dept. of the Navy, Bureau of Yards and Docks, April, 1951. (Prepared by The William F. Clapp Laboratories, Inc.)

2*bv*. *Wood Preservatives, U.S. Dept. Agr., Forest Prods. Lab.*, R149, rev. November, 1949.

2*bw*. *Results of Experiments on the Effectiveness of Various Preservatives in Protecting Wood against Marine-borer Attack, U.S. Dept. of Agr., Forest Prods. Lab.*, D1773, June, 1950.

2*bx*. R. H. Mann, "The Fire Resistance and Slow Burning Characteristics of Pressure Creosoted Piles and Timbers in Substructures of Piers and Wharves," *Proc. Am. Wood-Preservers' Assoc.*, April, 1950, pp. 339–348.

2*by*. W. F. Clapp, "Adequate Inspection—A Must for Marine Piles," paper prepared for Annual Meeting of American-Wood-Preservers' Association, April, 1951.

2*bz*. "Construction without Disruption," *Eng. News-Record*, vol. 157, No. 25, Dec. 20, 1956, pp. 38–40.

2*ca*. W. D. Keeney, R. H. Mann, and C. M. Burpee, *Pressure Treated Timber Foundation Piles*, American Wood Preservers Institute, Chicago, 1955.

2*cb*. P. W. Roberts, "What Woods Make Good Pier Fenders?" *Eng. News-Record*, vol. 160, No. 26, June 26, 1958, pp. 40, 41.

2*cc*. R. J. Menzies and R. Turner, "The Distribution and Importance of Marine Wood Borers in the United States," *Symposium on Wood for Marine Use and Its Protection from Marine Organisms, ASTM*, Los Angeles, Calif., Sept. 17, 1956, pp. 3–21.

2*cd*. H. Vind, H. Hochman, J. Muraoka, and J. Casey, "Relationship between Limnoria Species and Service Life of Creosoted Piling," *Symposium on Wood for Marine Use and Its Protection from Marine Organisms, ASTM*, Los Angeles, Calif., Sept. 17, 1956, pp. 35–50.

2*ce*. R. P. Woods, "Resistance of Timbers to Marine Borers—A Survey of Some Recent Experimental Work," *Dock and Harbour Authority*, vol. 38, No. 441, pp. 101–104. (Includes historical research bibliography.)

2*cf*. R. H. Colley, "Marine Biology and Commercial Treating," *Wood Preserving News*, vol. 35, No. 12, December, 1957, pp. 6–8.

2cg. A. P. Richards, "How to be Sure of Treated Wood Piling," *Eng. News-Record*, vol. 161, No. 4, July 24, 1958, pp. 51–53.

2ch. "Air-driven Axe Cuts Pile-pointing Time," *Eng. News-Record*, vol. 151, No. 21, Nov. 19, 1953, pp. 75, 76.

2ci. C. M. Burpee, "Pressure Preserved Wood for Permanent Structures," *J. Structural Div.*, No. ST 7, *Proc. ASCE, Paper* 1841, Part 1, vol. 84, November, 1958, pp. 1–10.

2cj. *Deterioration of Structures of Timber, Metal and Concrete Exposed to the Action of Sea-water*, Nineteenth Report of the Committee of the Institution of Civil Engineers, 2d ed., London, 1951.

2ck. D. B. Anderson, "Long Life Promised for Wooden Pilings in Marine Locations," *Southern Power and Ind.*, vol. 78, No. 10, October, 1960, p. 75.

2cl. A. C. Oliver, *Use of X-rays in the Detection of Teredo in Wood*, Timber Development *Assoc.*, London, 1960.

2cm. P. W. Roberts, "What Woods Make Good Pier Fenders?" *Eng. News-Record*, vol. 160, No. 26, June, 1958, pp. 40, 41.

2cn. "Bald Cypress Requires Preservative Treatment," *Wood Preserving News*, vol. 38, No. 10, October, 1960, pp. 14, 16.

2co. V. G. McClurg, "Working Loads for Timber Piles," *Wood Preserving News*, vol. 38, No. 9, September, 1960, pp. 4, 7, 8, 10, 12, 14, 15, 18, 20.

STEEL AND IRON PILES, INCLUDING CORROSION AND PRESERVATION*

3*a*. G. G. Greulich, "Steel Piles for Subaqueous Work," *Civil Eng.*, vol. 12, No. 10, October, 1942, p. 566.

3*b*. *Bitumastic Bull.* 19, Wailes Dove-Hermiston Corp., 1942.

3*c*. "Corrosion in Soils," *Letter Circular LC*-689, U.S. Dept. of Commerce, National Bureau of Standards, Apr. 22, 1942.

3*d*. A. H. Thomas, "Iron and Steel Corrosion Traced to Bacterial Action," *Eng. News-Record,* vol. 129, No. 11, Sept. 10, 1942, p. 105.

3*e*. T. D. Beckwith, "The Bacterial Corrosion of Iron and Steel," *J. AWWA*, vol. 33, No. 1, January, 1941, p. 147.

3*f*. E. L. Durkee, "The Design of a Steel-pile Pier," *Civil Eng.*, vol. 7, No. 3, March, 1937, pp. 187–191.

3*g*. T. J. Barnett, "Features of Steel-pile Pier Construction," *Civil Eng.*, vol. 7, No. 3, March, 1937, pp. 191–194.

3*h*. J. G. Mason and A. L. Ogle, "Steel-pile Foundations in Nebraska," *Civil Eng.*, vol. 2, No. 9, September, 1932, pp. 553–557.

3*i*. J. S. Unger, "Steel in Sea Water Examined After 31 Years Exposure," *Eng. News-Record,* vol. 111, Nov. 16, 1933, pp. 593–594.

3*j*. "Paper Permits Piling Salvage," *Eng. News-Record,* vol. 129, No. 19, Nov. 5, 1942, p. 654.

3*k*. *Deterioration of Structures in Sea-Water,* Report of the Committee of the Institution of Civil Engineers, London, 1920, and Annual Interim Reports through 1938.

3*l*. "Special Rig Pulls Twisted Sheetpiles," *Eng. News-Record,* vol. 133, No. 26, Dec. 28, 1944, pp. 820–822.

3*m*. R. Trench, "Cast-iron Screw—Driven to Rock," *Eng. News-Record,* vol. 133, No. 26, Dec. 28, 1944, pp. 830–838.

3*n*. R. C. Manning, "Long Life of Structural Steel Piles," *Can. Engr.*, July 7, 1931, p. 25.

3*o*. F. N. Speller, *Corrosion, Causes and Prevention: An Engineering Problem,* 3d ed., McGraw-Hill Book Company, Inc., 1951.

3*p*. "Long Steel Piles Spliced by Welding in Rotating Jigs," *Construction Methods,* vol. 28, No. 8, August, 1946, pp. 96–98.

3*q*. "Outdoor Laboratory Provides Information on Effects of Marine Corrosion," *Civil Eng.*, vol. 17, No. 10, October, 1947, pp. 32–34, 78.

3*r*. "Screw Piles Support Turkish Pier," *Eng. News-Record,* vol. 140, No. 2, Jan. 8, 1948, p. 99.

3*s*. "Concrete Jacketing H-piles in Soft Mud," *Eng. News-Record,* vol. 140, No. 6, Feb. 5, 1948, p. 179.

3*t*. *Research Report No. 1—Investigation of the Strength of the Connection Between a Concrete Cap and the Embedded End of a Steel H-pile,* State of Ohio, Department of Highways, Dec. 1, 1947.

* Also see items 36, 74, 85, 97, 103, 2*ab*, 2*ax*, 5*aq*, 5*as*, and 11*m*. Also see section on Pipe Piles.

3*u*. C. G. Cappel, "Steel H-piles and Pipe Bents Support Deep-water Drilling Platform," *Civil Eng.*, vol. 18, No. 7, July, 1948, pp. 46–48.

3*v*. "Steel Sheet Piling," *Steel Products Manual, Section* 5, American Iron and Steel Institute, New York, June, 1948.

3*w*. A. Casagrande, "The Pile Foundation for the New John Hancock Building in Boston," *J. Boston Soc. Civil Engrs.*, vol. 34, No. 4, October, 1947, pp. 297–315.

3*x*. B. G. White, "The Construction of Military Ports in Gareloch and Loch Ryan," *Civil Eng. and Pub. Works Rev.*, vol. 44, No. 514, April, 1949, pp. 212.

3*y*. "Huge Guide-traveler Speeds Driving of 65 Miles of Sheetpiling," *Eng. News-Record*, vol. 143, No. 19, Nov. 10, 1949, pp. 39–41.

3*z*. P. Leimdörfer, "Some View on the Selection of Steel Sheet Piling," paper in Sec. I, *Communication* 1 *of the* 17*th Intern. Nav. Congress*, Lisbon, 1949.

3*aa*. E. P. Muntz, "Steel Rail Piles Replace Concrete Piles," *Eng. J.*, vol. 30, No. 4, April, 1947, pp. 161–164.

3*ab*. D. T. Cremans, "El Paso Rail Cut Cuts Traffic Problem," *Eng. News-Record*, vol. 145, No. 10, Sept. 7, 1950, pp. 46–48.

3*ac*. "Bridge Foundation is Built from Top Down," *Eng. News-Record*, vol. 146, No. 2, Jan. 11, 1951, pp. 39–41.

3*ad*. G. G. Greulich, "Extracted Steel H-Piles Found in Good Condition," *Eng. News-Record*, vol. 145, No. 8, Aug. 24, 1950, pp. 41–42.

3*ae*. H. J. Schaufele, "Erosion and Corrosion on Marine Structures, Elwood, California," *Coastal Engineering, Proc. First Conf. on Coastal Eng.*, Long Beach, Calif., 1951, pp. 326–334.

3*af*. *Evaluation of Steel Beam Piling*, Final Report, ARF M516, Armour Research Foundation of Illinois Institute of Technology, Chicago, Ill., for United States Steel Corp., December, 1954.

3*ag*. "Fireproof Dock Serves Refiner," *Eng. News-Record*, vol. 53, No. 17, Oct. 21, 1954, pp. 34, 36, 37.

3*ah*. B. C. Lattin, "Steel Piling in Sea Water," *Consulting Engr.*, vol. 10, No. 6, June, 1958, pp. 78–83.

3*ai*. "Cathodic Protection Installation for Kwinana Refinery Jetty and Water Intake Flume," *Civil Eng. and Public Works Rev.*, vol. 50, No. 589, July, 1955, pp. 769, 770.

3*aj*. H. F. Cornick, "Corrosion and Preservation of Iron and Steel—A Review of Recent Technical Researches," *Dock and Harbour Authority*, vol. 33, No. 387, January, 1953, pp. 282–285, and No. 388, February, 1953, pp. 312–315. (Contains bibliographies.)

3*ak*. H. H. Uhlig, *The Corrosion Handbook*, John Wiley & Sons, Inc., New York, 1948.

3*al*. L. T. Ryan, "Cathodic Protection of Steel Piled Wharves—Description of Installations at Lae, New Guinea," *Dock and Harbour Authority*, vol. 35, No. 412, February, 1955, pp. 303–307, and No. 413, March, 1955, pp. 329–331, 334.

3*am*. J. C. Hudson, "The Corrosion of Iron and Steel and Its Prevention—With Special Reference to Harbour and Dock Installations," *Dock and Harbour Authority*, vol. 34, No. 392, June, 1953, pp. 47–53; No. 393, July, 1953, pp. 81–84; No. 394, August, 1953, pp. 111–116; No. 395, September, 1953, pp. 143–146. (Contains bibliography.)

3*an*. E. P. and G. L. Doremus, "Cathodic Protection of Fourteen Offshore Drilling Platforms," *Corrosion*, vol. 6, No. 7, July, 1950, pp. 216–224.

3*ao*. L. M. Applegate, "Cathodic Protection in Military Construction," *Military Engr.*, vol. 49, No. 329, May–June, 1959, pp. 188–191.
3*ap*. K. A. Spencer, "Cathodic Protection of Ships and Marine Structures," *Dock and Harbour Authority*, vol. 38, No. 447, January, 1958, pp. 307–312, and No. 448, February, 1958, pp. 361–363. (Contain bibliographies.)
3*aq*. A. C. Rayner, "Life of Steel Sheet Pile Structures in Atlantic Coastal States," *Coastal Engineering, Proc. Third Conf. on Coastal Eng.*, Cambridge, Mass., 1952, pp. 209–216.
3*ar*. C. R. Johnson, "Integration of Corrosion Control in Pier Substructures," *Corrosion*, vol. 12, No. 4, April, 1956, pp. 19–22.
3*as*. H. A. Humble, "The Cathodic Protection of Steel Piling in Sea Water," *Corrosion*, vol. 3, No. 9, September, 1949, pp. 292–302.
3*at*. R. M. Robinson, "Controlling Corrosion of Offshore Platforms," *Corrosion*. vol. 14, No. 11, November, 1958, pp. 93, 94, 96.
3*au*. W. R. Shepherd, "Cathodic Protection of Steel Sheet Piling on the River Tyne," *Dock and Harbour Authority*, vol. 39, No. 459, January, 1959, pp. 290–292.
3*av*. C. W. N. McGowan, Paper, *Proc. Eighteenth Intern. Navigation Congr.*, Sect. 2, Commun. 1, Rome, 1953, pp. 57–86.
3*aw*. *U.S.S. Steel H-piles*, United States Steel Corporation, Pittsburgh, Pa. (1961).

COMPOSITE PILES

4*a*. H. D. Dewell, "Experience in Driving and Testing Composite Piles," *Civil Eng.*, vol. 8, No. 9, September, 1938, pp. 574–577.

4*b*. "Super Rig Drives Super Piles," *Construction Methods*, vol. 31, No. 4, April, 1949, pp. 34–37.

4*c*. W. E. Dean, "Four Lanes across Tampa Bay at Low Cost," *Eng. News-Record*, vol. 157, No. 12, Sept. 20, 1956, pp. 42–48.

CONCRETE PILES INCLUDING DETERIORATION AND PRESERVATION*

5a. *Concrete Piles*, Portland Cement Association.

5b. *Recommended Practice and Standard Specifications for Concrete and Reinforced Concrete*, Report of Joint Committee on Standard Specifications for Concrete and Reinforced Concrete, Affiliated Committees of ACI, AIA, AREA, ASCE, ASTM, and PCA, June, 1940.

5c. J. H. Thornley, "Navy Warehouse Foundation on Bottom-bearing Piles with Concrete Tips," *Civil Eng.*, vol. 12, No. 7, July, 1942, pp. 372–373.

5d. N. A. Melick, "Cast-in-place Concrete Pile Work in Quaking Clay," *Eng. News-Record*, vol. 102, No. 21, May 23, 1929, pp. 823–826.

5e. O. J. West, "Cored-out Pipe Piles in Sticky Clay," *Eng. News-Record*, vol. 101, Nov. 29, 1928, pp. 799–800.

5f. F. J. Converse, "Cast-in-place Short Piles Show High Test Results," *Eng. News-Record*, vol. 115, Dec. 19, 1935, pp. 842–844.

5g. K. A. Reeder, "Boulder Transmission Line Utilizes Poured-in-place Foundations," *Civil Eng.*, vol. 11, No. 12, December, 1941, pp. 722–725.

5h. B. H. Johnson, "Precast Concrete Piles for Residential Foundations," *Civil Eng.*, vol. 11, No. 9, September, 1941, pp. 549–550.

5i. "Tests on Concrete Piles in Sea Water Undertaken by Navy," *Eng. News-Record*, vol. 97, No. 10, Sept. 2, 1926, pp. 372–373.

5j. "Bituminized Concrete Casing for Piles in Sea Water," *Eng. News-Record*, vol. 114, No. 4, Jan. 24, 1935, pp. 121–124.

5k. W. R. Shettel, "Asphalt-impregnated Slabs of Concrete in Salt Water," *Eng. News-Record*, vol. 110, No. 4, Jan. 26, 1933, pp. 108–109.

5l. G. F. Nicholson, "Protecting Reinforced Concrete Against Deterioration by Sea Water," *Civil Eng.*, vol. 1, No. 14, November, 1931, pp. 1243–1248.

5m. J. J. Collins, "New Type Sea Wall Built for Subsiding Lake Shore in Venezuela," *Eng. News-Record*, vol. 114, No. 12, Mar. 21, 1935, pp. 405–408.

5n. "Splicing Long Concrete Piles," *Concrete*, vol. 53, No. 4, April, 1945, pp. 2–3.

5o. "Simple Forms Cut Pile Casting Costs," *Construction Methods*, vol. 28, No. 6, June, 1946, pp. 86–88, 159.

5p. "Pre-cast and In Situ Piling," *Civil Eng. and Pub. Works Rev.*, vol. 30, December, 1935, p. 378.

5q. "Sheetpiles Cast on Concrete Platform for Kansas City Floodwall," *Construction Methods*, vol. 28, No. 9, September, 1946, pp. 96–99.

5r. "Driving 3,850 Piles for 60-ton Loads," *Eng. News-Record*, vol. 137, No. 22, Nov. 28, 1946, pp. 722–723; and vol. 137, No. 24, Dec. 12, 1946, p. 779.

5s. R. Glossop and I. S. Greeves, "A New Form of Bored Pile," *Concrete and Constructional Engineering*, vol. 41, No. 12, December, 1946, pp. 344–351.

5t. "L & N Replaces Long Timber Trestle," *Concrete for Railways*, vol. 11, No. 2, pp. 1–11.

* Also see items 27, 85, *1i*, *2ab*, *2ax*, *3k*, *3ax*, and *12r*.

5*u*. H. Caudill, "High Capacity Piles on Chicago Hardpan," *Eng. News-Record*, vol. 139, No. 10, Sept. 4, 1947, pp. 315–317.

5*v*. R. R. Shoemaker, "Concrete-filled Steel Piles Support Wharf at Long Beach, Calif.," *Eng. News-Record*, vol. 139, No. 10, Sept. 4, 1947, pp. 328–331.

5*w*. "Fitting Foundations to Variable Ground," *Eng. News-Record*, vol. 140, No. 6, Feb. 5, 1948, pp. 194–197.

5*x*. T. E. Stanton, "Durability of Concrete Exposed to Sea Water and Alkali Soils—California Experience," *J. ACI*, vol. 19, No. 9, May, 1948, pp. 821–847.

5*y*. "Electrolysis in Reinforced Concrete," *Engineering*, vol. 166, No. 4319, Nov. 5, 1948, pp. 453–4.

5*z*. H. W. Hunt, "Thin-shelled Cast-in-place Concrete Piles Driven 102 Ft," *Civil Eng.*, vol. 19, No. 3, March, 1949, pp. 38–39.

5*aa*. "Bridge on Santa Fe," *Concrete for Railways*, vol. 13, No. 1, pp. 13–15.

5*ab*. "First Rails across the Tigris," *Eng. News-Record*, vol. 142, No. 24, June 16, 1949, pp. 18–19.

5*ac*. F. H. Fay, paper in Sec. II, *Communication* 2 *of the* 17*th Intern. Nav. Congress*, Lisbon, 1949.

5*ad*. *Building Code Requirements for Reinforced Concrete*, American Concrete Institute Standard (ACI 318–56).

5*ae*. "New Method of Pile Driving," *New Eng. Construction*, vol. 14, No. 7, December, 1949, pp. 46, 68–69.

5*af*. E. A. F. Huber, "Oplegging, transport en hijsen van heipalen van gewapend beton," *Cement*, No. 15–16, 1950, pp. 320–327, 337.

5*ag*. "New Tool for Pile-driving—Pneumatic Mandrel Gets Test," *Eng. News-Record*, vol. 146, No. 2, Jan. 11, 1951, pp. 35–36.

5*ah*. H. K. Cook, "Protection of Concrete Piles," *Military Engr.*, vol. 43, No. 293, May–June, 1951, pp. 183–185.

5*ai*. S. J. Parrett, "First Prestressed Piles Carry Tank Platform," *Eng. News-Record*, vol. 147, No. 1, July 5, 1951, pp. 38, 39.

5*aj*. "Concrete in Sulphate-bearing Clays and Ground Waters," *Civil Eng. and Public Works Rev.*, vol. 46, No. 542, August, 1951, pp. 598, 599, and No. 543, September, 1951, pp. 683, 684.

5*ak*. N. L. Liver, E. C. Mardorf, and J. C. King, "Development and Application of Intrusion Grout Mixed-in-place Piles," *The Prepakt Reporter*, March–April, 1954, pp. 1–11.

5*al*. L. A. Gerdine, "Mixed-in-place Piles for Cut-off Wall," *The Prepakt Reporter*, January, 1957, p. 5.

5*am*. H. T. Perez, "Dual-purpose Piles Cut Building Costs," *Construction Methods and Equipment*, vol. 36, No. 5, May, 1954, pp. 54–57.

5*an*. R. H. Mann, "Jackets of Creosoted Wood Protect Concrete Piles," *Eng. News-Record*, vol. 158, No. 7, Feb. 14, 1957, p. 41.

5*ao*. ———, "Wooden Jackets Save Concrete Piles," *Wood Preserving News*, vol. 35, No. 1, January, 1957, pp. 12–14.

5*ap*. *Concrete Manual*, U.S. Dept. of the Interior, Bureau of Reclamation, Denver, Colo.

5*aq*. H. K. Cook, Paper, *Proc. Eighteenth Intern. Navigation Congr.*, Sect. 2, Commun. 1, Rome, 1953, pp. 15–39. (Includes extensive bibliography on deterioration and protection of marine structures. Also contains other articles on deterioration of marine wood and steel piles.)

5*ar*. C. M. Wakeman, E. V. Dockweiler, H. E. Stover, and L. L. Whiteneck, "Use of Concrete in Marine Environments," *J. ACI*, Title 54-46, vol. 29, No. 10, April,

1958, pp. 841–856; also discussions in vol. 30, No. 6, Part 2, December, 1958, pp. 1309–1346. (Both include extensive bibliographies.)

5*as*. L. E. Stiffler, Jr., "The Effects of Sea Water on Waterfront Structures," *Dock and Harbour Authority*, vol. 38, No. 443, September, 1957, pp. 181, 182, and No. 444, October, 1957, pp. 217–220.

5*at*. Herbert Brauner, "Prestressed Concrete Units of the Petaluma Creek Bridge at Black Point, California," *J. Prestressed Concrete Inst.*, vol. 3, No. 1, June, 1958, pp. 59–62.

5*au*. Ben C. Gerwick, Jr., "Torsion in Concrete Piles during Driving," *J. Prestressed Concrete Inst.*, vol. 4, No. 1, June, 1959, pp. 58–63.

5*av*. E. A. L. Smith, "Tension in Concrete Piles during Driving," *J. Prestressed Concrete Inst.*, vol. 5, No. 1, March, 1960, pp. 35–40.

5*aw*. T. Y. Lin, "Design Procedures for Pretensioning Prestressed Concrete Bearing Piles and Sheet Piles," Ben C. Gerwick, Inc., Precast-Prestressed Concrete Div., San Francisco, Calif., October, 1956.

5*ax*. H. K. Preston, *Practical Prestressed Concrete*, McGraw-Hill Book Company, Inc., New York, 1960.

5*ay*. R. L. Peck, "105 Miles of Concrete Piling," *Modern Concrete*, vol. 23, No. 1, May, 1959, pp. 46–51, 64.

5*az*. *Specification for Pre-tensioned Prestressed Concrete*, SPC-100-57T Prestressed Concrete Institute.

5*ba*. ACI-ASCE Joint Committee, "Tentative Recommendations for Prestressed Concrete," *J. ACI*, vol. 7, No. 29, January, 1958, pp. 545–578; also *J. Structural Div.*, No. ST 1, *Proc. ASCE, Paper* 1519, vol. 84, January, 1958, pp. 1519-1, 1519-39.

5*bb*. W. E. Dean, "Prestressed Concrete Sheetpiles for Bulkheads and Retaining Walls," *Civil Eng.*, vol. 30, No. 4, April, 1960, pp. 68–71.

5*bc*. Myers Van Buren, "Piles Double as Columns of Tollway Structures," *Civil Eng.*, vol. 28, No. 6, June, 1958, pp. 48–50.

5*bd*. W. E. Dean, "Prestressed Concrete—Difficulties Overcome in Florida Bridge Practice," *Civil Eng.*, vol. 27, No. 6, June, 1957, pp. 60–63.

5*be*. M. M. Upson, "Prestressing Warrants Study to Simplify Construction Methods," *Civil Eng.*, vol. 23, No. 1, January, 1953, pp. 2–5.

5*bf*. "Driving Tests Show Prestressed Piles Will Take More Punishment," *Eng. News-Record*, vol. 157, No. 6, Aug. 9, 1956, pp. 33, 34.

5*bg*. H. K. Preston, "Proper Use of Wire and Strand in Prestressed Concrete," *J. Prestressed Concrete Inst.*, vol. 2, No. 3, December, 1957, pp. 18–20.

5*bh*. W. E. Dean, "Peculiarities and Shortcomings of Prestressed Concrete in Florida Bridge Practice," *J. Prestressed Concrete Inst.*, vol. 2, No. 1, June, 1957, pp. 32–37.

5*bi*. *Highway Engineering Report No.* 3-60, Armco Drainage & Metal Products, Inc.

PIPE PILES*

6*a*. "Extra Long Cylinder Piles Carry Column Loads to Rock," *Eng. News-Record*, vol. 109, No. 1, July 7, 1932, pp. 5–7.

6*b*. H. T. Immerman, "Concrete Filled Steel Cylinders for Foundations," *General Building Contractor*, vol. 1, No. 2, pp. 31–36.

6*c*. "Pipe Piles Cleaned Out by Jetting Through Hood," *Construction Methods*, vol. 30, No. 6, June, 1948, pp. 104–105.

6*d*. M. P. Anderson, "Sea-going Construction Plant Drives Huge Pipe-piles to Close Tolerances," *Civil Eng.*, vol. 19, No. 5, May, 1949, pp. 30–33.

6*e*. "Here's a New Type of Foundation for Marine Structures," *Eng. News-Record*, vol. 147, No. 9, Aug. 30, 1951, pp. 41–43.

6*f*. Russell Barnard, *Pipe Piles for Bridges and Buildings*, Armco Drainage and Metal Products, Inc., Middletown, Ohio, 1956.

* Also see items 82 and 5*e*. Also see section on Steel and Iron Piles, including Corrosion and Preservation.

CAISSON-TYPE PILES

7a. R. E. White, "Heavy Foundations Drilled into Rock," *Civil Eng.*, vol. 13, No. 1, January, 1943, pp. 19–22.

7b. C. B. Spencer, "Drilled-In Caissons with Heavy H-cores Carry 16-story Navy Warehouse," *Eng. News-Record*, vol. 127, No. 13, Sept. 25, 1941, pp. 431–434.

7c. G. B. Sowers, "Through 130 Ft of Plastic Clay to Rock," *Eng. News-Record*, vol. 131, No. 21, Nov. 18, 1943, pp. 107–110.

7d. G. B. Sowers, "Open Steel Caissons for Cast-in-place Piles," *Eng. News-Record*, vol. 127, No. 21, Nov. 20, 1941, p. 717.

7e. H. A. Christie, "Boring Machine Digs Wells for Concrete Piers," *Eng. News-Record*, vol. 109, July 28, 1932, p. 105.

7f. J. W. Beretta, "Drilled, Reamed Holes for Bridge Foundation Piles," *Eng. News-Record*, vol. 118, Feb. 11, 1937, pp. 217–218.

7g. C. B. Spencer, "Deep Caissons Support River-side Power Plant," *Civil Eng.*, vol. 14, No. 2, February, 1944, pp. 67–68.

7h. C. G. Cappel, "Steel Caissons Support New Orleans Dock," *Civil Eng.*, vol. 13, No. 10, October, 1943, pp. 459–461.

7i. E. J. Cleary, "Prefabricated Caisson Foundations for Submarine Oil Wells," *Eng. News-Record*, vol. 137, No. 26, Dec. 26, 1946, pp. 859–867.

7j. "Big Bell-bottom Caissons Will Support Detroit's Second Civic Center Building," *Eng. News-Record*, vol. 147, No. 23, Dec. 6, 1951, pp. 44–46.

7k. R. E. White, "Grouted-stone Fills Caissons," *Construction Methods and Equipment*, vol. 40, No. 12, December, 1958, pp. 73, 74, 76.

CAISSONS

8*a*. F. W. Skinner, "Construction Methods Used on Cleveland Terminal," *Eng. and Contr.*, vol. 66, No. 7, July, 1927, pp. 297–300.

8*b*. "Concrete Lagging Used for Deep-well Caissons," *Eng. News-Record*, vol. 109, No. 19, Nov. 10, 1932, p. 563.

8*c*. J. Feld, "Precast Concrete Pit Sheeting," *J. ACI*, vol. 16, No. 5, April, 1945, pp. 441–451 (*ACI Proc.*, vol. 41).

8*d*. *Handbook of Drainage and Construction Products*, Armco Drainage & Metal Products, Inc., Middletown, Ohio, 1958.

8*e*. "Mobile Rig Digs Deep Caissons," *Construction Methods and Equipment*, vol. 39, No. 8, August, 1957, pp. 58–61.

SAND PILES*

9*a*. "Cellular Cofferdam Serves as Wall of Ship Construction Basins," *Eng. News-Record*, vol. 131, No. 17, Oct. 21, 1943, pp. 588–594.

9*b*. W. W. Moore, "Pacific Coast Quay Founded on Deep Clay," *Civil Eng.*, vol. 14, No. 12, December, 1944, pp. 515–517.

* Also see items 25, 10*r*, 10*x*, 10*y*, and 10*ab*.

SOIL STRENGTHENING

10*a*. J. D. Lewin, "Grouting with Chemicals," *Eng. News-Record*, vol. 123, No. 7, Aug. 17, 1939, pp. 221–222.

10*b*. K. Terzaghi, *Proc. Intern. Conf. Soil Mech. and Foundation Eng., Paper M*-6, Cambridge, Mass., vol. 3, 1936, pp. 180–182.

10*c*. K. Endell and U. Hoffmann, "Electrochemical Hardening of Clay Soils," *Proc. Intern. Conf. Soil Mech. and Foundation Eng.*, Cambridge, Mass., vol. 1, 1936, pp. 273–275.

10*d*. "Grouting Checks Vibration," *Eng. News-Record*, vol. 129, No. 17, Oct. 22, 1942, pp. 95–96.

10*e*. J. R. Pfeiffer, "A New Method of Impermeabilizing and Improving the Physical Properties of Pervious Subsoils by Injecting Bituminous Emulsions," *Proc. Intern. Conf. Soil Mech. and Foundation Eng.*, Cambridge, Mass., vol. 1, 1936, pp. 263–266.

10*f*. *Eng. News-Record*, vol. 131, No. 17, Oct. 21, 1943, p. 623.

10*g*. *Dowflakes*, vol. 6, No. 1, January, 1944.

10*h*. "Chemical Injections Solidify Soil to Stop Foundation Settlement," *Construction Methods*, vol. 27, No. 12, December, 1945, pp. 78, 79, 156, 158, 160, 162.

10*i*. D. E. Moran and F. O. Dufour, "Increasing the Bearing Power of Clay Soil," *Eng. News-Record*, vol. 108, No. 20, May 19, 1932, p. 726.

10*j*. J. W. Kelly and B. D. Keatts, "Two Special Methods of Restoring and Strengthening Masonry Structures," *J. ACI*, vol. 17, No. 4, February, 1946, pp. 289–304.

10*k*. "Building Sites Stabilized with Grout," *Eng. News-Record*, vol. 134, No. 2, Jan. 11, 1945, pp. 68–69.

10*l*. C. M. Riedel, "Chemical Soil Solidification and Chemical Sealing of Leaking Concrete," *J. AWWA*, vol. 37, No. 9, September, 1945, pp. 849–862.

10*m*. C. M. Riedel, "The Joosten Chemical Soil Solidification Process," *Improvement Bull.*, June 4, 1943.

10*n*. C. M. Riedel, "Chemical Joint Sealing and Soil Solidification," *Eng. News-Record*, vol. 127, No. 7, Aug. 14, 1941, pp. 222–225.

10*o*. H. Japp, "New Methods of Excavating Quicksands," *Engineer*, vol. 156, No. 4059, Oct. 27, 1933, pp. 398–402; and vol. 156, No. 4060, Nov. 3, 1933, pp. 422–425.

10*p*. F. L. Kassel, "Soil Consolidation by Electrolysis," *Civil Eng. and Pub. Works Rev.*, vol. 41, No. 479, May, 1946, pp. 187–188.

10*q*. H. E. Berger, "Well Points Used to Compact Sand Fill," *Eng. News-Record*, vol. 137, No. 2, July 11, 1946, pp. 44–45.

10*r*. "Vertical Sand Drains Speed Consolidation of Soft Water-bearing Subsoil," *Eng. News-Record*, vol. 138, No. 10, March 6, 1947, pp. 361, 386–389.

10*s*. A. H. D. Markwick and A. F. Dobson, "Application of Electrosmosis to Soil Drainage," *Engineering*, vol. 163, No. 4230, Feb. 7, 1947, pp. 121–122.

10*t*. D. J. Maclean and D. W. Rolfe, "Soil Drainage by an Electrical Method," *Civil Eng. and Pub. Works Rev.*, vol. 40, No. 464, 1945, p. 34, and *Road Abstracts*, vol. 12, No. 207, 1945.

10*u*. L. Casagrande, "Electrical Drainage of Fine-grained Soils," *Strasse*, vol. 8 (12/20), 1941, p. 3241, and *Road Abstracts*, vol. 9, No. 216, 1942.

10*v*. H. R. Reynolds and P. Protopapadakis, "Practical Problems in Soil Mechanics, Part XIV—Artificial Cementation and Ground Water Lowering," *Civil Eng. and Pub. Works Rev.*, vol. 42, No. 490, April, 1947, pp. 158–161.

10*w*. N. J. Cochrane, "Analysis of the Position in Soil Mechanics," *Engineering*, vol. 164, No. 4257, Aug. 29, 1947, pp. 213–214.

10*x*. *German Roads and Soil Stabilization*, British Intelligence Objectives Sub-Committee Final Report No. 917, Item No. 33, H. M. Stationery Office, London.

10*y*. *Report on Soil Compaction* by E. E. Halmas, issued by Parsons, Brinckerhoff, Hogan & Macdonald, New York, 1945.

10*z*. A. Machis, "Experimental Observations on Grouting Sands and Gravels," *Proc. ASCE*, vol. 72, No. 9, September, 1946, pp. 1203–1227, and subsequent discussions.

10*aa*. N. S. Bateman, "New Piers Save Old Bridges," *Eng. News-Record*, vol. 140, No. 4, Jan. 22, 1948, pp. 113–117.

10*ab*. J. C. Carpenter, "Stabilize Muck in Puerto Rico," *Eng. News-Record*, vol. 141, No. 4, July 22, 1948, pp. 52–54.

10*ac*. B. Fruhauf, "Wet Vibration Puts Strength in Sand," *Eng. News-Record*, vol. 142, No. 25, June 23, 1949, pp. 60–62.

10*ad*. A. B. Bridgwater, "Electro-kinetic Phenomena in Soils," *Civil Eng. and Pub. Works Rev.*, vol. 45, 1950: Part I, No. 526, April, pp. 234–236; Part II, No. 527, May, pp. 313–315; Part III, No. 528, June, pp. 385–387; Part IV, No. 528, July, pp. 451–452 (including bibliography).

10*ae*. H. E. Mussche and J. C. Waddington, "Applications of the Freezing Process to Civil Engineering Works," *Inst. Civil Engrs, Works Construction Div.*, May 14, 1946.

10*af*. H. E. Mussche, "A Blast Furnace Skip Pit Constructed by the Freezing Process," *Civil Eng. and Pub. Works Rev.*, vol. 45, No. 527, May, 1950, pp. 301–303 (including bibliography).

10*ag*. A. J. S. Taunton, "Dry Ice Creates Frozen Cofferdams for Bridge Pier Repair," *Civil Eng.*, vol. 20, No. 8, August, 1950, pp. 24–26, 85.

10*ah*. W. Schaad and R. Haefeli, "Electro-kinetic Phenomena and their Application to Soil Mechanics," *Technical Translation TT*-61, National Research Council of Canada, Feb. 25, 1948.

10*ai*. C. M. Riedel, "Chemicals Stop Cofferdam Leaks," *Civil Eng.*, vol. 21, No. 4, April, 1951, pp. 23–24.

10*aj*. G. Remenieras, "Application de l'électroosmose à l'exécution de certains travaux en terrains aquifères," *Houille Blanche*, special number A/1949, pp. 393–404. ("Application of Electroosmosis to the Execution of Certain Works on Water-bearing Soils," *U.S. Geol. Survey, Transl.* 21, 1951.)

10*ak*. "Soil Stabilization by a Thermal Treatment," *Civil Eng. and Public Works Rev.*, vol. 50, No. 594, December, 1955. (Review of article by I. M. Litvinov in *Stroitelnaya Promishlennost*, October, 1955.)

10*al*. R. Loughney, "Electricity Stiffens Clay Fivefold for Electric Plant Excavation," *Construction Methods and Equipment*, vol. 36, No. 8, August, 1954, pp. 70, 71, 74, 78, 82.

10*am*. "Electric Curtain Stabilizes Wet Ground," *Construction Methods and Equipment*, vol. 35, No. 4, April, 1953, pp. 52–58.

10*an*. Y. Poisson, "Traitement d'argile plastique par le courant électrique," *Travaux*, No. 164, 1948, pp. 386, 387. ("Electrical Treatment of Plastic Clay," *U.S. Geol. Survey, Transl.* 24, 1951.)

10*ao*. F. F. Kravath, "Soil Stabilization by Electroosmosis," *Military Engr.*, vol. 46, No. 312, July–August, 1954, pp. 273–275.

10*ap*. J. P. Gnaedinger, "Soil Stabilization by Injection Techniques," *Ind. Eng. Chem.*, vol. 47, November, 1955, pp. 2249–2253.

10*aq*. G. E. Johnson, "Stabilization of Soil by Silt Injection Method," *Proc. ASCE*, vol. 79, Separate No. 323, November, 1953.

10*ar*. "Grouting of Foundation Sands and Gravels," *Corps of Engineers, Waterways Expt. Sta., Tech. Mem.* 3-408, Vicksburg, Miss., June, 1955. (Includes bibliography.)

10*as*. C. F. Kollbruner and C. Blatter, "Injections—Stabilization and Densification of Pervious Soils, Fissured Rock, Porous Walls, Concrete, etc., *Private Soc. Soil Research and Soil Mech., Rept.* 4, Zurich, June, 1941.

10*at*. *The Use of AM-9,* American Cyanamid Co., New York, 1957.

10*au*. "Electricity Dries Wet Silt," *Eng. News-Record,* vol. 150, No. 18, Apr. 30, 1953, pp. 36, 37.

10*av*. R. H. Karol, "Development of a New Chemical Grout," *Proc. ASTM*, vol. 57, 1957, pp. 1219–1232.

10*aw*. L. Casagrande, "Electro-osmotic Stabilization of Soils," *J. Boston Soc. Civil Engrs.*, vol. 39, No. 1, January, 1952, pp. 51–83.

10*ax*. R. H. Karol, *Soils and Soil Engineering,* Prentice-Hall, Englewood, N.J., 1960.

10*ay*. P. A. Wild, "Tower Foundations Compacted by Explosives," *Electrical World,* vol. 155, No. 2, Jan. 9, 1961, pp. 36–38, 66.

10*az*. "Foundation Soil Compacted by Explosives," *New England Construction,* vol. 25, No. 13, Nov. 21, 1960, pp. 24, 25.

10*ba*. *A Unique and Exclusive Controlled Blasting Process, Tailored to Individual Conditions by Experienced Engineers,* Soil Compactors, Inc., Tampa, Fla.

PILE CATALOGUES

11*a*. *Steel H-beam Bearing Piles*, United States Steel Corp.

11*b*. *Typical Installations—Test Data—Details of Bethlehem H-Piling*, Booklet 99-B, Bethlehem Steel Co.

11*c*. *The B.S.P. Pocket Book*, British Steel Piling Co., Ltd.

11*d*. *Typical Installations of Bethlehem Steel Sheet Piling*, Booklet 127, Bethlehem Steel Co.

11*e*. *Steel Sheeting Piling*, Cat. 151-E, Bethlehem Steel Co.

11*f*. *U.S.S. Steel Sheet Piling*, United States Steel Corp.

11*g*. *Steel Sheet Piling*, Weirton Steel Co.

11*h*. *Inland Steel Sheet Piling and Accessories*, Inland Steel Co.

11*i*. *Z-type Steel Sheet Piling*, Bethlehem Steel Co.

11*j*. *Corrugated Steel Sheet Piling*, Caine Steel Co.

11*k*. *Armco Metal Sheeting*, Armco Drainage Products Ass'n.

11*l*. *Armco Spiral Welded Pipe for Pipe Piles, Pile Shells, and Caissons*, The American Rolling Mill Co.

11*m*. *Seamless Steel Pipe Piles—Design Manual No.* 27, National Tube Co.

11*n*. *Monotube Piles*, Cat. 91, Union Metal Mfg. Co.

11*o*. *Raymond Piles*, Cat. S-58, Raymond Concrete Pile Co.

11*p*. *Compressed Concrete Piles*, Western Foundation Corp.

11*q*. *Cased Concrete Piles*, Western Foundation Corp.

11*r*. *Compressed Concrete Piles*, MacArthur Concrete Pile Corp.

11*s*. *Raymond Composite Piles*, Raymond Concrete Pile Co.

11*t*. *Foundations—Underpinning*, Spencer, White & Prentis, Inc.

11*u*. *Circular Pre-cast Concrete Piles*, Seisel Construction Co.

11*v*. *Massey Reinforced Concrete Piles*, Massey Concrete Products Co.

11*w*. *Franki Piles* and *Pieux Franki*, Franki Compressed Pile Co., Ltd.

11*x*. *Franki Foundation Company*, Franki Foundation Co.

11*y*. *Foundations and Underpinning*, Underpinning & Foundation Co., Inc.

11*z*. *Drilled-In Caissons*, Drilled-In Caisson Corp.

11*aa*. *Buda-Hubron Earth Drill*, *Bull.* 1019-*B*, The Buda Co.

11*ab*. *Bulletin*, The Presscrete Co., Inc.

11*ac*. *The Prestcore Vibrationless Concrete Piling System*, The British Steel Piling Co., Ltd.

11*ad*. *Vibro Concrete Piles*, The British Steel Piling Co., Ltd.

11*ae*. *List of Algoma Sections Rolled on Mills*, Algoma Steel Corp., Ltd.

11*af*. *Palplanches métalliques*, La Société Lorraine des Aciéries de Rombas.

11*ag*. *Palplanches métalliques type "Senelle,"* Société Métallurgique de Senelle—Maubeuge.

11*ah*. *Union-Flachprofil*, Dortmund-Hoerder-Hüttenverein A. G.

11*ai*. *California Earth Boring Machine*, California Welding and Blacksmith Shop, Inc.

11*aj*. *Cut Drilling Costs with a California Earth Boring Machine*, California Welding and Blacksmith Shop, Inc.

11*ak*. "The Intrusion-Prepakt Encasement Method for the Repair of Timber Piles," *The Prepakt Reporter*, Intrusion-Prepakt, Inc., April, 1950.

11*al*. *Western Concrete Piles and Caissons*, Western Foundation Corp.

11*am*. *Grey Beams*, Differdange, Grand-Duchy of Luxembourg (United States agents, Ameurope Metals, Inc., New York).

11*an*. *Larssen Steel Sheet Piling*, Dortmund-Hoerder Hüttenverein, A.G. (United States agents, Kurt Orban Co., Inc., New York).

11*ao*. *Columeta* (Terres Rouge and Belval Sheet Piling), Belval Works, Comptois Métallurgique Luxembourgeois, Luxembourg. (U.S. Agents, Amerlux Steel Products Corp., New York.)

11*ap*. *Arbed-Belval Steel Sheet Piling*. (U.S. Agents, Amerlux Steel Products Corp., New York.)

11*aq*. *Soil Compaction by Vibroflotation*, Vibroflotation Foundation Co., Pittsburgh, Pa.

11*ar*. *Bored Piling*, The Cementation Co., Ltd., London.

11*as*. *Rombas Steel Sheet Piling* (Larssen, Lackawanna, Rombas), Union Sidérurgique Lorraine, Paris.

11*at*. *Peiner Kastenspunwand Handbuch*, Hüttenwerke Ilsede-Peiner, Peine, Hanover, Germany, 2d ed., 1955.

11*au*. *Raymond Cylinder Piles of Prestressed Concrete*, Cat. CP-3, Raymond Concrete Pile Co.

11*av*. *The Prepakt Reporter*, October-November, 1958, or "I-P Cast-in-place Piling Services," *Intrusion-Prepakt, Inc. Spec. Rept.* 102, Cleveland, Ohio, 1958.

11*aw*. *Armco Foundation Pipe*, Armco Drainage & Metal Products, Inc., Welded Pipe Sales Div.

11*ax*. *Guild Foundation Piles*, C. L. Guild Construction Co., Inc.

PILE-DRIVING-EQUIPMENT CATALOGUES*

12*a*. *Drop Pile Hammers, Bull.* 66-F, Vulcan Iron Works.
12*b*. *Single-acting Pile Hammers, Bull.* 68-E, Vulcan Iron Works.
12*c*. *Double-acting Pile Hammers and Extractors, Bulls.* 58R and 58A, McKiernan-Terry Corp.
12*d*. *Portable Pile Hammer, Bull.* 30, Vulcan Iron Works.
12*e*. *Differential-acting Pile Hammer—Open Type, Bull.* 70-E, Vulcan Iron Works.
12*f*. *Pile-driving Accessories, Bulls.* 30, 40, 45 and 67, Vulcan Iron Works.
12*g*. *Union Pile Hammers, Cat.* 184, Union Iron Works.
12*h*. *Pile Hammer Leads, Bull.* 66R, McKiernan-Terry Corp.
12*i*. *Single-acting Pile Hammers, Bull.* 69, McKiernan-Terry Corp.
12*j*. *Concrete Piles and Caissons,* Western Foundation Corp.
12*k*. *Link-Belt Speeder Diesel Pile Hammer,* Cat. 2582, Link-Belt Co.
12*l*. *Equipment and Methods for Sand Drains, Bull.* 61, McKiernan-Terry Corp. (Bibliography.)
12*m*. *Diesel Pile Hammers, Bull.* 67, McKiernan-Terry Corp.
12*n*. *Delmag Diesel Pile Hammers,* The Foundation Equipment Corp., Newcomerstown, Ohio.
12*o*. *Pile Extractors, Bull.* 71-D, Vulcan Iron Works.
12*p*. *Pile Driving Equipment,* British Steel Piling Co., Ltd.
12*q*. Associated Pipe and Fitting Co., Inc. (Pile Fittings for Pipe, Shell, Timber, and Composite Piles).
12*r*. T. Y. Lin and W. J. Talbot, Jr., *Design Procedures for Pretensioned Prestressed Concrete Bearing Piles and Sheet Piles,* prepared for Ben C. Gerwick, Inc., Precast-Prestressed Div., San Francisco, Calif.
12*s*. *Raykin Fender Buffers,* The General Tire and Rubber Co.

* Also see item 11*c*.

INDEX

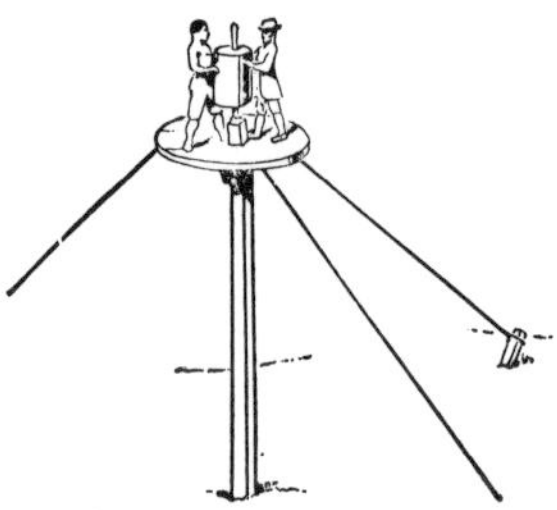

A hand ram

Driving pile with maul

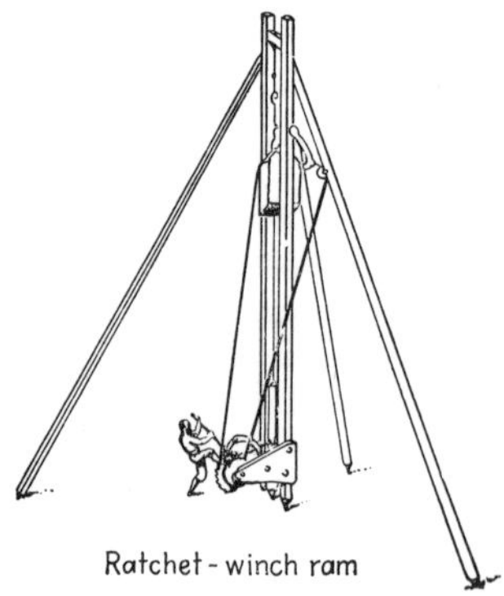

Ratchet-winch ram

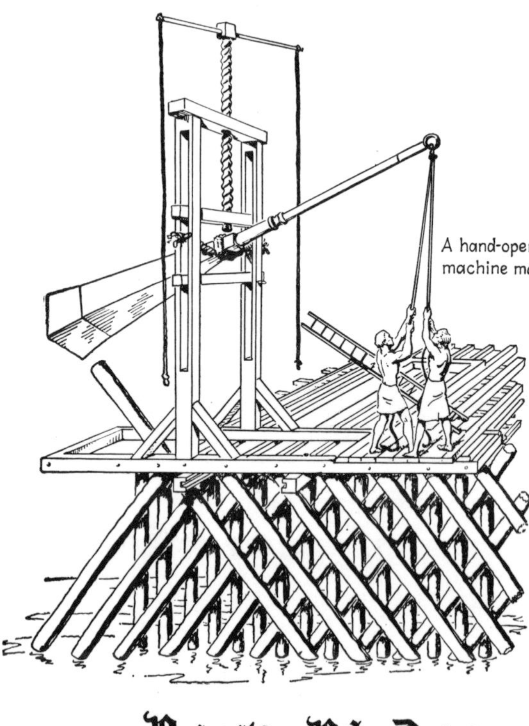

A hand-operated machine maul

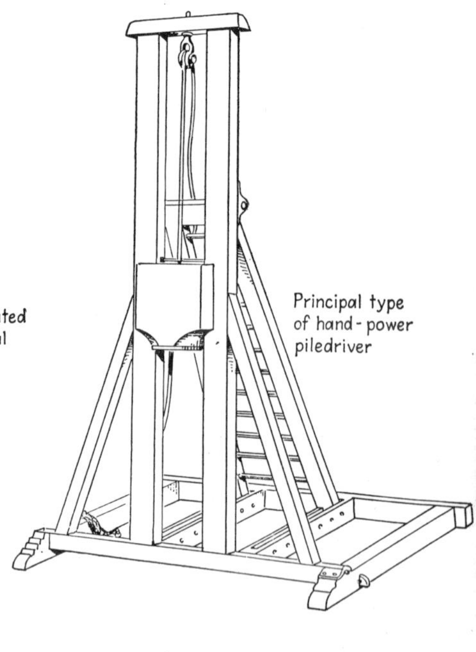

Principal type of hand-power piledriver

Primitive Pile Driving and Early Ram Drivers